Lecture Notes in Computer Science 13004

More information about this subseries at http://www.springer.com/series/7407

Martijn Mes · Eduardo Lalla-Ruiz ·
Stefan Voß (Eds.)

Computational Logistics

12th International Conference, ICCL 2021
Enschede, The Netherlands, September 27–29, 2021
Proceedings

Editors
Martijn Mes
IEBIS
University of Twente
Enschede, Overijssel, The Netherlands

Eduardo Lalla-Ruiz
IEBIS
University of Twente
Enschede, Overijssel, The Netherlands

Stefan Voß
IWI-Institute of Information Systems
University of Hamburg
Hamburg, Germany

ISSN 0302-9743 ISSN 1611-3349 (electronic)
Lecture Notes in Computer Science
ISBN 978-3-030-87671-5 ISBN 978-3-030-87672-2 (eBook)
https://doi.org/10.1007/978-3-030-87672-2

LNCS Sublibrary: SL1 – Theoretical Computer Science and General Issues

This Springer imprint is published by the registered company Springer Nature Switzerland AG
The registered company address is: Gewerbestrasse 11, 6330 Cham, Switzerland

Preface

Throughout the last decades, the increasing volume of information and operational workload in logistics caused a sharp interest in the automation of physical and informational logistical processes. Companies, institutions, and logistics stakeholders considering this aspect can react more efficiently to changes and disturbances resulting in more accurate planning, extending customer and product individualization while, in many cases, reducing operating costs. This resulted in advances in several logistics sectors, such as maritime shipping, multi-modal transport, urban logistics, warehousing, and inventory management. Computational logistics, as the driver between decision making and operations, has become a key component for economic and industrial growth.

Computational logistics covers the management of logistics' activities and tasks through the joint use of information and communication technologies and advanced decision support and optimization techniques. It is applied in several areas, e.g., the flow and storage of goods and services as well as the flow of related information. In this context, modeling and algorithmic approaches are developed, verified, and applied for planning and executing complex logistics tasks, e.g., for finding the most efficient routing plan and schedule to transport passengers or distribute goods. The models and algorithms are integrated with computing technologies, not only to get satisfactory results in reasonable times but also to exploit interactivity with the decision maker through visual interfaces, and to extract knowledge from data to improve future decision making. This promotes the joint effort of practitioners and scholars for better understanding and solving the logistics problems at hand.

The International Conference on Computational Logistics (ICCL) is a forum where recent advances in the computational logistics research area are presented and discussed. This volume offers a selection of 42 peer-reviewed papers out of the 111 contributions submitted to the this year's edition of ICCL, held virtually at the University of Twente, Enschede (The Netherlands), during September 27–29, 2021. The papers show various directions of importance in computational logistics, classified into five topic areas reflecting the interest of researchers and practitioners in this field. The papers in this volume are grouped accordingly:

1. **Maritime and Port Logistics**
   Maritime and port logistics is the backbone of global supply chains and international trade. The performance and functioning of its related activities are remarkably influenced by the quality of its planning and management. In ICCL 2021, the contributions that fall into this category relate to, among other things, berth allocation, ship routing, bulk logistics, simulation and proactive approaches, and various real-world maritime applications.
2. **Supply Chain and Production Management**
   The management of supply chains (SCs) and production covers different relevant logistics operations such as warehousing, workforce management, lot-sizing,

inventory management, and information sharing. The works included in this category pursue the efficient organization and management of the diverse resources and operations involved in such a way that the production, flow, and storage of products is as efficient as possible. Contributions related to all above-mentioned components, such as warehousing and inventory management, production scheduling, lot-sizing, and other SC-related topics fall into this category.

3. **Urban Transport and Collaborative Logistics**
   The progress in urban transport and collaborative logistics as well as the development of (smart) cities and regions require current systems to be adapted and updated to cope with changes that involve new transportation means, such as drones, the sharing of logistics resources, and collaboration among different logistics operations. The papers in this category relate to a diverse range of topics, such as car- and ride-sharing, drone-assisted delivery, self-coordination of vehicles, and micro-transit services.
4. **Routing, Dispatching, and Scheduling**
   The routing, dispatching, and scheduling of logistics resources constitute an important challenge in real-world transport and logistics activities. Due to numerous specific real-world features, there is a strong necessity for modeling and developing efficient solutions as well as formalizing cases that foster advancements in this area. The papers in this category address, among other things, green pickup and delivery, rerouting and dispatching operations, and service and tour planning approaches.
5. **Air Logistics and Multi-Modal Transport**
   Traditionally, the majority of studies presented at ICCL focus on maritime and road transport. However, nowadays there is an increasing interest in air logistics due to the necessity to operate more efficiently and sustainably. Furthermore, attention is given to logistics problems involving a combination of transportation means, leading to multi-modal transport, where at least two different transport modes are used (e.g., air, water, road, or rail). Thus, the papers that appear in this category relate to a range of topics concerning air logistics and multi-modal transport, such as aircraft routing, gate scheduling, cargo packing, multi-modal transport, and physical internet analysis.

ICCL 2021 was the 12th edition of this conference series, following the earlier ones held in Shanghai, China (2010, 2012), Hamburg, Germany (2011), Copenhagen, Denmark (2013), Valparaiso, Chile (2014), Delft, The Netherlands (2015), Lisbon, Portugal (2016), Southampton, UK (2017), Salerno, Italy (2018), Barranquilla, Colombia (2019), and Enschede, The Netherlands (2020). The editors thank all the authors for their contributions as well as the Program Committee and reviewers for their invaluable support and feedback. Finally, we would like to express our gratitude to Julia Bachale for her helpful support and assistance during the preparation of the conference. We trust that the present volume supports the continued advances within computational logistics and inspires all participants and readers to its fullest extent.

September 2021

Martijn Mes
Eduardo Lalla-Ruiz
Stefan Voß

Organization

Program Committee

Panagiotis Angeloudis	Imperial College London, UK
Tolga Bektas	University of Liverpool, UK
Francesco Carrabs	University of Salerno, Italy
Carlos Castro	Universidad Federico de Santa María, Chile
Raffaele Cerulli	University of Salerno, Italy
Joachim Daduna	Berlin School of Economics and Law, Germany
Christopher Expósito-Izquierdo	University of La Laguna, Spain
Yingjie Fan	Leiden University, The Netherlands
Elena Fernández	Universidad de Cádiz, Spain
Monica Gentili	University of Louisville, USA
Rosa González-Ramírez	Universidad de Los Andes, Chile
Hans-Dietrich Haasis	University of Bremen, Germany
Richard Hartl	University of Vienna, Austria
Geir Hasle	SINTEF Digital, Norway
Wouter van Heeswijk	University of Twente, The Netherlands
Leonard Heilig	University of Hamburg, Germany
Alessandro Hill	California Polytechnic State University, USA
Jan Hoffmann	UNCTAD, Switzerland
Manuel Iori	University of Modena and Reggio Emilia, Italy
Jiangang Jin	Shanghai Jiao Tong University, China
Raka Jovanovic	Qatar Environment and Energy Research Institute, Qatar
Herbert Kopfer	University of Bremen, Germany
René de Koster	Erasmus University Rotterdam, The Netherlands
Ioannis Lagoudis	University of Piraeus, Greece
Eduardo Lalla-Ruiz (Chair)	University of Twente, The Netherlands
Jasmine Siu Lee	LamNanyang Technological University, Singapore
Gilbert Laporte	HEC Montréal, Canada
Janny Leung	University of Macau, China
Israel López-Plata	University of La Laguna, Spain
Dirk Mattfeld	TU Braunschweig, Germany
Frank Meisel	University of Kiel, Germany
Gonzalo Mejía	Universidad de La Sabana, Colombia
Belén Melián-Batista	Universidad de La Laguna, Spain
Martijn Mes (Chair)	University of Twente, The Netherlands
José Marcos Moreno-Vega	Universidad de La Laguna, Spain
Adriana Moros-Daza	Universidad del Norte, Colombia

Rudy Negenborn	Delft University of Technology, The Netherlands
Dario Pacino	Technical University of Denmark, Denmark
Julia Pahl	University of Southern Denmark, Denmark
Carlos Paternina-Arboleda	Universidad del Norte, Colombia
Mario Ruthmair	University of Vienna, Austria
Dirk Sackmann	Hochschule Merseburg, Germany
Juan José Salazar González	Universidad de La Laguna, Spain
Frederik Schulte	Delft University of Technology, The Netherlands
Marco Schutten	University of Twente, The Netherlands
Xiaoning Shi	University of Hamburg, Germany
Douglas Smith	University of Missouri–St. Louis, USA
Maria Grazia Speranza	University of Brescia, Italy
Shunji Tanaka	Kyoto University, Japan
Kevin Tierney	Bielefeld University, Germany
Thierry Vanelslander	University of Antwerp, Belgium
Stefan Voß (Chair)	University of Hamburg, Germany

Additional Reviewers

Fabian Akkerman
Adina Aldea
Thiago Alves De Queiroz
Lorena Bearzotti
Breno Beirigo
Beatrice Bolsi
Matteo Brunetti
Giovanni Campuzano
Rafael Carmona-Benitez
Fabio D'Andreagiovanni
Alan Dávila de León
Oskar Eikenbroek
Alejandro Fernández-Gil
Jose García Conejeros
Rogier Harmelink
Xiaohuan Lyu
Meead Mansoursamaei
Bernardo Martin-Iradi
Javier Maturana-Ross
Mahmoud Moradi
Mirko Mucciairni
João Nabais
Yaxu Niu
Dennis Prak
Peter Shobayo
Engin Topan
Noemi Van Meir
Robert van Steenbergen
Matthias Volk
Daniel Wetzel
Jeffrey Willems
Vahid Yazdanpanah
Jun Ye
Jingjing Yu

Contents

Maritime and Port Logistics

Supply Chain and Production Management

Maritime and Port Logistics

An Integrated Planning, Scheduling, Yard Allocation and Berth Allocation Problem in Bulk Ports: Model and Heuristics

João Luiz Marques de Andrade[1](✉) and Gustavo Campos Menezes[2]

[1] Graduate Program in Mathematical and Computational Modeling, Federal Center for Technological Education of Minas Gerais, Belo Horizonte, MG, Brazil
[2] Department of Electronics and Computing, Federal Center for Technological Education of Minas Gerais, Belo Horizonte, MG, Brazil
gustavo@cefetmg.br

Abstract. Integrating operational and logistic processes is fundamental to ensure a port terminal's efficient and productive operation. This article deals with the integration of planning, scheduling, yard allocation, and berth allocation in dry bulk export port terminals. The integrated problem consists of planning and sequencing the flow of products between arrival at the terminal and the berths, allocating the products to the storage yards, and determining the sequence, berthing time, and position of each vessel. A mixed-integer linear programming formulation is proposed, connecting the problems and incorporating tidal time windows and non-preemptive scheduling. To solve the integrated problem more efficiently, we developed an algorithm based on a combination of a diving heuristic with limited backtracking, two relax-and-fix heuristics, a local branching heuristic, a rolling horizon heuristic, and a variable-fixing strategy. The mathematical formulation and proposed algorithm are tested and validated with large-scale instances. The computational results show that the proposed algorithm is effective in finding strong upper bounds.

Keywords: Integrated planning · Scheduling · Berth allocation · Yard allocation · Matheuristics

1 Introduction

Port terminals are an essential and strategic part of the global supply chain. They manage large volumes of products between land transport and sea vessels, and the performance of their operations directly affects the entire chain. For this reason, it is fundamental that various logistics operations and processes at the terminals are carried out with the best possible efficiency. Based on the previous literature, the benefits of using Operations Research and its methods in improving the performance of the activities practiced by port terminals are well known.

Port terminals are complex facilities that have several strongly related operational and logistical problems. For example, the equipment schedule is directly

M. Mes et al. (Eds.): ICCL 2021, LNCS 13004, pp. 3–20, 2021.
https://doi.org/10.1007/978-3-030-87672-2_1

related to both the allocation of products in the stockyards and the loading of the vessels because the equipment interferes in when and where a product is stacked and removed in the yards and when and where the vessel should be loaded. Similarly, the storage yards and the allocation of ships in the berths affect the use of equipment. Therefore, optimizing operations individually, ignoring the relationships between them, can generate solutions for the terminal with low-quality [17].

Motivated by the importance of integrating operations at port terminals, this paper has as its contribution a mathematical formulation that integrates the problems of planning, sequencing, yard allocation, and berth allocation in dry bulk exporting port terminals. It is well established the difficulties of solvers in solving complex models; for this reason, an algorithm has been developed that combines a diving heuristic with backtracking, two relax-and-fix heuristics adapted for the formulation, local branching heuristic, and a rolling horizon heuristic. Heuristics are applied to obtain integer solutions for specific sets of variables, and in the end, generate a good quality feasible integer solution to the problem. Computational experiments with large-scale instances are conducted to validate the formulation and test the performance of the proposed algorithm. The computational results prove the effectiveness of the algorithm and the model.

The remainder of the paper is structured as follows: Sect. 2 provides a literature review. Section 3 introduces the problem through a detailed description. The mathematical model is proposed in Sect. 4. Section 5 presents a solution approach. Results and computational experiments are analyzed in Sect. 6. Finally, conclusions and future research are addressed in Sect. 7.

2 Literature Review

The high complexity of management and heterogeneous processes of port terminals provides a fertile field to apply operational research and its methods. In recent years, the operational problems of port terminals have received attention from the academic community, and consequently, significant progress has been made. Most contributions are associated with optimizing container terminals, while bulk port terminals have received relatively little attention from researchers.

A part of the bulk terminal optimization literature focuses on individual problems, while the other part is focused on integrated problems. Barros et al. [1], Umang et al. [15], Ernst et al. [5] and Cheimanoff et al. [2] are papers that focus on the individual problems of berth allocation, and Hu and Yao [8], Unsal [16] investigate the individual problems of reclaimer scheduling. Robenek et al. [11], Tang et al. [14], Menezes et al. [9,10], Unsal and Oguz [17] and Rocha de Paula et al. [12] study integrated approaches to bulk terminal operational problems. Robenek et al. [11] extend the work of Umang et al. [15] by integrating the berth allocation problem with the yard assignment problem. The authors consider several realistic assumptions, such as dynamic ship arrival, cargo handling capacity, storage location restrictions based on the type of cargo, and congestion constraints. Tang et al. [14] study an integrated storage space allocation and ship scheduling problem to achieve better yard utilization and reduce product losses and transportation costs. The authors of this paper develop a yard management method based on dividing large yards into smaller areas. Unsal and Oguz [17] propose integrating the problems of berth allocation, reclaimer scheduling, and yard allocation. The problem incorporates the operational problems and constraints of the tidal window, multiple stocking pads, non-crossover of reclaimers, and vessel and berth size. Recently, Rocha de Paula et al. [12] has developed a genetic algorithm to maximize the efficiency of a coal export terminal. The article studies an optimization method that programs the arrival of coal at the terminal, determines the periods of stacking and recovery of the coal loads, and programs the arrival and departure times of the vessels. Other lines of research in the area of bulk terminals are represented by the works of Dávila de Leon et al. [3,4]. Dávila de Leon et al. [3] develops machine learning-based system for supporting berthing operations in bulk ports, and Dávila de Leon et al. [4] proposes a simulation-optimization in bulk berth scheduling.

Menezes et al. [9,10] study a production planning and scheduling problem in bulk cargo terminal. The problem considers planning and sequencing the flow of products between supply, storage, and demand nodes, minimizing the operational costs. The main highlights of the papers are the new mathematical formulation and solution approaches. In Menezes et al. [9], a hierarchical heuristic is presented, while Menezes et al. [10] developed a branch-and-price algorithm.

The formulation presented in the current paper is based on that of Menezes et al. [9,10]. In this paper, the integrated problem of Menezes et al. [9,10] is extended with the berth allocation problem. In addition, scheduling constraints are reformulated to consider a non-preemptive scheduling. The main contributions of this paper are a mathematical model for the integrated planning, scheduling, yard allocation, and berth allocation problem and a solution algorithm with new versions of matheuristics.

3 Problem Description

The problem investigated in this paper considers a dry bulk export terminal. The port complex as a whole can be represented by a set of three subsystems: reception, stockyards, and berth. Figure 1 provides an overview of the port terminal with the three subsystems.

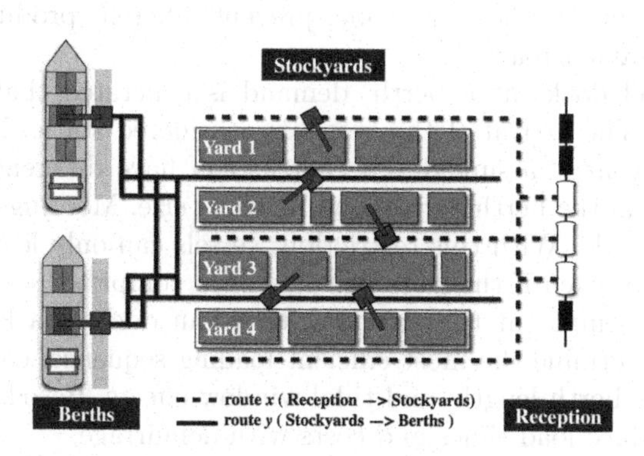

Fig. 1. Port complex and its three subsystems.

Products are transported between subsystems via routes x and y. Routes x stock the reception products in the stockyards, and the routes y transport the products from the yards to load the ships at the berths. These routes are responsible for transporting quantities (lots) of products and define the various options of paths along which the products can be transported. Thus, the goal is to define the quantity and destination of each product, and in addition, to determine which routes will transport them.

Routes are a combination of equipment previously defined. Each equipment has a predefined transportation capacity per hour. The lowest capacity equipment on the route defines the capacity of the route. Given equipment can be shared by more than one route. Thus, if a pair of routes share any equipment, these routes cannot be active simultaneously. The goal is to activate and schedule the best routes that can work simultaneously to transport the products. Figure 2 shows an example of two routes that share a stacker.

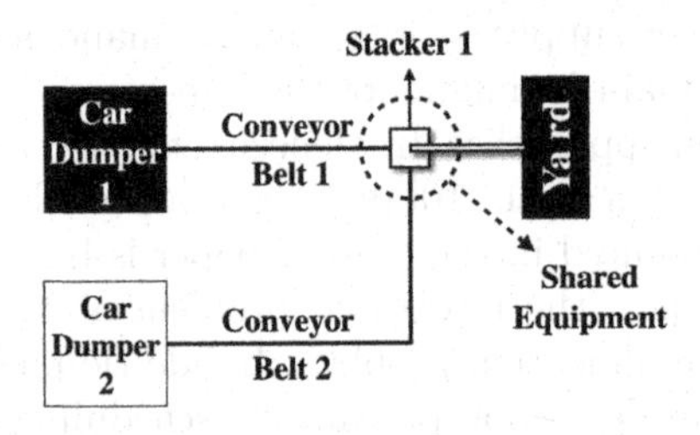

Fig. 2. Examples of routes with shared equipment.

The product flows arriving in the reception subsystem are directly transferred to a stock area in the stockyards via routes x. The stockyards subsystem consists of large product storage areas. Each storage area is further subdivided into smaller subareas, named subareas. The capacity limit of each subarea is an input data of the problem. Free space is maintained between the subareas to avoid contamination at the stacks of products. Each subarea can only store one product at a time. Whenever a subarea is empty, and a new product is assigned, it is considered a subarea cleanup cost. The goal is to allocate the best product in each subarea, and at the same time, prevent different products from being allocated in the same subarea.

When a vessel docks at a berth, demand is generated that must be met by the routes y. The arrival of the vessel is considered static. In this study, a discrete berth layout is assumed. The berths can have different sizes, so each vessel must dock in the berth corresponding to its size. Moorings are limited by the effects of the tide. With this restriction, vessels can only leave the berth at a high tide period, even if the shipment has been completed earlier. Moreover, as vessels arrived empty at the terminal, they can dock at a low tide period. The goal is to determine the most efficient loading sequence, considering stock product quantity, berth length, and tidal window effects. Based on this goal, it is intended to reduce load times and costs with demurrage.

4 Model Formulation

This section presents the mathematical formulation for the integrated planning, scheduling, yard allocation, and berth allocation problem in dry bulk export terminals. In this formulation, the time horizon is divided into T periods. The periods must have a time of fewer than 12 h to consider the high tide effects. The vessels enter at the beginning of each period and leave only at the end of each period. The notation used in the mathematical model is provided in Table 1.

Table 1. Notation used in the formulation.

Sets	
T	Set of periods
I^t	Set of microperiods available for period $t \in T$
P	Set of products
R^x	Set of routes (Reception/Stockyard)
R^y	Set of routes (Stockyard/Berths)
R	Set of all available routes ($R = R^x \cup R^y$)
S	Set of storage subareas
R^x_s	Subset of routes x that arrive in the subarea $s \in S$
R^y_s	Subset of routes y departing from the subarea $s \in S$
M	Set of equipment
R^x_m	Subset of routes x that use equipment $m \in M$
R^y_m	Subset of routes y that use equipment $m \in M$
B	Set of berths
N	Set of vessels
E	Pairs of routes that share at least one piece of equipment to transport products
Parameters	
O_{pt}	Supply (in ton) of product p at the beginning of period t
K_{np}	Amount of cargo (in ton) of product p for vessel n
l^s_{pt}	Storage capacity of subarea $s \in S$ for product p in period t
b_m	Capacity of equipment m
j^m_t	Available time (in hours) for the use of equipment m in period t
c^{rx}	Capacity (in tons/hour) of route $r \in R^x$
c^{ry}	Capacity (in tons/hour) of route $r \in R^y$
α_{pt}	Penalty for not meeting the supply at the reception of product p in period t
β_{np}	Penalty for not meeting the demand (cargo) of product p of vessel n
$\gamma^s_{pp't}$	Preparation cost associated with replacing product p by product p' in subarea s at period t
σ^r	Maintenance cost of using route $r \in R$
ϕ_n	Penalty on the loading time of vessel n at the terminal
lb_b	Length (in meters) of berth b
lv_n	Length (in meters) of vessel n
K_{max}	Constant value equal to the highest load
H_t	Maximum duration (in hours) of each period t
Π_t	Number of microperiods of each period t
μ_{it}	duration of each microperiod i in period t
T_{max}	Constant equal to the total number of periods

(continued)

Table 1. (*continued*)

Variables	
x^r_{pt}	Time (in hours) used by routes $r \in R^x$ to transport product p from reception to the stockyard in period t
y^r_{pt}	Time (in hours) used by routes $r \in R^y$ to transport product p from the stockyard to the berths in period t
d^b_{npt}	Amount of product p loaded on vessel n allocated in berth b in period t
IR_{pt}	Represents the amount of product p in the reception subsystem that was not delivered at the end of period t
IV_{np}	Represents the amount of product p that was not loaded from vessel n
f^s_{pt}	Binary, equals 1 if subarea s is allocated for product p in period t, 0 otherwise
$Sf^s_{pp't}$	Has a value of 1 when product p has been replaced with product p' at period t, 0 otherwise. This replacement can occur only when the amount of product p in subarea s has been exhausted in the preceding period $t-1$
e^s_{pt}	Amount of product p stored at subarea s in period t
v^b_{nt}	Binary, equals 1 if vessel n is moored to berth b in period t, 0 otherwise
w^b_n	Binary, equals 1 if the berth b is assigned to the vessel n, 0 otherwise
τ_n	Loading time of vessel n
q^r_{pit}	Binary, equals 1 if given route r is activated to transport the product p in the microperiod $i \in I^t$ in period t

The mixed-integer linear programming (MILP) model for the integrated planning, scheduling, yard allocation and berth allocation problem in dry bulk ports can be formulated as follows:

$$\begin{aligned}\textbf{min} \quad f = & \sum_{p\in P}\sum_{t\in T}\alpha_{pt}IR_{pt} + \sum_{n\in N}\sum_{p\in P}\beta_{np}IV_{np} + \sum_{s\in S}\sum_{p\in P}\sum_{p'\in P}\sum_{t\in T}\gamma^s_{pt}Sf^{pp't}_s \\ & + \sum_{p\in P}\sum_{t\in T}\sum_{r\in R^x}\sigma^r(c^{rx}x^r_{pt}) + \sum_{p\in P}\sum_{t\in T}\sum_{r\in R^y}\sigma^r(c^{ry}y^r_{pt}) + \sum_{n\in N}\phi_n\tau_n\end{aligned} \tag{1}$$

subject to

$$\sum_{r\in R^x} c^{rx}x^r_{pt} - IR_{p(t-1)} + IR_{pt} = O_{pt} \qquad \forall p \in P, \forall t \in T \tag{2}$$

$$\sum_{r\in R^y} c^{ry}y^r_{pt} = \sum_{n\in N} d^b_{npt} \qquad \forall b \in B, \forall p \in P, \forall t \in T \tag{3}$$

$$\sum_{b\in B}\sum_{t\in T} d^b_{npt} + IV_{np} = K_{np} \qquad \forall n \in N, \forall p \in P \tag{4}$$

$$\sum_{n\in(N\cup 0)} v^b_{nt} = 1 \qquad \forall b \in B, \forall t \in T \tag{5}$$

$$\sum_{p\in P} d^b_{npt} \leq K_{max}v^b_{nt} \qquad \forall b \in B, \forall n \in N, \forall t \in T \tag{6}$$

$$\sum_{m=1}^{t-1} v_{nm}^{b} - v_{n(t-1)}^{b}t + v_{nt}^{b}t \leq t \qquad \forall b \in B, \forall n \in N, \forall t \in T : t > 1 \quad (7)$$

$$\sum_{t \in T} v_{nt}^{b} = 0 \qquad \forall b \in B, \forall n \in N : (lb_b < lv_n) \quad (8)$$

$$\sum_{b \in B} \sum_{t \in T} v_{nt}^{b} = \tau_n \qquad \forall n \in N \quad (9)$$

$$\sum_{b \in B} w_n^{b} = 1 \qquad \forall n \in N \quad (10)$$

$$\sum_{t \in T} v_{nt}^{b} \leq T_{max} w_n^{b} \qquad \forall b \in B, \forall n \in N \quad (11)$$

$$e_{p(t+1)}^{s} = e_{pt}^{s} + \sum_{r \in R_s^x} c^{rx} x_{pt}^{r} - \sum_{r \in R_s^y} c^{ry} y_{pt}^{r} \qquad \forall s \in S, \forall p \in P, \forall t \in T \quad (12)$$

$$e_{pt}^{s} \geq \sum_{r \in R_s^y} c^{ry} y_{pt}^{r} \qquad \forall s \in S, \forall p \in P, \forall t \in T \quad (13)$$

$$e_{pt}^{s} \leq l_{pt}^{s} \qquad \forall s \in S, \forall p \in P, \forall t \in T \quad (14)$$

$$\sum_{p \in P} f_{pt}^{s} = 1 \qquad \forall s \in S, \forall t \in T \quad (15)$$

$$l_{pt}^{s} f_{pt}^{s} - e_{pt}^{s} \geq 0 \qquad \forall s \in S, \forall t \in T \quad (16)$$

$$l_{pt}^{s} f_{pt}^{s} - \sum_{r \in R_s^x} x_{pt}^{r} \geq 0 \qquad \forall s \in S, \forall t \in T \quad (17)$$

$$Sf_{pp't}^{s} \geq f_{p(t-1)}^{s} + f_{p't}^{s} - 1 \qquad \forall s \in S, \forall t \in T, \forall p \in P, \forall p' \in P, p \neq p' \quad (18)$$

$$\sum_{p \in P} \left(\sum_{r \in R_m^x} c^{rx} x_{pt}^{r} + \sum_{r \in R_m^y} c^{ry} y_{pt}^{r} \right) \leq j_t^{m} b^{m} \qquad \forall m \in M, \forall t \in T \quad (19)$$

$$x_{pt}^{r} \leq \sum_{i \in I^t} q_{pit}^{r} \mu_{it} \qquad \forall p \in P, \forall t \in T, \forall r \in R^x \quad (20)$$

$$y_{pt}^{r} \leq \sum_{i \in I^t} q_{pit}^{r} \mu_{it} \qquad \forall p \in P, \forall t \in T, \forall r \in R^y \quad (21)$$

$$q_{pit}^{r} + q_{pit}^{r'} \leq 1 \qquad \forall i \in I^t, \forall p \in P, \forall (r, r' \in E) \forall t \in T \quad (22)$$

$$\sum_{k=1}^{i-1} q_{pkt}^{r} - q_{p(i-1)t}^{r} i + q_{pit}^{r} i \leq i \qquad \forall r \in R, \forall p \in P, \forall t \in T, \forall i \in I^t : i > 1 \quad (23)$$

The objective function seeks to minimize the penalty of not meeting the supply of products at the reception subsystem (first term), the penalty of not meeting the demand of vessels (second term), the product exchange costs in the subareas (third term), the cost of using routes x and y to transport products (fourth and fifth terms) and the penalties on the loading time of vessels (sixth term).

Constraints (2) are related to meeting the product supply of the reception subsystem. Constraints (3) control the vessel loadings in the berth subsystem. Constraints (4) control the demand meeting of the vessels. Constraints (5) ensure that once a berth is assigned to a vessel, it can no longer be used by any other vessel in this same period. Note that the vessel 0 ($n = 0$) represents that the b berth is empty at period t. Constraints (6) ensure that a vessel can be loaded just when it is berthed at the berth. Constraints (3), (5) and (6) are complementary and ensure that vessels are loaded only within the periods they are allocated to a berth. Constraints (7) ensure no interruption in the vessel loading operation ([1]). Constraints (8) impose vessels with longer berth lengths from berthing ([17]). Constraints (9) are responsible for summing up all periods used to load each vessel. Constraints (10) and (11) control the allocation of vessel in berths.

Constraints (12) control the input and output of products in the storage subareas. Constraints (13) impose that the routes y transport only the products that are in stock at the beginning of the period. Constraints (14) define the storage capacity of each subarea. Constraints (15)–(17) are responsible for controlling the allocation of products to subareas. Constraints (15) impose that only one product can be allocated to a subarea in the period. Note that the product 0 ($p = 0$) represents that the subarea s is empty at period t. Constraints (18) control the replacement of products in each subarea.

Constraints (19) ensure that no equipment will have its capacity exceeded. Constraints (20)–(23) refer to scheduling. To perform scheduling, each period t is divided into microperiods (μ_{it}). Where $i \in I^t$ and I^t represents the total of microperiods available for period t. The duration of each microperiod is fixed and given by the following expression: $\mu_{it} = H_t/\Pi_t$. Each microperiod will contain only the routes that can be activated simultaneously, in other words, only routes that have no conflicts in their equipment. This condition is guaranteed by the constraints (22). The sums of the constraints (20) and (21) represent the microperiods available for each route. Variables x^r_{en} and y^r_{en} can be interpreted as transportation tasks because their values represent the hours used to transport the products between the subsystems. However, to enable the transportation of products on these routes, it is necessary to allocate some microperiods. Restrictions (23) ensure that microperiods are assigned consecutively on active routes, thus avoiding pre-emption of tasks ([1]).

$$d^b_{npt}, x^r_{pt}, y^r_{pt}, e^s_{pt}, IR_{pt}, IV_{np}, \tau_n \geq 0 \qquad \forall r \in R, \forall b \in B, \forall n \in N, \forall s \in S, \forall p \in P, \forall i \in I^t, \forall t \in T \tag{24}$$

$$f^s_{pt}, v^b_{nt}, w^b_n, q^r_{pit} \in \{0, 1\} \qquad \forall r \in R, \forall b \in B, \forall n \in N, \forall s \in S, \forall p \in P, \forall t \in T, \forall i \in I^t \tag{25}$$

$$0 \leq S^s_{pp't} \leq 1 \qquad \forall s \in S, \forall p \in P, \forall p' \in (P \cup 0), p \neq p', \forall i \in I^t, \forall t \in T \tag{26}$$

Constraint (24)–(26) determines the domains of variables.

5 Solution Approach

Solving MILP model formulation with large-scale instances via a solver, such as CPLEX, is a substantially tricky task and may even be unfeasible. Due to a large number of constraints and integer variables. Intending to circumvent these limitations and obtain good solutions, we developed an algorithm that combines matheuristics with a variable-fixing strategy, named the relax-solve-and-fix heuristic (RSFH). The strategy adopted by the RSFH consists of obtaining a feasible integer solution for a set of binary variables through a matheuristic or a combination of them, fixing this solution, repeating this procedure with other specific groups of binary variables until a feasible integer solution is obtained.

5.1 Diving Heuristic with Limited Backtracking

The sequence in which the variables are fixed is of critical importance, as this can result in a good quality solution or even an infeasible solution. The first variables to be set are the w_n^b (responsible for only assigning one berth to each vessel) to restrict the solution space of the remaining variables without generating infeasible solutions. The feasible integer solution for the variables w_n^b is obtained through a diving heuristic with limited backtracking (DHLB). The DHLB is based on the works of Harvey and Ginsberg [7], and Sadykov et al. [13]. The DHLB strategy consists of performing a limited number of depth-first searches in the solution tree and using the best feasible integer solution found.

DHLB is divided into two phases. In the first phase, the initial node is obtained with the solution of the relaxed model (all binary variables are relaxed). From this initial node, a depth-first search will be performed with the intention of finding a feasible integer solution. The depth is limited by the input parameter (*maxdepth*). A rounding strategy is adopted to generate the branch to the left of the initial node, which sets the fractional variables with a value higher than the parameter Δ to 1. If no variable can be fixed, the value of Δ is reduced by the parameter ϵ until at least one variable is fixed. After fixing some variables, the relaxed model is solved again via CPLEX, and the fixing procedure is repeated until an integer solution is obtained. With some variables fixed at 1, the relaxed model is solved again via CPLEX, and the fixing procedure is repeated to create a new node. Repeat the entire process until an integer solution is obtained. If you reach the last node and the solution obtained is still not an integer, the previous fractional variables with the value closest to 1 should be fixed.

In the second phase of DHLB, new branches are generated to the right at each node created in the first phase. In these new branches, the variables that are set to 1 in the branch to the left (in phase 1) are at this phase set to 0. The other variables that were relaxed should remain the same. Figure 3 illustrates part of a tree generated by DHLB and exemplifies the new branches on the right, which w^a represents the set of variables fixed to 1 in the first branch of phase 1 and w^b the group of variables fixed in the next branch in the same phase. Another detail of Fig. 3 is the nodes highlighted in black, representing the first phase, and the nodes in white, representing the second phase (backtrackings).

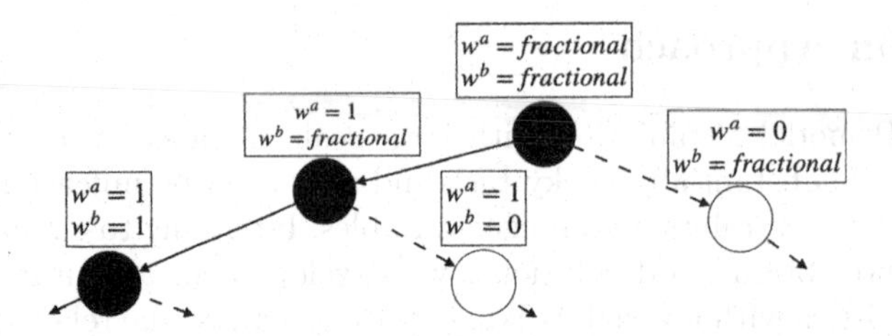

Fig. 3. Example of enumeration tree of the diving heuristic with backtracking.

With the specific variables fixed at zero, the same variable-fixing procedure performed in phase 1 is repeated, intending to find a new feasible integer solution for the variables w_n^b. After ending the depth-first search with the branch to the right of the initial node, the same procedure should be repeated for the second node generated in the first phase. This procedure of new right-hand branching (backtracking) is repeated for all nodes generated in the first phase of the heuristic.

The solution obtained at the end of the first phase is defined as the best solution. If a new integer solution is generated during right-hand branching and has a lower value f, this is the new best solution. If the value f of a solution generated during some backtracking in phase 2 is greater than the best solution, the search for that node is terminated, and a new depth-first search starts at the next node generated in the first phase. Therefore, at the end of the second phase of the heuristic, the best solution of the variables w_n^b is fixed.

5.2 Relax-and-fix Heuristic for Variables v_{nt}^b

The next step of RSFH, after vessels are allocated to a berth through the solution of DHLB, consists in assigning in which periods vessels are loaded (v_{nt}^b). To obtain an integer solution for the variables v_{nt}^b, first a relax-and-fix heuristic specific to that set of variables (HRF-V) is implemented, and then a local branching heuristic.

The strategy of HRF-V consists of fixing the variables by period, following the criterion of fixing the variables with the value closest to 1, then checking the quantity loaded by the vessel, and fixing the next period if the minimum loading percentage is not reached.

One input to the HRF-V is the vector V_{max}. This vector is used to limit the number of periods assigned to the vessels. The vector has a value for each available vessel in the planning horizon. These values are associated to an average of periods that each vessel has to load, and are calculated by the expression: $\bar{t} = \left\lfloor \frac{|B| \times |T|}{|N|} \right\rfloor$.

HRF-V is divided into iterations, with the number of iterations equal to the number of periods. Each iteration has two variable-fixing procedures. First, the relaxed model is solved via CPLEX, and then the first iteration is started. At the beginning of the iteration, the first variable fixing procedure occurs. This procedure consists in fixing at 1 the variables v_{nt}^{b} that has the highest value in each berth in period 1, thus allocating only one vessel in each berth in the iteration period. After the variables are fixed, the relaxed model is solved again by the solver, and the second variable fixing procedure is started.

In the second procedure for fixing the variables v_{nt}^{b}, each vessel that was set to 1 in the first procedure has the amount of product loaded by the period of the iteration verified. If the quantity loaded by the particular vessel is less than 70% of the total cargo and the number of periods assigned to the respective vessel is less than the maximum (given by V_{max}), the variable v_{nt}^{b} associated with period 2 is set to 1. For example, if ship 2 has the highest value among the ships allocated to berth 3 in period 1, the variable v_{21}^{3} is set to 1 (first procedure). After the model is solved, if the amount of product loaded by the vessel in period 1 is less than 70% and the number of periods given by the vector V_{max} is equal or greater than 2, the variable v_{22}^{3} is assigned to 1; otherwise, it is not assigned. Once the second variable-fixing procedure is finished in iteration 1, the second iteration begins.

In the second and subsequent iterations, the same procedures as in iteration 1 are repeated, i.e., the first procedure, the solution of the relaxed model via CPLEX, and the second variable-fixing procedure. All variables set to 1, whether set in the first procedure of the current period t or set in the second step of period $t-1$, are verified and fixed in the second variable fixing procedure of period t. In the previous example, the variables v_{21}^{3} and v_{22}^{3} of vessel 2 allocated to berth 3 were fixed at iteration 1 (period 1). In iteration 2 (period 2), no variable associated with berth three is analyzed in the first fixing procedure because v_{22}^{3} is already fixed at 1; however, in the second procedure, the loading of vessel two is checked, as also of the other vessels. Note that the second variable fixing procedure is not performed in the last period.

5.3 Local Branching Heuristic

A local branching matheuristic is implemented to refine the greedy solution given by HRF-V. Local branching is a technique introduced by Fischetti and Lodi [6] to solve integer problems. The local branching algorithm developed in this work is presented in Algorithm 1.

Algorithm 1: *Local Branching*

```
  Input: s', k_0, t_limit.
1 s* ⟵ s'
2 k ⟵ k_0
3 it ⟵ 0
4 while it ≤ it_max do
5     add the constraint Δ(s*, s') ≤ k to the model;
6     s' ⟵ solveModel(t_limit)
7     if value f of the feasible solution s' is less than feasible solution s* then
8         remove the constraint Δ(s*, s') ≤ k to the model;
9         add the constraint Δ(s*, s') ≥ k + 1 to the model;
10        s* ⟵ s'
11    else
12        remove the constraint Δ(s*, s') ≤ k to the model;
13        k = k + 1
14    it = it + 1
```

Initially, the solution s' obtained previously in the relax-and-fix heuristic is used as a reference solution (s^*). Then a local branching constraint of the form $\Delta(s^*, s') \leq k$ is added to the model, and then the CPLEX solver is run until it finds an optimal solution or reaches the execution time limit (t_{limit}). If this new solution (s') has a lower cost than the solution s^*, the last local branching constraint of the form $\Delta(s^*, s') \leq k$ is removed, and a local branching constraint of the form $\Delta(s^*, s') \geq k+1$ is added to the formulation.This procedure means that a branch from the right-hand side of the local branching enumeration tree will be explored, and the value of k returns to the minimum distance. If the solution s' has a higher cost than the solution s^*, the last local branching constraint of the form $\Delta(s^*, s') \leq k$ is removed, and the value of the parameter k is incremented by 1, extending the distance. Subsequently, to either of the two cases, a new local branching constraint of the form $\Delta(s^*, s') \leq k$ based on the new solution found, or on the new value of k, is added to the formulation. This procedure is repeated until the maximum iteration is reached. Both local branching constraints used in Algorithm 1 remain the same as proposed in Fischetti and Lodi [6].

5.4 Relax-and-Fix Heuristic for the Variables f^s_{pt}

With the integer solutions of the variables w^b_n and v^b_{nt} already defined, the next step of RSFH is obtain a feasible integer solution for the variables f^s_{pt} (allocate the products to the subareas). First, a relax-and-fix heuristic specific to the variables f^s_{pt} is applied to provide an initial solution, then the same local branching algorithm is applied to the variables v^b_{nt}, but with different parameters.

The relax-and-fix heuristic specific for the variables f^s_{pt} (HRF-F) is divided into two parts. The first part starts by solving the model with the variables f^s_{pt} relaxed via CPLEX. Using the solution generated by the solver, the variables f^s_{pt} that have a value greater than the parameter α is set to 1. Next, the parameter α is decremented by the value of 0.05.

These three procedures are repeated in a loop until a feasible integer solution is found, or while the value of α is greater than 0.5. This limiting value is justified by the possible case of two or more variables f^s_{pt} having a higher value α in only one subarea. If fixed these variables, it would originate an infeasible solution because it would assign more than one product in only one subarea and not respecting the constraints (15).

If an integer solution is not obtained yet, the second part of HRF-F begins. This part consists of converting to binary the variables f^s_{pt} that have not yet been fixed, solving the relaxed model via CPLEX, and fixing the generated solution.

The feasible integer solution found by HRF-F is used as the incumbent solution for the local branching heuristic.

5.5 Rolling Horizon Heuristic

The integer solution for the variables f^s_{pt} obtained via a local branching heuristic is fixed but may be modified. The integer solution for the variables f^s_{pt} and q^r_{pit} (responsible for scheduling the routes and preventing them from sharing equipment) is obtained in an integrated manner using a rolling horizon heuristic (RHH).

The main idea of RHH is to partition the planning horizon into smaller subproblems and then consecutively solve each of the subproblems. This strategy is intended to reduce computational effort since only one subproblem will present binary variables, while the subproblems that have been solved present fixed variables, and the following subproblems the variables are relaxed.

In this work, each subproblem is a period t. The strategy of RHH consists of repeating the following procedure for each t period: converting only the f^s_{pit} and q^r_{pit} variables of the t period to binary, solving the model via solver until an optimal solution is obtained, and fixing this solution. Note that while the variables f^s_{pt} and q^r_{pit} are binary of period t, in period $t+1$ the variables f^s_{pt} are fixed following the solution obtained by local branching and the variables q^r_{pit} are relaxed. Furthermore, the variables f^s_{pt} and q^r_{pit} in the period $t-1$ are fixed according to RHH's solution. With the completion of this last heuristic a feasible integer solution to the integrated problem is obtained.

6 Computational Results

To validate the proposed mathematical formulation and test the performance of the proposed algorithm, we elaborate a set of instances capable of simulating several characteristics of a dry bulk terminal. The goal is to generate instances that simulate configurations in relative proximity to real cases and stress the solution approach used as much as possible. Some of the parameters used are detailed in the Table 2.

Table 2. Data used to generate the instances.

Parameters	Description
Stockyard	The storage area has been divided into 50 subareas, where each subarea has a limit capacity of 100,000 tons of storage
Equipment	Five car dumpers, five reclaimers, four stackers/reclaimers (equipment that performs both tasks), five stackers, and six ship loaders. At the terminal in question, there are approximately 50 km of conveyor belts. Each route uses one belt segment
Mooring berths	The terminal has three berths. In which the berths have the lengths (lb_b) of 335, 300, 240 m, respectively
Vessel length	The length of the vessels (lv_n) is divided into three cases: 10% are vessels longer than 315 m, 40% are vessels between 275 and 290 m, and 50% are vessels shorter than 240 m
α_{pt}	0.5 (half monetary units) for all products and periods
β_{np}	1 (one monetary units) for all products and vessels
ϕ_n	10 (ten monetary units) for all for vessels
$\gamma^s_{pp't}$	1 (one monetary units) for any product exchange in all periods and subareas
σ^r	Based on the following formula: 0.01 (monetary unit) $\times$ length of route r

For the experiments described next, the capacity of the equipment is varied between 12,000 and 16,000 ton/h, 150 routes are considered, and the initial amount of product in the stockyard is set at 30%. About the supply amounts and shipments, they were generated randomly but are large scale with proximity to real data from bulk port terminals in Brazil. The maximum duration of each period (H_t) is 11 h, to respect the high tide restrictions and to consider a reasonable time for berthing and undocking vessels. The duration of each microperiod (μ_{it}) is one hour.

The input parameters for DHLB are as follows: $maxdepth = 5$, $\Delta = 0.95$ and $\epsilon = 0.05$. Through these parameters, it is intended to fix a reasonable amount of variables in each node so that consequently, the search tree is not relatively long and the computational time is reduced, but still generate good integer solutions for the variables w^b_{nt}. The input parameters for local branching heuristics for the variables v^b_{nt} are as follows: $it_{max} = 5$, $t_{limit} = [180, 540]$ seconds and $k_0 = \left\lfloor \frac{|T|}{2} \right\rfloor$. Whereas, the input parameters for local branching heuristics for the variables f^s_{en} are as follows: $it_{max} = 2$, $t_{limit} = [900, 1200]$ seconds and $k_0 = |T|$. These values for the parameters were obtained after numerous tests, and it was noted that with them, there was a more significant benefit between computational time and the upper bound found. The variation in the parameter values in both cases is associated with the number of types of products handled.

All computational experiments are performed using CPLEX 12.9 on a computer with a 3.50 GHz processor and 32 GB RAM. The time limit is set as six-hour (21600 s).

In the following table, the column *instance* refers to the instance number and specification. For example, instance 01-05T05P03V, represents instance number 1 and a time horizon of 5 periods (05T), 5 different products being handled (05P), and a queue with 5 available vessels (5V). The columns f and f_{ub_1} exhibit the best upper bounds obtained by the CPLEX solver and the proposed algorithm, respectively. The column f_{ub_2} presents the upper bounds obtained with the proposed algorithm without using the local branching heuristic. Column f_{lb} provides the lower bound for the MILP model (linear relaxation). Column GAP_1 gives the GAP between the best lower bound and upper bound obtained via CPLEX solver. Column GAP_2 gives the relative GAP between the upper bounds generated by the CPLEX solver and the developed algorithm, given $GAP_2 = ((f_{ub_1}/f) - 1)100$. Lastly, the columns t_{cplex} and t_{heu} are the computational times, expressed in seconds.

Table 3 presents the computational results for instances with variations in the number of vessels, types of products handled, and periods. The instances are separated into three groups, representing three congestion situations at the

Table 3. Computational results for a set of instances with variation in the number of available vessels.

Set	Instance	CPLEX				RSFH			
		f_{lb}	f	t_{cplex}	GAP_1	f_{ub_1}	t_{heu}	GAP_2	f_{ub_2}
1	01-05T05P03V	52.19	54.49	839	0.00	54.49	276	0,00	84.13
	02-05T10P03V	51.29	55.29	2456	0.00	56.41	1170	2,03	95.88
	03-10T05P08V	125.63	139.62	T.L.	4.33	146.37	2702	4,84	229.08
	04-10T10P08V	127.42	154.04	T.L.	9.40	153.33	4272	−0,46	2.40E+04
	05-15T05P12V	196.54	237.34	T.L.	10.70	241.00	3558	1,54	391.89
	06-15T10P12V	199.57	–	–	–	266.55	7777	–	5.92E+04
2	07-05T05P05V	75.66	77.30	658	0.00	78.60	267	1,68	105.87
	08-05T10P05V	75.85	81.38	2851	0.00	82.19	1416	0,98	146.91
	09-10T05P10V	155.66	172.25	T.L.	4.56	177.98	2245	3,33	251.01
	10-10T10P10V	156.70	218.25	T.L.	23.62	182.55	4926	−16,36	1.90E+04
	11-15T05P15V	234.06	273.20	T.L.	11.02	284.45	4028	4,12	1062.04
	12-15T10P15V	232.67	–	–	–	310.58	9599	–	7.46E+04
3	13-05T05P07V	101.89	103.26	626	0.00	103.67	166	0,39	157.30
	14-05T10P07V	103.49	110.77	6794	0.00	114.04	1389	2,95	194.08
	15-10T05P12V	180.50	206.90	T.L.	7.14	208.21	2827	0,64	302.63
	16-10T10P12V	183.81	282.52	T.L.	31.18	241.66	4609	−14,46	4.69E+04
	17-15T05P17V	255.66	311.47	T.L.	13.38	294.76	5487	−5,36	457.73
	18-15T10P17V	254.09	–	–	–	351.46	10220	–	1.05E+05

(-) represents instances for which the solver could not solve the integrated problem due to insufficient memory.

T.L. refers that instance is terminated by the time limit.

terminal. This congestion is associated with the number of ships that must be loaded and the number of periods in the time horizon. Sets 1, 2 and 3, are low, medium, and high congestion, respectively.

The results presented in the table demonstrate the difficulty of the CPLEX solver in finding good solutions for the MILP model. In the medium-scale instances, the solver could only find an upper bound within the 6-hour test limits. In the larger scale instances, CPLEX was unable to produce any significant results due to insufficient memory.

With the RSFH, it was possible to obtain good upper bounds for all instances, and in all solutions, all offers and demands were met. In the instances that the solver obtained an optimal solution (ins. 01, 02, 07, 08, 13, and 14), the algorithm obtained upper bounds with GAP little than 2%. In particular, instance 01, in which the RSFH obtained optimal solution in one-third of the computational time of solver. When comparing the solutions between the two methods, using the columns GAP_2, t_{cplex} and t_{heu}, the solver obtain a better upper bound than the algorithm in 55% of the instances; however, the computational time of the solver is significantly different compared to the developed algorithm. Note that in 23% of the instances, the algorithm generated an upper bound with a value of f better than found by the solver. The upper bounds of column f_{ub_2} highlight the importance of the local branching heuristic in refining the greedy solutions of the relax-and-fix heuristics.

7 Conclusions

This paper proposes an integrated model for the integrated planning, scheduling, yard allocation, and berth allocation problem for bulk cargo port terminals and an RSFH that combines new and specific versions of matheuristics into a variable-fixing strategy.

The proposed formulation was validated with instances based on parameters with proximity to real cases. As expected, the CPLEX solver presented significant difficulties in solving the MILP model, solving only the smaller instances. Meanwhile, the computational results show that the RSFH performed satisfactorily. The good upper bounds obtained result from the efficient combination of the proposed heuristics, especially the local branching heuristic and the variable-fixing strategy.

Based on the scale of the integrated problem and the formulation, future work focuses on improving the mathematical formulation and applying a decomposition method to obtain better, and eventually optimal, solutions more efficiently.

Acknowledgments. This research is supported by the following institutions: Brazilian National Research Council (CNPq), Coordination of Superior Level Staff Improvement (CAPES) and Federal Center for Technological Education of Minas Gerais (CEFET-MG).

References

1. Barros, V.H., Costa, T.S., Oliveira, A.C., Lorena, L.A.: Model and heuristic for berth allocation in tidal bulk ports with stock level constraints. Comput. Ind. Eng. **60**(4), 606–613 (2011). https://doi.org/10.1016/j.cie.2010.12.018
2. Cheimanoff, N., Fontane, F., Kitri, M.N., Tchernev, N.: A reduced vns based approach for the dynamic continuous berth allocation problem in bulk terminals with tidal constraints. Expert Syst. Appl. **168**,(2021). https://doi.org/10.1016/j.eswa.2020.114215
3. de León, A.D., Lalla-Ruiz, E., Melián-Batista, B., Marcos Moreno-Vega, J.: A machine learning-based system for berth scheduling at bulk terminals. Expert Syst. Appl. **87**, 170–182 (2017). https://doi.org/10.1016/j.eswa.2017.06.010
4. de León, A.D., Lalla-Ruiz, E., Melián-Batista, B., Moreno-Vega, J.M.: A simulation-optimization framework for enhancing robustness in bulk berth scheduling. Eng. Appl. Artif. Intell. **103**,(2021). https://doi.org/10.1016/j.engappai.2021.104276
5. Ernst, A.T., Oğuz, C., Singh, G., Taherkhani, G.: Mathematical models for the berth allocation problem in dry bulk terminals. J. Sched. 1–15 (2017). https://doi.org/10.1007/s10951-017-0510-8
6. Fischetti, M., Lodi, A.: Local branching. Math. Program. **98**, 23–47 (2003). https://doi.org/10.1007/s10107-003-0395-5
7. Harvey, W.D., Ginsberg, M.L.: Limited discrepancy search. In: IJCAI (1995)
8. Hu, D., Yao, Z.: Stacker-reclaimer scheduling in a dry bulk terminal. Int. J. Comput. Integr. Manuf. **25**(11), 1047–1058 (2012). https://doi.org/10.1080/0951192X.2012.684707
9. Menezes, G.C., Mateus, G.R., Ravetti, M.G.: A hierarchical approach to solve a production planning and scheduling problem in bulk cargo terminal. Comput. Ind. Eng. **97**, 1–14 (2016). https://doi.org/10.1016/j.cie.2016.04.007
10. Menezes, G.C., Mateus, G.R., Ravetti, M.G.: A branch and price algorithm to solve the integrated production planning and scheduling in bulk ports. Eur. J. Oper. Res. **258**(3), 926–937 (2017). https://doi.org/10.1016/j.ejor.2016.08.073
11. Robenek, T., Umang, N., Bierlaire, M., Ropke, S.: A branch-and-price algorithm to solve the integrated berth allocation and yard assignment problem in bulk ports. Eur. J. Oper. Res. **235**(2), 399–411 (2014). https://doi.org/10.1016/j.ejor.2013.08.015
12. Rocha de Paula, M., Boland, N., Ernst, A.T., Mendes, A., Savelsbergh, M.: Throughput optimisation in a coal export system with multiple terminals and shared resources. Comput. Ind. Eng. **134**, 37–51 (2019). https://doi.org/10.1016/j.cie.2019.05.021
13. Sadykov, R., Vanderbeck, F., Pessoa, A., Tahiri, I., Uchoa, E.: Primal heuristics for branch and price: the assets of diving methods. INFORMS J. Comput. **31**(2), 251–267 (2019). https://doi.org/10.1287/ijoc.2018.0822
14. Tang, L., Sun, D., Liu, J.: Integrated storage space allocation and ship scheduling problem in bulk cargo terminals. IIE Trans. **48**(5), 428–439 (2016). https://doi.org/10.1080/0740817X.2015.1063791

15. Umang, N., Bierlaire, M., Vacca, I.: Exact and heuristic methods to solve the berth allocation problem in bulk ports. Transp. Res. Part E: Logistics Transp. Rev. **54** (07 2013). https://doi.org/10.1016/j.tre.2013.03.003
16. Unsal, O.: Reclaimer scheduling in dry bulk terminals. IEEE Access 1 (05 2020). https://doi.org/10.1109/ACCESS.2020.2997739
17. Unsal, O., Oğuz, C.: An exact algorithm for integrated planning of operations in dry bulk terminals. Transp. Res. Part E: Logistics Transp. Rev. **126**, 103–121 (2019). https://doi.org/10.1016/j.tre.2019.03.018

Simulation of an AIS System for the Port of Hamburg

Pierre Bouchard[1(✉)], Adriana Moros-Daza[2(✉)], and Stefan Voß[1(✉)]

[1] Institute of Information Systems, University of Hamburg, 20146 Hamburg, Germany
pierre@bouchard.de, stefan.voss@uni-hamburg.de
[2] Universidad del Norte, Barranquilla, Colombia
amoros@uninorte.edu.co

Abstract. This paper shows that the prediction of vessel arrival times with AIS (Automatic Identification System) is increasing the number of vessels a port can handle without additional superstructure. The Port of Hamburg is used as a case study to show the difference between the as-is situation and one with the integrated information system. The simulation shows improvements with two different risk levels to prove the concept. The simulation uses simplified versions of an algorithm that assigns vessels to free berths without disrupting the normal terminal usage. It was possible to clear up to 44% more ships each day just with an additional system that utilises already existing data for achieving more efficiency within the port.

Keywords: Smart port · AIS · Berth allocation

1 Introduction

Logistics is necessary for almost every product, whether for manufacturing or sales. Parts for the final product need to be transported to the last production facility, and finished goods need to be transported to the consumer. Either way, ports are important gateways in this respect. The Port of Hamburg as an example is responsible for 9.3 M. TEU (Twenty-foot Equivalent Unit) in 2019 [20], thus, making it essential for many companies in and around its metropolitan area. Above-average growth of the Chinese, Indian and East European markets has caused a shift and most significant growth of cargo flows [8], which forces the Port of Hamburg to be more efficient to remain globally significant.

The spatial constraints of the port [8] do not allow for significant expansions along the river Elbe. Therefore, the Port of Hamburg is highly reliant on innovations and information systems. This holds for the past (see, e.g., [23]) as well as for today (see, e.g., [10]). In the paper "IT-Governance in the Port of the Future" [2] we already showcased the potential use and advantages of integrating AIS data for better predictions of arrival times of the incoming vessels. Having an AIS onboard is a regulation by the IMO (International Maritime Organisation)

M. Mes et al. (Eds.): ICCL 2021, LNCS 13004, pp. 21–35, 2021.
https://doi.org/10.1007/978-3-030-87672-2_2

for every container vessel [16]. Therefore, every vessel on its way to the Port of Hamburg will be recognised by this system. Additionally, the port has another advantage, the river Elbe: Due to strict regulations regarding speed and limitations by draught at specific tide levels, the earliest arrival time can precisely be calculated [2]. This information can be used mainly in two different ways. On the one hand, more vessels get cleared with the same infra- and superstructure without expanding the port spatially. On the other hand, customer satisfaction and hinterland transport can be improved.

In this paper we show the feasibility of this concept in a simplified way. Thus, no advanced algorithm is used for the berth allocation.[1] Therefore, the simulation uses a minimalistic allocation principle, which showcases an implemented advanced system's possibilities. Every advanced algorithm will allocate the incoming vessels more efficiently and further improve the Port of Hamburg's efficiency. Additionally, this paper shows how many vessels could be added to the overall schedule with the simplified principle. Common risks are implemented into the simulation to illustrate the productivity of the concept with more realistic conditions.

2 Problem Description

Hamburg's port is the gateway to the Baltic Sea Region; containers from the economically strongly growing East European countries are combined at the Port of Hamburg for cheaper transportation worldwide. The other way around, goods, especially from Asia, get prepared at the port for further transportation into the Baltic Sea Region. This rising number of containers forces the Port of Hamburg to grow and handle more containers to compete with other ports. Due to the spatial constraints, it needs to be more efficient, especially with the limited moorings [8].

On average, every vessel waits one hour and 45 min at the Port of Hamburg before it gets cleared [12]. Besides the obvious issue of the cost factor for the shipping companies, the Port of Hamburg has not yet enough possibilities for shore power for the vessels. Thus, the vessels have to run on their fuel to generate power [7]. In 2017, 39% of all nitric oxides in Hamburg were caused by vessel traffic [19]. This number is mostly caused by the container ships on their way entering or leaving the port. However, by an average retention time of 13 h and 49 minutes, 12.7% of the vessels' nitric oxide emissions of the container vessels are produced by waiting [12].

Therefore, this paper takes a closer look at the terminal reservations and berth allocation. Normally, shipping companies reserve berths for their vessels in advance, especially seagoing vessels that place their reservation up to two months before they arrive. Logically, those reservations are not precise, and vessels often have a delay or arrive too early (as shown in Fig. 1). 44% of the arriving vessels have a schedule deviation of 30 min or more. Those vessels are the interesting ones for the system because of a noticeable impact on the port

[1] For references stating the state-of-the-art over time see, e.g., [1,15,17].

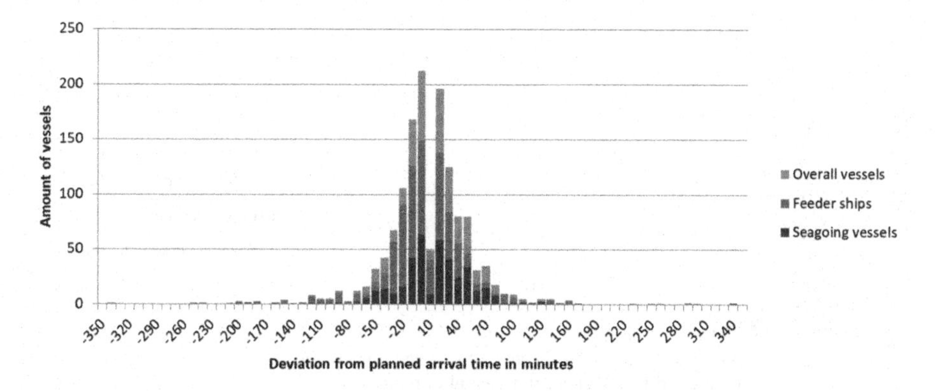

Fig. 1. Frequency distribution of the deviation from the to-be to the as-is arrival time [12]

as well as other vessels. With this, 24% are delayed, whereas almost 20% are too early. It could happen that the vessels were delayed for up to five-and-a-half hours [12].

Due to the fast-changing nature of the trade market, berth allocation needs a fast and reliable solution. It would also be preferable to use existing infrastructure and processed data without restructuring the berth reservation process completely. Thus the expense is very low, and the regulations are not affected as heavily.

3 The Concept

The idea is based on the precise forecast of the earliest arrival time of every incoming vessel. This knowledge is utilised to use reserved berths of vessels that delay clearing of one or more other vessels. Additionally, the system can provide a precise arrival time more than eight hours in advance. Therefore, time slots that have been too short for clearing additional vessels between two existing reservations can now be removed. This makes it possible to communicate the possibility for schedule changes to the waiting or incoming ships.

42.7% of all arriving vessels can be cleared in under five hours (as shown in Fig. 2). As an example, one of those vessels that have already arrived at the port or arrive soon could easily fit into a time slot of a vessel that should arrive in four hours and is two hours delayed. Before that, the vessel could not be cleared earlier because it would disrupt the reservation of the incoming vessel by one hour [12]. Additionally, the time slot of this vessel now got accessible by bringing the reservation to the front. This previously reserved time slot can be used for other vessels or make up for possible changes in the terminal schedule due to the delay of the original vessel. Most of the time, the knowledge of the earliest arrival time eight hours in advance is sufficient to communicate the possibility for change to the responsible persons before these container vessels have even

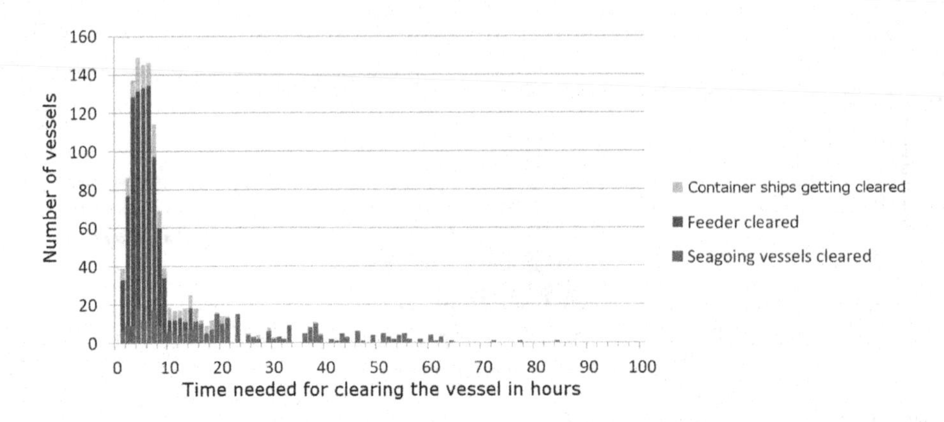

Fig. 2. Distribution of time needed to clear the vessels [12]

arrived at the port. Therefore, the container ships with a changed schedule can immediately head to the terminal without the need to stop at the waiting berth, which leads to a better structured and thus safer port.

The information regarding the arrival times can be used mainly in two different ways: On the one hand, as already described, there are advantages on the seaside of the port. Vessels already at the Port of Hamburg because they have arrived too early or vessels like feeder ships that have a fast way to travel before arriving and therefore are flexible in their schedule benefit from the system. On the other hand, the hinterland transportation and the terminal benefit, too. These correlations are not simulated but important for understanding the benefits of the core concept. In the Port of Hamburg, every truck driver has to register the container collection four hours in advance to collect a container. Thus, the earliest time to collect a container is four hours after it has arrived at the port [14]. With the knowledge of the arrival times of the vessels, the truck driver could get the opportunity to collect the container right after it is cleared and released from customs. Additionally, enhanced planning capabilities for the haulage companies and truck drivers due to the knowledge of the arrival time will reduce traffic in the port area and enhance customer satisfaction due to better arrival predictions in the hinterland transportation. The improvement of the container collection process, additionally, results in a major improvement for the port itself. If every truck driver collected the container immediately after arrival, the port would need 8% less storage capacity. This would also affect the overall efficiency of the terminal and the gantry cranes [2].

4 Discrete Event Simulation Model

This section introduces the model used to simulate using AIS data for rearranging booked time slots without disrupting the planned schedules of container vessels. This simulation uses real-life data from the HHLA [12]; therefore, just

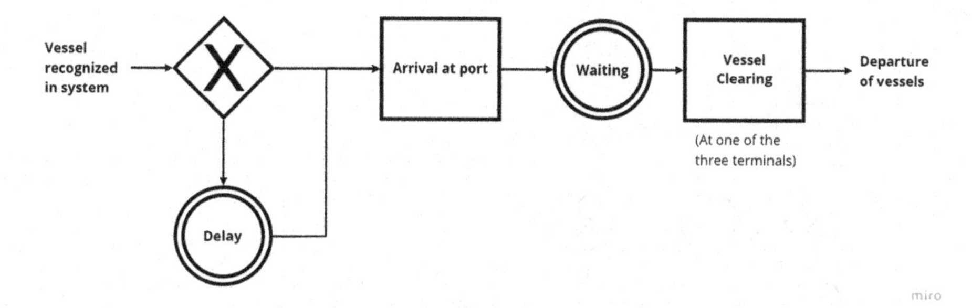

Fig. 3. Abstracted model of the arriving vessels without an AIS-system integrated

three of four terminals owned by the HHLA will be modelled. Additionally, the algorithms responsible for the rescheduling will be simplified. Thus, the outcomes of the simulation will possibly be worse than in reality. However, the results can showcase the system's possibilities for a real-life scenario due to the real-life input data. Thereby, the simplified algorithm will make up for not simulated processes like the time needed by the captain of a container vessel to decide whether to use the schedule change or not.

The model shown in Fig. 3 describes the voyage of a container vessel without using any advantages of the AIS. This type of vessel is called from now on type one. These vessels arrive on-time or delayed. As seen in Fig. 1, many of those ships are seagoing vessels. The type one vessels also often need more time to get cleared due to their size, and thus a higher amount of containers need to be cleared.

The type one vessels have no potential for getting cleared earlier due to their arrival time. Thus the process of arriving at the port displays the current situation at the Port of Hamburg. First, the AIS signature of the vessel gets recognised up to 24 h in advance by the multi-purpose platform HVCC (Hamburg Vessel Coordination Center) at the Port of Hamburg [25]. The incoming vessel may have a delay, which does not affect any other component of the port. After arrival, the ship has to wait for varying periods of time before it can be cleared and depart afterwards.

The new vessels added in Fig. 4 are subsequently called type two and are the interesting ones for this concept. This type of vessels will or have arrived too early at the port or have no reservation because of schedule changes and fully reserved terminal berths. Thus, they will or have arrived at the port without the possibility for immediate clearing due to reserved berths. At first sight, this could be feeder ships and seagoing vessels, as seen in Fig. 1. However, most of them will be feeder ships because these vessels need less time for clearing. As seen in Fig. 2 and as already described in the example in the previous paragraph, the vessels with a higher chance of fitting in a new schedule gap are those with short clearing times.

Every time a type one vessel gets recognised by the AIS, the new system is looking for incoming or waiting type two vessels. Thereby, the list of waiting and

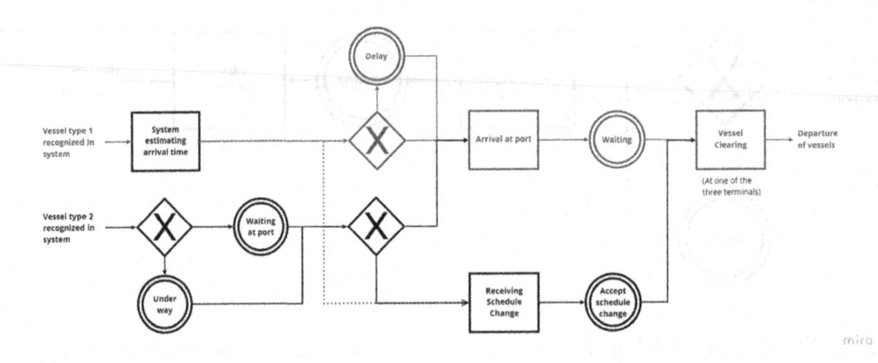

Fig. 4. Abstracted model of the arriving vessels with an AIS-system integrated

arriving vessels get checked whether any of the type two vessels can come and be cleared before the type one vessel is arriving. Besides, the algorithm previously checks whether all the terminals are in use and, if so, declines the inquiry. Thus, the simple but effective algorithm checks if the terminal can clear one vessel of type two before the type one vessel that has reserved the berth arrives at the port. If a waiting or arriving vessel with a reservation is waiting for too long, they will leave the system. These vessels would then be treated as a type one vessel in reality but are neglected in this simulation to keep the complexity low.

4.1 Validation

To prove that such a model will work accurately in a real-life scenario, such as implemented at the Port of Hamburg, historical data need to be used for verification. For the validation, the confidence interval method is used for comparing the historical data from the case study, shown in the columns "Intervals" and the modelled data shown in the "As is" columns (cf. Table 1).

Table 1. Input data validation

	As is		Intervals		
	Average	Half width	Minimum	Average case study	Maximum
Vessels Cleared (per day)	17.94	2.36	13.66	16.02	18.38
Time needed to clear vessels in h	9.44	1.40	8.71	10.11	11.51
Delay in hours	0.04	< 0.00	-3	-0.05	2.5

For the historical data, a sailing list from the HHLA was used. To validate the first model, shown in Fig. 3, three different pieces of information needed to be confirmed. As shown in Table 1, the first data used as input for the simulation is the number of vessels that get cleared daily. For an improved efficiency of the port, this number has risen after using the AIS system. The second and third piece of information (data) are the time each vessel will stay at a terminal to get cleared and the deviation from the planned arrival time as seen in Fig. 1.

For verification, the simulated incoming data needs to be represented within a 95% confidence interval of the real-life data from the HHLA. The number of cleared vessels and the time needed to clear the ships at the terminal are verified, as seen in Table 1. Thus these properties of the Port of Hamburg can be adequately depicted. However, the delay in hours has a slight variation. Due to the limited display of the half-width, there is no possibility to verify the 95% like above. However, the arrival-time deviation with the exact numbers of the real-life data and the simulated data is 32.6 s. Due to the widespread arrival time deviations of up to 5 h, this paper treats the delay as verified.

4.2 Analysis of Scenarios

To demonstrate the impact of the AIS system at the Port of Hamburg, it is necessary to highlight the difference between the real implemented system and the simulated system. Due to the simplified modelling of the AIS system, an implemented system will be more efficient and achieve better results.

Modeled System. In the modeled system, the type one vessels get recognised from the AIS systems and are checked with respect to two parameters: First, whether a type two vessel is arriving or already waiting at the port and second if one of the type two vessels can be cleared within the time the type one vessel is coming. After that, the vessel type one continues its journey until it arrives at its free and reserved berth. If a vessel type two was detected that could arrive at the berth, get cleared and leave the berth before the vessel type one arrives, a signal is sent to this vessel to get cleared at the terminal.

To demonstrate the effectiveness of the simulated AIS system, several simulation runs were started. As shown in Fig. 5, every round, an additional vessel per day was fed into the simulation. Then it was checked if the system can handle an additional vessel. Thereby, the maximum number of vessels the simulation could handle was 25 additional type two vessels. The simulation stopped when 150 vessels were simulated simultaneously. Hereby, a simulated peak was recorded at the simulation with 26 additional vessels. The simulation covers a period of ten years with 1000 runs per additional type two vessel. While the arrival of the type one vessels was simulated with the help of the "Segelliste" from the HHLA, the arrivals of the type two vessels were randomly distributed and were the reason for the simulation to stop. Additionally, the simulation is event-based, thus waiting for type two vessels is not able to exist by themselves, for example, when it is time for their reservation, but rather had to wait for a type one vessel to trigger the selection procedure before they can leave the queue.

Nevertheless, this simulation shows the capability of the simulated AIS system. The system handled up to 139% more vessels and thus up to 43.2 instead of 18.2 vessels a day. The time a type one vessel has to wait has increased by

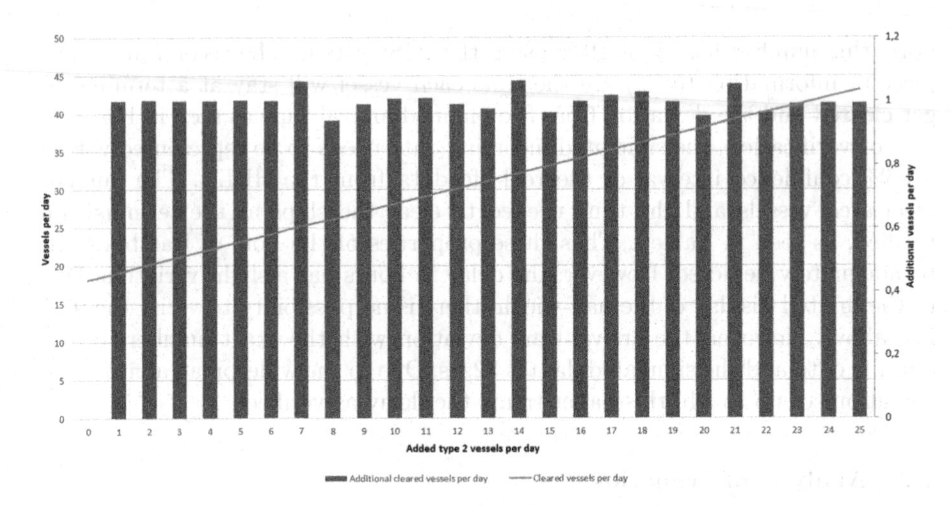

Fig. 5. Effectiveness of the simulated AIS system

15.5% or 10 min from zero to 25 additional type two vessels. Due to the random allocation of the terminals for the vessels in the simulation, it is possible that too many vessels had a reserved berth, for example, at terminal one, while terminal two and three were almost empty. Resulting is a simulated waiting time for each type one vessel of one hour and five minutes. The simulated waiting time for the type one vessels with an additional 25 type two vessels increased to one hour and sixteen minutes.

Real-life berth allocation systems like those in the Port of Hamburg are far more advanced and would only let the vessels reserve a berth at a free terminal (or even a free dedicated terminal due to the planning situation, respectively). But the change of this parameter is an indicator of the quality of an AIS system. With a deviation of ten minutes, the additional type two vessels have a negligible impact on the average residence time of 23.36 h for the type one vessels in the simulation.

Real-Life System. In the implemented and finished system, two algorithms are used for the type two vessel's berth allocation. The first one calculates the arrival time of every incoming container ship. Therefore, the AIS data, data from the harbour and terminals, weather and tidal data will be used to calculate the estimated arrival times while considering the spatial constraints of the river Elbe. The algorithms from Fancello et al. [6] and El Mekkaoui et al. [5] can be used as a basis for the software development to achieve a high-quality arrival estimation. These estimations were given in the simulations by the historical data to work correctly. The second algorithm is the more interesting one. This algorithm is used to allocate free berths if possible. Therefore, the arriving estimations from the first algorithm and data from the terminal operating system, for the time needed to clear the type two vessels, are required. This information needs to be precise, complete, correct and updated in a high frequency to allow smooth handling of the type one vessels. With this information, it is checked by the system whether it is possible to clear a type two-vessel with a sufficient time buffer. The new schedule can afterwards be communicated to the (captain of the) type two-vessel [2].

Comparison of the Systems. The significant difference between the simulation and the real-life system is the algorithm for the berth allocation. The real-life algorithm will check the incoming vessels for the best usage of the free berth or if the vessels are from haulage companies with a "premium agreement". Thus, not the first vessel that would fit into the slot will be allocated, but rather the vessel or vessel combination that would fit the best for the time slot is allocated to the berth. Thus, as mentioned before, an implemented system with advanced algorithms will be more capable than the simulation.

4.3 Risk Management

The port of Hamburg functions as a hub for container transport into and out of the Baltic Sea region. Additionally, many companies in Hamburg and the hinterland depend upon the port. Many goods need to pass the terminals of the port daily to arrive at various enterprises and factories, thus making the Port of Hamburg a bottleneck [8]. Simultaneously, different critical industries, like nutrition, health are interdependent and dependent on the Port of Hamburg. For example in 2020 when the world was hit by the global pandemic COVID-19 [21], the German government alone imported masks worth six billion from China [24]. Thus, making harbours and airports directly responsible for the health of the German population. The port counts to the critical infrastructure of Hamburg and thus needs special protection, risk and crisis management [4].

Every stock company has an obligation by the German law §91 Abs.2 AktG: "to take appropriate measures, particularly establishing a monitoring system, to ensure the timely identification of developments that might place the continued existence of the Company at risk" [3]. Thus, the companies that count as critical infrastructure are forced by law to implement functioning risk management. To support these companies, the German Federal Ministry of the Interior provided a risk and crisis management concept, consisting of five phases: Preliminary Planning, Risk Analysis, Preventive Measures, Crisis Management and Evaluation. For this work, the parts of the first phase and the second phase are primarily interesting.

Preliminary Planning. The first phase primarily discusses recommendations for personnel change and change in responsibilities, but also includes strategic protection objectives. Thereby, holistic and process independent protection goals are formulated. These aims are subordinated and help for the next phases.

The system described in this paper has an advantage, as it is an additional system just used for optimisation purposes. It aims to provide suggestions for a possible schedule change for vessels that arrive too early or without a reservation. Thus, in an event of a complete failure of this system, the port itself and the terminals could be operated just by using the normal scheduling system of the port. Nevertheless, the strategic protection objective is the "preservation of the terminals' capability to clear vessels".

Risk Analysis. To comply with the strategic protection aims, risks are identified, structured, objectified and stored in the second phase. Thereby, the risks can be grouped in two different sets, but due to the separation of this system and the main harbour system, only risks associated with the system introduced in this paper are going to be considered. Thus, risks that are only related to the port system and therefore managed by the port authority and the terminal operators are not handled in this paper. However, if those risks occurring from the main port's systems can, for example, change schedules of vessels and thus change the parameters for the AIS system, they will be considered in the risk analysis.

The simplest way to preserve the terminal's capability to clear vessels is to not disrupt the type one vessel's schedule. Thus, even when the additional system for the type two vessel's berth allocation would break down, the main system must not depend on the additional system to run smoothly. Nevertheless, wrong information or cyber-attacks could be able to damage the main port system's scheduling.

There are three main categories and points of attack capable of disrupting the schedule of the type one vessels and therefore the terminals' capability to clear vessels:

- Wrong arrival time estimation (ETA, estimated time of arrival)
- Wrong estimation of the time a vessel needs for mooring, clearing, etc.
- False calculation of the possible schedule changes

If the estimations either of the type one or the type two vessels are wrong the system is going to rearrange the schedules in a wrong way and thus may imply to block needed terminal space. Additionally, the system itself or other mistakes happening at the port have the possibility to stop the terminals from working properly. These risks can have several different reasons shown in Table 2. For this more precise risk analysis, the archive of the HHLA was analysed to achieve a more precise overview of the risks for the port in Hamburg [11].

Due to the automation of the system, human behaviour while using the system for rearranging the schedules will not be considered.

To simulate the risks, there are three different possibilities: Failure at the port, failure within the AIS system and wrong estimations of arrival time and time needed for arriving, clearing and departing of the type one and two container ships.

Table 2. Risk list

Kind of risk	Potential intensity	Possible impact
Weather		
Changed wind direction and speed	Average wind speed of 5.9 m/s per day [9]	Change in arrival time of vessels
Storms	Delayed vessel arrivals [13]	Delay of vessel clearing and not enough storage space for containers
Poor visibility	Thick fog or heavy snowfall can limit the visual range [18]	Delay of vessel clearing and delay of arrival times from vessels
Cold spell	Energy Blackout [22]	Vessels cannot be cleared for up to 2 days and information system failure
Wrong data delivered		
Wrong or missing reservation data	Type 1 vessel arrives at the port without the system knowing beforehand	The terminal of type 1 vessel may be occupied, despite the reservation of the type 1 vessel
Wrong AIS data	Vessels arrive too early or too late	The terminal of type 1 vessel may be occupied, despite the reservation of the type 1 vessel
Shorter duration for clearing the vessel by the harbour system estimated	Clearing the vessels takes longer than planned	- When type 1 vessel takes longer: Type 2 vessel may be sent to the wrong terminal - When type 2 vessel takes longer: Reserved berth may be occupied
Mistakes by rearranging the vessels		
Mistakes by communicating information to the type 2 vessels, pilots and terminal	Delay before or while clearing the vessel	Type 2 vessel cannot be cleared in the remaining time without disrupting the schedule of the type 1 vessel
Human behaviour		
Mistakes while driving and docking vessel	Vessel or terminal damaged	Vessel or terminal needs to be repaired, loss of a terminal, schedules for this terminal need to be replanned
Mistakes when clearing the vessels	- Wrong containers cleared - Vessel or terminal damaged	- Reloading the container, clearing vessels takes longer - Vessel or terminal needs to be repaired, loss of a terminal, schedule for this terminal need to be replanned
Cyber attacks		
System hacked and information changed	Changed information of ETA, or communication of wrong schedule to type two vessels	Too many vessels at the port, that have to wait
System break down	Vessels of type two cannot enter without reservation	Vessels that arrive too early have to wait for their reservation

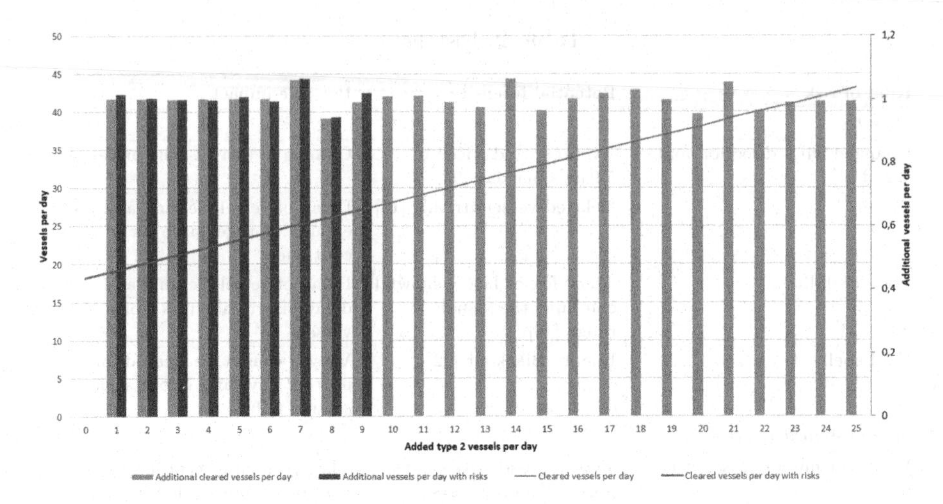

Fig. 6. Effectiveness of the simulated AIS system simulated with risks

As seen in Fig. 6, the simulation was not able to handle more than nine additional vessels. Hereby the extremes were decisive. As already mentioned the simulation had a maximum number of 150 entities in parallel in the simulation. With risks that stop clearing the vessels for two days or hacking attacks that disguise several type two vessel arrivals and send them as a bunch to the terminals, the number of parallel vessels rises. Table 3 supports this hypothesis. It shows the difference in the number of container ships waiting in a simulation at the terminal with and without risks. The average in both cases is pretty similar and small and shows the system's capability to work without disrupting the terminals' workflow. The maximum shown is the highest number of vessels over a period of ten years and 1000 simulation runs that had to wait in front of the terminal without getting cleared. Here, a significant difference can be seen between the simulations with and without simulated risks.

However, the simulated risks on average do not have a significant impact. The main reason for the high maximum values is the risks that were not counteracted by any measures. Additionally, the random terminal allocation for the vessels affected the capabilities of the simulation. Thus, a higher number of vessels can be achieved with good risk management and algorithms that allocate the type one and two vessels in a more efficient way. Additionally, to operate this additional system separate from the main port's system lowers the negative effects and risks for the type one vessels and the strategic protection objective: "Preservation of the terminals' capability to clear vessels".

Table 3. Vessels waiting at the terminals before clearing

Additional type 2 vessels per day	Number waiting at terminals before cleared		Number waiting at terminals before cleared with risks	
	Average	Maximum	Average	Maximum
1	0.00	2.67	0.04	46.67
2	0.00	3.33	0.05	51.67
3	0.00	3.66	0.05	54
4	0.00	4.3	0.05	53.67
5	0.02	4	0.05	59.67
6	0.00	5	0.06	62.67
7	0.01	5	0.06	62.33
8	0.01	6	0.06	61.67
9	0.01	5.7	0.06	61.67

5 Conclusion

This paper followed up with the subject of using AIS data to enable a more efficient berth allocation process without disrupting existing procedures. While using this data, vessels that arrive too early or without reservation at the port could be cleared without disrupting the vessels' berths if a reservation exists.

To prove the concept, a simple discrete event simulation model was built to simulate the port under normal conditions and with ascending numbers of additional vessels per day that had no reservation. Hereby real-life data from the Port of Hamburg's largest terminal operator was used for the container ships that arrive with reservations. For a better understanding of the problem, risks were introduced to conceive a more realistic model. This led to a unique perspective of the system. While it worked well for up to nine additional vessels, extreme situations led to failure at ten additional vessels per day. But these extremes are due to the limitation of the simulation not using advanced arrival estimation algorithms nor berth allocation systems that are already in use within the port. Thus these extremes could easily be reduced by a well-composed system added to the already existing berth allocation system of the terminal/port. Therefore, an even higher capacity utilisation rate for the vessels and the port capacity can be achieved with advanced algorithms for this add-on system.

This study provides more insight on the topic, proves the concept and provides more quantitative information for the possibilities of this topic. Additionally, this paper constitutes a good basis for further research and thus can be used to study the possibilities at different ports around the world under different conditions.

The limitations of this study are the simulation itself, while it creates a good foundation for further research and proves the concept, a study under real-world conditions with algorithms for the estimation of the earliest arrival time and the assignment of berths. The use of simple algorithms also was the reason that only one and only the first fitting additional vessel, per vessel that had reserved a berth, was able to be chosen from the system. Thus the results with advanced algorithms can be superior.

Future research will achieve more information on this topic with scenarios for different harbours around the globe. Additionally, it is proposed to work on more versatile case studies with real ports to indicate the real-life possibilities of using AIS data for advance berth allocation.

References

1. Bierwirth, C., Meisel, F.: A follow-up survey of berth allocation and quay crane scheduling problems in container terminals. Eur. J. Oper. Res. **244**(3), 675–689 (2015)
2. Bouchard, P., Moros-Daza, A., Voß, S.: IT-governance in the port of the future. IWI, Institute of Information Systems, University of Hamburg, Technical Report (2021)
3. Bundesministerium der Justiz und für Verbraucherschutz: Aktiengesetz §91 Organisation. Buchführung Abs. 2. (no date). https://www.gesetze-im-internet.de/aktg/__91.html. Accessed 30 Apr 2021
4. Bundesministerium des Innern: Schutz kritischer Infrastrukturen - Risiko- und Krisenmanagement. Leitfaden für Unternehmen und Behörden 2 (2011). https://www.bmi.bund.de/SharedDocs/downloads/DE/publikationen/themen/bevoelkerungsschutz/kritis-leitfaden.html. Accessed 22 Jul 2021
5. El Mekkaoui, S., Benabbou, L., Berrado, A.: Predicting ships estimated time of arrival based on AIS data. In: Proceedings of the 13th International Conference on Intelligent Systems: Theories and Applications (SITA 2020), pp. Article 6, 1–6. Association for Computing Machinery, New York (2020). https://doi.org/10.1145/3419604.3419768
6. Fancello, G., Pani, C., Pisano, M., Serra, P., Zuddas, P., Fadda, P.: Prediction of arrival times and human resources allocation for container terminal. Marit. Econ. Logistics **13**, 142–173 (2011). https://doi.org/10.1057/mel.2011.3
7. Fischer, M.: Keine PowerPacs zur Stromversorgung im Hamburger Hafen (2019). https://www.welt.de/regionales/hamburg/article194713295/Keine-PowerPacs-zur-Stromversorgung-im-Hamburger-Hafen. Accessed 22 Jul 2021
8. Freie und Hansestadt Hamburg - Behörde für Wirtschaft, Verkehr und Innovation: Der Hafenentwicklungsplan bis 2025 https://www.buergerschaft-hh.de/ParlDok/dokument/38168/hafenentwicklungsplan-%E2%80%9Ehamburg-h%C3%A4lt-kurs-%E2%80%93-der-hafenentwicklungsplan-bis-2025%E2%80%9C.pdf. Accessed 22 Jul 2021
9. Hamburg.de: Messdaten: Windgeschwindigkeit, https://luft.hamburg.de/clp/windgeschwindigkeit/clp1/. Accessed 3 May 2021
10. Heilig, L., Voß, S.: Information systems in seaports: a categorization and overview. Inf. Technol. Manage. **18**(3), 179–201 (2016). https://doi.org/10.1007/s10799-016-0269-1
11. HHLA: News archiv. https://hhla.de/medien/pressemitteilungen/archiv-pressemitteilungen. Accessed 06 May 2021

12. HHLA: Schiffsabfertigung: Segelliste. https://coast.hhla.de/report?id=Standard-Report-Segelliste. Accessed 01 Feb 2021
13. HHLA: Unwetterfolgen beeinträchtigen Betrieb auf Terminalanlagen im Hamburger Hafen. https://hhla.de/medien/pressemitteilungen/detailansicht/unwetterfolgen-beeintraechtigen-betrieb-auf-terminalanlagen-im-hamburger-hafen. Accessed 3 May 2021
14. HHLA: LKW-Transportvormeldung Benutzerhandbuch. Version 1.0.5 (2018). https://hhla.de/fileadmin/download/HHLA-Handbuch-LKW-Vormeldung-Slotbuchung-2018-04-30.pdf. Accessed 22 Jul 2021
15. Hoffarth, L., Voß, S.: Liegeplatzdisposition auf einem Container Terminal — Ansätze zur Entwicklung eines entscheidungsunterstützenden Systems. In: Operations Research Proceedings 1993. ORP, vol. 1993, pp. 89–95. Springer, Heidelberg (1994). https://doi.org/10.1007/978-3-642-78910-6_28
16. IMO: IMO identification number schemes. http://www.imo.org/en/OurWork/MSAS/Pages/IMO-identification-number-scheme.aspx. Accessed 08 Mar 2021
17. Kramer, A., Lalla-Ruiz, E., Iori, M., Voß, S.: Novel formulations and modeling enhancements for the dynamic berth allocation problem. Eur. J. Oper. Res. **278**, 170–185 (2019). https://doi.org/10.1016/j.ejor.2019.03.036
18. Kraus, M.: Wetter Hamburg versinkt in der Nebel-Suppe (2015). https://www.mopo.de/hamburg/wetter-hamburg-versinkt-in-der-nebel-suppe-23087862. Accessed 03 May 2021
19. NABU: Da liegt was in der Luft - Messungen des NABU zeigen hohe Luftschadstoffbelastung durch Schiffe. https://www.nabu.de/umwelt-und-ressourcen/verkehr/schifffahrt/messungen/16819.html. Accessed 16 Mar 2021
20. Port of Hamburg: Port of Hamburg Statistics. https://www.hafen-hamburg.de/de/statistiken/containerumschlag. Accessed 03 Mar 2021
21. Robert Koch Institute: Journal of health monitoring s11/2020 (2020). https://www.rki.de/DE/Content/Gesundheitsmonitoring/JoHM/2020/JoHM_Inhalt_20_S11.html. Accessed 29 Apr 2021
22. Sorge, N.: Warum der Stromausfall bisher ausfällt (2012). https://www.manager-magazin.de/unternehmen/energie/a-813570.html. Accessed 03 May 2021
23. Voß, S., Böse, J.: Innovationsentscheidungen bei logistischen Dienstleistern - Praktische Erfahrungen in der Seeverkehrswirtschaft. In: Dangelmaier, W., Felser, W. (eds.) Das reagible Unternehmen, pp. 253–282. HNI, Paderborn (2000)
24. Wallenfels, M.: Bundesbürger tragen meist Schutzmasken Made in China (2021). https://www.aerztezeitung.de/Wirtschaft/Bundesbuerger-tragen-meist-Schutzmasken-made-in-China-418066.html. Accessed 30 Apr 2021
25. Wärtsilä: The future of shipping. https://www.wartsila.com/marine/white-paper/the-future-of-shipping. Accessed 22 Jul 2021

Designing the Hydrogen Supply Chain for Maritime transportation in Norway

Šárka Štádlerová(✉) and Peter Schütz

Department of Industrial Economics and Technology Management, Norwegian University of Science and Technology, 7491 Trondheim, Norway
{sarka.stadlerova,peter.schutz}@ntnu.no,
https://www.ntnu.edu/

Abstract. We study the problem of locating hydrogen facilities for the maritime transportation sector in Norway. We present a multi-period model with capacity expansion to obtain optimal investment and expansion decisions and to choose optimal production quantities and distribution solutions. The objective is to minimize the sum of investment, expansion, production, and distribution costs while satisfying the demand in each period. Hydrogen production costs are subject to economies of scale which causes non-linearity in the objective function. We model long-term investment and expansion costs separately from short-term production costs. The short-term production costs depend on the installed capacity and production quantities. We analyze two models that differ in investment decision flexibility and two demand scenarios: demand only from the maritime sector and demand from the whole transportation sector in Norway. The results show that the scenario with higher demand does not lead to a higher number of built facilities due to the economies of scale. The model with higher flexibility leads to higher capacity utilization in the first periods and thus significantly lower production costs. The results further indicate that the initial demand is too low to build a steam methane reforming facility, instead only electrolysis facilities are built in both scenarios and both models.

Keywords: Facility location · Capacity expansion · Hydrogen supply chain

1 Introduction

Emission reduction in the transportation sector is a crucial step in order to meet the emission targets set in the Paris agreement on climate change. In 2015, the Norwegian parliament decided that CO_2 emissions must be decreased by at least

This work was performed within MoZEES, a Norwegian Center for Environment-friendly Energy Research (FME), co-sponsored by the Research Council of Norway (project number 257653) and 40 partners from research, industry and the public sector.

M. Mes et al. (Eds.): ICCL 2021, LNCS 13004, pp. 36–50, 2021.
https://doi.org/10.1007/978-3-030-87672-2_3

40% (compared to 1990) towards 2030 in an attempt to reach the targets of the Paris agreement. As a consequence of this ambitious decision, fossil fuels have to be replaced by alternative zero-emission fuels. The use of hydrogen fuel cells is considered as one way to decarbonize the transport sector and to decrease the emission of greenhouse gases (GHG) [12].

In 2017, the transport sector in Norway was responsible for emitting 15.8 mill. tons CO_2, accounting for 23% of all CO_2 emissions [1]. CO_2 emissions from domestic inland water and coastal transport in Norway accounted for 8.7% of emissions from the transport sector in 2018. Introducing zero-emission fuels such as hydrogen in maritime transportation can therefore considerably reduce emissions of CO_2. However, limited experience with hydrogen as fuel and uncertainty about hydrogen availability may affect the smoothness of the transition to hydrogen fuels [22]. One way to create an initial demand for hydrogen is to require that high-speed passenger ferries and car ferries have to use hydrogen as fuel when public transport contracts are renewed. In general, demand for hydrogen is expected to increase in the years to come and the production infrastructure has to adjust to this growth [11]. As such, the infrastructure needed to cover demand from the maritime sector can help ensuring a stable hydrogen supply also for other transportation sectors in Norway [12].

The two most relevant hydrogen production technologies for Norway are electrolysis (EL) and steam methane reforming with carbon capture (SMR+) [14]. While electrolysis is a more profitable technology in small-scale production (50–5,000 Nm^3/h), SMR+ is more favourable when producing large quantities of hydrogen (50,000–100,000 Nm^3/h). Scaling up the production results in lower average costs, leading to economies of scale. This property is significant for SMR+, but it also applies to electrolysis [20]. Figure 1 shows the economies of scale in the long-term hydrogen cost function. Note that the cost-axis uses a logarithmic scale.

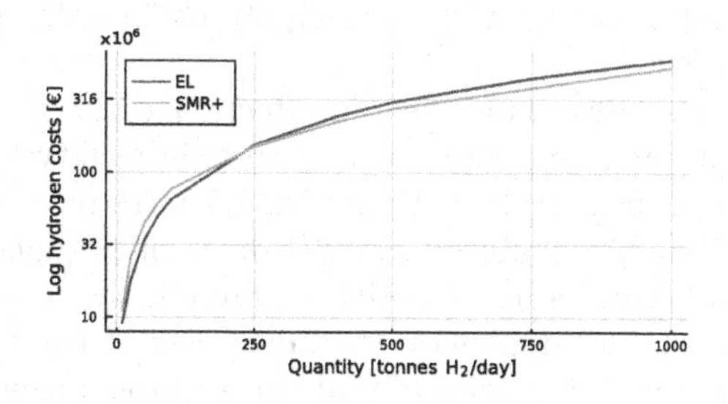

Fig. 1. Long-term hydrogen costs

In this paper, we study the problem of how to design the hydrogen supply chain for maritime transportation in Norway. The problem consists of investment and expansion decisions, production quantities, and distribution solutions. It belongs to the category of facility location problems with capacity expansion. An early review of pioneering papers dealing with capacity expansion can be found in [21]. Shulman [31] and Dias et al. [10] study a multi-period plant location problem with discrete expansion where a plant is modelled as a set of facilities in the same location. Capacity expansion is achieved by building an additional facility and the facility size must be chosen from a finite set of capacities. The production costs are defined for each facility and depend only on facility type and quantity produced in the facility. Behmardi and Lee [5] study a multi-period multi-commodity capacitated facility location problem with capacity expansion and relocation. The modelling approach differs from previous papers as Behmardi and Lee [5] work with dummy locations to relocate capacity. The dummy locations are used for modelling purposes to shift the capacity. Customers can only be served from real facilities. Torres et al. [32] present a comparison of multi-period facility location problems with growing demand where opening and closing decisions are allowed at any time during the planning horizon. Jena et al. [17] introduce a multi-period facility location model with a capacity expansion, reduction, and the option to temporarily close the facility. In their work, capacity expansion is modelled by the modification of existing facilities. Jena et al. [18] present a facility location problem with modular capacities where capacity expansion, as well as partial closing and reopening, are allowed. An extension of their model is published in [19] where also facility relocation is allowed. Castro et al. [6] present a large-scale capacitated multi-period facility location model where a set of capacitated facilities is progressively built during the planning horizon and simultaneously a maximum amount of operating facilities in each period is specified.

Facility location and supply chain design problems with a focus on hydrogen infrastructure are discussed in [3], [24] and [13]. In the work by Almansoori and Shah [3], a multi-period hydrogen supply chain for Great Britain is studied. However, in their work, expansion is not allowed. Myklebust et al. [24] present a case study from Germany and study the impact of demand and input costs on the optimal technology choice. Han et al. [13] present a different approach where an optimization model for the hydrogen supply chain with given production capacities is considered.

Economies of scale cause non-linear production costs. Several approaches for how to incorporate non-linear production costs in facility location problems have been published in the literature. Holmberg [15] introduces a piecewise linear staircase cost function that enables to model different production costs at different capacity levels. Correia and Captivo [7] present the modular capacitated facility location model and emphasize the advantage of the modular formulation as it enables to take economies of scale into consideration. They separate investment and operational costs and provide different unit operational costs for each facility size. Van den Broek et al. [34] study facility location problem with non-linear, non-convex, and non-concave objective function. They follow the idea of non-linear costs depending on installed capacity as presented in [7] however, they introduce a linear staircase cost approximation. The approach presented in [34] can capture economies as well as diseconomies of scale.

For more examples of facility location and supply chain design see the excellent reviews by Melo et al. [23], and Arabani et al. [4]. Review on multi-period facility location problems can be found in [26].

In this paper, we investigate the impact of demand and decision flexibility on the optimal design of the hydrogen infrastructure for maritime transportation in Norway. In particular, we study where to locate hydrogen production facilities, which capacity and production technology to install, and which period to choose for investment and expansion.

We distinguish between long-term costs and short-term costs. Long-term costs consist of investment and expansion costs, while the short-term costs are given as production costs, representing capital expenses (CAPEX) and operational expenses (OPEX) respectively.

The investment and expansion represent the long-term decision because a built facility cannot be closed down during the planning horizon. The short-term production costs depend on installed capacity and its utilization. We allow the production rate to deviate from the installed capacity, allowing for a more flexible production schedule. However, deviating from the installed capacity leads to increasing unit costs [29]. We carry out our analysis using two models and two demand scenarios. In the first model, opening new facilities is allowed during the whole planning horizon, while in the second model, opening facilities is restricted to the first period. In the first demand scenario, we assume demand only from the maritime sector, while in the second scenario, demand from the whole transportation sector in Norway is considered.

The remainder of this paper is organized as follows: in Sect. 2, we provide a mathematical formulation of the dynamic facility location problem with capacity expansion. Case description and computational results are discussed in Sects. 3 and 4, respectively. Conclusion is presented in Sect. 5.

2 The Mathematical Programming Model

We formulate our problem as a multi-period facility location problem with capacity expansion. The goal is to determine the optimal strategy for opening and expanding hydrogen production facilities such that demand is satisfied. Closing facilities is not allowed. The objective is to minimize the discounted sum of investment and expansion costs, production costs, and distribution costs.

We provide two models for our multi-period facility location problem with non-linear objective function and capacity expansion. In the first model, investing in a new facility is allowed in each period, while in the second model, the initial investment can only be made in the first period. In both models, capacity expansion is allowed for each facility once during the planning horizon, and technology change is not permitted. We assume that the cost functions are independent of selected locations and investment time. Each technology is characterized by its own cost function. However, the general properties described in Subsect. 2.1 apply to both considered technologies. The mathematical formulation is then presented in Subsect. 2.2.

2.1 Modelling Approach

We model investment decisions as a choice from a discrete set of available capacities similar to [7]. Capacity expansion here means modifying an existing facility and is modelled as a discrete jump between available capacities. This approach is also used in [18].

To model the cost of investing, expanding, and operating facilities, we separate the long-term investment and expansion costs from the short-term production costs. Each installed capacity has its own short-term production cost function. We model the short-term production costs as a piecewise linear, convex function. This is similar to the approach presented in [30]. From the point of view of short-term production costs, higher utilization of smaller capacity is always more favourable than smaller utilization of higher installed capacity.

Expanding capacity implies an additional investment as well as switching over to a new short-term production cost function. Figure 2a illustrates our approach for modelling the expansion of facilities. Let Q_k be the initially installed capacity and C_k the corresponding investment costs. The expansion costs of expanding from capacity Q_k to capacity Q_l are denoted as E_{kl}. As $C_k + E_{kl} > C_l$. Investing in a smaller facility and expanding to a larger capacity is more expensive than opening the bigger facility right away.

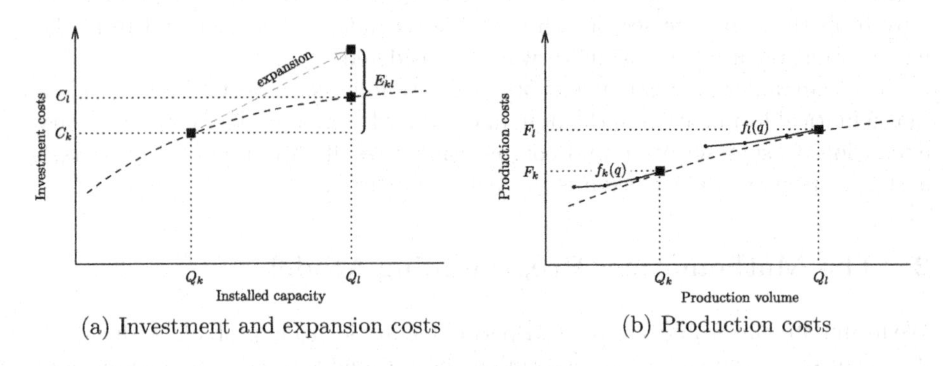

(a) Investment and expansion costs (b) Production costs

Fig. 2. Short-term and long-term costs

Due to separating the long-term investment and expansion costs from the short-term production costs, expansion implies moving from one short-term production cost function to another. An example of this can be seen in Fig. 2b. Before expanding the facility from capacity Q_k to capacity Q_l, the production cost function $f_k(q)$ applies, whereas function $f_l(q)$ is valid after the expansion has taken place.

2.2 Mathematical Formulation

Let us first introduce the following notation:

Sets

$\mathcal{B}$ Set of breakpoints of the short-term cost function
$\mathcal{I}$ Set of possible facility locations
$\mathcal{J}$ Set of customer ports
$\mathcal{K}$ Set of available discrete capacities
$\mathcal{P}$ Set of periods
$\mathcal{T}$ Set of available production technologies

Parameters and coefficients

C_{ikt} investment costs in location i, for point k of capacity function, and technology t;
D_{jp} demand in port j in period p;
E_{klt} costs of expansion from capacity in point k to capacity in point l for technology t;
F_{bkt} costs at breakpoint b of the short-term cost function given for capacity k and for technology t;
L_{ijp} 1 if demand at location j can be served from facility i in period p, 0 otherwise;
Q_{bkt} production volume at breakpoint b of the short-term cost function, for capacity point k and technology t;
T_{ijp} transportation costs from facility i to customer j in period p;
y_{iklt0} initial facility variable;
δ_p discount factor in period p;
τ_p length of time period p in years;

Decision variables

x_{ijp} amount of customer demand at location j satisfied from facility i in period p;
y_{ikltp} 1 if facility is opened in location i in period p, with originally installed capacity k, operated capacity l, and technology t, 0 otherwise;
μ_{biltp} weight of breakpoint b at location i for capacity point k and technology t in period p.

We present a multi-period model where investment and expansion decisions are allowed during the whole planning horizon. The changes in formulation needed for the first-period model are presented at the end of this section. The problem is given as:

$$\min \sum_{i\in\mathscr{I}}\sum_{k\in\mathscr{K}}\sum_{l\in\{l\geq k:l\in\mathscr{K}\}}\sum_{t\in\mathscr{T}}\sum_{p\in\mathscr{P}}\delta_p C_{ikt}\left(y_{ikltp}-y_{iklt(p-1)}\right)+$$
$$\sum_{i\in\mathscr{I}}\sum_{k\in\mathscr{K}}\sum_{l\in\{l>k:l\in\mathscr{K}\}}\sum_{t\in\mathscr{T}}\sum_{p\in\mathscr{P}}\delta_p E_{klt}(y_{ikltp}-y_{iklt(p-1)})+$$
$$\sum_{i\in\mathscr{I}}\sum_{j\in\mathscr{J}}\sum_{p\in\mathscr{P}}\delta_p\tau_p T_{ijp}x_{ijp}+ \tag{1}$$
$$\sum_{b\in\mathscr{B}}\sum_{i\in\mathscr{I}}\sum_{l\in\mathscr{K}}\sum_{t\in\mathscr{T}}\sum_{p\in\mathscr{P}}\delta_p\tau_p F_{blt}\mu_{biltp},$$

subject to:

$$\sum_{k\in\mathscr{K}}\sum_{l\in\{l\geq k:l\in\mathscr{K}\}}\sum_{t\in\mathscr{T}} y_{ikltp}\leq 1, \qquad p\in\mathscr{P}, \tag{2}$$

$$\sum_{l\in\{l\geq k:l\in\mathscr{K}\}} y_{ikltp}\geq \sum_{l\in\{l\geq k:l\in\mathscr{K}\}} y_{iklt(p-1)}, \qquad i\in\mathscr{I},k\in\mathscr{K},t\in\mathscr{T},p\in\mathscr{P}, \tag{3}$$

$$y_{ikltp}-y_{iklt(p-1)}\geq 0, \quad i\in\mathscr{I},k\in\mathscr{K},l\in\{l>k:l\in\mathscr{K}\},t\in\mathscr{T},p\in\mathscr{P}, \tag{4}$$

$$\sum_{b\in\mathscr{B}}\mu_{biltp}=\sum_{k\in\mathscr{K}} y_{ikltp}, \qquad i\in\mathscr{I},l\in\mathscr{K},t\in\mathscr{T},p\in\mathscr{P}, \tag{5}$$

$$\sum_{j\in\mathscr{J}} x_{ijp}=\sum_{b\in\mathscr{B}}\sum_{l\in\mathscr{K}}\sum_{t\in\mathscr{T}} Q_{blt}\mu_{biltp}, \qquad i\in\mathscr{I},p\in\mathscr{P}, \tag{6}$$

$$\sum_{i\in\mathscr{I}} x_{ijp}=D_{jp}, \qquad j\in\mathscr{J},p\in\mathscr{P}, \tag{7}$$

$$x_{ijp}\leq L_{ijp}D_{ip}, \qquad i\in\mathscr{I},j\in\mathscr{J},p\in\mathscr{P}, \tag{8}$$

$$y_{ikltp}\in\{0,1\}, \qquad i\in\mathscr{I},k\in\mathscr{K},l\in\{l\geq k:l\in\mathscr{K}\}t\in\mathscr{T},p\in\mathscr{K}, \tag{9}$$

$$x_{ijp}\geq 0, \qquad i\in\mathscr{I},j\in\mathscr{J},p\in\mathscr{P}, \tag{10}$$

$$\mu_{biltp}\geq 0, \qquad b\in\mathscr{B},i\in\mathscr{I},k\in\mathscr{K},t\in\mathscr{T},p\in\mathscr{P}. \tag{11}$$

The objective function (1) is the discounted sum of investment costs, expansion costs, distribution costs, and production costs. Restrictions (2) guarantee that only one facility can be opened at the given location. Constraints (3) ensure that a facility can expand but cannot be closed. Capacity expansion is allowed only once during the planning horizon. The variable y_{ikltp} contains information about the initially installed capacity k as well as the capacity l at which it is currently operated. After expansion, the operated capacity l is higher than the installed capacity k. Inequalities (4) ensure that capacity index l can change only once. Equations (5) ensure that production is allocated only to opened facilities and that the short-term production cost function depends on operated capacity. Equations (6) express the requirement that the whole production has to be distributed to customers. Equations (7) ensure demand satisfaction, while constraints (8) specify if customer j can be served from facility i. Restrictions (9)–(11) are the binary and non-negativity requirements.

In our second model, a facility can only be opened in the first period. Expansion is still allowed in later periods. In this model, constraint (12) replaces constraint (3):

$$\sum_{l \in \{l \geq k : l \in \mathscr{K}\}} y_{ikltp} = y_{ikkt1}, \qquad i \in \mathscr{I}, k \in \mathscr{K}, t \in \mathscr{T}, p \in \mathscr{P}. \tag{12}$$

The rest of the model is identical to the first model.

3 Case Study

In this section, we present the input data for the problem of designing the Norwegian hydrogen supply chain for maritime transportation. We include 17 candidate locations for hydrogen facilities on the Norwegian west coast. The candidate locations for hydrogen production are obtained from the interactive map set up by Ocean Hyway Cluster [28].

We consider two hydrogen production technologies: EL and SMR+. We approximate the facility capacity by 8 discrete points for EL and 7 points for SMR+. The discrete points are given in Table 1. We use the same discretization of capacity for both technologies, but we do not consider SMR+ for the smallest capacity. In Table 1, we provide facility investment costs and production costs per kilogram at the discrete capacity points. Note that with decreasing utilization, the production costs per unit increase. [33]

Table 1. Investment and production costs for EL and SMR+ at discrete capacity points

Discrete capacity	1	2	3	4	5	6	7	8
Capacity [tonnes/day]	0.6	3.1	6.2	12.2	30.3	61.0	151.5	304.9
Investment EL [mill. €]	1.4	6.0	11.2	20.5	46.5	87.2	197.7	371.5
Investment SMR+ [mill. €]	–	23.9	39.9	65.2	127.7	204.3	402.1	709.2
Production EL [€/kg]	1.95	1.61	1.53	1.45	1.43	1.42	1.40	1.38
Production SMR+ [€/kg]	–	1.91	1.61	1.42	1.28	1.18	1.04	1.00

The production rate for an EL facility can vary between 20 – 100% of the installed capacity [25]. We define a piecewise linear, convex short-term production costs for each discrete capacity. We approximate the short-term production costs by a piecewise linear function with breakpoints at 20%, 50%, 80% and 100%

of installed production quantity. For simplification, we use the same production rates for SMR+. We use the model by Jakobsen and Åtland [16] for calculating investment and short-term production costs for electrolysis and SMR+.

We calculate the expansion costs as the difference between the investment costs of opening two facilities with different capacities plus an additional mark-up. We assume the mark-up for expansion to be 10% of the difference in investment costs.

We derive the costs of distributing one kilogram of hydrogen for one kilometer for distances up to 800 km from [9]. To obtain the costs for distributing up to 1000 km, we extrapolate the distribution cost function. The distribution costs per kilometer and kilogram hydrogen are then valid for the appropriate interval as shown in Table 2. If a customer is located in the same municipality as a facility, we assume zero distribution costs. We set the distance limit between production facility and customer to 1000 km. Hydrogen distribution over 1000 km is suitable for pipelines. However, pipelines are not considered relevant for Norway [8].

Table 2. Hydrogen distribution costs in [€/km/kg H_2]

Distance [km]	1–50	51–100	101–200	201–400	401–800	801–1000
Costs	0.00498	0.00426	0.00390	0.00372	0.00363	0.00360

We use two demand scenarios where hydrogen demand is increasing during the planning horizon (see Fig. 3). In the maritime sector, demand moderately increases until period 11. In period 11, the coastal route Bergen-Kirkenes starts to operate on hydrogen fuels which causes a significant increase in demand. Until period 3, there is no difference between the two demand scenarios. In the whole transportation sector, the main demand growth is in periods 4 and 9 which corresponds to years 2025 and 2030. These dates represent two strategic phases for hydrogen transition in heavy transport and long-distance bus transport [11].

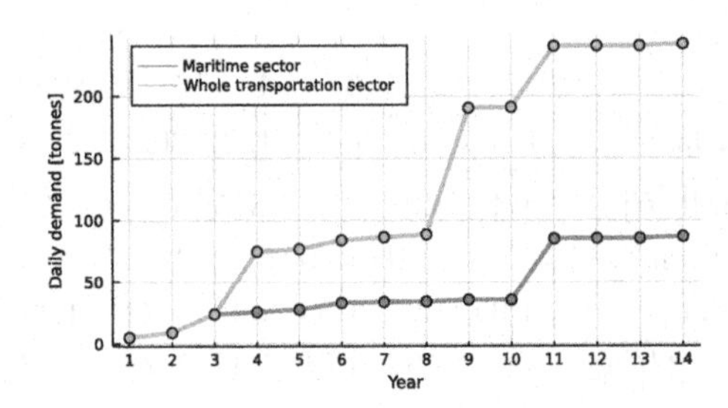

Fig. 3. Development of hydrogen demand during the planning horizon

- Maritime: high-speed passenger ferries, car ferries, and coastal route Bergen-Kirkenes, [2] and [27]

- All transportation: maritime sector plus road traffic and railway sector, [11]

Aarskog and Danebergs [2] and Ocean Hyway Cluster [27] present high-speed passenger ferry and car ferry routes that are relevant for hydrogen fuel as well as their bunkering locations. They list 51 relevant customer locations for the maritime sector and assume that new contracts for public transportation services will require a zero-emission solution and that hydrogen will be selected as fuel. For the whole transportation sector, the list of customers is extended to 70 locations and consists of bunkering ports and several inland locations relevant for hydrogen consumption in road traffic and the railway sector.

In our case, we assume the discounting interest rate to be zero. Thus, the discount factor δ_p is equal to one in each period.

4 Computational Results

The model is implemented in Mosel and solved with Xpress Optmizer Version 36.01.10. All calculations were run on a laptop with a Intel(R) Core(TM) i7-10510U CPU @ 1.80 GHz processor and 16 GB RAM.

A summary of the main results of both demand scenarios and both models can be found in Table 3. We provide the main characteristics of the built infrastructure as the number of built facilities and the number of expansions. Total capacity and average size refer to the installed capacity and average facility size in the last period. The total costs represent the sum of investment, expansion, production, and distribution costs. The average hydrogen costs are calculated over the entire planning horizon average. Note that the chosen technology is electrolysis in all cases.

Table 3. Hydrogen infrastructure characteristics.

Demand scenario	Maritime		All transportation	
Investment decision	First-period	Multi-period	First-period	Multi-period
Built facilities #	12	13	13	13
Expansion #	9	2	10	4
Total capacity [tonnes/day]	87.2	87.2	262.5	274.0
Average size [tonnes/day]	7.2	6.7	20.2	21.1
Total cost [mill. €]	606.0	578.0	1658.7	1594.1
Average hydrogen costs [€/kg]	2.73	2.61	2.53	2.43

Comparing the maritime sector and the whole transportation sector (all transportation), the installed capacity significantly increases in the scenario with higher demand, but not the number of built facilities. In the maritime sector, using the first-period model, the number of built facilities is 12. In all other cases, the number of built facilities is 13. As a result, the average facility size in the last period is almost three times higher in the scenario for the whole transportation sector comparing to the scenario for the maritime sector. The results further show that the expansion option is more often used in the first-period model as the number of expansion is 9 and 10 for the maritime and the whole transportation demand scenario, respectively. For the first-period model, expansion is the only way how increase capacity and so it leads to a higher number of expansions compared to the multi-period model which enables to build facilities later during the planning horizon.

Table 3 further indicates that the capacity utilization is better in the scenario for the maritime sector where the installed capacity is only slightly higher than demand in the last period. The installed capacity is 87.2 tonnes per day for both models and demanded hydrogen amount is 86.9 tonnes per day. In the whole transportation sector scenario, the infrastructure can daily provide 20 or 37 tonnes of hydrogen more than is the demanded amount for the first-period and the multi-period model, respectively.

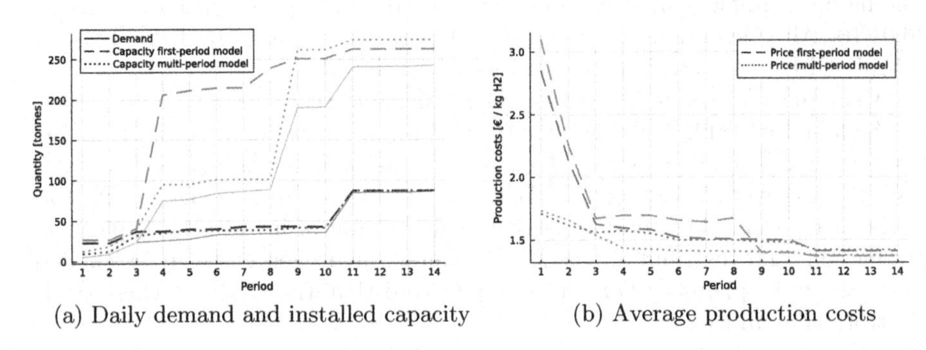

(a) Daily demand and installed capacity (b) Average production costs

Fig. 4. Illustration of installed capacity and average production costs in each period for both demand scenarios and both models. Blue lines refer to the maritime scenario and orange lines to the whole transportation sector. (Color figure online)

Figure 4a provides an overview of installed capacity during the planning horizon. The capacity difference between installed capacity and demand is generally low in the maritime scenario independently of the used model. In the whole transportation sector scenario, the first-period model expands in the period 4 and then the installed capacity is 2.7 times higher than the demand. In the multi-period model, the significant increase in capacity comes in period 9 where three of the four expansion in this scenario are performed and then the increase in capacity is significantly higher than the increase in demand. However, the difference is much lower than in the first-period model and the low capacity utilization affects only periods 9 and 10. The reason is that expansion is allowed only once so the expansion is performed directly to the target size. Figure 4a also shows that from period 11 onwards, demand remains constant and the installed capacity is just slightly higher than demand because the investment and expansion decision aimed to satisfy this target value of demand. In addition, the choice of capacities is limited by the discrete available capacities. With our choice of discrete capacities, the lower demand in the maritime scenario can be satisfied with low excess capacity. When larger capacities are needed, it becomes more difficult to successively build the capacity in line with growing demand because differences between adjacent capacities are increasing.

Figure 4b shows the average hydrogen production costs. In the first three periods, the multi-period model performs significantly better because it allows to build only a few facilities with high utilization in the first periods and to build more later when demand increases. This advantage of the multi-period model also leads to lower total costs and about 5% lower average hydrogen costs than the first-period model.

Figure 4 as a whole further illustrates the economies of scale, as increasing demand leads to lower unit production costs. We can see an exception in the first-period model in the scenario for the whole transportation sector. The average production costs in period 4 and 5 are higher than the costs in period 3 and then again the costs increases in period 8. Due to the capacity expansion in period 4 and 8 (see Fig. 4a), the increase in capacity is significantly higher than the demand growth. The capacity utilization is low, and the unit production costs increase.

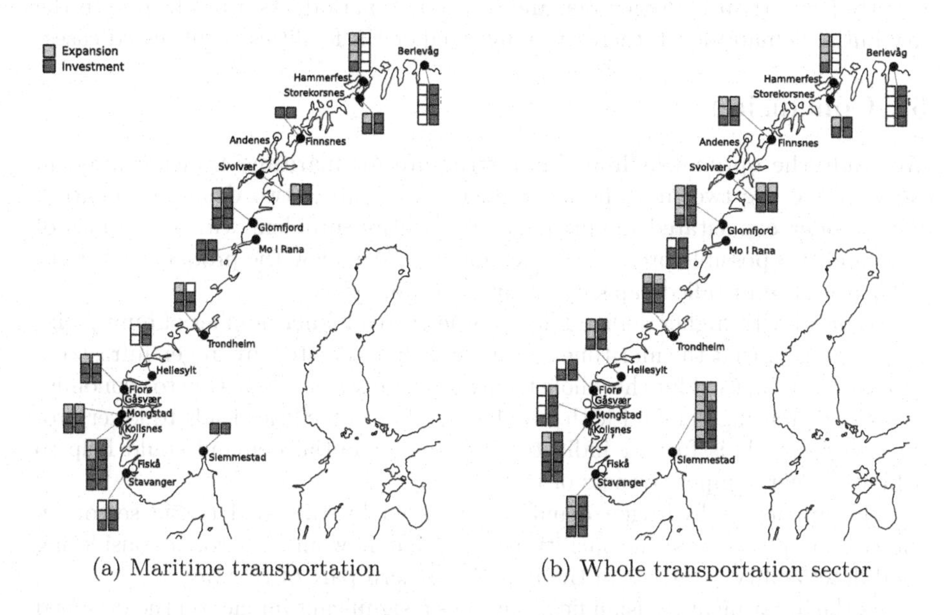

(a) Maritime transportation (b) Whole transportation sector

Fig. 5. Investment and expansion structure of opened hydrogen facilities. The column height corresponds to the installed discrete capacity. Left columns represent the first-period model and right column represent the multi-period model.

The optimal investments in opening and expanding facilities for both models and both demand scenarios are illustrated in Fig. 5. Figure 5a shows the hydrogen production infrastructure for maritime transportation and Fig. 5b shows the infrastructure when the whole transportation sector is considered. The blue boxes denote the discrete capacity that was originally invested in, and the green boxes represent the additional discrete expansion capacity. Comparing the Figs. 5a and 5b, there is no big difference in the infrastructure design in the northern part of Norway because most of the demand in that region comes from the maritime sector. The main difference and also the highest density of opened facilities is in the southern part of Norway. In the maritime scenario in the first-period model (left column), the facilities in Mongstad and Florø are larger than in the whole transportation sector scenario even if the demand in the whole transportation sector is higher and the basic demand from the maritime sector is the same. In the whole transportation scenario, the facility in Slemmestad expands already in period 4 to the target size (see Fig. 4) and helps to satisfy the demand on the west coast.

The demand in the first periods is very low. Because of that, the infrastructure has to be successively built to satisfy demand from the first period. Later, when demand increases, there are already several smaller facilities that still have to be used and satisfy a part of this demand. The remaining requested hydrogen amount is not large enough to build a new SMR+ facility. An SMR+ facility is favourable for quantities higher than 210 tonnes hydrogen daily which is just slightly lower than hydrogen demand in the last period. As a result, due to the low initial demand level, there are built smaller EL facilities in all tested cases.

5 Conclusion

We study the optimal hydrogen infrastructure for maritime transportation in Norway. We use two multi-period models and analyze two demand scenarios. We consider capacitated modular facility location problem with economies of scale and two possible production technologies. We allow the production rate to differ from the installed capacity for both technologies.

Scenario with higher demand does not lead to a higher number of built facilities suggesting that the maritime sector can help to create a hydrogen infrastructure that can be used for the whole transportation sector later. Due to economies of scale, increasing demand with a stable number of facilities leads to lower production costs. This further indicates that higher initial demand could help to achieve higher competitiveness of hydrogen.

The impact of hydrogen demand generated by the road traffic sector on the size of the Slemmestad facility reflects that it would be worth considering candidate facility locations in the inland southern part of Norway.

As the investment decision flexibility has a significant impact on the designed infrastructure, a natural extension of this work is to allow facility closing and technology change during the planning horizon.

The infrastructure design and overall costs highly depend on the demand scenario. An extension of this work is to introduce uncertain demand and thus several demand scenarios and construct a stochastic optimization model. It will be also interesting to analyze the technology choice and the cost structure if we consider uncertainty in costs.

Considering international maritime transportation, ships may purchase fuel in foreign countries. It may increase the uncertainty in demand and lead to pressure on the hydrogen price in Norway. A complex model where the impact of international hydrogen purchasing on national hydrogen demand and hydrogen price is studied is subject to future work.

References

1. Aarskog, F.G., Danebergs, J., Strømgren, T., Ulleberg, Ø.: Energy and cost analysis of a hydrogen driven high speed passenger ferry. Int. Shipbuild. Progr **67**(1), 97–123 (2020)
2. Aarskog, F.G., Danebergs, J.: Estimation of energy demand in the Norwegian high-speed passenger ferry sector towards 2030. IFE/E-2020/003, Halden, Norway (2020)
3. Almansoori, A., Shah, N.: Design and operation of a future hydrogen supply chain: multi-period model. Int. J. Hydrogen Energy **34**(19), 7883–7897 (2009)
4. Arabani, A.B., Farahani, R.Z.: Facility location dynamics: an overview of classifications and applications. Comput. Ind. Eng. **62**(1), 408–420 (2012)
5. Behmardi, B., Lee, S.: Dynamic multi-commodity capacitated facility location problem in supply chain. In: Proceedings of the 2008 Industrial Engineering Research Conference, pp. 1914–1919. Institute of Industrial and Systems Engineers (IISE) (2008)
6. Castro, J., Nasini, S., Saldanha-da-Gama, F.: A cutting-plane approach for large-scale capacitated multi-period facility location using a specialized interior-point method. Math. Program., 411–444 (2016). https://doi.org/10.1007/s10107-016-1067-6
7. Correia, I., Captivo, M.E.: A Lagrangean heuristic for a modular capacitated location problem. Ann. Oper. Res. **122**(1), 141–161 (2003)
8. Damman, S., Sandberg, E., Rosenberg, E., Pisciella, P., Johansen, U.: Largescale hydrogen production in Norway - possible transition pathways towards 2050. SINTEF Rapport 2020–00179, Trondheim, Norway (2020)
9. Danebergs, J., Aarskog, F.G.: Future compressed hydrogen infrastructure for the domestic maritime sector. IFE/E-2020/006, Halden, Norway (2020)
10. Dias, J., Captivo, M.E., Clímaco, J.: Dynamic location problems with discrete expansion and reduction sizes of available capacities. Investigação Operacional **27**(2), 107–130 (2007)
11. DNV GL: Produksjon og bruk av hydrogen i Norge. Rapport 2019–0039, Oslo, Norway, (in Norwegian) (2019)
12. Fridstrøm, L., et al.: Decarbonization of transport, a position paper prepared by FME MoZEES and FME CenSES (2018). ISBN 978-82-93198-25-3
13. Han, J.H., Ryu, J.H., Lee, I.B.: Modeling the operation of hydrogen supply networks considering facility location. Int. J. Hydrogen Energy **37**(6), 5328–5346 (2012)
14. Hirth, M., et al.: Norwegian future value chains for liquid hydrogen: NCE Maritime CleanTech, Report liquid hydrogen 2019, Stord, Norway (2019)
15. Holmberg, K.: Solving the staircase cost facility location problem with decomposition and piecewise linearization. Eur. J. Oper. Res. **75**(1), 41–61 (1994)
16. Jakobsen, D., Åtland, V.: Concepts for large scale hydrogen production. Master's thesis, Department of Energy and Process Engineering, NTNU, Trondheim, Norway (2016)

17. Jena, S.D., Cordeau, J.F., Gendron, B.: Dynamic facility location with generalized modular capacities. Transp. Sci. **49**(3), 484–499 (2015)
18. Jena, S.D., Cordeau, J.F., Gendron, B.: Solving a dynamic facility location problem with partial closing and reopening. Comput. Oper. Res. **67**, 143–154 (2016)
19. Jena, S.D., Cordeau, J.F., Gendron, B.: Lagrangian heuristics for large-scale dynamic facility location with generalized modular capacities. INFORMS J. Comput. **29**(3), 388–404 (2017)
20. Keipi, T., Tolvanen, H., Konttinen, J.: Economic analysis of hydrogen production by methane thermal decomposition: comparison to competing technologies. Energy Conv. Manag. **159**, 264–273 (2018)
21. Luss, H.: Operations research and capacity expansion problems: a survey. Oper. Res. **30**(5), 907–947 (1982)
22. Mäkitie, T., Hanson, J., Steen, M., Hansen, T., Andersen, A.D.: The sectoral interdependencies of low-carbon innovations in sustainability transitions. FME NTRANS Working paper 01/20, Trondheim, Norway (2020)
23. Melo, M.T., Nickel, S., Saldanha-Da-Gama, F.: Facility location and supply chain management-a review. Eur. J. Oper. Res. **196**(2), 401–412 (2009)
24. Myklebust, J., Holth, C., Tøftum, L.E.S., Tomasgard, A.: Optimizing investments for hydrogen infrastructure in the transport sector. In: Techno-economic modelling of value chains based on natural gas:with consideration of CO_2 emissions, pp. 27–70. Doctoral thesis at NTNU 2010:83, Department of Industrial Economics and Technology Management, Trondheim, Norway (2010)
25. NEL Hydrogen: Efficient electrolysers for hydrogen production (2015). http://wpstatic.idium.no/www.nel-hydrogen.com/2015/03/Efficient_Electrolysers_for_Hydrogen_Production.pdf/, Accessed 05 Feb 2021
26. Nickel, S., Saldanha-da-Gama, F.: Multi-period facility location. In: Laporte, G., Nickel, S., Saldanha da Gama, F. (eds.) Location Science, pp. 303–326. Springer, Cham (2019). https://doi.org/10.1007/978-3-030-32177-2_11
27. Ocean Hyway Cluster: 2030 hydrogen demand in the Norwegian domestic maritime sector. OHC HyInfra project, Workpackage C: Mapping future hydrogen demand (2020)
28. Ocean Hyway Cluster: Interactive map - potential maritime hydrogen in Norway. OHC HyInfra project, Workpackage C: Mapping future hydrogen demand (2020)
29. Schütz, P.: Managing uncertainty and flexibility in supply chain optimization, : Doctoral thesis at NTNU 2009:89. Department of Industrial Economics and Technology Management, Trondheim, Norway (2009)
30. Schütz, P., Stougie, L., Tomasgard, A.: Stochastic facility location with general long-run costs and convex short-run costs. Comput. Oper. Res. **35**(9), 2988–3000 (2008)
31. Shulman, A.: An algorithm for solving dynamic capacitated plant location problems with discrete expansion sizes. Oper. Res. **39**(3), 423–436 (1991)
32. Torres-Soto, J.E., Üster, H.: Dynamic-demand capacitated facility location problems with and without relocation. Int. J. Prod. Res. **49**(13), 3979–4005 (2011)
33. Ulleberg, Ø., Hancke, R.: Techno-economic calculations of small-scale hydrogen supply systems for zero emission transport in Norway. Int. J. Hydrogen Energy **45**(2), 1201–1211 (2020)
34. Van den Broek, J., Schütz, P., Stougie, L., Tomasgard, A.: Location of slaughterhouses under economies of scale. Eur. J. Oper. Res. **175**(2), 740–750 (2006)

Destination Prediction of Oil Tankers Using Graph Abstractions and Recurrent Neural Networks

Búgvi Benjamin Magnussen[1(✉)], Nikolaj Bläser[1], Rune Møller Jensen[1], and Kenneth Ylänen[2]

[1] IT -University of Copenhagen, Rued Langgaards Vej 7, 2300 Copenhagen, Denmark
bugvibenjamin@magnussen.email, rmj@itu.dk
[2] Torm, Tuborg Havnevej 18, 2900 Hellerup, Denmark
koyl@torm.com

Abstract. Predicting the destination of vessels in the maritime industry is a problem that has seen sustained research over the last few years fuelled by an increase in the availability of Automatic Identification System (AIS) data. The problem is inherently difficult due to the nature of the maritime domain. In this paper, we focus on a subset of the maritime industry - the oil transportation business - which complicates the problem of destination prediction further, as the oil transportation market is highly dynamic. We propose a novel model, inspired by research on destination prediction and anomaly detection, for predicting the destination port- and region of oil tankers. In particular, our approach utilises a graph abstraction for aggregation of global oil tanker traffic and feature engineering, and Recurrent Neural Network models for the final port- or region destination prediction. Our experiments show promising results with the final model obtaining an accuracy score of 41% and 87.1% on a destination port- and region basis respectively. While some related works obtain higher accuracy results - notably 97% port destination prediction accuracy - the results are not directly comparable, as no related literature found deals with the problem of predicting oil tanker destination on a global scale specifically.

Keywords: Oil tankers · Automatic identification system · Destination prediction · Graph abstraction · Recurrent neural networks · Deep learning

1 Introduction

Destination prediction of sea vessels is an area of research that has started to significantly grow in popularity during recent years, arguably correlated with the increase in available data in the maritime industry. Research has been conducted for various destination prediction applications [4,9]. However, what all work regarding destination prediction tends to agree upon, is that reliable

M. Mes et al. (Eds.): ICCL 2021, LNCS 13004, pp. 51–65, 2021.
https://doi.org/10.1007/978-3-030-87672-2_4

behavioural prediction is difficult. The reasons for this are numerous. In particular, behavioural prediction is difficult due to the fact that sea vessels move in a continuous space, and because sea vessels are strongly affected by their surrounding environment, which includes weather, currents and seasons. Factors related to the oil transportation business further complicate the problem. Specifically, the oil transportation business is characterised by the spot-market and short-term contracts which implies that oil tankers are constantly assigned to new contracts. As a consequence, the nature of oil shipping is much more dynamic compared to for example cargo shipping, where vessels often are assigned to long-term contracts, and thus follow more well-defined patterns. In this paper, we propose a novel model for predicting the future trajectory as well as the destination port and region of oil tankers on a global scale. We combine ideas from literature related to both destination prediction and anomaly detection [5,9,13]. Using Automatic Identification System (AIS) data, we construct a graph abstraction of global oil tanker traffic and use it to generate sequences that are used to train a Recurrent Neural Network deep learning model to predict the destination of oil tankers. In particular, we show that destination prediction is possible with 87.1% accuracy on the regional level.

The rest of this paper is organised as follows. First, the problem statement is outlined in Sect. 2. Then, related work is presented in Sect. 3, followed by a detailed outline of the proposed model in Sect. 4. Next, we present our results and discuss our experiments in Sect. 5. Finally, concluding remarks are delivered in Sect. 6.

2 Problem Statement

Companies operating in the oil transportation business are engaged in the tramp trade which means that vessels sail without published schedules. Further, these companies naturally do not share information with each other regarding the strategic positioning of their fleet and destinations of their vessels on voyages. As a consequence of these factors, situations often arise in which there is an excessive supply of oil transport vessels in certain regions, while there is an excessive demand in others. Committing a vessel to a voyage can thus in the worst case be a waste of time and resources, as competitors might reach the destination first and thus saturate the demand. As such, being able to conduct an accurate forecast of future destinations of competitors' oil tankers can lead to a competitive advantage.

Developing a model capable of predicting the destination of an oil tanker requires historical data. Oil tankers are obliged to transmit positional data using the "Automatic Identification System" (AIS). AIS is used worldwide and provides automatic positioning via small transponders installed on sea vessels. Data is emitted by these transponders periodically and is referred to as AIS data. It contains information such as latitude, longitude and ship type [3]. The data accessible to us is an AIS dataset provided by TORM A/S. It was collected from 4112 oil tankers, gathered in the time span $01-02-2019$ to $31-12-2020$, and contains 43948646 data points and 16 features (VesselID, Longitude, Latitude, Movement_Date_Time, MoveStatus, Speed, Heading, Destination, Destination-Tidied, ETA, CargoInfo, Beam, CallSign, Draught, Length ShipType). We propose a two step model, inspired by related research in destination prediction [5,9] and anomaly detection [13], to bridge the gap between the available AIS data and the ability to make predictions regarding the future destination of oil tankers on a global scale. Thus, given an oil tanker, when making a destination prediction, the input to the model is the recorded partial trajectory of the oil tanker since it last departed a port. The output of the proposed model is a probability distribution of future trajectories of an oil tanker, including the destination. The purpose of the model is to support human decision making, and thus a probability distribution is valuable, as it allows the model to express uncertainty, which in turn allows a human to decide when the model can be trusted, i.e., when the model assigns a high probability to a single trajectory/destination. In this paper, we consider destinations on two levels: destination port and destination region.[1] Here, predicting the destination region is arguably more valuable, as naturally, strategical positioning of a fleet will usually take place on a regional level.

The ability to predict the future positioning of competitors' oil tankers should lead to an improvement in the ability to strategically position an oil tanker fleet. The potential competitive benefits of this are numerous. In general, the implications are that oil tankers would less frequently embark on non-profitable journeys. This is due to the fact that prior to sending an oil tanker on a journey to a particular destination to meet demand, one could examine the current state of competitors' oil tankers, predict their destination using the model, and based on this information, determine whether demand would be saturated before the oil tanker in question would be able to arrive at the destination[2]. This would have a direct impact on an oil-shipping company's bottom-line, as it would lead to reduced costs in terms of crew salaries and other operating costs such as fuel consumption. Further, it can also directly increase profitability. By reducing the number of non-profitable journeys that oil tankers undertake, they are free to embark on profitable ones. Finally, in addition to the competitive benefits, forecasting oil tankers could also benefit the oil supply chain as a whole. Positioning a fleet strategically based on an accurate forecast of the global oil tanker population would result in an overall better distribution of oil tankers, and as a result, this could lead to an improved service for oil-providers and end-consumers.

3 Related Work

Oleh et al. [4] propose a method for predicting port destinations of all types of vessels in the Mediterranean Sea based on ensemble techniques and different variants of Decision Trees. They report a port prediction accuracy of 97%. While this result is impressive, it is not directly comparable to our work. The scale is

[1] The regions are defined by human experts.

[2] Estimated time of arrival is also an important aspect of this scenario. However, this is outside the scope of this paper.

regional and not global, i.e., only the Mediterranean Sea is targeted for prediction. Further, there are no details provided as to the nature of the data used, and it could thus stem from vessels of multiple types, notably container vessels, which exhibit much more predictable behaviour. Finally, the proposed model outputs a single prediction and not a probability distribution, which implies that it is not capable of expressing uncertainty and hence not suitable for our purposes.

Nguyen et al. [9] propose a method that utilises a sequence-to-sequence Recurrent Neural Network model to predict the trajectories of sea vessels, including their destination, in the Mediterranean Sea. Given a stream of AIS data-points, the spatial features (longitude/latitude) are mapped to a grid with evenly sized grid cells, each uniquely identified by an identifier (id). The grid cell ids are then fed as input to a Recurrent Neural Network (RNN) model. The authors also discuss how other features, such as ship length, can be incorporated. The authors report that the model is effective at predicting trajectories with a log perplexity score [2] of 1.44 (lower is better). Note that this method does take the historical trajectory of a vessel into account. However, their work is also not directly applicable to our problem, as they only target the Mediterranean Sea for prediction. Further, the nature of the data is unknown, i.e., it could stem from multiple types of vessels. For global destination prediction, the proposed approach would arguably be inefficient (see Sect. 4.5 for why).

Finally, it is worth highlighting a method proposed by Varlamis et al. [13]. Concretely, the authors do not propose a destination prediction model, but rather, an anomaly detection system where a graph abstraction is constructed using raw AIS data. The vertices of the graph abstraction are defined by clusters of AIS data-points where vessels are either standing still or performing significant turns. The edges of the graph abstraction are defined by aggregations of AIS data segmented into sub-trajectories between the graph vertices. The authors then propose that anomalous vessel behaviour can be identified by using the graph abstraction as a descriptor of what "normal" vessel behaviour is, and observing how much a vessel's trajectory deviates from the graph abstraction. A key idea of our work is to apply this graph abstraction to trajectory and destination prediction (see Sect. 4).

While the related work presented in this section each achieved excellent results, none of it is directly applicable to the problem we are faced with in this paper - global destination prediction of oil tankers. Further, as alluded to throughout this section, there are various reasons for why the related work cannot directly be used as a solution to the problem we are faced with. Thus, this highlights the need for a novel model designed specifically for global destination prediction of oil tankers.

4 Solution Approach

The proposed model can be split into two parts. The objective of the first part is to extract discrete port-to-port trajectories, as well as relevant associated constant features[3] from raw AIS data. This is achieved by building a multi-step data pipeline. The objective of the second part is to build a Recurrent Neural Network capable of predicting the future trajectory of a vessel, including the destination, given the processed discrete trajectory data. Figure 1 provides an overview of the model.

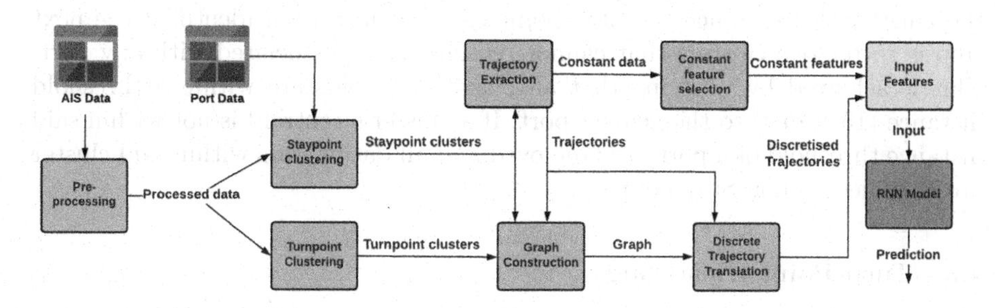

Fig. 1. The overall solution approach. The blue boxes represent the first part of the model and can be described as *feature engineering*. (Color figure online)

4.1 Pre-processing

In the pre-processing stage the speed and bearing rate[4] of a vessel are derived for each AIS data-point. To calculate said features, AIS data points for each vessel respectively are sorted by time and considered pairwise. Both features are needed for later stages in the proposed model. We also perform some basic data cleaning by removing duplicates and removing data points where features contain nonsensical values, e.g., a Heading value larger than 360.

4.2 Stay-Point Clustering

The next step in the proposed model is to perform *Stay-point* clustering. A Stay-point, is an AIS data-point where the vessel emitting the data-point is moving very slowly or is stationary. The purpose of Stay-point clustering is to identify clusters around the globe where vessels can be considered to be in port. These clusters are then used for the construction of the graph abstraction outlined in Sect. 4.5. Given the raw AIS data, Stay-points are identified by using the speed feature computed during the pre-processing stage. Specifically, data-points with a speed of less than 1 knot are classified as Stay-points, which is a commonly used threshold for the purpose [13]. Next, a clustering algorithm called DBSCAN [10] is used to cluster the identified Stay-points. The choice of DBSCAN is deliberate. DBSCAN can identify arbitrarily shaped clusters, which implies that no assumption has to be made with respect to the shape of the

[3] Constant features are features that do not change during the duration of a vessel's port-to-port voyage. For example, the *length* of a vessel.

[4] The bearing rate is the rate of turn of a vessel.

Stay-point clusters. The major limitation of DBSCAN is that it is restricted to finding clusters of similar density. However, in the case of finding Stay-point clusters this is less of an issue, as it can arguably be assumed that Stay-point clusters associated with frequently[5] visited ports have a similar density. The hyper-parameters used have been found empirically, albeit with some inspiration drawn from Chengkai et al. [17], by visually examining the convex polygons that the clusters define. Once the Stay-point clusters have been identified the next step is to remove clusters that cannot reasonably be associated with any port. This is achieved by verifying that all cluster centroids are within a threshold distance (15 *miles*) to the nearest port. If a cluster's centroid is not within said distance threshold of a port, it is removed and all data points within said cluster are classified as non Stay-points.

4.3 Turn-Point Clustering

The next step is to perform *Turn-point* clustering. A Turn-point, is an AIS data-point where the vessel emitting the data-point is performing a turn. The purpose of Turn-point clustering is to identify data-points which indicate that an important event is occurring during a vessel's journey, i.e., where the vessel performs a turn, indicating a change in course. These clusters, like Stay-point clusters, are also used to construct the graph abstraction which is described in Sect. 4.5. The Turn-point clustering process is highly similar to the Stay-point clustering process described previously. Given the raw AIS data, Turn-points are identified by using the bearing rate feature computed during the pre-processing phase (see Sect. 4.1). Concretely, data-points that are not Stay-points and that have a bearing rate of 0.3 or more degrees per minutes are labelled as Turn-points. Next, the data is split into Turn points "close" to shore and Turn points "far" from shore. The rationale behind performing this split is that clusters close to shore are assumed to have different characteristics than clusters far from shore, and should therefore be clustered differently. Similarly to Stay-points, Turn-points "close" to shore are clustered using DBSCAN. For Turn-points "far" from shore, a clustering algorithm called OPTICS [1] is used. OPTICS can cluster arbitrarily shaped clusters of varying density. This is particularly useful for identifying Turn-point clusters far from shore, as it can be hypothesised that they vary quite a bit in both shape and density.

4.4 Trajectory Extraction

The AIS data at this point contains labels indicating which, if any, Stay-point or Turn-point cluster a data-point has been assigned to. In order to extract port-to-port trajectories, the AIS data is grouped by the Vessel id attribute, sorted by time and then split into trajectories. The following example illustrates how

[5] Less-frequently visited ports are less interesting as they are a form of outlier in the dataset.

two port-to-port trajectories are extracted from a time-sorted sequence of AIS data-points emitted from one vessel.

$$data = \langle \hat{1}, \hat{2}, 3, 4, 5, 6, \hat{7}, \hat{8}, 9, 10, 11, 12, 13, \hat{14} \rangle$$
$$\downarrow$$
$$t_1 = \langle \hat{2}, 3, 4, 5, 6, \hat{7} \rangle \; t_2 = \langle \hat{8}, 9, 10, 11, 12, 13, \hat{14} \rangle,$$

where numbers with "hats" indicate data points classified as Stay-points. In this particular example the vessel starts out with being in-port $\langle \hat{1}, \hat{2} \rangle$, sets out on a voyage $\langle 3, 4, 5, 6 \rangle$, arrives at a port $\langle \hat{7}, \hat{8} \rangle$, sets out on a new voyage $\langle 9, 10, 11, 12, 13 \rangle$ and arrives at the final port $\langle \hat{14} \rangle$. Thus, two port-to-port trajectories are extracted from this example. Once all vessels have been processed, the result is a new dataset containing port-to-port trajectories from all vessels in the original AIS dataset.

4.5 Graph Abstraction

The extracted trajectory data could in principle already be used as an input feature for a sequential prediction model. However, as mentioned by Nguyen et al. [9], raw AIS data is, due to varying transmission frequencies, not well suited for this type of prediction. A much more promising approach is to discretize AIS trajectories into a format that eliminates the frequency problem. As mentioned in Sect. 3, one way of doing so is to split the coordinate space into a spatial grid and map each trajectory to a sequence of visited grid cells. This approach however arguably also has flaws. An even spaced grid does not take swimlanes, harbour areas, or landmass into account, and as such, there is a loss of information when used for discretization. Furthermore, while good results have been achieved for smaller regions [9], using a grid-cell approach for the whole globe would be less efficient, as the number of cells (state space) needed in order to represent trajectories sufficiently accurately grows proportionally with the size of the coordinate space. To address the aforementioned problems, we propose a graph-based method for discretizing trajectories. Specifically, our idea is to create a graph-abstraction in which vertices represent areas of interest and edges represent swimlanes connecting these areas. A trajectory can then be discretized into a sequence of traversed vertices or, alternatively, a sequence of traversed edges. This approach not only solves the frequency problem of raw AIS data, but also avoids the loss of information of the grid-based approach. Furthermore, note that compared to grid cells, a graph abstraction is more scalable, as the size of the graph does not depend on the size of the coordinate space.

Vertices. Vertices represent sea-areas in which vessels perform manoeuvres of interest. There are two types of areas that are of particular interest. One, areas in which vessels typically stand still (harbour areas). And two, areas in which vessels typically perform turn manoeuvres. Vertices can thus be split into two types, Stay-point vertices and Turn-point vertices. Stay-point vertices are constructed using the clustered Stay-point AIS data. For each Stay-point cluster (and thus harbour), a convex polygon is created, tightly encapsulating the data-points belonging to the cluster. Each polygon thus defines the sea-area of a harbour. The polygons are then slightly enlarged using a buffer, and overlapping polygons are merged. This is done to reduce the number of polygons. Note that this also implies that multiple ports will be mapped to the same polygon. Next, for each polygon, a Stay-point vertex is created representing the area defined by the polygon. Finally, the polygon itself as well as the names of the ports represented by the polygon are stored within the corresponding Stay-point vertex. Similarly, Turn-point vertices are constructed using the clustered turn-point AIS data. The only difference to Stay-point vertices, is that Turn-point vertices with overlapping polygons are not merged. Instead, the intersection is removed from one of the overlapping polygons. See Fig. 2 for an illustration.

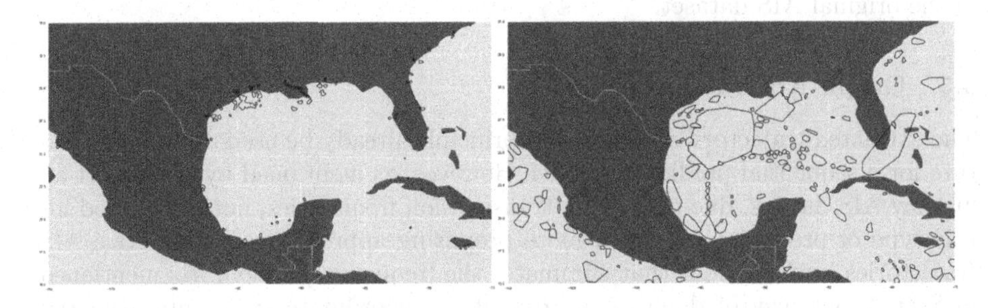

Fig. 2. The Gulf of Mexico. The image to the left shows Stay-point polygons and the image to the right shows Turn-point polygons.

Edges. Edges are added to the graph by splitting each vessel trajectory into several sub-trajectories. This is done by pairwise iterating through data-points in a trajectory and creating a sub-trajectory whenever the line between the pair of points intersects with a vertex. The edges are then added to the graph. Note that currently, edges of the graph are not used in further stages of the proposed methodology. See Sect. 6 for future plans.

Region Mapping. Since a part of the objective of this paper is to predict the destination region of a vessel, information is added to the graph in regards to which region any given Stay-point vertex belongs to.

4.6 Discrete Trajectory Translation

Port-to-port trajectories are discretised by iterating through the AIS data-points of each trajectory and checking for intersection with vertices (polygons). Whenever a line formed by a pair of consecutive points within a trajectory intersects with a vertex, the id of that vertex is added to a sequence of vertex ids. The result is thus a sequence of vertices traversed by a vessel throughout its port-to-port journey. See Fig. 3 for an illustration.

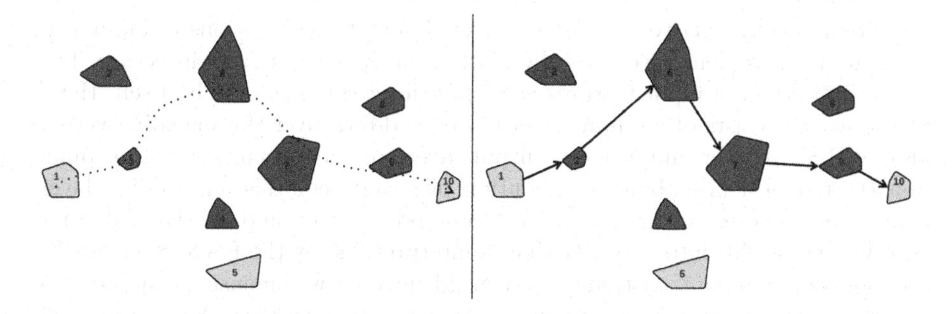

Fig. 3. Red polygons represent Stay-point vertices and green polygons represent Turn-point vertices. The left side shows a continuous trajectory and the right side shows the same trajectory after it is discretized using the graph. The resulting sequence of vertex ids is $\{1, 3, 6, 7, 9, 10\}$ (Color figure online)

4.7 Constant Feature Extraction

Apart from spatial features used to construct the aforementioned sequences, AIS data also contains other features that potentially are useful for predicting the future destination and trajectory of a vessel. These features will be referred to as *constant features*, as they should not change during the port-to-port voyage of a vessel. Relevant constant features include the vessel id, the length of a vessel, the draught of a vessel, the departure time of a vessel (month, day, hour), the previous departure port and information related to the cargo of a vessel. Combining sequential and constant features yields the following format for each data-point.

$$X = \{[x^0, ..x^{t-1}, x^t], id, length, draught, month, day, hour, prev\_dep, cargo\_info\},$$

where $[x^0, ..x^{t-1}, x^t]$ is the discretized trajectory (sequence of vertices).

4.8 Recurrent Neural Network Model

We propose a sequence-to-one RNN for predicting the future trajectory of an oil tanker. An RNN [11] is a type of Neural Network (NN) designed to detect patterns in sequential data. What allows an RNN to detect sequential patterns is the ability to remember previous inputs. That is, outputs are not only determined by the current input, but also by what has been observed in the past. The sequence-to-one RNN is thus able to model problems of the following form:

$$X = \{x^0, x^1 ... \; x^{t-1}, x^t\}, \; Y = \{x^{t+1}\}$$

where the output is generated not only based on the most recent observation x^t, but based on a sequence of previous observations. In our case, the input is a partial port-to-port trajectory (sequence of vertex ids) and the output is the most likely next vertex (id) a vessel intersects. Several different variations of RNNs have been developed that share the same underlying idea, but address some of the weaknesses of the original approach. One of the most popular variations is known as *Long short-term memory* (LSTM) [7]).

Network Architecture. In order to test different combinations of input features, we have implemented several variations of the network's architecture. The first combination of input features is simply the *vertex sequence* by itself. Here, the network consists of an input layer, an embedding layer (for encoding vertex ids), an LSTM layer and a dense output layer with a softmax activation function (to model a probability distribution over vertices). The embedding layer is an alternative to the commonly used one-hot-vectors, and is responsible for encoding vertex ids into a format that is interpretable by the RNN. Specifically, the embedding layer transforms a vertex id into a low dimensional dense vector, representing the vertex and its relation to other vertices. For subsequent input feature combinations major adaptions to the architecture of the network are required. This is due to the fact that we now start to mix features of different types. While the vertex sequence is a sequential feature, the other constant features are either categorical or continuous. Inspired by Yuan et al. [15], we propose a *hybrid* recurrent neural network architecture for modelling multi-type input features. Specifically, the architecture is split into two parts, referred to as the *feature learning* part and the *target learning* part. Note that several different architecture modifications in terms of number of neurons, activation functions and number of layers have been tested. The best performing architecture (when including all features) is shown in Fig. 4.

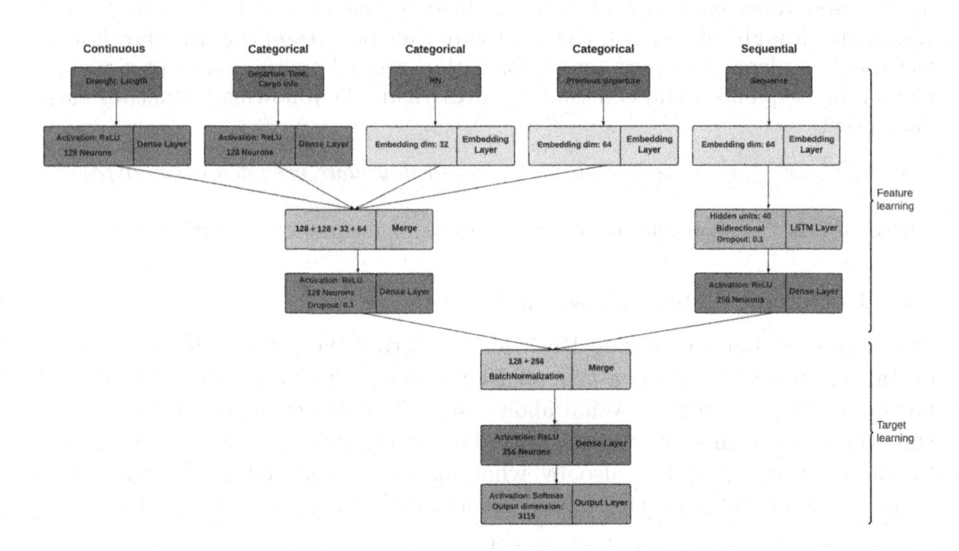

Fig. 4. The best performing architecture of the RNN model.

The *vessel id* and *previous departure port* are high-dimensional categorical input features each forwarded to an embedding layer. The network uses dropout [12] to reduce overfitting. A dropout of 0.1 is applied within the LSTM layer and after merging constant features. Furthermore, note that *batch normalization* [8] (BN) is applied after merging the sequential branch with the constant feature branches. Typically, BN is applied to *deep convolutional neural networks* to improve the rate of convergence. Applying it to our RNN model interestingly has the same effect. Please note that the RNN model described in this Section is implemented using Tensorflow[6].

Decoding. The RNN by itself is designed to predict a single vertex and not the entire remaining port-to-port trajectory of a vessel. To predict the future trajectory (sequence of vertices), as well as the future destination, we use the decoding algorithm beam-search, as described by Zhang et al. [16]. Summarised briefly, the decoder uses the RNN to iteratively generate tokens (in this case vertex ids) until an end-of-sequence (eos) token is generated. Given that the objective is to predict a port-to-port trajectory, the decoding process continues until a Stay-point vertex id is generated. Thus, every Stay-point vertex of the graph represents an eos token. Furthermore, note that a hyper-parameter *beam width* can be specified, which represents the number of likely trajectories the decoder should generate. Each returned trajectory includes a score, which indicates the likelihood of the trajectory relative to other trajectories. The most likely trajectory is thus the one with the highest score. The destination region and port are then derived from the predicted trajectory.

5 Experiments and Results

Prior to conducting our experiments, we split the data into training, test and validation datasets as indicated by Fig. 5.

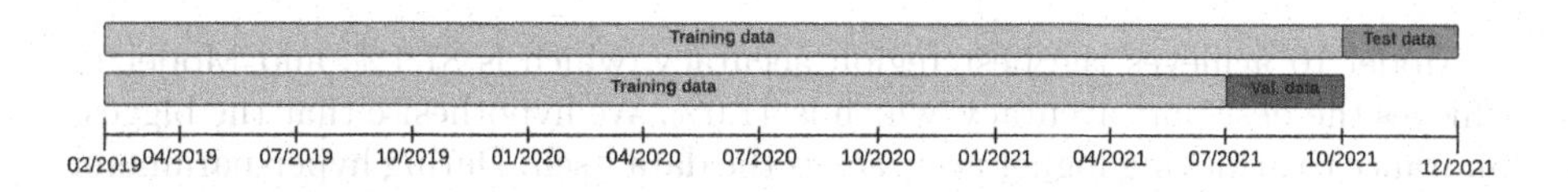

Fig. 5. An illustration of how the AIS data is divided.

For testing, the training dataset is extended by the validation dataset. We use the training data to construct the graph abstraction and train the RNN model. Note that usually, prior to splitting the dataset into training, test, and validation, the data is shuffled to improve the model's ability to generalise. However, due to the fact that the purpose of the RNN model is to predict future vessel destinations based on historical vessel data, the data is not shuffled, but split in chronological order as this gives a better indication of how the RNN model would perform when deployed. The data is then shuffled after the chronological split.

The RNN model uses a vertex-level loss at training. The loss is measured using *Categorical Cross-Entropy*, the optimiser used is *Adam* with a learning rate

[6] https://www.tensorflow.org/.

of 0.001 and the batch size is set to 32. Hyper-parameters of the RNN model, including the size of the network, are tuned using a trial-and-error approach. The accuracy of the trained RNN model is measured using halfway port-to-port trajectories. The measured metrics are next vertex, destination port and destination region accuracy. As mentioned previously, several different variations of the RNN model are trained in order to test different combinations of input features. The results are shown in Table 1.

Table 1. Results for different feature combinations.

	1	2	3	4	5	6	7	8	9	10	11	12
Sequence	✓	✓	✓	✓	✓	✓	✓	✓	✓	✓	✓	✓
Vessel id		✓						✓			✓	✓
Length			✓						✓	✓	✓	✓
Draught				✓					✓	✓	✓	✓
Depart. time					✓					✓	✓	✓
Prev. Depart						✓		✓	✓	✓	✓	✓
Cargo Info							✓	✓	✓			✓
Val. node	64.2	64.3	65.5	65.6	65.0	64.4	64.4	64.9	66.0	65.7	65.8	65.2
Val region	80.7	80.1	81.1	81.5	80.5	80.8	78.8	80.2	84.9	85.1	83.5	83.3
Val. port	35.1	35.1	37.0	37.2	36.4	35.1	33.1	37.0	37.3	37.1	40.5	40.4
Test node	64.5	64.8	65.8	65.9	65.3	64.7	64.5	65.2	66.4	66.0	65.8	65.8
Test region	81.0	81.0	81.7	82.1	81.1	80.9	78.1	83.1	86.5	87.1	84.3	84.0
Test port	35.4	36.1	37.4	37.5	36.7	35.9	33.3	38.2	38.0	38.0	40.9	41.0

Model 10 achieves the best region accuracy, which is 87.1%, and Model 12 achieves the best port accuracy, which is 41.0%. We hypothesise that the biggest hindrance to achieving higher accuracy is the data itself. During hyper-parameter tuning, we observed that the RNN model has a high tendency to over-fit the training data. Small increments in the complexity of the RNN model (for instance an increase in the number of neurons in a single layer), already causes the model to over-fit significantly. Arguably, the chronological split between the training and test data is a major cause for over-fitting. If the distribution of the test data differs from the distribution of the training data, the RNN model will not generalise well. It is hypothesised that this is indeed the case due to the COVID-19 pandemic, which affected global oil shipping during 2020 [6]. The training data, which spans from 2019 to 2020, is only partially affected by COVID-19, whereas the validation and test data exclusively consist of AIS data from the end of 2020, which is when the pandemic was at its worst. More data, or data ranging over a different time span, would thus likely give better results. Furthermore, as seen in Table 1, the test accuracy appears to be significantly higher than the validation accuracy, which could be due to the difference in size of the training data used for validation and testing respectively (see Fig. 5). Hence, this further supports the argument that more data would improve the accuracy of the RNN model.

Visualisation of Predictions. The trajectories predicted by the model can be plotted on a map and are thus easy to interpret. See Fig. 6.

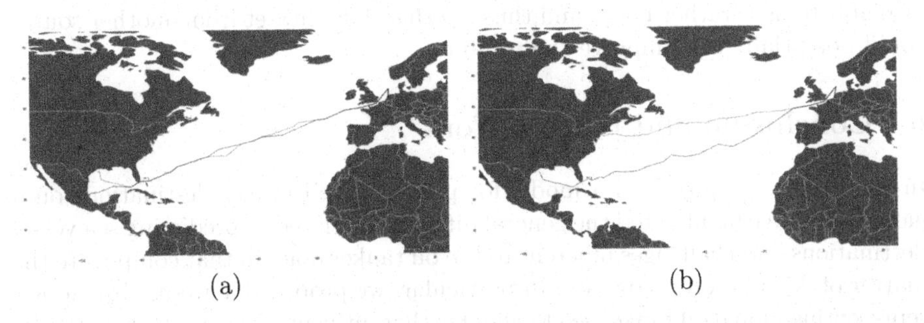

Fig. 6. The black line indicates the partial trajectory given as input to the RNN model. The red line indicates the prediction made by the RNN model and the green line the actual trajectory. (Color figure online)

As seen in the figure, the model is able to consistently produce realistic trajectories. However, even though the predictions are realistic, they can still substantially deviate from the actual trajectory of a vessel (as seen in example *(b)* of Fig. 6). Furthermore, note that predicting only a single trajectory does not serve as a good basis for decision making, as there is no indication in regards to how confident the RNN model is in its prediction. Instead, a much better approach would be to return a probability distribution over possible future trajectories. This is achieved by using a heatmap, which is extracted from the trajectories returned by the beam-search decoder. Figure 7 shows a heatmap over predicted trajectories for the partial trajectory shown in example *(b)* of Fig. 6.

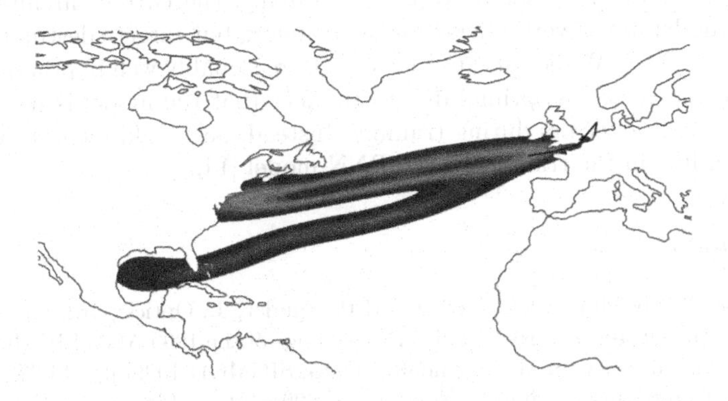

Fig. 7. A plotted heatmap using a beam width of 35.

While our results show promise our experiments also highlight the fact that making trajectory- and destination prediction in the oil transportation domain is difficult. Thus, further work is needed in order to improve the results, in particularly at the port destination prediction level. Finally, in addition to the aforementioned issues with the data, our experiments indicate that the data available to us is rather noisy, and thus another AIS dataset from another source could potentially also improve the results.

6 Conclusion and Future Work

In this paper, we proposed a model for predicting oil tanker destinations on a global scale. We highlighted the general difficulties related to predicting sea vessel destinations and challenges inherent to the oil tanker domain that complicate the matter of destination prediction. In particular, we proposed a graph abstraction representing global oil tanker sea traffic. Further, utilising this graph abstraction, sea vessel port-to-port trajectories are discretised into sequences that are then used to train a Recurrent Neural Network model for destination prediction on a port and regional level. We presented experiments that show promising results for both port- and region destination prediction, thereby highlighting the potential of the proposed model. While the destination prediction accuracy at the port level is quite low, at the regional level, it is quite high. Further, because the model outputs a probability distribution, as of now, it can arguably be used to aid human decision making. Thus, deploying the model should already lead to an improvement in the ability to strategically position a tanker fleet.

We consider this work as a first step and there are numerous potential improvements that can be made. For instance, currently port-to-port trajectories are converted into discrete sequences by checking for intersection with the vertices of the graph abstraction. An alternative to this would be to utilise the edges of the graph to discretise the port-to-port trajectories into a sequence of edge mappings. Using the edges in this way arguably allows for more details to be captured by the discrete sequences. Finally, the current architecture of the RNN model uses a vertex-level loss at training, but targets destination-level accuracy at testing. While this approach works reasonably well in practice, it can be argued that it is sub-optimal due to the fact that the model is not exposed to destination-level errors during training. Instead, one could incorporate beam search decoding in the training of the RNN model [14].

References

1. Ankerst, M., Breunig, M.M., Kriegel, H.P., Sander, J.: Optics: Ordering points to identify the clustering structure. In: Proceedings of the 1999 ACM SIGMOD International Conference on Management of Data, SIGMOD 1999, pp. 49–60. Association for Computing Machinery, New York (1999). https://doi.org/10.1145/304182.304187
2. Arora, K.: Contrastive perplexity: A new evaluation metric for sentence level language models. CoRR abs/1601.00248 (2016). http://arxiv.org/abs/1601.00248
3. BigOceanData: The definitive ais handbook. http://www.marineinsight.com/wp-content/uploads/2016/11/AiS-Whitepaper.pdf

4. Bodunov, O., Schmidt, F., Martin, A., Brito, A., Fetzer, C.: Grand challenge: Real-time destination and ETA prediction for maritime traffic. CoRR abs/1810.05567 (2018). http://arxiv.org/abs/1810.05567
5. Capobianco, S., Millefiori, L.M., Forti, N., Braca, P., Willett, P.: Deep learning methods for vessel trajectory prediction based on recurrent neural networks. CoRR abs/2101.02486 (2021). https://arxiv.org/abs/2101.02486
6. Ghosh, S.: Why the oil tanker business boomed during covid-19 pandemic? (February 2021). https://www.marineinsight.com/know-more/oil-tanker-business-boomed-during-covid-19-pandemic/
7. Hochreiter, S., Schmidhuber, J.: Long short-term memory. Neural Comput. **9**(8), 1735–1780 (1997). https://doi.org/10.1162/neco.1997.9.8.1735
8. Ioffe, S., Szegedy, C.: Batch normalization: Accelerating deep network training by reducing internal covariate shift. CoRR abs/1502.03167 (2015). http://arxiv.org/abs/1502.03167
9. Nguyen, D.D., Le Van, C., Ali, M.I.: Vessel trajectory prediction using sequence-to-sequence models over spatial grid. In: Proceedings of the 12th ACM International Conference on Distributed and Event-Based Systems, DEBS 2018, pp. 258–261. Association for Computing Machinery, New York (2018). https://doi.org/10.1145/3210284.3219775
10. Ram, A., Sunita, J., Jalal, A., Manoj, K.: A density based algorithm for discovering density varied clusters in large spatial databases. Int. J. Comput. Appl. **3** (06 2010). https://doi.org/10.5120/739-1038
11. Rumelhart, D.E., Hinton, G.E., Williams, R.J.: Learning representations by back-propagating errors. Nature **323**, 533–536 (1986). https://doi.org/10.1038/323533a0
12. Srivastava, N., Hinton, G., Krizhevsky, A., Sutskever, I., Salakhutdinov, R.: Dropout: a simple way to prevent neural networks from overfitting. J. Mach. Learn. Res. **15**(56), 1929–1958 (2014). http://jmlr.org/papers/v15/srivastava14a.html
13. Varlamis, I., Tserpes, K., Etemad, M., Júnior, A.S., Matwin, S.: A network abstraction of multi-vessel trajectory data for detecting anomalies. In: Papotti, P. (ed.) Proceedings of the Workshops of the EDBT/ICDT 2019 Joint Conference, EDBT/ICDT 2019, Lisbon, Portugal, March 26, 2019. CEUR Workshop Proceedings, vol. 2322. CEUR-WS.org (2019). http://ceur-ws.org/Vol-2322/BMDA_5.pdf
14. Wiseman, S., Rush, A.M.: Sequence-to-sequence learning as beam-search optimization. CoRR abs/1606.02960 (2016). http://arxiv.org/abs/1606.02960
15. Yuan, Z., Jiang, Y., Li, J., Huang, H.: Hybrid-dnns: Hybrid deep neural networks for mixed inputs. CoRR abs/2005.08419 (2020). https://arxiv.org/abs/2005.08419
16. Zhang, A., Lipton, Z.C., Li, M., Smola, A.J.: Dive into deep learning. CoRR abs/2106.11342 (2021). https://arxiv.org/abs/2106.11342
17. Zhang, C., et al.: Ais data driven general vessel destination prediction: a random forest based approach. Transp. Res. Part C: Emerg. Technol. **118**, 102729 (2020). https://doi.org/10.1016/j.trc.2020.102729, http://www.sciencedirect.com/science/article/pii/S0968090X20306446

Scheduling Drillships in Offshore Activities

Rafael Gardel Azzariti Brasil[1], Marco Aurélio de Mesquita[1], Dario Ikuo Miyake[1], Tiago Montanher[2], and Débora P. Ronconi[1](✉)

[1] Department of Production Engineering, University of São Paulo, Av. Luciano Gualberto, 1380, Cidade Universitária, São Paulo, SP 05508-010, Brazil
dronconi@usp.br

[2] Faculty of Mathematics, University of Vienna, Oskar-Morgenstern-Platz 1, 1090 Vienna, Austria

Abstract. This paper addresses the scheduling of offshore oil well construction using drilling vessels. Drilling costs constitute a substantial part of the total development costs for an offshore field, thus planning the efficient use of drilling rigs is crucial to ensure economic feasibility of oil and gas exploration and production (E&P) projects. The objective of this study is to minimize the completion time of all operations involved in the development of subsea wells considering the availability of the drilling rigs. These activities are drilling and completion of the well, and maintenance activities. Technical constraints and availability of the drilling vessels and release dates and precedence constraints of the activities are considered. In addition, vessel eligibility restrictions are respected. A mixed integer linear programming model was developed considering the goals and constraints above. Numerical experiments using instances based on real-world situations show adequate behavior, which demonstrates that it faithfully represents the situation portrayed and can be used, combined with more advanced optimization techniques, to achieve better results.

Keywords: Oil wells · Offshore activities · Mixed integer linear programming

1 Introduction

Despite the transition on course for the development of lower carbon sources of energy and the severe impacts on the global demand for fuels caused by the Covid-19 pandemic, the supply of oil and gas tends to continue playing an important role in global energy systems in the coming decades [9,30]. Even in a scenario of accelerating investments in renewable sources of energy, it will be necessary to keep heavy investments in the exploration and production (E&P) projects to avoid an imbalance of supply that may cause significant increases of prices and risk of supply disruptions [9,16].

M. Mes et al. (Eds.): ICCL 2021, LNCS 13004, pp. 66–81, 2021.
https://doi.org/10.1007/978-3-030-87672-2_5

The E&P activities constitute the upstream segment of the oil and gas industry which include onshore and offshore drilling. Nowadays, more than a quarter of the oil and gas supply is produced offshore and a relevant share of new reserves additions is expected to keep coming from fields underneath the ocean. However, offshore drilling is more expensive than onshore drilling. The drilling costs can be estimated in different ways (e.g. based on functional category, time dependency, variable or fixed cost classification) encompassing several components [12,17]. A common way is to consider that to a great extent, they will depend on the drilling time, which includes the non-productive time, and the daily rig rate [2,14]. In values of 2010, for deepwater wells, offshore rig rates per day could be six to eight times more expensive than for onshore wells. On its turn, the number of days will be a function of the well depth. Hence, while usual wells up to 20,000 ft requires 70 to 80 days, deeper wells up to 32,000 ft may need up to 150 days. Daily rig rate will vary according to the rig type. As of March 2021, the average day rate of semisubmersible drilling units and drillships to operate in water depth greater than 7,500 ft was, respectively, about USD 125,000 and USD 180,000 much higher than for jackup rigs used to operate in water depths up to 500 ft [13]. This, along with the greater distance from shore, contributes to make deepwater projects require more investment compared to onshore or shallow water developments. It is worth noting that numerous other factors may influence the efficiency of offshore drillings, such as geologic conditions, purpose of the well (i.e. exploratory, development, etc.), trajectory, water depth, hole diameter, weather, and operator's experience, among others [12,14]. Even so, propelled by advancements in E&P technologies and efficiency improvement efforts, new fields in deepwater areas have accounted for around half of discovered oil and gas resources over the last ten years [15]. Nowadays, much of deepwater oil production is concentrated in Angola, Brazil, Nigeria and United States [15]. In the coming years, it is expected that Brazil will remain as one of main drivers of drilling and well services expenditure to expand deepwater E&P activities [24]. To a great extent, this dynamism has been promoted by the opening of the Brazilian sector of oil and gas to allow the participation of foreign and other local players, besides Petrobras - Brazil's state controlled oil company -, in exploration and production activities. Since then, several multinational oil companies were granted rights to act as operators in order to explore and produce in the pre-salt fields [5]. As the drilling costs can make up some 25 to 35% of the total development costs for an offshore field [28], for these operators, it is imperative to seek the most efficient use of the drilling units they hire for their E&P projects given that by the end of 2019, the greatest share of the Brazilian oil (96%) and natural gas (81%) proved reserves was located in offshore areas [22].

Given the growth in the Brazilian offshore oil production in the past years and, especially, the contribution of wells from the pre-salt layer to this outcome, studies intended to characterize and improve this process are very relevant. Following this trend, our research study focus on the scheduling problem of oil well development activities. These activities can be the drilling, the completion (end phase) of the wells, as well as the maintenance activities. Technical constraints of each drilling rig, the ready times of the vessels, the release dates of the activities, and the precedence constraints among the activities are considered. Moreover, eligibility constraints set by the company are considered. A MILP model considering the constraints that structure this complex situation is proposed and evaluated using instances based on real-world situations. The considered objective is the minimization of the completion time of all activities considering the availability of the oil drilling rigs. This optimization criterion, known in the job scheduling literature as makespan, usually implies high resource utilization. As one of the main components of the well development cost is the cost of hiring the drilling ships, reducing their idleness is a relevant point for the optimization of this process.

Considering the characteristics of the problem under study, we present below studies conducted on the ship routing and scheduling problem which addresses the very issue of maximizing the efficiency of critical resources in offshore operations. Christiansen et al. [6] presented a survey approaching the subject of ship routing and scheduling in recent years. The authors highlighted the relevance of this theme for global development, presenting the increase in the fleet and the cargo to be transported, from 1980 to 2010. Hennig et al. [10] addressed a problem of routing and scheduling heterogeneous transportation ships with pickup and delivery of crude oil. Characteristics of this problem were the loading capacities at the ports, fractional loads, and the time windows for both pickup and delivery. A path flow model, which uses pre-generated ship routes, has been introduced to minimize fuel costs and port charges. Hennig et al. [11] presented a simpler and more compact model, in addition to detailing the pre-generated ship routing procedure. Another work that addresses the transportation of crude oil of various types can be found in Nishi and Izuno [23]. The authors addressed the problem of routing and scheduling of ships with fractional deliveries aiming to minimize distances and obeying the capacity of tankers, which collect oil in various points of the world and distribute it to some customers. The authors proposed a mixed integer linear programming (MILP) model and applied an iterative heuristic based on the column generation procedure. Lee and Kim [18] addressed the routing problem with fractional deliveries, time windows, and heterogeneous fleet in the context of a steel manufacturing company. In the considered problem there were two types of ships: owned, which can perform multiple pickups and multiple deliveries on a route; and contracted, which perform point-to-point activities. A MILP model was presented for the problem as well as a heuristic algorithm, based on Adaptive Large Neighborhood Search.

Assis and Camponogara [8] dealt with the problem of relieving ships, which perform the transport between the oil platforms, which are located offshore and the onshore terminal. A MILP model based on a graph in which the vertices represent the terminal and the platforms was presented. Route feasibility, number of ships to be used, inventory and cargo transfer balance constraints are established. In addition, two heuristic methods were applied: rolling-horizon and relax-and-fix. A MILP approach in the oil industry was also proposed by

Lin et al. [19] in a problem of scheduling a fleet of marine ships to unload crude oil into tankers. This type of activity is often related to the transfer of crude oil from a discharging tanker to smaller ships to make the tanker lighter. Stanzani et al. [27] addressed a real-world multi-ship routing and scheduling problem with inventory constraints that arise in crude oil gathering and delivery operations from multiple offshore platforms to coastal terminals. MILP models were presented to handle small to moderate cases; while a matheuristic was proposed to handle larger cases. A comprehensive analysis of the costs involved in abandoning wells that have reached their maturity stage was conducted in [1]. Additionally, the author proposes MILP models to represent the strategic, tactical and operational planning of this process, which involves the routing and scheduling of ships with high daily allocation costs.

Addressing the area of scheduling of Pipe Laying Support Vessels (PLSV), Bassi and Ronconi [4] tackled the problem of connecting subsea oil wells to offshore platforms through the use of the outsourced fleet of PLSVs. A MILP model was developed considering several features of the problem, among them the increase of the production curve using injection wells to combat the natural decline of well production over time. The objective considered was the maximization of the oil production curve. Also related to the area of scheduling PLSVs, [7,21,26] conducted studies making an analogy between routing activities with scheduling activities in parallel machines. In the most recent research [7] the goal was to reschedule identical PLSVs in order to minimize the impacts caused by operational interruptions. A MILP model was proposed for the routing problem that starts from previously created activity blocks (voyage) composed of setup times, activities to be performed and the return navigation to the loading base. The only navigation considered separately is between the port and the exploratory region. Additionally, a method based on the Iterated Local Search (ILS) meta-heuristic was proposed.

Considering oil well development activities, Pereira et al. [25] optimized the use of PLSVs and drilling rigs using the GRASP (Greedy Randomized Adaptive Search Procedure) metaheuristic. The objective was to maximize the oil production within a given time horizon. Moura et al. [20] addressed the same problem considered in [25], considering however the time for resource displacement. GRASP was also considered as the solving method. Bassi et al. [3] tackled the problem of planning and scheduling a fleet of rigs to be used for well drilling or for maintenance activities with uncertain service times. The authors developed a procedure based on simulation and optimization strategies to generate the expected solutions.

Through the literature review it can be observed that a large portion of the researches presents a mathematical model to represent the ship routing and scheduling problem in different contexts. However, few authors that approached the scheduling of oil well development activities presented MILP formulations. We point out that the representation of this problem as an integer linear programming model makes it possible to define the problem precisely, as well as to apply exact or heuristic techniques that are based on the model, such as matheuristics and relax-and-fix strategies.

This paper is organized as follows. Section 2 describes details of the problem while Sect. 3 presents the proposed mixed integer linear programming model. Section 4 reports numerical experiments with a set of small-sized instances based on real-world data. Section 5 presents the final considerations and next steps suggested for this study.

2 Problem Description

An offshore oil well life cycle can be summarized in the following phases: prospecting, construction, interconnection, production, maintenance, and closure. The initial phases of prospecting, construction, and interconnection are managed by an oil well project manager and involve a wide range of activities, which need to be carefully managed.

In the first stage, the characteristics of the well and its construction are defined, and risk and economic analyses are carried out before the project is approved. Once approved, the project moves on to the construction stage, consisting of a series of activities that can change greatly from one project to another. In the next phase, interconnection, the well is connected to a pipeline network and is ready to go into production. In the production phase, the well requires maintenance activities and, at the end of its useful life, the well is deactivated (abandoned). In all mentioned activities, except in production, a critical resource is specialized vessels, such as, drillships and pipelaying ships, used in the construction and interconnection phases of the wells, respectively. The drillships are also used in maintenance activities, whether scheduled or emergency.

This article addresses the scheduling of drillships used in the construction of offshore oil wells in a maritime oil field. The scheduling for this operation considers two main input data: *i*) a portfolio of approved projects and *ii*) a fleet of drillships in operation. The drillships scheduling is dynamic, as the vessels are in continuous operation while new projects are released. Thus, the scheduling must consider when each drillship will be available for a new operation, called "ready time". The projects also have an "earliest start date" since they require a range of materials and equipment for the execution of the operations in addition to the vessels. The availability dates for these materials and equipment are planned by the supply area and, based on these dates, the operation's earliest start date is defined, which will be called "release date." Once we know the ships' availability and the projects' release dates, the new projects may be scheduled.

Considering the construction stage, we have different activities that are previously planned by the engineering area. From the scheduling point of view, this activity will be divided into only two stages: drilling and completion. Completion consists of constructive activities that, after drilling, prepare the well for production. It includes finishing the well, installing equipment, among other activities. Ideally, drilling and completion activities should be carried out in sequence and without interruption, using the same drillship. However, due to contingencies of the operation, there are cases where there is a time gap between the execution of these activities, and the ship is moved from one geographical position to another. Thus, in the scheduling model, it is considered that each construction project consists exactly of two and sequential activities.

The proposed model aims to provide a schedule of the drilling vessels' activities that considers, in addition to the constraints of availability of the resources and the activities themselves, the time of displacement of the vessels from one geographical position to another. It also considers that the ships are different in terms of efficiency and that not all drillships are eligible for each of the projects. Based on the characteristics of the rigs, the engineering determines the time required to carry out drilling and completion activities for each of the eligible rigs, as well as the travel time between two construction sites.

It is noteworthy that in this study, we consider only two activities in the construction schedules. However, extending the model to consider cases with more activities and more general precedence relations is possible. Our choice is adequate for the case at hand and does not significantly compromise the computational efficiency of running the model. Another highlight is that, as it was developed, considering the initial availability of the rigs, the model meets the needs of rescheduling the operations, which could be triggered either by the launch of new projects or by uncertainty in the execution of offshore well construction projects.

3 Mathematical Model

The mixed integer programming model presented next is based on the model presented by [29] to minimize the completion time of the last task in a system with parallel heterogeneous machines with sequence-dependent setup times. In the proposed MILP model, the sequence-dependent setup times represent the travel times of the vessels from one position to another. The processing times of the tasks correspond to the execution times of the drilling and completion activities, which are dependent on the allocated drillship.

Let m be the number of rigs and n the number of activities. There is at least one rig i that can perform an activity j, so that we have: p_{ij} $(i=1,\ldots,m;$ $j=1,\ldots,n)$ the processing time of an activity j using rig i. A dummy activity 0 will be used to represent the starting point of each rig. Let t_{ijk} $(i=1,\ldots,m,$ $j=1,\ldots,n;$ $k=1,\ldots,n)$ be the travel time of rig i, after having performed activity j, to perform activity k. At the beginning of the schedule, each rig i may be at a particular location where an activity k is located, in which case $t_{i0k}=0$. The continuous variables C_{ij} $(i=1,\ldots,m;$ $j=1,\ldots,n)$ represent the completion time of an activity j using rig i. The following model aims to minimize the completion time of the last activity in the system, known in the job shop scheduling literature as makespan. Table 1 summarizes the notation used in the model.

Table 1. Notation of the model.

Sets	
$M = \{1, \ldots, m\}$	Set of rigs
$N = \{1, \ldots, n\}$	Set of activities
PR	Pairs of activities (u, v), where u must be finished before v can start to be processed
Parameters	
p_{ij}	Processing time of activity j using the rig i; $i \in M, j \in N$
t_{ijk}	Travel time of rig i, after execute activity j, to execute the activity k; $i \in M, j, k \in N$
V, W	Big numbers
r_i	Ready time of rig i; $i \in I$
F_i	End of contract of rig i; $i \in M$
$\min_k^{activ}$	Lower bound for the beginning of activity k (release date); $k \in N$
$\max_k^{activ}$	Upper bound for the beginning of activity k; $k \in N$
a_k^i	Binary constant whose value is 1 if the rig i is able to perform activity k; $k \in N$
Decision variables	
x_{ijk}	1, if activity j precedes activity k on rig i; 0, otherwise
C_i^{dril}	Completion time of the service of rig i;
C_{ij}	Completion time of activity j on rig i
y_{ik}	Auxiliary binary variable associated to rig i and the activity k
p'_k	Processing time of the activity k
C_k^{activ}	Completion time of activity k

MILP Model:

$$\text{Minimize } c_{max} \tag{1}$$

subject to

$$c_{max} \geq c_{i,j}, \quad i \in \mathcal{M}, j \in \mathcal{N}, \tag{2}$$

$$\sum_{i \in \mathcal{M}} \sum_{\substack{j \in \{0\} \cup \mathcal{N} \\ j \neq k}} x_{ijk} = 1, \quad k \in \mathcal{N}, \tag{3}$$

$$\sum_{i \in \mathcal{M}} \sum_{\substack{k \in \mathcal{N} \\ j \neq k}} x_{ijk} \leq 1, \quad j \in \mathcal{N}, \tag{4}$$

$$\sum_{k \in \mathcal{N}} x_{i0k} \leq 1, \quad i \in \mathcal{M}, \tag{5}$$

$$\sum_{\substack{h \in \{0\} \cup \mathcal{N} \\ h \neq k, h \neq j}} x_{ihj} \geq x_{ijk}, \quad j, k \in \mathcal{N}, j \neq k, i \in \mathcal{M}, \tag{6}$$

$$c_{ik} + V(1 - x_{ijk}) \geq c_{ij} + t_{ijk} + p_{ik}, \quad j \in \{0\} \cup \mathcal{N}, k \in \mathcal{N}, j \neq k, i \in \mathcal{M}, \tag{7}$$

$$c_{i0} = r_i, \quad i \in \mathcal{M}, \tag{8}$$

$$c_i^{drill} \geq c_{ij}, \quad i \in \mathcal{M}, j \in \mathcal{N}, \tag{9}$$

$$c_i^{drill} \leq F_i, \quad i \in \mathcal{M}, \tag{10}$$

$$p'_k = \sum_{i \in \mathcal{M}} \sum_{\substack{j \in \{0\} \cup \mathcal{N} \\ j \neq k}} p_{ik} x_{ijk}, \quad k \in \mathcal{N}, \tag{11}$$

$$c_k^{activ} = \sum_{i \in \mathcal{M}} c_{ik}, \quad k \in \mathcal{N}, \tag{12}$$

$$c_{ik} \leq W y_{ik}, \quad k \in \mathcal{N}, i \in \mathcal{M}, \tag{13}$$

$$1 - \sum_{\substack{j \in \{0\} \cup \mathcal{N} \\ j \neq k}} x_{ijk} \leq W(1 - y_{ik}), \quad i \in \mathcal{M}, k \in \mathcal{N}, \tag{14}$$

$$c_k^{activ} - p'_k \geq \min_k^{activ}, \quad k \in \mathcal{N}, \tag{15}$$

$$c_k^{activ} - p'_k \leq \max_k^{activ}, \quad k \in \mathcal{N}, \tag{16}$$

$$c_v^{activ} - p'_v \geq c_u^{activ}, \quad (u, v) \in PR, \tag{17}$$

$$\sum_{j \in \{0\} \cup \mathcal{N}} x_{ijk} \leq a_k^i, \quad k \in \mathcal{N}, i \in \mathcal{M}, \tag{18}$$

$$\sum_{k \in N} p'_k \leq \sum_{i \in \mathcal{M}} (F_i - r_i), \tag{19}$$

$$c_{ij} \geq 0, \quad j \in \mathcal{N}, i \in \mathcal{M}, \tag{20}$$

$$x_{ijk} \in \{0, 1\}, \quad j \in 0 \cup \mathcal{N}, \quad k \in \mathcal{N}, j \neq k, i \in \mathcal{M}, \tag{21}$$

$$y_{ik} \in \{0, 1\}, \quad k \in \mathcal{N}, i \in \mathcal{M}. \tag{22}$$

The objective function (1) aims to minimize the completion time of the last activity in the system (makespan). Constraints (2) establish the minimum value of the makespan. The constraint set (3) ensures that each activity is assigned to exactly one rig and has exactly one predecessor activity. Constraints (4) set the maximum number of successors of each activity to one. Constraints (5) limit the number of successors of the fictitious activity 0 to at most one per rig. The set of constraints (6) ensures that if activity j is allocated to a rig i, a predecessor h of this activity must already exist on the same rig. Constraints (7) aim to establish that if an activity k is allocated to vessel i after activity j, i.e., $x_{ijk} = 1$, its completion time, C_{ik}, has to be greater than or equal to the completion time of activity j plus the travel time between j and k and the processing time of k. Constraints (8) guarantee that vessel i can only start its activities after the ready time r_i, while constraints (9) state that a rig will only finish its trip after the completion of all activities assigned to this rig. In addition, the set of restrictions (10) states that rigs must complete their activities within the period in which they were contracted. Constraints (11) calculate the processing time of activity k considering that the activity will be executed by only one of the available rigs. With the same consideration, constraints (12) determine the completion time of activity k. Constraints (13) and (14) establish, using the auxiliary binary variable y_{ik}, that if an activity k is not allocated to a drilling ship i, the variable corresponding to its completion time on this vessel, C_{ik}, is equal to zero. Constraints (15-16) set the minimum and maximum limits for the beginning of activity k, while constraints (17) ensure that the precedence relationships between activities are respected. The constraint set (18) restricts the use of the vessels for certain activities. A parameter a_k^i is provided to check which ship can be used for each of the activities to be performed. Constraints (19) state that the sum of the activity processing times must be equal to or less than the total time available for vessels activities. Constraints (20), (21), and (22) indicate the domain of decision variables.

4 Numerical Experiments

Aiming to evaluate the performance of the proposed MILP model, several instances based on real-world data from the exploration and production projects of a major oil company that operates in Brazil were solved. Three different types of offshore activities performed by drillships were considered: drilling, completion and workover activities. Table 2 presents the characteristics of the vessels. Each drillship is identified by a number ("Drillship") and has an available time window ("Available From" and "Available Until" columns). The following columns specify the starting location of the drillship ("Pos. X" and "Pos. Y") and its speed ("Speed"). The last three columns specify the standard times for drilling, completion, and workover activities in days ("Drill", "Comp.", and "Workover"). Table 3 describes the activities that must be scheduled. The first two columns identify the activity and its type ("Activity" and "Type"). The fourth column ("Prec.") identifies the preceding activity if any (otherwise it is zero-filled). The activities belong to a project ("Proj.") that has an associated geographic location ("Pos. X" and "Pos. Y"). Finally, activities also have a time window defined by an earliest start time and a latest start time ("EST" and "LST") in days from time zero. Figure 1 shows the geographical distribution of the drillships and activities considered in this experiment. It is worth noting that the data were changed due to a confidentiality request by the oil company that provided access to the considered situations.

Table 2. Drillships: characteristics and technical limitations.

Drillship	Available from	Available until	Pos. X	Pos. Y	Speed	Drill	Comp.	Workover
1	31	857	100.00	392.56	406	36	33	24
2	12	973	750.00	100.00	451	43	23	17
3	28	472	303.44	366.08	240	60	34	35
4	13	496	349.28	372.24	293	64	29	13
5	0	364	301.43	100.00	226	46	31	25
6	20	364	500.00	400.00	226	61	28	17
7	26	218	600.00	50.00	257	45	24	38

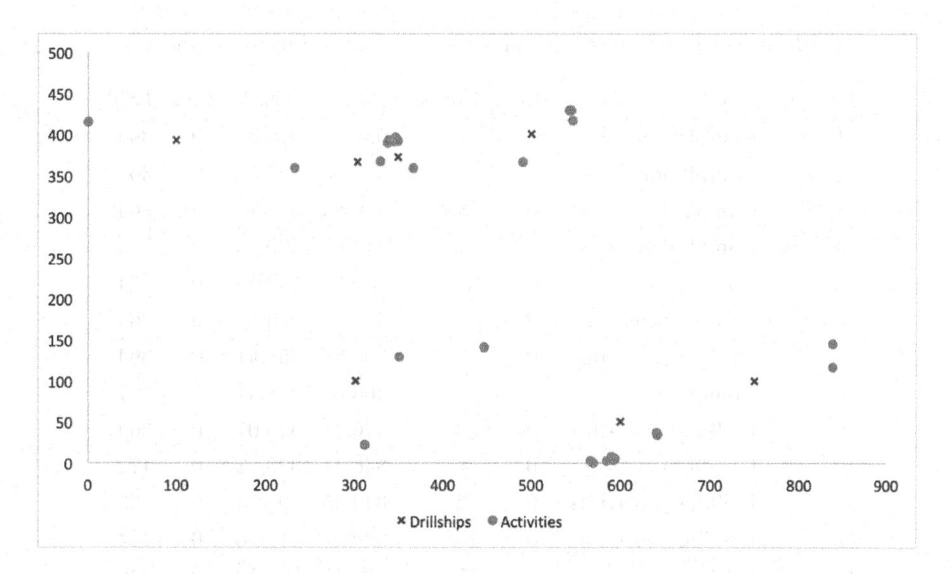

Fig. 1. Geographical positions of the drillships and activities of the experiments.

The dataset of vessels (Table 2) and activities (Table 3) allows composing different test instances. In order to illustrate a solution of the focused problem Fig. 2 shows the Gantt chart of the schedule achieved for the problem instance made of the first five drillships and the first twelve activities (5×12 instance).

Next, we aim to evaluate the performance of the proposed method in a set of small-sized instances in order to obtain its optimal solution. In this set the number of drillships varying between 2 and 7 and the number of activities varying from 10 to 13. Numerical experiments were run on a Huawei Matebook 13 computer with an AMD Ryzen 5 3500U 2.1 GHz and 8 Gb of RAM memory. The solver GUROBI 9.0, with all its default parameters, was able to find feasible solutions for all evaluated instances. A CPU time limit of one hour per instance was imposed. Table 4 presents the results obtained. The first column shows the dimension of each instance, while the next three columns indicate the core dimensions of the associated MILP model. The following columns show the objective function value, the running time required to solve each instance, and the column named *Gap* (%) indicates the percentage difference between the incumbent solution and the lower bound determined by the solver.

Table 3. Offshore activities: characteristics and technical limitations.

Activity	Type	Prec.	Project	Pos. X	Pos. Y	EST	LST
1	Completion	0	1	0.00	415.60	0	364
2	Completion	0	2	542.95	429.59	0	364
3	Completion	0	3	546.86	416.95	0	364
4	Completion	0	4	344.83	390.83	0	62
5	Completion	0	5	339.06	392.55	0	364
6	Drilling procedure	0	6	545.27	429.58	0	364
7	Drilling procedure	0	7	366.68	358.41	0	364
8	Completion	7	7	366.68	358.41	0	101
9	Drilling procedure	0	8	346.41	396.07	0	364
10	Completion	9	8	346.41	396.07	0	172
11	Drilling procedure	0	9	349.45	391.74	0	23
12	Drilling procedure	0	10	0.00	415.60	0	364
13	Drilling procedure	0	11	337.15	389.59	0	240
14	Completion	13	11	337.15	389.59	0	364
15	Drilling procedure	0	12	490.10	366.40	0	138
16	Drilling procedure	0	13	233.03	358.59	0	31
17	Completion	0	14	570.03	0.00	14	259
18	Workover	0	15	567.05	2.59	30	110
19	Drilling procedure	0	16	58517	2.90	29	183
20	Drilling procedure	0	17	591.46	5.62	7	56
21	Completion	20	17	591.46	5.62	20	103
22	Drilling procedure	0	18	592.29	3.20	24	104
23	Completion	22	18	592.29	3.20	34	102
24	Drilling procedure	0	19	311.39	21.91	0	1750
25	Completion	24	19	311.39	21.91	0	714
26	Workover	0	20	640.44	36.87	25	34
27	Drilling procedure	0	21	641.15	34.01	22	178
28	Completion	27	21	641.15	34.01	32	240
29	Completion	0	22	350.85	128.82	26	364
30	Completion	0	23	328.71	367.17	11	364
31	Completion	0	24	838.99	116.39	18	364
32	Completion	0	25	838.99	144.74	19	364
33	Drilling procedure	0	26	447.21	141.02	5	364
34	Completion	33	26	447.21	141.02	23	364
35	Drilling procedure	0	27	594.31	5.35	0	364
36	Completion	35	27	594.31	5.35	0	364
37	Workover	0	28	590.57	7.36	0	4

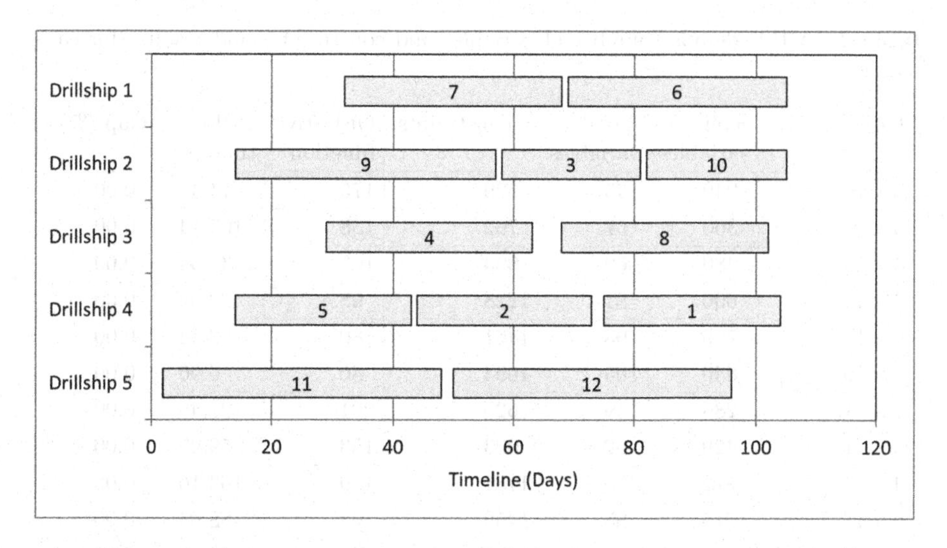

Fig. 2. Gantt chart for an instance with 5 drillships and 12 activities with the following values of the completion times: $C_{1,7} = 68$, $C_{1,6} = 105$, $C_{2,9} = 57$, $C_{2,3} = 81$, $C_{2,10} = 105$, $C_{3,4} = 63$, $C_{3,8} = 102$, $C_{4,5} = 43$, $C_{4,2} = 73$, $C_{4,1} = 104$, $C_{5,11} = 48$, $C_{5,12} = 96$.

Analyzing the figures in Table 4, it can be seen that the solver was able to achieve the optimal solution within the allowed runtime in almost 80% of the instances. For the remaining instances, the average difference was about 13%. As expected, the solutions found by the solver for the tighter instances (higher rate of activities per probe) contain results (instant of termination of the last activity) equal to or worse than the results for the looser instances. To illustrate the results achieved with the model, Fig. 2 shows an optimal schedule for an instance with five drilling ships and twelve activities.

In order to illustrate the effect of considering drilling and well completion activities as separate activities, the instance with five drillships and twelve activities was adapted. In this case, activities 7 and 8 were considered as a single activity at the same location. The same adaptation was performed with activities 9 and 10. As expected, the quantities of binary and real variables, 600 and 81, respectively, were reduced (see Table 4), as was the resolution time (14.50 s). However, on the other hand, the final completion time of all system activities, 111 days, increased by approximately 6%.

Table 4. MILP details (number of variables and constraints) and results (for each instance.

Drill × Activ.	#binary variables	#real variables	#constraints	Objective function	CPU time (s)	Gap (%)
2 × 10	240	45	529	175	14.44	0.00
3 × 10	360	57	762	138	107.14	0.00
4 × 10	480	69	995	105	31.54	0.00
5 × 10	600	81	1228	98	22.92	0.00
6 × 10	720	93	1461	80	8.44	0.00
7 × 10	840	105	1694	80	6.96	0.00
2 × 11	286	49	625	201	21.46	0.00
3 × 11	429	62	903	153	59.97	0.00
4 × 11	572	75	1181	129	163.10	0.00
5 × 11	715	88	1459	103	88.04	0.00
6 × 11	858	101	1737	98	115.40	0.00
7 × 11	1001	114	2015	80	13.07	0.00
2 × 12	336	53	729	212	37.21	0.00
3 × 12	504	67	1056	171	470.81	0.00
4 × 12	672	81	1383	138	765.27	0.00
5 × 12	840	95	1710	105	523.84	0.00
6 × 12	1008	109	2037	101	268.19	0.00
7 × 12	1176	123	2364	95	284.76	0.00
2 × 13	390	57	841	225	352.59	0.00
3 × 13	585	72	1221	179	3600.17	10.61
4 × 13	780	87	1601	145	3600.04	24.04
5 × 13	975	102	1981	124	3608.05	21.03
6 × 13	1170	117	2361	105	3600.05	7.62
7 × 13	1365	132	2741	100	3600.09	3.00

5 Conclusion and Further Research

This article dealt with the scheduling of offshore oil well construction using drillships. Considering that the costs of this operation constitute a substantial part of the total costs of developing an offshore field, the efficient scheduling of drilling rigs is crucial for the economic viability of oil exploration and production projects. Following this premise, this study aimed to minimize the completion time of a portfolio of subsea oil well development projects, considering the restrictions of precedence between activities and other operational restrictions such as eligibility, readiness, and contracts of rigs and readiness and deadlines for activities.

A mixed integer linear programming model was developed, considering the practical constraints of the problem addressed. Numerical experiments using examples based on real-world situations demonstrate that the model faithfully represents the described operation and can be used to achieve better schedules combined with more advanced optimization techniques. These numerical experiments showed consistent results, showing a reduction in the completion time of activities, as expected, with a larger number of drillships available. This indicates that with a larger number of vessels, despite the operational restrictions, more activities can occur in parallel, allowing for the anticipation of the start of production of producing wells and the increase in production volume in the analyzed period. It is noteworthy that the model presented can be easily adapted to consider different objective functions, such as minimizing the sum of the completion times or minimizing the total tardiness of the projects' completion based on the projects' due date.

As a future research topic, we suggest developing heuristic procedures to tackle larger instances. However, since our target problem has many limitations, preparing an advanced procedure is a big challenge because the solution can easily lose its feasibility characteristics.

Acknowledgments. This study was financially supported by the National Agency for Petroleum, Natural Gas and Biofuels (ANP) through the clauses for funding of Research, Development and Innovation (R, D&I) investments established by the Resolution no. 50/2015 and by PETROBRAS under grant TC 5900.0112830.19.9. The authors are grateful for the collaboration of Pedro H. M. S. Sousa and Marco V. S. Nóbrega from PETROBRAS-CENPES and to the Brazilian funding agencies CAPES, CNPq and FAPESP that partially supported this research.

References

1. Aarlott, M.M.: Cost Analysis of Plug and Abandonment Operations on the Norwegian Continental Shelf. Master's thesis, NTNU (2016)
2. Amado, L.: Field cases evaluations, chap. 12. In: Reservoir Exploration and Appraisal, pp. 53–156. Gulf Professional Publishing, Oxford (2013)
3. Bassi, H.V., Ferreira Filho, V.J.M., Bahiense, L.: Planning and scheduling a fleet of rigs using simulation-optimization. Comput. Ind. Eng. **63**(4), 1074–1088 (2012)
4. Bassi, S., Ronconi, D.P.: Optimization of the use of critical resources in the development of offshore oil fields. In: Gervasi, O., et al. (eds.) ICCSA 2020. LNCS, vol. 12249, pp. 391–405. Springer, Cham (2020). https://doi.org/10.1007/978-3-030-58799-4_29
5. Brito, T.C.: The growing relevance of the PPSA in the Brazilian exploration and production arena and its main challenges. J. World Energ. Law Bus. **12**(2), 156–168 (2019)
6. Christiansen, M., Fagerholt, K., Nygreen, B., Ronen, D.: Ship routing and scheduling in the new millennium. Eur. J. Oper. Res. **228**(3), 467–483 (2013)
7. Cunha, V., Santos, I., Pessoa, L., Hamacher, S.: An ILS heuristic for the ship scheduling problem: application in the oil industry. Int. Trans. Oper. Res. **27**(1), 197–218 (2020)

8. de Assis, L.S., Camponogara, E.: A MILP model for planning the trips of dynamic positioned tankers with variable travel time. Transp. Res. Part E Logist. Transp. Rev. **93**, 372–388 (2016)
9. Empresa de Pesquisa Energética: Oil price forecasts 2021–2030 (February 2021). https://www.epe.gov.br/sites-en/publicacoes-dados-abertos/publicacoes/Paginas/Special-Report-Oil-Price-Forecasts-2021-2030.aspx
10. Hennig, F., et al.: Maritime crude oil transportation - a split pickup and split delivery problem. Eur. J. Oper. Res. **218**(3), 764–774 (2012)
11. Hennig, F., Nygreen, B., Furman, K.C., Song, J.: Alternative approaches to the crude oil tanker routing and scheduling problem with split pickup and split delivery. Eur. J. Oper. Res. **243**(1), 41–51 (2015)
12. Hossain, M.E.: Drilling costs estimation for hydrocarbon wells. J. Sustain. Energ. Eng. **3**(1), 3–32 (2015)
13. IHS Markit: Petrodata offshore rig day rate trends (April 2021). https://ihsmarkit.com/products/oil-gas-drilling-rigs-offshore-day-rates.html
14. Ikwan, U., Egba, A.N., Dosunmu, A., Iledare, W.: Comparative analysis of drilling cost used for petroleum economics in the North Sea, Gulf of Mexico and Niger Delta regions. In: Society of Petroleum Engineers (eds.) SPE Nigeria Annual International Conference and Exhibition, Lagos, Nigeria, vol. 2, pp. 1179–1190 (2016)
15. International Energy Agency: Offshore energy outlook (May 2018). https://www.iea.org/reports/offshore-energy-outlook-2018
16. International Energy Agency: Oil 2021 - analysis and forecast to 2026 (March 2021). https://www.iea.org/reports/oil-2021
17. Kaiser, M.J.: Modeling the time and cost to drill an offshore well. Energy **34**(9), 1097–1112 (2009)
18. Lee, J., Kim, B.I.: Industrial ship routing problem with split delivery and two types of vessels. Exp. Syst. Appl. **42**(22), 9012–9023 (2015)
19. Lin, X., Chajakis, E.D., Floudas, C.A.: Scheduling of tanker lightering via a novel continuous-time optimization framework. Ind. Eng. Chem. Res. **42**(20), 4441–4451 (2003)
20. Moura, A.V., Pereira, R.A., De Souza, C.C.: Scheduling activities at oil wells with resource displacement. Int. Trans. Oper. Res. **15**(6), 659–683 (2008)
21. Moura, V.C.: Programação de frota de embarcações de lançamento de dutos. Master's thesis, Universidade de São Paulo (2012)
22. National Agency for Petroleum: Natural Gas and Biofuels: oil, natural gas and biofuels statistical yearbook 2020 (2020). http://www.anp.gov.br/publicacoes/anuario-estatistico/statistical-yearbook-2020
23. Nishi, T., Izuno, T.: Column generation heuristics for ship routing and scheduling problems in crude oil transportation with split deliveries. Comput. Chem. Eng. **60**, 329–338 (2014)
24. Offshore Magazine: Brazil, Guyana to boost offshore drilling market recovery (2021). https://www.offshore-mag.com/drilling-completion/article/14199897/brazil-guyana-to-boost-offshore-drilling-market-recovery
25. Pereira, R.A., Moura, A.V., de Souza, C.C.: Comparative experiments with GRASP and constraint programming for the oil well drilling problem. In: Nikoletseas, S.E. (ed.) WEA 2005. LNCS, vol. 3503, pp. 328–340. Springer, Heidelberg (2005). https://doi.org/10.1007/11427186_29
26. Queiroz, M.M., Mendes, A.B.: Heuristic approach for solving a pipe layer fleet scheduling problem. In: Sustainable Maritime Transportation and Exploitation of Sea Resources, pp. pp. 1073–1080. Taylor & Francis Group, London (2012)

27. de Lorena Stanzani, A., Pureza, V., Morabito, R., da Silva, B.J.V., Yamashita, D., Ribas, P.C.: Optimizing multiship routing and scheduling with constraints on inventory levels in a Brazilian oil company. Int. Trans. Oper. Res. **25**(4), 1163–1198 (2018)
28. Thakkar, A., Raval, A., Chandra, S., Shah, M., Sircar, A.: A comprehensive review of the application of nano-silica in oil well cementing. Petroleum **6**(2), 123–129 (2020)
29. Vallada, E., Ruiz, R.: A genetic algorithm for the unrelated parallel machine scheduling problem with sequence dependent setup times. Eur. J. Oper. Res. **211**(3), 612–622 (2011)
30. World Bank Group: Commodity markets outlook: causes and consequences of metal price shocks (April 2021). https://thedocs.worldbank.org/en/doc/c5de1ea3b3276cf54e7a1dff4e95362b-0350012021/original/CMO-April-2021.pdf

Solving a Real-Life Tramp Ship Routing and Scheduling Problem with Speed Profiles

Lucas Louzada, Rafael Martinelli(✉), and Victor Abu-Marrul

Departamento de Engenharia Industrial, Pontifícia Universidade Católica do Rio de Janeiro (PUC-Rio), Rio de Janeiro, Brazil
lucas.louzada@aluno.puc-rio.br, martinelli@puc-rio.br, victor.cunha@tecgraf.puc-rio.br

Abstract. Shipowners seek to increase their profits by optimizing the operation of their available fleet, increasing its capacity, and reducing costs while meeting the customers' demands. Ship routing stands out as a relevant topic of study, especially for tramp shipping companies, due to the high competitivity on this market, which highlights the importance of providing reliable and price competitive services. This work presents a mixed-integer programming formulation to maximize the profit of an actual tramp shipping company. The studied problem considers pick-up-and-delivery for different cargoes, partial contract orders, time window restrictions, heterogeneous fleet, cargo split, varying navigation speeds, and guaranteed transit time terms. We perform computational experiments with real data from the studied company, comparing the achieved solutions with those developed by the company following their current planning process. The mathematical formulation improves the existing solutions in all tested cases with total costs up to 7% smaller, including fuel, port, and ship's operational costs.

Keywords: Ship routing · Mixed-integer programming · Tramp shipping · Speed profiles

1 Introduction

The United Nations Conference on Trade and Development (UNCTAD), in its 2018 annual review [30], highlights that maritime modal transports around 80% of the total volume of the global trade of goods, reaching more than 10.7 billion tons of cargoes transported in 2017. The review by UNCTAD indicates that, since seaborne trade plays such an essential role in the international trade of goods and the global economy overall, the global fleet of ships and the industries related to it can be considered the backbone of global trading. The review also states that the industry of maritime transportation is volatile, being directly impacted by changes in geopolitics, economics, and international trade policies, affecting its profitability. Moreover, the imbalance between the availability and

M. Mes et al. (Eds.): ICCL 2021, LNCS 13004, pp. 82–96, 2021.
https://doi.org/10.1007/978-3-030-87672-2_6

demand of ships combined with the low global economic growth affected the maritime transport companies. The reduction in demand for freight combined with the increase in fuel prices (the highest variable cost in operating a ship) leads ship operators to seek strategies to increase profits and reduce costs [31].

Maritime transportation is usually classified as: liner, industrial, and tramp shipping [21]. Liner shipping companies plan their fleet to follow a strict itinerary, similar to a bus line. Thus, the cargo shippers must adjust their demand for cargo transportation to the available schedule [7]. Industrial shipping companies control their fleet and decide how to schedule it to transport their cargo, which is more common in vertically integrated businesses, such as oil, chemical, and mining companies [7,27]. Finally, tramp shipping companies operate similarly to a taxi service, i.e., the ships are planned according to a demand for cargo transportation [7]. On the one hand, this shipping model works better when the demand for cargo transportation increases; on the other hand, it is less effective for the operating companies when the demand decreases. For this reason, contracts of affreightment are usually fixed between cargo shippers and ship owners or operators, in which a certain amount of cargo is transported between defined ports within an agreed period. Freight is paid on the transported cargo unit, and ship operators aim to maximize profit by time unit.

As highlighted by Fagerholt [9], the ship scheduling directly affects the operating costs for ship owners. And, these schedules are usually manually defined according to expert knowledge within the companies. Ship routing and scheduling problems have been increasingly studied by academics, with most works focusing on industrial shipping with recent growth in tramp shipping planning problems. The current increase in tramp shipping is because companies decide to focus on their core business, outsourcing their cargo transportation by contracting independent ship owners [26].

In this work, we deal with a ship scheduling problem related to a real tramp shipping company. The studied company is investing in several initiatives to optimize the schedule of its fleet, improving the service level for its customers, and increasing the flexibility to fix spot cargoes. Spot cargoes can be defined as irregular, sporadic cargoes, usually available for prompt loading and covered by a contract associated with a specific shipment [33]. We propose a Mixed-Integer Programming (MIP) formulation to solve the tramp ship routing and scheduling problem for the studied company to minimize the operating costs and increasing the service level towards customers. Real data from the company is considered, such as orders, ports, cargo quantity, vessel capacity, and contractual restrictions, enabling the mathematical formulation to generate optimal routing solutions suitable to the actual problem. To the best of our knowledge, no attempt has been made to consider all these aspects simultaneously. The vessel planners within the company analyzed the generated solutions and validated them for later implementation with customers. The mathematical formulation currently works as a decision support tool for the ship planners within the company. Before developing the model, we performed two steps to allow applying it to a real-life problem described as following: process mapping to define the scope of the problem among with the vessel cargo lifting, port, and contractual restrictions (such as arrival windows at load ports and transit time); data gathering regarding orders, ships, distances, and the actual company's planning. The company defines a plan for periods of 30 to 45 days. After implementing the formulation, we conducted experiments with data regarding six months demands.

This work is organized into six sections, including this introduction. A literature review related to tramp ship routing and scheduling problems is presented in Sect. 2. Section 3 describes the studied problem, while Sect. 4 presents the developed MIP formulation. In Sect. 5, computational experiments are performed. The results are presented and compared to real solutions provided by the company planners. Finally, Sect. 6 concludes the paper and includes some recommendations for future studies.

2 Literature Review

Al-Khayyal and Hwang [1] define that cargo routing problems are often restrictive since they consider the cargo conditions, either due to the aspects of the loading and discharging ports (such as capacity and productivity) or due to the arrival windows and transit times required for the ships to reach their destinations. To Christiansen et al. [6], the concept of scheduling includes the temporal aspects in a ship routing problem, i.e., when planners must consider time-marked events regarding the vessels' voyages. Both works consider ship routing and scheduling problems of high importance for the tramp and industrial shipping companies. Christiansen et al. [6] define 'cargo' as an individual number of products to be collected in a specific port, transported, and discharged in a particular delivery port. Usually, an arrival window is imposed at the loading port and, potentially also at the discharging port. The ship operator controls a heterogeneous fleet that is available for transporting the cargo demands. Several reasons might prevent ships from being compatible with loading certain cargoes. Christiansen et al. [6] mention draft constraints at the loading and discharging ports, for instance. Depending on the ship's capacity and the total amount of cargo transported within a certain period, a ship could load multiple cargoes on a voyage. For major bulk cargoes, such as grains and iron ore, it is typical for the entire cargo on board of a ship to belong to only one shipper with only one loading port for a given voyage. Minor bulks and liquids can be jointly loaded within the vessels.

The work of Flood [12], one of the first studies regarding ship routing, deals with a homogeneous military fleet that transports fuel. The author considered equal ship operational and port costs focusing on re-positioning ballast voyages to minimize the operating costs. Brown et al. [5] address a tanker fleet routing problem considering variable speeds and time windows. This work is a significant reference for problems including speeds as variables in the routing decision-making process [11,13,24,32].

Fagerholt and Christiansen [10] apply a dynamic programming algorithm to solve a traveling salesman problem similar to a ship routing problem. Andersson et al. [2] studies a collection and delivery problem with arrival windows at the loading port and orders split. The authors propose a methodology that generates routes using an exact method to use in two network flow models. Regarding split cargo shipments, Stålhane et al. [28] use a Branch-and-Cut algorithm for solving the problem. Vilhelmsen et al. [31] introduces a column generation approach to solve full shipment problems limited to one cargo on-board, integrating and optimizing fuel bunkering.

Regarding the use of heuristics and metaheuristics, Homsi et al. [16] highlights some works using different approaches in this context for solving ship routing and scheduling problems as follows: multi-start local search by Brønmo et al. [4], unified tabu search by Korsvik and Fagerholt [18], large neighborhood search by Korsvik et al. [20] and Hemmati et al. [14], and hybrid genetic search by Borthen et al. [3]. Homsi et al. [16] also use a hybrid genetic search combining it with a Branch-and-Cut algorithm.

Several works deal with tramp ship routing and scheduling problems, although none of them deals with break-bulk cargo. We refer the reader to the works of Andersson et al. [2] and Kang et al. [17] for a review about these problems. We also highlight the works of Stålhane et al. [29], and Hemmati et al. [15], in which shipowner companies and charterer vessel are considered. Based on our literature review, we note a continued interest in tramp shipping problems. Different aspects are considered in the literature, such as heterogeneous fleets, arrival windows for loading and discharging, pickup and delivery, transit time, and cargo lifting capacity. Nevertheless, to the best of our knowledge, no attempt has been made to consider all these aspects simultaneously addressing a real-life tramp shipping problem.

3 Problem Description

The addressed tramp ship scheduling problem considers a real-life problem of a company that operates a permanent fleet of over 90 ships. However, we focus our approach on the fleet that operates routes between the East Coast of South America and Asian ports, with four cargo loading ports in Brazil and 12 discharging ports across China, South Korea, and Japan. The company plans its fleet for horizons of 30 to 45 days. The available space within the ships is commercialized through long-term Contracts of Affreightment (CoAs) or spot fixtures (one-time or few-times loading in a short period). Spot cargoes with fixtures before the foreseeable scheduling period for the respective route are equivalent to affreightment contracts.

The company aims at accomplishing all contracted cargo transportation orders, including spot, minimizing the total operational cost. Each order is identified by the demanding customer, the loading port, the discharging port, the total cargo volume to be transported, the arrival window at the loading port, and the maximum transit time between the ports. All these information and conditions are defined by contract. Orders can be served by more than one ship due to the limited capacity of the vessels. After loading the cargo, all ships proceed to the port of Singapore for bunkering, due to its lower fuel price, to further continue their voyage to the discharging port. Thus, the fuel price is given by the average fuel price at Singapore port considering the planning horizon. Vessels can operate under three possible speed profiles (full speed, economical speed, and super economical speed), independently if it transports cargo or not. As highlighted by Psaraftis [25], the possibility to adjust a ship's speed allows the ship operator to reduce costs by avoiding idleness at the port.

Whenever a ship anchors or berths, known as a port call, port fixed or variable fees are charged (the latter depending on the length of stay in the port). Fixed costs represent 98% of the total cost of a port call [22]. Thus, we only consider fixed costs for the port calls. The company's fleet is heterogeneous, and each ship type has a daily operational cost (USD/day), regardless of the route performed, which covers crew, maintenance, certification, and insurance costs. Larger and older ships tend to have higher daily operational costs when compared to smaller or newer ones. Vessels should be available for loading the cargo within a pre-defined time window agreed with the client. Moreover, the company must accomplish all CoAs. Each ship has a different number of cranes, equipment capacity, cargo hold dimensions, and crane operation visibility. We consider average loading productivity (tons/day) for each port, which will affect the length of stay in the port of each order. Due to the fleet's heterogeneity, not all ships can load in all ports due to their width or length. Furthermore, some ports have arrival and departure draft restrictions. Draft is the distance between the waterline and the ship's keel. The more cargo a ship has on board, the deeper a ship will be and closer to reaching a port's maximum allowable draft. A ship's capacity at a given port is calculated considering the local draft restriction and water density. All ships in the fleet have a maximum cargo intake associated with these restrictions.

4 MIP Formulation

We propose a MIP formulation to model and solve the real-life tramp ship scheduling problem described in the previous section. Table 1 shows the sets considered by the MIP formulation, while Tables 2 and 3 presents the parameters and variables, respectively. The formulation uses six decision variables. The binary variable x_{ijk} assumes 1 if a ship k travels from the loading port i to the discharging port j, and 0, otherwise. Following the same idea, the binary variable s_{ijkv} defines the speed profile s that a ship k will follow while navigating between ports i and j. The parameter S_{ks} gives the actual speed (in knots) of vessel k when navigating under profile s. As mentioned previously, three speed profiles can be considered: full speed, economical speed, and super economical speed. The idea is to reduce costs by controlling the navigation speeds to accomplish the required time windows for the ship's arrival in the loading ports. The third binary variable, y_{rk}, assumes 1 if order r is addressed by ship k, and 0, otherwise. The continuous variable z_{ikr} indicates the cargo volume of order r loaded at port i by vessel k. The continuous variable w_{ik} indicates the total cargo on board of ship k after it departs from port i. This variable is important to control the maximum cargo a ship can take at each port. The last continuous variable, t_{ik}, identifies the arrival time of ship k at loading port i that must respect the allowed time window $[LE_r, LB_r]$, if ship k addresses order r.

Table 1. Formulation sets

Name	Description
N	Ports, indexed by i and j
$Nc \subseteq N$	Loading ports
$Nd \subseteq N$	Discharging ports
K	Ships, indexed by k
R	Orders, indexed by r
$R_i \subseteq R$	Subset of orders to be loaded in port i
V	Speed profiles, indexed by s: 1 (full speed), 2 (economical speed) and 3 (super economical speed)

Table 2. Formulation parameters

Name	Description
O_k	Starting point of ship k
Z_k	Starting day of ship k
D_{ij}	Distance, in nautical miles, between ports i and j
Q_{ik}	Maximum cargo intake of ship k at port i
c	Port of Singapore
F	Artificial port where all ships end
CD_k	Daily operational cost in USD of ship k
CP_i	Cost of port i
PR_{ik}	Productivity (tons per day) of ship k at port i
S_{ks}	Speed (in knots) of ship k under speed profile s
CC_{ks}	Daily fuel consumption (in tons) of ship k at speed profile s
LE_r	Starting time of arrival window at the loading port of order r
LB_r	Final time of arrival window at loading port of order r
TT_r	Maximum transit time of order r
BP	Fuel price in USD per ton
Ol_r	Loading port of order r
Od_r	Discharging port of order r
Oq_r	Total cargo of order r

Table 3. Formulation decision variables

Name	Type	Description
x_{ijk}	Binary	1, if ship k travels from port i to port j. 0, otherwise
y_{rk}	Binary	1, if order r is carried on ship k. 0, otherwise
s_{ijkv}	Binary	1, if ship k travels from port i to port j at speed v. 0, otherwise
z_{ikr}	Continuous	Cargo volume carried from order r loaded or discharged at port i on board ship k
w_{ik}	Continuous	Cargo volume remaining on board ship k after calling port i
t_{ik}	Continuous	Arrival time for ship k at port i

The developed mathematical formulation extends the one proposed by Korsvik and Fagerholt [19], with transit time constraints and speed profiles decision variables included fitting the actual problem addressed. The objective function, shown in Eq. (1), aims at minimizing the total operating cost of the company's fleet, considering fuel consumption costs, port fees costs, and ship operational costs. The mathematical formulation, including the objective function and the respective constraints, is shown in the following:

$$\min \underbrace{\sum_{k\in K}\sum_{i\in N}\sum_{j\in N}\sum_{v\in V} S_{kv}CC_{kv}BPs_{ijkv}}_{\text{Fuel cost}} + \underbrace{\sum_{k\in K}\sum_{i\in N}\sum_{j\in N} CP_i x_{ijk}}_{\text{Port cost}} + \underbrace{\sum_{k\in K} CD_k\left(t_{Fk} - Z_k\right)}_{\text{Ship cost}} \quad (1)$$

subject to

$$\sum_{i\in N} x_{ijk} \leq 1 \qquad \forall k \in K, j \in N \quad (2)$$

$$\sum_{j\in N} x_{ijk} \leq 1 \qquad \forall k \in K, i \in N \quad (3)$$

$$\sum_{i\in N} x_{ijk} - \sum_{i\in N} x_{jik} = 0 \qquad \forall k \in K, j \in N\backslash\{O_k, F\} \quad (4)$$

$$\sum_{k\in K}\sum_{j\in N} x_{ijk} \geq 1 \qquad \forall i \in N \quad (5)$$

$$\sum_{r\in R_i} z_{ikr} \leq w_{ik} \leq Q_{ik} \qquad \forall k \in K, i \in Nc \quad (6)$$

$$\sum_{r\in R_i} z_{ikr} + w_{ik} \leq Q_{ik} \qquad \forall k \in K, i \in Nd \quad (7)$$

$$\sum_{k\in K} y_{rk} \geq 1 \qquad \forall r \in R \quad (8)$$

$$\sum_{i\in Nc\cup\{c\}} x_{Ol_r jk} \geq y_{rk} \qquad \forall k \in K, r \in R \quad (9)$$

$$\sum_{i \in Nd \cup \{F\}} x_{Od_r jk} \geq y_{rk} \qquad \forall k \in K, r \in R \tag{10}$$

$$\sum_{k \in K} z_{Ol_r kr} = Oq_r \qquad \forall r \in R \tag{11}$$

$$z_{Ol_r kr} \leq Oq_r y_{rk} \qquad \forall k \in K, r \in R \tag{12}$$

$$\sum_{k \in K} z_{Od_r kr} = Oq_r \qquad \forall r \in R \tag{13}$$

$$z_{Od_r kr} \leq Oq_r y_{rk} \qquad \forall k \in K, r \in R \tag{14}$$

$$z_{Ol_r kr} = z_{Od_r kr} \qquad \forall k \in K, r \in R \tag{15}$$

$$w_{ij} = \left(w_{ik} + \sum_{r \in R_j} z_{jkr} \right) x_{ijk} \qquad \forall k \in K, i \in Nc, j \in Nc \cup \{c\} \tag{16}$$

$$w_{ij} = \left(w_{ik} - \sum_{r \in R_j} z_{jkr} \right) x_{ijk} \qquad \forall k \in K, i \in Nd \cup \{c, F\}, j \in Nd \tag{17}$$

$$\sum_{v \in V} s_{ijkv} = x_{ijk} \qquad \forall k \in K, i \in N, j \in N \tag{18}$$

$$t_{O_k k} \geq Z_k \qquad \forall k \in K \tag{19}$$

$$t_{jk} \geq \left(t_{ik} + \sum_{v \in V} \frac{s_{ijkv} D_{ij}}{24 S_{kv}} + \sum_{r \in Ri} \frac{z_{ikr}}{PR_{ik}} \right) x_{ijk} \qquad \forall k \in K, i \in N, j \in N \tag{20}$$

$$LB_r y_{rk} \leq t_{Ol_r k} \leq LE_r y_{rk} \qquad \forall k \in K, r \in R \tag{21}$$

$$t_{Od_r k} \leq \left(t_{Ol_r k} + \sum_{r \in Ri} \frac{z_{Ol_r kr}}{PR_{Ol_r k}} + TT_r \right) y_{rk} \qquad \forall k \in K, r \in R \tag{22}$$

$$x_{ijk} + x_{jik} = 1 \qquad \forall k \in K, i \in N, j \in N \tag{23}$$

$$\sum_{j \in Nc} x_{O_k jk} = 1 \qquad \forall k \in K \tag{24}$$

$$\sum_{i \in Nc} x_{ick} = 1 \qquad \forall k \in K \tag{25}$$

$$\sum_{j \in Nd} x_{cjk} = 1 \qquad \forall k \in K \tag{26}$$

$$x_{ijk} = 0 \qquad \forall k \in K, i \in Nc, j \in Nd \tag{27}$$

$$\sum_{i \in Nd} x_{iFk} = 1 \qquad \forall k \in K \tag{28}$$

$$\sum_{i \in Nc} x_{iFk} = 0 \qquad \forall k \in K \tag{29}$$

$$w_{O_k k} = 0 \qquad \forall k \in K \tag{30}$$

$$w_{Fk} = 0 \qquad \forall k \in K \tag{31}$$

$$x_{ijk} \in \{0, 1\} \qquad \forall k \in K, i \in N, j \in N \tag{32}$$

$$y_{kr} \in \{0, 1\} \qquad \forall k \in K, r \in R \tag{33}$$

$$s_{ijkv} \in \{0,1\} \qquad \forall k \in K, i \in N, j \in N, v \in V \tag{34}$$

$$z_{ikr} \geq 0 \qquad \forall k \in K, i \in N, r \in R \tag{35}$$

$$w_{ik} \geq 0 \qquad \forall k \in K, i \in N \tag{36}$$

$$t_{ik} \geq 0 \qquad \forall k \in K, i \in N \tag{37}$$

Constraints (2) and (3) guarantee that each ship only attends a port call once, whether for loading or discharging. With the former referring to departing ports, and the latter to arriving ports. Constraints (4) guarantee the flow conservation of the ships. Constraints (5) forces all ports calls (orders) to be addressed. Constraints (6) make sure the amount of cargo after ship k departs port i respects the ship's maximum intake (tons) at this same port, also ensuring that the volume w_{ik} (remaining on board of the ship after departing from port i) is superior to the amount of cargo loaded at that port. Constraints (7) do the same but for the discharging ports. Constraints (8) guarantee that every order is addressed at least by one ship. Constraints (9) and (10) force ship k to visit the loading and discharging ports of an addressed order r. Constraints (11) and (12) guarantee that the cargo loaded within a ship k respects the total amount of cargo regarding an addressed order r, while constraints (13) and (14) do the same but for discharging ports. Constraints (15) ensure that the amount loaded is equal to the amount discharged if a ship k addresses an order r. Constraints (16) and (17) compute variables w_{ik}. We linearize these constraints using a large value M [23], but we present it in its quadratic form for a better understanding. Constraints (18) connect the speed profile variables with the routing variables. Constraints (19) compute the period in which each ship is available to start its route. Constraints (20) compute the arrival time of ship k at port i, considering the departure from the previous port visited and the traveling times between the ports. We also linearize these constraints using a large value M [23]. Constraints (21) guarantee the arrival at the loading port according to the customer's declaration or contractual terms if a ship k addresses an order r. Constraints (22) force the transit time between the loading port i and the discharging port j to respect the maximum transit time TT_r of an order r. Constraints (23) force every ship departing to a port different from the previous one. Constraints (24) guarantee that each ship route starts from its starting port. Constraints (25) and (26) force all ships to perform a call at Singapore port, following the company's policy. Constraints (28) force all routes to end at the artificial node F. Constraints (29) remove arcs between loading ports and the artificial node F. Constraints (30) and (31) force all ships to be empty at the beginning and at the end of its routes, respectively. Finally, constraints (32) to (37) set the variables' domains.

5 Computational Experiments

This section presents the experiments conducted on a set of six real-life instances from the studied tramp shipping company. Each instance represents one specific period planned by the company, which comprises 30 to 45 days, with numerous available ships and several orders to serve. Each order is associated with a volume of cargo to transport. The identifier of each instance with the total number of orders and ships, total cargo volume (in metric tons), percentage cargo volume, and bunker price (in dollars) are shown in Table 4. The cargo volume percentage indicates the percentage of the available capacity among the vessels that needs to be used to accommodate the total cargo volume. Higher values limit the model's decision to allocate cargo on ships. All experiments were performed on a computer with an Intel Core i7-7600U with a 2.80 Ghz CPU and 8.00 GB of RAM running Windows 10. The proposed MIP formulation was coded in Julia language v.0.19.2 using the JuMP library [8], and the model was solved by Gurobi 8.1 solver.

Table 4. Instances' details

Instance	Orders	Ships	Total cargo Volume	Cargo volume Percentage	Bunker Price
1	14	3	135,200	86%	370.00
2	17	6	335,589	93%	412.00
3	13	6	300,646	97%	432.50
4	19	7	379,486	95%	427.00
5	24	8	429,088	94%	427.00
6	13	6	243,577	80%	349.00

We compare all results with the actual ship schedules developed by the company's planners. Three shipping planners validated our results regarding the operational viability and the solution cost calculation: the current tramp ship planner within the company, who has five years of experience, the commercial manager, and a ship planner expert, with more than twenty-five years working with ship scheduling problems in the company. According to the experts' validation, the proposed approach can be used as an actual decision support tool within the company due to the feasibility and quality of the solutions provided in a computational time that respects their process. Usually, when a new demand arrives, the company takes 24 to 48 h to respond if they can accomplish it due to their limited capacity. Moreover, the company defines when the order will be addressed and how much will be charged for the cargo transportation. Thus, we limited the model execution to 24 h. Anyway, none of the instances took 24 h to run, the faster ones took less than 3 h, while the harder one took almost 16 h to reach optimality. Table 5 depicts the percentage deviation of our solutions in relation to the ones provided by the company regarding the six real-life instances described above in terms of the objective function elements (fuel cost, port cost, ship operation cost). We use the deviation to present the results due to confidential issues regarding the real costs of the solutions. In the company's solution, the average contribution of the fuel costs to the total objective function value is 34%, the port costs 8% and the operating costs 58%. In the obtained solution, these contributions change to 31%, 8%, and 60%, respectively.

Table 5. Detailed results regarding the objective function elements.

Instance	1	2	3	4	5	6
Total fuel costs	−7%	−15%	−17%	−22%	−6%	−1%
Total port costs	−5%	−29%	6%	67%	−22%	−21%
Total ship operating costs	−7%	6%	0%	−2%	−2%	−1%
Total costs	−7%	−4%	−6%	−5%	−5%	−3%

One can note that the MIP formulation reduces the total operating costs in all instances, with the largest one being Instance 1 with costs 7% smaller when compared to the solution provided by the company experts. It is interesting to see that, even when some costs increases (see for example the increase of 67% in the port costs of Instance 4), the total costs are at least 3% smaller. In the context of high fuel prices, the mathematical model focused on defining more loading port calls to reduce discharging port calls, reducing the distance traveled, and consequently, the fuel consumption. In addition, the optimization of the ships' speed allowed the model to efficiently reduce the fuel consumption while accomplishing the time windows for the addressed orders.

In Table 6, we detail the obtained solutions considering several aspects, highlighting the deviation from the MIP formulation solutions to the company's solutions in terms of the number of loading port calls, number of discharging port calls, the total number of port calls, capacity occupation, total days of the ships' operation, nautical miles traveled, total days of navigation, and total fuel consumption. One can note a more significant reduction in the indicators that affect fuel costs, such as fuel consumption and nautical miles traveled due to the high fuel prices, as highlighted previously. Note that in some cases, the model increased the total number of port calls (Instances 3 and 4), the total number of ships' operating days (Instance 2), or even the total number of navigation days (Instances 2, 3 and 5). However, those decisions led to better solutions in terms of total costs, with a reduced capacity occupation variance. An interesting aspect of this analysis is highlighting the model capacity of modifying the solutions according to the real aspects of each scenario. This makes the approach flexible and capable of providing solutions for different conditions of operational demands and costs, in many cases making decisions that a company planner would never make. Usually, programmers follow fixed rules with low flexibility and inadequate response to different input data characteristics.

Some interesting points were discussed with the company experts while validating and comparing the MIP formulation solutions with their solutions. Regarding Instance 1, the solution of the mathematical formulation led to a paradigm change for the planners by defining the largest ship available to perform

Table 6. Detailed results regarding other solution aspects.

Instance	1	2	3	4	5	6
Loading port calls	−13%	−25%	27%	36%	6%	−17%
Discharging port calls	0%	−27%	−13%	39%	−18%	0%
Total port calls	−5%	−26%	4%	38%	−8%	−8%
Capacity occupation	−2%	0%	−2%	0%	−2%	0%
Ships' operating days	−6%	6%	0%	−4%	−1%	−1%
Nautical miles traveled	0%	−6%	−1%	−9%	−2%	−46%
Sailing days	−6%	1%	1%	−8%	1%	−2%
Fuel consumption	−7%	−18%	−3%	−10%	−6%	−1%

the shortest route, leading to more balanced routes and better use of the vessel's capacities. Planners were used to allocate larger ships on more extensive routes to always navigate to more distant destinations. Concerning Instance 4, the mathematical model defined the final ports of the routes of two ships in the southern region of China, a region generally with low demand for cargo loading. However, according to planners, the study of this strategy seems to be interesting as it could reduce navigation costs due to its location and the easiness of attending any demanding port in China. The results and the discussions led the company to start using the MIP formulation as a decision support tool to help ship planners.

6 Conclusion and Future Works

Given the scenario that a large part of international trade takes place through maritime transport, arises the opportunity for tramp shipping companies to meet cargo transportation demands worldwide. Tramp shipowners have faced, in recent years, difficulties in increasing freight rates, among several reasons, due to the asymmetry between the availability of vessels and cargo demand, highlighting the importance of seeking operational costs reduction. The optimization of ship routing and scheduling, usually done manually by experienced professionals in these organizations, is increasingly being explored by academic literature, helping ship owners enhance their competitivity.

In this work, we studied a real-life tramp ship scheduling problem to address a set of cargo transportation orders to be loaded on the East Coast of South America and discharged in Asian ports, minimizing the total operating costs of the company's fleet. We consider navigation speeds as a problem's decision to accomplish the available loading time windows for the vessels in the ports.

The study has an essential role in addressing a real problem and in developing an approach that improves the current scheduling methodology used in the studied company. According to the validation within the company, our approach serves as a decision support tool for ship planners. We conducted experiments on six instances with actual data from the company with different scenarios of vessel availability and demands for cargo transportation. The results show a clear advantage of using the proposed approach to improve solutions in all instances, with costs up to 7% smaller. Due to the good results achieved, the company expects to expand its applicability and capture more spot cargo fixtures.

Throughout the development of the work, we noticed some innovation opportunities within the studied company. The application of approaches explored in the academic literature to ship routing and scheduling problems can help to reduce costs and improve service levels within the company. Furthermore, details of the company's operation can be further explored by the academic community to support new researches. Below, we highlight some opportunities for future research:

- Include specific constraints for spot parcels, as an optional decision to be taken during the planning horizon.
- Investigate the relationship between speed and fuel consumption as continuous functions instead of using discrete levels for these variables.
- Consider sustainability indicators in the problem, such as reducing the social and environmental impact.
- Analyze longer planning horizons with up to one year.
- Include other trade routes incorporating strategic (allocation of ships on the lines) and tactical (ports, sequence, and volumes of each ship) decisions.
- Consider uncertainties on the problem, such as seasonal weather events which affect productivity and congestion in ports, variation in fuel prices and daily operating cost, associated with market indexes for each line and type of ship.

Acknowledgments. This study was financed in part by PUC-Rio, by the Coordenação de Aperfeiçoamento de Pessoal de Nível Superior - Brasil (CAPES) - Finance Code 001, by the Conselho Nacional de Desenvolvimento Científico e Tecnológico (CNPq) under grant 315361/2020-4 and by Fundação de Amparo à Pesquisa do Estado do Rio de Janeiro (FAPERJ) under grant E-26/010.002232/2019. The financial support is gratefully acknowledged.

References

1. Al-Khayyal, F., Hwang, S.J.: Inventory constrained maritime routing and scheduling for multi-commodity liquid bulk, part i: applications and model. Eur. J. Oper. Res. **176**(1), 106–130 (2007)
2. Andersson, H., Christiansen, M., Fagerholt, K.: The maritime pickup and delivery problem with time windows and split loads. INFOR: Inf. Syst. Oper. Res. **49**(2), 79–91 (2011)
3. Borthen, T., Loennechen, H., Wang, X., Fagerholt, K., Vidal, T.: A genetic search-based heuristic for a fleet size and periodic routing problem with application to offshore supply planning. EURO J. Transp. Logist. **7**(2), 121–150 (2017). https://doi.org/10.1007/s13676-017-0111-x
4. Brønmo, G., Christiansen, M., Nygreen, B.: Ship routing and scheduling with flexible cargo sizes. J. Oper. Res. Soc. **58**(9), 1167–1177 (2007)

5. Brown, G.G., Graves, G.W., Ronen, D.: Scheduling ocean transportation of crude oil. Manage. Sci. **33**(3), 335–346 (1987)
6. Christiansen, M., Fagerholt, K., Nygreen, B., Ronen, D.: Ship routing and scheduling in the new millennium. Eur. J. Oper. Res. **228**(3), 467–483 (2013)
7. Christiansen, M., Fagerholt, K., Ronen, D.: Ship routing and scheduling: status and perspectives. Transp. Sci. **38**(1), 1–18 (2004)
8. Dunning, I., Huchette, J., Lubin, M.: Jump: a modeling language for mathematical optimization. SIAM Rev. **59**(2), 295–320 (2017)
9. Fagerholt, K.: A computer-based decision support system for vessel fleet scheduling-experience and future research. Decis. Support Syst. **37**(1), 35–47 (2004)
10. Fagerholt, K., Christiansen, M.: A travelling salesman problem with allocation, time window and precedence constraints-an application to ship scheduling. Int. Trans. Oper. Res. **7**(3), 231–244 (2000)
11. Fagerholt, K., Christiansen, M., Hvattum, L.M., Johnsen, T.A., Vabø, T.J.: A decision support methodology for strategic planning in maritime transportation. Omega **38**(6), 465–474 (2010)
12. Flood, M.M.: Application of transportation theory to scheduling a military tanker fleet. J. Oper. Res. Soc. Am. **2**(2), 150–162 (1954)
13. Gatica, R.A., Miranda, P.A.: Special issue on Latin-American research: a time based discretization approach for ship routing and scheduling with variable speed. Netw. Spat. Econ. **11**(3), 465–485 (2011)
14. Hemmati, A., Hvattum, L.M., Fagerholt, K., Norstad, I.: Benchmark suite for industrial and tramp ship routing and scheduling problems. INFOR: Inf. Syst. Oper. Res. **52**(1), 28–38 (2014)
15. Hemmati, A., Stålhane, M., Hvattum, L.M., Andersson, H.: An effective heuristic for solving a combined cargo and inventory routing problem in tramp shipping. Comput. Oper. Res. **64**, 274–282 (2015)
16. Homsi, G., Martinelli, R., Vidal, T., Fagerholt, K.: Industrial and tramp ship routing problems: closing the gap for real-scale instances. Eur. J. Oper. Res. **283**(3), 972–990 (2020)
17. Kang, K., Zhang, W., Guo, L., Ma, T.: Research on ship routing and deployment mode for a bulk. In: 2012 19th Annual Conference Proceedings of the International Conference on Management Science & Engineering, pp. 1832–1837. IEEE (2012)
18. Korsvik, J.E., Fagerholt, K.: A tabu search heuristic for ship routing and scheduling with flexible cargo quantities. J. Heuristics **16**(2), 117–137 (2010)
19. Korsvik, J.E., Fagerholt, K., Laporte, G.: A tabu search heuristic for ship routing and scheduling. J. Oper. Res. Soc. **61**(4), 594–603 (2010)
20. Korsvik, J.E., Fagerholt, K., Laporte, G.: A large neighbourhood search heuristic for ship routing and scheduling with split loads. Comput. Oper. Res. **38**(2), 474–483 (2011)
21. Lawrence, S.A.: International Sea Transport: The Years Ahead. Lexington Books (1972)
22. Meersman, H., Strandenes, S.P., Van de Voorde, E.: Port pricing: principles, structure and models. NHH Department of Economics Discussion Paper (14) (2014)
23. Miller, C.E., Tucker, A.W., Zemlin, R.A.: Integer programming formulation of traveling salesman problems. J. ACM (JACM) **7**(4), 326–329 (1960)
24. Norstad, I., Fagerholt, K., Laporte, G.: Tramp ship routing and scheduling with speed optimization. Transp. Res. Part C Emerg. Technol. **19**(5), 853–865 (2011)
25. Psaraftis, H.N.: Ship routing and scheduling: the cart before the horse conjecture. Marit. Econ. Logist. **21**(1), 111–124 (2019)

26. Rowbotham, M.: Introduction to Marine Cargo Management. Informa Law from Routledge (2014)
27. Sanghikian, N., Martinelli, R., Abu-Marrul, V.: A hybrid VNS for the multi-product maritime inventory routing problem. In: Mladenovic, N., Sleptchenko, A., Sifaleras, A., Omar, M. (eds.) ICVNS 2021. LNCS, vol. 12559, pp. 111–122. Springer, Cham (2021). https://doi.org/10.1007/978-3-030-69625-2_9
28. Stålhane, M., Andersson, H., Christiansen, M., Cordeau, J.F., Desaulniers, G.: A branch-price-and-cut method for a ship routing and scheduling problem with split loads. Comput. Operat. Res. **39**(12), 3361–3375 (2012)
29. Stålhane, M., Andersson, H., Christiansen, M., Fagerholt, K.: Vendor managed inventory in tramp shipping. Omega **47**, 60–72 (2014)
30. UNCTAD: Review of Maritime Transport. United Nations Conference on Trade and Development, New York and Geneva, vol. e.18 ii.d, 5 edn. (2018). https://unctad.org/en/PublicationsLibrary/rmt2018_en.pdf. Accessed 25 May 2020
31. Vilhelmsen, C., Lusby, R.M., Larsen, J.: Tramp ship routing and scheduling with voyage separation requirements. OR Spectrum **39**(4), 913–943 (2017). https://doi.org/10.1007/s00291-017-0480-4
32. Wen, M., Ropke, S., Petersen, H.L., Larsen, R., Madsen, O.B.: Full-shipload tramp ship routing and scheduling with variable speeds. Comput. Oper. Res. **70**, 1–8 (2016)
33. Yu, B., Wang, K., Wang, C., Yao, B.: Ship scheduling problems in tramp shipping considering static and spot cargoes. Int. J. Shipping Transp. Logist. **9**(4), 391–416 (2017)

Optimizing Maritime Preparedness Under Uncertainty – Locating Tugboats Along the Norwegian Coast

Julie Louise Musæus, Håkon Nøstvik, Henrik Andersson, and Peter Schütz[✉]

Department of Industrial Economics and Technology Management, Norwegian University of Science and Technology, Trondheim, Norway
{julielm,haakonmn}@stud.ntnu.no
{henrik.andersson,peter.schuetz}@ntnu.no

Abstract. We study the strategic problem of locating tugboats along the Norwegian coast to optimize maritime preparedness. The problem is formulated as a two-stage stochastic program. In the first stage, we locate the tugboats such that nominal coverage requirements are satisfied, whereas we deploy the located tugboats in the second stage in order to assist vessels in distress. The objective is to minimize the sum of the costs of publicly operated tugboats in the emergency towing service and the expected penalty costs due to insufficient preparedness. We solve the problem using Sample Average Approximation in combination with a self-developed heuristic. Our results indicate that we can achieve a sufficient preparedness level with six tugboats.

Keywords: Maritime preparedness · Set covering problem · Uncertainty

1 Introduction

A safe and reliable maritime transportation system is of great economic importance to Norway: Approximately 83% of all international cargoes (in tonnes) and about 44% of all domestic cargoes (in tonne-kilometres) are carried on board ships [11]. In addition, overall ship traffic (in kilometres) along the coast of Norway is predicted to increase by 41% until 2040. However, maritime transportation is risky and the number of accidents is also expected to increase by as much as 30% during the same period [15].

A recent example of the risk in maritime transportation is the incident involving the MV Eemslift Hendrika that was abandoned off the west coast of Norway in April 2021. The ship drifted for 2 days before tugboats were able to secure the ship and avert grounding of the ship. Up to 350 tonnes of heavy fuel oil and up to 50 tonnes of diesel fuel could have caused severe environmental pollution in the area if the ship's hull had been damaged [23].

M. Mes et al. (Eds.): ICCL 2021, LNCS 13004, pp. 97–111, 2021.
https://doi.org/10.1007/978-3-030-87672-2_7

To assist drifting vessels and prevent them from grounding or colliding in coastal waters, Norway has established an emergency towing service [4]. The ambition is to be able to reach and assist a drifting vessel with the necessary tugboats to prevent an accident from happening [18]. The tactical problem of how to dynamically position these tugboats according to current vessel traffic has been extensively studied, see e. g. [2,3,5] or [6]. However, these approaches all focus on the allocation of existing resources and consider the number of available tugboats as given. In this paper, we consider the strategic problem of determining both the optimal location and number of tugboats to achieve a satisfying maritime preparedness level.

A common characteristic of emergency response planning is that the location decisions have to be made before the actual emergency is known. Approaches considering uncertainty are therefore common in the literature. Some recent and relevant examples of this literature are: [16] discusses the problem of locating ambulances to serve random road crashes. [19] uses a maximal covering formulation to evaluate the effectiveness of different fleets when responding to emergencies. [14] introduces a robust formulation of the set covering problem for locating emergency service facilities. In [1], two variants of the probabilistic set covering problem are discussed: the first variant considers a problem with uncertainty regarding whether a selected set can cover an item, while the second variant aims at maximizing the minimum probability that a selected set can cover all items. For a more thorough overview over covering models in the literature see [10]. See also [13] for an overview of the literature on covering models used for emergency response planning.

We propose a two-stage stochastic programming model for the strategic Tugboat Location Problem (TLP) under uncertainty that covers nominal preparedness requirements in the first stage and uses the second-stage decisions to evaluate the quality of the first-stage decisions by calculating the expected costs in case accidents cannot be prevented. To solve the model, we use Sample Average Approximation (SAA) combined with a self-developed heuristic.

The remainder of this article is structured in the following way: The problem is presented in more detail in Sect. 2, before the mathematical model is introduced in Sect. 3. Subsequently, the solution scheme is presented in Sect. 4 and followed by a description of our case study and computational results in Sect. 5. Lastly, we conclude in Sect. 6.

2 Problem Description

In this section, we present the TLP in more detail. First, we introduce the problem before discussing the uncertainty in the problem and its consequences.

2.1 Problem Structure

The overall goal of the strategic TLP is to determine the optimal fleet of tugboats in the emergency towing service and to locate the tugboats in ports such that they can respond to vessels in need of assistance. As the decision which tugboats to locate where has to be made before the location of the vessels in need of assistance is known, this gives rise to a two-stage structure of the problem where the first-stage decisions include the number and type of required tugboats and which ports to locate them in. The second-stage decisions are made after a vessel calls for assistance and cover which tugboat(s) to deploy to the incident site to assist the vessel.

The first-stage decisions have to satisfy dimensioning criteria for Norway's emergency towing service. The dimensioning criteria are derived from dimensioning incidents that represent the assistance needs of high-risk vessels such as large oil tankers [17]. The maximum time to reach the vessel is determined by the vessel's drifting time, whereas the tugboats' time to assist the vessel consists of mobilization and sailing time as well as hook-up time. In addition, we include a safety margin to enforce even stricter requirements for high-risk areas and vessels. We formulate these criteria as minimal covering requirements for maritime preparedness, ensuring that a sufficient number of tugboats are able to reach any vessel in any area within the required time.

Once an incident becomes known, tugboats are deployed to assist the vessel in distress. However, due to weather conditions at the time of the incident, not only drifting speed of the vessel but also sailing speed of the tugboat and hook-up time can be different from the nominal values used in the first stage. It is therefore possible that tugboats arrive too late and an accident occurs. In this case, a penalty cost is incurred.

Each located tugboat has a chartering and operating cost associated with it and can provide a certain bollard pull. The combined bollard pull of all deployed tugboats has to be higher than the required bollard pull to assist the vessel in distress. If multiple tugboats are deployed, hook-up to the vessel will start once all tugboats have arrived at the scene.

Some ports in Norway serve as bases for a large number of privately operated vessels with tugboat capabilities. When the number of these tugboats is sufficiently large, we assume that a small number can contribute to maritime preparedness. We therefore account for them when considering the minimal covering requirements. However, the cost of these tugboats does not become part of the objective function as they are not paid for by the emergency towing service.

The objective is to minimize the sum of the costs of publicly operated tugboats in the emergency towing service and expected penalty costs due to insufficient preparedness. Incident-dependent operational costs are not included in the model as these are usually paid for by the vessel in need of assistance.

Note that the probability of a vessel starting to drift is small. This implies that the probability of two or more drifting incidents happening simultaneously is extremely low, and these incidents are therefore neglected in this paper. This is a common simplification in preparedness planning, see e.g. [2] and [20].

2.2 Geography, Vessel Types and Tugboat Types

We divide the Norwegian coast into geographical zones, as illustrated in Fig. 1. Every zone is described by its seabed characteristic, ecosystem vulnerability, weather conditions and traffic density. Each zone also contains an impact point and a set of one or more corridors where vessels are sailing. A zone is defined from the coastline to the corridor located farthest away from the shore. When a vessel encounters a problem leading to loss of propulsion, it will start to drift from the center of its corridor towards the coast. If no tugboats are able to assist the drifting vessel before it reaches the impact point, an accident happens. The impact point is located in the middle of each zone's baseline.

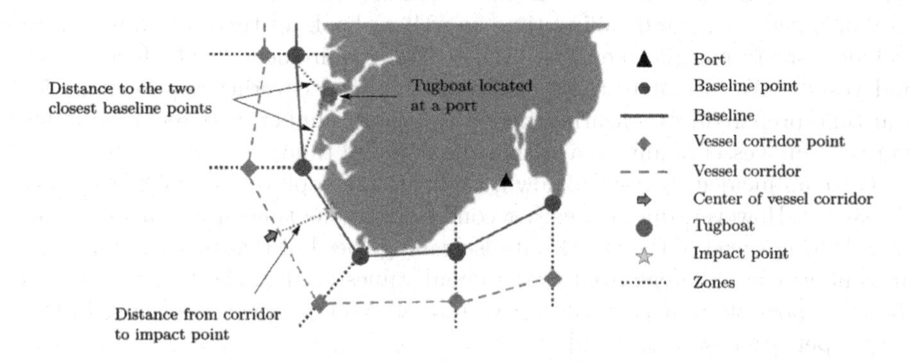

Fig. 1. Representation of the Norwegian coastline.

We categorize vessels by vessel type. A vessel type is defined by its size, cargo type and sailing corridor. All vessel belonging to a given vessel type are considered to be identical. Tugboats are grouped by tugboat type. Each tugboat type is described by its bollard pull and maximal sailing speed, with all tugboats belonging to a tugboat type having the same properties.

2.3 Uncertain Parameters

We define date and location of an incident, characteristics of the involved vessel and wind speed (as a proxy for weather conditions) as uncertain parameters affecting the potential consequences of the incident and the lateness of the tugboats deployed to assist the vessel in distress. See Fig. 2 for an illustration of how these parameters relate to each other and the penalty cost. Both consequence, lateness and penalty cost are explained in more detail below.

Consequence. Consequences result from an accident when the deployed tugboats are not able to assist a drifting vessel before it reaches the impact point. The consequences depend on vessel and seabed characteristics as well as the

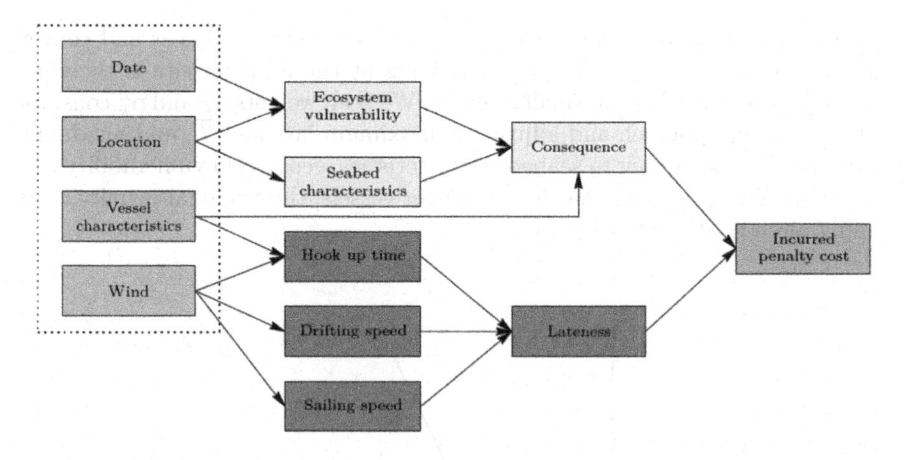

Fig. 2. Uncertain parameters and their relationship to the incurred penalty cost.

vulnerability of the ecosystem. The vessel characteristics, and cargo type in particular, influence the expected consequence of an accident. Incidents involving passenger vessels put human lives at stake whereas spills from oil tankers have higher consequences in more vulnerable ecosystems. The ecosystem vulnerability is changing throughout the year and differs across areas. We assume that grounding on a rocky seabed will cause a greater fracture in the vessel's hull than a sandy seabed, thus leading to a greater spill.

Lateness. By lateness we describe the tugboat's ability to assist the vessel in time. Drifting speed increases with wind speed and determines how much time the tugboats have to reach and assist the vessel in distress. The hook-up time is the time is takes to connect the tugboat's tugline to the vessel. We assume that the hook-up time increases with higher wind speed and heavier vessels. The tugboat's sailing speed determines how long it takes to sail from port to the incident site. Sailing speed is negatively correlated with wind speed.

Lateness is positive if the sum of the tugboat's mobilization time, sailing time and hook-up time is larger than the vessel's drifting time. In the other case, the accident is avoided and lateness is zero.

Penalty Cost. We introduce a fictitious penalty cost function depending on lateness to penalize accidents. The penalty cost reflects that it might not always be possible to prevent accidents from happening, but that is often beneficial to have vessels quickly at the site of the accident to reduce its consequences, for example through search-and-rescue operations or oil-spill mitigation. By adjusting the penalty cost function, e.g. for accidents involving high-risk vessels or in high-risk areas, a policy maker can analyze how different expected costs of accidents influence the optimal number and location of the tugboats.

We have chosen to model the penalty cost as a piecewise linear and convex function, see Fig. 3. Large lateness in arriving at the accident site is therefore penalized more severely than small lateness. We keep weights α_k^t and α_k^c constant in our modeling approach and adjust the maximum lateness, $\overline{T}$, and maximum penalty cost, $\overline{C}$, according to seabed characteristics, ecosystem vulnerability and vessel characteristics. Note that for passenger vessels, the maximal penalty costs $\overline{C}$ apply already at a lateness of 0.

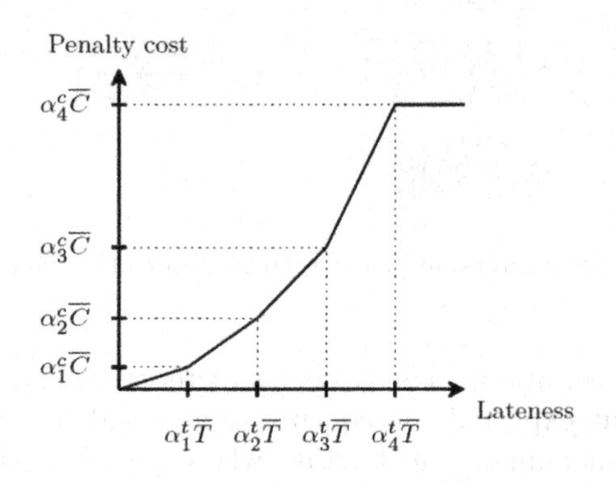

Fig. 3. Penalty cost as a function of lateness.

3 Model Formulation

We formulate the problem of optimizing maritime preparedness as a two-stage stochastic programming problem. The objective is to minimize the sum of the costs of tugboats in the emergency towing service and the expected penalty costs resulting from accidents.

Let us first introduce the following notation for our problem formulation:

Sets

$\mathcal{B}$ Set of tugboat types.
$\mathcal{K}$ Set of breakpoints of the penalty cost function.
$\mathcal{P}$ Set of ports.
$\mathcal{S}$ Set of scenarios.
$\mathcal{V}$ Set of vessel types.
$\mathcal{Z}$ Set of zones along the coast.

Parameters

A_{bpvz} 1 if tugboat type b in port p is close enough to vessel type v in zone z, 0 otherwise, $b \in \mathcal{B}, p \in \mathcal{P}, v \in \mathcal{V}, z \in \mathcal{Z}$.
C_b^B Cost of chartering and operating a tugboat of type b, $b \in \mathcal{B}$.
$\overline{C}^s$ Maximal penalty cost in scenario s, $s \in \mathcal{S}$.

F_b Bollard pull of tugboat type b, $b \in \mathcal{B}$.
$\hat{F}_b^s$ Effective bollard pull of tugboat type b in scenario s, $b \in \mathcal{B}, s \in \mathcal{S}$.
Q_v Number of tugboats required by vessel type v, given the minimal covering requirement, $v \in \mathcal{V}$.
$\hat{Q}^s$ Number of tugboats required, depending on the characteristics of the drifting vessel in scenario s, $s \in \mathcal{S}$.
T_{bp}^s Lateness of tugboat type b in port p in scenario s, $b \in \mathcal{B}, p \in \mathcal{P}, s \in \mathcal{S}$.
$\overline{T}^s$ Maximal lateness in scenario s, $s \in \mathcal{S}$.
W_v Total bollard pull required by vessel type v, given the minimal covering requirement, $v \in \mathcal{V}$.
$\hat{W}^s$ Total effective bollard pull required, depending on the characteristics of the drifting vessel in scenario s, $s \in \mathcal{S}$.
X_{bpz} 1 if a privately operated tugboat of type b in port p has zone z within its operating range, 0 otherwise, $b \in \mathcal{B}, p \in \mathcal{P}, z \in \mathcal{Z}$.
$\hat{X}_{bp}^s$ 1 if the range of privately operated tugboat type b in port p includes the drifting vessel in scenario s, 0 otherwise, $b \in \mathcal{B}, p \in \mathcal{P}, s \in \mathcal{S}$.
α_k^c Breakpoint k's weight of the maximal penalty cost, $k \in \mathcal{K}$.
α_k^t Breakpoint k's weight of the maximal lateness, $k \in \mathcal{K}$.
p^s Probability of scenario s, $s \in \mathcal{S}$.

Decision variables

u_{bp}^s 1 if tugboat type b located in port p is deployed in scenario s, 0 otherwise, $b \in \mathcal{B}, p \in \mathcal{P}, s \in \mathcal{S}$.
t^s Maximum lateness in scenario s, $s \in \mathcal{S}$.
x_{bp} 1 if tugboat type b is located in port p, 0 otherwise, $b \in \mathcal{B}, p \in \mathcal{P}$.
y_{bvz} 1 if tugboat type b is able to rescue vessel type v in zone z, given the minimal cover requirement, 0 otherwise, $b \in \mathcal{B}, v \in \mathcal{V}, z \in \mathcal{Z}$.
μ_k^s Weight of breakpoint k in scenario s, $k \in \mathcal{K}, s \in \mathcal{S}$.

In the first stage of the two-stage stochastic programming problem, the number of tugboats and their locations are decided upon. In the second-stage, tugboats are deployed in order to prevent a specific drifting accident. The resulting model formulation is given below:

$$\min \sum_{b \in \mathcal{B}} \sum_{p \in \mathcal{P}} C_b^B x_{bp} + \sum_{s \in \mathcal{S}} p^s \mathcal{Q}^s(x) \tag{1}$$

subject to

$$\sum_{p \in \mathcal{P}} A_{bpvz} \cdot (x_{bp} + X_{bpz}) \geq y_{bvz} \qquad b \in \mathcal{B}, v \in \mathcal{V}, z \in \mathcal{Z}, \tag{2}$$

$$\sum_{b \in \mathcal{B}} y_{bvz} \geq Q_v \qquad v \in \mathcal{V}, z \in \mathcal{Z}, \tag{3}$$

$$\sum_{b \in \mathcal{B}} F_b \cdot y_{bvz} \geq W_v \qquad v \in \mathcal{V}, z \in \mathcal{Z},, \tag{4}$$

$$x_{bp} \in \{0,1\} \qquad b \in \mathcal{B}, p \in \mathcal{P}, \tag{5}$$

$$y_{bvz} \in \{0,1\} \qquad b \in \mathcal{B}, v \in \mathcal{V}, z \in \mathcal{Z}, \tag{6}$$

where $\mathcal{Q}^s(x)$ is the solution to the following second-stage problem:

$$\mathcal{Q}^s(x) = \min \sum_{k \in \mathcal{K}} \alpha_k^c \overline{C}^s \mu_k^s \tag{7}$$

subject to

$$u_{bp}^s \leq x_{bp} + \hat{X}_{bp}^s \qquad b \in \mathcal{B}, p \in \mathcal{P}, s \in \mathcal{S}, \tag{8}$$

$$\sum_{b \in \mathcal{B}} \sum_{p \in \mathcal{P}} u_{bp}^s \geq \hat{Q}^s \qquad s \in \mathcal{S}, \tag{9}$$

$$\sum_{b \in \mathcal{B}} \sum_{p \in \mathcal{P}} \hat{F}_b^s \cdot u_{bp}^s \geq \hat{W}^s \qquad s \in \mathcal{S}, \tag{10}$$

$$t^s \geq T_{bp}^s u_{bp}^s \qquad b \in \mathcal{B}, p \in \mathcal{P}, s \in \mathcal{S}, \tag{11}$$

$$t^s = \sum_{k \in \mathcal{K}} \alpha_k^t \overline{T}^s \mu_k^s \qquad s \in \mathcal{S}, \tag{12}$$

$$\sum_{k \in \mathcal{K}} \mu_k^s = 1 \qquad s \in \mathcal{S}, \tag{13}$$

$$\mu_k^s \geq 0, \text{ SOS2} \qquad k \in \mathcal{K}, s \in \mathcal{S}, \tag{14}$$

$$u_{bp}^s \in \{0, 1\} \qquad b \in \mathcal{B}, p \in \mathcal{P}, s \in \mathcal{S}. \tag{15}$$

The first-stage problem (1)–(6) is formulated as a covering problem, satisfying the dimensioning criteria for planning maritime preparedness. The first-stage objective (1) minimizes the sum of the tugboats' chartering and operating costs and the expected penalty costs. Constraints (2) ensure that tugboat type b located in port p has to be close enough to zone z to rescue a vessel of type v. Constraints (3) make sure that the number of available tugboats for rescuing vessel type v in zone z exceeds the required number of tugboats. Constraints (4) ensure that the combined bollard pull of the available tugboats is sufficiently large. Constraints (5) and (6) are the binary restrictions on x_{bp} and y_{bvz}, respectively.

The second-stage problem is given by (7)–(15). The second-stage objective (7) minimizes the penalty costs of responding to an incident. Constraints (8) ensure that a tugboat type b has to be located in port p in order to be deployed in scenario s. Restrictions (9) and (10) ensure that a sufficient number of tugboats with a sufficiently large combined effective bollard pull are deployed to assist the drifting vessel. Constraints (11) determine the lateness for a given scenario. Lateness for a scenario is defined by the lateness of the last deployed tugboat to reach the drifting vessel. Constraints (12) link the scenario lateness to the penalty cost in the objective function (7) through weights μ_k^s on the breakpoints of the penalty cost function. Constraints (13) in combination with constraints (14) define variable μ_k^s as a special ordered set of type 2 (SOS2), see e.g. [24] for more details. The binary requirements on variable u_{bp}^s are imposed through constraints (15).

4 Solution Approach

We first introduce the scenario generation procedure, before we briefly present our solution approach. Our approach combines Sample Average Approximation to determine a lower bound to the problem with a heuristic for finding upper bounds.

4.1 Scenario Generation

A scenario is mainly characterized by 3 groups of uncertain parameters: location, date and incident type. The location of the incident affects the seabed characteristics and, together with the date of the incident, the vulnerability of the ecosystem. The date also provides the weather conditions for the given location and wind speed in particular. The incident type specifies the vessel characteristics such as vessel type and size as well as its cargo.

The date of the incident is drawn randomly from the datasets of available wind data. We use historical weather data from the ERA-Interim dataset [7], made available by the European Centre for Medium-Range Weather Forecasts (ECMWF). One might assume that locations with extreme weather conditions are more likely to experience an incident. However, human error and technical conditions of the vessel actually have a greater impact on the probability of an incident [9]. We therefore sample date and location of an incident independent of each other. The probability distribution for the location of an incident is based on the accidents statistics provided by the Norwegian Maritime Authority. The incident type is sampled based on historical marine traffic information from the Norwegian Coastal Administration.

4.2 Sample Average Approximation for Estimating a Lower Bound

A common challenge in stochastic programming is that the problems often become computationally intractable when using an appropriately large number of scenarios. We therefore apply Sample Average Approximation (SAA) to obtain a lower bound estimate for our problem [12]. SAA is based on the idea that it is often easier to solve multiple smaller optimization problems than one single large problem.

By solving M problems with N scenarios each, we can derive a statistical lower bound for our problem. Note that the scenario trees for the M problems are generated independent of each other and that the number of scenarios N used to solve the problem is small compared with the true problem size. See [12,21] or [22] for detailed descriptions of the algorithm.

4.3 Search Heuristic for Determining an Upper Bound

To calculate an upper bound for the problem, we need to evaluate a feasible solution over a reference sample of N'_{ref} scenarios. The reference sample is sampled independently from the $M \cdot N$ scenarios of the SAA-problems and usually $N'_{\text{ref}} \gg N$.

The solutions for the different SAA problems can be used as candidate solutions for calculating an upper bound, but they may be infeasible for the reference sample. We therefore use a search heuristic that generates feasible solutions in the neighbourhood of the solutions provided by the SAA problems. A solution's neighbourhood is defined as the closest surrounding ports and tugboat types. For example, if a tugboat of type B is located in port 13, its neighbourhood are tugboat types A–C located in ports 12–14. We group the different unique solutions from the SAA problems according to the number of used tugboats and then generate the neighbourhood solutions for the best solutions of each group. These solutions are then checked for feasibility and infeasible solutions are discarded. The remaining feasible solutions are then evaluated in the reference sample.

However, in many cases it is impractical to evaluate all solutions in the reference sample N'_{ref} as the computational time could be huge. We therefore apply a stepwise approach to evaluating feasible solutions: All feasible candidate solutions are evaluated over a relatively small sample N'_1. Solutions that perform poorly in N'_1 are assumed to be bad and are discarded, while the K_1 best solutions are evaluated a second time. In the second evaluation, the sample size is larger, $N'_2 \approx 100 \cdot N'_1$, thus providing a better estimate of the true objective function value. The K_2 best solutions in N'_2 are evaluated a third time, now in the reference sample N'_{ref}. The best objective function value from the reference sample is then used as an upper bound estimate.

Note that our heuristic can only search a neighbourhood where all solutions use a predefined number of tugboats. We therefore denote the upper bound for solutions using n tugboats UB_n with the corresponding solution $x^{(n)}$.

5 Computational Results

In this section we present our computational study of locating tugboats along the Norwegian coast. We first introduce the input data and the problem instance in Sect. 5.1, before presenting our results in Sect. 5.2.

5.1 Problem Instance

This section discusses how the parameters used in the model are calculated and estimated. First the geography, vessel and tugboat characteristics are introduced. Then penalty costs are presented before the calculations of the time-related parameters are provided.

Geography and Ports. The Norwegian coastline is divided into 38 zones of approximately equal width. The coastline itself is represented as a set of line segments, referred to as the baseline. The impact point is located in the middle of the baseline and its seabed characteristics and ecosystem vulnerability are representative for the entire zone.

The seabed in a zone is categorized as either sandy, rocky or cliffy. The ecosystem vulnerability is classified as low, medium or high, depending on the season [8]. Both of these factors have an impact on the penalty cost function described below.

Only ports satisfying the International Ship and Port Facility Security (ISPS) Code are regarded as sufficiently large to be used for tugboat operations. Among the 640 ISPS ports in Norway, 30 ports close to relatively large densely populated areas are chosen as possible base locations for tugboats.

Vessel Types and Tugboat Characteristics. A vessel type is categorized by cargo type as either "passenger", "oil", or "other", sailing in the inner or outer corridor depending on cargo type. The size of the vessels is classified as either "light" or "heavy", with a required effective bollard pull of 45 tonnes and 95 tonnes, respectively.

A tugboat type is defined by its bollard pull, sailing speed and chartering and operating cost. We define four tugboat types, A–D, as shown in Table 1. The types represent different real-world tugboats, from smaller harbour tugs to larger tugboats used in the offshore oil & gas industry. Effective bollard pull and sailing speed are adjusted for wind speed according to a linear reduction function. The maximal reduction occurs when the wind speed is 28 m/s, corresponding to violent storm on the Beaufort wind force scale, reducing bollard pull and sailing speed by 20% and 30% respectively.

Table 1. Overview of tugboat types used in the model.

Tugboat type	A	B	C	D
Maximal sailing speed [knots]	15	14	13	12
Bollard pull [tonnes]	200	150	100	80
Chartering and operating cost [MNOK/year]	51	36	33	18

We include two tugboats of type B located in Bergen and Oslo in our analysis to reflect the availability of privately operated tugboats in those areas.

Generation of Penalty Cost Functions. The penalty costs for not being able to provide assistance in time depend on the severity of the accident. The expected spillage in case of an accident depends on the given seabed characteristics. We assume no leakage if the seabed is classified as sandy, only leakage from the fuel tank if rocky and leakage from the fuel and oil tanks if cliffy.

Human lives are generally considered more important than environmental damage. Therefore, we regard all incidents involving "passenger" vessels as equally bad and worse than any other accidents. Consequently, neither $\overline{C}$ nor $\overline{T}$ depend on seabed characteristics or ecosystem vulnerability.

Vessels with cargo type "oil" spill from both the fuel tank and the oil tank, depending on the seabed characteristics at the impact point. The consequences in case of an accident are determined by spill size (with heavy vessels spilling more oil than light vessels) and ecosystem vulnerability. In case the vessel is grounding on a sandy seabed, no fuel or oil spill is expected. Thus, the penalty cost is in this case independent of ecosystem vulnerability.

Vessels with cargo type "other" can by definition only spill fuel, as they do not have an oil tank. This means that the consequences of grounding on a cliffy and rocky seabed are the same, resulting in the same penalty cost function as long as the ecosystem vulnerability does not change.

We model the penalty cost function as an approximated quadratic function using five breakpoints $(\alpha^t \overline{T}, \alpha^c \overline{C})$, where the weights are chosen as $\alpha^t = (0\ \frac{1}{4}\ \frac{2}{4}\ \frac{3}{4}\ 1)$ and $\alpha^c = (0\ \frac{1}{20}\ \frac{1}{4}\ \frac{5}{9}\ 1)$. The actual values for maximal penalty cost $\overline{C}$ and maximal lateness $\overline{T}$ depend on the combination of cargo type, seabed characteristics and ecosystem vulnerability at the time of the accident and are therefore calculated for each scenario. The values used in the model range from 100 to 10 000 MNOK for $\overline{C}$ and 0 to 10 h for $\overline{T}$.

Preprocessing A_{bpvz} and T^s_{bp}. The incidence matrix $\mathbf{A}$ is used in the first-stage problem to identify which tugboats at which ports can contribute to satisfying the covering requirements. Its values are mainly determined by the nominal dimensioning criteria for maritime preparedness. We calculate the nominal time for tugboat assistance as the sum of a tugboat b's mobilization and sailing time from port p to zone z, T^B_{bpz}, the hook-up time T^H_{vz} for connecting the tugboat to the vessel v in zone z and a safety margin T^S_{vz} dependent on vessel v and zone z. The value of the incidence matrix is then set based on whether the nominal time for tugboat assistance is less than vessel v's nominal drifting time to the impact point in zone z, T^V_{vz}. So, if (16) holds, we set $A_{bpvz} = 1$ and $A_{bpvz} = 0$ otherwise.

$$T^B_{bpz} + T^H_{vz} + T^S_{vz} \leq T^V_{vz} \qquad v \in \mathcal{V}, b \in \mathcal{B}, p \in \mathcal{P}, z \in \mathcal{Z}. \qquad (16)$$

Matrix $\mathbf{T}$ is needed in the second-stage problem and depends on the realization of the uncertain parameters. Its values, T^s_{bp}, represent the lateness of tugboat b located in port p in scenario s. Lateness is non-negative and cannot be larger than the maximal lateness $\overline{T}$. As each scenario s only considers a single incident, the location, date and vessel type are known. We can therefore preprocess lateness for different combinations of tugboat type b and port p as the difference between the time needed by tugboat type b in port p to reach the drifting vessel in scenario s, $T^B_{bp}(s)$, the corresponding hook-up time $T^H(s)$, and the vessel's actual drifting time, $T^V(s)$. Hence, the values of matrix $\mathbf{T}$ are determined by (17):

$$T^s_{bp} = \max\left\{0, \min\left\{\overline{T}, T^B_{bp}(s) + T^H(s) - T^V(s)\right\}\right\} \quad b \in \mathcal{B}, p \in \mathcal{P}, s \in \mathcal{S}. \quad (17)$$

5.2 Case Study

We use the SAA method, solving $M = 20$ problems with $N = 60$ scenarios each. The resulting lower bound estimate LB is 179.8 with a corresponding standard deviation σ_{LB} of 2.67%. We observe that the minimum number of tugboats satisfying the minimal covering requirement is 6. To not exclude possibly better solutions with more tugboats, we use a search heuristic to generate neighbourhood solutions for the solutions containing 6, 7 and 8 tugboats, respectively.

For each neighbourhood, the solutions are evaluated using $N_1' = 500$, $K_1 = 500$, $N_2' = 50\,000$ and $K_2 = 20$. The best solutions are then evaluated in the reference sample N_{ref}' containing 700 000 scenarios. These numbers are chosen based on experience from initial testing. Table 2 shows that the best upper bound estimate found by the search heuristic is $UB^{(6)} = 245.3$ for a solution with 6 tugboats, resulting in an optimality gap of approx. 36%. Furthermore, Table 2 indicates that solutions with 7 and 8 tugboats perform considerably worse when evaluated in the reference sample, even though the neighbourhoods are ~100 times larger.

Table 2. Upper bound estimates found by the search heuristic and the corresponding neighbourhood size. Values for UB in MNOK.

Solution	Upper bound		Neighborhood size	
	$UB^{(n)}$	$\sigma_{UB^{(n)}}$	All solutions	Feasible solutions
$x^{(6)}$	245.300	0.03%	41 472	3 888
$x^{(7)}$	257.431	0.05%	6 531 264	315 792
$x^{(8)}$	257.055	0.04%	6 398 388	537 408

To provide a better estimate on the lower bound, we solve the SAA problems again, but force the number of tugboats in the solution to be either 6, 7 or 8. The lower bounds and optimality gaps from these constrained SAA problems are shown in Table 3.

Table 3. Results for the constrained SAA problems. The upper bounds are found by the search heuristic and the confidence interval is at 90%. Values for LB and UB in MNOK.

Solution	Lower bound		Upper bound		Optimality gap	
	$LB^{(n)}$	$\sigma_{LB^{(n)}}$	$UB^{(n)}$	$\sigma_{UB^{(n)}}$	*Estimate*	Confidence interval
$x^{(6)}$	245.113	8.03%	245.300	0.03%	0.08%	10.36%
$x^{(7)}$	202.255	5.97%	257.431	0.05%	21.44%	27.44%
$x^{(8)}$	189.308	1.92%	257.055	0.04%	26.35%	28.17%

Table 3 shows that the optimality gap estimate from the constrained problem with 6 tugboats is less than 0.1%. The optimality gap estimates for the constrained problem with 7 and 8 tugboats are still above 20%. However, due to the lower upper bound for the 6 tugboat solution and the stable performance of the tested solutions in the evaluation procedure, we believe that the optimal solution consists of 6 tugboats with an objective value of approx. 245 MNOK.

6 Conclusions

The goal of this work is to develop an optimization framework that can support the design of maritime preparedness systems. In our case, we want to compose the tugboat fleet and locate the tugboats along the Norwegian coast such that they can assist vessels in distress before accidents happen. We therefore formulate a two-stage stochastic programming model with the objective to find the optimal trade-off between the costs of tugboats in the emergency towing service and expected penalty costs from accidents that could not be prevented.

Our results indicate that six tugboats are sufficient to satisfy the dimensioning criteria for maritime preparedness along the Norwegian coast. Restricting the original problem, we can show that our approach is capable of identifying near-optimal locations and tugboat types for the six tugboats.

Areas of future research include studying how a more detailed description of the Norwegian coastline, i. e. a larger problem instance, affects runtime and solution quality of the approach presented in this paper. It should also be possible to use a formulation based on facility location models to design the maritime preparedness system. In addition, further work is needed to better understand how different penalty costs impact the optimal number and location of the tugboats.

References

1. Ahmed, S., Papageorgiou, D.J.: Probabilistic set covering with correlations. Oper. Res. **61**(2), 438–452 (2013)
2. Assimizele, B.: Models and algorithms for optimal dynamic allocation of patrol tugs to oil tankers along the northern Norwegian coast. PhD theses in Logistics 2017:1, Molde University College, Molde, Norway (2017)
3. Assimizele, B., Royset, J.O., Bye, R.T., Oppen, J.: Preventing environmental disasters from grounding accidents: a case study of tugboat positioning along the Norwegian coast. J. Oper. Res. Soc. **69**(11), 1773–1792 (2018)
4. Berg, T.E., Selvik, Ø., Jordheim, O.K.: Norwegian emergence towing service - past - present and future. TransNav, Int. J. Mar. Navig. Saf. Sea Transp. **14**(1), 83–88 (2020)
5. Bye, R.T.: A receding horizon genetic algorithm for dynamic resource allocation: a case study on optimal positioning of tugs. In: Madani, K., Dourado Correia, A., Rosa, A., Filipe, J. (eds.) Computational Intelligence. SCI, vol. 399, pp. 131–147. Springer, Heidelberg (2012). https://doi.org/10.1007/978-3-642-27534-0_9
6. Bye, R.T., Schaathun, H.G.: A simulation study of evaluation heuristics for tug fleet optimisation algorithms. In: de Werra, D., Parlier, G.H., Vitoriano, B. (eds.) ICORES 2015. CCIS, vol. 577, pp. 165–190. Springer, Cham (2015). https://doi.org/10.1007/978-3-319-27680-9_11
7. Dee, D.P., et al.: The ERA-Interim reanalysis: configuration and performance of the data assimilation system. Q. J. R. Meteorol. Soc. **137**(656), 553–597 (2011)

8. DNV: Miljørisiko ved akutt oljeforurensning fra skipstrafikken langs kysten av Fastlands-Norge for 2008 og progonoser for 2025. Technical report, 2011–0850 (2011), http://www.kystverket.no/contentassets/d6d1509b3b5b46f4b3d58628e99c7437/miljorapport-10.10.2011.pdf. Accessed: 10 Jun 2021 (in Norwegian)
9. DNV GL: Årsaksanalyse av grunnstøtinger og kollisjoner i norske farvann. Technical report, 2014–1332 Rev. C, Høvik, Norge (2015). https://www.kystverket.no/contentassets/f056df3c875140aa98ef49a25cc082c6/3_arsaksanalyse.pdf. Accessed: 1 Jun 2021 (in Norwegian)
10. Farahani, R.Z., Asgari, N., Heidari, N., Hosseininia, M., Goh, M.: Covering problems in facility location: a review. Comput. Ind. Eng. **62**(1), 368–407 (2012)
11. Farstad, E., Flotve, B.L., Haukås, K.: Transportytelser i Norge 1946–2019. TØI rapport 1808/2020, Transportøkonomisk institutt, Oslo, Norway (2020). in Norwegian
12. Kleywegt, A., Shapiro, A., Homem-de-Mello, T.: The sample average approximation method for stochastic discrete optimization. SIAM J. Optim. **12**(2), 479–502 (2001)
13. Li, X., Zhao, Z., Zhu, X., Wyatt, T.: Covering models and optimization techniques for emergency response facility location and planning: a review. Math. Methods Oper. Res. **74**, 281–310 (2011)
14. Lutter, P., Degel, D:, Büsing, C., Koster, A.M.C.A., Werners, B.: Improved handling of uncertainty and robustness in set covering problems. Eur. J. Oper. Res. **263**(1), 35–49 (2017)
15. Meld. St. 35 (2015–2016): På rett kurs - forebyggende sjøsikkerhet og beredskap mot akutt forurensing. Det Kongelige Samferdselsdepartementet, Oslo, Norway. https://www.regjeringen.no/no/aktuelt/venter-vekst-i-skipstrafikken-nye-tiltak-skal-pa-plass/id2502911/. Accessed 5 Jun 2021 (in Norwegian)
16. Mohri, S.S., Haghshenas, H.: An ambulance location problem for covering inherently rare and random road crashes. Comput. Ind. Eng. **251**, 106937 (2021)
17. Norwegian Coastal Administration: Nasjonal slepeberedskap. Rapport fra arbeidsgruppe. https://www.regjeringen.no/globalassets/upload/kilde/fkd/prm/2006/0006/ddd/pdfv/271859-2005-00047_vedlegg_1_samlet_rapport-endelig.pdf. Accessed 8 Jun 2021 (in Norwegian)
18. Norwegian Coastal Administration: Konseptvalgutredning nasjonal slepebåtberedskap. https://www.kystverket.no/globalassets/beredskap/slepeberedskap/kvu-slepeberedskap.pdf. Accessed 9 Jun 2021 (in Norwegian)
19. Pettersen, S., Fagerholt, K., Asbjørnslett, B.: Evaluating fleet effectiveness in tactical emergency response missions using a maximal covering formulation. Naval Eng. J. **131**(1), 65–82 (2019)
20. Psaraftis, H.N., Tharakan, G.G., Ceder, A.: Optimal response to oil spills: the strategic decision case. Oper. Res. **347**(2), 203–217 (1986)
21. Santoso, T., Ahmed, S., Goetschalckx, M., Shapiro, A.: A stochastic programming approach for supply chain network design under uncertainty. Eur. J. Oper. Res. **167**(1), 96–115 (2005)
22. Schütz, P., Tomasgard, A., Ahmed, S.: Supply chain design under uncertainty using sample average approximation and dual decomposition. Eur. J. Oper. Res. **199**(2), 409–419 (2005)
23. Wikipedia: MV Eemslift Hendrika (2015). https://en.wikipedia.org/wiki/MV_Eemslift_Hendrika_(2015). Accessed 19 May 2021
24. Williams, H.P.: Model Building in Mathematical Programming, 5th edn. Wiley, Chichester (2013)

Supply Chain and Production Management

Layout-Agnostic Order-Batching Optimization

Johan Oxenstierna[1,2(✉)], Jacek Malec[1], and Volker Krueger[1]

[1] Department of Computer Science, Lund University, Box 118, SE-221 00 Lund, Sweden
{johan.oxenstierna,jacek.malec,volker.krueger}@cs.lth.se
[2] Kairos Logic AB, Lund, Sweden
https://rss.cs.lth.se/

Abstract. Order-batching is an important methodology in warehouse material handling. This paper addresses three identified shortcomings in the current literature on order-batching optimization. The first concerns the overly large dependence on conventional warehouse layouts. The second is a lack of proposed optimization methods capable of producing approximate solutions in minimal computational time. The third is a scarcity of benchmark datasets, which are necessary for data-driven performance evaluation. This paper introduces an optimization algorithm, SBI, capable of generating reasonably strong solutions to order-batching problems for any warehouse layout at great speed. On an existing benchmark dataset for a conventional layout, Foodmart, results show that the algorithm on average used 6.9% computational time and 105.8% travel cost relative to the state of the art. New benchmark instances and proposed solutions for various layouts and problem settings were shared on a public repository.

Keywords: Order-batching problem · Order picking · Discrete optimization

1 Introduction

There are many optimizable processes within warehouse operations. One of these is *order-picking*, which refers to the retrieval of shipment orders, where each order contains one or several products (items stored in the warehouse) [23]. As much as 55% of all operating expenses in a warehouse are allocated for order-picking [21]. A common method with which to conduct order-picking is *order-batching*, where each picker (vehicle) is set to pick a batch of one or more orders [37]. Within optimization literature order-batching is known as the Order-Batching Problem (OBP) [15] or the Joint Order-Batching and Picker Routing Problem (JOBPRP) [39]. The Picker Routing Problem is the Traveling Salesman Problem (TSP) [33] applied in warehouses (henceforth the Picker Routing Problem is referred to as

Supported by the Wallenberg Autonomous Systems Program.

M. Mes et al. (Eds.): ICCL 2021, LNCS 13004, pp. 115–129, 2021.
https://doi.org/10.1007/978-3-030-87672-2_8

TSP). Most of the literature assesses quality of batches based on travel cost estimation while calling the problem an OBP (without necessarily incorporating *picker routing* in the term), and this paper follows this convention. The OBP is usually formulated as a special version of the more well known Vehicle Routing Problem (VRP) [13]. While the general objective in the OBP is the same as in the VRP, i.e., to assign a set of vehicles to visit a set of locations at minimum travel cost, the literature on the OBP includes two distinguishing features:

- *Order-integrity*: In the OBP products of one order cannot be picked by more than one vehicle [18] whereas in the VRP this constraint is not used (orders are not defined in the VRP) [13].
- *Obstacle-layout*: As far as we are aware, all previous work on the OBP requires a certain form of obstacle layout in the warehouse (the *conventional layout*) (e.g. [5,24,38]). The conventional layout means that warehouse racks are placed in Manhattan style blocks with parallel aisles and cross-aisles (see Fig. 1a). The VRP does not have this requirement.

We are not aware of any reference in the literature which suggests a proportion of conventional versus unconventional layouts in the warehousing domain. Figure 1 includes examples of unconventional layouts used in industry. We see an overly large reliance on conventional layouts as a shortcoming in research on OBP optimization.

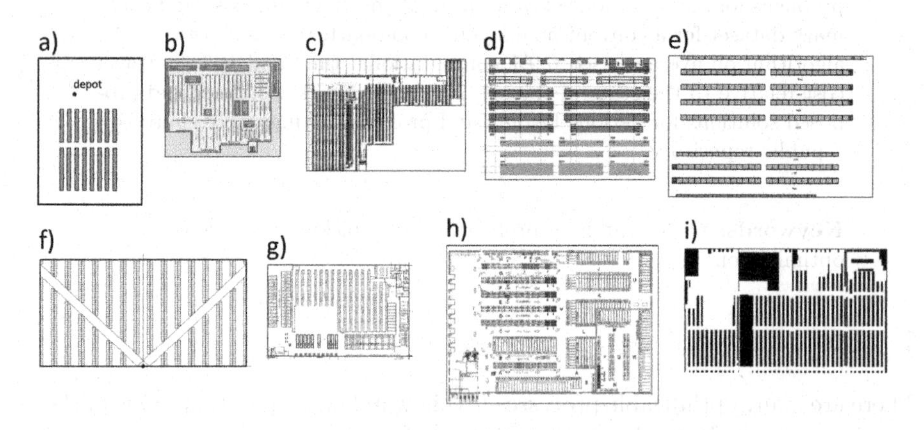

Fig. 1. Examples of warehouse layouts. All except a) are unconventional.

A second identified shortcoming concerns the subject of an OBP optimization module's required computational time, versus the ease with which it can be integrated with a Warehouse Management System (WMS). The WMS manages the overall operation of a warehouse and there is a complex interaction between processes such as order-picking, delivery scheduling, quality assurance, location tracking, packing, verification, shipping, replenishment, yard management, labor management etc., and time margins are usually tight [4]. The WMS gets orders for picking dynamically during the workday. If a subset of these orders are sent to an optimization module, which will select some of them to be picked by a vehicle, it is therefore preferable, from an optimization point of view, to have this selection and corresponding picking tour computed before new orders have arrived to the warehouse. The simplest form of integration is by a synchronous request/response cycle between the WMS and the optimization module, instead of an asynchronous setup where the WMS first sends a request for optimization followed by the collection of a response at a later time (when the original request may already be obsolete). Synchronous request/response is only preferable if optimization can be completed within a few seconds. This paper showcases the kind of OBP optimization performance achievable in such a short time.

A third identified shortcoming is a scarcity of publicly shared benchmark datasets on the OBP. These types of datasets are crucial to allow for experiment reproducibility and peer collaboration.

Our contributions are as follows:

1. The introduction of an optimization algorithm, SingleBatchIterated (SBI), with capability of producing fast approximate solutions to the OBP, irrespective of warehouse layout. SBI's performance is evaluated against the state of the art on Foodmart, a publicly available dataset, which models a warehouse with a conventional layout. The evaluation concerns distance minimization as well as computational time.
2. The introduction of a publicly shared OBP dataset with six types of warehouse layouts and 203 test-instances. Optimization results using SBI and various settings are included in each instance.

2 Literature Review

OBP's for warehouses with conventional layouts have been formulated using integer programming (e.g. [39]) or set-partitioning (e.g. [15]). The conventional layout appears in these formulations as required input parameters such as "number of aisles" [5], intra-aisle distance [7], the cross-distance between two consecutive aisles [15] or number of vertices in the subaisle [39]. Some authors have requested investigations into more layouts than the conventional layout [14,16,18,28]. One benefit of this generalization is that more problem scenarios within logistics could be explored as OBP's. One drawback is that it is very challenging to reduce the OBP solution space without taking advantage of regularities in the layout [39].

Authors often discuss OBP optimization with regard to two fundamental components: 1. Order to vehicle assignments. 2. Solving the TSP's needed to visit all products in the proposed order assignments. The two components can either be optimized jointly [39] or in separate phases [3,38]. The TSP component is often optimized using linear time S-shape or Largest-gap heuristics [18,35] which are specifically designed for conventional layouts. The order to vehicle assignment component is often optimized using so called *proximity batching*, which heuristically ensures vehicles are assigned orders whose pick products are located close together [15]. Sharp & Gibson [37] propose First-Come-First-Served

(FCFS), Space Filling Curve (SFC) and Sequential Minimal Distance (SMD) heuristics to ensure closeness between batched products. Rosenwein [36] proposes Minimum Additional Aisle (MAA) and Centre of Gravity (COG) heuristics. Ho et al. [20] propose 25 different heuristics to initialize and then add one order at a time to a batch until vehicle capacity is exceeded. These heuristics are sometimes collectively referred to as *seed heuristics* or *seed algorithms* [22]. Another heuristic optimization method for order to vehicle assignment is the so called Clark & Wright (C&W) savings algorithm [5]. In this algorithm travel cost to pick individual orders are first estimated and then compared against the cost required to pick larger collections of orders. This algorithm is known to produce batches with less travel cost than seed algorithms, while the computational effort is 100–200 times greater [22].

The OBP optimization objective can be stated as minimizing the sum of all TSP solution costs needed to pick all products (henceforth referred to as *minisum*) [5,6] or to minimize the maximum TSP solution costs (*minimax*) [18]. Solution cost is mostly expressed in terms of distance or time. The latter is more complex but also more realistic to work with as it involves predicting vehicle velocities, time to search for and pick items on shelves etc.

There is a broad array of focus areas in the literature on OBP optimization, reflecting different types of warehouse models and constraints. Chew & Tang [11], for example, examine the relationship between the travel cost of a vehicle, number of available vehicles and where products are stored in the warehouse. The latter is an optimization problem on its own called the Storage Location Assignment Problem (SLAP) [9]. It is rarely studied in conjunction with batching although there is a clear interdependence [18,29]. If there are different origin and destination locations for vehicles the OBP is said to be a multi-depot or Dial-A-Ride Problem (DARP) [18]. A basic multi-depot example is whenever vehicles are set to drop off their picked orders at one pre-designated location, then move to another pre-designated location to collect empty boxes, i.e. orders that have not been picked yet, before moving out to collect a new batch. If all products that need to be picked are assumed to be known apriori the OBP is said to be *static* or *off-line* as opposed to *dynamic* or *on-line* [40]. Proposed optimization programs for the OBP versions described above include integer and mixed integer [5], dynamic programming [6], data mining [10], clustering [24] and meta-heuristics such as Tabu Search [19], Ant Colony Optimization [26] and Genetic Algorithms [8].

Computational time used for OBP optimization and its relevance within warehouse operations is a topic rarely discussed in the literature. Some authors set timeouts for optimization but these are only arbitrarily defined to simulate a "tolerable" time horizon [24]. The largest test-instance results with 30–5000 orders in Briant et al. [6], were achieved after optimization was set to run between 30 min to 2 h. Briant et al. do not discuss whether a WMS provider would be interested in allocating 30 min for generating 6 optimized batches out of 30 unassigned orders.

Two examples of OBP benchmark datasets are Foodmart [39] and HappyChic [6], which are designed for static OBP's and two conventional layouts. The vast majority of benchmarking in OBP research is not carried out on public datasets, but instead on a described model/simulation of a warehouse with a conventional layout. For comparison, in the related research domain on the Vehicle Routing Problem (VRP), there are several widely used benchmark datasets which researchers use to evaluate optimization performance, including the Solomon, Christofides, Taillard, Augerat et al., Fisher and Kilby instances [27,30–32]. A commonly used data format for VRP instances is TSPLIB [17]. We have extended on TSPLIB to introduce new OBP instances in the experimental part of this paper (Sect. 5).

3 Preliminaries

In this section we define all relevant terms and parameters that will be needed for the reminder of the paper. For better readability, we keep the definitions on an intuitive level and use mathematical precision only where necessary.

A batch b is defined as a set of orders from customers, selected out of a set of unassigned orders. The unassigned orders are denoted O and the set of all possible batches is denoted B. Each order contains a set of products and each product has a volume and weight. A batch is picked by a vehicle, m, selected out of a set of available vehicles, $m \in M$. The vehicle's capacities are expressed in number of orders, weight and volume. Each product has a location in the warehouse. The union of all locations in a batch b is retrievable with a function loc^b. The sequence of location visits a vehicle follows to pick a batch (including an origin and a destination location for the vehicle) is computable with a function T^b. Note T^b gives a solution to a Traveling Salesman Problem (TSP). The distance of T^b is computable with a function D^b. The D^b function makes use of a distance matrix which contains the shortest distance between all locations in a given warehouse (without crossing obstacles). The distance matrix is assumed pre-computed. For a presentation of the digitization steps followed to produce it see [34].

The optimization objective of the minisum OBP [5,15,19] is to assign batches to vehicles such that the distance required to pick the orders is minimized, while not breaking any of the following constraints:

1. Each unassigned order is assigned to exactly one vehicle (order-integrity).
2. Each product location in each order assigned to a vehicle must be visited at least once.
3. Capacities of vehicles may never be exceeded.

The proposed optimization algorithm (Sect. 4) makes use of an optimization module which optimizes a more tractable form of the OBP, the so called *single batch* OBP. The optimization objective of the single batch OBP is to find a single batch b with the minimal batch distance. Constraints 2 and 3 still apply for the single batch version of the problem. The following additional constraint is added:

4. The number of orders in a single batch must be as large as possible.

Without this last constraint i.e. the maximization of number of orders, a single batch optimization algorithm would always create a batch with just a single order. This is because the minimal batch distance is always achieved if the batch is made up of just a single order. Note that it is possible to define constraint 4 as both a constraint and an objective. Constraint 4 is delimited from including weight and volume of products in the maximization since this would necessitate decision making over whether weigth, volume or number of orders is more important. Both above models are concerned with *static* OBP's (Sect. 2) i.e. ones where all unassigned orders can be batched at any time.

4 Optimization Algorithm

In this section we will introduce the SingleBatchIterated (SBI) optimization algorithm, which produces an approximate solution to the minisum OBP. Internally it makes use of the *SingleBatch* algorithm, which produces an approximate solution to the single batch OBP (Sect. 3). SingleBatch is shown in the lower rectangular box in Fig. 2. It is used to produce single batches and corresponding picking tours iteratively until there are no more unassigned orders left.

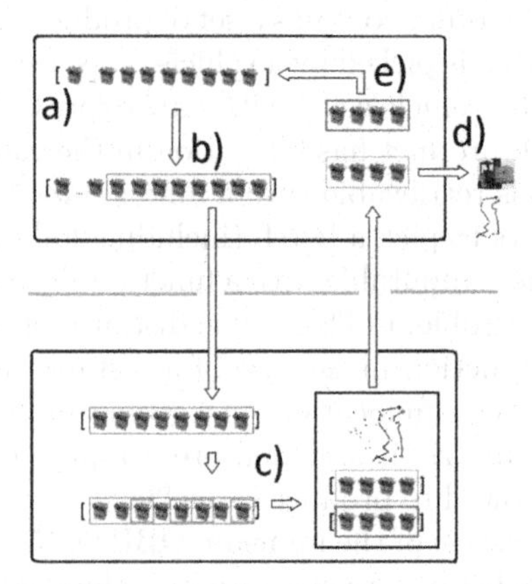

Fig. 2. Flowchart showing SingleBatchIterated (SBI). First unassigned orders are processed according to priority and a subset of orders is then sent to the SingleBatch optimization algorithm (lower box), which produces a single batch and the TSP solution required to pick that batch. The algorithm runs until all unassigned orders have been batched. (Color figure online)

A vehicle is first selected from a set of vehicles (a) and a subset of unassigned orders from the set of all unassigned orders is selected (b). This subset selection is done to reduce the amount of computational time needed for the subsequent optimization. A single batch b as well as a TSP solution for that batch are then computed using the SingleBatch optimization algorithm (c). The distance of the TSP solution is added to the total cost of the OBP solution (initialized as

0). The selected vehicle is dispatched to pick batch b (d). The orders in b are removed from the set of unassigned orders (e). The steps in Fig. 2 correspond to Algorithm 1 shown below.

Algorithm 1: SBI

```
cost ← 0
while O do
    m ← select_vehicle(M)
    O_s ← select_subset(O)
    b ← single_batch(O_s, m, D)
    cost = cost + D(b)
end
```

Algorithm 2: SingleBatch

```
single_batch(O_s, m, D)
// Phase 1
b_ord ← seed_algorithm(O_s, m, D)
// Phase 2
b_tour ← solve_tsp(b, D)
return b
```

The SingleBatch algorithm, Algorithm 2, takes a subset of unassigned orders O_s, a vehicle m and the distance matrix D as input. Order selection using one of two seed algorithms is used to initialize a batch and assign orders to it (b_{ord}) until vehicle capacity runs out. A tour to pick the batch (b_{tour}) is computed using the Concorde TSP solver (details for Concorde are beyond the scope of this paper, for details see [2,12]). The SingleBatch function returns the batch (including the orders and the tour).

SBI requires that there are enough vehicles to batch all orders. This delimitation is used because the vehicle selection part is handled by the Warehouse Management System (WMS) in the intended industrial application (the WMS takes over the full handling of the upper rectangle in Fig. 1, i.e., it decides when and which vehicles should be assigned a batch).

The purpose of the SingleBatch seed algorithm (inside Algorithm 2) is to return a batch of orders that allows subsequent TSP optimization (for the locations in that batch) to result in a short distance. One way to achieve a batch selection quickly is to use heuristics such as Sequential Minimal Distance (SMD) [37] or Centre of Gravity (COG) [36], and these are tested and compared in the experimental part of this paper. SMD and COG can be used to output a scalar value ("distance") that estimates the distance that would be achieved if the locations of the products of two orders were used to formulate and optimize a TSP. The "seed algorithm" works sequentially by adding an order at a time to a sequence of assigned orders (i.e. the single batch). The "seed order" denotes the order which was last added to the sequence (while the sequence is being populated). SMD and COG are used to search for an unassigned order, with a low "distance" to the seed order, to add next. The first order in the sequence can for example be selected randomly [37]. In SingleBatch's seed algorithm it is instead selected as the order with the least sequential minimal distance (SMD) or the shortest distance to the centre of gravity (COG) (depending on which is used). To enable this the vehicle origin location is used as a first seed placeholder. Using SMD or COG for the first order selection is motivated by the SingleBatch optimization objective which states that the distance of the batch should be as short as possible regardless of how many orders end up in that batch.

SMD is computed using the following:

$$SMD(s,o) = \sum_{i \in s} \min_{j \in o} |d_{ij}|, \quad o \in O, o \notin b, s \in b \tag{1}$$

where d_{ij} is the distance between product i in order s (the seed) and product j in unassigned order o. $SMD(s, o)$ is then calculated as the sum of these minimal distances d_{ij}. Sharp & Gibson [37] present a way in which to compute d_{ij} in the conventional layout scenario. For the unconventional layout scenario it is given as $d_{ij} \in D$ (D is the shortest paths distance matrix, assumed pre-computed).

The COG heuristic was introduced by Rosenwein [36] and is for a single order given as:

$$COG(o) = \frac{1}{|o|} \sum_{p \in o} a_p \tag{2}$$

where a_p denotes the location of the product, and $|o|$ is the number of products in the order. The COG of two orders is given by the Manhattan distance between two order COGs: $COG(s, o) = |COG(s) - COG(o)|$ where s and o denote the seed order and an unassigned order, respectively. Note this version of COG does not make use of distance matrix D and hence does not take the warehouse layout into account.

Once the order with the least SMD or COG has been found it is added to the batch and set as the new seed. New orders are then added in the same way until vehicle capacity is full or there are no more unassigned orders left.

5 Experiments

In this section we first discuss the datasets used i.e. Foodmart and the new test instances generated. Then,

1. we discuss OBP results using our SBI approach on the datasets in terms of distance minimization, as well as computational times.
2. we compare results using a seed algorithm running either the SMD or COG heuristics.
3. we compare computational times required by the seed algorithm and the TSP solver.

5.1 Datasets

Foodmart. Foodmart contains test-instances for static OBP's and a conventional layout. It was introduced by Valle et al. [39] and includes 135 test-instances with up to 50 unassigned orders and 7 larger testing-instances with 50 to 5000 orders. The layout has 3 cross-aisles and a maximum of 8 aisles (see Fig. 1a)). There is only a single origin and destination location.

In Foodmart each vehicle carries 8 bins, where each bin has a volume capacity of "40 V". Each product has a volume ranging from 1 to 40 V, and if an order contains products whose sum of V's exceeds 40, or exceeds the volume left in any of the 8 bins, the order may be split between different bins on the same vehicle. This way to formulate vehicle capacity is specific to Foodmart. There are many possible alternatives, e.g. maximum number of orders [25], products [5], volume [8] or weight [24]). The number of available vehicles is unlimited in Foodmart.

Presented results for Foodmart in [6,39] include optimal OBP results for 130 test-instances where the number of orders to be batched varies between 5–100. These instances can therefore be used to evaluate our approach against optimal results on conventional layouts. We believe the gap between SBI's results and optimal results can be used as an estimate of how far away from optimality SBI's results are on unconventional layouts.

Generated Test-Instances. Six different types of warehouse layouts on a 80×80 grid were first generated with the following name-tags: "No obstacles", "conventional layout with 3 cross-aisles and 12 aisles", "1 single rack", "12 racks", "NR1" and "NR2" (see Fig. 4). "NR" stands for non-regular. The unconventional layouts were chosen as simplified representations of real examples seen in the industry (see Fig. 1).

Using the generated layouts, 203 test-instances on a modified TSPLIB format were then generated (30–40 instances for each layout)[1]. The modifications made to the TSPLIB are described in a text file in the provided link. For simplicity vehicle capacity is the same for all vehicles and only expressed in number of orders (between 2–30) in these instances, and experiments involving more capacity types (e.g. volume, weight, number of products, Foodmart type bins and combinations of capacities and/or vehicle types) are left for future work. The number of vehicles in the instances is set as the ceiling of number of unassigned orders divided by the vehicle number of orders capacity (denoted k^M): $|M| = \lceil \frac{|O|}{k^M} \rceil$. Concerning where the products are placed in the warehouse (see Sect. 2 for an explanation for why this is relevant in OBP's), either 1, 2 or 4 rectangular storage assignment zones are used. These zones are placed anywhere on the grid and are generated in two steps: First a random x, y storage zone centroid coordinate within the 80×80 grid is generated. Then storage locations for products (for each order in the generated instance) are generated such that the Manhattan distance between the product location and the storage zone centroid coordinate do not exceed a specified distance[2]. Each of the six layout types has a differing origin and destination location where vehicles start and end their tours.

[1] https://github.com/johanoxenstierna/OBP_instances.

[2] it is called "min_distance_to_slotting_zone" and can be found in a specs JSON in each instance.

5.2 Experimental Results

Since the vehicles in Foodmart use bins into which orders are placed the SingleBatch algorithm was first adapted to be able to handle that particular capacity type. To be exact, the modification was conducted within the call to the "seed_algorithm" function in Algorithm 2. Three modifications were made: 1. The batch object b was modified to include a key-value dictionary "bins" with 8 enumerated keys and corresponding values to keep track of how much volume has been taken up in each bin. 2. A function $check\_candidate\_order(b, o)$ (inside "seed_algorithm"), which checks if a candidate order can be added to a batch without breaking constraints, was modified to find the bin which, if the order is added to it, comes as close as possible to the 40 volume capacity without breaking it. 3. If there exists such a bin its key is returned, the order is added to the batch and the given bin is updated with the added volume. If the order cannot be added to any bin in this batch it is excluded and added to a different batch at a later stage. Only SMD was used as order selection heuristic for the Foodmart experiment.

The OBP experimental results are summarized in Table 1 (Foodmart) and Table 2 (generated instances). On the Foodmart instances an average of 105.8% distance and 6.9% computational time was achieved relative to reported optimal results in [6,38]. The result shows that fast approximate OBP optimization can be accomplished with a relatively small penalty in added distance.

Concerning the comparison of the SMD and COG heuristics, results only concern the generated instances (since COG was not used on Foodmart). It was found that, on the 203 generated OBP instances, SBI with SMD yielded solutions with 97.9% distance and 131% computational time, relative to SBI with COG. Within SingleBatch, the seed algorithm on average used only 7.3% of the total computational time versus the TSP solver Concorde's 92.7%. On average, the seed algorithm requires 0.05–0.1 s to construct a batch using SMD, whereas Concorde requires anywhere between 0.001–3 s to solve a batch TSP, depending on various factors such as number of product locations in the batch (see Fig. 3).

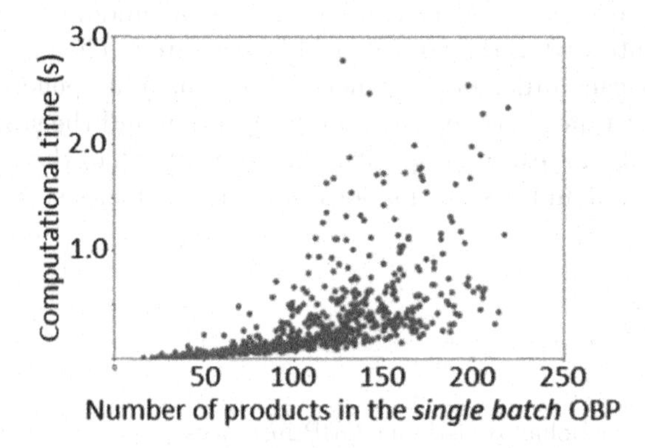

Fig. 3. CPU-time (y-axis) of the SingleBatch algorithm versus number of products in the single batch OBP's (x-axis) (this figure excludes results on Foodmart and "NR1").

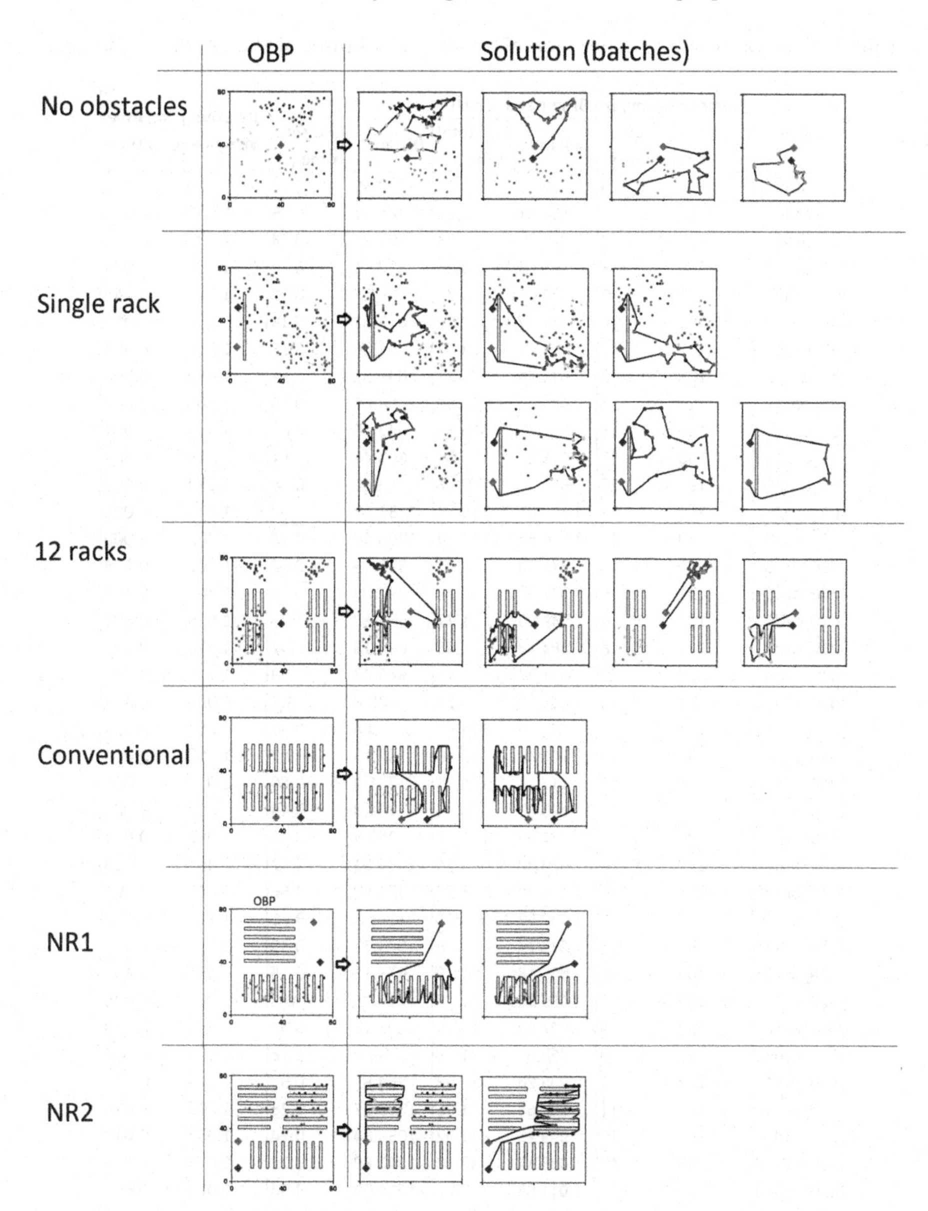

Fig. 4. Six examples of instances (one for each layout type) and solutions (from top left to bottom right) using SBI with the SMD heuristic. The larger red and blue dots are the origin and destination locations for vehicles. Each smaller dot denotes a product and their color denotes the order which the product belongs to. (Color figure online)

Table 1. This table shows experimental results on a subset of the Foodmart dataset

Name	Number of products	Number of orders	Briant et al. (2020)		SBI		Relative distance	Relative CPU time
			Distance	CPU total time (s)	Distance	CPU total time (s)		
d5_ord5	59	5	348.59	1.15	348.59	0.21	1.000	0.183
d5_ord6	67	6	364.81	1.11	364.81	0.05	1.000	0.045
d5_ord7	74	7	374.81	1.84	374.81	0.14	1.000	0.076
d5_ord8	81	8	503.75	3.23	503.9	0.17	1.000	0.053
d5_ord9	88	9	539.61	4.27	561.5	0.18	1.041	0.042
d5_ord10	95	10	581.42	10.89	611.61	0.18	1.052	0.017
d5_ord11	102	11	611.63	53.67	645.6	0.18	1.056	0.003
d5_ord12	109	12	613.40	45.27	649.51	0.18	1.059	0.004
d5_ord13	116	13	623.39	75.86	685.8	0.19	1.100	0.003
d5_ord14	123	14	637.73	116.19	647.46	0.08	1.015	0.001
d5_ord15	130	15	653.42	166.54	663.46	0.1	1.015	0.001
d5_ord20	165	20	862.10	1596.8	960	0.13	1.114	0.000
d5_ord25	200	25	1 083.85	7246.5	1233.22	0.37	1.138	0.000
d5_ord30	230	30	1 000.53	7304.03	1208.61	0.28	1.208	0.000
d10_ord5	80	5	371.09	0.87	371.09	0.09	1.000	0.103
d10_ord6	91	6	377.09	1.23	377.09	0.2	1.000	0.163
d10_ord7	102	7	549.80	1.77	565.61	1.23	1.029	0.695
d10_ord8	113	8	584.18	1.43	648.18	0.42	1.110	0.294
d10_ord9	124	9	637.37	7.15	693.47	0.58	1.088	0.081
d10_ord10	134	10	661.80	16.23	685.8	1.2	1.036	0.074
d10_ord11	144	11	692.43	114.83	729.8	0.41	1.054	0.004
d10_ord12	154	12	703.42	182.96	752.18	0.7	1.069	0.004
d10_ord13	164	13	712.24	124.13	761.9	0.59	1.070	0.005
d10_ord14	174	14	723.83	182.6	761.8	0.48	1.052	0.003
d10_ord15	183	15	881.69	189.19	958.4	0.46	1.087	0.002
d10_ord20	223	20	976.32	955.79	1064.61	3.62	1.090	0.004
d10_ord25	258	25	1 187.30	2677.58	1355.42	3.66	1.142	0.001
d10_ord30	293	30	1 159.58	7305.76	1295.8	3.71	1.117	0.001
d20_ord5	91	5	573.77	0.66	573.77	0.33	1.000	0.500
d20_ord6	105	6	656.18	1.19	678.18	0.16	1.034	0.134
d20_ord7	119	7	689.80	8	714.18	0.2	1.035	0.025
d20_ord8	132	8	697.80	10.54	698.18	0.94	1.001	0.089
d20_ord9	145	9	726.80	41.18	751.8	1.13	1.034	0.027
d20_ord10	158	10	905.26	37.92	964.86	0.51	1.066	0.013
d20_ord11	171	11	980.51	11.65	1009.27	0.28	1.029	0.024
d20_ord12	183	12	999.25	41.92	1056.84	0.74	1.058	0.018
d20_ord13	195	13	1 008.12	82.72	1040.89	0.73	1.033	0.009
d20_ord14	207	14	1 011.08	70.6	1084.99	3.56	1.073	0.050
d20_ord15	219	15	1 025.46	87.81	1096.99	3.57	1.070	0.041
d20_ord20	279	20	1 332.26	619.56	1435.98	0.51	1.078	0.001
d20_ord25	334	25	1 602.86	3288.5	1738.87	0.76	1.085	0.000
d20_ord30	389	30	1 843.13	7271.35	2025.12	2.74	1.099	0.000

CPU used: Intel Core i7-4710MQ 2.5 GZ 4 cores, 8GB of memory

Table 2. This table summarizes the experimental results on 14 types of instances (Foodmart can be seen in the lowest row).

Name	Number of products avg.	Layout	Storage zones	SMD avg. distance	SMD avg. CPU time (s)	COG avg. distance	COG avg. CPU time (s)
s1	88	None	1	516	0.30	516	0.33
s2	118	None	2	380	0.41	380	0.39
s2_1o	111	1 rack	2	369	0.65	369	0.58
s2_12o	110	12 racks	2	377	0.43	380	0.39
s2_12o40	42	12 racks	2	290	0.03	293	0.12
s2_12o120	122	12 racks	2	421	0.59	422	0.55
s2_12o200	117	12 racks	2	395	0.55	409	0.49
s1_3CAo	66	Conv.	1	618	0.24	621	0.32
s4	83	None	4	840	0.47	852	0.38
s2_3CAo	131	Conv.	2	652	0.56	683	0.42
s2_NR1o	254	NR1	2	1233	1.88	1295	1.47
s2_NR1o	146	NR1	2	1001	0.91	1015	0.72
s2_NR2o	140	NR2	2	650	0.93	651	0.68
Foodmart	173	Conv.	-	903	0.91	-	-

CPU used: Intel Core i7-4710MQ 2.5 GZ 4 cores, 8GB of memory

No attempt was made to infer how features such as layout, storage zones and depot locations affect the computational times shown in Fig. 3. Concorde has a high degree of internal variance when it comes to computational time [1,2,12]. It would therefore require a large number of OBP test instances to make this type of inference.

6 Conclusion

This paper introduced an optimization algorithm, SingleBatchIterated (SBI), capable of producing strong approximate solutions to the OBP at minimal computational time for both conventional and unconventional warehouse layouts. The algorithm was evaluated on the Foodmart benchmark dataset, where it showed that OBP solutions could be obtained at great speed and with a relatively low penalty in added distance compared to optimal results. Additionally, a new OBP dataset with several types of layouts, depot locations and storage zone settings was introduced. Proposed solutions using SBI were uploaded together with visualizations of the new instances.

The vast majority of computational time in SBI was allocated to TSP solving rather than order selection. Results show that this is mostly due to the TSP solver Concorde, which has a high internal variance in terms of computational time. Instead of replacing Concorde with a TSP optimizer which is more stable with regard to computational time, it is deemed more relevant to allocate more computational time at the order selection phase. As Fig. 3 and Table 2 show, most OBP instances were optimized in well under 1 s, which allows for more optimization in many scenarios. One alternative could be to add the *savings algorithm* (Sect. 2) as an alternative for order selection and to use it if there are relatively few products in the batch. Further work on dataset generation is also needed, especially for OBP instances involving dynamicity and more vehicle capacity options.

References

1. Applegate, D., Cook, W., Dash, S., Rohe, A.: Solution of a min-max vehicle routing problem. INFORMS J. Comput. **14**, 132–143 (2002)
2. Applegate, D.L., Bixby, R.E., Chvatal, V., Cook, W.J.: The Traveling Salesman Problem: A Computational Study. Princeton University Press, Princeton (2006)
3. Azadnia, A., Taheri, S., Ghadimi, P., Samanm, M., Wong, K.: Order batching in warehouses by minimizing total tardiness: a hybrid approach of weighted association rule mining and genetic algorithms. Sci. World J. **2013** (2013). Article ID 246578 . https://doi.org/10.1155/2013/246578
4. Bartholdi, J., Hackman, S.: Warehouse and distribution science Release 0.98 (2019)
5. Bozer, Y.A., Kile, J.W.: Order batching in walk-and-pick order picking systems. Int. J. Prod. Res. **46**(7), 1887–1909 (2008)
6. Briant, O., Cambazard, H., Cattaruzza, D., Catusse, N., Ladier, A.L., Ogier, M.: An efficient and general approach for the joint order batching and picker routing problem. Eur. J. Oper. Res. **285**(2), 497–512 (2020)
7. Bu, M., Cattaruzza, D., Ogier, M., Semet, F.: A Two-Phase Approach for an Integrated Order Batching and Picker Routing Problem, pp. 3–18 (2019)
8. Cergibozan, C., Tasan, A.: Genetic algorithm based approaches to solve the order batching problem and a case study in a distribution center. J. Intell. Manuf. 1–13 (2020). https://doi.org/10.1007/s10845-020-01653-3
9. Charris, E., Rojas-Reyes, J., Montoya-Torres, J.: The storage location assignment problem: a literature review. Int. J. Ind. Eng. Comput. **10**, 199–224 (2018)
10. Chen, M.C., Wu, H.P.: An association-based clustering approach to order batching considering customer demand patterns. Omega **33**(4), 333–343 (2005)
11. Chew, E.P., Tang, L.C.: Travel time analysis for general item location assignment in a rectangular warehouse. Eur. J. Oper. Res. **112**(3), 582–597 (1999)
12. Cook, W.: Concorde TSP Solver (2020). http://www.math.uwaterloo.ca/tsp/concorde/index.html
13. Cordeau, J.F., Laporte, G., Savelsbergh, M., Vigo, D.: Vehicle routing. Transp. Handb. Oper. Res. Manage. Sci. **14**, 195–224 (2007)
14. Fumi, A., Scarabotti, L., Schiraldi, M.: The effect of slot-code optimization in warehouse order picking. Int. J. Bus. Manage. **5**, 5–20 (2013)
15. Gademann, N., Velde, V.D.S.: Order batching to minimize total travel time in a parallel-aisle warehouse. IIE Trans. **37**(1), 63–75 (2005)
16. Gue, K.R., Meller, R.D.: Aisle configurations for unit-load warehouses. IIE Trans. **41**(3), 171–182 (2009)
17. Hahsler, M., Kurt, H.: TSP - infrastructure for the traveling salesperson problem. J. Stat. Softw. **2**, 1–21 (2007)
18. Henn, S.: Algorithms for online order batching in an order picking warehouse. Comput. Oper. Res. **39**(11), 2549–2563 (2012)
19. Henn, S., Wscher, G.: Tabu search heuristics for the order batching problem in manual order picking systems. Eur. J. Oper. Res. **222**(3), 484–494 (2012)
20. Ho, Y.C., Su, T.S., Shi, Z.B.: Order-batching methods for an order-picking warehouse with two cross aisles. Comput. Ind. Eng. **55**(2), 321–347 (2008)

21. Jiang, X., Zhou, Y., Zhang, Y., Sun, L., Hu, X.: Order batching and sequencing problem under the pick-and-sort strategy in online supermarkets. Procedia Comput. Sci. **126**, 1985–1993 (2018)
22. Koster, M.B.M.D., Poort, E.S.V.d., Wolters, M.: Efficient orderbatching methods in warehouses. Int. J. Prod. Res. **37**(7), 1479–1504 (1999)
23. Koster, R.D., Le-Duc, T., Roodbergen, K.J.: Design and control of warehouse order picking: a literature review. Eur. J. Oper. Res. **182**(2), 481–501 (2007)
24. Kulak, O., Sahin, Y., Taner, M.E.: Joint order batching and picker routing in single and multiple-cross-aisle warehouses using cluster-based tabu search algorithms. Flex. Serv. Manuf. J. **24**(1), 52–80 (2012)
25. Le-Duc, T., Koster, R.M.B.M.D.: Travel time estimation and order batching in a 2-block warehouse. Eur. J. Ope. Res. **176**(1), 374–388 (2007)
26. Li, J., Huang, R., Dai, J.B.: Joint optimisation of order batching and picker routing in the online retailers warehouse in China. Int. J. Prod. Res. **55**(2), 447–461 (2017)
27. Mańdziuk, J., Świechowski, M.: UCT in capacitated vehicle routing problem with traffic jams. Inf. Sci. **406–407**, 42–56 (2017)
28. Masae, M., Glock, C.H., Grosse, E.H.: Order picker routing in warehouses: a systematic literature review. Int. J. Prod. Econ. **224**, 107564 (2020)
29. Nieuwenhuyse, I., De Koster, R., Colpaert, J.: Order batching in multi-server pick-and-sort warehouses. Katholieke Universiteit Leuven, Open Access publications from Katholieke Universiteit Leuven (2007)
30. Okulewicz, M., Mańdziuk, J.: The impact of particular components of the PSO-based algorithm solving the dynamic vehicle routing problem. Appl. Soft Comput. **58**, 586–604 (2017)
31. Pillac, V., Gendreau, M., Guret, C., Medaglia, A.L.: A review of dynamic vehicle routing problems. Eur. J. Oper. Res. **225**(1), 1–11 (2013)
32. Psaraftis, H., Wen, M., Kontovas, C.: Dynamic vehicle routing problems: three decades and counting. Networks **67**, 3–31 (2015)
33. Ratliff, H., Rosenthal, A.: Order-picking in a rectangular warehouse: a solvable case of the traveling salesman problem. Oper. Res. **31**, 507–521 (1983)
34. Rensburg, L.J.V.: Artificial intelligence for warehouse picking optimization - an NP-hard problem. Master's thesis, Uppsala University (2019)
35. Roodbergen, K.J., Koster, R.: Routing methods for warehouses with multiple cross aisles. Int. J. Prod. Res. **39**(9), 1865–1883 (2001)
36. Rosenwein, M.B.: A comparison of heuristics for the problem of batching orders for warehouse selection. Int. J. Prod. Res. **34**, 657–664 (1996)
37. Sharp, G., Gibson, D.: Order batching procedures. Eur. J. Oper. Res. **58**, 57–67 (1992)
38. Valle, C., Beasley, B.: Order batching using an approximation for the distance travelled by pickers. Eur. J. Oper. Res. **284**, 460–484 (2019)
39. Valle, C., Beasley, J.E., da Cunha, A.S.: Optimally solving the joint order batching and picker routing problem. Eur. J. Oper. Res. **262**(3), 817–834 (2017)
40. Yu, M., Koster, R.BMd.: The impact of order batching and picking area zoning on order picking system performance. Eur. J. Oper. Res. **198**(2), 480–490 (2009)

Automated Negotiation for Supply Chain Finance

Alexandra Fiedler(✉) and Dirk Sackmann

Faculty of Business Administration and Information Sciences,
University of Applied Sciences Merseburg, Merseburg, Germany
alexandra.fiedler@hs-merseburg.de
https://www.hs-merseburg.de

Abstract. The growing importance of supply chain finance and the possibility of digital descriptions for goods and services increase the urgency of providing sophisticated solutions for automating negotiations in this area. Multi-agent systems technology plays an essential role in this regard. This paper highlights the specifics of automated negotiations and describes financial supply chain actors as agents. It also describes the complexity of possible supply chain finance solutions. A scenario for automated decision making for the best financing option is explained and the negotiation flow of a multi-agent system implemented in Java Agent Development Framework is demonstrated.

The negotiations, in the form of an auction, are aimed at minimizing the capital costs of the supply chain. Here, it is important to weigh up whether internal financing via an investor within the supply chain or external financing via the capital market is more advantageous. The different roles of the supply chain finance agents capital demander, investor and capital market are described in detail. The use of so-called HelperAgents for the negotiating participants capital demander and investor as negotiating agents within a supply chain finance negotiation protocol is also explained.

Keywords: Automated negotiation · Supply chain finance · Multi-agent system

1 Introduction

Automation and robotization are also making their way into the world of negotiations. This is made possible by the use of so-called BOTs. While in a "normal" negotiation people interact with each other, in the case of an automated negotiation the process is controlled by a BOT. According to the term, a BOT is

---

The work is supported by the federal–state funding initiative "Innovative Hochschule", funded by the Federal Ministry of Education and Research and the Federal State of Saxony-Anhalt. It is carried out within the joint project "TransInno_LSA - Transfer and innovation service in the federal state of Saxony-Anhalt" (grant number 03IHS013).

M. Mes et al. (Eds.): ICCL 2021, LNCS 13004, pp. 130–141, 2021.
https://doi.org/10.1007/978-3-030-87672-2_9

a computer program that automatically processes procedures. In the case of a negotiation BOT, the program takes over and controls the negotiation and interacts with the user on the other side: clarifications about options are made, proposals and counterproposals are made, and agreements are reached without the need for human intervention on either side. Due to the structure of communication and decision variants, this form of negotiation can be seen as a kind of hybrid of a usual negotiation and an auction. The automation of the negotiation process leads to a purely transactional orientation of the negotiation. The relationship level is reduced to the technical interaction, the focus is on result optimization.

In a multi-agent environment, such BOTs can be implemented in the form of negotiating agents. These are autonomous and therefore did not require a human during the negotiation. However, the specific requirements must be defined by a human before the actual negotiation begins. The process of decision making through automated negotiation can quickly become complex, so techniques such as artificial intelligence [10,17], game theory [13,18], or evolutionary programming [5,14] are needed and explored.

This paper firstly provides an overview of automated negotiation and supply chain finance (SCF) and secondly presents a multi-agent approach to identify the best financing option in the supply chain. In particular, the DFHelperAgent developed in JADE is discussed and its use in the specially developed negotiation protocol is explained.

2 Automated Negotiation

An automated negotiation is understood as an iterative communication and decision-making process between at least two agents (persons or their representatives) who cannot fulfill their goals through unilateral actions and exchange offers and arguments to reach a consensus [3]. In addition to the understanding of the term, there are also physical characteristics into which negotiations can be differentiated:

- Number of participants (cf. Fig. 1) *bilateral:* The negotiation is limited to one supplier and one buyer. If deception and the withholding of information are neglected as negotiation strategies, this is a simple scenario. However, if these behaviors are taken into account, complexity can increase significantly, making automation difficult [22].
  multilateral: One-sided multilateral negotiations are usually automated in the form of auctions [19] regardless of whether there is one supplier and several demanders or vice versa. If multiple demanders negotiate with multiple suppliers, it is a two-sided multilateral negotiation. Such complex situations require appropriate rules, as considered by Wooldridge [27], among others.
- Number of attributes (cf. Fig. 2)
  In auctions, as a special case of a negotiation it is often only about the price [15]. All other attributes of the object of negotiation are not part of the process

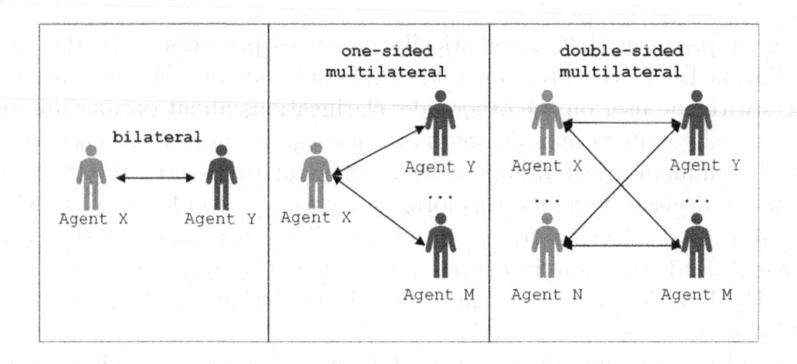

Fig. 1. Negotiation characteristics according to number of participants

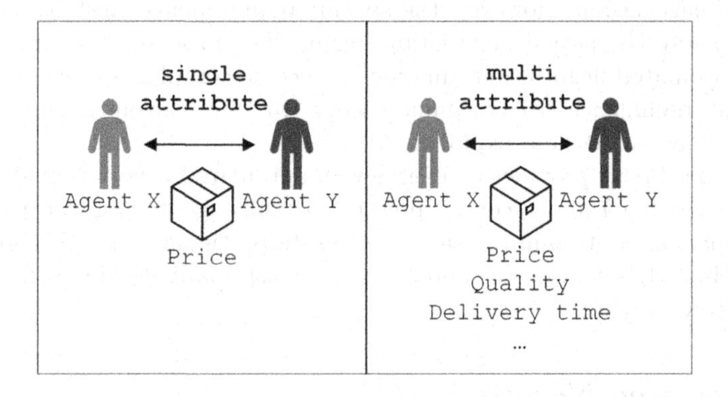

Fig. 2. Negotiation characteristics according to number of attributes

and must be agreed upon separately. In the business environment, however, negotiation is often about more than just price. The need to consider several attributes simultaneously is difficult to automate as research shows [2,25].
- With and without mediator
  In a negotiation without a mediator, communication takes place exclusively between the negotiating parties, i.e., directly. With the use of mediators (also called brokers or intermediaries), an intermediary role is introduced between the negotiating parties, who can also serve to control the market process [6].
- Number of negotiation items (cf. Fig. 3)
  In *non-combinatorial* negotiations, a decision is made only on a single or several similar negotiation items within the negotiation process. In contrast, in *combinatorial* negotiations, the negotiation takes place over a bundle (package) of different negotiation items. The automation of combinatorial negotiations is complex, so this is usually done in the form of bidding or auction packages to obtain a combined product package [4].

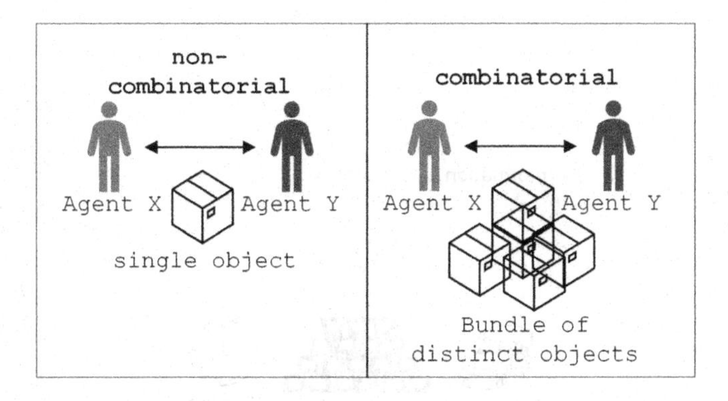

Fig. 3. Negotiation characteristics according to number of negotiation items

Because of the previously documented multiple possible manifestations of negotiation, fully automated models require strong structuring so that software agents can negotiate autonomously [3]. They are considered result-oriented because they are used to generate a negotiation outcome. The technology used for automation must be able to reflect this.

3 Supply Chain Finance

The globalization of trade has contributed to the development of the concept of supply chain finance (SCF). Within a supply chain, suppliers strive to receive their payments as early as possible, while buyers extend their payment terms as far as possible. SCF provides solutions to get a grip on this dilemma and thus offers opportunities for all stakeholders involved. Pfohl and Gomm [21] formulate that SCF is about which objects within a supply chain are financed by which actor and under which conditions. In the meantime, many SCF solutions have been developed in this area of tension [11,20,23,28].

The basis of functioning solutions is the exchange of information between the actors. The higher the level of information sharing, the closer the cooperation between upstream and downstream parts of the supply chain or between companies [24]. The efficiency of information sharing depends on the technological development of information systems and the improvement of transmission technology [12]. Nevertheless, the relationship between capital demanders (agent) and capital providers (principal) will be characterized by information asymmetry, partly because of incomplete information and partly because of information inequality. It can be assumed that the capital demander has more information about the investment project than the investor. If financing is provided, the cost of capital will be determined not least by the resulting risk to the donor. Donors will install monitoring and control mechanisms (agency costs) to protect and monitor the investment. In the case of internal supply chain (SC) financing, although commonality between principal and agent can be assumed with respect to SC objectives (e.g., lowering the SC's cost of capital), the pursuit of

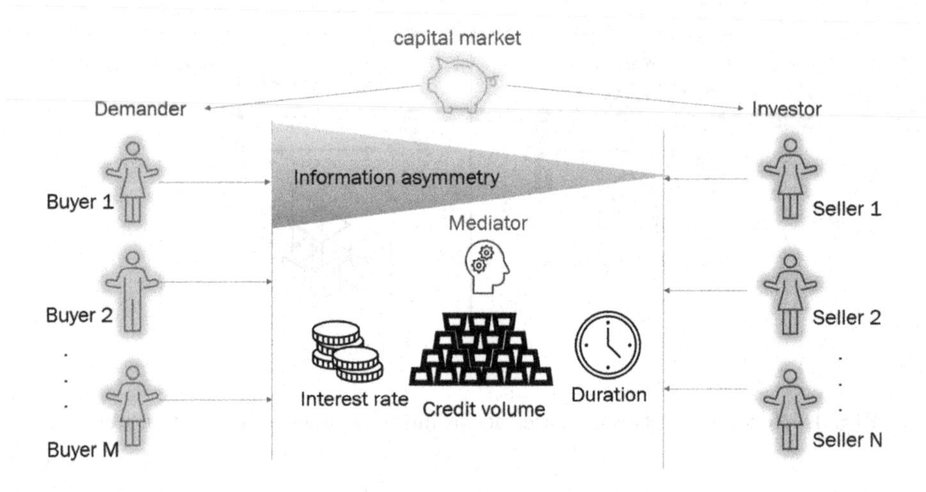

Fig. 4. Complex SCF negotiation

self-interest (e.g., maximizing profit from the investment) cannot be ruled out. The higher the degree of integration of principal and agent within the SC, the more likely there is to be a relationship of trust and the better the exchange of information, which minimizes the occurrence of information problems.

As Fig. 4 illustrates, negotiations in the context of SCF can take the possible forms of multilateral (there are multiple capital demanders and investors in the supply chain) and multiattribute (negotiations e.g. on cost of capital, loan amount and duration). In addition, intermediaries in the form of banks, platform providers, FinTechs or similar are often used. A multi-agent environment provides an infrastructure that specifies the communication and interaction of agents in such a way.

4 Multi-agent Systems

Multi-agent systems (MAS) are populations of agents that either work together towards a goal or work against each other [26]. The focus is on interaction, i.e. coordination and planning mechanisms that allow the different agents to solve a problem. MAS can be divided into the classes of problem solving, simulation, construction of artificial worlds, robotics and program design according to the type of use [7]. The approach presented here for SCF is used for problem solving (making funding decisions). Problem solving in the broadest sense of the word refers to all processes in which software agents solve tasks that are useful for humans. In problem solving with agents, one can distinguish the three subcases of distributed problem solving, solving distributed problems, and distributed techniques for problem solving [7,16].

Distributed problem solving is necessary when the task to be solved is complex and cannot be solved by one agent alone, but rather requires multiple specialized agents that complement each other in the solution. The task itself is not distributed, so the agents must collaborate appropriately in solving it. When solving distributed problems, an additional aspect is that the problem itself is inherently distributed. Accordingly, problem solving is also distributed. Distributed problems are typically encountered in the analysis, identification, troubleshooting, and control of physically distributed systems where a centralized overview is difficult (e.g., power grids). Distributed techniques for problem solving are used for problems that, in principle, a single agent can solve. Neither the scope nor the expertise is distributed. However, the distributed approach to the problem can lead to simpler solutions (e.g., assembling parts).

In the context of this thesis, the Java Agent Development Framework (JADE) [1] serves as the environment for the development of agent-oriented software. JADE is an open-source platform for the development of agent-based applications with the following characteristics: it is based on Java and benefits from third-party libraries, it is written based on the FIPA standard (Fipa, 2010), it supports the simulation of distributed systems, and it has a graphical interface for the design of MAS. To regulate communication, it uses the agent communication language (FIPA-ACL), a language based on speech act theory that allows agents to represent the communicative acts or purpose of a message (e.g., inform, request, reject). In addition, JADE has the Agent Management System (AMS), the Directory Facilitator (DF), and the Message Transport Service (MTS). The AMS entity is responsible for managing operations such as agent creation and deletion. The DF is responsible for providing yellow page services to other agents. It manages the full list of agents that have public services to help agents find services in the platform. Finally, the MTS is a service responsible for transporting and delivering ACL messages between agents on the platform [9].

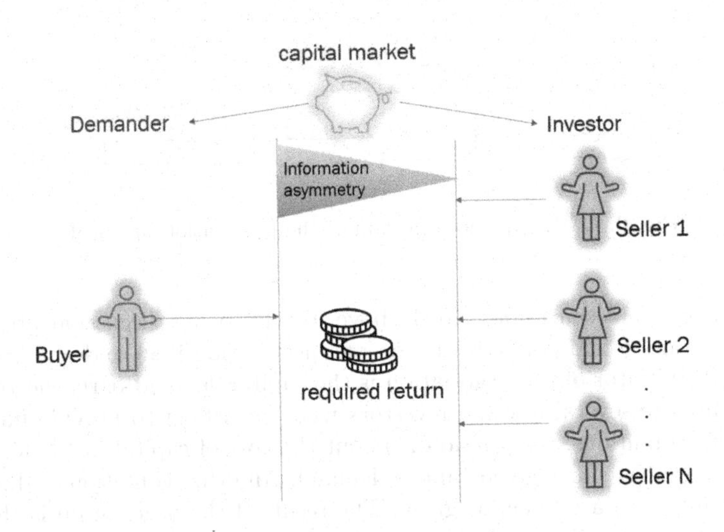

Fig. 5. One-to-many SCF scenario

5 An Illustrative Example

For the implementation of a SCF solution by MAS, the primary and supporting actors [21] must be represented in the form of agents. The roles of capital demander, internal investor and external investor must be represented.

The following process is to be mapped: For a project to be financed, an actor in the supply chain requires capital. It is possible to finance either externally via the capital market or via investors within the SC. These also finance via debt capital. Due to their position in the SC, they may have better conditions and are also interested in investing in an SC project in order to strengthen relationships and gain trust.

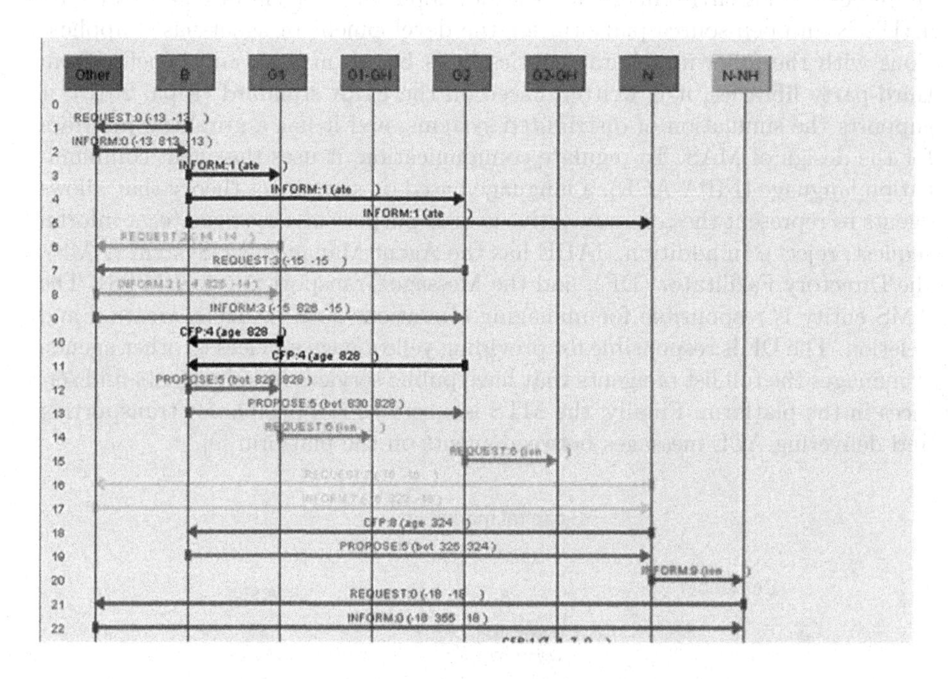

Fig. 6. UML sequence diagram of the negotiation protocol

Based on the scenario described above, the following discussion presents a MAS for a one-to-many negotiation on one negotiation item and one attribute (cf. Fig. 5). Initiator of the negotiation is the capital demand party, negotiation participants are several internal investors who are willing to provide financing at different conditions. Negotiation is about the cost of capital in the form of a return requirement for a given capital demand. All other conditions (duration, loan amount, etc.) are taken as given. The result of the negotiation is the best internal financing option. Subsequently, the demand agent checks whether the internal financing is more favorable than the external one and decides accordingly. The roles of the agents used are described below (cf. Table 1).

The classes are connected via helper agents, so-called DFHelper. This class is created when an InvestorAgent (participant) or a DemanderAgent (initiator) is initialized. Its purpose is to provide the agents with information relevant to the bidding process, as well as the termination of agents. The InvestorAgent and DemanderAgent classes use an evolution of the FIPA Iterated Contract Net Protocol [8] and are closely comparable to reverse English auctions.

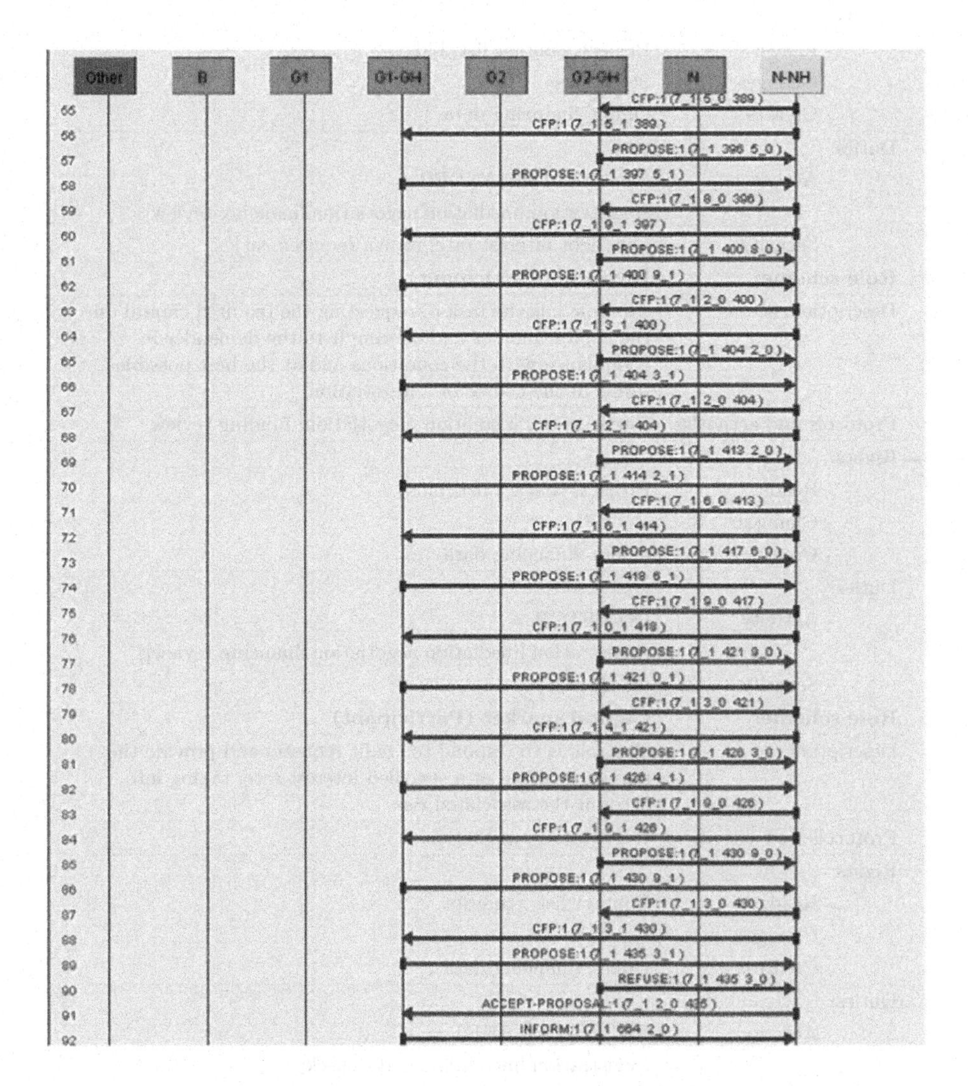

Fig. 7. UML sequence diagram of the auction

Table 1. Description of the agent roles

Role scheme:	Capital demander (Initiator)
Description:	The role has the task of meeting the desired capital requirements in compliance with the conditions and on the best possible terms in the context of a negotiation
Protocols and activities:	Registration, mediation, negotiation, funding review
Rights:	
- Reads:	Project, contact list, bids
- Changes:	Start bids
- Creates:	Offers, financing data
Duties:	
- Activity:	CAPITALDEMANDER = (registration.[mediation.negotiation.financing review])
- Security:	min{debt interest rate; return requirement}
Role scheme:	**Investor (Participant)**
Description:	The role has the task of requesting the required capital on the capital market and offering it to the demander in compliance with the conditions and at the best possible terms in the course of a negotiation
Protocols and activities:	Registration, mediation, negotiation, funding review
Rights:	
- Reads:	Project, contact list, bids
- Changes:	–
- Creates:	Offers, financing data
Duties:	
- Activity:	INVESTOR = (registration.[mediation.negotiation.financing review])
- Security:	max{profit}
Role scheme:	**Capital market (Participant)**
Description:	The role is to respond to credit requests and provide the required capital at a specified interest rate, taking into account the associated risk
Protocols and activities:	Registration, mediation
Rights:	
- Reads:	Contact list, requests
- Changes:	–
- Creates:	Offers, financing data
Duties:	
- Activity:	CAPITALMARKET = (registration.[mediation.credit check])
- Security:	max{profit}

Figure 6 is a section of a UML sequence diagram of the negotiation protocol, recorded with the sniffer agent provided by JADE. It shows the communication of each agent in a negotiation with a CapitalMarketAgent (B), a DemanderAgent (N), his helper (N-NH), and two InvestorAgents (G1, G2) and their helpers (G1–GH, G2–GH). After the agents are registered, G1, G2 and N ask B for debt capital. Within the negotiation protocol this is done by the message type "call for proposal" (cfp) in lines 10, 11 and 18. In lines 12, 13 and 19, the CapitalMarketAgent responds with the message type "propose". B submits corresponding offers, the helpers are generated, and the starting bid and the respective bid lower bounds are calculated.

The actual negotiation process starts (cf. Fig. 7). N-NH as helper of the auction initiator requests with the message "cfp" all investor helpers to submit bids (for example in lines 56 from N-NH to G1–GH and 59 from N-NH to G2–GH). In addition, the necessary requirements of the auction are transmitted with this message. Participants respond with a bid through the message type "propose" (for example, G1–GH submits a bid in line 58 and G2–GH submits a bid in line 61) or decline to bid further through the message "refuse" (in line 90, G2–GH drops out). If no new bids are made, N-NH can accept the offer after checking the financing options via the message "accept-proposal" or reject it with the message "reject-proposal". Here, in line 91, the offer from investor G1 is accepted.

Subsequently, the DemanderHelper N-NH informs the DemanderAgent N (line 93) and the contracts are concluded. The investor G1 accepts the offer (line 98) of the CapitalMarketAgent (B) and lends capital to the demander at the conditions negotiated in the auction (line 99). The process is completed as soon as the funds have been repaid with corresponding interest premiums (cf. Fig. 8, lines 100–103).

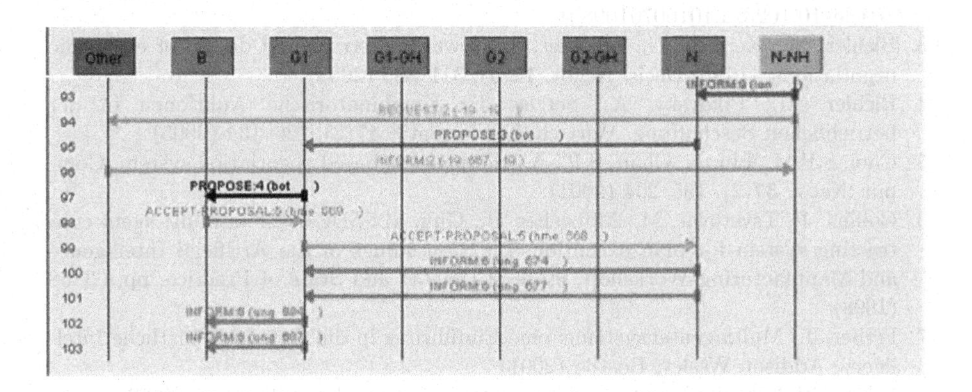

Fig. 8. UML sequence diagram of the contract closures

6 Conclusion and Outlook

This paper has provided a brief overview of automated negotiation. It outlines the possible characteristics of such negotiations in terms of the number of participants, attributes, and items of negotiation, as well as the involvement of mediators.

Supply chain finance is presented as an application area for automated negotiation. In particular, the importance of information exchange between the involved actors is pointed out.

After introducing multi-agent technology in particular as a technique for problem solving and describing the Java Agent Development Framework as the development environment used, the demonstration is based on a scenario from SCF. The agents and their roles, which are used, are described in detail.

The successful implementation of an automated negotiation of the best funding option for a project in a supply chain is demonstrated using a UML sequence diagram. In summary, automated negotiation using agents has been shown to be of interest in the field of SCF and requires further investigation.

References

1. Bellifemine, F.L., Caire, G., Greenwood, D.: Developing Multi-Agent Systems with JADE. John Wiley and Sons, Hoboken (2007)
2. Bichler, M.: Trading financial derivatives on the Web–an approach towards automating negotiations on OTC markets. Inf. Syst. Front. **1**(4), 401–414 (2000). https://doi.org/10.1023/A:1010070109801, https://link.springer.com/article/10.1023/a:1010070109801
3. Bichler, M., Kersten, G., Strecker, S.: Towards a structured design of electronic negotiations. Group Decis. Negot. **12**(4), 311–335 (2003)
4. Bichler, M., Pikovsky, A., Setzer, T.: Kombinatorische Auktionen in der betrieblichen Beschaffung. Wirtschaftsinformatik **47**(2), 126–134 (2005)
5. Choi, S.P.M., Liu, J., Chan, S.P.: A genetic agent-based negotiation system. Comput. Netw. **37**(2), 195–204 (2001)
6. Collins, J., Tsvetovat, M., Mobasher, B., Gini, M.: MAGNET: a multi-agent contracting system for plan execution. In: Proceedings of the Artificial Intelligence and Manufacturing Workshop: State of the Art and State of Practice, pp. 63–68 (1998)
7. Ferber, J.: Multiagentensysteme: eine Einführung in die verteilte künstliche Intelligenz. Addison-Wesley, Boston (2001)
8. FIPA: FIPA iterated contract net interaction protocol specification (2000)
9. FIPA: Foundation for Intelligent Physical Agents (2020). http://www.fipa.org/
10. Gerding, E.H., van Bragt, D.D.B., La Poutré, J.A.: Scientific Approaches and Techniques for Negotiation A Game Theoretic and Artificial Intelligence Perspective. Centrum voor Wiskunde en Informatica (2000)
11. Hausladen, I., Dachsel, B.: Supply Chain Finance im Überblick. WiSt - Wirtschaftswissenschaftliches Studium **47**(2–3), 4–11 (2018). https://doi.org/10.15358/0340-1650-2018-2-3-4

12. Hu, J., Erdogan, B., Jiang, K., Bauer, T.N., Liu, S.: Leader humility and team creativity: the role of team information sharing, psychological safety, and power distance. J. Appl. Psychol. **103**(3), 313–323 (2018). https://doi.org/10.1037/apl0000277
13. Jennings, N.R., Faratin, P., Lomuscio, A.R., Parsons, S., Sierra, C., Wooldridge, M.: Automated negotiation: prospects, methods and challenges. Int. J. Group Dec. Negotiation **10**(2), 199–215 (2001)
14. de Jonge, D., Sierra, C.: GANGSTER: an automated negotiator applying genetic algorithms. In: Recent Advances in Agent-based Complex Automated Negotiation, pp. 225–234. Springer, Cham (2016). https://doi.org/10.1007/978-3-319-30307-9_14
15. Kehagias, D.D., Symeonidis, A.L., Mitkas, P.A.: Designing pricing mechanisms for autonomous agents based on bid-forecasting. Electron. Mark. **15**(1), 53–62 (2005)
16. Klügl, F.: Multiagentensysteme. In: Görz, G., Schneeberger, J. (eds.) Handbuch der künstlichen Intelligenz, pp. 527–556. Walter de Gruyter (2012)
17. Li, C., Giampapa, J., Sycara-Cyranski, K.: A review of research literature on bilateral negotiations. Technical report (2003)
18. Liang, Y., Yuan, Y.: Co-evolutionary stability in the alternating-offer negotiation. In: IEEE Conference on Cybernetics and Intelligent Systems, 2008. IEEE, Piscataway, NJ (2008). https://doi.org/10.1109/iccis.2008.4670896
19. Lomuscio, A.R., Wooldridge, M., Jennings, N.R.: A classification scheme for negotiation in electronic commerce. Group Decis. Negot. **12**(1), 31–56 (2003). https://doi.org/10.1023/A:1022232410606
20. Marak, Z., Pillai, D.: Factors, outcome, and the solutions of supply chain finance: review and the future directions. J. Risk Finan. Manage. **12**(1), 3 (2019). https://doi.org/10.3390/jrfm12010003, https://www.mdpi.com/1911-8074/12/1/3
21. Pfohl, H.-C., Gomm, M.: Supply chain finance: optimizing financial flows in supply chains. Logistics Res. **1**, 149–161 (2009). https://doi.org/10.1007/s12159-009-0020-y
22. Rebstock, M.: Elektronische Unterstützung und Automatisierung von Verhandlungen. Wirtschaftsinformatik **43**(6), 609–619 (2001)
23. Taschner, A., Charifzadeh, M.: Supply chain finance. In: Management Accounting in Supply Chains, pp. 121–151. Springer, Wiesbaden (2020). https://doi.org/10.1007/978-3-658-28597-5_6
24. Tchamyou, V.S., Asongu, S.A.: Information sharing and financial sector development in Africa. J. Afr. Bus. **18**(1), 24–49 (2017). https://doi.org/10.1080/15228916.2016.1216233
25. Teuteberg, F.: Experimental evaluation of a model for multilateral negotiation with fuzzy preferences on an agent-based marketplace. Electron. Mark. **13**(1), 21–32 (2003)
26. Veit, D.J.: Matchmaking in Electronic Markets: An Agent-Based Approach towards Matchmaking in Electronic Negotiations, Lecture notes in computer science, vol. 2882. Springer, Berlin and Heidelberg (2003). https://doi.org/10.1007/b94069
27. Wooldridge, M., Bussmann, S., Klosterberg, M.: Production sequencing as negotiation. In: Proceedings of the First International Conference on the Practical Application of Intelligent Agents and Multi-Agent Technology (PAAM-96), pp. 709–726 (1996)
28. Zhao, L., Huchzermeier, A.: Supply chain finance: integrating operations and finance in global supply chains. In: EURO Advanced Tutorials on Operational Research. Springer, Cham, Switzerland (2018). https://doi.org/10.1007/978-3-319-76663-8_6

Production Scheduling with Stock- and Staff-Related Restrictions

Carlo S. Sartori(✉), Vinícius Gandra, Hatice Çalık, and Pieter Smet

Department of Computer Science, CODeS, Leuven.AI, KU Leuven, Leuven, Belgium
{carlo.sartori,vinicius.gandramartinssantos,hatice.calik, pieter.smet}@kuleuven.be

Abstract. Effective production scheduling allows manufacturing companies to be flexible and well-adjusted to varying customer demand. In practice, production scheduling decisions are subject to several complex constraints which emerge from staff working hours and skills, delivery schedules, stock capacities, machine maintenance and machine setup. This paper introduces a novel production scheduling problem based on the real-world case of a manufacturing company in Belgium. Given a set of customer requests which may only be delivered together on one of the provided potential shipment days, the problem is to select a subset of these requests and schedule the production of the required item quantities subject to the aforementioned restrictions. All decisions must be taken for a time horizon of several days, leading to a complex problem where there may not be enough resources to serve all requests. We provide an integer programming formulation of this novel problem which is capable of solving small-scale instances to proven optimality. In order to efficiently solve large-scale instances, we develop a metaheuristic algorithm. A computational study with instances generated from real-world data indicates that the metaheuristic can quickly produce high-quality solutions, even for cases comprising several days, requests and limited stock capacities. We also conduct a sensitivity analysis concerning characteristics of the schedules and instances, the results of which can be exploited to increase production capacity and revenue.

Keywords: Production scheduling · Stock levels · Integer programming · Metaheuristic

1 Introduction

Due to increased global competitiveness and market uncertainty, manufacturing companies have become increasingly flexible to meet varying demand for their products. Typically, production lines have finite capacities and demand cannot be met by simply increasing production rates. Instead, careful lot-sizing decisions must be made to determine how much of each product is produced and how much stock is maintained. In practice, these decisions are subject to a variety of constraints, including staff-related restrictions and delivery schedules.

M. Mes et al. (Eds.): ICCL 2021, LNCS 13004, pp. 142–162, 2021.
https://doi.org/10.1007/978-3-030-87672-2_10

Our work is motivated by a problem faced by a manufacturing company in Belgium. The production environment is equipped with two non-identical machines capable of producing multiple item types under the supervision of human operators. Each machine can only produce one item type at a time given that the necessary configuration (setup) for an item type is conducted before production begins. The item types and quantities demanded by a set of customers are fixed, but the date to deliver the entire demand of each customer must be selected from multiple options provided by the customer. If resources are not sufficient to satisfy all customer demands, the lost sales are reflected as a penalty cost. Machines require maintenance at regular intervals. Maintenance and certain setups can only be performed by a skilled operator who is available during a specific time slot each day. In order to increase production capacity, the two machines can operate in parallel. They can also operate with overtime or during night shifts, but each of these options incurs additional costs. Another type of cost is incurred when the daily safety stocks are not maintained, meaning that the stock level of an item type drops below a minimum desired threshold. Moreover, a predetermined maximum stock level may not be exceeded under any circumstances for any item type.

Given all these machine-, operator- and stock-related restrictions, a solution for the problem involves generating a production schedule at minimum cost for a finite period of time, which comprises a number of days subdivided into several time blocks. We refer to this problem as the *Production Scheduling Problem with stock- and staff-related restrictions* (PSP). Before providing a detailed description of the PSP in Sect. 2, we will briefly review some related problems to position the PSP in the literature.

The literature most closely related to the PSP is that of *Lot-sizing and Scheduling Problems* (LSPs): a class of problems with many variants for the planning of production schedules. A review concerning standard models proposed for several LSPs was provided by [3]. Additionally, due to its numerous side constraints, the PSP can also be positioned within the literature of LSPs with *Secondary Resources* (SRs) [12]. SRs are subcategorized as one of two types: disjunctive or cumulative. A disjunctive SR can only be employed once at a time, whereas a cumulative SR is one for which the total accumulated value restricts the solution in some way. The PSP contains SRs of both types. Similar to the study conducted by [11], certain operations can only be performed in the presence of a skilled operator. This operator is a disjunctive SR who cannot perform multiple duties in parallel. Meanwhile, daily stock capacities and the stocks themselves are cumulative SRs since although they constrain the solutions to the problem, they can be used to satisfy multiple requests.

More specifically, the PSP can be considered a variant of the *Discrete Lot-sizing and Scheduling Problem (DLSP)* [5]. This is one of the classic LSPs as per the classification provided by [3]. There are two characteristics that distinguish the DLSP from other LSPs [4]. First, it considers both macro and micro periods where macro periods are formed by a sequence of micro periods. Other LSPs only consider the macro scale. Second, the DLSP assumes all-or-nothing production:

only one item type may be produced by a machine during a micro period while running at full capacity. In the PSP, we have both macro and micro periods: days and blocks. Furthermore, in each block, a machine is either idle or entirely dedicated to a single type of operation (maintenance, setup or production of one item type). One key difference, however, is that while the DLSP considers a holding cost per item in stock, the PSP assumes no holding costs but instead enforces a maximum stock level for each item type.

Another key challenge in the PSP is the order selection and scheduling, for which we refer interested readers to [10] for a thorough review of the topic. Note that order selection and scheduling coupled with sequence-dependent setup times becomes significantly more challenging to solve since the production order can impact the amount of unproductive time introduced depending on the setups required for each machine. A recent problem variant which combines both characteristics was studied by [9].

The PSP differs from the aforementioned DLSP and other LSP variants in three main ways which in combination with one another makes the problem very challenging. First, demand and due dates are not entirely fixed and should be decided. Second, daily production capacities can be increased via parallel production on machines, overtime and night shift production, however all of these options incur additional costs. Third, setup or maintenance operations can be spread over multiple non-continuous blocks, between which the dedicated operator might be assigned other duties, such as the setup or maintenance of another machine. The demand selection and multiple due date options associated with the PSP resemble the order selection and scheduling with leadtime flexibility considered by [2] for a single-machine system. However, the PSP remains distinct due to the second and third aforementioned characteristics.

The remainder of this paper is organized as follows. Section 2 formally introduces the PSP. Section 3 details a *Late Acceptance Hill-Climbing* (LAHC) heuristic [1] we designed for the PSP. A recent successful application of LAHC in combination with exact methods to solve a variant of the LSP [6] encouraged us to select this method. Section 4 goes on to provide a computational study and introduce new instances, while Sect. 5 concludes the paper.

2 Notation and Problem Description

The PSP is modeled over an ordered set $D = \{1, \ldots, |D|\}$ of days, each of which begins at 05:00 and has a duration of 24 h. Each day is decomposed into a set $B = \{1, \ldots, b_{\mathrm{n}}\}$ of time *blocks* of equal length as depicted in Fig. 1. Figure 1 also highlights specific blocks that define intervals during which certain operations can take place. A task to be carried out on a machine takes an integer number of blocks to be completed and cannot take less than one block. Throughout the remainder of this paper, time is expressed in terms of number of blocks.

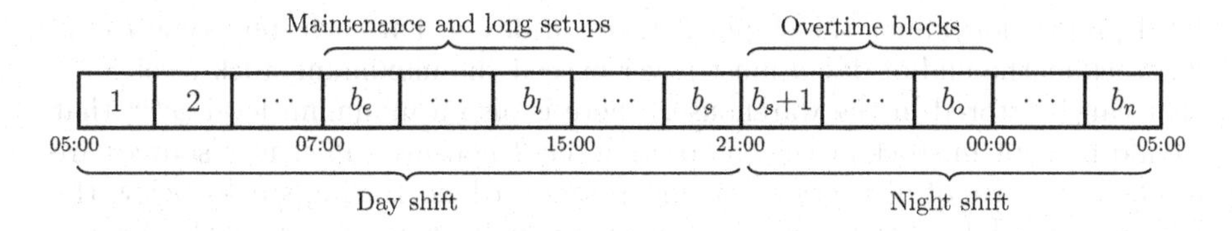

Fig. 1. Representation of one day as a set of time blocks.

A day begins with block 1 and ends with block b_n, after which the following day begins. The factory is open every day between blocks 1 and b_s. Blocks $1, \ldots, b_s$ constitute the *day shift*, whereas blocks $b_s+1, \ldots, b_n$ constitute the *night shift*. The night shift may be used at an additional cost in one of two different ways. First, blocks $b_s+1, \ldots, b_o$ may be individually scheduled as *overtime* and incur a cost p_o per block. Overtime may be scheduled any day and as often as necessary. Alternatively, a *full night shift* may be scheduled at a cost p_n per day, thereby enabling all night blocks $b_s+1, \ldots, b_n$ to be used. The organization in question requires full night shifts to be scheduled for at least d_n consecutive days. We refer to this as the *minimum consecutive night shifts* constraint. It is not possible to schedule overtime and a full night shift on the same day.

There are two machines available to produce a set I of different item types. Let $M = \{m_1, m_2\}$ denote the set of machines and $e_i^m \geq 0$ represent the total number of type $i \in I$ items that machine $m \in M$ produces per block. If machine m cannot produce items of type i, then $e_i^m = 0$.

Machines can operate simultaneously, which we will henceforth refer to as *parallel operation*. Parallel operation occurs in a day $d \in D$ and a block $b \in B$ when no machines are idle during this time slot. Parallel operation during the day shift incurs an additional cost p_p for that day because the company must hire an additional outsourced external worker. During night shifts, however, machines can operate in parallel without additional costs because night shifts already include the outsourced external worker in their cost.

Machines operate to fulfill a set R of customer requests, where each $r \in R$ requires a quantity q_r^i of item type $i \in I$ to be produced. Let c_i be the sales price of each item type $i \in I$, then $c(r) = \sum_{i \in I} q_r^i c_i$ denotes the revenue obtained from request $r \in R$ when it is fulfilled. Each request r has a set of allowed shipping days $D_r \subseteq D$ which are specified by the customer. A request r can only be shipped on day $d \in D_r$ if all the necessary quantities of items are available at the end of day d (partial shipments are not permitted). Items do not necessarily need to be produced during day d and may instead be taken from the available stock. If a request cannot be shipped by its latest feasible day, it is considered *unserved*.

The allowed shipping days are also crucial with respect to maintaining feasible stock levels. Shipments are scheduled by the end of each day (after block b_n). After shipment, but before the start of the next day, stock levels must be updated. The stock on day $d \in D$ corresponds to the stock on day $d-1$ plus the total production minus the shipped items on day d. For each item type $i \in I$, its stock at the end of day d must never exceed the maximum stock level $S_i^{\max}$ that can be stored in the warehouse. There is also a minimum level $S_i^{\min}$ that should be kept in stock at the end of each day to ensure sufficient resources are available in case of disruptions or unexpected orders. Having stocks below the minimum stock level is allowed, but incurs a penalty p_s at the end of the day per unit below the minimum. Naturally, the stock of an item can never be negative.

For each (m, d, b) tuple such that $m \in M$, $d \in D$ and $b \in B$, we assign one of the following four tasks: (i) production of an item type, (ii) machine setup from one item type to another, (iii) maintenance and (iv) idle. If a task requires more than one block of time, then multiple blocks must be scheduled. Tasks with multiple blocks are not necessarily continuous. Idle and maintenance/setup blocks may be scheduled in between other setup/maintenance tasks. When night shift blocks are not employed in the schedule, they are all set to idle.

The type of item produced on a machine can be changed with sequence-dependent setup times that require s_{ij}^m blocks to change the configuration of machine m from production of item type $i \in I$ to type $j \in I$. These setups are classified as either short (U) or long (L). A short setup $(i, j) \in U,\ i, j \in I$ may be carried out at any time. Meanwhile, a long setup $(i, j) \in L,\ i, j \in I$ is only allowed during the range of blocks $[b_e, b_l]$ due to staff-related constraints. Additionally, maintenance must be scheduled for each machine $m \in M$ at most $g_{\max}^m$ days apart and each maintenance takes f^m blocks. No production is possible while a machine is undergoing maintenance or setup. Long setups and maintenance can only be scheduled during the range $[b_e, b_l]$ and are never allowed in parallel with another maintenance or long setup.

Since production is supposed to be a continuous process, we need to take into account certain information concerning schedules from the previous time horizon. This historic information includes: the last item type for which each machine was configured, the number of days since maintenance was performed for each machine, the stocked quantity of each item, and the number of consecutive night shifts scheduled at the end of the preceding scheduling period. The last of these parameters provides us with the number of *mandatory night shifts* $h_n \geq 0$: the number of night shifts that must be scheduled at the beginning of the time horizon in order to comply with the minimum consecutive night shifts constraint.

The primary decision is to select when and which requests to produce and ship. This not only involves deciding on which date/time, on which machine and in which order to produce items, but also how many items of each type to produce, and how many items to take from the stock. Additionally, we must decide when to schedule overtime or night shifts to extend production capacity. All decisions must also account for the constraints related to the limited shipping days, stock levels, machine maintenance and limited setup times.

A solution is sought which minimizes the revenue loss from unserved requests, additional personnel costs (overtime, night shift and parallel operations) and the penalties incurred by stock deficits. Let X be the set of all feasible solutions to the PSP and $f(s)$ be the cost of solution $s \in X$. The aim is to find the minimum cost solution $s^* = \arg\min_{s \in X} f(s)$. This cost is defined by five components. First, the cost $c(r)$ of all unserved requests $r \in R$ (or the revenue loss). Second, a cost p_p for each day $d \in D$ that contains at least one parallel operation. Third, a cost p_n for each day $d \in D$ that a night shift has been scheduled. This cost is incurred even if all night shift blocks are completely idle, as long as day d is part of a minimum consecutive night shifts sequence. Fourth, a cost p_o for each block $b \in B$ of overtime scheduled for each day $d \in D$. Finally, a penalty p_s is incurred per unit of item $i \in I$ below its minimum stock level $S_i^{\min}$ at the end of every day $d \in D$.

In order to formally define the problem, an Integer Linear Programming (ILP) formulation is provided in Appendix 5. This ILP formulation is also used to assess the performance of the heuristic approach proposed in Sect. 3.

3 A Heuristic Approach

Preliminary experiments revealed that the ILP formulation can only provide high-quality solutions for very small instances within a time limit of one hour. Thus, a tailored algorithm is required in order to produce high-quality solutions for large-scale PSP instances within reasonable processing times. In this paper, we employ a heuristic which improves an initial solution through insertion and removal of requests in order to efficiently explore the PSP's solution space.

The proposed heuristic to solve the PSP adapts LAHC [1], which is a simple metaheuristic framework that requires only a single parameter: the length of the list containing previous solution costs. LAHC has recently been employed to solve a variant of the LSP and achieved high-quality results [6], demonstrating that it is also a good choice for this class of problems. Algorithm 1 outlines the LAHC algorithm employed to solve the PSP.

Algorithm 1: Late Acceptance Hill-Climbing (LAHC).

```
1  Input: Instance of the PSP, parameters M_iter, α, L_s, γ, β_roll, β_night, η;
2  s ← initialSolution();
3  F[k] ← +∞, k = 1,...,L_s;
4  iter, idle ← 0;
5  while iter < M_iter and idle < ⌊αM_iter⌋ do
6  |   s' ← buildNewSolution(s, iter, M_iter, γ, β_night, η);
7  |   if f(s') ≥ f(s) then
8  |   |   idle ← idle + 1;
9  |   else
10 |   |   idle ← 0;
11 |   k ← iter mod L_s;
12 |   if f(s') < F[k] or f(s') ≤ f(s) then s ← s';
13 |   if f(s) < F[k] then F[k] ← f(s);
14 |   if idle = ⌊αβ_roll M_iter⌋ then s ← s*;
15 |   iter ← iter + 1;
16 return s*
```

Algorithm 1 begins by constructing an initial feasible solution using the procedure which will be detailed in Sect. 3.1, followed by initializing LAHC's fitness array F and counter variables *iter* and *idle* (lines 2–4). The main loop of the algorithm (lines 5–18) is iterated over until either M_{iter} iterations have been reached or no improvement has been observed for $\lfloor \alpha M_{\text{iter}} \rfloor$ iterations. Both M_{iter} and $\alpha \in [0, 1]$ are parameters of the LAHC.

In every iteration of the LAHC, a new solution s' is generated using the current solution s and the procedure which will be described in Sect. 3.2 (line 6). Then, lines 7–15 update the idle iteration counter, the current solution s and the fitness array F according to the original strategy introduced by [1]. Note that F allows a worsening solution s' to be accepted whenever its total cost is less than the cost of the solution L_s iterations earlier, where L_s is the size of LAHC's list of solution costs. To increase intensification of the search, we included a rollback procedure so that when the number of idle iterations reaches a specified percentage of its maximum value, the current solution s is replaced with s^* (lines 16–17). Here, $\beta_{\text{roll}} \in [0, 1]$ is also a parameter. The first rollback is always allowed. Any additional rollbacks can only occur if s^* is improved after the preceding rollback. The best solution s^* generated over all iterations is returned (line 19).

3.1 Initial Solution

Initial solution s is constructed by first initializing the blocks for every day and every machine as idle. Next, the required maintenance is scheduled for each machine with as many days in-between as possible while avoiding parallel maintenance blocks. These initial maintenance days are fixed throughout the entirety of LAHC's execution. After these steps, solution s should contain a feasible maintenance schedule for all machines, otherwise the instance is considered infeasible. Algorithm 2 outlines the steps to build the initial solution.

Algorithm 2: initialSolution()

```
1 s ← idleSolution();
2 s ← scheduleMaintenance(s);
3 O_r ← true, ∀r ∈ R ;                    // Initially all requests are available
4 while ∃ r ∈ R : O_r = true do
5     c^max ← max c(r) : O_r = true;
6     select request y : O_y = true with probability 0.8 · c(y)/c^max;
7     s ← insertRequestBestPosition(s, y, true);
8     O_y ← false;
9 return s;
```

Once a feasible maintenance schedule is generated, requests are inserted into s along with the necessary production and setup blocks. These insertions are performed in a greedy-randomized manner (lines 4–8 of Algorithm 2). While there remain available requests to be inserted, one request $r \in R$ is selected at random with probability $0.8c(r)/c^{\max}$, where $c(r)$ denotes the profit for shipping request r and $c^{\max}$ is the value of the most profitable request among all currently available ones (lines 5–6). The selected request r is inserted into its best shipment day $d^* \in D_r$ using the procedure described in Sect. 3.3 (line 7). Once r is inserted into s or once no feasible day remains into which r can be inserted, the request is marked as *processed* so that it is no longer considered for insertion in the construction phase (line 8). Request insertion is repeated until all requests have been marked *processed.*

3.2 New Solution Generation

A new solution s' is constructed from the current solution s via a series of modifications. Algorithm 3 outlines the steps to perform these changes. The algorithm removes requests at random from s', as well as unnecessary production blocks arising from such removal, employing the procedure described in Sect. 3.4 (line 7 of Algorithm 3). Note that by removing requests and unecessary production blocks, more space or *slack* is created to later reinsert requests and therefore more effectively explore the solution space of the PSP. The algorithm reinserts these requests in a random order on their best possible shipping dates using the heuristic outlined in Sect. 3.3 (line 8). The number of removed requests is selected uniformly from $[1, \max\{20, \lfloor\gamma|R|\rfloor\}]$, where $\gamma \in [0,1]$ is a parameter of the LAHC. We limit the removal to a maximum of 20 requests to account for very large instances. In the case that not all of the removed requests are successfully reinserted, the remaining unserved requests incur a penalty.

Algorithm 3: buildNewSolution(s, iter, $M_{\text{iter}}, \gamma, \beta_{\text{night}}, \eta$)

```
1  Input: Solution s, num. of iterations iter, parameters M_iter, γ, β_night, η;
2  s' ← s;
3  if iter > M_iter β_night and iter mod η = 0 then
4  |   s' ← removeUnusedOvertimeAndNightShift(s');
5  |   s' ← insertRandomOvertimeOrNightShift(s');
6  y ← rand(1, min{20, γ|R|});
7  s' ← randomRequestRemoval(s', y);
8  s' ← bestRequestInsertion(s');
9  s' ← fixStockBelowMinimum(s');
10 return s';
```

Once the number of iterations in Algorithm 1 reaches $\lfloor\beta_{\text{night}} M_{\text{iter}}\rfloor$, new solutions are permitted to employ overtime or additional night shifts. Here, value $\beta_{\text{night}} \in [0,1]$ is another parameter of the LAHC. Overtime and night shifts are modified as follows (lines 3–5). First, unused but active overtime and night shifts are removed, followed by the activation of new overtime or night shift blocks. The procedures for (de)activating overtime and night shift blocks are detailed in Sect. 3.5. These (de)activations are only executed every η iterations of the LAHC so that the algorithm has sufficient time to make best use of the new overtime or night shift blocks. The value of η is parameterized as well. The decision to postpone the activation of night shifts and overtime to later iterations in the LAHC's execution is not arbitrary. In practice, night shifts and overtime are deemed undesired by both employers and employees. Therefore, avoiding their use forces the LAHC to produce solutions without featuring them.

When all requests have been fulfilled in s', a procedure to increase stock levels by scheduling production blocks is employed (line 9). This procedure examines each day $d \in D$ for the items $i \in I$ that have stocks below their minimum level. It then attempts to schedule production blocks for these items on days $d' \leq d$ to reduce stock penalties. Production blocks on days $d' < d$ are only added if they do not result in a solution exceeding maximum stock levels.

3.3 Request Insertion Heuristic

A shipping day must be determined for each request while respecting stock levels and only considering the allowed shipping days for that request D_r. Our proposed request insertion heuristic inserts requests into each allowed shipping day. The shipping day which yields the best result is then selected. When attempting to ship a request r on a given day d, production of all requested items q_r^i $(i \in I)$ is scheduled in such a way that the maximum possible number of items is produced from day 0 until day d. When the produced items are not enough to serve the request, the shortfall of production is compensated by preexisting stock. By producing as many items as possible for each inserted request, solutions have less chance of violating minimum stock levels and stocks may be preserved to help serve requests with greater demand.

Algorithm 4 outlines the overall framework of the request insertion heuristic for a single request r. The insertion of parallel production is optional and given as a parameter. Production of items is inserted backwards, beginning from the shipping day d back until the first day of the time horizon (lines 8–11). This mechanism aims to maintain production as close to the shipment date as possible, which decreases the chance of violating maximum stock levels. If a solution producing all q_r items is not found, the remaining items to complete request r are removed from stocks (lines 12–15). In this case, minimum stock levels can be violated. After all permitted shipping days have been checked, the best solution is returned (line 19).

The *insertProduction* method receives a set of item types and their respective quantities RP to be produced on a given day d'. For every $i \in I$ where $RP_i \geq 1$, the method attempts to insert a total of $b = \lceil * \rceil RP_i/e_i^m$ production blocks on any machine $m \in M$ where $e_i^m > 0$. The method ends as soon as either all production has been scheduled or once it is not possible to insert any more production on that day. The items and machines are iterated over sequentially.

The production of b blocks of item type i is then scheduled on machine m on day d' in accordance with one of two possible methods.

The first method is used when machine m on day d' has no production of item i. In this scenario, production of item i is inserted along with the necessary setups into the first sequence of idle blocks on day d' where it can be fit. If the production of b blocks is not possible, the same method is called to insert $b-1$ blocks on the same day and on the same machine. The second method handles the insertion of production of item i on a day and machine that is already producing at least one block of i. This method increments the sequence of production blocks of i by b blocks. Tasks already scheduled on day d' are pulled back or pushed forward, replacing idle blocks. The resulting day is feasible if every non-idle task remains scheduled and no conflict is found.

Algorithm 4: insertRequestBestPosition(s, r, bp)

```
1  Input: Solution s, Request r, boolean bp to enable parallel production;
2  s* ← s;
3  foreach d ∈ D_r do
4      s' ← s;
5      RP ← q_r^i ∀i ∈ I ;                     // Remaining production of each item
6      IP ← ∅ ∀i ∈ I ;                         // Inserted production of each item
7      d' ← d;
8      while d' ≠ 0 or RP ≠ ∅ do
9          insertProduction(s', d', RP, IP, bp);
10         RP ← RP \ IP;
11         d' ← d' − 1;
12     if RP ≠ ∅ then
13         possible ← RemoveFromStock(s', RP, d);
14         if possible = False then go to next d;
15     s' ← Schedule shipping of request r on day d;
16     if f(s') < f(s*) then
17         s* ← s';
18 return s*;
```

3.4 Request Removal Heuristic

Given a current solution and a set of requests $\overline{R}$ to be removed from the schedule, the request removal heuristic starts by removing the scheduled shipping days of all requests in $\overline{R}$. Stock levels for every item and day are then recalculated. In this step, solutions often have a large number of items being produced which are not shipped, possibly violating maximum stock levels.

In order to remove unnecessary production and correct stock level violations, a removal slack SLK is calculated for each day and item type. SLK expresses how many production tasks of item type i can be removed from day d without violating S_i^{min} and is calculated as $SLK_{[d][i]} = \min(stock_{[d]} - S_i^{min}, SLK_{[d+1][i]})$.

SLK is used as an upper bound and the request removal heuristic continues by removing as many production tasks as possible for each day and item type. Removing a large number of production tasks results in a partial solution with more idle blocks, providing additional flexibility for request insertion. When removing production tasks, setups for certain item types become obsolete as those items are no longer being produced, hence these setups are also removed.

3.5 Overtime and Night Shift Heuristics

Once the number of iterations performed by Algorithm 1 exceeds the threshold defined in Sect. 3.2, non-mandatory night shifts and overtime blocks are (de)activated in the schedule. For a particular day $d \in D$, (de)activation of night shifts and overtime is performed by the use of *Boolean flags* indicating whether a night shift (alternatively overtime) is active for a day d. Overtime and night shift flags cannot be simultaneously active during the same day d.

Overtime and Night Shift Removal: Before the insertion of overtime or night shifts, the algorithm first removes all unused blocks. For every day d with active overtime but for which all blocks in $[b_{s+1}, b_o]$ contain idle tasks, the overtime flag is deactivated. For night shifts, a sequence of at least d_n consecutive days with night shifts is extracted from the solution (if such a sequence exists). All empty night shifts – those with only idle blocks in both machines – in this sequence are deactivated from either the beginning, the end, or the middle of the sequence so long as the minimum consecutive night shift constraint is respected. Note that mandatory night shifts are never removed and are always available for use.

Overtime and Night Shift Insertion: After the removal of unused overtime and night shifts, new insertions are performed. The algorithm chooses with uniform probability one of the three following insertion methods:

(I1) Overtime insertion: a number $\delta \in [1, |D|]$ is selected with uniform probability. Then, the procedure iterates over all days D in a random order, activating overtime flags for days without any active flags. This continues until either δ overtime blocks have been activated or all days have been checked.

(I2) Earliest night shift insertion: this can only be executed if the instance contains mandatory night shifts. The algorithm selects with uniform probability a number $\delta \in [1, d_n - 1]$. Then, starting from the first day without a mandatory night shift, the algorithm activates night shifts for the next δ days in the time horizon. Any of the δ days which already contains an activated night shift is counted as an activation. In the case that a day contains overtime, it is deactivated and a night shift is activated in its place.

(I3) Random night shift insertion: a day $d_1 \in [h_n, |D|]$ is selected at random. Then, for all days $d_1, \ldots, d_1 + d_n$, night shifts are activated. Any active overtime is replaced with a night shift.

For both (I2) and (I3), feasibility with respect to the minimum consecutive night shift constraint is maintained at all times.

4 Computational Study

In order to provide some managerial insights on certain PSP characteristics and analyze the performance of the ILP model as well as the LAHC heuristic, this section presents the results obtained from a computational study on the PSP.

All experiments were conducted on a computer with Intel Xeon E5-2660 at 2.6 GHz and 160 GB of RAM running Ubuntu 20.04 LTS. The LAHC was implemented in C++, compiled using g++ 9.3 and executed in single-thread mode. The ILP was implemented using the C++ API of Gurobi 9 and it was run for up to one hour per instance, using maximum eight threads. Based on the company's requests, we set the maximum execution time of the LAHC to ten minutes.

4.1 New Instance Sets

In order to run experiments using the proposed algorithm and stimulate further research regarding the PSP, instances were generated based on real-world data. The instances, solutions and a validator are publicly available at an online repository [8]. The company that inspired this work provided us with a set of items (I) and machines (M), minimum and maximum stock levels per item ($S_i^{\min}$ and $S_i^{\max}$), a set of short/long setups and their duration (U, L and s_{ij}^m), maintenance durations and frequencies per machine (f^m and $g_{\max}^m$) and time restrictions for long setup and maintenance (bounded by the block indexes $b_{e,l,s,o,n}$). The efficiency of machines per item (e_i^m) is identical for both machines with the exception of certain items which cannot be produced by machine m_2, in which case $e_i^{m_1} > 0$ and $e_i^{m_2} = 0$. This data was considered standard and left unaltered for every instance. The company also provided the cost of items (c_i), overtime (p_o), night shift (p_n) and parallel production (p_p). For privacy reasons, these values were converted into proportional values. Finally, a month's worth of customer requests were provided and used as the basis for generating multiple instances.

Consider $AvgI = \frac{\sum_{r \in R} \sum_{i \in I} q_r^i}{|D|}$, which is the average number of items requested per day in a given time horizon of $|D|$ days. The company generates their production schedule for a time horizon of 10 days considering blocks of 60 min and an $AvgI$ up to one million in high-demand weeks. Based on the provided data, we generated two instance sets corresponding to periods of low and high production demands. For every instance in the low-demand and high-demand sets, $AvgI$ = 500,000 and 1,000,000, respectively. The high-demand set corresponds to a busy scenario for the company and can therefore be considered realistic in terms of size. Each benchmark set contains 18 instances. Each instance is named in accordance with its three primary attributes: $|D|$, $|R|$ and b_{dur} (the length of each block in minutes). For example, the high-demand instance H_10_15_30 has a time horizon of 10 days, 15 requests (with an average of 1,000,000 items per day) and blocks of 30 min.

Given $AvgI$ and these three primary attributes, the remaining attributes of each instance were generated as follows. The minimum number of consecutive days with night shifts d_n was selected with uniform probability from

$[2, min(10, |D| * 0.5)]$. For each request, the permitted shipping days $D_r \subseteq D$ are either every day, every two days or every five days with selection probabilities 0.4, 0.4 and 0.2, respectively. Each request $r \in R$ comprises of at most three different items selected with uniform probability from I. The demand for each request r is selected from $[0.8, 1.1]$ of the average item demand per request $(\frac{AvgI*|D|}{|R|})$ and divided randomly among the items comprising request r. Note that two instances with the same $|D|$ and $|R|$ are identical in every aspect, except for b_{dur}. Moreover, instances from the same instance set have the same $AvgI$ despite the varying number of requests $|R|$.

4.2 LAHC Parameters

Parameters for LAHC were obtained by tuning the algorithm with the irace package [7]. Tuning was performed for the two instance types, resulting in a low-demand and a high-demand parameter set. This is not arbitrary as schedules for low- and high-demand instances differ considerably. In each tunning, irace was given a budget of 5,000 runs and six randomly selected instances as the training set. We fixed LAHC's maximum number of iterations to $M_{\text{iter}} = 20,000$ and the idle rate $\alpha = 0.2$. The best parameter sets reported by irace are provided in the format $\{L_s, \gamma, \beta_{\text{roll}}, \beta_{\text{night}}, \eta\}$. For the low-demand instances parameter values were $\{2000, 0.60, 0.56, 0.62, 20\}$, whereas for the high-demand instances they were $\{2000, 0.64, 0.84, 0.01, 60\}$.

4.3 Results

Table 1 provides the results of the ILP and those obtained by the LAHC. The results concerning instances in the same set with the same $|D|$ and $|R|$ but different block sizes are aggregated into a single row. For example, row L_10_15 provides the aggregated results for instances L_10_15_15, L_10_15_30 and L_10_15_60. Moreover, since the LAHC is an inherently stochastic algorithm, we ran it ten times per instance with different seeds for the random number generator. Therefore, the cells associated with LAHC correspond to the averages or minimums of 30 runs: 10 runs per instance for 3 different block sizes. Similarly, for the ILP, each cell corresponds to the average or minimum of 3 runs: 1 run per instance for 3 different block sizes. The online repository [8] provides a complete table with detailed results to each individual instance.

In these experiments, the ILP reached the one-hour time limit for all instances except L_10_15_30 and L_10_15_60 (solved in 1800 and 800 s, respectively). Therefore, the ILP columns in Table 1 report only the upper bounds (the values of the best solutions found) and lower bounds provided by the solver, but not the computation times. More specifically, columns UB_{min}, UB_{avg} and LB_{avg} report the minimum upper bound, average upper bound and average lower bound, respectively. For the LAHC, column BKS reports the best-known solution value (cost). The next columns report the average values for the solution cost (S_{avg}), execution time in seconds (Time_{avg}), standard deviation of the solution costs (SD_{avg}), number of blocks used for overtime (OT_{avg}), number of days with night shifts (NS_{avg}), number of days with parallel tasks (PD_{avg}) and number of unserved requests (UR_{avg}).

Table 1. ILP and LAHC results.

Instance	ILP			LAHC							
	UB_{min}	UB_{avg}	LB_{avg}	BKS	S_{avg}	$Time_{avg}$	SD_{avg}	OT_{avg}	NS_{avg}	PD_{avg}	UR_{avg}
L_10_15	900.00	143,438.03	900.00	985.00	1,074.50	341.93	34.17	4.63	0.00	0.53	0.00
L_10_25	120.00	212,243.26	0.00	258.75	348.06	259.74	42.20	7.82	0.00	1.43	0.00
L_20_15	331,921.48	522,131.40	0.00	307.50	431.10	522.36	48.38	16.89	0.00	0.37	0.00
L_20_25	683,697.80	703,534.73	0.00	595.00	801.96	525.58	76.02	8.17	0.00	4.30	0.00
L_40_50	1,538,832.12	1,539,432.12	0.00	1,780.00	2,044.94	601.65	97.98	6.04	0.00	14.87	0.00
L_40_100	1,578,473.24	1,579,463.24	0.00	1,920.00	2,282.71	602.96	76.14	2.72	0.00	18.33	0.00
H_10_15	70,580.86	394,387.34	1,594.74	3,782.50	3,963.75	228.38	49.52	0.40	6.83	8.13	0.00
H_10_25	64,098.45	317,605.53	15,649.56	62,604.85	84,184.75	185.93	10,383.02	2.75	8.80	6.00	2.53
H_20_15	1,497,758.35	1,621,849.64	76,390.43	233,532.05	269,658.30	547.06	37,532.99	4.93	15.23	12.50	2.37
H_20_25	1,281,182.11	1,533,486.47	2,855.43	7,637.50	20,550.81	577.50	8,581.04	0.19	11.83	15.17	0.23
H_40_50	3,481,699.59	3,481,999.59	0.00	282,959.48	493,129.05	601.41	66,928.92	64.87	10.30	35.70	6.47
H_40_100	3,324,599.38	3,331,056.53	1,800.00	13,128.75	59,202.84	603.47	67,203.97	8.81	27.17	34.37	1.33

The first thing to note from Table 1 is that the ILP was able to find the best solution for instances L_10_15 and L_10_25 when comparing the minimum values UB_{min} (ILP) and BKS (LAHC) for these instances. However, when considering the average across all three block sizes, LAHC obtains far lower solution costs in under five minutes. For both low- and high-demand instance types, LAHC finds solutions with costs far lower than the ILP's as the instances become larger. Such differences in solution quality are due to the size of the ILP model, which becomes significantly large and experiences significant difficulty to solve even for a time horizon of just 20 days. Indeed, the ILP was unable to produce any feasible solution for instances of 40 days and blocks of 15 min.

In terms of the lower bounds produced by the ILP, the LB_{avg} for low-demand instances is always the trivial bound considering only mandatory night shifts and maintenance without any production. Meanwhile, for high-demand instances the ILP improved the lower bound for those with time horizons of less than 40 days, sometimes even by a large margin (for example instance H_20_15 for which the trivial LB is 0). This may occur due to the fact that in high-demand instances, machine occupancy rates are high enough for the ILP to prove that lower solution values are impossible, whereas with low occupancy rates this is harder to prove since more blocks are likely idle and could be used to avoid parallel tasks or stock penalties. Because block usage depends on several other factors, the model requires longer execution times to improve lower bounds.

LAHC's standard deviation is low for low-demand instances, whereas for high-demand instances the observed variation increases significantly. This high standard deviation is possible given the different number of unserved requests which incur large penalties. Column UR_{avg} shows that while all requests were served for the low-demand instances, in the high-demand instances a number of requests remained unserved, increasing the solution cost. For example, on average 97% of the total cost of solutions for H_40_50 instances is due to unserved requests. Meanwhile, solutions for H_40_100 are only penalized in the 15-minute block set where unserved requests account for 50% of the total cost on average, but for blocks of 30 and 60 min all requests are served and so no penalty is incurred. These results indicate how the difficulty of solving the problem increases as the search space expands.

Out of the total hours available for overtime and total number of days, the low-demand instance set employs on average 15% of overtime hours and 21% of days with parallel tasks. These are the two main components that incur costs in the low-demand instances as no night shifts are employed and the penalty per item under minimum stock levels is on average responsible for only 4% of the total solution cost. Meanwhile, high-demand instances employs on average 64% of the available night shifts, 13% of overtime hours and 76% of days with parallel tasks. For high-demand instances with 40 days the usage of night shifts, overtime and parallel tasks may reach as high as 79%, 65% and 90%, respectively. Items bellow the minimum stock levels were also successfully avoided in the high-demand instances, representing 0.5% of the total solution cost.

Further analyses are performed concerning the impact of block lengths and the relaxation of different constraints. Table 2(a) provides the solution gaps for each block size and instance set. For a block size $b_{dur} \in \{15, 30, 60\}$, gap_{BKS} of b_{dur} is calculated by $\frac{BKS(b_{dur}) - \text{min}_{\text{BKS}}}{\text{min}_{\text{BKS}}}$, where $BKS(b_{dur})$ is the best solution found for b_{dur} and $\text{min}_{\text{BKS}} = \min_{t \in \{15,30,60\}} BKS(t)$. Similarly, gap_{avg} is calculated between the average solution of b_{dur} and the minimum average solution of all block sizes. The average processing time is reported by time_{avg}. Solutions with blocks of 30 min often perform better than the other two block lengths for both instance sets. Instances with blocks of 60 min have the quickest processing times and obtain the second best gap. In contrast, when scheduling blocks of 15 min, the search space is much larger and this results in longer processing times, fewer iterations and worse solution values. To evaluate statistically significant differences, the pairwise T-test was performed with a confidence level of 95%. Although using blocks of 30 min resulted in the best solutions, no statistically significant difference was found when comparing the results of the three block sizes.

Table 2(b) provides the gap between the average solutions produced by the LAHC for the original instance sets and those obtained when relaxing one of the following PSP constraints: (i) shipment day, meaning requests may be shipped on any day; (ii) time windows for long setup and maintenance, meaning these two tasks may be performed during any block and; (iii) maximum stock levels, where S_i^{max} is doubled for every item. These constraints were selected because we consider them to be the most constraining. For the low-demand instance set, statistical tests were performed using the Wilcoxon signed-rank test and demonstrated significant differences between LAHC's results and those obtained by the shipment day and time window relaxations. While relaxing shipment days improves the solutions, relaxed time windows counter-intuitively resulted in a worsening of solution quality. The reasons behind these results are twofold. First, time window relaxation increases the number of blocks to be considered for maintenance and long setups and increases the size of the search space significantly. Second, results show an increase of days with parallel tasks and total number of used overtime blocks, while a decrease on items under the minimum stock is observed. Therefore, the flexibility given by the relaxed time windows enables more production of items to be scheduled, which is considered a priority of the request insertion heuristic (to insert as many production tasks as possible and take as few items as possible from stock). To remedy this behavior one option would be to calibrate the algorithm and give it more time to insert production while prohibiting parallel tasks and overtime blocks.

Table 2. Sensitivity analysis.

b_{dur}	Low			High		
	15	30	60	15	30	60
gap_{BKS}	13.39	0.00	13.77	19.51	4.32	7.58
gap_{avg}	11.29	0.06	13.57	254.71	5.26	6.46
time_{avg}	572.78	483.93	370.40	521.14	458.01	392.73

(a) Impact of block size

Inst. set	Ship any day	Maintenance any time	Double max stock
Low	−5.00	13.13	−3.11
High	−3.48	−16.68	−33.91

(b) Gap_{avg} with relaxations

When considering high-demand instances, the Wilcoxon signed-rank test suggests statistically significant differences for all relaxations. The results indicate that improvements may be obtained by increasing the shipping frequency, doubling the stock capacity or hiring more skilled workers for maintenance and long setups. Indeed, the results demonstrate that doubling stock capacity would bring the largest profits, although one should also consider the construction or rental costs incurred by doing so. Interestingly, increasing the stock capacity by more than 100% did not present significant gains for the considered instances.

5 Conclusion

This paper introduced a real-world production scheduling problem with stock- and staff-related restrictions. To serve a profitable selection of available customer requests within a given time horizon, production, setup and maintenance tasks must be scheduled in blocks of time of predetermined lengths. In addition to an integer programming formulation of the problem, this paper also designed a heuristic algorithm with local search moves based on the insertion and removal of requests. Given the originality of the problem, and thus the lack of benchmark instances in the literature, and with an aim to stimulate future research on the subject, a set of instances was derived from real-world data provided by the company which inspired this research.

A computational study demonstrated the efficacy of the proposed metaheuristic in producing high-quality schedules in quick processing times, even for the more challenging scenarios. Moreover, experiments were carried out to understand the impact of varying demand and block size, along with a sensitivity analysis concerning constraints regarding stock capacities, request shipment days and task time windows. This analysis suggested that companies confronted with similar situations ought to consider operational changes regarding limited shipping days, maintenance windows and stock limits. All of these changes should be exploited to increase revenue. However, we foresee that a broader analysis concerning the trade-off between the gain from such changes and the costs associated with making them represents a crucial consideration which ought to be explored by future research. Furthermore, additional studies could be conducted considering the following extensions: technician scheduling, which would result in flexible times for long setups maintenance; more than two machines, which should be considered along with the scheduling of multiple operators so as to allow parallel operations; and a dynamic version of the problem where requests are not known *a priori*.

Acknowledgments. Research supported by KU Leuven (C2 C24/17/012) and 'Data-driven logistics' (FWO-S007318N). Editorial consultation provided by Luke Connolly (KU Leuven).

Appendix A Integer Linear Programming Formulation

To model the PSP as an ILP we introduce some additional notation.

- T: set of all blocks in the scheduling horizon: $T = \{1, \ldots, |B|, |B| + 1, \ldots, 2|B|, 2|B| + 1, \ldots, |D||B|\}$
- $T^M \subset T$: set of all maintenance blocks.
- $T^N \subset T$: set of all night-shift blocks.
- $T^O \subset T^N$: set of all overtime blocks.
- $T_r \subset T$: set of all shipping blocks for request $r \in R$.
- $d(t) \in D$: the day index of block $t \in T$: $d(t) = 1$ for $t = 1, \ldots, |B|$; $d(t) = 2$ for $t = |B| + 1, \ldots, 2|B|$ ...
- h_n: the index of the last day with a night shift pushed from the previous scheduling horizon.
- g_0^m: at the beginning of current scheduling horizon, the number of days passed without a maintenance for machine $m \in M$ since the last maintenance in the previous scheduling horizon.
- b_n^d: the last block of day $d \in D$.

Additionally, a set of decision variables is used.

- $\eta_d = 1$ if there is a night shift on day $d \in D$, 0 otherwise.
- $\pi_d = 1$ if there is parallel processing of machines on day $d \in D$, 0 otherwise.
- $\theta_t = 1$ if block $t \in T^O$ is used as overtime, 0 otherwise.
- $y_t^m = 1$ if machine $m \in M$ is idle during block $t \in T$, 0 otherwise.
- $\mu_t^m = 1$ if $m \in M$ is under maintenance during block $t \in T$, 0 otherwise.
- $\gamma_r^t = 1$ if request $r \in R$ is fulfilled by shipping at $t \in T_r$, 0 otherwise.

- $z_{ti}^m = 1$ if $m \in M$ is producing item $i \in I$ during block $t \in T$, 0 otherwise.
- Δ_{id} is the stock level of item $i \in I$ at the end of day $d \in D$. Δ_{i0} is the initial stock of i.
- ϕ_{id} is the stock deficit of $i \in I$ at the end of day $d \in D$.
- $\tau_d^m = 1$ if $m \in M$ undergoes a maintenance on day $d \in D$, 0 otherwise.
- $v_{tt'}^m = 1$ if during block $t \in T^M$, machine $m \in M$ is occupied by a long setup to be finished during block $t' \in T^M$, 0 otherwise.
- $w_{ij}^{mt'} = 1$ if during block $t' \in T^M$, machine $m \in M$ is occupied and finished a long setup from item $i \in I$ to item $j \in I$, 0 otherwise.
- $\psi_{tt'}^m = 1$ if during block $t \in T$, machine $m \in M$ is occupied by a short setup to be finished during block $t' \in T$, 0 otherwise.
- $u_{ij}^{mt} = 1$ if during block $t \in T$, machine $m \in M$ is occupied and finished a short setup from item $i \in I$ to item $j \in I$, 0 otherwise.
- $\rho_{ti}^m = 1$ if machine $m \in M$ is set-up to produce item $i \in I$ during block $t \in T$, 0 otherwise ($\rho_{0i}^m = 1$ if the initial configuration of machine m is for item i).

The following is an integer programming formulation for the PSP.

$$\min \sum_{r \in R} \sum_{t \in T_r} c(r)(1-\gamma_r^t) + \sum_{t \in T^O} p_o \theta_t + \sum_{d \in D} (p_p \pi_d + p_n \eta_d + \sum_{i \in I} p_s \phi_{id}) \tag{1}$$

$$\text{s.t. } \theta_t + \eta_{d(t)} \leq 1, \qquad \forall t \in T^O, \tag{2}$$

$$\theta_{t+1} \leq \theta_t, \qquad \forall t \in T^O, \tag{3}$$

$$\theta_t + \eta_{d(t)} + y_t^m \geq 1, \qquad \forall t \in T^O, m \in M \tag{4}$$

$$\eta_{d(t)} + y_t^m \geq 1, \qquad \forall t \in T^N \setminus T^O, m \in M \tag{5}$$

$$(d_n - 1)\eta_d \leq \sum_{d'=d+1}^{d+d_n-1} \eta_{d'} + (d_n - 1)\eta_{d-1}, \qquad \forall d : |D| - d_n \geq d > h_n \tag{6}$$

$$(d_n - 1)\eta_d \leq \sum_{d'=d-d_n+1}^{d-1} \eta_{d'} + (d_n - 1)\eta_{d+1}, \qquad \forall d \geq \max\{h_n + 1, d_n\} \tag{7}$$

$$y_t^m + \sum_{i \in I} z_{ti}^m + \sum_{t' \in T : t' \geq t} \psi_{tt'}^m + \sum_{t' \in T^M : t' \geq t} v_{tt'}^m + \mu_t^m = 1, \qquad \forall t \in T^M, m \in M \tag{8}$$

$$y_t^m + \sum_{i \in I} z_{ti}^m + \sum_{t' \in T : t' \geq t} \psi_{tt'}^m = 1, \qquad \forall t \in T \setminus T^M, m \in M \tag{9}$$

$$\sum_{m \in M} (1 - y_t^m) \leq (|M| - 1)\pi_d + 1, \qquad \forall d \in D, t \in T \setminus T^N : d(t) = d \tag{10}$$

$$\Delta_{id} = \Delta_{id-1} + \sum_{t \in T : d(t) = d} \sum_{m \in M} e_i^m z_{ti}^m - \sum_{r \in R} q_r^i \gamma_r^{t'}, \qquad \forall d \in D, i \in I, t' = b_n^d \tag{11}$$

$$\phi_{id} \geq S_i^{min} - \Delta_{id}, \qquad \forall d \in D, i \in I \tag{12}$$

$$\sum_{t \in T_r} \gamma_r^{t'} \leq 1, \qquad \forall r \in R, i \in I \tag{13}$$

$$u_{ij}^{mt} \leq \psi_{tt}^m, \qquad \forall t \in T, m \in M, (i,j) \in U \tag{14}$$

$$(s_{ij}^m - 1)u_{ij}^{mt'} \leq \sum_{t \in T : d(t) = d(t'), t < t'} \psi_{tt'}^m, \qquad \forall t' \in T, m \in M, (i,j) \in U \tag{15}$$

$$\sum_{t^3 \in T : t^3 \geq t, t^3 \neq t^2} \psi_{tt^3}^m + \sum_{t^3 \in T^M : t^3 \geq t} v_{tt^3}^m + \sum_{i \in I} z_{ti}^m + \psi_{t^1 t^2}^m \leq 1, \qquad \forall t, t^1, t^2 \in T : t^1 \leq t \leq t^2, m \in M \tag{16}$$

$$\sum_{t \in T: d(t)=d} u_{ij}^{mt} \leq 1, \qquad \forall d \in D, m \in M, (i,j) \in U \tag{17}$$

$$w_{ij}^{mt} \leq v_{tt}^{m}, \qquad \forall t \in T^M, m \in M, (i,j) \in L \tag{18}$$

$$(s_{ij}^{m} - 1) w_{ij}^{mt'} \leq \sum_{t \in T^M : d(t)=d(t')}^{t'-1} v_{tt'}^{m}, \qquad \forall t' \in T^M, m \in M, (i,j) \in L \tag{19}$$

$$\sum_{t^3 \in T^M : t^3 \geq t, t^3 \neq t^2} v_{tt^3}^{m} + \sum_{t^3 \in T: t^3 \geq t} \psi_{tt^3}^{m} + \sum_{i \in I} z_{ti}^{m} + v_{t^1 t^2}^{m} \leq 1, \qquad \forall t, t^1, t^2 \in T^M : t^1 \leq t \leq t^2, m \in M \tag{20}$$

$$\sum_{t \in T: d(t)=d} w_{ij}^{mt} \leq 1, \qquad \forall d \in D, m \in M, (i,j) \in L \tag{21}$$

$$\sum_{m \in M} \sum_{t' \in T^M : t' \geq t} v_{tt'}^{m} + \sum_{m \in M} \mu_t^m \leq 1, \qquad \forall t \in T^M \tag{22}$$

$$\sum_{t: d(t)=d} \mu_t^m = f^m \tau_d^m, \qquad \forall d \in D, m \in M \tag{23}$$

$$\sum_{d=0}^{g_{max}^m - d_0^m} \tau_d^m \geq 1, \qquad \forall m \in M \tag{24}$$

$$\sum_{d'=d-g_{max}^m}^{d} \tau_d^m \geq 1, \qquad \forall d \in D : d \geq g_{max}^m, m \in M \tag{25}$$

$$\mu_{t^1}^m + \mu_{t^2}^m + \sum_{i \in I} z_{ti}^m \leq 2, \qquad \forall t, t^1, t^2 \in T^M, m \in M, \; t^1 \leq t \leq t^2, d(t^1) = d(t^2) \tag{26}$$

$$z_{ti}^m \leq \rho_{ti}^m, \qquad \forall t \in T, m \in M, i \in I \tag{27}$$

$$\sum_{i \in I} \rho_{ti}^m \leq 1, \qquad \forall t \in T, m \in M \tag{28}$$

$$\rho_{ti}^m \leq \rho_{(t-1)i}^m + \sum_{j:(j,i) \in U} u_{ji}^{mt-1} + \sum_{j:(j,i) \in L} w_{ji}^{mt-1}, \qquad \forall t \in T^M, m \in M, i \in I \tag{29}$$

$$\rho_{ti}^m \leq \rho_{(t-1)i}^m + \sum_{j:(j,i) \in U} u_{ji}^{mt-1}, \qquad \forall t \in T \setminus T^M, m \in M, i \in I \tag{30}$$

$$\sum_{j:(i,j) \in U} u_{ij}^{mt} + \sum_{j:(i,j) \in L} w_{ij}^{mt} \leq \rho_{(t-1)i}^m, \qquad \forall t \in T^M, m \in M, i \in I \tag{31}$$

$$\sum_{j:(i,j) \in U} u_{ij}^{mt} \leq \rho_{(t-1)i}^m, \qquad \forall t \in T \setminus T^M, m \in M, i \in I \tag{32}$$

$$\Delta_{id} \leq S_i^{max}, \qquad \forall d \in D, i \in I \tag{33}$$

$$\phi_{id}, \Delta_{id} \geq 0, \qquad \forall d \in D, i \in I \tag{34}$$

$$\eta_d, \pi_d \in \{0,1\}, \qquad \forall d \in D \tag{35}$$

$$\tau_d^m \in \{0,1\}, \qquad \forall d \in D, m \in M \tag{36}$$

$$\theta_t \in \{0,1\}, \qquad \forall t \in T^O \tag{37}$$

$$y_t^m \in \{0,1\}, \qquad \forall t \in T \tag{38}$$

$$\mu_t^m \in \{0,1\}, \qquad \forall t \in T^M \tag{39}$$

$$z_{ti}^m, \rho_{ti}^m \in \{0,1\}, \qquad \forall t \in T, i \in I, m \in M \tag{40}$$

$$u_{ij}^{mt} \in \{0,1\}, \qquad \forall t \in T, (i,j) \in U, m \in M \tag{41}$$

$$w_{ij}^{mt} \in \{0,1\}, \qquad \forall t \in T^M, (i,j) \in L, m \in M \tag{42}$$

$$v_{tt'}^{m} \in \{0,1\}, \qquad \forall t, t' \in T^M : t' \geq t, m \in M \tag{43}$$

$$\gamma_r^t \in \{0,1\}, \qquad \forall r \in R, t \in T_r \tag{44}$$

Objective function (1) minimizes the sum of revenue loss from unserved requests, additional personnel costs (overtime, night shift and parallel operations) and penalties incurred by stock deficits. Constraints (2) ensure that if a block is used for overtime then there is no night shift that day and vice versa (meaning no overtime is possible if there is a night shift on a certain day). Constraints (3) forbid using isolated overtime blocks, in other words: overtime in a block is possible only if the previous overtime block is also used, except for the first overtime block, which can be used without preceding overtime blocks. Constraints (4) and (5) enforce the machines to be idle during night-shift blocks if there is no overtime or night shift used that day. Constraints (6) ensure that for a certain day with no night shift on the preceding day, night shifts are only allowed if there is a night shift on every single one of the following $d_n - 1$ consecutive days. Similarly, Constraints (7) ensure that for a certain day with no night shift on the following day that night shifts are only allowed if there is a night shift on every single one of the preceding $d_n - 1$ consecutive days. Constraints (8) and (9) ensure that during any block, a machine is either idle or occupied by a single operation, namely: setup (short or long), maintenance or production. Constraints (10) enforce a parallel processing penalty if more than one machine is not idle. Constraints (11) are the inventory (stock) balance constraints. Constraints (12) retrieve the daily stock deficit per item, if there is any. If the inventory level is greater than the minimum stock requirement, the deficit variable assumes a value of zero thanks to the objective function (1) and binary restrictions (34). By Constraints (13), at most one shipping is conducted per request. Constraints (14)–(16) ensure that no production is performed in between the blocks of the same short setup. Constraints (18)–(20) ensure that no production is performed in between the blocks of the same long setup. Constraints (17) and (21) ensure that only one type of setup is scheduled per day. Constraints (22) ensure that at most one long setup or maintenance takes place during any single block. Constraints (23)–(26) ensure that maintenance blocks are assigned with the required frequency while ensuring that no production is conducted in between the blocks of a maintenance. Constraints (27)–(30) guarantee that production of an item is only possible if the machine has the right configuration, which is validated by a previous block either with the identical configuration or via a completed setup to that item. Maximum stock levels are respected thanks to Constraints (33). Finally, (34)–(44) are nonnegativiy and binary restrictions.

References

1. Burke, E.K., Bykov, Y.: The late acceptance hill-climbing heuristic. Eur. J. Oper. Res. **258**(1), 70–78 (2017)

2. Charnsirisakskul, K., Griffin, P.M., Keskinocak, P.: Order selection and scheduling with leadtime flexibility. IIE Trans. **36**(7), 697–707 (2004)
3. Copil, K., Wörbelauer, M., Meyr, H., Tempelmeier, H.: Simultaneous lotsizing and scheduling problems: a classification and review of models. OR Spectrum **39**(1), 1–64 (2016). https://doi.org/10.1007/s00291-015-0429-4
4. Drexl, A., Kimms, A.: Lot sizing and scheduling - survey and extensions. Eur. J. Oper. Res. **99**(2), 221–235 (1997)
5. Fleischmann, B.: The discrete lot-sizing and scheduling problem. Eur. J. Oper. Res. **44**(3), 337–348 (1990)
6. Goerler, A., Lalla-Ruiz, E., Voß, S.: Late acceptance hill-climbing matheuristic for the general lot sizing and scheduling problem with rich constraints. Algorithms **13**(6), 138 (2020)
7. López-Ibáñez, M., Dubois-Lacoste, J., Pérez Cáceres, L., Birattari, M., Stützle, T.: The irace package: iterated racing for automatic algorithm configuration. Oper. Res. Perspect. **3**, 43–58 (2016)
8. Sartori, C.S., Gandra, V., Çalik, H., Smet, P.: Instances for production scheduling with stock- and staff-related restrictions. Mendeley Data, V2 at http://dx.doi.org/10.17632/rpbv622wyd.2 (2021). Accessed 26 July 2021
9. Silva, Y.L.T., Subramanian, A., Pessoa, A.A.: Exact and heuristic algorithms for order acceptance and scheduling with sequence-dependent setup times. Comput. Oper. Res. **90**, 142–160 (2018)
10. Slotnick, S.A.: Order acceptance and scheduling: a taxonomy and review. Eur. J. Oper. Res. **212**(1), 1–11 (2011)
11. Tempelmeier, H., Copil, K.: Capacitated lot sizing with parallel machines, sequence-dependent setups, and a common setup operator. OR Spectrum **38**(4), 819–847 (2015). https://doi.org/10.1007/s00291-015-0410-2
12. Wörbelauer, M., Meyr, H., Almada-Lobo, B.: Simultaneous lotsizing and scheduling considering secondary resources: a general model, literature review and classification. OR Spectrum **41**(1), 1–43 (2018). https://doi.org/10.1007/s00291-018-0536-0

Chances of Interpretable Transfer Learning for Human Activity Recognition in Warehousing

Michael Kirchhof[1], Lena Schmid[1](✉), Christopher Reining[2], Michael ten Hompel[2], and Markus Pauly[1]

[1] Department of Statistics, TU Dortmund University, 44221 Dortmund, Germany
lena.schmid@tu-dortmund.de

[2] Chair of Material Handling and Warehousing, TU Dortmund University, 44221 Dortmund, Germany

Abstract. Human activity recognition evolves around classifying and analyzing workers' actions quantitatively using convolutional neural networks on the time-series data provided by inertial measurement units and motion capture systems. However, this requires expensive training datasets since each warehouse scenario has slightly different settings and activities of interest. Here, transfer learning promises to shift the knowledge a deep learning method gained on existing reference data to new target data. We benchmark interpretable and non-interpretable transfer learning for human activity recognition on the LARa order-picking dataset with AndyLab and RealDisp as domain-related and domain-foreign reference datasets. We find that interpretable transfer learning via the recently proposed probabilistic rule stacking learner, which does not require any labeled data on the target dataset, is possible if the labels are sufficiently semantically related. The success depends on the proximity of the reference and target domains and labels. Non-interpretable transfer learning via fine-tuning can be applied even if there is a major domain-shift between the datasets and reduces the amount of labeled data required on the target dataset.

Keywords: Domain-shift · Few-shot learning · Interpretability · Logistics · Multi-label classification · Time-series · Zero-shot learning

1 Introduction and Related Work

Manual processes in warehouses make up more than half of their total operating expenses [12,17]. Consequently, human activities need to be quantitatively determinable to allow for their assessment and improvement in regards to economics and ergonomics [5]. Detailed information on the occurrence and duration of activities is crucial to draw conclusions on how to enhance warehouse layout and employee performance. In fact, it is seen as a managerial failure not to account for human characteristics when planning warehousing activities [11].

M. Mes et al. (Eds.): ICCL 2021, LNCS 13004, pp. 163–177, 2021.
https://doi.org/10.1007/978-3-030-87672-2_11

Due to advancements in sensor technology and data processing, IT-supported approaches for automated recognition of human activities receive increasing attention [6]. This leads to the emerging research field of Human Activity Recognition (HAR) that is highly relevant for logistics.

Typically, body-worn sensors such as inertial measurement units (IMUs) are deployed for data gathering in industrial settings [25]. These low-power devices are cheap, highly reliable, non-invasive and easy-to-use as they are not affected by occlusion. They do not portray human identities as in the case of videos. A so-called classifier is capable of automatically recognizing human activities, which are referred to as classes or labels in this context. In the past decade, methods of deep learning, in particular neural networks, have become the state of the art for HAR [25]. For training such a classifier, a set of labelled examples is necessary. The creation of a dataset that comprises the relevant activity labels remains a challenging task as recorded data needs to be annotated.

Manual annotation of multi-channel times series data recorded with IMUs is time-consuming and expensive [33]. The effort scales with the amount of data to annotate. The intra- and inter-class variability of human motion and the influence of the employees' physique necessitate a high quantity of observations from different subjects for classifier training [24]. For example, annotating the OPPORTUNITY dataset took 7–10 h per 30 min of video [28]. In [10], it took 26 min to annotate 1 min of order-picking activities from a video that was synchronized with an IMU data stream. Reining et al. state that annotating the LARa dataset [23] took 85 min per 2 min of recorded material [26]. The research group was further able to halve the time consumption by semi-automated annotation procedures [3]. However, their approach is restricted to recordings in laboratory environments as an marker-based motion capturing system is deployed for reference. This technology may not be available in many cases. As a result, recording new data for each application scenario of HAR in warehousing still entails substantial expenditures.

This contribution investigates to what extent already existing datasets can be utilized for HAR in different warehousing scenarios, even though the datasets do not originate from the same scenario. This is feasible with the help of transfer learning techniques. Here, the goal is to connect the labels of an existing dataset to the possibly different labels of a related target dataset. For example, a combination of labels from a posture dataset can be utilized to describe the labels of an ergonomics dataset. Transfer learning can be conducted both with interpretable and non-interpretable means. In the former case, different HAR classes are connected with rules or structures that are semantically interpretable. Thus, besides being self-learned on examples, they can also be given by experts without requiring any examples, known as zero-shot classification [32]. In particular, recent approaches use knowledge graphs [20], attribute-class structures [2] and logical rules to connect the classes [9]. We focus on the recently proposed probabilistic rule stacking learner (pRSL) [15] that exploits logic rules in a probabilistic framework for the classification task. Non-interpretable transfer learning usually evolves around fine-tuning of neural networks [35]. Here, a neural network is trained on a HAR dataset to recognize a set of classes, and for that it has to learn a way of summarizing the raw IMU data into more compact form, its internal latent representation space, which is usually a black-box. This representation space is then kept as starting point to learn the classes of a new dataset. Thereby, it needs only a few labeled examples on the new dataset, known as few-shot learning [7].

In this work, we focus on localizing the border between zero-shot interpretable transfer learning and few-shot non-interpretable transfer learning in HAR. Hence, we apply pRSL and fine-tuning to both a transfer from ergonomics to logistics and to a transfer from sports to logistics. The idea to use transfer learning in intra-logistics is motivated by the variety of emerging HAR-datasets, that share activities and labels resembling warehousing activities [23]. Making full use of existing data sources extends a classifier's application potential in warehousing without requiring cumbersome recording and annotating sessions. Thus, the effort for taking new warehousing scenarios into account is reduced as less new data needs to be recorded and annotated.

The paper is structured as follows: The ideas behind the applied pRSL transfer learner and its training are explained in Sect. 2. Section 3 introduces the LARa dataset for which we seek transfer learning and two anotated datasets from which we like to gain knowledge for this task. In Sect. 4, pRSL is applied and its results are analyzed. The paper closes with a discussion and an outlook.

2 Methods

2.1 tCNN

We choose temporal convolutional neural networks (tCNNs) [29] as deep learning method to cast the IMU and motion capture (MoCap) data into predictions on the labels as it proved successful in previous HAR problems [22]. Its main idea is to learn convolution filters along the time dimension of the input time-series. The tCNN framework has two variants: one for IMUs, called tCNN-IMU, and one for MoCap data, called tCNN-MoCap. Apart from the dimension of the input data, the frameworks differ in that tCNN-IMU applies one convolutional neural network (CNN) per IMU before concatenating them together in a later step, while tCNN-MoCap analyzes the whole body posture in one CNN.

2.2 pRSL

pRSL was recently proposed by the same authors. As the method contains a plethora of technical ideas we refer to [15] for its explicit definition and all detailed technicalities. Instead, we focus on describing its interpretable transfer learning ideas along Fig. 1 by means of the following illustrative example: Suppose we have three sensor set-ups, each analyzed by an individual machine learning model: An RGB-camera C_1 can distinguish locomotion w of a person from a standing position n. Furthermore, workers wear surface electromyography

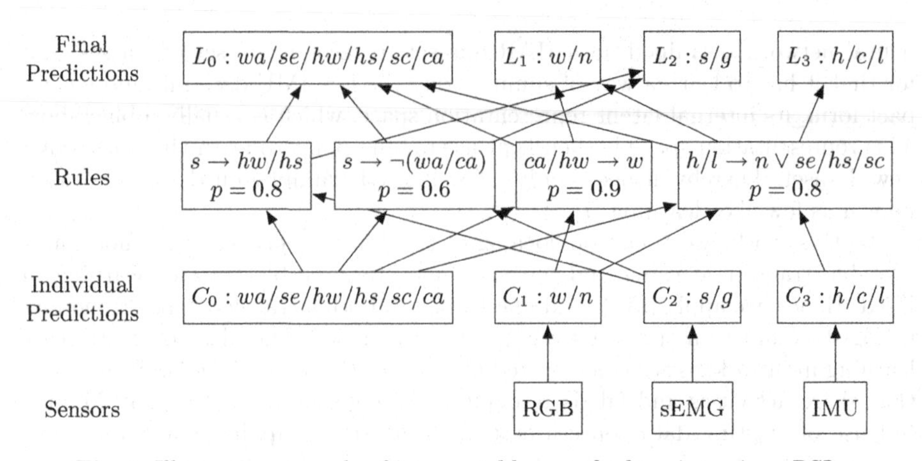

Fig. 1. Illustrative example of interpretable transfer learning using pRSL.

(sEMG) sensors attached to their forearms C_2 that can detect handling activities s and the lack thereof g. Also, a set of IMUs attached to the hands and belt C_3, each including accelerometers and gyroscopes, observe in what direction an activity is being performed - upwards h, centered c or downwards l.

So far, data has been recorded in a conventional 'person-to-stock' order picking system. Employees walk from a base to the retrieval locations according to their order, pick packages and bring them back to the base where they are consolidated. The conventional activity classes C_0 would be, e.g., walking wa, searching se, handling ha, scanning sc, and carrying ca. In this scenario, the classes may be well recognizable using C_2 or C_3.

Now, the classifier shall be deployed in a two-stage picking scenario in which the first stage is a pick-to-belt system. Here, the employees tend to put the packages on the conveyor belt during locomotion as the handling is far easier than taking or placing an item from a box or a shelf. The classifier trained in the first scenario cannot properly tell walking, carrying and handling apart since the corresponding motion pattern overlap, rendering the assessment of human activities in the second scenario impossible. The idea rises to subdivide the handling class ha in handling while walking hw and handling while standing hs. But recognizing these new classes would require recording and annotating new data. To avoid this or to reduce the dataset creation effort, we can make use of the camera and the following rules:

- Handling in any direction is likely to happen when sEMG sensors detect activity. This is true for both handling while walking and handling while standing. This can be expressed as the logical formula $s \rightarrow hw/hs$ with a probability of $p = 0.8$. Similarly, walking and carrying are less likely, when sEMG sensors detect activity, that is $s \rightarrow \neg(wa/ca)$ with $p = 0.6$.
- carrying or walking while handling are far more likely, when the camera detects locomotion, i.e. $ca/hw \rightarrow w$ with $p = 0.9$.

- Locomotion and the corresponding classes are less likely, when the IMUs recognize handling in upwards or downwards direction, i.e. $h/l \rightarrow n \vee se/hs/sc$ with $p = 0.8$. This is because employees rarely walk in bent over position or with the arms above their head.

As visualized in Fig. 1, pRSL uses these rules to interlink the previously independent $C_0, C_1, C_2,$ and C_3, where especially C_0 has no specially trained classifier and thus no prediction yet. After consolidating all information, it returns a coherent prediction on all labels denoted as $L_0 - L_3$. In particular, L_0 is now predicted based on the states of $L_1, L_2,$ and L_3.

2.3 Fine-tuning

Fine-tuning [35] is a non-interpretable transfer learning technique where the tCNN is first trained on a given reference dataset in order to learn weights of lower layers that embed the input data into a space suitable for classification. Then, the final classification layer is removed from the tCNN and replaced with a layer for the target dataset's classification. The weights of this layer are then trained using the available annotated target data. All other layers are trained as well, but due to the bottleneck of provided target data, they need to rely on the pre-training performed on the reference dataset.

3 Datasets

We used the LARa [23] logistics dataset along with one domain-foreign and one domain-related HAR dataset to study to what extent transfer learning via pRSL is feasible. They were selected from the 61 HAR datasets surveyed in [23] using the following inclusion criteria:

1. The sensor data must be compatible with LARa.
2. The dataset must comprise several participants.
3. The domain-related dataset's labels must be related to LARa, and the domain-foreign dataset's labels must describe diverse human motion in sufficient detail.

For Criterion 1, LARa offers MoCap and IMU data. Although IMUs can be simulated at arbitrary positions using the MoCap data, [29] report poor performance using such simulated data. Hence, for Criterion 1, datasets had to have IMUs at positions compatible with those in LARa, or offer MoCap data. Criterion 3 was the biggest filter as most datasets only included general locomotion classes (e.g. standing, walking).

Finally, we selected RealDisp as domain-foreign and AndyLab as domain-related datasets. They are described along with LARa below.

LARa. Our target dataset, LARa, includes 14 workers performing different tasks in a logistics settings. As labels, six mutually exclusive activity categories describe the workers' current action (e.g. pushing a cart, handling items on a centered level) and 17 binary attributes describe these actions in finer detail (e.g. stepping, holding a bulky object), some of which are mutually exclusive. In this work, we exclude the activity categories and focus on the attributes as we are mainly interested in linking these to attributes of other datasets. As input data, the dataset comprises 5 IMUs measuring tri-axial linear and angular acceleration 100 Hz, positioned at the forearms, calves, and back, and 39 MoCap markers measuring 3D-locations 200 Hz, positioned according to the Vicon Full-Body Scheme [31]. We therefore correspond to these datasets as LARa-IMU and LARa-MoCap.

The preprocessing of IMUs and MoCap data was the same as explained for the next two datasets. We follow the original paper's train-validation-test split, where 8 subjects are used for training and 3 are used for validating and testing each. Note that for IMUs, this split reduces to 4 subjects for training and 2 for validating and testing each, because not all subjects were recorded with IMUs. This also means that a direct comparison between IMU and MoCap data is not possible.

RealDisp. In RealDisp [4], 17 subjects perform different sports exercises, making it the domain-foreign dataset. It was designed to provide measurements for both ideally placed and misplaced IMUs. However, we restricted on the ideally placed IMUs. The dataset is labeled with 33 mutually exclusive categories giving the current sports exercise (e.g. waist-bends, high-knees running). 9 IMUs positioned at the forearms, upper arms, calves, thighs, and back measure tri-axial linear and angular acceleration as well as the magnetic field and orientation 50 Hz serve as input variables.

To make it compatible with LARa, the IMUs at the upper arms and thighs were discarded along with the measurements of the magnetic field and orientation. Further, LARa's IMUs were downsampled 50 Hz. Following [29], each IMU channel was standardized to have zero mean and unit variance as learned on the training data. There is no train-validation-test split in the literature. As no potential stratification data on the subjects is reported and they all performed the same routines, they were randomly assigned into 11 train, 3 validation and 3 test subjects.

AndyLab. AndyLab [21] serves as domain-related dataset, containing 13 workers performing industrial tasks such as screwing or carrying packages in different experiment setups. These tasks are labeled in 3 groups, not with the primary goal of activity recognition, but for ergonomics assessment following [30]: They describe the general posture in 6 mutually exclusive categories (e.g. sitting, standing), an ergonomics-oriented posture in 5 mutually exclusive categories (e.g. overhead work, forward bending) and the performed action in 8 mutually exclusive categories (e.g. picking, carrying). The input data includes a sensor glove, an IMU suit and MoCap markers measuring the same 39 3D-locations as in LARa 120 Hz as well as two additional markers on each foot.

We only used the 39 3D-position markers and downsampled the measurements in LARa and AndyLab to their biggest common factor 40 Hz. The further preprocessing followed [29]: To prevent the models from memorizing which spatial locations belong to which actions, all coordinates were given in relation to the subject's approximate center of body mass. This is defined as the middle of the right and left posterior inferior iliac markers. Each coordinate channel was normalized to the interval [0,1] using their minimum and maximum values in the training data.

In 76.74% of AndyLab's measurements, the position of at least one marker is missing. The dataset's authors note that this was often caused by workers interacting with objects that covered the marker. Thus, the data is not missing completely at random and can not simply be left out. As creating a human-motion-aware imputation model, e.g. via expectation-maximization, exceeds this work's scope, we applied linear interpolation in the time dimension.

[27] provide a train-validation-test split for AndyLab. However, this split is unbalanced: E.g., two of the three test persons are left-handed whereas only right-handed persons are seen during training and validation. Also, the distributions of categories is unbalanced, e.g. the category kneeling is seen for three seconds during training, but for 2.7 min during testing. Hence, we constructed a stratified train-validation-test split with 7 persons for training, 3 for validating, and 3 for testing. The split had to fulfill the following constraints:

1. At least one left-handed person must be both in the train and in the test data.
2. At least one male and one female worker must be both in the train, in the validation, and in the test data.
3. The two experiment setups must be at least once both in the train, in the validation, and in the test data.
4. Each category must be seen for at least one minute during training.

From the splits that fulfilled these criteria, the split with the most similar conditions during training, validating, and testing was selected, as measured by the χ^2 statistic [1] between the category distributions of train, validation, and test data. The chosen split, along with its category distributions, is reported in the online appendix [16].

4 Application

In this section, we use predictions on previously learned categories of reference datasets in order to classify new categories on the LARa data. We frame this as a zero-shot problem where we aim to predict novel labels without having seen examples for them. This task poses a strong concept shift: The classifiers trained on reference data are not only applied to new, slightly differently distributed LARa data, but also the labels of LARa are far off the labels of the reference datasets. E.g. in RealDisp, the classifier's predictions on (by design incorrectly) detected sports activities on the LARa data were used to find out which work in logistics a worker performs. This is a harder task than attribute-based descriptions of new classes and also fine-tuning, since, besides the concept shift, pRSL did not have access to the high-dimensional tCNN-embedding of the data, but only to the low-dimensional classifier beliefs on each category.

4.1 Experimental Setup

Manual Rules. To perform a zero-shot transfer, interpretable rules that link the reference dataset's labels to those of the target dataset need to be provided to pRSL. To this, we directly linked combinations of AndyLab's labels to combinations of LARa's labels as detailed in the online appendix. For RealDisp, this is infeasible since RealDisp's sport activities cannot be directly linked to LARa's logistics labels. Hence, we inserted a mid-layer of latent labels that describe which body part is in motion in each sports or logistics action, as visualized in Fig. 2. The construction of these rules, as derivated from the datasets' documentations, is also further explained in the online appendix.

Self-learned Rules. Besides using pRSL to apply designed rules, we also trained it on LARa to find rules on its own. This allows a better insight into whether the manually designed rules exploit the information stored in the reference labels, or whether better a better interpretable transfer is possible beyond the above manual rules. pRSL was trained to find rules between the reference and target labels using 20% of the available data, with a budget of 500 batches of 20 samples each or a maximum training time of 44 h on ten cores of an Intel Xeon E5-2640v4@2.40GHz CPU and 8GB RAM. Note that this used pRSL's ability to train when parts of the ground truths, here those for RealDisp and AndyLab, are missing. The ideal number of rules K was found by applying a grid search on validation data over $K \in \{10, 20, \ldots, 100\}$.

Fine-tuning. In defining manual rules, none of LARa's data was used. Hence, as a comparison to the manual rules, we applied fine-tuning [35] with 1% of LARa's data, selected at random from all available one-second long samples. As a comparison to the learned rules, we performed fine-tuning using 20% of the LARa data. As an additional baseline for the zero-shot experiments, a classifier that predicts labels in each XOR group or each binary label at random is provided.

Performance Metrics. We chose the metrics from [15] and [29] to quantify the performance of the resulting transfer learned classifiers. As such, we report the F1-Score for each label on LARa, as well as a weighted average (w-F1). The weights for the average are proportional to how often each label is present in the test data, as reported in the online appendix. As additional summary metrics over all labels, we report the hamming loss, that is the average number

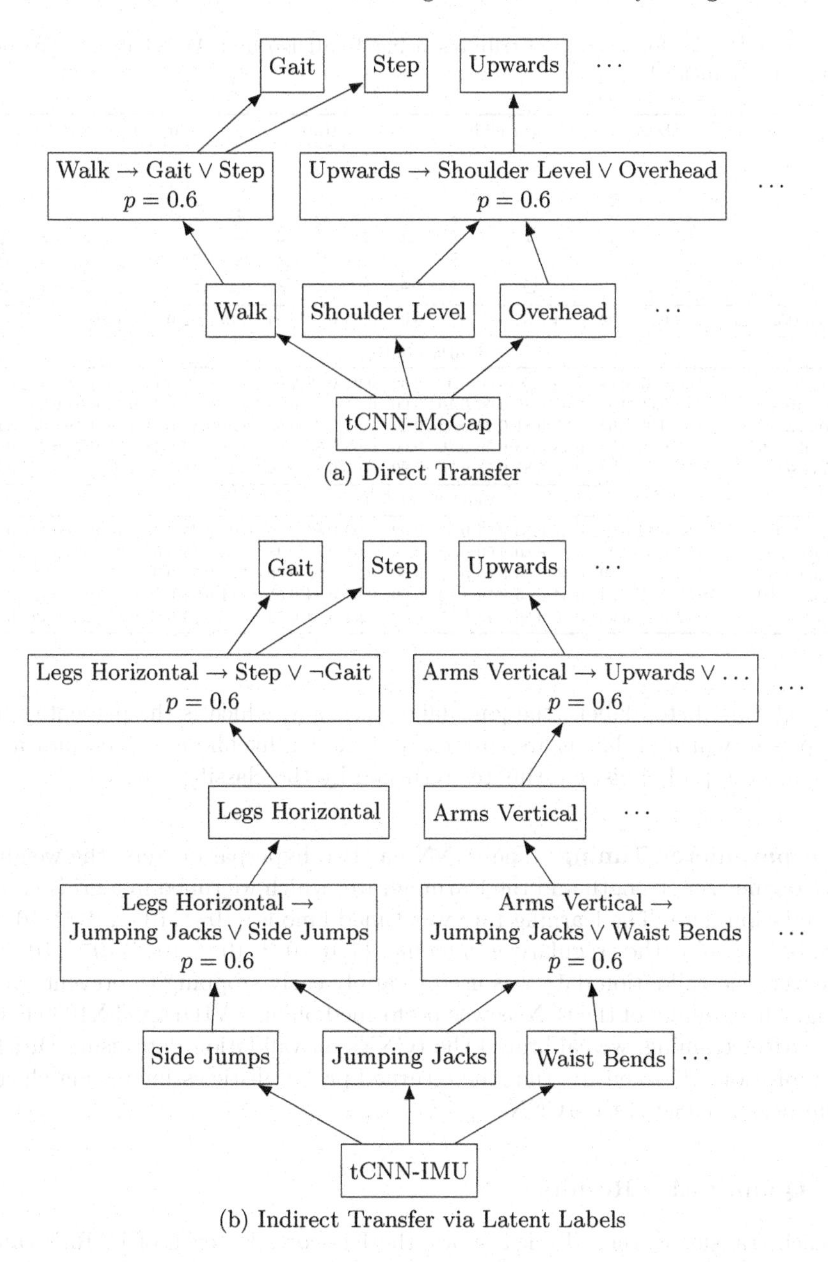

(a) Direct Transfer

(b) Indirect Transfer via Latent Labels

Fig. 2. Connection of reference labels to target labels via rules. (a) shows the direct transfer in AndyLab where labels are domain-related (b) shows the indirect transfer in RealDisp where the domain-foreign reference labels are first abstracted to limb movement labels and then connected to the target labels.

Table 1. F1-score for LARa's attributes using RealDisp and AndyLab as reference datasets for transfer learning.

	Legs			Upper Body					Arms			Item Pose						
Method	Gait Cycle	Step	Standing Still	Upwards	Centered	Downwards	No Motion	Torso Rotation	Right Hand	Left Hand	No Hand	Bulky Unit	Handy Unit	Utility	Cart	Computer	No Item	Weighted Average
Random (Baseline)	0.28	0.38	0.32	0.14	0.35	0.08	0.24	0.02	0.63	0.58	0.24	0.19	0.22	0.16	0.14	0.07	0.13	0.38
RealDisp (IMU)																		
Manual Rules	0.00	0.47	0.38	0.00	0.56	0.00	0.31	0.02	0.91	0.82	0.00	0.19	0.22	0.17	0.15	0.07	0.12	0.48
Learned Rules	0.10	0.39	0.56	0.00	0.77	0.00	0.10	0.00	0.91	0.82	0.00	0.08	0.52	0.00	0.16	0.00	0.23	0.52
Fine-tuning 1%	0.67	0.35	0.63	0.35	0.81	0.40	0.66	0.00	0.94	0.85	0.53	0.11	0.57	0.31	0.68	0.28	0.67	0.66
Fine-tuning 20%	0.75	0.60	0.72	0.48	0.85	0.59	0.76	0.00	0.95	0.87	0.60	0.41	0.60	0.48	0.78	0.69	0.72	0.75
Full Dataset	0.75	0.59	0.72	0.50	0.85	0.53	0.78	0.00	0.95	0.87	0.66	0.47	0.61	0.48	0.83	0.71	0.73	0.76
AndyLab (MoCap)																		
Manual Rules	0.58	0.42	0.08	0.32	0.53	0.54	0.34	0.02	0.93	0.84	0.00	0.00	0.00	0.19	0.18	0.07	0.14	0.48
Learned Rules	0.51	0.54	0.48	0.14	0.80	0.00	0.59	0.00	0.91	0.80	0.48	0.13	0.55	0.00	0.25	0.00	0.43	0.61
Fine-tuning 1%	0.66	0.24	0.60	0.16	0.80	0.58	0.58	0.00	0.92	0.82	0.00	0.23	0.56	0.05	0.47	0.36	0.08	0.60
Fine-tuning 20%	0.76	0.37	0.64	0.64	0.85	0.72	0.74	0.00	0.93	0.84	0.55	0.14	0.54	0.46	0.71	0.74	0.72	0.70
Full Dataset	0.77	0.47	0.68	0.65	0.86	0.73	0.75	0.00	0.94	0.84	0.59	0.25	0.55	0.47	0.72	0.72	0.72	0.72

of wrongly predicted labels, the joint-label accuracy, which is the percentage of examples in which all labels are correct, and the log-likelihood, which indicates how good the probabilistic estimates returned by the classifier are.

Hyperparameter Tuning. The tCNN has two hyperparameters, the weight-decay regularizer strength and the learning rate, which we tuned in a grid-search on validation data. The learning rate was tuned among $\{10^{-1}, 10^{-2}, 10^{-3}, 10^{-4}, 10^{-5}, 10^{-6}\}$ and the regularizer among $\{1, 0, 10^{-1}, 10^{-2}, 10^{-3}, 10^{-4}, 10^{-5}\}$. Moreover, the validation data was used to apply early stopping to prevent overfitting. The training of the tCNNs was performed on an NVIDIA GTX1060 6GB GPU. After training, we calibrated the tCNNs on validation data using Dirichlet calibration [19] to ensure that the returned probabilistic estimates match the actual performance of the tCNNs.

4.2 Quantitative Results

For each transfer learner, Table 1 shows the F1-scores for each of LARa's categories and Table 2 reports the summary metrics across all categories.

On RealDisp, at first glance, it appears that the manual rules outperformed the random baseline both in terms of summary metrics and the F1-scores for labels where rules were provided. However, closer inspection reveals that the manual rules lead to constant or purely random predictions for all labels, which results in F1-scores of 0 and equal scores as the random baseline, respectively. Hence, the performance is limited in comparison to the 1% fine-tuning. The

Table 2. Quality metrics for transfer learning using RealDisp and AndyLab as reference datasets. Best result per reference dataset are printed in bold.

	Hamming loss	Weighted F1	Log-likelihood	Accuracy
Random (baseline)	0.607	0.484	−7.05	0.001
RealDisp (IMU)				
Manual rules	0.448	0.484	−6.203	0.008
Learned rules	0.315	0.523	−4.448	0.076
Fine-tuning 1%	0.241	0.664	−3.138	0.154
Fine-tuning 20%	0.189	0.751	−2.239	0.278
Full dataset	**0.184**	**0.759**	**−2.152**	**0.284**
AndyLab (MoCap)				
Manual rules	0.435	0.480	−6.419	0.007
Learned rules	0.295	0.607	−4.436	0.085
Fine-tuning 1%	0.284	0.596	−3.707	0.108
Fine-tuning 20%	0.224	0.700	−2.739	0.210
Full dataset	**0.212**	**0.723**	**−2.529**	**0.229**

learned rules improved over manual rules in all summary metrics by 8% to 850%, but could not outperform the 20% fine-tuning. Comparing the fine-tuning results given different amounts of data reveals diminishing returns: While the summary metrics increased by 13% to 81% when comparing 1% to 20% fine-tuning, the gap between 20% and the full dataset is just 1% to 4%.

Manual rules did not suffer from constant predictions on AndyLab. For LARa labels that were connected to (sets of) semantically equivalent labels on AndyLab, e.g. gait cycle or upwards and downwards work, the F1-score is competitive with or outperforms that of 1% fine-tuning. Learned rules lead to an increased F1-score in 9 of the 17 labels against manual rules, showing that further relations between the labels were found and utilized. We provide a qualitative analysis of these relations in Sect. 4.3. The diminishing returns of fine-tuning are also present on AndyLab: Using 20% of the data improves the summary metrics by 17% to 94% compared to using 1%, while adding the remaining 80% increases them by 3% to 9%.

4.3 Qualitative Analysis

To analyze the semantic meaning of learned rules, we randomly selected three out of the 90 rules learned on AndyLab in the previous Section and showcase them in Fig. 3. The figure reveals that pRSL did not only model within-LARa dependencies, but transferred the information provided in the AndyLab labels, since both AndyLab and LARa labels are present in each rule.

The first rule considers work, where the operator applies some fine manipulation using both hands. The second revolves around actions where the worker

(Stand) ∧ (Bulky Unit / Computer / No Item) → Right Hand ∨ Left Hand ∨ (Step / Standing Still)

(Standing Still / Step) ∧ (Centered / Downwards) → (Kneeling / Lying) ∨ (Handy / Utility / No Item) ∨ Bent Forward / Bent Strongly Forward)

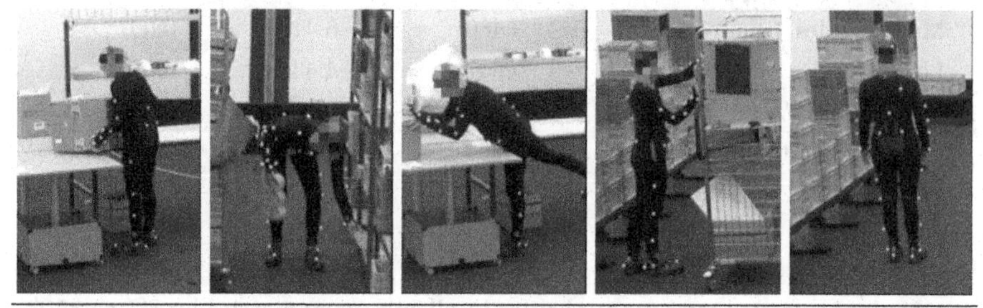

(Cart / No Item) ∧ (No Motion) ∧ (Walking / Standing) ∧ (No Hand) → (No Torso Rotation)

Fig. 3. Three randomly selected self-learned rules from the AndyLab transfer learning task along with five examples where the rule has a high impact on the classification decision, as measured by the L1 distance to the prediction without the rule.

is leaning forward while standing to handle a utility or bulky item. The third is harder to interpret, though appears to refer to situations in which the worker carries items or pushes the cart with otherwise resting hands. Here, it is interesting to see that pRSL used the torso rotation category, which nearly never occurred in the data. This also happened in other rules not shown here with categories like kneeling or lying. Aside from possibly being a fragment from training, this might have been used to reduce the rule strength since "No Torso Rotation" applies almost always.

4.4 Discussion

In summary, the fact that the performance of manual and learned rules is more favorable on AndyLab than on RealDisp (see Table 2) indicates that label-based transfer learning requires closely related domains and labels. The, by construction wrongly, detected sports categories and the coarse-grained latent labels in RealDisp proved to be insufficient information for transfer learning. In contrast, using the higher dimensional and more adjustable embedding space in fine-tuning outperformed the label-based transfer learning on both datasets. For fine-tuning, there was a notable decrease in returns when training with more data. However, it should be noted that this does not justify recording less data in general since the data was sampled at random from the whole dataset, and hence was scattered across all recordings.

Finally, the examples from Sect. 4.3 show that pRSL's label-based transfer learning can yield interpretable rules in real world environments, though we stress that the interpretation strategy is still unpolished

5 Conclusion

The results of our experiments showed that an interpretable label-based transfer achieves a competitive zero-shot detection but requires the labels of the reference and target dataset to be semantically related. When moving towards few-shot learning, non-interpretable fine-tuning allowed transfer learning even under a big domain-shift from sports to logistics. Moreover, fine–tuned learners, even with only a few annotated examples on the target dataset, quickly outperformed learned zero-shots. First results confirmed that this transfer might help to reduce the amount of required labeled target data, which is a main cost factor in deploying deep learning techniques to novel warehousing scenarios.

6 Outlook

To make more precise statements about the potential of interpretable or non-interpretable transfer learning with multi-channel times series data, the methodology should be applied to solve related problems for other data sets, experimental set-ups and even other domains, e.g. a (back) transfer in sports, production management etc., in subsequent research.

Moreover, beyond classification many logistics tasks are related to predicting quantitative outcomes, i.e. regressions tasks. Considering the above-mentioned advantages, it is tempting to exploit the applicability of different transfer learning techniques in logistics more exhaustively in the future, e.g. beside warehousing [18] also for transport planning [8,14] or predictive maintenance [13,34].

References

1. Agresti, A.: An Introduction to Categorical Data Analysis. John Wiley, Hoboken (2018)
2. Atzmon, Y., Chechik, G.: Probabilistic AND-OR attribute grouping for zero-shot learning. In: Conference on Uncertainty in Artificial Intelligence (2018)
3. Avsar, H., Altermann, E., Reining, C., Rueda, F.M., Fink, G.A., ten Hompel, M.: Benchmarking annotation procedures for multi-channel time series HAR dataset. In: 2021 IEEE International Conference on Pervasive Computing and Communications Workshops and other Affiliated Events, pp. 453–458 (2021)
4. Banos, O., Toth, M.A., Damas, M., Pomares, H., Rojas, I.: Dealing with the effects of sensor displacement in wearable activity recognition. Sensors **14**(6), 9995–10023 (2014)
5. Calzavara, M., Glock, C.H., Grosse, E.H., Persona, A., Sgarbossa, F.: Analysis of economic and ergonomic performance measures of different rack layouts in an order picking warehouse. Comput. Ind. Eng. **111**, 527–536 (2017)
6. Chen, C., Jafari, R., Kehtarnavaz, N.: A survey of depth and inertial sensor fusion for human action recognition. Multimed. Tools Appl. **76**(3), 4405–4425 (2017)
7. Cheng, H.T., Sun, F.T., Griss, M., Davis, P., Li, J., You, D.: NuActiv: recognizing unseen new activities using semantic attribute-based learning. In: 11th Annual Conference on Mobile Systems, Applications, and Services, pp. 361–374 (2013)
8. Daduna, J.R.: Automated and autonomous driving in freight transport - opportunities and limitations. In: Lalla-Ruiz, E., Mes, M., Voß, S. (eds.) ICCL 2020. LNCS, vol. 12433, pp. 457–475. Springer, Cham (2020). https://doi.org/10.1007/978-3-030-59747-4_30
9. Ding, N., Deng, J., Murphy, K.P., Neven, H.: probabilistic label relation graphs with Ising models. In: Proceedings of the 2015 IEEE International Conference on Computer Vision, pp. 1161–1169 (2015)
10. Feldhorst, S., Aniol, S., ten Hompel, M.: Human Activity Recognition in der Kommissionierung - Charakterisierung des Kommissionierprozesses als Ausgangsbasis für die Methodenentwicklung. Logistics J. **2016**(10) (2016)
11. Grosse, E.H., Calzavara, M., Glock, C.H., Sgarbossa, F.: Incorporating human factors into decision support models for production and logistics: current state of research. IFAC-PapersOnLine **50**(1), 6900–6905 (2017)
12. Grosse, E.H., Glock, C.H., Neumann, W.P.: Human factors in order picking system design: a content analysis. IFAC-PapersOnLine **48**(3), 320–325 (2015)
13. Guo, L., Lei, Y., Xing, S., Yan, T., Li, N.: Deep convolutional transfer learning network: a new method for intelligent fault diagnosis of machines with unlabeled data. IEEE Trans. Industr. Electron. **66**(9), 7316–7325 (2018)
14. Huang, H., Pouls, M., Meyer, A., Pauly, M.: Travel time prediction using tree-based ensembles. In: Lalla-Ruiz, E., Mes, M., Voß, S. (eds.) ICCL 2020. LNCS, vol. 12433, pp. 412–427. Springer, Cham (2020). https://doi.org/10.1007/978-3-030-59747-4_27
15. Kirchhof, M., Schmid, L., Reining, C., ten Hompel, M., Pauly, M.: pRSL: interpretable multi-label stacking by learning probabilistic rules. In: Uncertainty in Artificial Intelligence. PMLR (2021). (in press)
16. Kirchhof, M.: GitHub repository for this article (2021). https://github.com/mkirchhof/rslAppl
17. de Koster, R., Le-Duc, T., Roodbergen, K.J.: Design and control of warehouse order picking: a literature review. Eur. J. Oper. Res. **182**(2), 481–501 (2007)

18. Krüger, A., Feldmann, F., Pauly, M., ten Hompel, M.: Einsatzmöglichkeiten maschineller Lernverfahren in einer dezentral organisierten Lagerverwaltung auf Basis intelligenter Behälter. Logistics J. Proc. **2020**(12) (2020)
19. Kull, M., Perello Nieto, M., Kängsepp, M., Silva Filho, T., Song, H., Flach, P.: Beyond temperature scaling: obtaining well-calibrated multi-class probabilities with Dirichlet calibration. Adv. Neural. Inf. Process. Syst. **32**, 12316–12326 (2019)
20. Liu, L., Zhou, T., Long, G., Jiang, J., Zhang, C.: Attribute propagation network for graph zero-shot learning. In: Proceedings of the AAAI Conference on Artificial Intelligence, vol. 34(04), pp. 4868–4875 (2020)
21. Maurice, P., et al.: Human movement and ergonomics: an industry-oriented dataset for collaborative robotics. Int. J. Robot. Res. **38**(14), 1529–1537 (2019)
22. Rueda, F.M., Grzeszick, R., Fink, G.A., Feldhorst, S., Ten Hompel, M.: Convolutional neural networks for human activity recognition using body-worn sensors. Informatics **5**(2), 26 (2018)
23. Niemann, F., et al.: LARa: creating a dataset for human activity recognition in logistics using semantic attributes. Sensors **20**(15), 4083 (2020)
24. Ordóñez, F., Roggen, D.: Deep convolutional and LSTM recurrent neural networks for multimodal wearable activity recognition. Sensors **16**(1), 115 (2016)
25. Reining, C., Niemann, F., Rueda, F.M., Fink, G.A., ten Hompel, M.: Human activity recognition for production and logistics - a systematic literature review. Information **10**(8), 245 (2019)
26. Reining, C., Rueda, F.M., Niemann, F., Fink, G.A., ten Hompel, M.: Annotation performance for multi-channel time series HAR dataset in logistics. In: 2020 IEEE International Conference on Pervasive Computing and Communications Workshops, pp. 1–6 (2020)
27. Ribeiro, P.M.S., Matos, A.C., Santos, P.H., Cardoso, J.S.: Machine learning improvements to human motion tracking with IMUs. Sensors **20**(21), 6383 (2020)
28. Roggen, D., et al.: Collecting complex activity datasets in highly rich networked sensor environments. In: Seventh International Conference on Networked Sensing Systems (INSS), pp. 233–240 (2010)
29. Rueda, F.M., Fink, G.: From human pose to on-body devices for human-activity recognition. In: 26th International Conference on Pattern Recognition (ICPR), pp. 10066–10073 (2021)
30. Schaub, K., Caragnano, G., Britzke, B., Bruder, R.: The European assembly worksheet. Theor. Issues Ergon. Sci. **14**(6), 616–639 (2013)
31. Vicon: Full Body Modeling with Plug-in Gate (2017). https://docs.vicon.com/display/Nexus26/Full+body+modeling+with+Plug-in+Gait. Accessed 16 Mar 2021
32. Xian, Y., Lampert, C.H., Schiele, B., Akata, Z.: Zero-shot learning - a comprehensive evaluation of the good, the bad and the ugly. IEEE Trans. Pattern Anal. Mach. Intell. **41**(9), 2251–2265 (2018)
33. Yordanova, K., et al.: Challenges in Annotation of useR Data for UbiquitOUs Systems: Results from the 1st ARDUOUS Workshop (2018). arXiv:1803.05843
34. Zhang, A., et al.: Transfer learning with deep recurrent neural networks for remaining useful life estimation. Appl. Sci. **8**(12), 2416 (2018)
35. Zhang, A., Lipton, Z.C., Li, M., Smola, A.J.: Dive into Deep Learning (2020)

A Multi-periodic Modelling Approach for Integrated Warehouse Design and Product Allocation

Martin Scheffler, Lisa Wesselink(✉), and Udo Buscher

Faculty of Business and Economics, TU Dresden, 01069 Dresden, Germany
{martin.scheffler,lisa.wesselink,udo.buscher}@tu-dresden.de
http://www.industrielles-managment.de

Abstract. Consumers today expect supermarkets to offer a wide variety of items at favourable prices and with consistent availability. These expectations lead to major capacity challenges for distribution centres responsible for replenishing goods since consumer demand fluctuates weekly. Therefore, we investigate an integrated warehouse design and product allocation problem for a distribution centre in the retail food industry. For this purpose, we formulate a multi-periodic mixed-integer program that reflects the flow of goods within the distribution centre and, thus, explicitly captures daily fluctuations in demand. The suitability of the approach is demonstrated using modified data from a distribution centre in Germany. The results show that a static approach violates the weekday-specific capacity restrictions and that only a multi-periodic approach can meet the requirements of practice. In analyzing the real-world case we selected, the trade-off between handling and transport costs reveals that the automated storage and retrieval system is fully utilized. Interestingly, it shows that, even considering 20,000 different items, the problem can be solved in seconds using the presented model formulation.

Keywords: Warehouse design · Product allocation · Integrated approach · Mixed integer programming · Decision support

1 Introduction

In recent decades, warehouses have evolved from temporary storage facilities to complex distribution centres as an essential part of the supply chain [2]. In addition to warehousing, one of the greatest planning challenges in food retailing is picking items for individual stores. For the most-efficient commissioning possible, several storage and handling concepts (referred to as flows) have been developed over the years to meet the different characteristics of individual items (e.g. throughput speeds, dimensions). Due to limited resources (e.g. technical limits of the equipment, number of picking locations), not every item can be assigned to the most cost-efficient flow through the warehouse. In addition, the

M. Mes et al. (Eds.): ICCL 2021, LNCS 13004, pp. 178–191, 2021.
https://doi.org/10.1007/978-3-030-87672-2_12

available space can be divided very flexibly between the individual flows, creating a highly complex planning problem that requires automated decision-making support. Both, capacity restriction and flexibility of the space use avoid the use of simple assignment heuristics for allocating products. A suitable approach for combining the allocation of space (warehouse design) and the allocation of products was presented by [7]; they considered one year as a single-period planning horizon and only space as a resource.

In practice product demand is subjected to weekday-dependent fluctuations. Thus, the main contribution of this paper is a new multi-periodic modelling approach that takes these fluctuations into consideration. Further, this approach enables us to integrate several technical restrictions and requirements related to the flow through an automated storage and retrieval system (ASRS). For practical application, a trade-off emerges from the transportation systems utilized to move products from the distribution centre to the markets. Picking in high-bay warehouses is traditionally done directly onto pallets (term refers to a manual, conventional pallet warehouse in the following), which are then transported to stores. In contrast, in automated systems which often consist of a high-bay warehouse, picking is partially conducted in other transport systems. In the selected case, this involves small containers that cause greater volume losses compared to transport on pallets. The result is that the usually lower handling or picking costs in automated storage are offset by higher transport costs to markets. Therefore, our last contribution considers managerial insights on interdependence caused by the cost structure occurring in practice.

The paper is structured as follows: Sect. 2 provides a brief literature review focussing on the combined consideration of warehouse design and product allocation; in Sect. 3, we describe the problem in detail; Sect. 4 introduces the mathematical problem formulation; and finally a real-world case is described and the multi-periodic approach tested against the single-period variant for evaluating the improvements. Further, the cost interdependence of transport costs is discussed in detail, followed by Sect. 6, which summarizes the paper and briefly discusses further research.

2 Literature Review

Technical evolutions of the past three decades have led to more integrated and automated warehousing systems, with an increased demand for in-depth analysis regarding, e.g. the planning of operations [8]. In this context, both storage and picking processes form the core of the warehouse and significantly affect performance and operating costs [5,15]. Therefore, the design of the warehouse has to be considered as a basis for the operational system (with the aim of obtaining high performance at low cost) [1,2,12]. The integrated approach determining size and to allocate items to different storage areas as seen in [7] and as deployed in this paper can be classified as 'infrastructure design'. It is important to distinguish this type of problem from the so-called storage-location assignment problem, which aims to determine a distinct location inside a storage type (i.e. flow) for an item [4]. A thorough review has been provided by, e.g. [10].

Similar to the problem introduced herein is the forward-reserve problem, which deals with the allocation of items and their quantity to the forward (efficient for order picking) and reserve (efficient for storing) areas while trading off the costs of order picking and internal replenishment as presented by [3] or [14]. Possible additional considerations include the compatibility of different items or item groups and three-dimensional slots, as examined by [6]. Also relevant in the context of this paper is the approach by [9], who focused on the available space in a warehouse with the option of renting additional space under item-specific shortage probabilities. Moreover, integrated approaches, as in [7,11,13], play a vital role for combining warehouse design and operations, e.g. by simultaneously determining the warehouse layout and its control policies such as the applied storage policy. Nevertheless, in contrast to the present paper, none of the above-mentioned contributions has taken into consideration multi-periodicity or a combination of automated and 'manual' storage areas for planning.

3 Problem Description

We consider a distribution centre in the retail food industry that consists of two storage areas: a high-bay storage (HBS) area and an ASRS. The physical dimension of the latter is fixed. The ground floor of the HBS area is used for forward storage acting as picking locations. It is possible to either pick directly from pallets (pallet picking, PP) or to set up sloping shelves (shelf picking, SP) to increase the number of available picking locations. The upper levels of the HBS are used as reserve storage, from which the replenishment of the items on the ground floor takes place. Figure 1 illustrates this; note that, in practice, there is more than one storage level above the ground-floor level.

None of the items require refrigeration or frozen storage. On the receiving side of the warehouse, items arrive from producers on pallets (i.e. on each pallet, there is only one type of item). Each type of item has to be assigned to one flow through the warehouse (i.e. ASRS, PP or SP). Depending on this decision, the items pass through different processes, thereby resulting in different costs. In the HBS (i.e. PP and SP), an incoming pallet has to be stored in the upper levels by a forklift first. For PP flow, as soon as the current pallet in the picking location of a particular item is empty, a new one is replenished from the upper levels. In the SP flow, no entire pallet can be replenished, only the quantity that fits onto the sloping shelf. In both flows, picking takes place onto mixed pallets for stores on the shipping side. In contrast, the items are first repacked from pallets into boxes for storage for the ASRS on the receiving side.

Each item must either be stored in and picked from the ASRS or requires storage and picking from a picking location in the HBS. This means that the capacity limitation here is not the storage space itself but the number of picking locations, which is dependent on the shares of PP and SP flows of the HBS. One pallet position in the PP flow can be substituted by several sloping shelves, whereby this decision on the distribution of the HBS area is also considered later in this work. Figure 2 illustrates the structure of the distribution centre and the

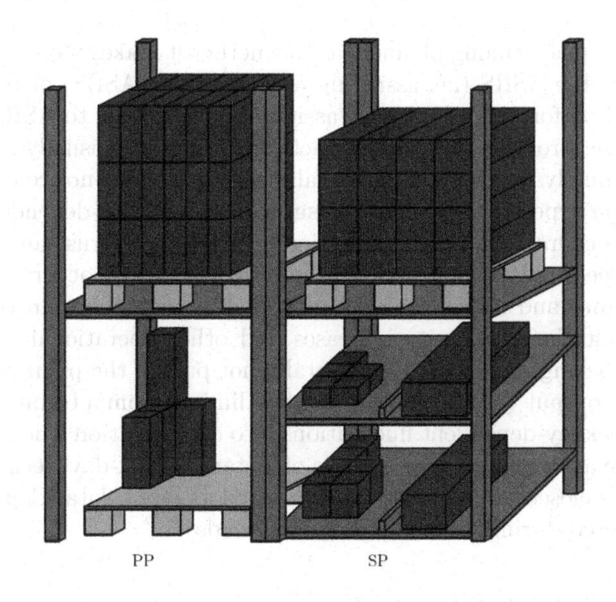

Fig. 1. Possible uses of the ground-floor level of the HBS

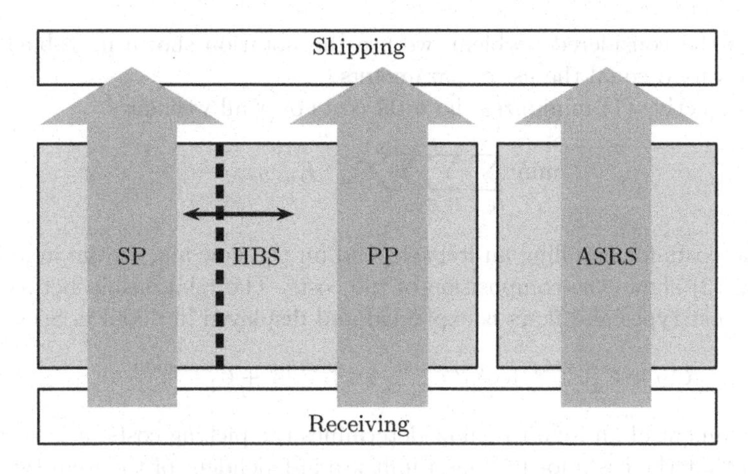

Fig. 2. Item flows in the distribution centre

three possible item flows. The dashed line represents the flexibility of area usage in HBS.

On the shipping side, picking using the ASRS involves moving items to small containers specially designed for the stores. Based on this, the (volume-based) transport costs to the stores must also be taken into account since the containers result in more volume loss than the pallets during transport by truck. Without transportation costs, the most-favourable allocation would be for all items in the ASRS. In the combined consideration of both types of costs, several items result in the lowest (total) costs if they are assigned to PP. In practice, frequent discussions take place among planners as to whether it makes more sense to only partially utilize the ASRS (i.e. assigning fewer items to ASRS) or to operate it at its technical performance limit (i.e. assigning many items to ASRS).

The planning problem occurs on a tactical level and is usually solved semi-annually to annually. In some cases, it is also solved at short notice to be able to assign new item types. Item demand is subject to weekday-dependent fluctuations. Over the course of a year, highly similar weekly patterns can be observed for about 40 weeks, while significant fluctuations in demand occur, for example, around Christmas and New Year celebrations. However, these fluctuations are managed by additional capacity increases and other operational actions (e.g. special cross-docking areas) and are generally not part of the planning problem. Since the daily output quantity of the ASRS is limited from a technical perspective, taking weekday-dependent fluctuations into consideration when planning is inevitable. The average output quantities of a standard six-day week (the distribution centre is closed on Sundays) are thus used as input data. Capacity limits must be respected during planning for each weekday.

4 Mathematical Formulation

To model the considered problem, we use the notation shown in Table 1. The input data used equal the list of parameters.

The objective (1) minimizes the total costs over all weekdays.

$$\min \sum_{t \in \mathcal{T}} \sum_{l \in \mathcal{L}} \sum_{a \in \mathcal{A}} C_{al} \cdot K_{at} \cdot x_{al}. \tag{1}$$

The total costs for handling an item depend on the flow and on the item itself. Equation (2) shows the composition of the costs. The relationship between the different cost types and flows is explained and displayed in detail in Sect. 5.3.

$$C_{al} = C_{al}^{storing} + C_{al}^{replenish} + C_{l}^{picking} + C_{l}^{\text{transport}} \cdot V_a \tag{2}$$

The assignment of an item to a flow determines the picking costs, it is therefore assumed that the costs for picking a unit are independent of the item type. We will refer to the sum of $C_{al}^{storing}$, $C_{al}^{replenish}$ and $C_{l}^{picking}$ as handling costs. The transportation costs $C_{l}^{\text{transport}} \cdot V_a$ depend on the volume of a unit (depending on the item type) and the flow, as described in Sect. 3.

Items that are stored in the ASRS must be repacked into boxes. Analogously to picking, the costs are independent of the item itself. If an item is stored in HBS, the incoming pallet has to be fork-lifted to the storage levels. The costs depend on the number of units on the incoming pallet.

$$C_{al}^{storing} = \begin{cases} C^{\text{repacking}} & \text{if } l = [\text{ASRS}] \\ C^{\text{fork-lifting}} \cdot \frac{1}{F_a} & \text{else.} \end{cases} \tag{3}$$

Table 1. Symbols

Sets	
$\mathcal{A}$	Set of all items a
$\mathcal{L}$	Set of all flows l; $\mathcal{L} = \{[ASRS], [PP], [SP]\}$
$\bar{\mathcal{L}}_a$	Set of all inappropriate flows for item a
$\mathcal{T}$	Set of all (work)days t
Parameter	
R	Number of picking locations in HBS, if all used for PP
U^{SP}	Number of picking locations resulting from replacing a PP zone with an SP zone
$M_{[ASRS]}$	Number of boxes available in ASRS
M_a	Max. unit quantity of item a that fits in one box
K_{at}	Average unit quantity of item a picked on day t
P_a	Average number of units picked in one pick
F_a	Number of units of item a on an incoming pallet
B_a	Number of units per replenishment of item a to a sloping shelf
G_a	Number of picking locations in HBS required for item a
V_a	Volume of one unit of item a
A^d	Max. working time on one day in hours
A^{2d}	Max. cumulated working time on two consecutive days in hours
A^w	Max. cumulated working time per week in hours
D^{ASRS}	Max. number of moved boxes per hour (ASRS)
$C_{al}^{storing}$	Costs for storing one unit of item a in storage l
$C^{repacking}$	Costs for repacking one unit in boxes for storage [ASRS]
$C^{fork-lifting}$	Costs for moving a pallet to the storage levels of HBS by a forklift
$C_{al}^{replenish}$	Costs for replenishing one unit of item a in storage l
$C_{[\mathrm{PP}]}^{replenish}$	Costs for replenishing one pallet to a PP zone
$C_{[\mathrm{SP}]}^{replenish}$	Costs for refilling the shelf of a SP zone
$C_l^{picking}$	Costs for picking one unit from storage l
$C_l^{transport}$	Transport costs of one dm^3 in storage l
C_{al}	Total costs for handling item a in storage l
Variables	
x_{al}	Binary, equals 1, if item a is assigned to storage l, otherwise 0
z_l	Continuous, share of storage places assigned to storage type l
$p_{[\mathrm{ASRS}]t}$	Continuous, resulting number of box movements in ASRS on day t (dependent auxiliary variable)

Similar considerations apply to replenishment. For PP, a complete pallet is moved to the picking location. For SP, only a specific number of units is replenished to the shelves. If the item is assigned to ASRS, replenishment is not necessary.

$$C_{al}^{\text{replenish}} = \begin{cases} 0 & \text{if } l = [\text{ASRS}] \\ C_{[\text{PP}]}^{\text{replenish}} \cdot \frac{1}{F_a} & \text{if } l = [\text{PP}] \\ C_{[\text{SP}]}^{\text{replenish}} \cdot \frac{1}{B_a} & \text{if } l = [\text{SP}]. \end{cases} \tag{4}$$

Constraints (5) and (6) model the item assignment and the warehouse design. Here, the model's similarity to the structure used by [7] becomes clear. The rest of the model differs significantly.

$$\sum_{l \in \mathcal{L}} x_{al} = 1 \qquad \forall \quad a \in \mathcal{A} \tag{5}$$

$$z_{[\text{PP}]} + z_{[\text{SP}]} \leq 1 \tag{6}$$

Constraint (7) avoids the assignment of some items to flows. For example, based on practicality, items such as edible oils or rice are stored in the ASRS reluctantly, as considerable cleaning is necessary in the event of damage or spillage. Because of the short computing times required to solve the model (see Sect. 5), sensitivity analyses can be used efficiently in practical planning for this purpose (i.e. determining the extra costs for restricting certain items from an ASRS).

$$x_{al} = 0 \qquad \forall \quad a \in \mathcal{A}, l \in \bar{\mathcal{L}}_a \tag{7}$$

Constraints (8) and (9) link variables, z, to the available space in the distribution centre. For some items, multiple picking locations are required in HBS for operational reasons.

$$\sum_{a \in \mathcal{A}} G_a \cdot x_{a[\text{PP}]} = R \cdot z_{[\text{PP}]}, \tag{8}$$

$$\sum_{a \in \mathcal{A}} G_a \cdot x_{a[\text{SP}]} = U^{\text{SP}} \cdot R \cdot z_{[\text{SP}]}, \tag{9}$$

Constraint (10) calculates the daily number of box movements in the ASRS. Constraint (11) limits this to the technical limit, which is subject to daily flexibility spread over the week. This is modelled by Constraints (12) and (13). Index $\hat{t}$ represents the day before t and is represented by $(t-1) \bmod |\mathcal{T}|$. For the parameters related to working time applies $2 \cdot A^d \geq A^{2d}$ and $6 \cdot A^d \geq A^w$ meaning that the maximum cumulated working time of two consecutive days (or one week) has to be less than or equal to the maximum working time of two (or six) individual days . This allows maintenance to be distributed flexibly over the week and also facilitates the handling of peaks.

$$\sum_{a \in \mathcal{A}} \frac{K_{at}}{P_a} \cdot x_{a[\text{ASRS}]} = p_{[\text{ASRS}]t} \qquad \forall \quad t \in \mathcal{T}. \tag{10}$$

$$p_{[\text{ASRS}]t} \leq A^d \cdot D^{\text{ASRS}} \qquad \forall \quad t \in \mathcal{T}, \tag{11}$$

$$p_{[\text{ASRS}]\hat{t}} + p_{[\text{ASRS}]t} \leq A^{2d} \cdot D^{\text{ASRS}} \qquad \forall \quad t \in \mathcal{T}, \tag{12}$$

$$\sum_{t \in \mathcal{T}} p_{[\text{ASRS}]t} \leq A^w \cdot D^{\text{ASRS}}, \tag{13}$$

A second capacity limit for the ASRS results from the number of boxes. Constraint (14) limits the number of required boxes to the number of available boxes. To ensure continuous operation, each item must be available in at least five boxes. The weekly maximum of outgoing units of each item a is shown by $K_a^{max} = \max_{t \in \mathcal{T}} K_{at}$.

$$\sum_{a \in \mathcal{A}} \max\left(5, \left\lceil \frac{K_a^{max}}{M_a} \right\rceil\right) \cdot x_{a[\text{ASRS}]} \leq M_{[\text{ASRS}]}. \tag{14}$$

Finally, Constraints (15)–(17) state the domains:

$$x_{al} \in \{0,1\} \qquad \forall \quad a \in \mathcal{A}, l \in \mathcal{L} \tag{15}$$

$$p_{[\text{ASRS}]t} \geq 0 \qquad \forall \quad t \in \mathcal{T} \tag{16}$$

$$0 \leq z_l \leq 1 \qquad \forall \quad l \in \mathcal{L} \setminus [\text{ASRS}] \tag{17}$$

The complete optimization problem is represented by min. (1) s.t. (5)–(17).

5 Computational Analysis

5.1 Experimental Design

The mathematical formulation is solved by Gurobi 9.1 using the C#-API on an Intel Core i7-3770 CPU and 8 GB RAM. We consider the real-world case of a distribution centre in Germany with almost 20,000 items. An initial analysis of the cost structure is visualized by the pie chart at the centre of Fig. 3. The share of items for which each of the three flows is the most economical is shown. Note that SP cannot be the least costly flow for any item because of the replenishment process. The three outer rings represent the assignment restrictions with respect to Constraint (7). The shares of items marked in white must not be assigned to the respective flow. For easy referencing of the shares, we have labelled them alphabetically. The interpretation of the figure can be well illustrated by (e): 9% of the items can be assigned to either ASRS or PP. An assignment to SP is prohibited (e.g. because the items are too large for the shelves). As another example, 52% of the items can be aggregated in (d), thereby implying that the lowest costs are incurred when these items are assigned to the ASRS. Nevertheless, an assignment to each of the flows (SP, PP or ASRS) is possible. By contrast, although the items in (c) can also be assigned to every flow, assignment to PP is the least costly. For all items in (e) applies $C_{a[ASRS]} \leq C_{a[PP]}$. The items in (a) and (f) could be ignored in the optimization process (combined with a corresponding capacity reduction), but they are included for clarity. Gurobi automatically removes the associated variables in a presolve step.

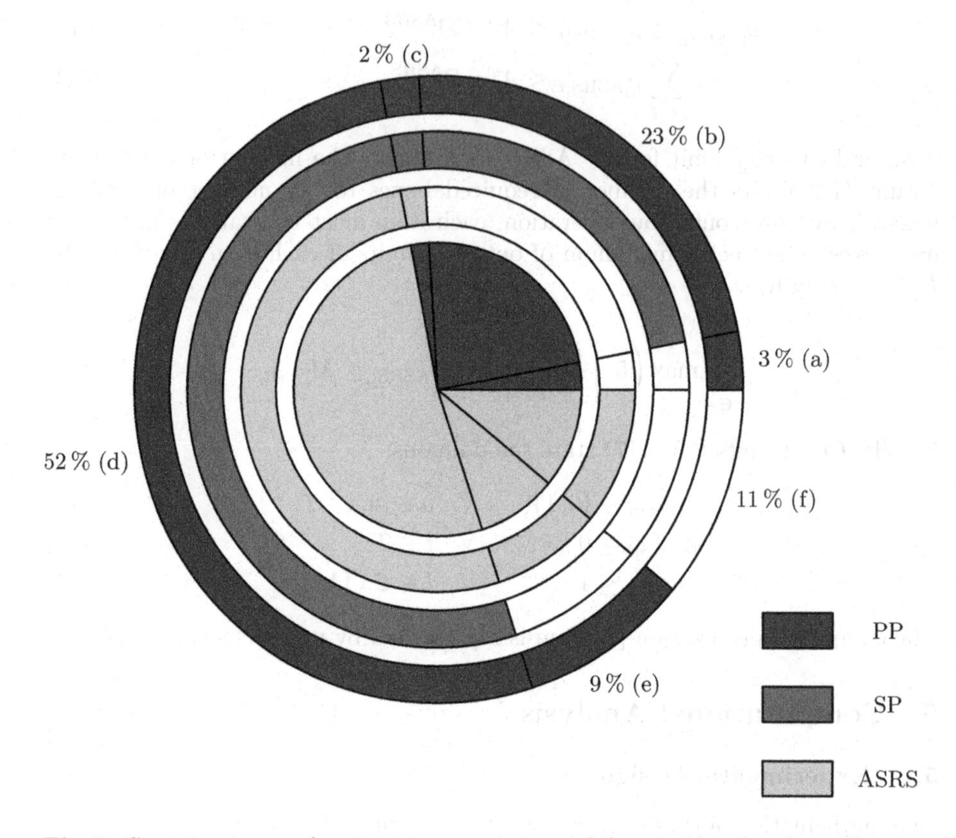

Fig. 3. Cost structure and assignment restrictions of the considered real-world case

We refer to solving the model with $\mathcal{T} = \{\text{Mo, Tu, We, Th, Fr, Sa}\}$ as using a multi-periodic approach. For comparison, we also solve the model with $|\mathcal{T}| = 1$ and refer to this as a single-periodic approach. Note that, in this case, Constraint (11) dominates (12) and (13) automatically. We use the average values from all weekdays as input data. This is the same as the single-periodic approach used by [7]. In a first step, we compare both variants. The resulting costs are considered in monetary units or cost points respectively. Thus, the objectives do not represent real cash flows but only reflect the ratio. For simplification, we have therefore omitted the specification of units. A second evaluation is performed for the capacity utilization of the ASRS with a special focus on the two different transport systems to markets (pallets and containers).

5.2 Single-Periodic Approach vs. Multi-periodic Approach

Solving the single-periodic version with a daily uptime of 24 h of the ASRS results in costs of 40,515.70. This is equal to 243,094.20 for a six-day week. Reducing the uptime to 20 h incurs almost identical daily (weekly) costs of 40,552.79 (243,316.74). Solving the multi-periodic version with a daily uptime limit of 24 h, a two-day limit of 40 h and a weekly limit of 120 h results in costs in the amount of 243,714.63. In all automatically generated solutions by solving the model, a maximum of seven items were assigned to SP. Therefore, SP is not visualized and discussed in detail in the following. Note that the fact that SP is almost never used in the optimal solutions does not mean that design Constraints (6), (8) and (9) should not be considered. This simply means that there are no capacity bottlenecks in the current product mix that would require more-extensive use of SP.

For a better understanding of the differences between the single-periodic and multi-periodic formulations, an illustrative example is provided in Table 2. By assuming one box movement per hour, the maximum technical limit of the ASRS is 24 box movements per day. The single-periodic approach is based on the average values, and it is clear that the limit is respected even if all items are assigned to the ASRS. In contrast, it becomes obvious that the technical limits will be exceeded on a daily basis on Monday, Tuesday and Wednesday. As a solution, only the following combined assignments of items to the ASRS are possible: {1;2}, {1;3}, {2;3}, {3;4}. Thus, it can be seen that the shortcomings of the static single-periodic model can only be overcome with a multi-periodic approach.

Table 2. Example of (average) unit throughput K_{at} and $\bar{K}_a$

Item	Mo	Tu	We	Th	Fr	Sa	AVG
1	15	9	9	5	3	1	7.00
2	5	13	5	5	5	5	6.17
3	5	7	1	0	0	0	2.17
4	19	12	10	5	3	1	8.33
Sum	44	41	25	15	11	7	23.67

In the following, we take another look at our practical example. Figure 4 shows the resulting number of picks for the ASRS and PP. For the ASRS, the number of picks equals the total daily number of box movements (bm), which is restricted to a technical limit determined by $24\,\text{h} \cdot 3200\,\text{bm/h}$; this is represented by the horizontal line.

In the solution of the single-periodic version with a limit of 24 h uptime, this technical limit is exceeded significantly on Monday and Tuesday. Thus, this solution is not feasible in practice. Even with the solution with an uptime limit

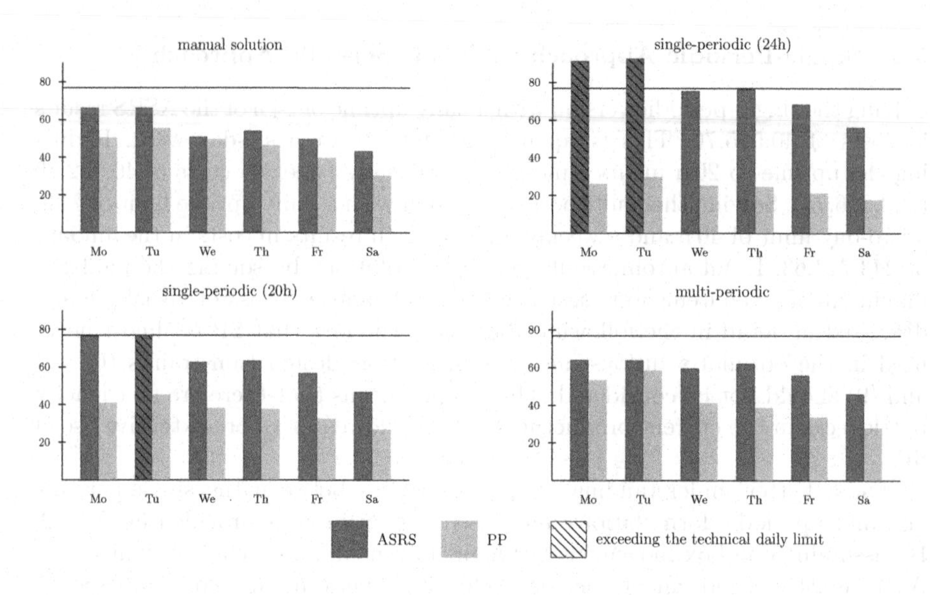

Fig. 4. Daily number of picks in thousands

of 20 h, the absolute technical limit of 24 uptime hours is exceeded on Tuesday. This solution leads to considerable capacity bottlenecks on Monday and Tuesday and is, therefore, also not feasible in practice. Of course, for both solutions, the respective limit is adhered to on a weekly average. For practical application, however, the day-to-day consideration of a complete week is unavoidable.

Since the cost difference between the single- and the multi-periodic version is less than 0.2%, it can be said that it is possible to keep the technical capacity limits of the ASRS at the same cost. Both variants can be solved in about one second; this is caused by the very strong linear relaxation of the problem (243,711.30 for the standard week). Compared to the capacity limits (i.e. the technical limit in the ASRS and the space limit in HBS), the impact of a single item is extremely small. This leads to the fact that, inherent in the LP solution, a large number of the variables already take integer values. Roughly speaking, only a very small number of items must be split in the LP solution; this makes solving with branch and bound strategies much easier. Therefore it can be assumed that no special heuristics need to be developed for problems of this kind.

The solution provided by the multi-periodic approach is visualized in Fig. 5. The average throughput speed ($\bar{K}_a = 1/T \cdot \sum_{t \in T} K_{at}$) is plotted against the item size (V_a). Small and slow-moving items are, preferably, assigned to the ASRS, whereas larger, slow-moving items are more likely to be found in PP. In the case of fast-moving items (right side of both graphs), it can be seen that, more often, they tend to be assigned to the ASRS regardless of volume. However, it should be noted that there are separate cross-docking areas in the distribution centre for extremely fast-moving items (e.g. beverages), which are generally not part of the optimization considered here.

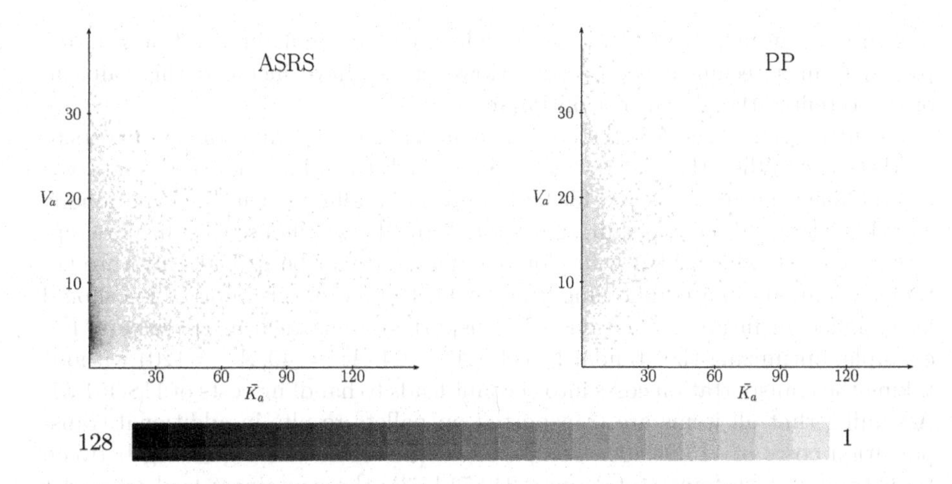

Fig. 5. Assigned number of items of (c), (d) and (e)

5.3 Cost Effects Depending on the Capacity Utilization of the ASRS

The general cost structure for the transportation related to the different transport systems to markets used in the HBS (pallet) and ASRS (container) is shown by $C^{\text{transport}}_{[\text{PP}]} = C^{\text{transport}}_{[\text{SP}]} \leq C^{\text{transport}}_{[\text{ASRS}]}$. This is in contrast to the handling costs shown by $C^{\text{handling}}_{al} = C^{\text{storing}}_{al} + C^{replenish}_{al} + C^{picking}_{l}$ with the following cost structure: $C^{\text{handling}}_{[\text{ASRS}]} \leq C^{\text{handling}}_{[\text{PP}]} \leq C^{\text{handling}}_{[\text{SP}]}$. In practice, these counteractive cost structures lead to intensive discussions about the trend in costs as the utilization of the ASRS increases. Therefore, Fig. 6 shows the cost trend and composition for increasing utilization of the ASRS.

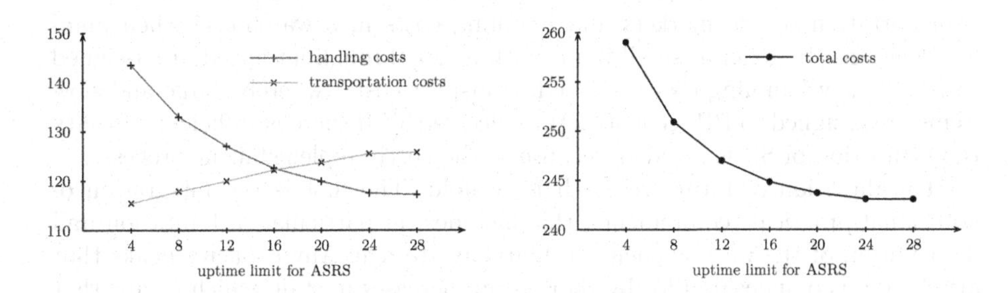

Fig. 6. Change in costs for increasing ASRS utilization

The data were generated by solving the multi-periodic version without considering Constraints (12) and (13) for an easier interpretation. The uptime limits correspond to an increase in the utilization of ASRS with a constant performance (D^{ASRS}). Except for the values for 28 h, these correspond to an uptime limit of 24 h with an increased performance, which cannot be maintained over a long period from a technical perspective. However, we have included this value in order to reflect the global cost optimum.

As utilization of the ASRS increases, handling costs fall and transport costs to markets rise. When the cost optimum is reached, the additional transport costs can no longer be compensated for by savings in handling when using containers. This knowledge about the shifting of costs from the warehouse to transport represents considerable added value for planners and can be gained only through the use of an automated planning process. Furthermore, this information should be included in future evaluations of transport systems such as containers. For example, optimizing the standard week ($A^d = 24, A^{2d} = 40, A^w = 120$) without taking the transportation costs into account leads to handling costs of 118,451.31. Assuming that all items are transported on pallets results in additional transportation costs of 113,969.09. Compared to the results distinguishing between both transport systems (total costs 243,714.63), the containers lead to a cost increase of 11,292.86, which corresponds to 4.9% of the total costs. Of course, the containers can offer benefits at other points in the supply chain (e.g. due to different costs occurring during handling when placing items in the store). Nevertheless, a concrete consideration for future deployment can be made, and the shift of costs between the sub-processes can be analyzed.

6 Summary

This paper presents a multi-periodic mixed-integer formulation for the integrated product assignment and warehouse design problem. In cooperation with a German food retailer, the formulation was developed, implemented, tested and evaluated in practice. Based on a real-world case, the inevitability of a multi-periodic approach was demonstrated. Further, several technical requirements for an ASRS are considered and integrated for the first time. From a managerial point of view it is of utmost importance to consider the relationship between costs for transportation to the markets and handling costs in a warehouse when making decisions. In general, small items with a low throughput speed are assigned to ASRS (low handling costs and high transport costs) whereby large and slow items are assigned to PP (cost structure vice versa). If there is sufficient capacity the utilization of SP is avoided because of the costly replenishment process.

The direction of future research is twofold. The first is the integration of additional practical requirements; this includes, in particular, a limitation and distribution of the work capacity in high-bay storage. Any capacity peaks that arise can be compensated for by short-term increases in staff, which means that these limitations have not yet been taken into account. With increasing staff shortages, capacity restrictions will also have to be expected here in the future. The second and somewhat more significant point is the fact that modelling has, thus far, been performed for only the single distribution centre under consideration, but in practice, not every distribution centre is equipped with an ASRS. To still be able to take advantage of the cost benefits of an ASRS, several centres can cooperate. This results in more-complex flows, which are reflected in a more complicated form in the capacity restrictions. It also offers the possibility of splitting product flows so that the capacities of both warehouses can be fully utilized. However, this makes modelling more difficult, and whether a model-based solution with state-of-the-art solvers is still possible without further effort requires further research and examination.

References

1. Baker, P., Canessa, M.: Warehouse design: a structured approach. Eur. J. Oper. Res. **193**(2), 425–436 (2009)
2. De Koster, R.B.M., Johnson, A.L., Roy, D.: Warehouse design and management. Int. J. Prod. Res. **55**(21), 6327–6330 (2017)
3. Gu, J., Goetschalckx, M., McGinnis, L.F.: Solving the forward-reserve allocation problem in warehouse order picking systems. J. Oper. Res. Soc. **61**(6), 1013–1021 (2010)
4. Gu, J., Goetschalckx, M., McGinnis, L.F.: Research on warehouse operation: a comprehensive review. Eur. J. Oper. Res. **177**(1), 1–21 (2007)
5. Gu, J., Goetschalckx, M., McGinnis, L.F.: Research on warehouse design and performance evaluation: a comprehensive review. Eur. J. Oper. Res. **203**(3), 539–549 (2010)
6. Guerriero, F., Pisacane, O., Rende, F.: Comparing heuristics for the product allocation problem in multi-level warehouses under compatibility constraints. Appl. Math. Model. **39**(23), 7375–7389 (2015)
7. Heragu, S.S., Du, L., Mantel, R.J., Schuur, P.C.: Mathematical model for warehouse design and product allocation. Int. J. Prod. Res. **43**(2), 327–338 (2005)
8. Kumar, S., Narkhede, B.E., Jain, K.: Revisiting the warehouse research through an evolutionary lens: a review from 1990 to 2019. Int. J. Prod. Res. 1–23 (2021)
9. Lee, M.K., Elsayed, E.: Optimization of warehouse storage capacity under a dedicated storage policy. Int. J. Prod. Res. **43**(9), 1785–1805 (2005)
10. Reyes, J., Solano-Charris, E., Montoya-Torres, J.: The storage location assignment problem: a literature review. Int. J. Ind. Eng. Comput. **10**(2), 199–224 (2019)
11. Roodbergen, K.J., Vis, I.F., Taylor, G.D., Jr.: Simultaneous determination of warehouse layout and control policies. Int. J. Prod. Res. **53**(11), 3306–3326 (2015)
12. Rouwenhorst, B., Reuter, B., Stockrahm, V., van Houtum, G.J., Mantel, R., Zijm, W.H.: Warehouse design and control: framework and literature review. Eur. J. Oper. Res. **122**(3), 515–533 (2000)
13. Strack, G., Pochet, Y.: An integrated model for warehouse and inventory planning. Eur. J. Oper. Res. **204**(1), 35–50 (2010)
14. Van den Berg, J.P., Sharp, G.P., Gademann, A., Pochet, Y.: Forward-reserve allocation in a warehouse with unit-load replenishments. Eur. J. Oper. Res. **111**(1), 98–113 (1998)
15. Van den Berg, J.P., Zijm, W.H.: Models for warehouse management: classification and examples. Int. J. Prod. Econ. **59**(1–3), 519–528 (1999)

New Valid Inequalities for a Multi-echelon Multi-item Lot-Sizing Problem with Returns and Lost Sales

Franco Quezada[1,2(✉)], Céline Gicquel[3], and Safia Kedad-Sidhoum[4]

[1] Sorbonne Université, LIP6, 75005 Paris, France
[2] Universidad de Santiago de Chile, LDSPS, Santiago, Chile
franco.quezada@usach.cl
[3] Université Paris Saclay, LISN, 91190 Gif-sur-Yvette, France
celine.gicquel@lri.fr
[4] CNAM, CEDRIC, 75003 Paris, France
safia.kedad_sidhoum@cnam.fr

Abstract. This work studies a multi-echelon multi-item lot-sizing problem with remanufacturing and lost sales. The problem is formulated as a mixed-integer linear program. A new family of valid inequalities taking advantage of the problem structure is introduced and used in a customized branch-and-cut algorithm. The provided numerical results show that the proposed algorithm outperforms both the generic branch-and-cut algorithm embedded in a standard-alone mathematical solver and a previously published customized branch-and-cut algorithm.

Keywords: Production planning · Lot-sizing · Remanufacturing · Mixed-integer linear programming · Valid inequalities

1 Introduction

Industrial companies face an increasing pressure from customers and governments to become more environmentally responsible and mitigate the environmental impact of their products. One way of achieving this objective is to remanufacture the products once they have reached their end of life. Remanufacturing is defined as a set of processes transforming used products into like-new products, mainly by rehabilitating damaged components. By reusing the materials and components embedded in used products, remanufacturing both contributes in reducing pollution emissions and natural resource consumption, making production processes more environment-friendly.

The present work considers a remanufacturing system involving three production echelons: disassembly of used products brought back by customers, refurbishing of the recovered parts and reassembly into like-new finished products. We

This work was partially funded by the National Agency for Research and Development (ANID)/Scholarship Program/DOCTORADO BECAS CHILE/2018 - 72190160.

M. Mes et al. (Eds.): ICCL 2021, LNCS 13004, pp. 192–207, 2021.
https://doi.org/10.1007/978-3-030-87672-2_13

aim at optimizing the production planning for the corresponding three-echelon system over a multi-period horizon. Within a remanufacturing context, production planning includes making decisions on the used products returned by customers, such as how much and when used products should be disassembled, refurbished or reassembled in order to build new or like-new products. The main objective is to meet customers' demand for the remanufactured products in the most cost-effective way.

We consider the case where the production on a machine requires setup operations such as machine calibration and incurs fixed setup costs. As a naive perception, to reduce these setup costs, production should be run using large lot sizes. However, this generates desynchronized patterns between the customers' demand and the production plan leading to costly high levels of inventory. Lot-sizing models thus aim at reaching the best possible trade-off between minimizing the setup costs and minimizing the inventory holding costs, taking into account both the customers' demand satisfaction and the practical limitations of the system. In the present work, we investigate the problem of reaching the best possible trade-off between setup and inventory holding costs within a remanufacturing environment and introduce an additional lost sales cost to be paid when the customers' demand is not satisfied on time. We thus study a 3-echelon lot-sizing problem with returns and lost sales.

Only a few works have addressed such multi-echelon production systems through exact solution approaches. A first attempt at tackling this difficulty can be found in [12]. Quezada et al. [12] considered the problem in a stochastic setting, taking into account uncertainties on the problem input parameters. They proposed a multi-stage stochastic approach based on the use of scenario trees. The problem was formulated as a MILP and solved through a new customized branch-and-cut algorithm. This algorithm relied on valid inequalities focused on strengthening the formulation of the single-echelon uncapacitated lot-sizing sub-problems embedded in the main problem. Although this approach was successful at providing near optimal solutions for small to medium size instances, some numerical difficulties were encountered to solve the larger instances. Intuitively, this difficulty might be partly due to the fact that the valid inequalities used to strengthen the formulation considered uncapacitated single-echelon sub-problems. They did not take into account the fact that, even if the production resources are assumed uncapacitated, the amount of products that can be processed on a resource at a given time period is limited among others by the amount of available used products returned up to this time period and by the yield of the disassembly process, i.e. by the proportion of disassembled parts that are in a sufficiently good state to be refurbished and reused in a remanufactured product. Hence, using valid inequalities taking into account this aspect of the problem might contribute in further strengthening its MILP formulation and decrease the computational effort needed to solve large-size instances.

To the best of our knowledge, the formulation of valid inequalities that explicitly take into account the impact of a limited returns quantity on the production plan has not yet been studied for a multi-echelon remanufacturing system. The present work aims at partially closing this gap by proposing new valid inequalities for this problem. However, in view of the theoretical and numerical difficulties encountered when developing new valid inequalities, we focus on the deterministic variant of the problem. Our contributions are thus twofold. First, we propose a new family of valid inequalities for the problem under study. These valid inequalities can be seen as an extension of valid inequalities previously known for the uncapacitated single-echelon lot-sizing problem with lost sales (the (k, U) inequalities first introduced in [8]) to take into account, at each echelon of the studied multi-echelon production system, the constraints on the production plan coming from the limited availability of the returns. We prove that these new valid inequalities are at least as strong as the previously known inequalities. Second, we develop a branch-and-cut algorithm based on the newly proposed valid inequalities and seek to assess its computational performance by comparing it with the one of a stand-alone mathematical programming solver and the one of a branch-and-cut algorithm based on previously known valid inequalities. The numerical results show the usefulness of the proposed inequalities at solving the problem under study.

The remainder of this paper is organized as follows. We first provide a brief overview of the related literature in Sect. 2. The problem description, together with its MILP formulation, are provided in Sect. 3. We then present the proposed new family of valid inequalities in Sect. 4. Computational results are summarized in Sect. 5. Finally, Sect. 6 gives a conclusion and some research perspectives.

2 Related Works

Throughout the last decade, several works sought to strengthen the MILP formulation of single-echelon lot-sizing problems involving remanufacturing, either through extended reformulations or through valid inequalities.

Helmrich et al. [13] discussed several MILP formulations of the uncapacitated single-item single-echelon lot-sizing problem with remanufacturing and introduced new valid inequalities by adapting the previously known (l, S, WW) proposed by [10] to their problem. The inequalities developed in [13] are based on the assumption that all returned products are either processed or kept in stock and do not consider the possibility that some of the returns may be discarded in case of an unbalance between returns quantity and demand. They can therefore not be directly used for the problem under study here. Similarly, [5] proposed a multi-commodity reformulation and a new set of valid inequalities for this problem. In particular, they further strengthened the (l, S, WW) inequalities presented in [13] by considering that the amount of finished products remanufactured in a given period t is limited by the cumulative quantity of returned products brought back up to t. Ali et al. [2] enriched the previous works by highlighting a theoretical property with regards to the equivalence of the shortest path and facility location reformulations. They also carried out a polyhedral analysis of a related sub-problem based on the single node fixed-charge network problem, proving the validity of several flow cover inequalities and their facet-defining conditions as well. Akartunali and Arulselvan [1] studied both the uncapacitated and capacitated variants of this single-item single-echelon lot-sizing problem. They showed that the uncapacitated problem cannot have a fully polynomial time approximation scheme (FPTAS) and provided a pseudo-polynomial algorithm to solve the problem. They also provided valid inequalities based on the flow-cover inequalities for the problem with a limited production capacity. Finally, we refer the reader to [4] for a recent survey on single-item lot-sizing problems with remanufacturing.

We note that all the above mentioned works focus on single-item single-echelon problems and do not consider the fact that remanufacturing may involve several processing steps, i.e. several production echelons, in order to transform the returned used products into like-new products. Moreover, these works consider hybrid manufacturing/remanufacturing systems and assume that, thanks to the presence of an uncapacitated manufacturing system, it will always be possible to satisfy the demand for finished products on time. In contrast, we investigate a pure remanufacturing system and consider that the demand will be lost in case the quantity and/or the quality of the returned products are insufficient to meet it on time. This means among others that the inequalities introduced in [5,13] and [2] are not valid for our problem and that new valid inequalities taking into account its specific features are needed.

3 Problem Description and Modeling

3.1 Production System

We consider a remanufacturing system comprising three main production echelons: disassembly, refurbishing and reassembly. We seek to plan the production activities in this system over a horizon comprising a discrete set $\mathcal{T} = \{1, .., T\}$ of periods. The system involves a set $\mathcal{I}$ of items. Among these ones, item $i = 0$ represents the used products returned by customers in limited quantities at each period. A used product is composed of I parts. Let α_i be the number of parts i embedded in a used product. The returned products are first disassembled to obtain a set $\mathcal{I}_r = \{1, ..., I\}$ of recoverable parts. Due to the usage state of the used products, some of the parts obtained during disassembly have to be discarded. In order to reflect the variations in the quality of the used products, i.e. the yield of the disassembly process, we let π_i^t denote the proportion of parts which will be recoverable at each time period t for each item $i = \{1, \ldots, I\}$. The recoverable parts are then refurbished on dedicated refurbishing processes. The set of $\mathcal{I}_s = \{I + 1, ..., 2I\}$ of serviceable parts obtained after refurbishing are reassembled into remanufactured products which have the same bill-of-material as the used products. These remanufactured products, indexed by $i = 2I + 1$, are used to satisfy the dynamic demand of customers.

The system comprises a set $\mathcal{P} = \{0, ..., I + 1\}$ of production processes: $p = 0$ corresponds to the disassembly process, $p \in \{1, ..., I\}$ corresponds to the process refurbishing the recoverable part indexed by p into the serviceable part indexed by $p+I$ and $p = I+1$ corresponds to the reassembly process. All these processes are assumed to be uncapacitated. However, the system might not be able to satisfy the customer demand on time due to part shortages if there are not enough used products returned by customers or if their quality is low. In this situation, the corresponding demand is lost incurring a high penalty cost to account for the loss of customer goodwill. Moreover, some used products and recoverable parts are allowed to be discarded. This option might be useful in case more used products are returned than what is needed to satisfy the demand for remanufactured products and in case there is a strong unbalance between the part-dependent disassembly yields leading to an unnecessary accumulation in inventory of the easy-to-recover parts.

All input parameters of the problem are time-dependent: r^t denotes the quantity of collected used products, d^t the customers' demand and π_i^t the proportion of recoverable parts $i \in \mathcal{I}_r$ obtained by disassembling one unit of returned product at period t. As for the costs, for each period t, we have the setup cost f_p^t for process $p \in \mathcal{P}$, the unit inventory cost h_i^t for part $i \in \mathcal{I}$, the unit lost-sales penalty cost l^t, the unit cost q_i^t for discarding item $i \in \mathcal{I}_r \cup \{0\}$ and the unit cost g^t for discarding the unrecoverable parts obtained while disassembling one unit of returned product (Fig. 1).

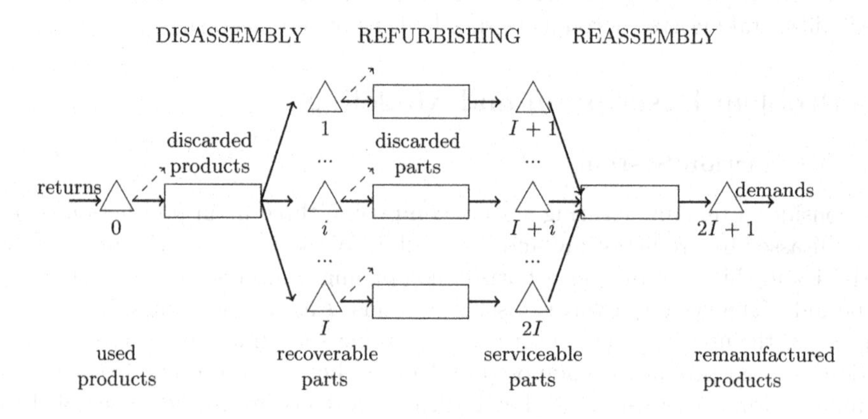

Fig. 1. Studied remanufacturing system

3.2 Natural Formulation

In order to build a mathematical model for the problem, we introduce the following decision variables at time period $t \in \mathcal{T}$: X_p^t the quantity of parts processed on process $p \in \mathcal{P}$, $Y_p^t \in \{0,1\}$ the setup variable for process $p \in \mathcal{P}$, S_i^t the inventory level of part $i \in \mathcal{I}$, Q_i^t the quantity of part $i \in \mathcal{I}_r \cup \{0\}$ discarded and L^t the lost sales of remanufactured products. This leads to the following MILP model.

$$\min \sum_{t \in \mathcal{T}} \Big(\sum_{p \in \mathcal{J}} f_p^t Y_p^t + \sum_{i \in \mathcal{I}} h_i^t S_i^t + l^t L^t + \sum_{i \in \mathcal{I}_r \cup \{0\}} q_i^t Q_i^t + g^t X_0^t \Big) \quad (1)$$

$$X_p^t \leq M_p^t Y_p^t \qquad \forall p \in \mathcal{J}, \forall t \in \mathcal{T} \quad (2)$$

$$S_0^t = S_0^{t-1} + r^t - X_0^t - Q_0^t \qquad \forall t \in \mathcal{T} \quad (3)$$

$$S_i^t = S_i^{t-1} + \pi_i^t \alpha_i X_0^t - X_i^t - Q_i^t \qquad \forall i \in \mathcal{I}_r, \forall t \in \mathcal{T} \quad (4)$$

$$S_i^t = S_i^{t-1} + X_{i-I}^t - \alpha_i X_{I+1}^t \qquad \forall i \in I_s, \forall t \in \mathcal{T} \quad (5)$$

$$S_{2I+1}^t = S_{2I+1}^{t-1} + X_{I+1}^t - d^t + L^t \qquad \forall t \in \mathcal{T} \quad (6)$$

$$S_i^0 = 0 \qquad \forall i \in \mathcal{I} \quad (7)$$

$$S_i^t \geq 0 \qquad \forall i \in \mathcal{I}, \forall t \in \mathcal{T} \quad (8)$$

$$X_p^t \geq 0, Y_p^t \in \{0,1\} \qquad \forall p \in \mathcal{J}, \forall t \in \mathcal{T} \quad (9)$$

The objective function (1) aims at minimizing the total remanufacturing cost over the whole planning horizon, i.e., the sum of the setup, inventory holding, lost sales and disposal costs. Constraints (2) link the production quantity variables to the setup variables. Constraints (3)–(6) are the inventory balance constraints. More specifically, Constraints (3) ensure that any returned product is either disassembled, discarded, or kept in stock. Constraints (4) guarantee that any item obtained from the disassembly process is either refurbished, discarded, or kept in stock. Constraints (5) ensure that any refurbished item is either used in the reassembly process or kept in stock. Constraints (6) ensure that any remanufactured/finished product is either used to satisfy the demand or kept in stock and that, if there is not enough remanufactured products to satisfy the demand, the unsatisfied demand is lost. Without loss of generality, we assume that the initial inventory, S_i^0, is set to 0 for each item $i \in \mathcal{I}$ (see Constraints (7)). Finally, Constraints (8)–(9) provide the domain of the decision variables.

Note that the value of each constant M_p^t can be set by using an upper bound on the quantity that can be processed on process p at each time period t. This quantity is limited by two elements: the availability of the used products already returned by customers and the future demand for remanufactured products. We thus have, for each period t:

- $M_0^t = \min \left\{ \sum_{1 \leq \kappa \leq t} r^\kappa, \frac{\sum_{t \leq \kappa \leq T} d^\kappa}{\min_{i=1,\ldots,I} \pi_i^t} \right\}$
- $M_p^t = \alpha_p \min \left\{ \sum_{1 \leq \kappa \leq t} r^\kappa \hat{\pi}_p^{\kappa,t}, \sum_{1 \leq \kappa \leq T} d^\kappa \right\}$, for $p \in \mathcal{I}_r$.
- $M_{I+1}^t = \min \left\{ \min_{p \in \mathcal{I}_r} \left\{ \sum_{1 \leq \kappa \leq t} r^\kappa \hat{\pi}_p^{\kappa,t} \right\}, \sum_{t \leq \kappa \leq T} d^\kappa \right\}$

where $\hat{\pi}_p^{\kappa,t} = \text{argmax}\{\pi_p^\theta, \theta = \kappa, \ldots, t\}$ denotes the maximum disassembly yield that can be obtained for recoverable item p over the time interval $[\kappa, t]$.

3.3 Echelon Stock Reformulation

We now provide a reformulation of the problem using the echelon stock concept [11]. The echelon stock of a product in a multi-echelon production system corresponds to the total quantity of the product held in inventory, either as such or as a component within its successors in the bill-of-material. We thus denote by E_i^t the echelon stock level of item $i \in \mathcal{I} \setminus \{0\}$ at the end of period t. Replacing variables S_i^t by variables E_i^t in Problem (1)–(9) leads to the following MILP formulation:

$$\min \sum_{t \in \mathcal{T}} \Big(\sum_{p \in \mathcal{J}} f_p^t Y_p^t + h_i^t S_0^t + \sum_{i \in \mathcal{I} \setminus \{0\}} eh_i^t E_i^t + \sum_{i \in \mathcal{I}_r \cup \{0\}} q_i^t Q_i^t + g_0^t X_0^t \Big) \tag{10}$$

$$X_p^t \leq M_p^t Y_p^t \qquad \forall p \in \mathcal{J}, \forall t \in \mathcal{T} \tag{11}$$

$$S_0^t = S_0^{t-1} + r^t - X_0^t - Q_0^t \qquad \forall t \in \mathcal{T} \tag{12}$$

$$E_i^t = E_i^{t-1} + \pi_i^t \alpha_i X_0^t - \alpha_i (d^t - L^t) - Q_i^t \qquad \forall i \in \mathcal{I}_r, \forall t \in \mathcal{T} \tag{13}$$

$$E_i^t = E_i^{t-1} + X_{i-I}^t - \alpha_i (d^t - L^t) \qquad \forall i \in I_s, \forall t \in \mathcal{T} \tag{14}$$

$$E_{2I+1}^t = E_{2I+1}^{t-1} + X_{I+1}^t - d^t + L^t \qquad \forall t \in \mathcal{T} \tag{15}$$

$$S_0^0 = 0 \tag{16}$$

$$E_i^0 = 0 \qquad \forall i \in \mathcal{I} \setminus \{0\} \tag{17}$$

$$E_i^t - E_{I+i}^t \geq 0 \qquad \forall i \in \mathcal{I}_r, \forall n \in \mathcal{T} \tag{18}$$

$$E_i^t - \alpha_i E_{2I+1}^t \geq 0 \qquad \forall i \in \mathcal{I}_s, \forall n \in \mathcal{T} \tag{19}$$

$$E_i^t \geq 0 \qquad \forall i \in \mathcal{I}, \forall t \in \mathcal{T} \tag{20}$$

$$S_0^t, L^t \geq 0 \qquad \forall t \in \mathcal{T} \tag{21}$$

$$X_p^t \geq 0, Y_p^t \in \{0, 1\} \qquad \forall p \in \mathcal{J}, \forall t \in \mathcal{T} \tag{22}$$

The objective function (10) aims at minimizing the total cost over the whole planning horizon. Constraints (11) link the production quantity variables to the setup variables. Constraints (12)–(15) are the inventory balance constraints. Constraints (12) use the classical inventory variables, whereas Constraints (13)–(15) make use of the echelon inventory variables. Constraints (16)–(17) translate the fact that the initial inventory of each item is assumed to be equal to 0. Constraints (18)–(19) ensure consistency between the echelon inventory at the different echelons of the bill-of-material and guarantee that the physical inventory of each product, S_i^t, remains non-negative for all $i \in \mathcal{I}$. Finally, Constraints (20)–(22) define the domain of the decision variables.

The use of the echelon stock reformulation (11)–(22) enables us to decompose the initial problem into a series of single-echelon sub-problems by relaxing the linking constraints (18)–(19). Each of these sub-problems is an uncapacitated single-echelon single-item lot-sizing problem with lost sales, whose formulation can be strengthened by the (k, U) valid inequalities proposed in [8]. We refer the reader to [12] for a detailed description of each subproblem and the single-echelon (k, U) inequalities applied to each subproblem. Nonetheless, this decomposition into single-echelon uncapacitated sub-problems overlooks the fact that the production on each process at a given period is limited by the amount of used products returned up to this period. Therefore, in what follows, we investigate a class of valid inequalities in which these aspects of the problem are explicitly considered.

4 Single-Echelon (ℓ, k, U) Inequalities

We now seek to strengthen the single-echelon (k, U) inequalities investigated in [8] and [12] by considering the limited quantity of returned products available at each time period in the system. The (ℓ, k, U) inequalities are defined as follows:

Proposition 1. *Let $0 \leq \ell \leq k \leq T$ be two periods of the planning horizon. Let $U \subseteq \{k+1, ..., T\}$ be a subset of periods and $t^* = \max\{\tau : \tau \in U\}$ be the last time period belonging to U.*
The following inequalities are valid for Problem (11)–(22):

$$S_0^{\ell} \hat{\pi}_i^{\ell,t^*} + \alpha_i^{-1} E_i^k + \sum_{k<t\leq t^*} \phi_i^t Y_0^t \geq \sum_{t\in U} (d^t - L^t) \qquad \forall i \in \mathcal{I}_r \tag{23}$$

$$S_0^{\ell} \hat{\pi}_i^{\ell,t^*} + \alpha_i^{-1}(E_i^{\ell} - E_{i+I}^{\ell}) + \alpha_i^{-1} E_{i+I}^k + \sum_{k<t\leq t^*} \phi_i^t Y_i^t \geq \sum_{t\in U} (d^t - L^t) \quad \forall i \in \mathcal{I}_r \tag{24}$$

$$S_0^{\ell} \hat{\pi}_i^{\ell,t^*} + (\alpha_i^{-1} E_i^{\ell} - E_{2I+I}^{\ell}) + E_{2I+1}^k + \sum_{k<t\leq t^*} \phi_i^t Y_{I+1}^t \geq \sum_{t\in U} (d^t - L^t) \quad \forall i \in \mathcal{I}_r \tag{25}$$

with $\phi_i^t = \min \left\{ \sum_{\ell<\nu\leq t} r^{\nu} \hat{\pi}_i^{\nu,t}, \sum_{\nu\in U: t\leq\nu} d^{\nu} \right\}$

Proof. Let (X, Y, S, E, L, Q) be a feasible solution of Problem (10)–(22). We show that this solution complies with inequalities (23) for any pair of periods (ℓ, k), any subset $U \subset \{k + 1, ..., T\}$ and any recoverable item $i \in \mathcal{I}_r$.

Let $\tau \in [k+1, t^*]$ be the first production period in which $\phi_i^\tau = \sum_{\nu \in U:\tau \leq \nu} d^\nu$.
By convention, $\tau = t^* + 1$ if there is no such period.
We have $\phi_i^\tau Y_0^\tau = \sum_{t \in U:\tau \leq t} d^t \geq \sum_{t \in U:\tau \leq t} (d^t - L^t)$.
We consider two cases.
- Case 1: there is no production on $p = 0$ over the interval $[k+1; \tau - 1]$

In this case, $Y_0^t = 0$ and $X_0^t = 0$ for all periods t in $[k+1, \tau - 1]$. As no disassembly occurs over $[k+1, \tau-1]$, all the recoverable items needed to satisfy the demand over this time interval, and in particular needed to satisfy $\sum_{t \in U; t \leq \tau - 1} (d^t - L^t)$, should already have been disassembled previously and be in stock at the end of period k. This gives $\alpha_i^{-1} E_i^k \geq \sum_{t \in U:t \leq \tau-1} (d^t - L^t)$. We thus have:

$$S_0^\ell \hat{\pi}_i^{\ell,t^*} + \alpha_i^{-1} E_i^k + \sum_{k < t \leq t^*} \phi_i^t Y_0^t \geq \alpha_i^{-1} E_i^k + \phi_i^\tau Y_0^\tau \geq \sum_{t \in U} (d^t - L^t)$$

Inequality (23) is thus valid in this case.
- Case 2: there is at least one production period on $p = 0$ over interval $[k+1; \tau-1]$

Let θ be the last period of production on $p = 0$ over the interval $[k+1; \tau-1]$. By definition of θ, we have: $\phi_i^\theta = \sum_{\ell < \nu \leq \theta} (r^\nu \hat{\pi}_i^{\nu,\theta})$.

By summing up the inventory balance constraints (13) over periods $k+1, \ldots, \tau - 1$ and using the fact that variables E_i^k and $Q_i^t, \forall t = k+1, \ldots, \tau - 1$, are non-negative, we have:

$$\alpha_i^{-1} E_i^k + \sum_{t=k+1}^{\tau-1} \pi_i^t X_0^t \geq \sum_{t=k+1}^{\tau-1} (d_t - L_t) \tag{26}$$

By definition of τ, θ and ℓ, we have:

$$\sum_{t=k+1}^{\tau-1} \pi_i^t X_0^t = \sum_{t=k+1}^{\theta} \pi_i^t X_0^t \leq \sum_{t=\ell+1}^{\theta} \pi_i^t X_0^t \tag{27}$$

This gives:

$$\alpha_i^{-1} E_i^k + \sum_{t=\ell+1}^{\theta} \pi_i^t X_0^t \alpha_i^{-1} \geq E_i^k + \sum_{t=k+1}^{\tau-1} \pi_i^t X_0^t \tag{28}$$

$$\geq \sum_{t=k+1}^{\tau-1} (d_t - L_t) \tag{29}$$

$$\geq \sum_{t \in U:t \leq \tau-1} (d_t - L_t) \tag{30}$$

We now compute an upper bound of $\sum_{t=\ell+1}^{\theta} \pi_i^t X_0^t$. This one is obtained by first computing the linear combination $\sum_{t=\ell+1}^{\theta} \left(\hat{\pi}_i^{t,\theta}\right) \times (12)_t$. This gives:

$$\sum_{t=\ell+1}^{\theta} \left(\hat{\pi}_i^{t,\theta}\right) S_0^t = \sum_{t=\ell+1}^{\theta} \left(\hat{\pi}_i^{t,\theta}\right) \left[S_0^{t-1} + r^t - X_0^t - Q_0^t\right] \quad (31)$$

By the non-negativity of variables Q_0^t and S_0^t and the fact that $\hat{\pi}_i^{t,\theta} \geq \hat{\pi}_i^{t+1,\theta}$, we have:

$$\sum_{t=\ell+1}^{\theta} \pi_i^t X_0^t \leq \sum_{t=\ell}^{\theta} \hat{\pi}_i^{t,\theta} X_0^t \quad (32)$$

$$\leq \sum_{t=\ell+1}^{\theta} \left(\hat{\pi}_i^{t,\theta}\right) S_0^{t-1} - \sum_{t=\ell+1}^{\theta} \left(\hat{\pi}_i^{t,\theta}\right) S_0^t + \sum_{t=\ell+1}^{\theta} \left(\hat{\pi}_i^{t,\theta}\right) r^t \quad (33)$$

$$\leq \left(\hat{\pi}_i^{\ell,\theta}\right) S_0^\ell + \sum_{t=\ell+1}^{\theta} \left(\hat{\pi}_i^{t,\theta}\right) r^t \quad (34)$$

$$\leq \left(\hat{\pi}_i^{\ell,t^*}\right) S_0^\ell + \phi_i^\theta Y_0^\theta \quad (35)$$

Replacing $\sum_{t=\ell}^{\theta} \pi_i^t X_0^t$ in Inequalities (30) by its upper bound provided by (35), we have:

$$\alpha_i^{-1} E_i^k + \left(\hat{\pi}_i^{\ell,t^*}\right) S_0^\ell + \phi_i^\theta Y_0^\theta \geq \sum_{t \in U: t \leq \tau - 1} (d_t - L_t) \quad (36)$$

Finally, we have:

$$S_0^\ell \hat{\pi}_i^{\ell,t^*} + \alpha_i^{-1} E_i^k + \sum_{k < t \leq t^*} \phi_i^t Y_0^t \geq S_0^\ell \hat{\pi}_i^{\ell,t^*} + \alpha_i^{-1} E_i^k + \phi_i^\theta Y_0^\theta + \phi_i^\tau Y_0^\tau$$

$$\geq \sum_{t \in U: t \leq \tau - 1} (d_t - L_t) + \sum_{t \in U: t \geq \tau} (d_t - L_t)$$

$$\geq \sum_{t \in U} (d_t - L_t)$$

This concludes the proof of validity for Inequality (23). The same arguments can be used to prove the validity of Inequalities (24) and (25). □

It is worth mentioning that the (k, U) inequalities used in [12] to strengthen the formulation (10)–(22) can be seen as a particular case of the more general family of (ℓ, k, U) inequalities (23)–(25) proposed in this work. Namely, by setting ℓ to 0 and by computing the value of ϕ^t without taking the returns into account (i.e. by setting ϕ^t to $\sum_{\nu \in U: t \leq \nu} d^\nu$), each (ℓ, k, U) inequality (23)–(25) becomes a (k, U) inequality.

Proposition 2. *The linear relaxation of formulation* (10)–(22) *strengthened by valid inequalities* (23)–(25) *is at least as tight as the linear relaxation strengthened by the* (k, U) *valid inequalities used in [12].*

Proof. Let P_{LR} be the linear relaxation of polyhedron given by inequalities (11)–(22), (23)–(25) and $\tilde{P}_{LR}$ be the linear relaxation of polyhedron given by inequalities (11)–(22) and the (k, U) inequalities. As any (k, U) inequality is a valid inequality (23)–(25) with $\phi^t = \sum\limits_{\nu \in U: t \leq \nu} d^\nu$ and $\ell = 0$, we have $P_{LR} \subseteq \tilde{P}_{LR}$. □

The main implication of Proposition 2 is that the lower bound obtained by strengthening the formulation (10)–(22) with the (ℓ, k, U) inequalities is at least as tight as the lower bound obtained while using the single-echelon (k, U) inequalities.

We now briefly discuss the resolution of the separation problem for the (ℓ, k, U) valid inequalities. Recall that this problem consists in finding an inequality (23)–(25) violated by a given solution $(\tilde{X}, \tilde{Y}, \tilde{S}, \tilde{E}, \tilde{L}, \tilde{Q})$ of the linear relaxation of Problem (11)–(22) or prove that no such inequality exists. In the present case, in order to find the most violated inequality among e.g. inequalities (23), we should find, for each period $k = 1, ..., T$, the set U of time periods and the period ℓ that maximize the difference between the right-hand and the left-hand side of the inequality. This is not trivial, in particular because the value of each coefficient ϕ_i^t simultaneously depends on U and ℓ. We thus consider a heuristic separation algorithm in our computational experiments. This one can be summarized as follows. For a given process p and time period k:

1. For each period $t = k+1, ..., T$, add t to U if $d^t(1 - \sum_{\tau=k+1}^{t} \tilde{Y}_p^\tau) - \tilde{L}^t > 0$.
2. For each period $\ell = 0, ..., k$,
   - compute the value of each coefficient $\phi_i^t = \min\left\{\sum_{\ell < \nu \leq t} r^\nu \hat{\pi}_i^{\nu,t}, \sum_{\nu \in U: t \leq \nu} d^\nu\right\}$
   - compute the left-hand side of the inequality (23) (resp. (24) and (25)).
3. Set ℓ to the period index which minimizes this left-hand side value.

This algorithm has a time complexity of $\mathcal{O}(T^2)$ as the computation of set U in step 1 and of coefficients ϕ_i in step 2 both require $\mathcal{O}(T^2)$ operations.

5 Computational Experiments

In this section, we focus on assessing the performance of the proposed valid inequalities when used within a customized branch-and-cut algorithm. We compare the performance of this algorithm with the one of the generic branch-and-cut algorithm embedded in a mathematical programming solver and the one of a branch-and-cut algorithm using single-echelon (k, U) inequalities.

5.1 Instance Generation

We considered two sets of instances: Set 1 instances involve $T = 25$ periods and $I = 10$ parts whereas Set 2 instances involve $T = 35$ periods and $I = 10$ parts. Within each set, the instances were randomly generated by adapting the procedure presented in [6]. More precisely, we considered four values of the setup-holding cost ratio $f/h \in \{600, 1200, 1800, 2400\}$, two values for the production-holding cost ratio $g/h \in \{2, 4\}$ and three values of the returns-demand quantity ratio $r/d \in \{1, 2, 3\}$. For each set and each possible combination of f/h, g/h, r/d, ten random instances were generated, resulting in a total of 480 instances.

For each instance, the value of each problem parameter was set as follows.

- Demand d^t was uniformly distributed in the interval $[0, 100]$ and the returns quantity r^t was uniformly distributed in the interval $[0.8(r/d)\bar{d}, 1.2(r/d)\bar{d}]$, where $\bar{d} = \frac{\sum d^t}{T}$ is the average demand per period.
- The proportion of recoverable parts π_i^t was uniformly distributed in the interval $[0.4, 0.6]$.
- The bill-of-materials coefficients $\alpha_i = \alpha_{i+I}, i \in \mathcal{I}_r$, were randomly generated following a discrete uniform distribution over $[1; 6]$ and we set $\alpha_0 = \alpha_{2I+1} = 1$.
- The holding cost h_0^t for the returned product $i = 0$ was fixed to 1. The holding cost h_i^t for each recoverable item $i \in \mathcal{I}_r$ was randomly generated following a discrete uniform distribution over interval $[2, 7]$. Similarly, the holding cost h_i^t for each serviceable item $i \in \mathcal{I}_s$ was randomly generated following a discrete uniform distribution over interval $[7, 12]$. Finally, in order to ensure non negative echelon costs, we set the value of the inventory holding cost for the remanufactured product, h_{2I+1}^t, to $\sum_{i=1}^{I} \alpha_i h_{I+i}^t + \epsilon$, where ϵ follows a discrete uniform distribution over interval $[80, 100]$.
- The production cost g^t was uniformly distributed in the interval $[0.8(g/h)\bar{h}, 1.2(g/h)\bar{h}]$, where $\bar{h} = \frac{\sum h_{2I+1}^t}{T}$ is the average holding cost.
- The setup cost f^t was uniformly distributed in the interval $[0.8(f/h)\bar{h}, 1.2(f/h)\bar{h}]$.
- Discarding costs were set to $q_i^t = 0.8\bar{h}_i^t$, where $\bar{h}_i^t = \frac{1}{T}\sum_{\kappa=t}^{T} h_i^\kappa$ The unit penalty cost for lost sales, l^n, was fixed to 10000 per

5.2 Results

We carried out extensive numerical experiments in order to assess the computational performance of the proposed valid inequalities. This was achieved by solving each considered instance using three alternatives branch-and-cut algorithms:

- *CPX*: the generic branch-and-cut algorithm embedded in CPLEX 12.8 using the echelon-stock formulation (11)–(22).
- (k, U): a customized branch-and-cut algorithm using the (k, U) inequalities to strengthen the echelon-stock formulation (11)–(22) similarly to what was done in [12].

- (ℓ, k, U): a customized branch-and-cut algorithm using the newly introduced (ℓ, k, U) inequalities to strengthen the echelon-stock formulation (11)–(22). This algorithm is based on the solver CPLEX 12.8. It generates inequalities of type (23)–(25) through a cutting-plane generation algorithm at the root node and at intermediate nodes of the branch-and-bound search tree using the UserConstraints callbacks provided by the solver.

All related linear programs and mixed-integer linear programs were solved using CPLEX 12.8 with the solver default settings. The algorithms were implemented in C++ using the Concert Technology environment. All tests were run on the computing infrastructure of the Laboratoire d'Informatique de Paris VI (LIP6), which consists of a cluster of Intel Xeon Processors X5690. We set the cluster to use two 3.46 GHz cores and 12 GB RAM to solve each instance. We imposed a time limit of 3600 s.

Table 1. Performance of CPLEX and branch-and-cut methods over instance in Set 1.

r/d	g/h	Method	R.LP$_{gap}$	R.MIP$_{gap}$	MIP$_{gap}$	C.Time	R.Time	T.Time
1	2	CPX	8.26	4.10	0.06	0.04	1.16	898.00
		(k, U)	6.08	3.66	0.05	0.92	2.93	1,011.79
		(ℓ, k, U)	4.17	3.55	0.03	0.97	2.35	792.64
	4	CPX	5.38	2.49	0.06	0.03	0.97	1,360.48
		(k, U)	3.95	2.33	0.05	0.50	1.47	1,088.78
		(ℓ, k, U)	2.84	2.22	0.04	0.85	1.98	1,056.29
2	2	CPX	40.86	11.28	0.97	0.03	1.81	2,570.62
		(k, U)	19.36	9.17	0.46	0.35	2.74	1,883.71
		(ℓ, k, U)	11.88	8.68	0.22	0.74	2.67	1,558.09
	4	CPX	37.79	10.71	2.29	0.03	1.59	3,280.23
		(k, U)	18.39	9.25	1.06	0.19	1.47	2,856.09
		(ℓ, k, U)	11.89	8.48	0.94	0.62	2.27	2,883.50
3	2	CPX	41.86	13.47	0.04	0.03	1.38	545.15
		(k, U)	20.60	9.68	0.02	0.29	1.96	296.65
		(ℓ, k, U)	14.16	9.11	0.02	0.48	1.91	250.52
	4	CPX	38.29	12.44	0.09	0.02	0.98	817.93
		(k, U)	20.81	9.71	0.02	0.15	1.00	458.41
		(ℓ, k, U)	15.36	8.99	0.06	0.41	1.57	522.70

The corresponding results are displayed in Table 1 for Set 1 instances and Table 2 for Set 2 instances. Column `Method` indicates the branch-and-cut algorithm used to solve the instances. Column R.LP$_{gap}$ reports the gap between the value of the linear relaxation strengthened by the corresponding valid inequalities and the best feasible solution found through the branch-and-bound search.

Table 2. Performance of CPLEX and branch-and-cut methods over instance in Set 2.

r/d	g/h	Method	R.LP$_{gap}$	R.MIP$_{gap}$	MIP$_{gap}$	C.Time	R.Time	T.Time
1	2	*CPX*	7.48	3.65	0.25	0.06	2.26	2,544.42
		(k, U)	5.09	3.08	0.25	8.08	13.32	2,300.22
		(ℓ, k, U)	3.49	2.95	0.11	7.59	11.07	1,764.24
	4	*CPX*	4.96	2.33	0.19	0.06	1.91	2,930.72
		(k, U)	3.31	2.00	0.11	3.66	6.58	2,529.28
		(ℓ, k, U)	2.40	1.91	0.10	5.09	7.95	2,319.77
2	2	*CPX*	44.73	12.89	6.76	0.06	4.43	3,599.07
		(k, U)	20.10	10.00	4.34	0.94	5.71	3,508.81
		(ℓ, k, U)	12.85	9.54	3.77	2.09	6.08	3,362.24
	4	*CPX*	43.15	12.54	7.29	0.06	3.57	3,599.07
		(k, U)	20.02	10.40	5.03	0.53	3.08	3,599.54
		(ℓ, k, U)	13.14	9.49	4.61	2.42	6.01	3,599.44
3	2	*CPX*	43.25	14.11	2.67	0.04	2.93	3,044.18
		(k, U)	18.97	9.48	0.55	0.87	4.26	1,855.28
		(ℓ, k, U)	13.38	9.11	0.64	1.40	4.30	1,614.35
	4	*CPX*	37.56	11.83	1.60	0.04	2.55	2,802.69
		(k, U)	16.82	8.30	0.19	0.49	2.30	1,035.84
		(ℓ, k, U)	12.67	7.86	0.20	1.21	3.67	1,029.84

For the CPX method, it reports the gap between the value of the initial linear relaxation and the best feasible solution found through the branch-and-bound search. Column `R.MIP`$_{gap}$ reports the gap between the lower bound at the root node (after the generation of CPLEX generic cutting planes) and the best feasible solution found through the branch-and-bound search. Column `MIP`$_{gap}$ reports the gap between the best lower and the best feasible solution found through the branch-and-bound search. The average CPU time for the cutting-plane generation of each method is reported in column `C.Time`, the CPU time spent at the root node in Column `R.Time` and the average total CPU time in Column `T.Time`. Note that each line corresponds to the average value of the corresponding 40 instances.

In general, we observe that the customized branch-and-cut algorithms based either on the (k, U) or on the (ℓ, k, U) inequalities outperform method CPX, providing solutions of better quality within shorter computation times. Specifically, the total computation time is reduced on average by 20% when using the branch-and-cut algorithm based on the (k, U) and by 26% when using the branch-and-cut algorithm based on the (ℓ, k, U) inequalities.

Regarding the relative performance of the (k, U) and (ℓ, k, U) inequalities, we note that the branch-and-cut algorithm based on the (ℓ, k, U) outperforms the algorithm based on (k, U) when the value of the demand-returns ratio is small, i.e. when $r/d \in \{1,2\}$. Thus, over the 160 instances corresponding to a value of r/d equal to 1, the total average computation time is reduced from1732 s when using (k, U) inequalities to1482 s when using (ℓ, k, U) inequalities. Similarly, over the 160 instances corresponding to a value of r/d equal to 2, the average MIP gap is reduced from 2.72% when using (k, U) inequalities to 2.38% when using (ℓ, k, U) inequalities. We note however that the relative performance of the proposed (ℓ, k, U) inequalities deteriorates for the instances corresponding to the largest considered value of the demand-returns ratio. Namely, when r/d is set to 3, the branch-and-cut algorithm based on the (k, U) inequalities provides a smaller MIP gap and/or smaller computation times. This might be explained by the fact that the corresponding instances involve a large amount of returned products so that the quantity processed on a resource at a given period is not (or at least to a lesser extent) limited by the availability of the returned products. This means that the proposed refinements in the expression of the valid inequalities are less relevant in this case.

6 Conclusion and Perspectives

We considered a lot-sizing problem aiming at planning production for a multi-echelon remanufacturing system. This problem can be formulated as a mixed-integer linear program. We focused on strengthening this formulation in order to be able to provide optimal or near-optimal solutions of this problem. Our main contribution is the development of a new set of valid inequalities which take into account, at each production echelon, the limitations on the produced quantities coming from the limited availability of the returned products. The results of our computational experiments show that a branch-and-cut algorithm based on these new valid inequalities performs well as compared to the generic branch-and-cut algorithm of CPLEX solver and to a previously published branch-and-cut algorithm based on less general valid inequalities.

A first possible research direction could be to develop an exact separation algorithm for the (ℓ, k, U) inequalities in the presented branch-and-cut algorithm as this may further improve their performance when used in a branch-and-cut algorithm. Moreover, it would also be worth studying whether valid inequalities previously proposed for capacitated lot-sizing problems (see e.g. [3,7,9]) might be useful to help solving our problem. Additional computational experiments are also needed to assess the size of the largest instances that may be solved with the proposed exact solution approach.

On a longer perspective, we could seek to extend the proposed valid inequalities to lot-sizing problems with returns involving complicating features such as a limited capacity, backlogging, safety stocks or minimum production levels. Finally, extending the proposed valid inequalities and the cutting-plane generation to solve the stochastic version of the problem studied in [12] is also worth investigating.

References

1. Akartunalı, K., Arulselvan, A.: Economic lot-sizing problem with remanufacturing option: complexity and algorithms. In: Pardalos, P.M., Conca, P., Giuffrida, G., Nicosia, G. (eds.) MOD 2016. LNCS, vol. 10122, pp. 132–143. Springer, Cham (2016). https://doi.org/10.1007/978-3-319-51469-7_11
2. Ali, S.A.S., Doostmohammadi, M., Akartunalı, K., van der Meer, R.: A theoretical and computational analysis of lot-sizing in remanufacturing with separate setups. Int. J. Prod. Econ. **203**, 276–285 (2018)
3. Bansal, M.: Facets for single module and multi-module capacitated lot-sizing problems without backlogging. Discret. Appl. Math. **255**, 117–141 (2019)
4. Brahimi, N., Absi, N., Dauzère-Pérès, S., Nordli, A.: Single-item dynamic lot-sizing problems: an updated survey. Eur. J. Oper. Res. **263**(3), 838–863 (2017)
5. Cunha, J.O., Konstantaras, I., Melo, R.A., Sifaleras, A.: On multi-item economic lot-sizing with remanufacturing and uncapacitated production. Appl. Math. Model. **50**, 772–780 (2017)
6. Guan, Y., Ahmed, S., Nemhauser, G.L.: Cutting planes for multistage stochastic integer programs. Oper. Res. **57**(2), 287–298 (2009)
7. Leung, J.M., Magnanti, T.L., Vachani, R.: Facets and algorithms for capacitated lot sizing. Math. Program. **45**(1), 331–359 (1989)
8. Loparic, M., Pochet, Y., Wolsey, L.A.: The uncapacitated lot-sizing problem with sales and safety stocks. Math. Program. **89**(3), 487–504 (2001)
9. Miller, A.J., Nemhauser, G.L., Savelsbergh, M.W.: On the capacitated lot-sizing and continuous 0–1 knapsack polyhedra. Eur. J. Oper. Res. **125**(2), 298–315 (2000)
10. Pochet, Y., Wolsey, L.A.: Polyhedra for lot-sizing with Wagner–whitin costs. Math. Program. **67**(1), 297–323 (1994)
11. Pochet, Y., Wolsey, L.A.: Production Planning by Mixed Integer Programming. Springer Science & Business Media (2006). https://doi.org/10.1007/0-387-33477-7
12. Quezada, F., Gicquel, C., Kedad-Sidhoum, S., Vu, D.Q.: A multi-stage stochastic integer programming approach for a multi-echelon lot-sizing problem with returns and lost sales. Comput. Oper. Res. **116**, 104865 (2020)
13. Retel Helmrich, M.J., Jans, R., van den Heuvel, W., Wagelmans, A.P.: Economic lot-sizing with remanufacturing: complexity and efficient formulations. IIE Trans. **46**(1), 67–86 (2014)

Interactive Multiobjective Optimization in Lot Sizing with Safety Stock and Safety Lead Time

Adhe Kania[1,3](✉), Juha Sipilä[2], Bekir Afsar[1], and Kaisa Miettinen[1]

[1] University of Jyvaskyla, Faculty of Information Technology, P.O. Box 35 (Agora), 40014 University of Jyvaskyla, Finland
adhe.a.kania@student.jyu.fi

[2] JAMK University of Applied Sciences, School of Technology, Jyvaskyla, Finland

[3] Institut Teknologi Bandung, Faculty of Mathematics and Natural Sciences, Jl Ganesha 10, Bandung 40132, Indonesia

Abstract. In this paper, we integrate a lot sizing problem with the problem of determining optimal values of safety stock and safety lead time. We propose a probability of product availability formula to assess the quality of safety lead time and a multiobjective optimization model as an integrated lot sizing problem. In the proposed model, we optimize six objectives simultaneously: minimizing purchasing cost, ordering cost, holding cost and, at the same time, maximizing cycle service level, probability of product availability and inventory turnover. To present the applicability of the proposed model, we consider a real case study with data from a manufacturing company and apply the interactive NAUTILUS Navigator method to support the decision maker from the company to find his most preferred solution. In this way, we demonstrate how the decision maker navigates without having to trade-off among the conflicting objectives and could find a solution that reflects his preference well.

Keywords: Inventory management · Uncertain demand · Uncertain lead time · Interactive decision making · NAUTILUS Navigator

1 Introduction

Lot sizing has emerged as one of the key factors for the effective supply chain management. The purpose of lot sizing is to determine an optimal order quantity that minimizes costs while satisfying demand. After Harris's economic order quantity concept for solving a simple lot sizing problem [10], there has been a dramatic increase in interest over the last century in developing lot sizing models to adapt to more complex situations [1,8].

Uncertainties complicate lot sizing problems. In fact, predicting the exact demand for future needs is challenging. Commonly, many companies hold a certain amount of stock, known as a safety stock (SS), as a buffer to cope when

M. Mes et al. (Eds.): ICCL 2021, LNCS 13004, pp. 208–221, 2021.
https://doi.org/10.1007/978-3-030-87672-2_14

demand exceeds the prediction [31]. Another source of uncertainty is the delivery lead time [24]. Companies usually have an agreement with their suppliers for the delivery time, but for many reasons, there can be delays. To overcome this issue, an additional time period, known as a safety lead time (SLT), is defined [31]. During the SLT period, companies keep their stocks available to satisfy the demand.

The problem of determining an optimal SS value has been studied by many researchers [9]. Various methods have been developed [26] to find an optimal value of SS that should be small enough to reduce costs while satisfying demand and guaranteeing a high service level. Most studies expand the cycle service level (CSL) formula [23] to adapt to various conditions. When lead time is uncertain, the CSL formula takes into account the average and standard deviation of the lead time [28]. On the other hand, the problem of finding an optimal SLT value is not as popular as the previous one [7]. In [12], inventory costs are minimized subject to a service level constraint to find an optimal SLT, and an optimization model based on Markov Chain is proposed in [6]. However, there is a lack of formula to control the quality of SLT.

The relationship between lot sizing problems with SS and SLT has been studied in [22]. Keeping stock for SS and SLT increases order quantity, which also increases the costs. Some researchers have studied lot sizing problems with uncertainty on demand and lead time [7]. However, they mostly use statistical tools to handle uncertainty in lot sizing models, but not simultaneously find SS or SLT. Some of them use simulation to find an optimal SS and SLT. There is a lack of integration of a lot sizing problem and problems of determining SS and SLT values in the literature. The problem of integrating lot sizing and SS determination is proposed in [18], but they consider SLT as the input value.

Lot sizing problems naturally include a conflict between minimizing costs and satisfying demand simultaneously. Additional problems of determining SS and SLT increase the conflict because holding more stock for SS and SLT makes the costs higher. For this reason, multiobjective optimization [19] is a good tool to solve lot sizing problems [2]. A multiobjective optimization problem has several mathematically incomparable solutions, called Pareto optimal solutions. Solving a multiobjective optimization problem can be understood as finding the most preferred solution for a decision maker (DM), who has expertise in the problem domain. Interactive methods [20] are regarded as promising because the solution process is iterative and they allow the DM to gain insight into the problem and change his/her preferences during the solution process, thanks to learning. So far, however, there have been a few studies applying interactive multiobjective optimization in lot sizing problems [29].

In this research, we consider a single item multi period lot sizing problem with uncertainty on demand and lead time. The main contributions of this paper are threefold. First, we propose a novel formula, named probability of product availability (PPA), for measuring the quality of SLT to handle unpredicted lead time. Second, we develop a multiobjective optimization model that determines the optimal lot sizes for each period and simultaneously finds the optimal values of SS and SLT. Last but not least, we support a DM to find the most preferred solution for the optimization model by applying an interactive NAUTILUS Navigator method [25].

The proposed multiobjective optimization model has six objectives to optimize simultaneously. Three of them are minimizing cost functions, i.e. purchasing cost (PC), ordering cost (OC), and holding cost (HC). We consider them separately to see trade-offs between objectives. The CSL is maximized to improve safety against demand uncertainty. We propose a PPA formula to assess the quality of SLT to buffer lead time uncertainty, which is maximized in the model. Lastly, inventory turnover (ITO) as the primary performance indicator for inventory management [27] is maximized to measure the effectiveness of this model in handling the inventory system.

Most lot sizing problems are difficult to solve because of their complexity [14]. In this paper, we use the interactive NAUTILUS Navigator method [25]. The strength of this method in handling computationally expensive problems meets the need of this kind of problem. Another strength of this method is allowing the DM to find his/her most preferred solution without sacrifices, which meets the needs of the DM. In this, the strategy is starting from a bad point and improving all objectives simultaneously. We use real data from a manufacturing company and a real DM to prove the validity of our proposed model. Finally, we support the DM to find the most satisfying solution for him by using this method.

The remainder of the paper is organized as follows. Section 2 reviews the basic concepts of multiobjective optimization and the NAUTILUS Navigator method. Then, the proposed multiobjective optimization model is presented in Sect. 3. In Sect. 4, the case study together with the real data from a manufacturing company is described, following by results and analysis of the decision making process using NAUTILUS Navigator. Finally, conclusions and discussions of possible extensions are presented.

2 Background on Multiobjective Optimization

In this section, we briefly review the basic concepts and definitions related to multiobjective optimization, followed by the NAUTILUS Navigator method.

2.1 Basic Concepts and Definitions

A multiobjective optimization problem can be formulated in the following form:

$$\begin{aligned} \text{minimize} \quad & \boldsymbol{f}(\boldsymbol{x}) = (f_1(\boldsymbol{x}), ..., f_k(\boldsymbol{x}))^T \\ \text{subject to} \quad & \boldsymbol{x} \in S, \end{aligned} \tag{1}$$

where $k \geq 2$ objective functions, $f_i : S \rightarrow \mathbb{R}$ for $1 \leq i \leq k$, are simultaneously optimized. The vector of decision variables $\boldsymbol{x} = (x_1, ..., x_n)^T$ belongs to the feasible region $S \subset \mathbb{R}^n$, which is formed by constraints. The image of the feasible region $Z = \boldsymbol{f}(S), Z \subset \mathbb{R}^k$ is called a feasible objective region, which is formed by the vectors of objective values $\boldsymbol{z} = \boldsymbol{f}(\boldsymbol{x}) = (f_1(\boldsymbol{x}), ..., f_k(\boldsymbol{x}))^T$, $\boldsymbol{z} \in Z, \boldsymbol{x} \in S$.

Because of the conflicting objectives, a multiobjective optimization problem (1) has several different solutions, called Pareto optimal solutions, which reflect the trade-offs among the conflicting objectives. A solution $\boldsymbol{z}^1 \in Z$ is said to dominate another solution $\boldsymbol{z}^2 \in Z$ if $z_i^1 \leq z_i^2$ for all $i = 1, ..., k$ and $z_j^1 < z_j^2$ for at least one $j = 1, ..., k$. A solution $\boldsymbol{z} \in Z$ is called a Pareto optimal solution if $\boldsymbol{z}$ is not dominated by any other solution. The lower and upper bounds of the Pareto optimal solutions are called an ideal point $\boldsymbol{z}^*$ and a nadir point $\boldsymbol{z}^{nad}$, respectively, which reflect the best and the worst values that each objective function in the Pareto optimal solutions can achieve.

Pareto optimal solutions are incomparable mathematically. Additional preference information from a DM is needed to identify the most preferred solution as the final solution. A DM is an expert who has a responsibility to make a decision in the problem domain, who is usually a supply chain manager in lot sizing. The preference information from the DM can be incorporated before the optimization process (a priori methods), after having generated a representative set of Pareto optimal solutions (a posteriori methods), or during an iterative optimization process (interactive methods) [19]. The advantages of interactive methods, which allow the DM to learn different aspects of the problem during the solution process and change their preferences during the solution process if desired, are the main reasons we chose this type of methods. Many interactive methods have been developed [20]. In this paper, we apply the NAUTILUS Navigator method [25] because of its ability in handling computationally expensive problems and the possibility to find the most preferred solution without trading-off. This is important since DMs sometimes get anchored around the initial solution and a trade-off free method avoids anchoring.

2.2 NAUTILUS Navigator

The NAUTILUS Navigator method combines the idea of NAUTILUS methods [21] to avoid trading-off and navigation ideas elaborated in [11]. Due to the fact that people do not respond similarly to losses and gains [15], trading-off among Pareto optimal solutions causes some decisional stress to the DM [17]. Motivated by this fact, NAUTILUS methods start from the worst possible objective function values and iteratively gain in all objectives without sacrificing any of the current values. Methods in the NAUTILUS family [21] differ in the way used to interact with the DM to find the final solution. NAUTILUS Navigator uses navigation to direct the movement from the worst starting point, which is the nadir point or any undesirable point provided by the DM, to a Pareto optimal solution as the final solution. In this process, the DM specifies a desirable value for each objective function, which are the components of a reference point, as a search direction to direct the movement towards desired Pareto optimal solutions. During the navigation process, the DM can change the reference point, the movement speed, or even go backwards if he/she wishes so.

To handle computationally expensive problems, a set of Pareto optimal solutions is generated before the interactive process starts. The generation may take time because of expensive functions, but it is done without involving the DM.

Any a posteriori methods can be used to generate a set of Pareto optimal solutions or a set that approximates Pareto optimal solutions. When involving the DM, the navigation process takes place using this set without solving the original computationally expensive problem. This allows showing real-time movement without waiting times to the DM. The detailed algorithm can be seen in [25].

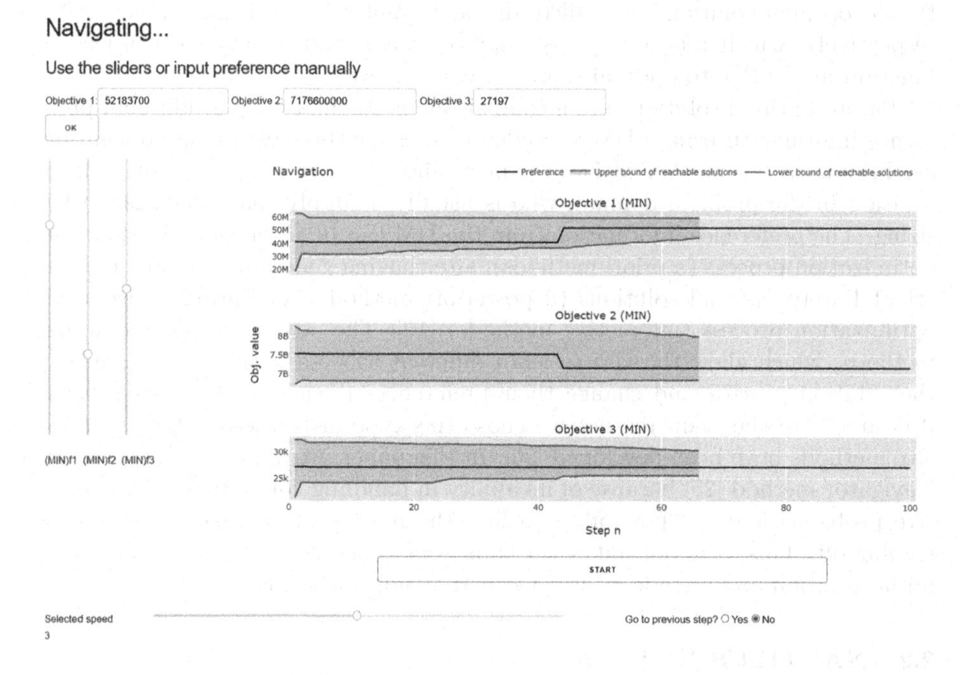

Fig. 1. GUI of the NAUTILUS Navigator method

A graphical user interface (GUI) is important for NAUTILUS Navigator to visualize the navigation process. Figure 1 shows the available GUI that can be freely downloaded from https://desdeo.it.jyu.fi. The DM provides his/her preferences using the sliders on the left side or inputs values in text boxes at the top. The green area in the graph shows the reachable ranges, which are the best and the worst objective function values, that each objective can reach from the current step without sacrifices in any objectives. Thus, the reachable ranges shrink when approaching Pareto optimal solutions. Whenever the DM wants to change his/her preference, he/she can stop the process and change the reference point. The black lines in the middle of the graphs show the positions of the components of the reference point. The DM is allowed to jump to any previous step using the radio button in the bottom right. He/she then needs to provide which step to go to and re-specify his/her preferences in order to define a new direction. The DM can navigate until he/she finds his/her most preferred Pareto optimal solution at the end of the solution process. In that case, the ranges shrink to a single point.

3 Problem Formulation

We study a lot sizing problem for a single item with a single supplier and in multiple time periods. We follow a periodic review policy, where orders are reviewed over discrete time periods $t = 1, ..., T$. The order quantity ($Q(t)$) is reviewed at the beginning of period t, and the order arrives after a stochastic lead time. The following assumptions are made throughout this paper.

1. The predicted demand during period t ($D(t)$) follows a normal distribution with a mean μ and a standard deviation σ. The demand in each period is independent of other periods.
2. The lead time follows a normal distribution with a mean L and a standard deviation s.
3. The price for purchasing one unit of item (p) is constant in all time periods and does not depend on the order quantity.
4. The cost for a single order is c without any capacity limit.
5. The cost of holding one unit of item (h) is constant throughout all time periods.
6. There is no backorder cost involved.
7. There is an agreement between the company and the supplier that the company must order with a minimum order quantity moq and it rounds up by a rounding value r. Therefore, the order can only be placed by following the formula $moq + a\,r$ for any integer $a \geq 0$.

3.1 Safety Stock and Safety Lead Time Formulation

As said, we focus on the lot sizing problem with uncertainty in demand and lead time. Many researchers have utilized a SS to protect against demand uncertainty and a SLT to handle lead time uncertainty [16]. A SS means keeping more stocks as a buffer against demand fluctuations. To control the amount of SS, the cycle service level (CSL) formula is applied [4]. CSL is the probability of not hitting a stockout in a replenishment time (RT). A RT is a time needed to refill the stock, that is from the arrival of one order to the arrival of the next one. We set $RT = 1 + SLT$ since we order in each period and prepare for late delivery in the SLT period. To prevent stockout during a RT, the difference between an actual demand (D^*_{RT}) and a predicted demand (D_{RT}) must be less than SS. We adopt the CSL formula for demand and lead time uncertainty [28] with our definition of RT, which can be written as follows:

$$\begin{aligned} CSL &= P(D^*_{RT} \leq D_{RT} + SS) \\ &= F(D_{RT} + SS, D_{RT}, \sigma_{RT}) = F\left(\frac{SS}{\sigma_{RT}}\right), \end{aligned} \tag{2}$$

where F is the standard normal distribution function and σ_{RT} is the standard deviation of demand during a RT, which can be formulated as $\sigma_{RT} = \sqrt{\sigma^2(1 + SLT) + \mu^2 s^2}$.

A SLT is assigned to handle unpredicted lead time. During the SLT period, the availability of stock to cover predicted and unpredicted demand must be guaranteed. Therefore, we consider an additional SLT period in the fill rate (FR) constraint to secure the availability of the stock during SLT to cover the predicted demand. A SLT period is also considered in CSL to buffer unpredicted demand during SLT. However, if the order arrives after the SLT period, the stockout may occur. Therefore, it is important to decide an optimal SLT value with a low possibility of having stockout. In this paper, we propose the probability of product availability (PPA) formula to measure the quality of SLT. PPA is defined as the probability of not having stockout because of the late delivery, which occurs when the actual order arrives during the period $L+SLT$. The PPA formula can be written as follows:

$$\begin{aligned} PPA &= P(\text{actual delivery time} \leq L + SLT) \\ &= F(L + SLT, L, s) = F\left(\frac{SLT}{s}\right). \end{aligned} \tag{3}$$

This formula can be used to find the SLT value by defining an appropriate PPA level.

3.2 Multiobjective Optimization Model

As said, we propose a multiobjective optimization model with six objectives, three to minimize and three to maximize. The main goal of this model is to find the order quantity of each period $(Q(t), t = 1, ..., t_n)$ together with SS and SLT values with the best balance between the objective functions. We define $I(t)$ as the inventory level at the end of period t where $I(t) = I(t-1)+Q(t-\lfloor L \rfloor)-D(t)$, and $Y(t)$ as the order indicator where $Y(t) = 1$ if the order is placed $(Q(t) > 0)$, otherwise $Y(t) = 0$. The proposed optimization model can be written as follows.

$$\text{min} \quad PC = \sum_t Q(t)\, p,\ OC = \sum_t Y(t)\, c,\ HC = \sum_t \frac{I(t-1) + I(t)}{2}\, h,$$

$$\text{max} \quad CSL\ (2),\ PPA\ (3),\ ITO = \sum_t \frac{D(t)}{(I(t-1) + I(t))/2},$$

$$\text{s.t.} \quad \frac{I(t-1) + \sum_{i=t-\lfloor L \rfloor}^{t} Q(i) - SS}{\sum_{j=t}^{t+\lfloor P \rfloor} D(j) + (P - \lfloor P \rfloor) D(\lceil P \rceil)} \geq 1, \text{ for } t = 1, ..., T, \tag{4}$$

$$Q(t) = Y(t)\,(moq + a\,r)\ , \text{ for any integer } a \geq 0 \text{ and } t = 1, ..., T, \tag{5}$$

$$SS \geq 0 \text{ and } SOT \geq 0. \tag{6}$$

Following the dynamic lot sizing problem [14,30], three types of cost are considered: PC, OC, and HC. We consider them separately to see the trade-offs. Minimizing PC implies minimizing HC, but OC has a trade-off with HC because ordering the same amounts of items many times makes OC higher and HC lower. For inventory management purposes, it is important to understand both HC and OC. In order to prevent partial optimization, which could be the case if only total costs were measured, it is important to separate them. When targeting at low HC only, one can be misled, as then there could be a temptation to order more often, resulting in higher OC.

We maximize CSL to prevent stockout because of the demand uncertainty and maximize PPA to avoid stockout due to late delivery. Keeping a high value of SS raises the CSL but PC and HC increase, which is a conflict as we need to maximize CSL but minimize OC and HC. Having a long SLT increases the PPA but decreases CSL with the same SS value. Then PPA has a trade-off with CSL, PC and HC. Maximizing ITO is our last objective function. To have a high ITO, the order must be as close to the demand as possible in order to hold less stock, which has a trade-off with OC. Furthermore, ITO has a trade-off with CSL and PPA as less stock is needed to have a high ITO, but CSL and PPA need more stock to have better safety in handling uncertainties.

FR represents customer service for an inventory control system. It is defined as the fraction of orders that are filled from stock [13]. It is an important indicator in daily operations. In the proposed model, FR is the first constraint (4) to fulfill the predicted demand. In each period, we guarantee that our stock (excluding SS) can satisfy the predicted demand. The consideration period for one order (P) in the periodic review policy is $1 + L$ [4], but an additional SLT period is also considered to ensure the stock availability during SLT. Thus, we set $P = 1 + L + SLT$. FR is a fraction between available stock without SS and the predicted demand during P. When FR is at least one, the stock availability to handle the predicted demand is guaranteed. Furthermore, we ensure that all orders follow the agreement of minimum order quantity and rounding value in constraint (5), while constraint (6) is defined to confine the lower bounds of SS and SLT.

4 Computational Results

We consider a case study from a manufacturing company to demonstrate the applicability of the proposed model. We apply the interactive NAUTILUS Navigator method to support the supply chain manager of the said company, acting as the DM, to find his most preferred solution without trading-off.

4.1 Information About the Case

We review a weekly single item lot sizing problem for 41 weeks. Thus, the optimization model has 43 integer decision variables, including weekly order quantities, SS and SLT. We received data of an item, which is a component of the company's product. The data is generated from the company's planning system. The data contains current inventory information for the item as well as a consumption projection according to the company's production plan. Based on the data, the price to purchase one unit of the item is €91.18, the cost for a single order is €200, and the cost of holding one unit of item is ten percent of the price annually. The lead time for this item is 6 weeks, with a standard deviation $s = 0.93$ days. The company has made a prediction for the weekly demand data based on its historical data, which varies with a mean $\mu = 116.22$ and a standard deviation $\sigma = 29.04$. The opening inventory is 312 units and the company has made previous orders for the next six weeks, which are $(48, 119, 120, 120, 48, 96)$. Based on the agreement between the company and the supplier, the company must place an order with a minimum of 48 units and round by 48 units.

As a request from the DM, bounds for SS and SLT were defined as additional constraints. The DM was only interested in SS values lower than μ and SLT values below four days. He also requested to see at least one day SLT or one day's worth of demand for SS, which is $\mu/5$. Furthermore, low ITO values below ten were not interesting for the DM.

As said, a GUI plays an important role in NAUTILUS Navigator. A few modifications of the available GUI were done in this research to make the GUI more useful for the DM in this case. The DM preferred to see the probability of product unavailability (PPU) rather than PPA. Thus, we switched to minimize $PPU = 1 - PPA$ in the fifth objective. Furthermore, the DM wanted to see the information of days of stock (DoS). DoS is an inventory performance indicator describing the number of days needed to sell an item. DoS is calculated as the number of days in one year (we use 254 working days) divided by ITO.

4.2 Computational Results

As described in Sect. 2.2, the starting point of the NAUTILUS Navigator method is a set of pre-generated solutions. As said, lot sizing problems are computationally expensive problems. Because of their complexity, many researchers use metaheuristic methods, like evolutionary algorithms, to solve various problems of lot sizing [14]. In this paper, we applied NSGA-III [5] by using the pymoo framework [3] because of its ability to solve constrained multiobjective optimization problems with integer variables. Evolutionary algorithms cannot guarantee Pareto optimality but can generate sets of solutions where no solution dominates the others.

Some strategies were needed to generate a large amount of nondominated solutions. Because a single run of NSGA-III was not able to generate enough solutions, we ran the algorithm several times with different initial populations. Furthermore, to get more solutions, various parameters of evolutionary operators were used that were available in the framework. Finally, all solutions were combined, dominated solutions were deleted, and 1503 nondominated solutions were obtained that approximate Pareto optimal solutions.

The DM started the navigation process by investigating the reachable ranges for the first step, which were represented by the ideal point and the nadir point initially derived from the set. With the bounds defined by the DM, the ideal point was $\boldsymbol{z}^* = (358\ 884.48,\ 1\ 000,\ 674.73,\ 0.9945,\ 0,\ 97.45)$ and the nadir point was $\boldsymbol{z}^{nad} = (367\ 637.76,\ 6\ 800,\ 4\ 782.04,\ 0.5,\ 0.5,\ 10.19)$ (remember that the fourth and sixth objectives are to be maximized and the others are to be minimized). Initially, the DM wanted to set the ideal point as the reference point to investigate how the navigation ran and which Pareto optimal solutions can be found if he wanted all the objectives to navigate towards their best values.

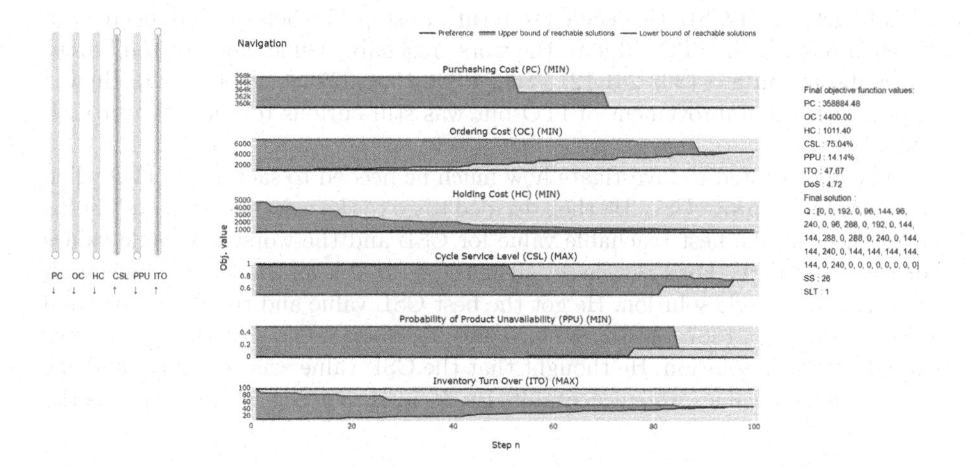

Fig. 2. A Pareto optimal solution for the ideal point as the reference point

Because of the trade-offs among the objectives, getting the best possible values for all objectives is naturally impossible, but, the DM navigated till the Pareto optimal solution $z = (358\,884.48, 4\,400, 1\,011.40, 0.7504, 0.1414, 47.67)$ was reached. Thus, the reachable range was finally a single point. Figure 2 shows this navigation. The DM analyzed that, in step 52, there was a significant decrease of the upper bound for the reachable CSL values to 0.8116, and the ITO reachable range shrunk with the upper bound 59.53. Because of this, the DM decided to go backwards to step 50 and provided new preferences.

The DM wanted to keep the ITO in the best value at this step, which was 59.53. He then set the components of the reference point for PC and OC to their worst values, and keep the other components as their best reachable values at this step. Therefore, the new reference point was (367 637.76, 6 800, 901.98, 0.9835, 0, 59.53). He let the navigation continue until the end to check the Pareto optimal solution that could be reached. The Pareto optimal solution obtained was $z = (363\,261.12, 6\,000, 1\,108.19, 0.8437, 0.0159, 39.25)$. He found the CSL value better but it was not satisfactory enough for him. He learned that the upper bound of the CSL's reachable values started to decrease at step 80. He then decided to return to this step to set a new reference point.

The DM navigated with different desired values of ITO to observe how much he needed to sacrifice in ITO to get better values for CSL. He returned to step 80 a few times with different desired values for ITO, but he only got 0.9041 as the best value for CSL. He decided to go further backwards to step 16 because the upper bound of ITO and HC in reachable values had a significant decrease after this step. He set all cost objectives in their worst reachable values, CSL and PPU in their best reachable values, and ITO=48. He let the reachable ranges shrink till the Pareto optimal solution $\boldsymbol{z} = (363\,261.12, 6\,400, 1\,183.94, 0.9366, 0.0159, 35.68)$. The DM found that the CSL value was not satisfactory enough.

The DM realized that CSL had a trade-off with PPU, and he needed to relax PPU to get better CSL. He decided to return to step 75 when the CSL decreased. He then relaxed the ITO value to the worst reachable value, and got the Pareto optimal solution $\boldsymbol{z} = (363\,261.12, 5\,800, 1\,066.10, 0.9272, 0.1414, 42.69)$. He was happy with the improvement of ITO but was still curious to find a better CSL value.

The DM wanted to investigate how much he needed to sacrifice in ITO when he desired to improve CSL. He then decided to go to the very first step and set his preferences at the best reachable value for CSL and the worst reachable values for costs and PPU. For ITO, he set 40 as the desired level. He let the navigation converge to a single solution. He got the best CSL value and the Pareto optimal solution was $\boldsymbol{z} = (367\,637.76, 5\,800, 1\,061.90, 0.9945, 0.5, 42.94)$. He was very happy with this solution. He thought that the CSL value was very good and the other objective values were acceptable. He decided to accept this solution as the final one.

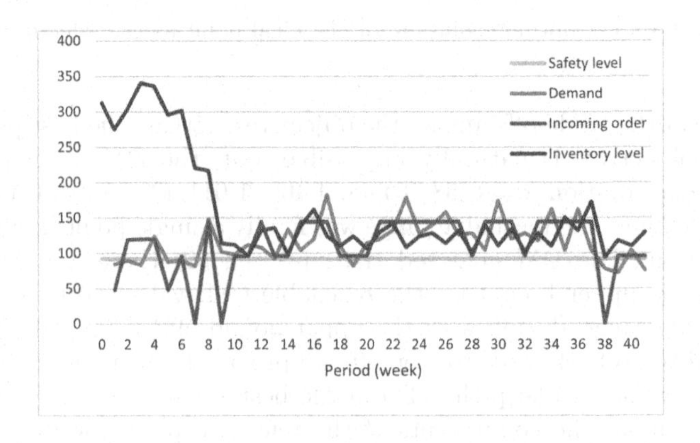

Fig. 3. The decision variables corresponding to the final solution

The decision variables corresponding to the final solution for order quantities can be seen in Fig. 3. The other decision variables were $SS = 92$ and $SLT = 0$. The green line in Fig. 3 shows the incoming order quantities for each week, which are the previously set order data for $t = 1, ..., 6$ and the optimized order quantities $Q(t - L)$ for $t = 7, ..., 41$. The inventory level in the blue line shows that during the first six weeks, which cannot be controlled by the model due to the lead time, the company had excess inventory. The inventory level then decreased and followed the demand quantity to have a higher ITO, which is a useful indicator for inventory management and planning purposes.

By deepening his understanding of the interdependencies between conflicting objectives, the DM learned a lot from his own area of responsibility as a supply chain manager and also gained the confidence to modify his original preferences. At first, he was not willing to sacrifice on any objectives, but during the decision making process, there was a growing awareness that not everything can be achieved, but sacrifices have to be made. These included, among other things, the CSL and ITO. However, in his day-to-day operations, ITO is a goal set by the company's top management. Therefore, deviating from this objective must be strongly justified to the management.

As a result of the learning process, the DM gained confidence in setting his preferences, and thus multiobjective optimization and NAUTILUS Navigator supported his understanding and ability to justify his decisions. The DM greatly appreciated the fact that as the decision making process progressed, he constantly saw the navigator's results and understanding of achieving objectives, which guided him in setting his preferences. The possibility to stop the process at any time and the feature to go backwards in the navigator, were, in his view, excellent opportunities to make decisions easily. The GUI of the navigation and the real-time updating of the results also supported his decision making. The navigator graphs and the sliders for setting the reference point were, in the DM's view, a clear advantage in support of decision making. The whole process was so instructive and professionally useful.

As can be seen in Fig. 3, the inventory level was significantly reduced from its original level. The DM commented that this is a typical example of decisions being made in the past "for the sake of certainty", where typically stock levels tend to rise. NAUTILUS Navigator as a method responded precisely to the need for decisions to be based on calculations rather than assumptions. The DM was pleased with the result of the objective function values, as well as the corresponding decision variables. Overall, the DM was satisfied with the results and operation of NAUTILUS Navigator and found an interactive method very suitable for learning. He is willing to adopt the method more widely for inventory planning and control, especially for critical items.

5 Conclusions

In this paper, we considered a single item multi period lot sizing problem in a periodic review policy under a stochastic environment on demand and lead time. We used a SS to handle uncertainty on demand and CSL to measure the quality of SS. To handle uncertainty on lead time, a SLT was used and we proposed the PPA formula to measure the quality of SLT. The aim of this paper was to integrate the lot sizing problem with the problem of determining the optimal values of SS and SLT. We developed a multiobjective optimization model to solve the integrated lot sizing problem. Six objectives were optimized simultaneously to find the optimal order quantity in each period and at the same time determine the optimal values of SS and SLT.

Real data from a manufacturing company was used to demonstrate the applicability and usefulness of the proposed model. A supply chain manager from the said company acted as the DM to draw managerial insights into the decision making process. The interactive NAUTILUS Navigator method was successfully applied to solve our integrated computationally expensive lot sizing problem. The DM appreciated the navigation process that allowed him to learn during the decision making process and find the most satisfying solution for him. He confirmed the validity of the solution and found it useful for his daily operation.

For future research, considering many items would present more computational challenges but meet the needs of real industrial problems. A company may have thousands of items that are impossible to consider separately. Another possible future research topic is to address the variation of price based on the order quantity, or integrating the model with the problem of determining minimum order quantity and rounding value.

Acknowledgements. This research was partly funded by LPDP, the Indonesian Endowment Fund for Education (grant number S-5302/LPDP.4/2020), and the Academy of Finland (grants 322221 and 311877). The research is related to the thematic research area DEMO (Decision Analytics utilizing Causal Models and Multiobjective Optimization, jyu.fi/demo) of the University of Jyväskylä.

References

1. Andriolo, A., Battini, D., Grubbström, R.W., Persona, A., Sgarbossa, F.: A century of evolution from Harris's basic lot size model: survey and research agenda. Int. J. Prod. Econ. **155**, 16–38 (2014)
2. Aslam, T., Amos, H.C.N.: Multi-objective optimization for supply chain management: a literature review and new development. In: 8th International Conference on Supply Chain Management and Informatio, pp. 1–8. IEEE (2010)
3. Blank, J., Deb, K.: Pymoo: multi-objective optimization in Python. IEEE Access **8**, 89497–89509 (2020)
4. Chopra, S., Meindl, P.: Supply chain management: strategy, planning, and operation. Pearson, 6 edn. (2016)
5. Deb, K., Jain, H.: An evolutionary many-objective optimization algorithm using reference-point-based nondominated sorting approach, Part I: Solving problems with box constraints. IEEE Trans. Evol. Comput. **18**(4), 577–601 (2014)
6. Dolgui, A., Ould-Louly, M.A.: A model for supply planning under lead time uncertainty. Int. J. Prod. Econ. **78**(2), 145–152 (2002)
7. Dolgui, A., Prodhon, C.: Supply planning under uncertainties in MRP environments: a state of the art. Annu. Rev. Control. **31**(2), 269–279 (2007)
8. Glock, C.H., Grosse, E.H., Ries, J.M.: The lot sizing problem: a tertiary study. Int. J. Prod. Econ. **155**, 39–51 (2014)
9. Gonçalves, J.N., Sameiro Carvalho, M., Cortez, P.: Operations research models and methods for safety stock determination: a review. Oper. Res. Perspectives **7**, 100164 (2020)
10. Harris, F.W.: How many parts to make at once. Factory, The Magazine of Management **10**, 135–136 (1913)
11. Hartikainen, M., Miettinen, K., Klamroth, K.: Interactive Nonconvex Pareto Navigator for multiobjective optimization. Eur. J. Oper. Res. **275**(1), 238–251 (2019)

12. Hegedus, M.G., Hopp, W.J.: Setting procurement safety lead-times for assembly systems. Int. J. Prod. Res. **39**(15), 3459–3478 (2001)
13. Hopp, W.J., Spearman, M.L.: Factory Physics. Waveland Press Inc, 3 edn. (2008)
14. Jans, R., Degraeve, Z.: Meta-heuristics for dynamic lot sizing: A review and comparison of solution approaches. Eur. J. Oper. Res. **177**(3), 1855–1875 (2007)
15. Kahneman, D., Tversky, A.: Prospect theory: an analysis of decision under risk. Econometrica **47**(2), 263–291 (1979)
16. van Kampen, T.J., van Donk, D.P., van der Zee, D.J.: Safety stock or safety lead time: coping with unreliability in demand and supply. Int. J. Prod. Res. **48**(24), 7463–7481 (2010)
17. Korhonen, P., Wallenius, J.: Behavioural issues in MCDM: neglected research questions. J. Multi-Criteria Decision Anal. **5**(3), 178–182 (1996)
18. Kumar, K., Aouam, T.: Integrated lot sizing and safety stock placement in a network of production facilities. Int. J. Prod. Econ. **195**, 74–95 (2018)
19. Miettinen, K.: Nonlinear Multiobjective Optimization. Kluwer Academic Publishers (1999)
20. Miettinen, K., Hakanen, J., Podkopaev, D.: Interactive nonlinear multiobjective optimization methods. In: Greco, S., Ehrgott, M., Figueira, J.R. (eds.) Multiple Criteria Decision Analysis. ISORMS, vol. 233, pp. 927–976. Springer, New York (2016)
21. Miettinen, K., Ruiz, F.: NAUTILUS framework: towards trade-off-free interaction in multiobjective optimization. J. Bus. Econ. **86**(1), 5–21 (2016)
22. Molinder, A.: Joint optimization of lot-sizes, safety stocks and safety lead times in an MRP system. Int. J. Prod. Res. **35**(4), 983–994 (1997)
23. New, C.: Safety stocks for requirements planning. Prod. Invent. Manag. **12**, 1–18 (1975)
24. Pahl, J., Voß, S., Woodruff, D.L.: Production planning with load dependent lead times: an update of research. Ann. Oper. Res. **153**, 297–345 (2007)
25. Ruiz, A.B., Ruiz, F., Miettinen, K., Delgado-Antequera, L., Ojalehto, V.: NAUTILUS Navigator: free search interactive multiobjective optimization without trading-off. J. Global Optim. **74**(2), 213–231 (2019)
26. Schmidt, M., Hartmann, W., Nyhuis, P.: Simulation based comparison of safety-stock calculation methods. CIRP Ann. **61**(1), 403–406 (2012)
27. Silver, E.A., Pyke, D.F., Thomas, D.J.: Inventory and production management in supply chains, 4 edn. CRC Press (2017)
28. Talluri, S., Cetin, K., Gardner, A.J.: Integrating demand and supply variability into safety stock evaluations. Int. J. Phys. Distrib. Logist. Manag. **34**(1), 62–69 (2004)
29. Torabi, S., Hassini, E.: An interactive possibilistic programming approach for multiple objective supply chain master planning. Fuzzy Sets Syst. **159**(2), 193–214 (2008)
30. Wagner, H.M., Whitin, T.M.: Dynamic version of the economic lot size model. Manage. Sci. **5**, 89–96 (1958)
31. Whybark, D.C., Williams, J.G.: Material requirements planning under uncertainty. Decision Sci. **7**(4), 595–606 (1976)

The Craft Beer Game and the Value of Information Sharing

Joshua Grassel[1], Alfred Craig Keller[1], Alessandro Hill[1(✉)], and Frederik Schulte[2]

[1] California Polytechnic State University, San Luis Obispo, CA 93407, USA
{jtgrasse,ackeller,ahill29}@calpoly.edu
[2] Delft University of Technology, Mekelweg 2, 2628 CD Delft, Netherlands
f.schulte@tudelft.nl

Abstract. The craft beer supply chain in the USA differs from the supply chain of macro breweries in its structure, handled volumes and product shelf-life. In this work, we study how these smaller craft breweries can benefit from transparency in their supply chain. We consider additional information sharing of orders and inventories at downstream nodes. The levels that we investigate grant the brewery incremental access to distributor, wholesaler, and retailer data. We show how this knowledge can be incorporated effectively into the brewery's production planning strategy. Extending the well-known beer game, we conduct a simulation study using real-world craft beer supply chain parameters and demand. We quantify the impact of information sharing on the craft brewery's sales, spoilage, and beer quality. Our model is designed to directly support the brewery when evaluating the value of downstream information and negotiating data purchases with brokers. Through a computational analysis, we show that the brewery's benefits increase almost linearly with every downstream node that it gets data from. Full transparency allows to halve the missed beer sales, and beer spoilage can even be reduced by 70% on average.

Keywords: Craft beer industry · Supply chain management · Information sharing · Production planning · Simulation

1 Introduction

The classic beer game of supply chain management (SCM) has received significant attention for demonstrating the bullwhip effect and the value of vertical collaboration in supply chains. With the rising popularity of craft beers, the craft beer game naturally emerges as a variant of the classic game with slightly altered rules and different insights to be gained for SCM (education). In contrast to the supply chains of the large breweries and many other supply chains, information in craft beer supply chains (CBSCs) is mostly not shared directly between the supply chain members but bought from data brokers [6]. Thus, the player of the

M. Mes et al. (Eds.): ICCL 2021, LNCS 13004, pp. 222–236, 2021.
https://doi.org/10.1007/978-3-030-87672-2_15

craft beer game has to take two decisions: how many units of stock to order and how much information to buy.

This, however, involves understanding a trade-off between the cost and the value of shared information. As an integral element of self-thinking supply chains [4], information sharing and its value for supply chains have been investigated considering different scenarios. Lee et al. [11], for instance, investigated the case of a simple two-stage supply chain and found that the manufacturer could reduce inventory costs with information sharing. Ouyang [15], on the other hand, showed that information sharing could improve the supply chain stability and mitigate the bullwhip effect. Nevertheless, the value of information sharing in CBSCs has not yet been investigated, and the impact of information sharing on important metrics of the sector such as average beer age, spoilage, and missed sales is not yet well understood. Existing CBSC literature rather focuses on sustainability factors of the supply chain [1,5] or develops models for the inherent supply uncertainty in these supply chains [19]. A general overview on the craft beer industry is given by Biano [2] considering the special craft requirements in terms of food quality. Further works on craft beer address, amongst other aspects, corporate social responsibility [9], sustainability objectives [8,14], economies of scale [20], and craft beer as a means of economic developments [12].

In this work, we model different levels of information sharing in the craft beer supply chain and develop a simulation approach based on real-world data to understand the value of information sharing concerning the CBSC metrics, average beer age, spoilage (in total and at each stage of the supply chain), and missed sales. We observe that all of the above metrics are significantly reduced when more information becomes available, and with more information, this reduction becomes more significant. Moreover, we find that the variability in the results over different scenarios is also clearly lowered with more information sharing. These findings and the presented approach generally help to understand the potential of information sharing platforms in CBSCs and related domains. On top of that, they help to improve the product quality in the considered supply chains.

2 The Craft Beer Supply Chain

The origins of beer can be traced as far back as 7000 BCE and the recipe has basically been the same to this day: water, malt, hops and yeast. While the ingredients have stayed constant over thousands of years, the US American Beer Industry has been relatively short with some notable occurrences have altered demand, manufacturing and supply chain operations. Most recently, the resurrection of US American Craft Breweries has changed the way that beer products must be managed within the supply chain. While most supply chain analyses are done under the model assumptions developed in the Beer Game [7], this model does not consider modern quality standards and beer style preferences.

We consider the Craft Beer Supply Chain (CBSC) in the USA. The CBSC is used by so-called craft breweries to supply end customers with craft beer.

The majority of the US beer industry is controlled by a small amount of macro-breweries, or breweries that produce over 6,000,000 barrels (9.5 MM hL) of beer, controlling over 87% of the market in 2020[1]. On the other side, there are thousands of craft breweries that provide the remaining volume[2]. Even though the craft beer segment has seen dramatic growth over the past 30 years, it still only controls a small portion of the market and has a strikingly different relationship with the other agents in the supply chain. The US American beer supply chain, also called the *Three Tier System* [13], consists of four echelons, and is illustrated in Fig. 1. The Three Tier System dates back to the 1920s and refers to a law that enforces breweries to sell their product to a distributor before being sold to wholesale and retail locations.

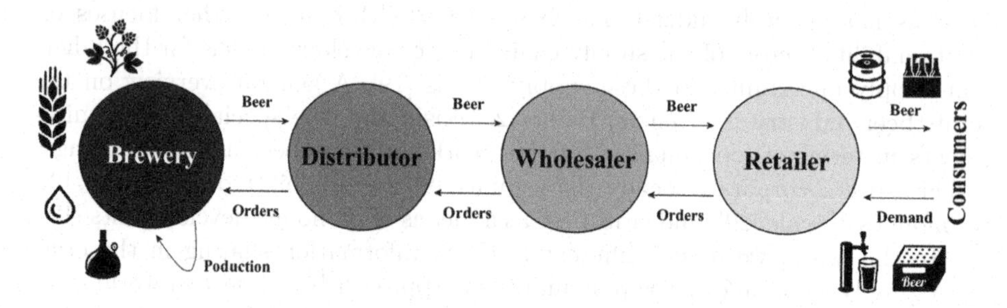

Fig. 1. The Three-Tier beer supply chain used by both macro breweries and craft breweries with its nodes (brewery, distributor, wholesaler, and retailer) down to the consumers.

The main contrasts between the regular beer supply chain and the CBSC are in ownership, information access and product shelf life. These differences directly affect the breweries' abilities to manage their production strategies efficiently, as well as control their storage and spoilage costs. In the following, we explain these differences in more detail.

Ownership and Information Access. Macro breweries' distributing centers exist around the country and are wholly owned by the breweries' themselves. This is a luxury that craft breweries rarely have. They rely on third party distributors for additional warehousing. These inventory transactions with external parties adds cost for the craft breweries.

Further, the Three-Tier Laws also dictate that breweries can have ownership of no more that 5% of the wholesalers that sell its products. Though, due to the sheer dominance that macro breweries have on the industry and benefits they can provide (e.g., vehicles), they have nearly full control on the wholesalers' operations for their products. These wholesalers are even barred from carrying other macro brewery products and are referred to as 'houses' for their connected

[1] https://www.brewersassociation.org/statistics-and-data/craft-brewer-definition.

[2] A craft brewery is considered a micro brewery when producing less than 15,000 barrels (ca. 17900 hL) per year.

macro brewery. This control includes ordering strategy decision making and information systems integration that allows full visibility of inventory movement. Ownership structure and information access are depicted in Fig. 2.

At the retail level, the macro breweries are on a more similar playing field with the craft breweries due to another set of laws called the Tied House Laws which penalize manufacturers for influencing retailers. Though, where macro and craft differ is in the access of information of direct consumer sales from the retailers themselves. The cost of this data through third party data brokers is in the six figures annually and while macros have no problem purchasing the data, it is nearly impossible for craft breweries to justify the cost.

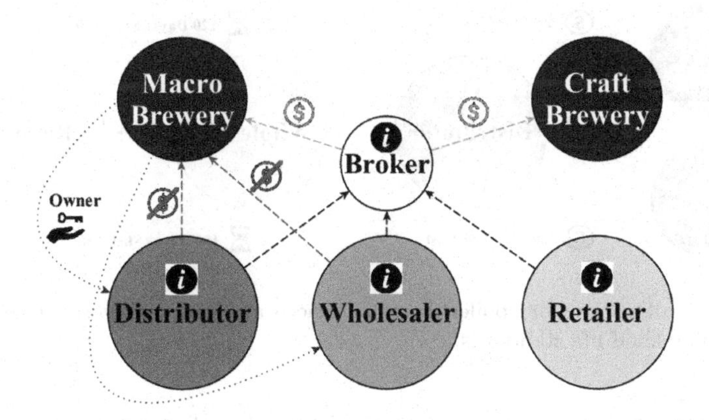

Fig. 2. Ownership relation, and paid and unpaid flow of information between supply chain members and data brokers.

Shelf Life. Quality standards within the US beer industry have changed dramatically since the 1960s, mainly due to the advancements in brewing technology and increased consumer product knowledge. These two changes have led to the development of and adherence to shelf-life standards for beer products. These standards designate how long products can age before they are deemed spoiled and discarded. Quality focused breweries and consumers are keen to minimizing the age of beer products as they make their way through the supply chain [16]. Aging speeds up with increasing exposure to oxygen and light. Beer becomes oxidized when it is introduced to oxygen and produces off-flavors described as cardboard or paper. Oxidation speed increases with heat and agitation, which are common with transportation. Hops are the primary cause of oxidation and thus beers with more hops are more susceptible to oxidation [10]. Beer becomes 'light-struck' when UV light reacts with hops producing a skunky smell. This flavor is evident in many popular beer brands that are found in green or clear glass bottles [3].

With increased visibility to information of their down line agents, macro breweries have the ability to control inventory levels at each agent and monitor the shelf-life of their products throughout. This ability becomes significantly more important for products that have shorter and gradually decreasing shelf-lives, which describes nearly all products produced by craft breweries. While keeping larger days on hand inventory at each agent, can help maximize the fulfillment rates to the consumer, this strategy can also lead to a longer aging of product at time of consumer purchase. Therefore, craft breweries' lower visibility to information of their down line agents, decreases the quality of their product and increases their cost.

The difference in shelf life standards and the corresponding cost factor are illustrated in Fig. 3.

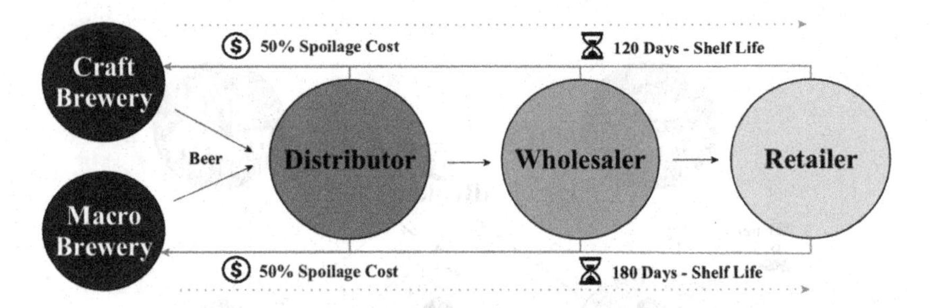

Fig. 3. Cost allocation for spoiled beer for macro and craft breweries based on the corresponding shelf life standards.

The predominant cost factors to be considered in the craft beer supply chain are as follows. *Missed Beer Sales*: The cost for the retailer of not fulfilling customer orders. *Beer Age*: The cost of providing a customer a beer product that is past its prime drinking age. *Beer Spoilage*: The cost of beer product that has been destroyed due to reaching its beer spoilage age. Note that beer spoilage can occur at any tier in the supply chain. The brewery is interested in minimizing all three metrics to reduce costs and increase product quality.

We assume sufficient transport capacity since shipping in the CBSC is done at lower quantities than for macro breweries. However, we account for shipping times. We do not consider costs related to transportation and inventory since they play a minor role in the CBSC.

The Craft Beer Game. The Beer Game is a well-known and heavily analyzed model that is used for applications, not limited to, but including classrooms, business management seminars and scientific research. The fundamental logic of the game is based on a multi-player system where product orders are placed upstream and fulfillment of these orders is completed downstream. Each player has its own strategies on how to fulfill these orders while considering their own limitations in storage capacity, transportation time and order quantities. The objective of the Beer Game is to come up with strategies at the player level to either minimize costs, maximize sales or a combination of the two.

In this work, we introduce the Craft Beer Game, which extends the Beer Game. In addition to the fundamental four steps (check deliveries, check orders, deliver beer, make order decision), we consider a new *acquire information* step. The latter is conducted in every round prior to making an order decision. The obtained information includes insights into the other players' operations. This is so important in the CBSC because of both, the strong competition in the craft beer industry, and the need to compete with larger breweries.

3 Information Sharing

Reordering strategies at the different tiers are key mechanisms in a supply chain. We take the perspective of the manufacturer, i.e., the brewery, which does not order but plan production instead. In practice, this operational planning step takes into account historical demands, current inventories and the planners' domain expertise. The main challenge is to predict future demands at the best possible level of accuracy. These demands are dependent on the next node's requirements which stem from the demand that it is facing.

There is a difference between information sharing and transparency. Information sharing leads to transparency. It can follow a mutual agreement between two supply chain members, resulting in collaboration. Or, it can be asymmetric, so that only one node obtains (partial) access to the node's data. We focus on the latter case and restrict ourselves to down-stream transparency resulting from up-stream sharing. Moreover, information can be obtained after involuntary disclosure, possibly through third-parties. That is, a node might not be willing to share information, but peripheral analysis could give insights into some of its operations. For example, shipping companies delivering products from the wholesaler to the retailer may provide insight into the corresponding order patterns. Cooperation between competitors, also called *coopetition*, happens when all involved parties expect benefits.

The information of interest in the CBSC can be categorized as order-related and inventory-related. Information is closely related to parameters that can be used to describe a supply chain. Both types can be static or dynamic (i.e., historical). Information can be stored in a centralized or a decentralized fashion. Currently, information is available from free data consolidators (e.g., VIP[3], GP-Analytics[4]) providing data-analytics services to distributors, wholesalers, and retailers. They commonly sell the data to breweries at a relatively high cost. They are information-sharing platforms to the distributors etc., and a data broker to the brewery.

We consider four information sharing models yielding different supply chain transparency from the manufacturer's (i.e., the brewery's) perspective.

I. Baseline: No transparency at all. The brewery has no insights in current or past downstream operations.

II. First-Level Transparency: The distributor's inventory and historical demand are known.

[3] https://www.mysoftwaresolutions.com/vip-analytics.
[4] https://www.gp-analytics.com.

III. Second-Level Transparency: In addition to distributor data access, the wholesaler's inventory and historical demand are known.
IV. Third-Level Transparency: First and second-level transparency is extended by retailers information.

Figure 4 illustrates the considered information sharing levels within the supply chain. Note that the brewery always has access to its own inventory levels and the current and historical orders submitted by the distributor. In this work, we do not consider information sharing with other nodes than the manufacturer, since we are interested in potential benefits for the brewery.

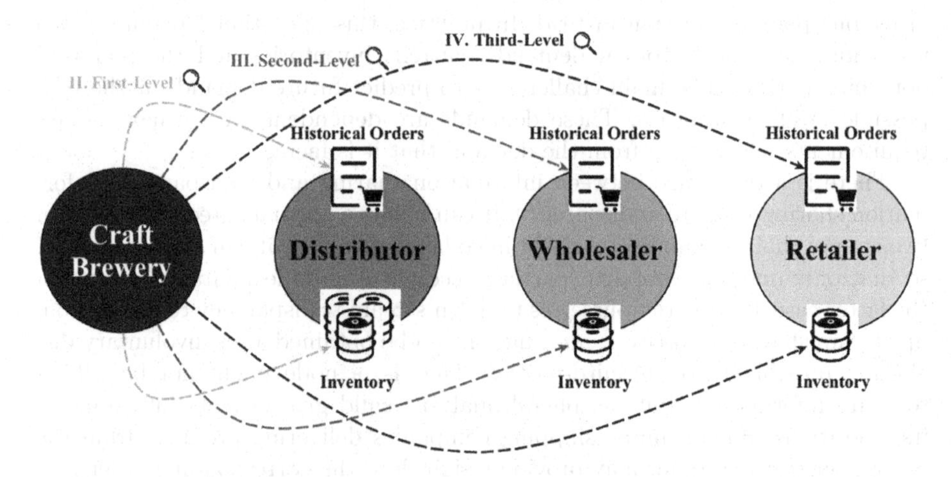

Fig. 4. The considered levels for information access for the brewery in the craft beer supply chain: No Sharing (I), Distributor (II), Distributor+Wholesaler (III), Distributor+Wholesaler+Retailer (IV).

4 Reordering and Production Planning

We first describe the base reorder strategy used at each node. Afterwards, we explain how the information sharing levels are used to adjust the brewery's beer production planning. All beer quantities in our model are measured using *case equivalents* (CEs), the standard measure utilized in the U.S. Beverage Wholesale industry. A CE is comparative to 24 cans of 12 fluid ounces.

4.1 Reorder Strategies

We build the reorder strategies for our Craft Beer Game based on the existing strategies for beer supply chains in [17,18]. Let $n \in \{1, .., 4\}$ denote the supply chain node from the brewery to the retailer (increasing from left to right). Each node places an order once per reorder cycle with node-dependent *cycle time* (CT_n) measured in days. The considered time periods t are the end times of the order cycles, which are different for the nodes. For every node n, reorders are executed at times in $\{0, 1 \cdot CT_n, 2 \cdot CT_n, 3 \cdot CT_n, \ldots\}$. In time period t at node n, the *suggested order quantity* ($SOQ_{n,t}$) in CEs is calculated as follows:

$$SOQ_{n,t} = max\{ED_{n,t} + AS_{n,t} + ASL_{n,t}, 0\} \tag{1}$$

The *expected demand* is defined as

$$ED_{n,t} = \theta \cdot INC_{n,t} + (1 - \theta) \cdot ED_{n,t-CT_n}$$

where $INC_{n,t}$ stands for the total *incoming orders* over the last cycle to node n (sent by node $n+1$), and $\theta \in [0, 1]$ is an expectation update The *adjusted supply* is defined as

$$AS_{n,t} = \alpha_S \cdot (DINV_{n,t} - INV_{n,t} + BL_{n,t})$$

where $DINV_{n,t}$ is the node's *desired inventory*; $INV_{n,t}$ is the actual *inventory*, including the beer that is currently being transported to the corresponding node; $BL_{n,t}$ the *backlog*; and $\alpha_s \geq 0$ is a fractional adjustment rate. The *Adjusted supply line* is defined as

$$ASL_{n,t} = \alpha_{SL} \cdot (-BL_{n-1,t})$$

where α_{SL} is a fractional adjustment rate. Note that $BL_{n-1,t}$ is known because node n knows both, what order was placed and how much of it was fulfilled by up-stream node $n-1$. This equation slightly differs from [17] since we use reorder cycle times that are longer than the summation of fulfillment and shipping times. This is common for a CBSC because of low volume and short transportation distances. At the wholesaler and the distributor, the $SOQ_{n,t}$ value will be rounded up to the next suitable batch size (see also Sect. 5), whereas the retailer precisely orders SOQ_t units. The reorder strategy defined above applies in particular to the brewery, where orders correspond to production orders. In Sect. 4.2, we will present a set of revised strategies that take into account the additional information that is being shared when planning the beer production.

4.2 Brewery Reordering with Different Information Levels

In the following, we incorporate the additional information available at the different sharing levels into the brewery's production planning. To this end, we suggest effective demand forecasting methods for all scenarios. Under additional information sharing using level L, the base strategy for the brewery node given in Eq. (1) extends as follows.

$$SOQ_{1,t} = \max\{ED_{1,t} + AS_{1,t} + ASL_{1,t} + ATS^L_{1,t}, 0\} \tag{2}$$

Here, the newly integrated level-dependent *adjusted total supply* is defined as

$$ATS^L_{1,t} = \alpha_{TS} \cdot (DTS^L_{1,t} - TS^L_{1,t}) \tag{3}$$

Similar to the adjusted supply $AS_{n,t}$, α_{TS} functions as a fractional adjustment rate. The *total supply* reflects the known actual amount of beer in all downstream node inventories, dependent on the information level:

$$TS^L_{1,t} = \begin{cases} 0 & \text{if } L = \text{I} \\ INV_{1,t} + INV_{2,t} & \text{if } L = \text{II} \\ INV_{1,t} + INV_{2,t} + INV_{3,t} & \text{if } L = \text{III} \\ INV_{1,t} + INV_{2,t} + INV_{3,t} + INV_{4,t} & \text{if } L = \text{IV} \end{cases} \tag{4}$$

We intentionally set the $TS^L_{1,t}$ to 0 (as we do for $DTS^L_{1,t}$) in the case that no additional information is available (Level I). Herewith, we ensure that the adjustment term $ATS^L_{1,t}$ cancels out in this information sharing level. The *desired total supply* corresponds to the desired amount of inventory contained in the entire supply chain, the strategy we use is defined as

$$DTS^L_{1,t} = \begin{cases} 0 & \text{if } L = \text{I} \\ \gamma_L \cdot \dfrac{INC_{2,t}}{CT_2} & \text{if } L = \text{II} \\ \gamma_L \cdot \dfrac{INC_{3,t}}{CT_3} & \text{if } L = \text{III} \\ \gamma_L \cdot \dfrac{INC_{4,t}}{CT_4} & \text{if } L = \text{IV} \end{cases} \tag{5}$$

where the parameter $\gamma_L \in \{1, 2, \ldots\}$ is used to specify the days of inventory that the brewery desires to be available in the supply chain down to the last node that it has data access to. A larger value typically results in an increased adjustment (see Eq. (3)) and yields overproduction. Conversely, reducing γ_L tends to decrease production. Note that for level I, this strategy reduces to the base strategy defined in Eq. (1), since no desired downstream inventories and total supply are included. $ATS^L_{1,t}$ can be negative, since it is used to adjust the production quantities.

We point out that adjusting the production has an impact on the brewery's objectives in two ways. If the adjustment is negative (i.e., less is produced than originally planned) then we will likely see less beer spoilage. If adjustment is positive (i.e., we produce more) then we expect to reduce the missed beer sales.

In the base production planning, the brewery is dependent on the distributor's estimation of downstream demand in form of the corresponding distributor orders. This might not be ideal for the brewery's objective to reduce the beer age. In information sharing levels II-IV, the brewery's planning can bypass the distributor's planning by adjusting according to an own real-data-based downstream demand estimation. Example: Consider the case that the distributor overestimates the future demand. If this is reflected in low retailer demand (that we have access to) then the brewery would adjust by reducing its production. Even if this way it is not possible to meet the distributors orders, we expect to avoid beer sitting in downstream inventories longer than needed.

Distributors are set up to hold large inventories with long shelf lives (Macro-Beer), but that causes a problem for Craft Breweries whose product has shorter shelf-life. Distributors are focused on selling the product within the spoilage window, which means they will focus on holding as much product, as possible as long as it does not spoil. Whereas, breweries want their product sold as fresh as possible, which would mean smaller inventories. Therefore, the Craft Breweries are self-regulating the supply chain by not filling every distributor order.

5 A Simulation Approach

We use simulation to quantify the impact of availability of downstream data for the brewery. The considered scenarios emanate from the information sharing models and the corresponding brewery production planning strategies introduced in Sect. 3 and Sect. 4. The used real and simulated market data is described in Sect. 5. We develop a hybrid agent-based and discrete-event simulation system to model the CBSC. We use AnyLogic[5] as simulation modelling system. In the following, we present our simulation approach including model logic, parameters, data, and metrics which is inspired by [17].

Model and Logic. We model every supply chain node (brewery, distributor, wholesaler, retailer, and customer) as a separate section of the agent flow logic. Orders (in CEs) are explicitly modelled as agents on all levels. They originate at distributor, wholesaler, retailer, and customer, and are terminated at the preceding node (see Fig. 5). Moreover, backorder agents originate at distributor, wholesaler, and retailer in case of insufficient inventory.

In Fig. 5, we illustrate the generic logic used to represent the supply chain nodes using the wholesaler. A recurring event causes an order agent to be generated at the order source with a quantity parameter defined by the node's reorder strategy. The order agent then functions as a container and picks up the desired quantity of CE agents from the previous node's inventory. The order and CE agents pass through a delay representing shipping, then the beer is dropped into this node's inventory while the order is disposed. In the case that an order attempts to pick up more beer than the previous node has, a new back order agent is put into a queue. The next time an order picks up from that node, it will attempt to pickup the new order quantity in addition to the quantity on back order.

The main events correspond to recurring customer demand (based on daily stochastic market data), reorders, beer production, backorders (dropped after one reorder cycle if not filled, which is common), and inventory quality control (spoilage check).

[5] www.anylogic.com.

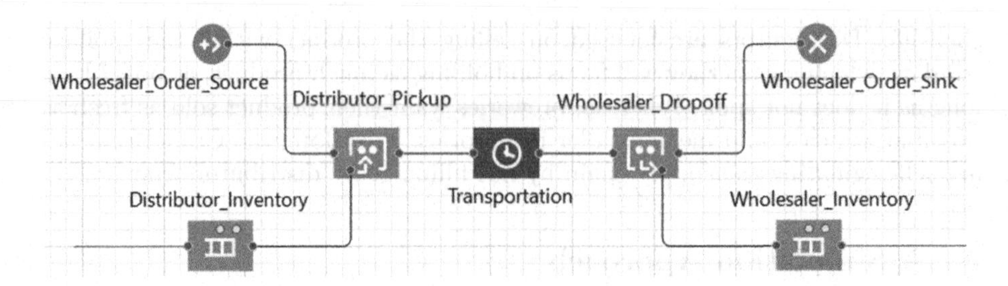

Fig. 5. The generic node logic with inventory queues, ordering, and shipping in our AnyLogic simulation model illustrated by the wholesaler.

Parameters. The following parameters are used to configure the CBSC. The beer production time is 21 days per batch (100 CE). Reorder cycle times are 1/14/14/7 days, and reorder quantities are rounded up to the next 100 (in CE) at the distributor and the wholesaler (not rounded at the retailer). The reorder parameters (used in Sect. 4) reflect industry standards: $\theta = 0.5$, $\alpha_S = 0.5$, $\alpha_{SL} = 1$, $\alpha_{TS} = 1$, $\gamma_{II} = 30$, $\gamma_{III} = 45$, $\gamma_{IV} = 60$. A desired inventory level for node n in time period t ($DINV_{n,t}$) is set to be a single order cycle's expected demand ($ED_{n,t}$). The retailer holds twice this volume. The brewery's production quantities are rounded up to the next 100 (in CE). The corresponding production limit is assumed to be 1000 CE; no limits at other nodes. The maximum beer age in number of days before being discarded during the inventory quality control, also called *hold days*, is 40 (brewery), 70 (distributor), 90 (wholesaler), and 120 (retailer). A spoilage check is performed on a daily base at every node. We set the transportation times to one day and do not incorporate capacity or cost. Moreover, we do not use inventory capacity limits.

Real-World Data. The customer demand is assumed to be stochastic. We base our experiments on real-world data from a craft beer brewery. Using four-year daily demand, we generate 19 randomly simulated time series. These are derived by a time series decomposition approach in which we detect the error distribution after subtracting linear trend and exponentially smoothed pattern. The historical demand data for the four-year period is illustrated on a monthly base in Fig. 6. Furthermore, the figure shows the simulated demands. A steady demand growth can be seen that is typical for early-stage craft beer breweries.

Key Performance Indicators. The performance of the CBSC is measured using the following metrics (see also Sect. 4).

1. Beer Sales: The relative missed beer sales in CE with respect to the overall customer demand.
2. Average Beer Age: The average number of days that a CE spends in the supply chain before reaching the end consumer.
3. Beer Spoilage: The relative beer in CE that is spoiled at any node due to an excess of shelf life.

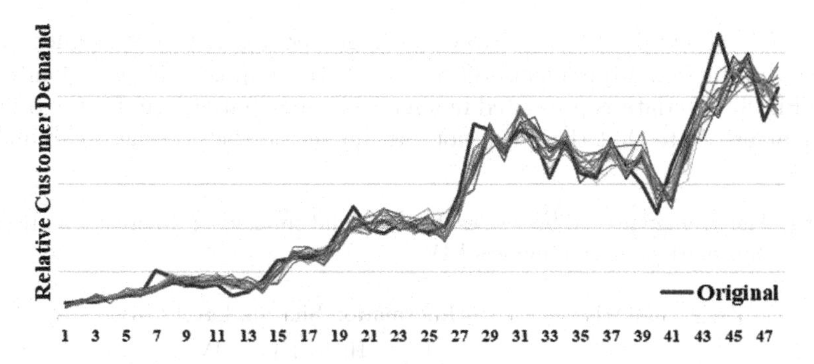

Fig. 6. Real-world time series and 19 simulations of customer beer demand data over the four-year horizon (monthly aggregation).

6 Computational Analysis

In this section, we present and analyze the results of our simulation study. We quantify the impact of the four different information sharing levels (Sect. 4) and the corresponding brewery planning strategies (Sect. 4) on the CBSC model described in Sects. 2 and 5. The used simulation data and parameters are described in Sect. 5.

We report our main results in Table 1 using the metrics introduced in Sect. 5: Missed Beer Sales, Beer Age, and Beer Spoilage. These numbers correspond to a breakdown of major cost factors associated to a craft brewery. Relative missed beer sales are given with respect to the overall customer demand. The spoilage at a node is compared to the overall beer volume that entered the node. The average beer age is calculated over the beer that is delivered to the end customer, not considering spoiled material. We recall that level I does not allow the brewery to look into the other nodes' operations at all.

We observe a significant reduction of missed beer sales when augmenting the information shared in the different levels. When allowing full transparency, the missed sales can almost be halved (4.9% $\rightarrow$ 2.5%). An even stronger impact can be seen in terms of beer spoilage. The overall spoilage can be reduced from 24.0% to 7.1%. The node-dependent breakdown confirms this gradual improvement. Brewery and distributor benefit the most since the corresponding detected spoiled beer reduces to 0.8% and 1.8%, respectively. However, the beer age remains consistently around 67%, indicating that the information levels do not help. This minimal effect on the beer age could be due to the fact that each node holds enough inventory to cover till their next shipment. Thus, the average beer age is rather correlated to the sum of days between shipments for each node, i.e., the reorder cycles. The distributions for missed beer sales, average beer age and beer spoilage are further described in Fig. 7. We observe some variation in missed sales but only small changes for spoilage and beer age. Overall, the missed beer sales range from 0.8% to 8.6%, whereas the average beer age is greater than 64.0% does not exceed 69.8%. Moreover, the standard deviation in all metrics decreases as more information becomes available: 1.9 $\rightarrow$ 1.3% (missed sales);

2.2 → 1.7% (spoilage); 1.2 → 0.7 days (average beer age). To better understand the beer production adjustments $ATS^L_{1,t}$ (Sect. 4), we illustrate the absolute values in Fig. 8. The data is presented in a monthly aggregated form for the original beer demand. Note that there is a notable impact on the production volumes.

Table 1. The average missed beer sales, beer age and spoilage at different supply chain nodes for information sharing models I-IV.

Metrics	Information Sharing Level			
	I	II	III	IV
Missed Sales (%)	4.9	4.5	3.5	2.5
Beer Age (∅)	67.7	66.0	66.2	68.1
Spoilage (%)				
Total	24.0	15.8	8.1	7.1
Brewery	13.4	6.0	0.5	0.8
Distributor	8.4	7.3	5.5	4.6
Wholesaler	3.2	3.0	2.5	1.8
Retail	0.00	0.00	0.00	0.00

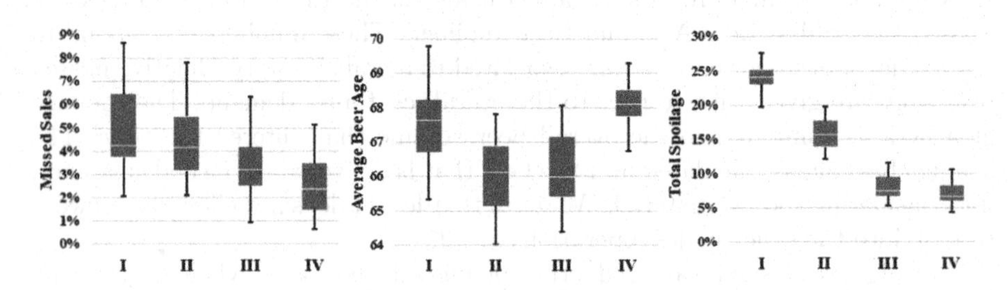

Fig. 7. The distributions of average relative missed beer sales (left), average beer age (center), and relative beer spoilage (right) observed over the simulation repetitions.

7 Conclusion

We studied the US American craft beer supply chain from the brewery's perspective in a beer game fashion. After defining its industry-specific properties, we developed practically relevant scenarios for how availability of down-stream information can be incorporated into production planning. Our main goal is to help the brewery's production planning regarding sales, product quality, and spoilage. We conducted a simulation study based on real-world craft beer data, in which we quantified the value of information sharing in the craft beer supply chain. We showed that the acquisition of downstream information from third-party brokers yields significant benefit. With every node for that the brewery

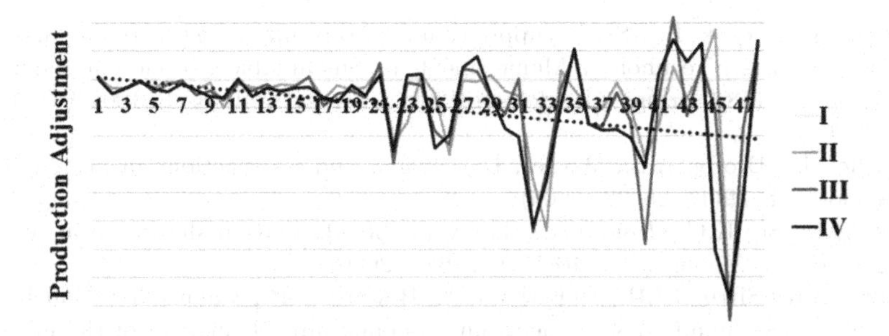

Fig. 8. Adjustment in beer production ($ATS^L_{1,t}$) at the brewery for the different information sharing levels (I-IV); shown for original demand date in monthly aggregation; level I indicates zero adjustment.

obtained data access, its planning improved near-linearly. In the case of complete supply chain transparency, the missed beer sales could be reduced by 50% on average. The costly beer spoilage could even be decreased by 70%.

From a managerial perspective, the developed approach can be used to support breweries when negotiating with data brokers. In addition, it can be used to evaluate collaboration opportunities with respect to information sharing in the platform economy. Based on these positive results, we suggest exploring further adaptation of the brewery's production planning strategy concerning demand forecasting and collaboration. Also, the investigation of interplay of production and reorder mechanisms could be of interest. Moreover, we see importance in an in-depth formalization and study of the generic Beer Game with information sharing.

References

1. Bahl, H.C., Gupta, J.N., Elzinga, K.G.: A framework for a sustainable craft beer supply chain. Int. J. Wine Bus. Res. Online **33**, 394–410 (2021)
2. Baiano, A.: Craft beer: an overview. Comprehensive Rev. Food Sci. Food Safety **20**(2), 1829–1856 (2021)
3. Burns, C.S., Heyerick, A., De Keukeleire, D., Forbes, M.D.E.: Mechanism for formation of the lightstruck flavor in beer revealed by time-resolved electron paramagnetic resonance. Chem. Eur. J. **7**(21), 4553–4561 (2001)
4. Calatayud, A., Mangan, J., Christopher, M.: The self-thinking supply chain. Supply Chain Manage. Int. J. **24**(1), 22–38 (2019)
5. Capitello, R., Todirica, I.C.: Concepts and practices of sustainable craft beer in Italy: a case study analysis. In: Case Studies in the Beer Sector, pp. 313–326. Elsevier (2021)
6. Clemons, E.K., Gao, G.G., Hitt, L.M.: When online reviews meet hyperdifferentiation: a study of the craft beer industry. J. Manag. Inf. Syst. **23**(2), 149–171 (2006)
7. Edali, M., Yasarcan, H.: A mathematical model of the beer game. J. Artif. Soc. Soc. Simul. **17**(4), 1–2 (2014)
8. Grunde, J., Li, S., Merl, R.: Craft Breweries and Sustainability: Challenges, Solutions, and Positive Impacts. Master's thesis, Blekinge Institute of Technology, Karlskrona, Sweden (2014)

9. Kawa, A., Łuczyk, I.: CSR in supply chains of brewing industry. In: Golińska, P., Kawa, A. (eds.) Technology Management for Sustainable Production and Logistics, pp. 97–118. Springer, Heidelberg (2015). https://doi.org/10.1007/978-3-642-33935-6_5
10. Kuchel, L., Brody, A.L., Wicker, L.: Oxygen and its reactions in beer. Packag. Technol. Sci. **19**(1), 25–32 (2006)
11. Lee, H.L., So, K.C., Tang, C.S.: The value of information sharing in a two-level supply chain. Manage. Sci. **46**(5), 626–643 (2000)
12. Miller, S.R., Sirrine, J.R., McFarland, A., Howard, P.H., Malone, T.: Craft beer as a means of economic development: an economic impact analysis of the michigan value chain. Beverages 5(2) (2019)
13. NABCA Research: The three-tier system: A modern view (2015). www.nabca.org/sites/default/files/assets/files/ThreeTierSystem_Mar2015.pdf
14. Ness, B.: Beyond the pale (ale): an exploration of the sustainability priorities and innovative measures in the craft beer sector. Sustainability **10**(11) (2018)
15. Ouyang, Y.: The effect of information sharing on supply chain stability and the bullwhip effect. Eur. J. Oper. Res. **182**(3), 1107–1121 (2007)
16. Stewart, G.: Beer shelf life and stability. In: Subramaniam, P. (ed.) The Stability and Shelf Life of Food, pp. 293–309. Woodhead Publishing Series in Food Science, Technology and Nutrition, Woodhead Publishing, 2nd edn. (2016)
17. Strozzi, F., Bosch, J., Zaldívar, J.: Beer game order policy optimization under changing customer demand. Decis. Support Syst. **42**(4), 2153–2163 (2007)
18. Thomsen, J.S., Mosekilde, E., Sterman, J.D.: Hyperchaotic phenomena in dynamic decision making. In: Mosekilde, E., Mosekilde, L. (eds.) Complexity, Chaos, and Biological Evolution. NAS, vol. 270, pp. 397–420. Springer, New York (1991). https://doi.org/10.1007/978-1-4684-7847-1_30
19. Warsing, D.P., Jr., Wangwatcharakul, W., King, R.E.: Computing base-stock levels for a two-stage supply chain with uncertain supply. Omega **89**, 92–109 (2019)
20. Wells, P.: Economies of scale versus small is beautiful: a business model approach based on architecture, principles and components in the beer industry. Organization Environ. **29**(1), 36–52 (2016)

Smarter Relationships? The Present and Future Scope of AI Application in Buyer-Supplier Relationships

Anna-Maria Nitsche[1,2](✉), Markus Burger[3], Julia Arlinghaus[4], Christian-Andreas Schumann[2], and Bogdan Franczyk[1,5]

[1] University of Leipzig, Augustusplatz 10, 04109 Leipzig, Germany
anna-maria.nitsche@uni-leipzig.de
[2] University of Applied Sciences Zwickau, Kornmarkt 1, 08056 Zwickau, Germany
[3] RWTH Aachen University, Templergraben 55, 52062 Aachen, Germany
[4] Otto von Guericke University, Universitätspl. 2, 39106 Magdeburg, Germany
[5] Wrocław University of Economics, Komandorska 118/120, 53-345 Wrocław, Poland

Abstract. The last decade has seen rapid developments in the area of artificial intelligence (AI). While research focuses on technical challenges and enablers of AI, the number of publications examining application approaches at the buyer-supplier interface is increasing. To accelerate the related discussion and to add clarity and richness to this fragmented research field, a systematic overview of the existing comprehensive body of literature is essential. We contribute to the academic debate by applying a combined systematic literature review with a text mining and machine learning-based literature review. Thus, we categorize and cluster different research streams and analyze the application of AI at the buyer-supplier interface. Subsequently, we identify gaps resulting from the comparison of the technology and the application domain and derive the main points of discussion from the literature. As a result, we present ten central questions outlining future requirements and research opportunities in the field of AI application at the buyer-supplier interface.

Keywords: Artificial Intelligence · Machine learning · Buyer-supplier relationship · Literature review

1 Introduction

Artificial intelligence (AI) is often portrayed as an important factor for the digitalization of future supply chain management (SCM). While the utilization of AI in SCM has been low for decades [1], research interest significantly increased within the context of development from technology-enabled to technology-centric SCM. At the interface between buyer and supplier, AI provides various benefits, for instance in terms of cost analysis and risk monitoring [2]. However, the practical implementation of AI and machine learning (ML) technologies at buyer-supplier interface is still scarce. Although the adaption rate has increased over the last years, still more than every second company has not

M. Mes et al. (Eds.): ICCL 2021, LNCS 13004, pp. 237–251, 2021.
https://doi.org/10.1007/978-3-030-87672-2_16

implemented AI or ML solutions within their procurement activities [3]. Further, current research regarding AI at the buyer-supplier interface focuses on specific entities, such as procurement [4] or on specific activities, such as risk management [2]. Consequently, researchers and practitioners face the situation of a significantly increased number of theoretical publications and a comparatively low rate of AI adoption at the practical buyer-supplier interface. As a result, potential benefits, like enhanced decision making and new business model opportunities [e.g. 2] remain untapped. This indicates a need to structure the current research for gaining clarity about dominating research streams and for identifying open questions for researchers and practitioners. Thus, the purpose of this article is to define ten central questions for future research by providing a systematic overview of the existing comprehensive body of literature of AI application at the buyer-supplier interface. In doing so, we categorize and cluster different research streams by conducting a text mining and machine learning literature review (MLR) combined with a systematic literature review (SLR). Compared to traditional SLRs, the applied combined approach can handle large quantities of publications to conduct comprehensive reviews while at the same time incorporating the benefits of manual reviews [5].

2 Theoretical Background

2.1 Industry 4.0 and AI in Buyer-Supplier Relationships

SCM and operations management are influenced by progressive digital developments such as the Internet of Things and AI [6]. AI as a disruptive technology is widely considered to be of growing importance for supply networks [6–9]. Many definitions exist due to the complexity of the field [10, 11], for example "the branch of computer science that is concerned with the automation of intelligent behavior" [12, p.1]. AI applies method sets including statistical learning and machine learning [10]. The terms AI and machine learning are often used interchangeably as different research communities use them with varying meanings [10, 11]. While supply networks and operations management constitute adequate data sources, it is argued that the potential of AI utilization in SCM has not yet been fully exploited and is likely to expand [1]. The complex nature of SCM goes along with a broad range of relationships between buyers and suppliers. Previous studies identified several ways of classifying in Buyer-Supplier Relationships (BSRs), e.g. in terms of competency [13], power relation [14] or degree of collaboration [15]. However, a positive impact on performance is not related to a special type of BSR but to the context of the specific supply chain [16]. While several business processes like supplier selection mark crucial milestones in every type of BSR, the depth of other business processes like supplier risk management depend on the BSR classification. Several researchers [17, 18] provide models for depicting and measuring business processes within BSRs. Within the development towards Industry 4.0 experts observe a consolidation of the supply base. While the relationships to key suppliers are getting even more intense, firms often refrain from collaboration with non-strategic suppliers [19]. In doing so, both the implementation of Industry 4.0 technologies as well as classical processes, like supplier selection or negotiations, are affected.

2.2 Research Gap

In this article we accelerate the related discussion and contribute to the research field of AI in SCM by addressing the following research gap: The increasing amount of available information that goes along with the ongoing implementation of Big Data solutions requires AI and ML approaches to analyze this information and to make it valuable for daily business users. Research currently contributes to this issue by theoretical foundations of AI, ML approaches and on knowledge representation and reasoning [4]. The mentioned technologies are considered amongst others within supply chain integration [20], procurement [4] and risk management literature [2]. Within the last years, the rising interest in the topic has led to an increasing number of publications [4]. However, the practical implementation of AI and ML technologies in firms is still scarce [3]. Consequently, a low adaption rate faces a broad and comprehensive scope of literature. Thus, the following questions regarding AI at buyer-supplier interface challenge researchers and practitioners:

- What are the main research areas?
- Where is the research focus?
- What are the open research gaps?
- What publications are suitable starting points, when examining the gaps?

The purpose of this article is to answer these questions by analyzing the present scope of AI research at buyer-supplier interface and exploring untapped potentials for future research scope. Thus, we present a systematic overview of the comprehensive body of literature of AI application at the buyer-supplier interface. We bring clarity and richness to this fragmented research field by clustering related publications, identifying potential research streams, and highlighting ten central questions for further research in terms of AI at buyer-supplier interface.

3 Research Design

This paper applies a MLR [21] combined with a SLR and elements of scientometric and bibliometric studies to provide the literature basis for the review (Fig. 1). Literature reviews provide a valuable source for policy, practice and academic research [22, 23] as they analyze and summarize the existing literature [24]. Traditional literature review forms include narrative reviews [25], systematic reviews [26] as well as scientometric and bibliometric studies [27–29]. Compared to traditional SLRs, the applied combined approach can handle large quantities of publications to conduct comprehensive reviews and incorporates the benefits of manual interventions, subjective experience, and knowledge-based decisions [5]. This paper consequently does not aim at a deeper understanding of individual research areas within the research field of AI application in BSRs, but rather at the identification, consolidation, and combination of the main research streams and at providing a bird's eye perspective of the thematic developments within the past few years at a higher level of abstraction. The SLR approach is based on the steps proposed by management and information systems research [26, 30, 31]. Following the definition of the review scope and aim, the database search (Scopus, Google

Scholar) is executed. As a search phrase we applied two segments: 1. related to AI ("Artificial Intelligence", "machine learning", "machine intelligence"), 2. BSRs in a wider sense ("buyer", "supplier", "seller", "supply chain", "purchasing" and "procurement"). Only English language publications from a business, computer science, decision science, or economics context are included. Data refers up to the search date March 2021. The resulting list of 1,936 titles is then checked for duplicates, subsequently eliminating 34 publications. The MLR is processed with the help of the open-source ML and data visualization software Orange [32] in terms of definition and analysis of clusters and for aggregating the relating research streams. The principal Orange workflow included in Fig. 1 applies the following steps for identifying the clusters: To prepare the text for further analysis, the widget *pre-processor* splits the text into smaller units, uses the Porter stemming algorithm to normalize the text and creates one-grams and two-grams to start the text mining process (n-grams are contiguous sequences of n items from a text). Next, the widget *bag of words* adds word counts for each data instance created in the previous step, thus rearranging the data into rows of entities (i.e., individual titles) and columns of features (i.e., every word that appears in all titles). The widget *distances* subsequently calculates Euclidian distances between the rows of the dataset as a basis for the *hierarchical clustering*. As a result, we identify 24 clusters of articles exhibiting certain similarities in terms of the titles.

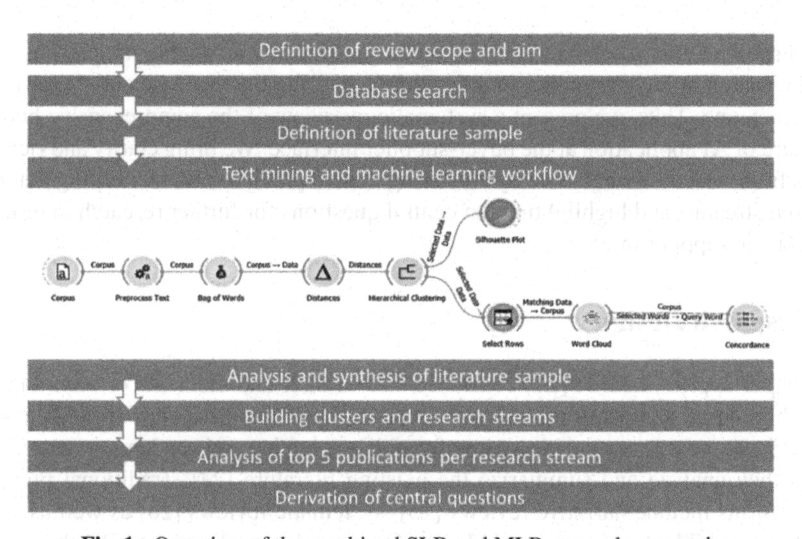

Fig. 1. Overview of the combined SLR and MLR research approach.

We further examine the identified clusters deeply by a three-step approach. First, we analyze buzzwords, characterizing the cluster based on their frequency and content within the cluster. Second, we aggregate thematically related clusters within research streams. Third, we manually analyze the five most influential publications (according to the average cite rate per article) of each research stream and thus, derive central questions to guide future research.

4 Findings and Discussion

This chapter presents the main findings of the comprehensive literature review and contains information regarding the main research streams examining AI approach at the buyer-supplier interface and their development over time, the research focus, and the prevalent methods within the research field.

What are the relevant research streams examining AI approaches at the buyer-supplier interface? We identified a total of 1,902 published articles on AI application in BSRs in the Scopus database and Google Scholar search engine. The hierarchical cluster analysis using Orange resulted in the identification of 24 clusters. We summarize these clusters within the following five main research streams as listed in Table 1:

Table 1. Research streams and cluster

Research stream	Cluster
AI technologies in general	C1 (AI – general, AI application in business and industry context), C2 (Learning, deep learning, reinforcement learning), C4 (Machine Learning – general, ML as application technique), C13 (Ethics of AI – Role of human and society), C15 (Multi-agent systems), C19 (Neural networks and demand forecasting), C21 (Fuzzy – logic, evaluation, model)
Industry 4.0	C6 (Internet of Things), C11 (Industry 4.0 general), C17 (Intelligent systems – CPS application for quality)
Data utilization	C3 (Data for machine learning), C5 (Data analysis – Big data, data intelligence, data mining)
Decision making	C9 (Decision making supported by AI), C23 (Decision-support systems), C24 (supplier selection based on fuzzy logic)
Procurement	C12 (e-purchasing, e-commerce), C14 (Procurement – electronic, public)

Further, several clusters can be summarized within *Others*. This category includes a fuzzy cluster with 474 articles as well as clusters concerning research approaches and methods. Each of the five main research streams aggregates thematic clusters that apparently refer to similar topics. Within the research streams, the associated clusters are sorted in ascending order. The cluster numbers indicate the closeness of the titles contained in the clusters as related titles are sorted into adjacent clusters. Most of the clusters seem clear regarding their thematic content and contain between 30 and 100 articles.

Figures 2 describes the research stream *AI technologies in general.* This is the oldest and largest one of the identified research streams, containing articles published in the 1970s. Hence, the most cited publications within this stream are textbooks, summing up the knowledge of decades [e.g. 33, 34]. Within this context AI technologies are analyzed, especially from a technical point of view. Applications at the buyer-supplier interface are examined within the context of smaller use cases, e.g., the application of neural networks in demand forecasting. Important technologies are deep learning and ML, neural networks, and fuzzy logic.

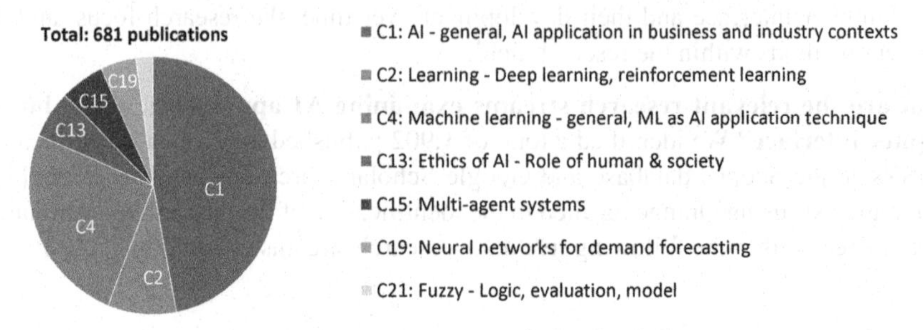

Fig. 2. AI technologies in general: Clusters and sizes.

All other research streams are shown in Fig. 3: In comparison to the stream *AI technologies in general*, the research stream *Industry 4.0* is a relatively young research stream that emerged in 2011 with the first announcement of this term [35]. Comparable approaches are described by the term (industrial) Internet of Things. The parallel use of this terms is reflected by the clusters C6 and C11 which are based on the respective buzzwords. While Industry 4.0 brings multi-disciplinary opportunities for operations management [36], it requires the integration of key suppliers with the help of digital technologies. The application of AI includes all kinds of related processes such as supplier selection [e.g. 37], data and information sharing [e.g. 38], planning, scheduling and forecasting [e.g. 39], transportation [e.g. 40], and inventory management [e.g. 41].

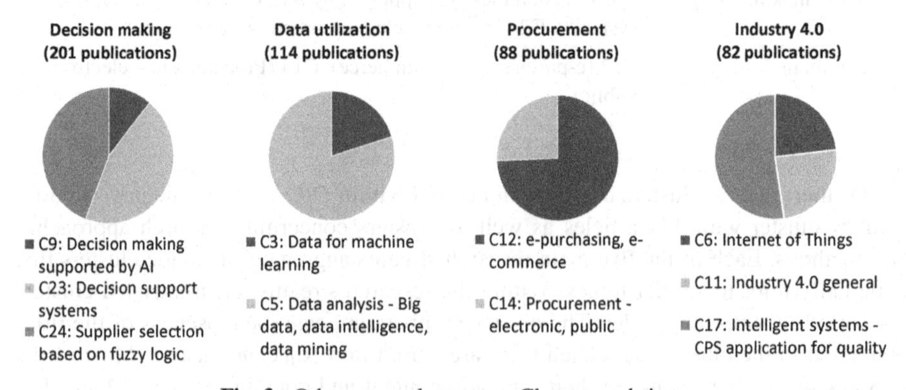

Fig. 3. Other research streams: Clusters and sizes.

As the average publication year within the research stream *Data utilization* is 2018, it can be assumed that there is a growing interest in this area. The increasing amount of available data goes along with the need to derive advantages from it. Cluster C5 covers the issues of generating these data within the context of data and process mining as well as the analysis of so-called Big Data. Cluster C4 puts a special emphasis on an iterative process of data capturing, data analysis and a continuing adaption of processes based on ML technologies. The research stream *Data utilization* covers all practice cases where huge data sets are available, like the analysis of market data in terms of supplier selection [42, 43] or price development [44]. The research stream *Decision-making* depicts how AI can support decision making at the buyer-supplier interface, especially by decision-support systems (DSS) that focus on the supplier selection process [45, 46]. In doing so, several AI technologies are applied to evaluate different selection criteria. The AI technologies fuzzy/ rough set theory, neural network, grey system theory, and genetic algorithm are identified as major ones [45]. Various research projects combine these technologies with multi-attribute decision making techniques like analytic hierarchy process or analytic network process [47, 48]. Further, research within this area deals with the factor 'human' by examining the risk-trust bias impact on the cognitive DSS performance [49] and how AI can be used to enhance (team) collaboration [50]. The research stream *Procurement* considers AI at the buyer-supplier interface especially from a buyer's point of view. First considerations concerning how AI can support purchasing have been made in the 1990s within the context of e-procurement [51]. E-procurement describes the support or replacement of paper or working intensive processes with the help of IT systems [52]. While the terms e-procurement, purchasing or commerce still characterize this research stream, research have developed away from e-procurement towards Procurement 4.0. Within this context, a steady flow of data and information between buyers and suppliers leads to a leap in productivity and performance [52]. AI is applied for operational, tactical, and strategic procurement in terms of supplier selection, cost analysis, and procurement strategy [4]. Several clusters of the category *Others* are mainly characterized by buzzwords and prominent research methods. The largest cluster covers optimization or simulation studies, followed by a cluster that mainly focuses on case studies. Hence, researchers tend to mostly develop AI concepts and validate them by single or multiple case studies. Both literature reviews and quantitative research appear to play a minor role.

Where is the research focus? Fig. 4 depicts the 24 identified clusters along the dimensions cite rate, and total cites. A publication's cite rate is calculated by dividing the total cites by the number of years that have passed since the article was published, thus avoiding disadvantaging more recent publications. The respective number of publications within a cluster is indicated by the size of the circle while the allocation to the different research streams is indicated through color coding. Thus, the relevance of the individual clusters is presented, and comparisons can be performed. Most of the clusters have an average cite rate of up to 5.0 and an average of total cites of up to 40. Thus, they form a rather homogenous group. Interestingly, the clusters *C4: ML – general, ML as AI application technique, C13: Ethics of AI – role of human & society* and *C1: AI – general, application in business and industry context* have comparably high average total cites and cite rates and take up a special position. This might be due to the popularity of the related buzzwords as these clusters are also primarily concerned with general aspects and effects of AI and less with the application of specific technologies. The importance of research into AI ethics and its impact on individuals and the society is shown by the substantially higher average cite rate of C13.

How did research develop over the past years? Figure 5 brings the identified clusters into a timely context by contrasting the average publication year with the year of the first publication. Thus, a cluster's age and development can be derived. Besides the fuzzy cluster C18, the clusters *C1: AI – general, application in business and industry context*, and *C4: ML – general, ML as AI application technique* can be seen as starting point and basis for further research activities. The high cite rates of both clusters (Fig. 4) underline their importance for subsequent AI research at buyer-supplier interface.

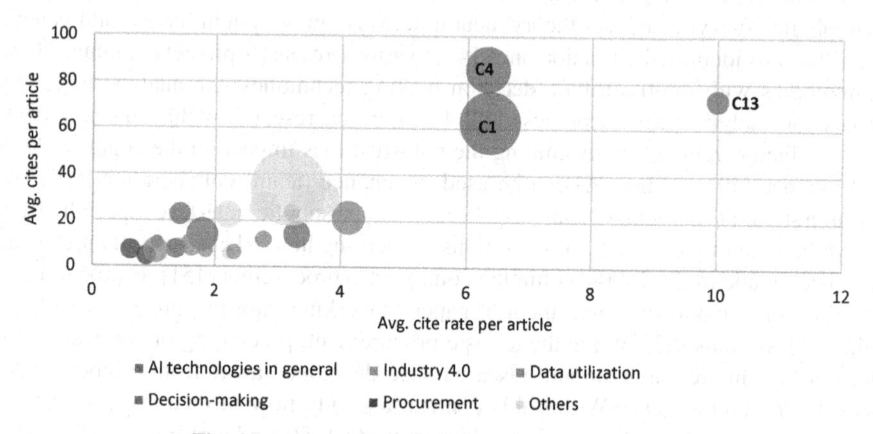

Fig. 4. Average cite rates and total cites per cluster.

Within the 1990s first AI approach have been concretized within the context of e-procurement (*C14: Procurement – public, electronic*). Further approaches refer to decision-making, with the help of decision-support systems, e.g., in terms of supplier selection, in the beginning of this century. Considering the year of the first publication and the average publication year, the research streams *Data utilization* and *Industry 4.0* are the youngest ones. Of course, the interrelationship between Industry 4.0 and creating value from data go hand in hand. Cluster *C11: Industry 4.0 general* can be found in the top right corner which highlights the novelty and growing relevance of this topic. The

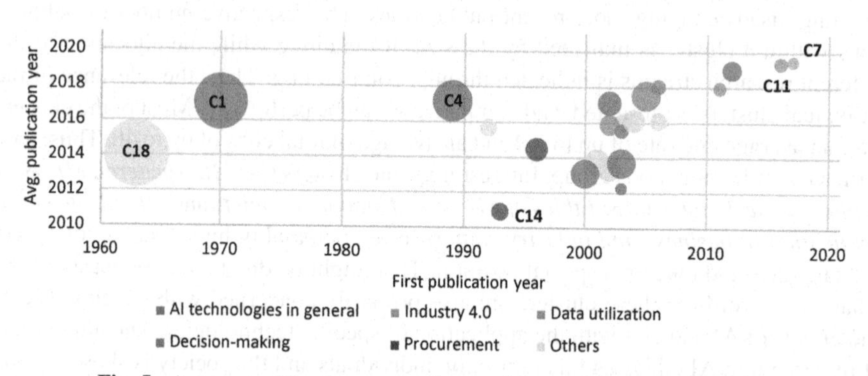

Fig. 5. Average publication year and year of first publication per cluster

placement of *C7: (Systematic) literature review* indicates the current need for resuming AI research at buyer-supplier interface and thus, strengthens or research goal.

Considering the publication date of each article, we remark a significant increase in the number of articles within the past two decades (Fig. 6). While the utilization of AI in the area of SCM has been low for decades [1], digitalization has led to a rapid development of AI and ML-related approaches in theory and practice as tremendous changes are expected in the logistics and SCM sector over the next years [6, 53]. Currently, the literature points towards a development from technology-enabled to technology-centric SCM as the advancing intertwinement of physical logistics with information and data drives the increased use of technology within SCM and logistics [54].

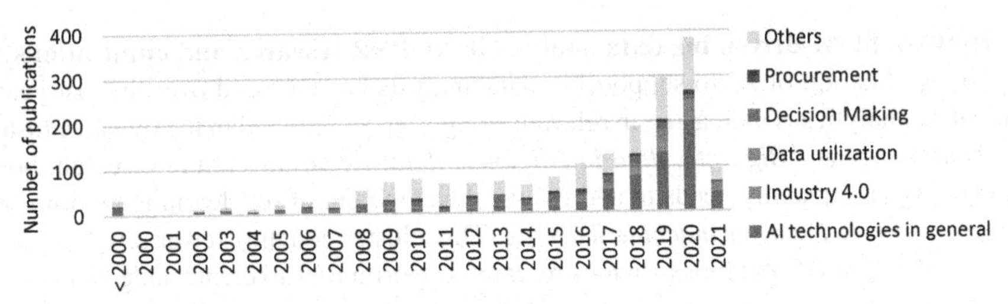

Fig. 6. Distribution of published articles during the years 2000 and 2021.

The total number of articles per year has been rising in all research streams. The share of articles of the streams *Industry 4.0* and *Data utilization* increased significantly over the last ten years. Thus, both streams can be assumed to play a crucial role for future AI research at the buyer-supplier interface. Despite the search date for this review (March 2021), already 96 publications are found for 2021 compared to 383 publications for 2020 and the total number of publications in 2021 is likely to surpass previous years.

5 Discussion of Future Research Directions

Based on the identified clusters and the subsequent in-depth analysis of the 25 most influential papers from the sample (Fig. 1), the underlying research themes and the synthesis of the findings are used to discuss ten central questions for future research in the field of AI application at the buyer-supplier interface:

1. **How can AI support business processes within a BSR?**

To be effective and easily accessible to supply chain managers and others, practice oriented BSR research needs to state what AI brings to the processes under consideration of performance aspects. Besides a research focus in AI-based decision support processes for supplier selection within the context of the *Decision-making* stream, there remains untapped potential for AI implementation in further processes like order management, risk management or supplier development. Research can be based on common models, depicting relevant business processes in a BSR [e.g. 17, 55]. Within this content the type of BSR plays a crucial role when evaluating costs and benefits of AI approaches. The key objectives of readers from practice backgrounds, such as staying up to date with methods and ensuring competitiveness [e.g. 45] and the transferability to other contexts [e.g. 56], should be addressed.

2. **Which research methods should be applied to further develop and examine the possibilities of AI application for BSRs?**
   Within the sample, the main research approaches included simulation, case examples, and mathematical and conceptual papers. The notable position of cluster *C7: (Systematic) literature review* in Fig. 5 highlight an increasing relevance of literature reviews [e.g. 36] which may be useful to gain a better understanding of this complex research field and to benefit from the aggregation of research findings.

3. **How could AI-driven big data analysis drive BSR research and applications?**
   The application of AI to support big data analysis has emerged over the last years while data analytics is increasing in relevance [e.g. 42], as depicted in the related Cluster *C5: Data analysis - Big data, data intelligence, data*: Regarding future research, both data quality and quantity needs to address the development of self-learning algorithms, as related to *C3: Data for machine learning*. This cluster contains fewer articles but is closely related to *C5* and thus provides research opportunities to create added value out of big data.

4. **How could the use of AI influence risk and security within supplier networks?**
   Due to the increasing technological centricity of SCM and BSRs, the potential impact of attacks and uncertainty is growing. This issue is especially considered in terms of the increasing interlinkage of systems, processes, and companies within the context of Industry 4.0 (*C6: Internet of Things, C11: Industry 4.0 general, C17: Intelligent systems – CPS application for quality*) Thus, aspects of risk assessment [e.g. 57] and security [e.g. 58] need to be incorporated in future research projects.

5. **Which areas of BSRs should be considered in terms of decision support?**
   The analysis of the decision-support related clusters *C9: Decision making supported by AI, C23: Decision support systems*, and *C24: Supplier selection based on fuzzy logic* highlights a focus on the final evaluation and selection process step at the buyer-supplier interface [e.g. 45, 46]. Thus, potential remains regarding the application of AI for an ongoing BSR. Some relevant areas could be the consideration of the dynamics characteristics of a long-term relationship [e.g. 59] and e-commerce [e.g. 60], and thus built a bridge to procurement research in *C12: e-purchasing, e-commerce*, and *C14: Procurement – electronic, public.*

6. **How could AI alleviate or intensify biased decision-making in BSRs?**
   Previous research has shown that cognitive-biases [61] negatively affect human decision-making processes in BSRs and consequently provide potential opportunities for technical support systems. Similarly, the impact of risk-trust-bias on the cognitive DSS performance should be considered [49] and contribute to the clusters *C9: Decision making supported by AI, C23: Decision support systems*, and *C24: Supplier selection based on fuzzy logic.*

7. **What could be possible effects of AI on personal relationships and trust?**
The investigation of social metrics needs to be increased as some more sensitive processes such as product development require a more intense relationship [e.g. 62] and might thus benefit from AI support. The manifold effects of AI on interpersonal interaction need to be examined. This question combines approaches concerning, amongst others decision-making and procurement literature.

8. **How could the interplay between human and artificial intelligence be shaped successfully and ethically?**
The omnipresence of AI and related technologies substantially changes our personal and working lives and thus also affects BSRs. Future research needs to address this area and can generate relevant insights as shown by some of the sample publications which, for example, discuss how humans can develop trust in AI [e.g. 63]. The high cite rate of cluster *C13: Ethics of AI – role of human & society* (see Fig. 4) underlines the major interest in this question.

9. **How could AI foster sustainability in BSRs?**
Social, economic, and ecological sustainability are key aims of SCM research and need to be incorporated within BSRs. The application of AI to foster strategic green and sustainable relationships in BSRs and more operative tasks such as waste elimination and decarbonization can be targeted. The term "sustainability" does not represent a distinct cluster but can be found in each research stream [e.g. 46, 64].

10. **How could enabling and future-oriented technologies advance BSR research?**
Despite the suitability of the application area of BSRs, the use of AI appears to stay behind its potential. Some AI technologies such as fuzzy logic are highly relevant within BSR research [e.g. 47]. Technologies that have not been used frequently but also the integration of hybrid techniques [e.g. 45] and the inclusion of emerging mobile technologies or cloud computing [e.g. 65] constitute interesting future directions. In general, a development towards technology-centric SCM can be observed, as is highlighted by research stream *Industry 4.0* which contains *C6*, *C12* and *C17* and drives considerations of platform economy and the Internet of Things.

6 Conclusion, Contributions and Limitations

This article provides critical insights into the present and future scope of AI research at buyer-supplier interface. We outline a systematic overview of the existing comprehensive body of literature by clustering related publications with a MLR approach and identify unexploited research topics. Thus, we theoretically contribute to this issue by facilitating the navigation in this complex and heterogenous research field and by presenting ten central questions as a clue on new research directions. This article will provide an insightful understanding of the scope of AI research at buyer-supplier interface and set impulses for relevant and practice-oriented future projects.

Findings of this review are limited by the choice methodology, including the selection of databases and search engines as well as by the selective literature search process.

The applied MLR is mainly limited by its overarching perspective instead of in-depth analysis. The selected sample of the five most influential publications of each research stream limits the transferability and generalizability of the findings. In addition, choosing general keywords such as AI might eliminate methods, as for example cluster analysis or artificial neural networks, which might not necessarily be connected or have previously been used in relation to this umbrella term. Furthermore, the in-depth analysis of the top five publications for each of the identified research streams reduces the generalizability and transferability of the conclusions.

References

1. Min, H.: Artificial intelligence in supply chain management: theory and applications. Int. J. Log. Res. Appl. **13**, 13–39 (2010)
2. Baryannis, G., Validi, S., Dani, S., Antoniou, G.: Supply chain risk management and artificial intelligence: state of the art and future research directions. Int. J. Prod. Res. **57**, 2179–2202 (2019)
3. Bode, C., Vollmer, M., Burkhart, D.: 2020 CPO Survey. University of Mannheim (2020)
4. Spreitzenbarth, J.M., Stuckenschmidt, H., Bode, C.: Methods of artificial intelligence in procurement: a conceptual literature review. In: International Purchasing and Supply Education and Research Association Conference (IPSERA) (2021)
5. Feng, L., Chiam, Y.K., Lo, S.K.: Text-mining techniques and tools for systematic literature reviews: a systematic literature review. In: Asia-Pacific Software Engineering Conference (APSEC), pp. 41–50 (2017)
6. Backhaus, A., et al.: Logistik 2020: Struktur- und Wertewandel als Herausforderung. Gipfel der Logistikweisen: Initiative zur Prognose der Entwicklung der Logistik in Deutschland (2020)
7. Koutsojannis, C., Sirmakessis, S. (eds.): Tools and Applications with Artificial Intelligence, Vol. 166 (2009)
8. Schill, K., Scholz-Reiter, B., Frommberger, L.: Preface: artificial intelligence and logistics. In: International Joint Conference on Artificial Intelligence (IJCAI), pp. 1–2. Universität Bremen, Universität Freiburg, (2011)
9. Anon: Digitalisierungsindex Mittelstand 2019/2020 - Der digitale Status Quo in deutschen Transport- und Logistikunternehmen Telekom (2019)
10. Kühl, N., Goutier, M., Hirt, R., Satzger, G.: Machine learning in artificial intelligence: towards a common understanding. In: Hawaii International Conference on System Sciences, Hawaii (2019)
11. Wang, W., Siau, K.: Artificial intelligence, machine learning, automation, robotics, future of work and future of humanity: A review and research agenda. J. Database Manage. (JDM) **30**, 61–79 (2019)
12. Luger, G.F.: Artificial intelligence: Structures and Strategies for Complex Problem Solving. Pearson Education, Inc. (2009)
13. Hvolby, H.H., Trienekens, J., Steger-Jensen, K.: Buyer–supplier relationships and planning solutions. Prod. Plann. Control **18**, 487–496 (2007)
14. Morsy, H.: Buyer-supplier relationships and power position: interchaning. Int. J. Supply Oper. Manage. **4**, 33–52 (2017)
15. Cooper, M.C., Gardner, J.T.: Building good business relationships: more than just partnering or strategic alliances? Int. J. Phys. Distrib. Logist. Manage. **23**, 15–26 (1993)
16. Håkansson, H., Persson, G.: Supply chain management: the logic of supply chains and networks. Int. J. Logistics Manage. **15**, 11–26 (2004)

17. Lambert, D.M., Schwieterman, M.A.: Supplier relationship management as a macro business process. Supply Chain Manage.: Int. J. (2012)
18. Mahdikhah, S., Messaadia, M., Baudry, D., Evans, R., Louis, A.: A business process modelling approach to improve OEM and supplier collaboration. J. Adv. Manag. Sci. **2**, 246–253 (2014)
19. Veile, J.W., Schmidt, M.-C., Müller, J.M., Voigt, K.-I.: Relationship follows technology! How Industry 4.0 reshapes future buyer-supplier relationships. J. Manuf. Technol. Manage. **31**, 977–997 (2020)
20. Tremblay, M.C.: Uncertainty in the information supply chain: integrating multiple health care data sources. In: Americas Conference on Information Systems (AMCIS) (2006)
21. Reuther, K.: A Systems Theory Perspective of Interconnected Influence Factors on Front-End Innovation: The Role of Organisational Structures. School of Business and Enterprise, vol. PhD Thesis. University of the West of Scotland, United Kingdom (2019)
22. Okoli, C., Schabram, K.: A guide to conducting a systematic literature review of information systems research. Working Papers on Information Systems. Sprouts (2010)
23. Petticrew, M., Roberts, H.: Systematic reviews in the social sciences: a practical guide. Blackwell, Malden, USA (2006)
24. Fink, A.: Conducting research literature reviews: From the internet to paper. Sage publications (2019)
25. Baker, J.D.: The purpose, process, and methods of writing a literature review. AORN J. **103**, 265–269 (2016)
26. Denyer, D., Tranfield, D.: Producing a systematic review. In: Bryman, B.a. (ed.) SAGE Handbook of Organizational Research Methods. SAGE Publications Ltd., London, England (2009)
27. Zupic, I., Čater, T.: Bibliometric methods in management and organization. Organ. Res. Methods **18**, 429–472 (2015)
28. Heilig, L., Voß, S.: A scientometric analysis of public transport research. J. Public Transp. **18**, 8 (2015)
29. Chellappandi, P., Vijayakumar, C.: Bibliometrics, scientometrics, webometrics/cybermetrics, informetrics and altmetrics–an emerging field in library and information science research. Shanlax Int. J. Educ. **7**, 5–8 (2018)
30. Tranfield, D., Denyer, D., Smart, P.: Towards a methodology for developing evidence-informed management knowledge by means of systematic review. Br. J. Manag. **14**, 207–222 (2003)
31. Schryen, G.: Writing qualitative is literature reviews—guidelines for synthesis, interpretation, and guidance of research. Commun. Assoc. Inf. Syst. **37**, 286–325 (2015)
32. Demsar, J., et al.: Orange: data mining toolbox in Python. J. Mach. Learn. Res. **14**, 2349–2353 (2013)
33. Murphy, R.R.: Introduction to Al robotics. The Mit Press, Cambridge, USA (2000)
34. Jang, J.-S.R., Sun, C.-T., Mizutani, E.: Neuro-fuzzy and soft computing-a computational approach to learning and machine intelligence [Book Review]. IEEE Trans. Autom. Control **42**, 1482–1484 (1997)
35. Kiel, D.: What do we know about "Industry 4.0" so far. Int. Assoc. Manage. Technol. **2**, 1–22 (2017)
36. Ivanov, D., Tang, C.S., Dolgui, A., Battini, D., Das, A.: Researchers' perspectives on Industry 4.0: multi-disciplinary analysis and opportunities for operations management. Int. J. Prod. Res., 1–24 (2020)
37. Nejma, M., Zair, F., Cherkaoui, A., Fourka, M.: Advanced supplier selection: a hybrid multi-agent negotiation protocol supporting supply chain dyadic collaboration. Decis. Sci. Lett. **8**, 175–192 (2019)

38. Dominguez, R., Cannella, S., Barbosa-Póvoa, A.P., Framinan, J.M.: OVAP: a strategy to implement partial information sharing among supply chain retailers. Trans. Res. Part E: Logistics Trans. Rev. **110**, 122–136 (2018)
39. Syntetos, A.A., Babai, Z., Boylan, J.E., Kolassa, S., Nikolopoulos, K.: Supply chain forecasting: theory, practice, their gap and the future. Eur. J. Oper. Res. **252**, 1–26 (2016)
40. Xu, S., Liu, Y., Chen, M.: Optimisation of partial collaborative transportation scheduling in supply chain management with 3PL using ACO. Expert Syst. Appl. **71**, 173–191 (2017)
41. Borade, A.B., Sweeney, E.: Decision support system for vendor managed inventory supply chain: a case study. Int. J. Prod. Res. **53**, 4789–4818 (2015)
42. Cavalcante, I.M., Frazzon, E.M., Forcellini, F.A., Ivanov, D.: A supervised machine learning approach to data-driven simulation of resilient supplier selection in digital manufacturing. Int. J. Inf. Manage. **49**, 86–97 (2019)
43. Fallahpour, A., Olugu, E.U., Musa, S.N., Khezrimotlagh, D., Wong, K.Y.: An integrated model for green supplier selection under fuzzy environment: application of data envelopment analysis and genetic programming approach. Neural Comput. Appl. **27**(3), 707–725 (2015). https://doi.org/10.1007/s00521-015-1890-3
44. Yang, R., et al.: Big data analytics for financial market volatility forecast based on support vector machine. Int. J. Inf. Manage. **50**, 452–462 (2020)
45. Chai, J., Liu, J.N.K., Ngai, E.W.T.: Application of decision-making techniques in supplier selection: a systematic review of literature. Expert Syst. Appl. **40**, 3872–3885 (2013)
46. Zimmer, K., Fröhling, M., Schultmann, F.: Sustainable supplier management–a review of models supporting sustainable supplier selection, monitoring and development. Int. J. Prod. Res. **54**, 1412–1442 (2016)
47. Kar, A.K.: A hybrid group decision support system for supplier selection using analytic hierarchy process, fuzzy set theory and neural network. Journal of Computational Science **6**, 23–33 (2015)
48. Pitchipoo, P., Venkumar, P., Rajakarunakaran, S.: Fuzzy hybrid decision model for supplier evaluation and selection. Int. J. Prod. Res. **51**, 3903–3919 (2013)
49. Lai, K., Oliveira, H.C., Hou, M., Yanushkevich, S.N., Shmerko, V.: Assessing risks of biases in cognitive decision support systems. In: 2020 28th European Signal Processing Conference (EUSIPCO), pp. 840–844. IEEE (2021)
50. Metcalf, L., Askay, D.A., Rosenberg, L.B.: Keeping humans in the loop: pooling knowledge through artificial swarm intelligence to improve business decision making. Calif. Manage. Rev. **61**, 84–109 (2019)
51. Goch, B., Julien, P., Lias, J.: Intelligent network element software procurement and delivery. IEEE International Conference on Communications, vol. 2, pp. 1189–1196. IEEE (1993)
52. Glas, A.H., Kleemann, F.C.: The impact of industry 4.0 on procurement and supply management: A conceptual and qualitative analysis. International Journal of Business and Management Invention 5, 55–66 (2016)
53. Junge, A.L., Verhoeven, P., Reipert, J., Mansfeld, M.: Pathway Of Digital Transformation In Logistics: Best Practice Concepts and Future Developments. Universitätsverlag TU Berlin, Berlin (2019)
54. Witten, P., Schmidt, C.: Globale Trends und die Konsequenzen für die Logistik der letzten Meile. In: Schröder, M., Wegner, K. (eds.) Logistik im Wandel der Zeit – Von der Produktionssteuerung zu vernetzten Supply Chains, pp. 303–319. Springer Gabler, Wiesbaden (2019)
55. Burger, M., Arlinghaus, J.: Digital Supplier Integration - The Impact of Buyer-Supplier Relationships on Industry 4.0 Transaction Maturity. Wissenschaftliches Symposium des BME, Online (2021)
56. Ben Othman, S., Zgaya, H., Dotoli, M., Hammadi, S.: An agent-based decision support system for resources' scheduling in emergency supply chains. Control. Eng. Pract. **59**, 27–43 (2017)

57. Tsang, Y.P., Choy, K.L., Wu, C.-H., Ho, G.T., Lam, C.H., Koo, P.: An Internet of Things (IoT)-based risk monitoring system for managing cold supply chain risks. Ind. Manage. Data Syst. (2018)
58. He, H., et al.: The security challenges in the IoT enabled cyber-physical systems and opportunities for evolutionary computing & other computational intelligence (2016)
59. Ferreira, L., Borenstein, D.: A fuzzy-Bayesian model for supplier selection. Expert Syst. Appl. **39**, 7834–7844 (2012)
60. Liu, G., et al.: Repeat buyer prediction for e-commerce. In: 22nd ACM SIGKDD International Conference on Knowledge Discovery and Data Mining, pp. 155–164 (2016)
61. Kahneman, D., Knetsch, J.L., Thaler, R.H.: Anomalies: the endowment effect, loss aversion, and status quo bias. J. Econ. Perspect. **5**, 193–206 (1991)
62. Santos, L.F.d.O.M., Osiro, L., Lima, R.H.P.: A model based on 2-tuple fuzzy linguistic representation and analytic hierarchy process for supplier segmentation using qualitative and quantitative criteria. Expert Syst. Appl. **79**, 53–64 (2017)
63. Andras, P., et al.: Trusting intelligent machines: deepening trust within socio-technical systems. IEEE Technol. Soc. Mag. **37**, 76–83 (2018)
64. Kannan, D., Mina, H., Nosrati-Abarghooee, S., Khosrojerdi, G.: Sustainable circular supplier selection: a novel hybrid approach. Sci. Total Environ. **722** (2020)
65. Ngai, E., Peng, S., Alexander, P., Moon, K.K.: Decision support and intelligent systems in the textile and apparel supply chain: an academic review of research articles. Expert Syst. Appl. **41**, 81–91 (2014)

The Effect of Order Batching on a Cyclical Order Picking System

Flora Maria Hofmann(✉) and Stephan Esterhuyse Visagie

Stellenbosch University, Stellenbosch 7600, South Africa
{fhofmann,svisagie}@sun.ac.za

Abstract. Order batching on a unidirectional cyclical picking system implemented at a prominent South African retailer is investigated. Four interdependent sub-problems are solved sequentially to optimise the entire system. These sub-problems are (a) the picking line assignment problem, (b) the stock keeping unit arrangement problem, (c) the system configuration problem, and (d) the order sequencing problem. The picking is performed in waves. The four sub-problems are viewed as decision tiers that must be solved to optimise each wave. The main objective is to minimise overall walking distance and thus reduce total picking time for a picking wave. Order batching is introduced to this picking system to explore its effect on total completion time. Orders are formed during the optimisation process and thus not known from the start. This also raises the question of where in the optimisation process to include order batching. Furthermore, the effect of increasing pick density to indirectly improve order batching is analysed. The combination of all solution approaches for each of the four decision tiers including the additional layer of order batching is evaluated. Three scenarios based on real-life historical data of the retailer are tested. The best solution approach is compared to a benchmark. The suggested batching approach saves up to 27.8% in total picking time.

Keywords: Order picking system optimisation · Assignment problem · Unidirectional cyclical picking line · Order batching · Complex logistics system

1 Problem Background

The picking activities in a real life distribution centre (DC) is considered in this study. The DC is owned by a large South African retailer (the Retailer). The Retailer serves about 2 500 stores in Southern Africa. It sells apparel and homeware to the low income population of South Africa. The main activity in the Retailer's DC is to transform bulk supply into customer (store) requirements. This process accounts for 50% to 65% of the cost of operation [4,10]. Due to the large number of stores and the economic situation in South Africa, the Retailer developed a unique picking system for their DCs. The Retailer adopted a central

M. Mes et al. (Eds.): ICCL 2021, LNCS 13004, pp. 252–268, 2021.
https://doi.org/10.1007/978-3-030-87672-2_17

planning approach to stock stores. This kept the cost of operation low for such a large number of stores. In this approach, planners assign stock, on a stock keeping unit (SKU) level, to stores centrally, rather than store managers ordering or requesting stock. The picking process that executes these central plans operates in picking waves. A picking wave has four sub-processes: (1) taking stock to a picking line, (2) locating stock on the picking line, (3) picking the orders, and (4) clearing away stock (if any is left). This implies that all store requirements for a SKU present in a wave are fulfilled during that wave. Similar SKUs, for example the different sizes of the same product, are grouped into distributions (DBNs). All SKUs in DBNs are schedule for picking in the same wave, once the DBNs including pick instructions are released to the DC [25].

The DC independently operates unidirectional cyclical picking lines in parallel. On each of these physical picking lines a wave of picking will be executed. In Fig. 1 this set up is displayed.

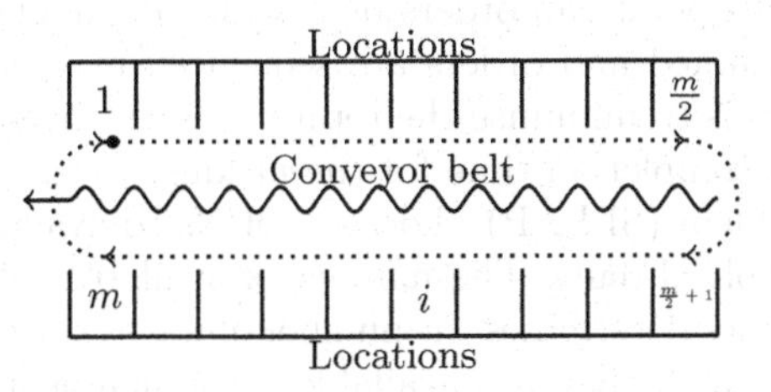

(a) A single picking line consisting of m locations. Source: [11].

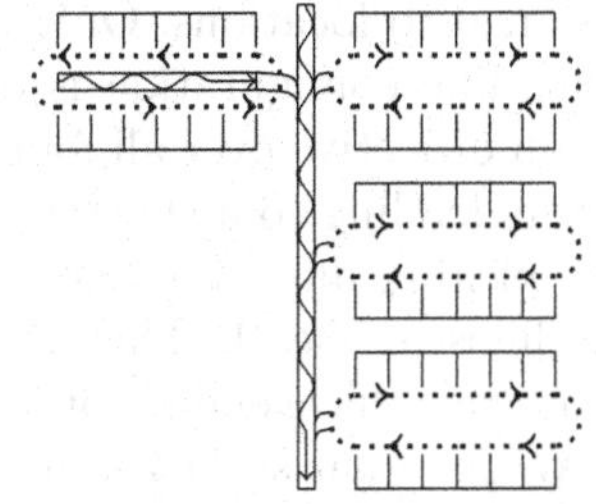

(b) A combination of picking-lines in the DC in Cape Town.

Fig. 1. The order picking system in the DC of the Retailer is represented schematically. An individual picking line (picking locations around a conveyor belt) is shown on the left. The dashed lines show how the pickers walk while picking. The picking line system on the right has one line around a conveyor belt and three more lines without conveyor belts.

The physical layout of a unidirectional cyclical picking line is schematically depicted in Fig. 1(a). It contains m locations. Typically, a picking line includes between 64 and 76 locations with the option of a conveyor belt. A unique location is assigned to each SKU during a wave of picking. At each location up to five pallets of stock can be stored avoiding replenishment during a picking wave. A voice recognition system (VRS) leads pickers onto their pick path. There are two configurations of the picking line, namely the U- and the Z-configuration. Pickers walk in a clockwise direction while picking from the locations in a U-configuration. Pickers are allowed to cross over the aisle to a picking location at the opposite pick face in a Z-configuration [16].

The U-configuration resembles a unidirectional carousel system described in literature if pickers are seen as static relative to moving SKUs. In carousels pickers remain at a fixed location while the carousel presents stock to them, thus accommodating only one picker per picking line [21]. In the U-configuration more pickers can be accommodated and for comparison purposes the number of pickers was set to eight pickers [15]. However, the major difference between these two systems are the presence of wave picking with deterministic orders that is present in the U-configuration [25].

The DC in Cape Town is schematically displayed in Fig. 1(b). One picking line has a conveyor belt in the middle, allowing only the U-configuration. The other three lines (organised as a picking module) do not have conveyor belts and thus allow both U- and Z-configurations. After picking, cartons are transported via a main conveyor belt connecting all picking lines to a dispatching area.

2 Problem Description

Managing the picking waves requires the solution of four sequential and dependent decision tiers. DBNs are allocated to picking lines in Tier 1. In Tier 2 SKUs are assigned to locations. Only at this point can orders be calculated. In Tier 3 a picking-line configuration is determined and orders are sequenced in Tier 4. The main objective over all four tiers is to minimise the total distance traversed by pickers leading to a shorter total completion time for the picking wave.

The picking line assignment problem (SPLAP) allocates DBNs to available picking lines in Tier 1. This tier is solved daily. The number of available (after previous waves have ended) picking lines determines the number of waves starting that day. As soon as DBNs are released, they are available for assignment to waves of picking [27]. Picking-line managers assign DBNs according to in-house rules. For example, they might try to minimise the number of different store departments (like girls, boys, women, men) per wave. The SKU arrangement problem (SAP) arranges SKUs on an available picking line during Tier 2. In Fig. 2, the interaction between Tier 1 and 2 is illustrated [27]. Once again picking line managers assign SKUs based on in-house rules, usually to spread out SKUs with high volumes evenly to avoid picker congestion.

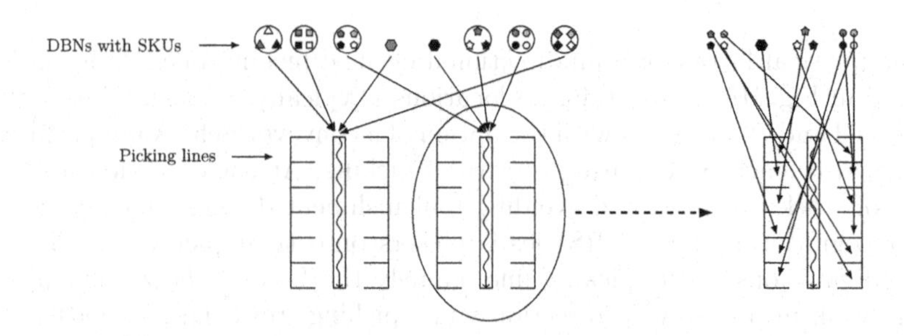

Fig. 2. The allocation of DBNs to picking lines during Tier 1 (left) and arrangement of SKUs on a picking line in Tier 2 (right). A circle depicts a DBN and each shade within a circle depicts a SKU. SKUs of the same DBN have different shades. Source: [24].

The system configuration problem (SCP) selects the configuration in which the picking line operates in Tier 3. Only picking lines that do not have a conveyor belt, as depicted in Fig. 3, have this selection. The configuration determines the logic of how pickers walk while picking. Figure 3(a) shows a U-configuration in which all pickers move in a cyclical clockwise direction from one location to another. Figure 3(b) depicts the Z-configuration in which a picker may move in a single direction, while switching between both pick faces when picking an order. A picker turns around at the end of the line and starts moving in the opposite direction. The picker stays on the same side (pick face) if the next required SKU on the same side is closer than the required SKU on the opposite side, but the picker will cross if the SKU on the opposite side is closer [16]. Managers stick to the in-house rule to use the Z-configuration if the pickers stop on average at less than a third of the locations. The U-configuration is used if stops are above this number.

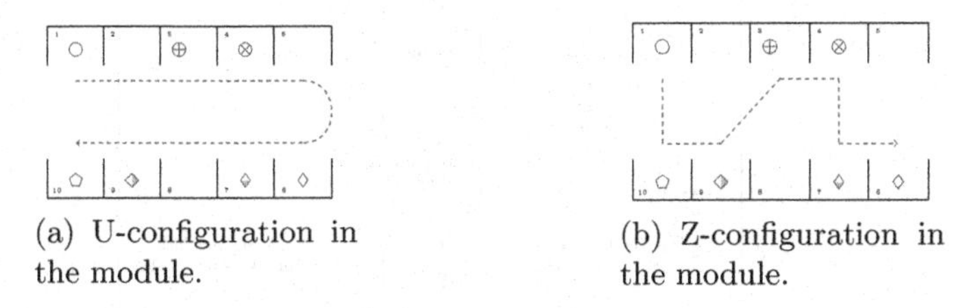

(a) U-configuration in the module.

(b) Z-configuration in the module.

Fig. 3. Picker walking (dashed lines) compared in representation of U- and Z-configuration. SKUs to be picked are illustrated by different shapes. An order is picked if a picker visits all those locations.

The order sequencing problem (OSP) sequences orders of a wave before it commences in Tier 4. Not all orders (often none) require all SKUs present in a wave and after completing the previous order a picker can get assigned to any order and start that order at any location [27]. Figure 4 displays the effect of order sequencing. The coloured, dashed lines illustrate the pick path of the same three orders, but in different sequences in a U-configuration. In the first case picking starts with yellow, followed by green and blue. As displayed in Fig. 4(a), this results in 27 locations passed. In Fig. 4(b) the blue order is picked first, followed by the yellow and green orders with total length of 24 locations. The objective is to find a sequence that minimise the total number of locations passed.

The information flow while solving all tiers is depicted in Fig. 5. Essential to optimising the picking system globally is the connection between all four decision tiers [27]. On top of this system, that is already studied, for example in [25–27], this paper introduces order batching in an attempt to further reduce the total distance walked by all pickers.

Order batching allows a single picker to pick all orders in a batch simultaneously to reduce walking distance [9]. The influence of order batching on

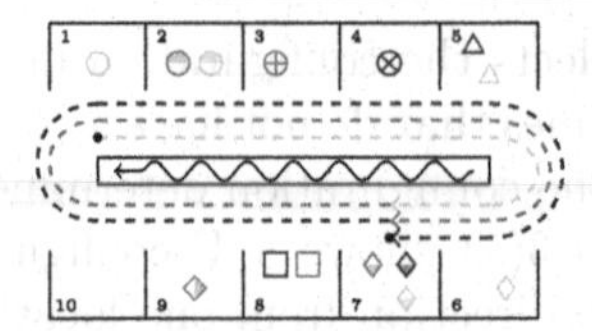

(a) Sequence 1: Yellow, green, blue.

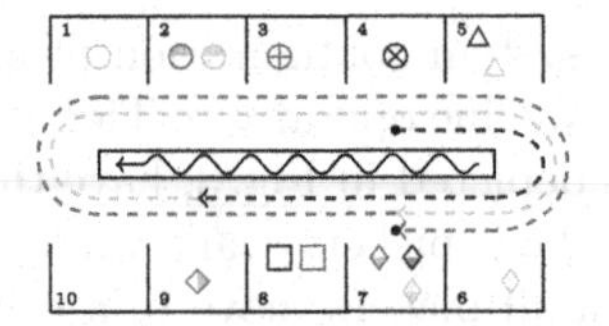

(b) Sequence 2: Blue, yellow, green.

Fig. 4. Tier 4 is represented by differences in order sequencing with the picker passing 27 locations in the first option and 24 locations in the second. (colour figure online)

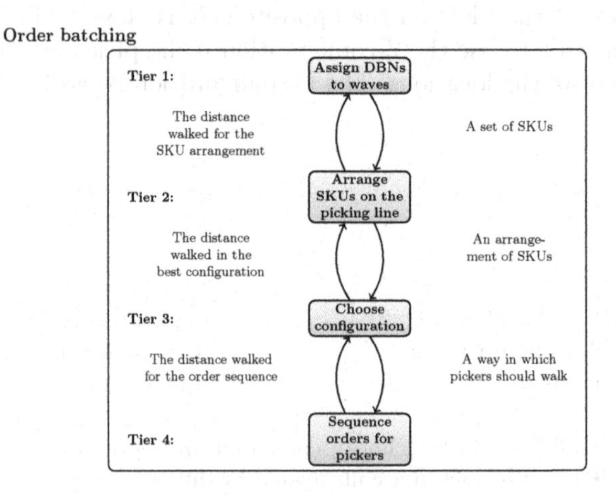

Fig. 5. The information flow between the four tiers represented schematically.

walking distance is depicted in Fig. 6. Walking distance is measured in *cycles traversed* [25]. Each time a picker completes a full cycle, it counts as one more cycle traversed. Consider a small example to illustrate the concept. In Fig. 6(a) the yellow order is picked first, followed by green, blue, and red. This results in four cycles traversed. In Fig. 6(b) yellow and red make up the orange batch, while green and blue forms the purple batch. This introduction of batching reduces the walking distance by 50% to only two cycles traversed.

The main objective of this paper is to use order batching to reduce the total completion time (by minimising walking distance). This will also lead to a reduction in total completion time of a wave of picking. After order batching metrics have been developed in [14,17], the associated picking time reduction through batching has been shown in [15], and the influence of batching on the configuration choice has been evaluated in [16], order batching seems to influence all four tiers directly or indirectly within the order picking system. Therefore, order batching is now added as another layer to the optimisation problem as depicted in Fig. 5. This paper thus aims at identifying a solution approach for

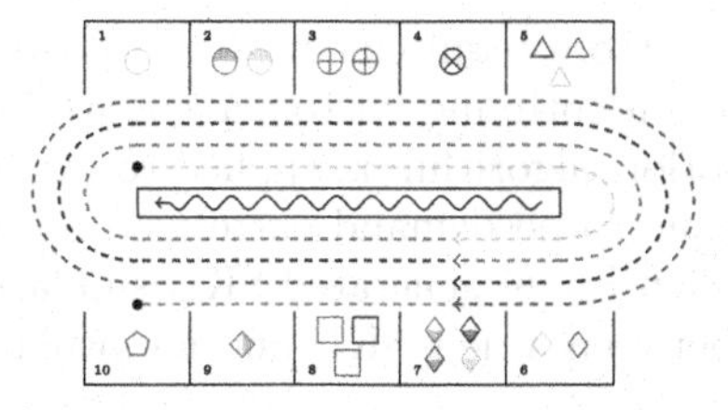

(a) A single order picking representation.

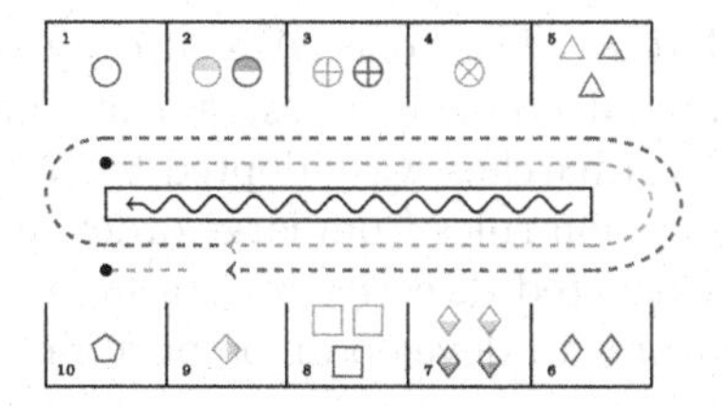

(b) A batch picking representation.

Fig. 6. Schematic representation comparing the walking distance between single and batch picking. (colour figure online)

each tier that supports order batching. The point when order batching should be included in the decision-making process will also be investigated.

A brief literature review follows in the next section. In Sect. 4 all four decision tiers with their possible solution approaches are presented. This is followed by testing combinations of solution approaches in Sect. 5. This is performed on real-life historical data that spans a full month of picking. A summary and possible ideas for future research conclude this study in the final section.

3 Literature

The three decision tiers in the unidirectional cyclical picking system have been addressed in literature, but without order batching [23,25–27]. The addition of another tier in which the configuration is chosen has also been investigated by [16]. A brief overview of the published research on each tier in this picking system will be discussed. The discussion will follow the four decision tiers.

Decision Tier 1: Each picking line functions independently allowing picking waves to be viewed as zones. Therefore, zone or modified non-zoned approaches can be tested for solving the SPLAP. A framework based on assigning SKUs in some order to locations was proposed [18,26]. DBNs may be assigned to picking waves in a greedy insertion procedure reducing the walking distance by minimising the sum of maximal SKU sizes per picking line. The *maximal SKU* is defined as the SKU that is needed by the largest number of stores. This number of stores serves as a lower bound for the number of cycles that must be traversed to pick all orders in a wave [25]. The cyclical layout and presence of multiple pickers is not included in this approach [26]. A correlation measure calculating the number of orders that require two specific SKUs for non-zoned picking systems was developed by [22]. A similarity measure with regards to SKU correlations, a stock turn coefficient, and a Jaccard statistic improving throughput at the DC was introduced by [5]. Simultaneously considering storage allocation and assignment, [1] suggested a top-down hierarchical solution procedure. Four desirability scores based on these studies, combined with the greedy insertion procedure, were developed by [23]. The authors recommended a desirability score that considers the number of stores needed by the candidate DBN that

includes at least one DBN already allocated. Store requirements were added to zone and non-zoned assignment approaches. Identifying customer connections to form batches was adopted by [7]. The apriori algorithm is applied to extract association rules from large order sets. The indicated demand patterns can then be extracted directly. A data mining approach by [8] generated SKU similarity measures to develop an association index between a new SKU and an available location.
Decision Tier 2: Minimising walking distance by arranging SKUs in the SAP was presented by [27], who introduced modified heuristic approaches from literature to meet layout requirements, namely the organ pipe arrangement, assigning SKUs to locations in a greedy sequential manner, placing SKUs close together if the probability of them appearing in the same order is high, and the classroom discipline heuristic [12,27,31,35,36]. However, order sequencing can reduce walking distance by 15% as opposed to a 1% improvement for Tier 2 in terms of optimisation possibility (Tier 4 far outweighs Tier 2). Therefore, a heuristic that is easily implementable is recommended for this tier [27]. Storage assignment in carousels was optimised with a branch-and-bound algorithm by [19], but order sequencing was not included in the model assumptions.
Decision Tier 3: The *pick density measure* d_{m} divides all picker stops by an approximation of distance travelled by pickers [16]. This serves as a guide of when to alternate between a Z- and U-configuration. This approach will be used to choose a configuration in this paper.
Decision Tier 4: Investigations on the OSP by [25] introduced a novel maximal cut approach, but the authors recommended a nearest end heuristic that provides comparable results with less computational effort. In this heuristic, the order with the nearest ending location is selected to be picked next, given the current location of a picker [3].

The solution approaches reviewed here are adopted by incorporating a batch of orders as a single big order to include order batching in the calculations. The system is optimised globally by finding the best combination of approaches over all four decision tiers resulting in the shortest total walking distance traversed by pickers for the batched orders.

4 Incorporating Order Batching in the Four Decision Tiers

In the following section the picking system's four decision tiers are analysed. Note that even though orders are not defined in the first tier yet, the effect of order batching is already investigated in this tier.

4.1 DBN Assignment to Picking Waves

DBNs are assigned to picking waves in Tier 1. The maximal SKU forms a lower bound on the number of pick cycles needed. With this in mind, the objective can be reformulated to minimising the maximal SKU [20]. At this stage, orders are not defined yet and thus order batching cannot be introduced here. However, walking distance can be reduced by selecting DBNs to increase pick density thus indirectly benefiting order batching even before orders are known [15,16].

Algorithm 1 describes the greedy insertion approach (GI) that is based on ranking all items that have not been assigned yet from best to worst possible allocation in decreasing order [34]. It is ensured that all DBNs are allocated by holding out DBNs that only need one location and assigning them in a second phase [26]. Furthermore, the picking density may increase if DBNs needed by a large number of stores are added first.

Algorithm 1: `Greedy insertion (GI)`

Input: A set of picking lines $\mathcal{L}$ – ordered by number of locations. A set of DBNs $\mathcal{D}$.
Output: Allocation of DBNs to picking lines.

1: **while** unassigned DBNs fit into l **do**
2: Assign DBN with largest maximal SKU to l.
3: **end while**

The desirability measure (DM) incorporating how many stores needed a particular DBN shows the best results amongst correlation approaches that minimise walking distance [24]. Therefore, the candidate DBN d is added to the set $\mathcal{D}_l$ of DBNs already allocated to picking line l, that has the most stores in common with it. All assigned DBNs are used to compute the correlation between this new DBN and the store set that needs the candidate DBN. The desirability score $\mathcal{A}(\mathcal{D}_l, d)$ is calculated as

$$\mathcal{A}(\mathcal{D}_l, d) = |\mathcal{G}_l \cap \mathcal{G}_d|, \tag{1}$$

with the set of stores $\mathcal{G}$ [24].

In Algorithm 2 this desirability score is added to the greedy insertion algorithm (GIDM) as the greedy insertion heuristic. If the same store set requires a DBN already assigned, the pick density may increase.

Algorithm 2: `GI with desirability measure (GIDM)`

Input: A set of picking lines $\mathcal{L}$ – ordered by number of locations. A set of DBNs $\mathcal{D}$ and a pre-allocated set $\mathcal{D}_l$ for picking line l.
Output: Allocation of DBNs to picking lines.

1: **while** unassigned DBNs fit into l **do**
2: Assign DBN with largest desirability score $\mathcal{A}(D_l, d)$ to l.
3: **end while**

SKUs are linked with customer requirements before DBNs are allocated to picking waves and can thus be used before waves are determined. The *apriori* algorithm [2] uncovers relationships to form batches in a pattern mining approach [7] and was extended further by [6]. Associations between DBNs are recognised with regards to *support*, *confidence*, and *lift*. Support is showing DBNs requested together by stores. Confidence describes the likelihood of requesting DBN 1 and 2. Correlation measures the degree to which requesting DBN 1 increases requesting DBN 2 [13]. Afterwards, a clustering procedure maximising the sum of the support values is applied [6]. Algorithm 3 describes the pattern mining approach (PMA). Until no significant association rules are generated any more, the support value is lowered dynamically and unassigned DBNs are added in a greedy manner. The pick density may increase if SKUs assigned to the same picking line are required by the same stores.

Algorithm 3: `Pattern mining (PMA)`

Input: A set of picking lines $\mathcal{L}$. A set of DBNs $\mathcal{D}$ together with the quantities going to each store on a SKU level.
Output: DBNs allocated to picking lines.

1: **while** unassigned DBNs fit into l **do**
2:     **while** The apriori algorithm generated association rules are not empty **do**
3:         Allocate DBN with the highest support value to picking line l.
4:         **if** Unassigned DBNs **then**
5:             The support value is lowered and the association rule is updated.
6:         **end if**
7:     **end while**
8:     The remaining DBNs are assigned in a greedy manner to l.
9: **end while**

A greedy random assignment approach (GRA) will be used to benchmark results. The proposed algorithms are also compared to historical assignments (HA) to evaluate if increasing the pick density in Tier 1 supports batching.

4.2 Arranging SKUs on a Picking Line

It has been shown that arranging the set of SKUs in Tier 2 is less than 1% and is outweighed by the gains in the final tier [27]. Therefore, quick approaches for this tier are recommended. Order-to-route closeness metrics might be considered, but cannot be applied because order routes have not been defined yet. Increasing pick density thus remains the objective during this stage.

SKU arrangement is determined by means of a greedy random (GRL) and greedy sequential (GSL) heuristic. The GRL assigns SKUs randomly to locations and the GSL arranges according to the maximal SKU or pick frequency. GRL provides a benchmark, while GSL may support batching later by assigning SKUs on high demand close to each other.

4.3 Configuration Selection

In Tier 3 a decision is made between the U- or Z-configuration for each wave of picking. This choice only influences batching metrics that use the actual picker path. A configuration should thus be selected before introducing order batching, because picker paths are not known at this point yet. Selecting a Z- or

U-configuration can be determined by the pick density measure d_{m} [16]. This measure is based on the number of (a) stops s, (b) locations m, and (c) the maximal SKU v. This measure is computed as

$$d_{\mathrm{m}} = \frac{s}{m \cdot v} \tag{2}$$

and will always lie between 0 and 1 [15].

Note that if the pick density is increased in Tier 1 and 2, it should make the selection of the U-configuration more likely. Tier 3 on the other hand directly influences selecting order batching metrics.

4.4 Order Sequencing Including Order Batching

In Tier 4 orders (or batches of orders) have to be sequenced. All information that is necessary to introduce order batching is available in this tier. Several metrics are available for consideration when batching orders.

The combination of stops metric and greedy random heuristic (RGR) [14] reduces walking distance the most and will thus also be used here. The number of non-overlapping stops n_{ij} between orders i and j is divided by the number of overlapping stops t_{ij} to calculate this metric. The matrix $\mathbf{R}$ with elements r_{ij} is computed by

$$r_{ij} = \frac{n_{ij}}{t_{ij}}, \quad \text{for all orders } \; i, j \text{ and } i \neq j. \tag{3}$$

Given a starting location, a *span* is the distance travelled to collect all items needed by that order [25]. This layout specific measurement is incorporated in the spans metric. The spans metric is combined with a greedy smallest entry heuristic (ZGS) [17]. All stops for orders i and j are included in the sets $\mathcal{S}_i$ and $\mathcal{S}_j$, respectively. Let the gap (in number of locations) between the ending and starting points of order i's minimum span be $\widetilde{P}_i^{\mathrm{min}}$. Let the number of locations that overlaps between P_i^{min} and P_j^{min} be $|P_i^{\mathrm{min}} \Cap P_j^{\mathrm{min}}|$. The greedy smallest entry heuristic (GS) searches globally through elements z_{ij} of matrix $\mathbf{Z}$ which can be computed with the formula

$$z_{ij} = (|\mathcal{S}_i| + |\mathcal{S}_j| - |\mathcal{S}_i \cap \mathcal{S}_j|) + (|\widetilde{P}_i^{\mathrm{min}}| + |\widetilde{P}_j^{\mathrm{min}}| - |\widetilde{P}_i^{\mathrm{min}} \Cap \widetilde{P}_j^{\mathrm{min}}|), \text{ with } i \neq j. \tag{4}$$

Finally, the nearest end heuristic (NE) is used to sequence the batches in Tier 4 because it is proven to perform well [25]. It is a path construction approach. Given the picker's current position, the batch with the nearest end point is added to the pick path. NE can be used to determine pick sequences in both U- and Z-configurations.

5 Results

Evaluating the picking system's performance (total completion time) globally can only be achieved by solving all four decision tiers together. This entails to sequentially group DBNs into picking waves (in Tier 1), arrange these waves/groups of SKUs on the available locations (in Tier 2), select a configuration (in Tier 3), batch the newly composed orders, and finally sequence batches (in Tier 4).

5.1 Data and Scenarios

The proposed combinations of approaches are evaluated using real data obtained from the Retailer's DC in Cape Town. The dataset runs over 27 work days and includes 45 picking waves. There are between two and four picking waves operating simultaneously. The 2 212 unique SKUs are incorporated in 1 206 unique DBNs. This master dataset is reorganised into 4 scenarios, because of different numbers of scheduled picking lines per day. This helps to evaluate picking waves that are not completed in a day and makes solution approaches more comparable. Data are cleaned if they contain less than 15 SKUs for example. As described in Table 1, a uniform set of instances is included in each scenario.

Table 1. Properties of the three test scenarios based on the DC in Cape Town.

Scenario	Number of			
	Lines	Days	DBNs	Waves
1	2	27	767	30
2	3	21	708	24
3	4	9	330	12

The numerical experiments were implemented in Python 3.6 [29]. It was executed on a Dell Optiplex 5050 running the Microsoft Windows 10 Enterprise 2016 LTSB operating system [28]. The computer has a 3.6 GHz Intel Core i7-7700 CPU, with a 1×8 GB 2400 MHz DDR4 RAM. Results were analysed in R [30].

Figure 7 describes all combinations of solution approaches for each tier. The abbreviation of the methods applied along the path of choices are used to describe an approach. For example, choosing GRA to solve the SPLAP, GRL to solve the SAP, a U-configuration, and RGR to form batches would configure the abbreviation GRA-GRL-U-RGR-NE which is derived from Fig. 7. The total completion time per wave is determined with a discrete event simulation (DES) making the different solution approaches comparable [15,16].

For Scenario 1, the sum of completion times measured in seconds for 30 picking waves is depicted in Fig. 8. With on average 1 h per picking wave, the configuration choice (Tier 3) influences completion time the most. The shortest time in the U-configuration is produced by ZGS, while RGR generates slightly lower times in the Z-configuration. As expected from previous studies, Tier 2 does not have a significant impact on completion time – influencing it less than 1%. GI generates the lowest time, which is followed by GIDM and PMA for Tier 1.

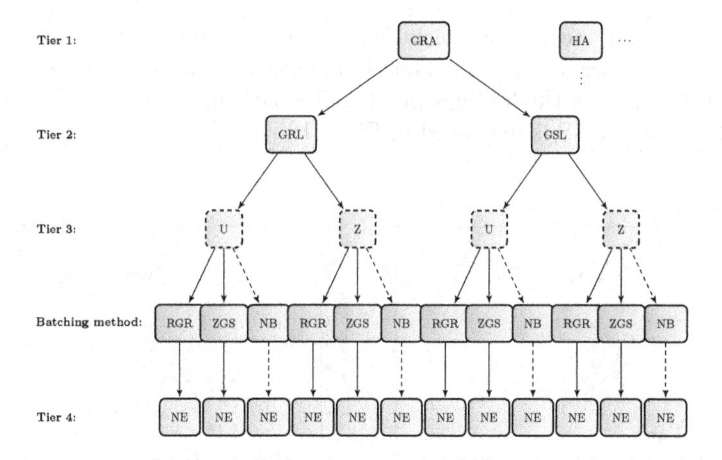

Fig. 7. Different optimisation combinations that include batching are depicted in a decision tree. Order batching is highlighted in blue, while newly introduced Tier 3 is illustrated with a dashed line. (colour figure online)

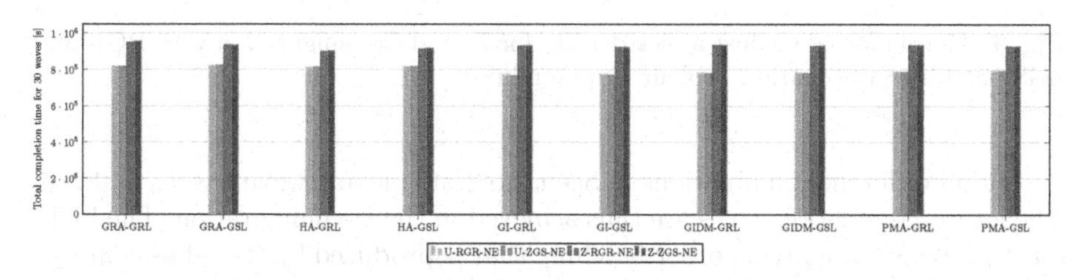

Fig. 8. All potential solution combination results for total completion time of Scenario 1. Tier 1 and 2 are displayed on the x-axis label. Tier 3, batching, and Tier 4 are described in the key. Completing all 30 waves for each combination is illustrated on the y-axis label.

For Scenario 2, Tier 3 contributes, with about 1.5 h difference, an even bigger influence on the completion time of waves. In Scenario 2, batching has the same effect as in the first scenario and in Tier 1 GI provided the shortest completion times, with GIDM in second place.

For Scenario 3, the third tier influences the total completion time, with about 2 hours per picking wave, the most. All other tiers produce similar results as in the first and second scenario.

The average pick density is computed after Tier 3 to investigate its interaction with order batching. Scatter plots of the pick density measure d_{m} vs the total completion time for Scenario 1 is provided in Fig. 9. The red dashed lines indicate a d_{m} of 0.22 and 0.38, respectively. The Z-configuration performs better on average at a d_{m} below 0.22, while the U-configuration outperformed Z for a d_{m} larger than 0.38 in this dataset [16]. Few picking lines perform better when applying the assignment approach of GRA and HA at $d_{\mathrm{m}} > 0.22$, since both approaches do not incorporate pick density. All average pick densities for GI, GIDM and PMA are above 0.38. Therefore, they produce lower times in the U-configuration, echoing the findings in [16]. The batching metric RGR is depicted in Fig. 9(a), while ZGS is displayed in Fig. 9(b).

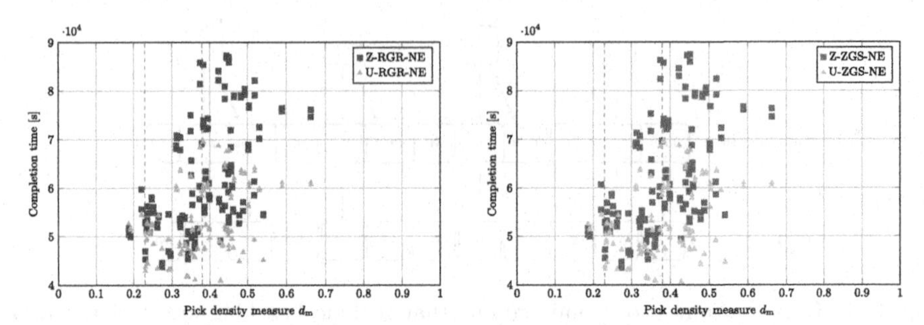

(a) Density and completion time using RGR in Z- and U-configuration.

(b) Density and completion time using ZGS in Z- and U-configuration.

Fig. 9. Times plotted against a pick density for Z- and U-configuration with RGR and ZGS batching in Scenario 1 (colour figure online).

For all different combinations in Scenario 2, no picking wave has $d_{\mathrm{m}} < 0.276$. The lowest completion times are thus achieved in the U-configuration. The highest d_{m} is 0.680 supporting order batching that is produced by the GI assignment approach.

For the third scenario, the lowest d_{m} is 0.283. Therefore, the lowest total completion times are achieved by the U-configuration. Again, the highest d_{m} of 0.681 was produced by the GI assignment.

Analysing the three scenarios, the preferred solution approach per decision tier can be selected. The GI assignment consistently increases pick density and should thus be chosen in Tier 1. The easily implementable solution approach GRL is suitable for Tier 2, since its influence on picking time is negligible. An increased pick density favours the U-configuration in Tier 3. The lowest completion time is generated applying batching metric ZGS. Tier 4 uses the NE heuristic to sequence orders. Therefore, the lowest results in overall picking time are generated by the GI-GRL-U-ZGS-NE combination. The time frame to solve all decision tiers as determined by the Retailer can incorporate this holistic solution approach.

A benchmark including historical assignment (HA) with GRL and a high enough pick density for the U-configuration, but without order batching (NB) is compared to GI-GRL-U-ZGS-NE and the completion times are illustrated for all scenarios in Fig. 10. A saving of 28.4% on picking time is achieved in the first scenario, in the second the saving is 27.9% and in the third 27.2%.

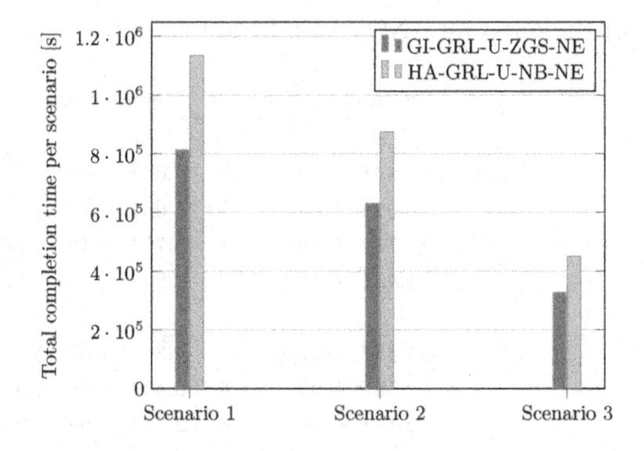

Fig. 10. Comparing the benchmark solution to the best performing solution approach over all tiers.

5.2 Statistical Analysis

With the help of inferential statistics the influence on total completion time per picking wave of Tier 1, Tier 2, Tier 3 and batching as the independent variables is investigated. Tier 4 is excluded in this analysis because batching is performed before Tier 4 is solved and thus does not influence batching. Per independent variable a one-way ANOVA including a Tukey's HSD *post-hoc* test and for all variables a fractional ANOVA is applied to the scenarios. Thereby, the statistical difference between the variables is investigated.

The ANOVA results, to check for statistical differences between variables, are displayed in Table 2. For Tiers 1 and 2 the differences are not significant, while batching, with $F(2, 1\ 557) = 117.2$, $p = 0.000$ is statistically significant. The difference between ZGS and NB is also significant (difference between the means is −21 418.8, with $p = 0.000$) as displayed in the Tukey's HSD *post-hoc* test. Furthermore, the Z- and U-configurations in Tier 3 are also statistically significant (difference between the means is 12 148.67, with $p = 0.000$).

The Tukey's HSD *post-hoc* test also concludes a significant difference in means between RGR and NB (−21 449.4, with $p = 0.000$) and between ZGS and NB (−21 418.8, with $p = 0.000$). Furthermore, the difference between the two configurations in Tier 3 (12 148.67, with $p = 0.000$) is also statistically significant [32].

In Table 2, the factorial ANOVA including Tier 1, Tier 2, Tier 3 and batching displays no significant influence through the interaction of all variables ($F(8, 1\ 500) = 0.002, p = 1.000$) [33].

The configuration choice seems to influence total completion times the most. Therefore, a strong focus on optimising the third tier and batching in an integrated approach of optimising the picking system is recommended.

Table 2. Results of the one-way ANOVA and the four-way ANOVA on variables.

	Sum of squares	*df*	Mean square	*F*	*p*
One-way ANOVA					
Tier 1	$3.85E+09$	4	$9.62E+08$	1.233	0.213
Tier 2	$2.08E+06$	1	2 081 219	0.003	0.955
Tier 3	$5.76E+10$	1	$5.76E+10$	77.32	0.000**
Batching	$1.59E+11$	2	$7.96E+10$	117.2	0.000**
Four-way ANOVA					
Tier 1 × Tier 2	$5.48E+07$	4	$1.37E+07$	0.021	0.999
Tier 1 × Tier 3	$2.58E+09$	4	$6.44E+08$	0.976	0.419
Tier 1 × Batching	$2.19E+09$	8	$2.74E+08$	0.414	0.913
Tier 2 × Tier 3	$9.15E+06$	1	$9.15E+06$	0.014	0.906
Tier 2 × Batching	$2.26E+07$	2	$1.13E+07$	0.017	0.983
Batching × Tier 3	$1.36E+09$	2	$6.80E+08$	1.031	0.357
Tier 1 × Tier 2 × Tier 3	$5.09E+07$	4	$1.27E+07$	0.019	0.999
Tier 1 × Tier 2 × Batching	$9.38E+06$	8	$1.17E+06$	0.002	1.000
Tier 1 × Batching × Tier 3	$2.46E+08$	8	$3.08E+07$	0.047	1.000
Tier 2 × Batching × Tier 3	$7.91E+07$	2	$3.96E+07$	0.060	0.942
Tier 1 × Tier 2 × Batching × Tier 3	$1.17E+07$	8	$1.47E+06$	0.002	1.000

Note: Two asterisks indicate significance at the 5% level or below.

6 Conclusion

This paper analyses the effects of order batching on a cyclical order picking system. With batching metrics for this system developed by [14,17], the effect of batching on completion times evaluated by [15], and the difference including batching on different configuration options investigated in [16], this paper concludes the optimisation of the unidirectional cyclical picking line by introducing batching to all decision tiers in an effort to optimise the system holistically.

The main objective of minimising total completion time (by minimising walking distance) is achieved by incorporating the important layer of order batching. On average 27.8% of total picking time can be saved using the algorithm sequence GI-GRL-U-ZGS-NE that includes batching and comparing it to the benchmark decision sequence HA-GRL-U-NB-NE as used in the historical data before the introduction of batching. These findings could be generalised for application on the unidirectional carousel for example.

The Z-configuration is outperformed by U if the pick density is high enough. The optimisation effort could focus on the U-configuration exclusively. Therefore, only three decision tiers would be in the centre of attention of future optimisation studies. Alternatively, for the Z-configuration a bidirectional option could improve its competitiveness. Pickers could pick items that are placed in the opposite direction of their current path. A bidirectional option could be introduced and compared in future studies.

References

1. Accorsi, R., Manzini, R., Bortolini, M.: A hierarchical procedure for storage allocation and assignment within an order-picking system. A case study. Int. J. Log. Res. Appl. **15**(6), 351–364 (2012)
2. Agrawal, R., Imieliński, T., Swami, A.: Mining association rules between sets of items in large databases. In: Proceedings of the 1993 ACM SIGMOD International Conference on Management of Data, pp. 207–216 (1993)
3. Bartholdi, J.J.I., Platzman, L.K.: Retrieval strategies for a carousel conveyor. IIE Trans. **18**(2), 166–173 (1986)
4. Van den Berg, J.P., Zijm, W.H.: Models for warehouse management: classification and examples. Int. J. Prod. Econ. **59**(1–3), 519–528 (1999)
5. Bindi, F., Manzini, R., Pareschi, A., Regattieri, A.: Similarity-based storage allocation rules in an order picking system: an application to the food service industry. Int. J. Log. Res. Appl. **12**(4), 233–247 (2009)
6. Chen, M.C., Huang, C.L., Chen, K.Y., Wu, H.P.: Aggregation of orders in distribution centers using data mining. Exp. Syst. Appl. **28**(3), 453–460 (2005)
7. Chen, M.C., Wu, H.P.: An association-based clustering approach to order batching considering customer demand patterns. Omega **33**(4), 333–343 (2005)
8. Chiang, D.M.H., Lin, C.P., Chen, M.C.: The adaptive approach for storage assignment by mining data of warehouse management system for distribution centres. Enterp. Inf. Syst. **5**(2), 219–234 (2011)
9. De Koster, M., Van der Poort, E.S., Wolters, M.: Efficient order batching methods in warehouses. Int. J. Prod. Res. **37**(7), 1479–1504 (1999)
10. De Koster, R., Le-Duc, T., Roodbergen, K.J.: Design and control of warehouse order picking: a literature review. Eur. J. Oper. Res. **182**(2), 481–501 (2007)
11. De Villiers, A.P.: Minimising the total travel distance to pick orders on a unidirectional picking line. Master's thesis, Stellenbosch University (2012)
12. Hagspihl, R., Visagie, S.E.: The number of pickers and stock-keeping unit arrangement on a unidirectional picking line. S. Afr. J. Ind. Eng. **25**(3), 169–183 (2014)
13. Han, J., Pei, J., Kamber, M.: Data Mining: Concepts and Techniques. Elsevier (2011)
14. Hofmann, F., Visagie, S.: Picking location metrics for order batching on a unidirectional cyclical picking line. ORiON **35**(2), 161–186 (2019)
15. Hofmann, F., Visagie, S.: The effect of order batching on a unidirectional picking line's completion time. Int. J. Logistics Syst. Manage. (2020, accepted to appear)
16. Hofmann, F., Visagie, S.: Configuration selection on a unidirectional cyclical picking line (2021, in the process of submission)
17. Hofmann, F., Visagie, S.: Route overlap metrics for order batching on a unidirectional cyclical picking line (2021, in the process of submission)
18. Kim, B.S., Smith, J.S.: Slotting methodology using correlated improvement for a zone-based carton picking distribution system. Comput. Ind. Eng. **62**(1), 286–295 (2012)
19. Kress, D., Boysen, N., Pesch, E.: Which items should be stored together? A basic partition problem to assign storage space in group-based storage systems. IISE Trans. **49**(1), 13–30 (2017)
20. Le Roux, G.J., Visagie, S.E.: A multi-objective approach to the assignment of stock keeping units to unidirectional picking lines. S. Afr. J. Ind. Eng. **28**(1), 190–209 (2017)

21. Litvak, N., Vlasiou, M.: A survey on performance analysis of warehouse carousel systems. Stat. Neerl. **64**(4), 401–447 (2010)
22. Manzini, R.: Correlated storage assignment in an order picking system. Int. J. Ind. Eng. Theor. Appl. Pract. **13**(4), 384–394 (2006)
23. Matthews, J., Visagie, S.E.: SKU assignment to unidirectional picking lines using correlations. ORiON **31**(2), 61–70 (2015)
24. Matthews, J.: SKU assignment in a multiple picking line order picking system. Ph.D. thesis. Stellenbosch University, Stellenbosch (2015)
25. Matthews, J., Visagie, S.E.: Order sequencing on a unidirectional cyclical picking line. Eur. J. Oper. Res. **231**(1), 79–87 (2013)
26. Matthews, J., Visagie, S.E.: Assignment of stock keeping units to parallel unidirectional picking. S. Afr. J. Ind. Eng. **26**(1), 235–251 (2015)
27. Matthews, J., Visagie, S.E.: SKU arrangement on a unidirectional picking line. Int. Trans. Oper. Res. **26**(1), 100–130 (2019)
28. Microsoft (2018). https://www.microsoft.com/
29. Python Software Foundation: Python 3.6 (2018). https://www.python.org/
30. R Core: The R Project for Statistical Computing (2019). https://www.r-project.org/
31. Stern, H.: Parts location and optimal picking rules for a carousel conveyor automatic storage and retrieval system. In: Proceedings of the 7th International Conference on Automation in Warehousing, pp. 185–193 (1986)
32. STHDA: One-way ANOVA Test in R (2020). http://www.sthda.com/english/wiki/one-way-anova-test-in-r#relaxing-the-homogeneity-of-variance-assumption
33. STHDA: Two-way ANOVA Test in R (2020). http://www.sthda.com/english/wiki/two-way-anova-test-in-r
34. Toth, P., Martello, S.: Knapsack Problems: Algorithms and Computer Implementations. Wiley, Hoboken (1990)
35. Vickson, R.G., Fujimoto, A.: Optimal storage locations in a carousel storage and retrieval system. Locat. Sci. **4**(4), 237–245 (1996)
36. Vickson, R., Lu, X.: Optimal product and server locations in one-dimensional storage racks. Eur. J. Oper. Res. **105**(1), 18–28 (1998)

Bi-objective Optimization for Joint Production Scheduling and Distribution Problem with Sustainability

Ece Yağmur and Saadettin Erhan Kesen(✉)

Konya Technical University, Selcuklu, Konya, Turkey
{ecyagmur,sekesen}@ktun.edu.tr

Abstract. This paper considers joint production and distribution planning problem with environmental factors. While the production phase of the problem consists of job shop production environment running under Just-In-Time (JIT) philosophy, the distribution phase involves a heterogeneous fleet of vehicles with regards to capacity and fuel consumption rate. Therefore, we tackle two well-known problems in Operations Research terminology which are called machine scheduling and vehicle routing problems. The joint problem is formulated as a bi-objective structure, the first of which is to minimize the maximum tardiness, the second of which aims to minimize the total amount of CO2 emitted by the vehicles. Orders are required to be consolidated to reduce the traveling time, distance, or cost. An increase in the vehicle capacity results in a higher possibility of consolidation, but in this case, the amount of CO2 emission that the vehicle emits into the air will also increase. Having shown that two objectives are conflicting in an illustrative example, we formulate the problem as a mixed integer programming (MIP) formulation and use an Augmented Epsilon Constraint Method (AUGMECON) for solving the bi-objective model. On randomly generated test instances, the applicability of the MIP model through the use of AUGMECON is reported.

Keywords: Joint production and distribution scheduling · Vehicle routing · Job shop · Sustainability · Heterogeneous fleet · Mixed integer model

1 Introduction

Joint decisions for production and distribution operations in supply chain management have been essential for many applications. Especially in make-to-order business, orders are ready for distribution as soon as their production is completed, since there is no reason to keep finished product in stock. As customers can customize their orders to meet their own needs in make-to-order businesses, production can start as soon as demand arrives. Therefore, the joint approach is vital for this business to respond to customer requests immediately.

In this study, we examine the joint production and distribution scheduling problem for job shop production environment in which each job undergoes multiple operations which need to be performed in different machines. In the distribution phase, we consider

M. Mes et al. (Eds.): ICCL 2021, LNCS 13004, pp. 269–281, 2021.
https://doi.org/10.1007/978-3-030-87672-2_18

a heterogeneous fleet that consists of different vehicle types in terms of capacity and fuel consumption rate. In addition, vehicles can be used more than once for delivery. We formulate the joint problem as a bi-objective structure. It is very important for customer satisfaction to deliver the orders by considering the specified due dates. So, the first objective of the problem is minimizing the maximum tardiness in order to meet required service level. In the globalizing world, the importance of environmental factors is on the rise. Besides the efficient use of resources, making environmentally friendly decisions is among the new goals of businesses. For this reason, the other objective of the problem is structured as the minimization of the total amount of CO_2 emission.

The main contributions of the paper can be summarized as follows:

- We examine one of the most complex production environments (i.e., job shop) which is only studied by Mohammadi et al. (2019) in the field of joint production and distribution scheduling.
- In the distribution phase, a fleet of limited heterogeneous vehicles with multiple use is studied, which is rarely studied in even vehicle routing literature.
- We also consider multiple objectives, one of which deals with environmental factors.

To the best of our knowledge, the considered problem is examined for the first time in terms of both production and distribution settings. It is obvious that the considered joint problem is NP-hard because standalone problems namely job shop scheduling and heterogeneous multiple tour vehicle routing belong to NP-hard problem class.

The remainder of this paper is constructed as follows: In Sect. 2, we present a literature survey of related works. In Sect. 3, we describe the problem with a formal definition and develop a Mixed Integer Programming (MIP) formulation for the problem. Following the problem description, we demonstrate an illustrative example to show the conflicting objectives. In Sect. 4, the application of AUGMECON method on the problem is given on the illustrative example. While computational experiments are given in Sect. 5, conclusions and future research directions are stated in Sect. 6.

2 Literature Survey

The coordination of supply chain functions is a small but growing subject for management science [1]. Many researchers have paid special emphasis to joint decisions for production and distribution functions at the operational level [2] and many different variants have been encountered relating to both production and distribution phases. When assessing the related literature according to the production environment, we can see that simpler production environments (single machine, parallel machine) are studied frequently. Scholz-Reiter et al. [3] is the first paper addressing the joint production and distribution scheduling problem where jobs having multiple operations. Meinecke and Scholz-Reiter [4] examine a real-world job shop problem with inventory stage between the production and distribution functions. Ramezanian et al. [5] compare two different delivery policies (direct delivery and delivery with routing) for permutation flow shop environment. Wang et al. [6] study three-stage hybrid flow shop scheduling problem, which also includes distribution decisions. Mohammadi et al. [7] address a case study from a furniture manufacturer with flexible job shop environment. More recently, Yağmur and Kesen [8] examine a permutation flow shop environment and develop a memetic algorithm for the joint problem.

According to [2], majority of the studies only aims to optimize a single objective, mainly focusing on cost minimization or service level maximization. While cost-related objectives may include production cost, distribution cost, and penalty costs, time-related objectives have usually been used for service level maximization. Additionally, very few researchers examined the joint problem with multiple objectives. Farahani et al. [9] investigate how the quality of perishable foods can be improved both cost and quality decay functions. Jamili et al. [10] optimize the customer's service level, measured as the mean delivery time and the total transportation cost. In the study of Ganji et al. [11], environmental objectives such as fuel consumption minimization and traditional objectives (i.e., cost) were considered together.

To conclude, this paper fills some gap in the literature as it tackles unstudied variant of joint production and distribution problem.

3 Problem Description

In this section, we first develop an MIP model for joint production and distribution scheduling problem and subsequently give an illustrative example for a clear and lucid account. For formal description, we consider a make-to-order production environment in which any order has multiple operations, each of which is performed on different machines to become a finished product. The job sequence on any machine and the machine sequence of any job can be different (i.e., job-shop). When processing of all orders in a particular batch completes, a vehicle with sufficient capacity delivers them to the related customers. There are three different types of vehicles in a fleet and any vehicle can be used more than once. Additionally, the amount of CO_2 emissions released into the air varies depending on the vehicle size.

3.1 MIP Formulation

Using the symbols given in Table 1, we propose a MIP formulation for the considered problem in this section.

$$min f_1 = T_{max} \tag{1}$$

$$min f_2 = \sum_{i \in N} \sum_{j \in N} \sum_{k \in V} \left(\rho_k^0 X_{ijk} + \frac{\rho_k^* - \rho_k^0}{Q_k} u_{ijk} \right) t_{ij} \tag{2}$$

Subject to

$$S_{im} - S_{jm} + HO_{ijm} \geq p_{jm} \qquad \forall (i,m) \rightarrow (j,m) \in DA \tag{3}$$

$$S_{jm} - S_{im} + H\left(1 - O_{ijm}\right) \geq p_{im} \qquad \forall (i,m) \rightarrow (j,m) \in DA \tag{4}$$

Table 1. Definitions of symbols used in MIP formulation

Symbol	Definition
Indices	
i, j	: Job
m, l	: Machine
k	: Vehicle
Sets	
N_C	: Set of jobs
N	: Set of nodes
M	: Set of machines
V	: Set of vehicles
DA	: Set of disjunctive arcs
CA	: Set of conjunctive arcs
LM	: Set of the machines on which the last operations of jobs are processed
Parameters	
d_i	: Demand of customer i
p_{im}	: Processing time of order i on machine m
t_{ij}	: Travel distance between customers i and j
dd_j	: Due date for order j
Q_k	: Capacity of vehicle k
H	: Sufficiently large number
ρ_k^0	: Fuel consumption rate of vehicle k (empty load)
ρ_k^*	: Fuel consumption rate of vehicle k (full load)
Decision Variables	
O_{ijm}	=1 if operation of order i is processed before operation of order j on machine m, 0 otherwise
X_{ijk}	=1 if vehicle k goes directly from node i to node j, 0 otherwise
W_{ijk}	=1 if vehicle k completes preceding tour with node i and starts succeeding tour with node j, 0 otherwise
u_{ijk}	: The total load of vehicle k when vehicle goes from customer i to customer j
S_{im}	: Production starting time of order i on machine m
C_{im}	: Production completion time of order i on machine m
B_i	: Production completion time of the last operation for job i

(*continued*)

Table 1. (*continued*)

Symbol	Definition
A_i	: Delivery time to customer i
Y_i	: Production completion time of a batch to which order i belongs
T_i	: Tardiness for customer i

$$O_{ijm} + O_{jim} = 1 \qquad \forall (i, m) \to (j, m) \in DA \tag{5}$$

$$S_{jm} - S_{jl} \geq p_{jl} \qquad \forall (j, l) \to (j, m) \in CA \tag{6}$$

$$C_{jm} - S_{jm} = p_{jm} \qquad \forall j \in N_C; m \in M \tag{7}$$

$$B_j = C_{jm} \qquad \forall j \in N_C; m \in LM \tag{8}$$

$$\sum_{k \in V} \sum_{i \in N_C} X_{ijk} = 1\partial \qquad \forall j \in N_C \tag{9}$$

$$\sum_{j \in N} X_{ijk} = \sum_{j \in N} X_{jik} \qquad \forall i \in N; k \in V \tag{10}$$

$$u_{iok} = 0 \qquad \forall i \in N_C; k \in V \tag{11}$$

$$\sum_{j \in N} u_{jik} - \sum_{j \in N} u_{ijk} = \sum_{j \in N} X_{ijk} d_i \qquad \forall i \in N_C; k \in V \tag{12}$$

$$u_{ijk} \leq (Q_k - d_i) X_{ijk} \qquad \forall i, j \in N; k \in V \tag{13}$$

$$\sum_{j \in N_C} W_{ijk} \leq X_{i0k} \qquad \forall i \in N_C; i \neq j; k \in V \tag{14}$$

$$\sum_{i \in N_C} W_{ijk} \leq X_{0jk} \qquad \forall j \in N_C; i \neq j; k \in V \tag{15}$$

$$\sum_{j \in N_C} X_{0jk} - \sum_{i \in N_C} \sum_{j \in N_C; i \neq j} W_{ijk} \leq 1 \qquad \forall k \in V \tag{16}$$

$$Y_i \geq B_i \qquad \forall i \in N_C \tag{17}$$

$$Y_j - Y_i \leq H\left(1 - \sum_{k \in V} X_{ijk} - \sum_{k \in V} X_{jik}\right) \qquad \forall i, j \in N_C; i \neq j \tag{18}$$

$$A_i - A_j + H\sum_{k \in V} X_{ijk} + (H - t_{ij} - t_{ji})\sum_{k \in V} X_{jik} \leq H - t_{ij} \qquad \forall i,j \in N_C; i \neq j \tag{19}$$

$$A_i - A_j + H\sum_{k \in V} W_{ijk} \leq H - t_{i0} - t_{0j} \qquad \forall i,j \in N_C; i \neq j \tag{20}$$

$$A_j \geq H\sum_{k \in V} W_{ijk} - H + Y_j + t_{0j} \qquad \forall i,j \in N_C; i \neq j \tag{21}$$

$$A_j \geq Y_j + t_{0j} - H\left(1 - X_{0jk} + \sum_{i \in N_C} W_{ijk}\right) \qquad \forall j \in N_C; i \neq j; k \in V \tag{22}$$

$$A_j \leq Y_j + t_{0j} + H\left(1 - X_{0jk} + \sum_{i \in N_C} W_{ijk}\right) \qquad \forall j \in N_C; i \neq j; k \in V \tag{23}$$

$$T_i \geq A_i - dd_i \qquad \forall i \in N_C \tag{24}$$

$$T_{max} \geq T_i \qquad \forall i \in N_C \tag{25}$$

$$u_{ijk} \geq 0 \qquad \forall i,j \in N; k \in V \tag{26}$$

$$A_i, S_{im}, C_{im}, Y_i, T_i \geq 0 \qquad \forall i,j \in N_C; m \in M \tag{27}$$

$$X_{ijk} \in \{0, 1\} \qquad \forall i,j \in N; k \in V \tag{28}$$

$$W_{ijk} \in \{0, 1\} \qquad \forall i,j \in N_C; k \in V \tag{29}$$

$$O_{ijm} \in \{0, 1\} \qquad \forall i,j \in N_C; m \in M \tag{30}$$

The first objective of the model which is defined in Eq. (1) minimize the maximum tardiness. The second objective shown in Eq. (2) minimize the total amount of CO_2 emission that the vehicles emit into the air. As seen from the Eq. (2) the amount of emission is a function of fuel consumption (for more details interested readers are referred to the works of Xiao et al. [12] and Zhang [13].

The studies in which fuel consumption function is handled in more detail by evaluating different parameters can be found in Bektaş and Laporte [14], Kirschstein and Meisel [15], Franceschetti et al. [16]. The reason why we choose the function determined by the traveling distance and the load of vehicles is that the integrated problem under consideration is already quite complex even under a single objective as it contains two NP-hard problems and we intend to simplify the second objective function.

While Eqs. (3) - (8) are related to the production constraints, remaining equations are related to distribution constraints. Equations (3) and (5) are called as disjunctive constraints which determines the sequence of jobs processed on the same machine. Equation (6) determines the production starting times of consecutive operations of each job by conjunctive arcs which reflect the precedence constraints (i.e., machine sequence for any job).

Equation (7) guarantees that the production completion time of any operation is equal to the sum of production starting time of this operation and processing time of this operation on destinated machine. Equation (8) guarantees that the production of any jobs completes at the time when production of the last operation of that job completes.

While Eq. (9) states that each customer is served exactly once with a particular vehicle, Eq. (10) provides the input and output balance of any nodes. Equations (11)–(13) are associated with sub-tour elimination and capacity constraint. Equation (11) guarantees that the amount of load on any vehicle when returning to the depot must be equal to zero. Equation (12) determines the total load of any vehicle on any arc. Equation (13) indicates the boundaries of the u_i variable which is used for sub-tour elimination. Equation (14) and Eq. (15) are the constraints that provide tour combination for any vehicle. Equation (16) ensures that the number of consecutive tour combinations of any vehicle is always less than or equal to the total number of tours of that vehicle. Equation (17) and Eq. (18) calculate the production finish time of each batch. Equation (19) determines the order delivery time (i.e., service time) for consecutive customers in the same tour. Equation (20) and Eq. (21) determine the order delivery time for the first customer in any tour following another tour. Equation (20) is used when the next production batch to be distributed by a vehicle is ready before the vehicle returns to the depot. Equation (21) is active when a vehicle returns to the depot before the next production batch to be distributed by the vehicle is ready. Equation (22) and Eq. (23) state the order delivery time for the first customer in the first tour of each vehicle. Equation (24) determines the tardiness in the case of late delivery of the orders, considering the due date predetermined by the customers. Equation (26) - Eq. (30) are the non-negative and 0–1 integer constraints of the variables in the model.

3.2 Illustrative Example

In this section, we introduce an illustrative example for clear understanding of the problem. The parameters of the problem which covers six customers, three machines and three heterogenous vehicles are given in Table 2.

The first column of Table 2 represents the depot node and customers. While second and third column defines the coordinates of each customer on the two-dimensional plane, the fourth column represents the demand of each customer. The processing times of each job on each machine are given in the following three columns. The machine sequence of each job is given in Machine sequence column. And finally, the last column defines the due date parameter of each customer. The vehicle capacities are set to 140, 210 and 280 for small, medium, and large vehicle types, respectively. While fuel consumption rates for empty load situation are defined as 0.2, 0.3 and 0.4; for full load situation are defined as 0.26, 0.39 and 0.52 for three vehicle types, respectively.

We use lexicographic optimization to form a payoff table for the illustrative example. According to lexicographic optimization, first, the first objective function is optimized, then, by adding the optimal solution obtained from the first optimization as a constraint

Table 2. The parameters of illustrative example

i	X	Y	d_i	p_{i1}	p_{i2}	p_{i3}	Machine sequence	dd_i
0	0	0	–	–	–	–	–	–
1	52	11	60	69	58	55	2,3,1	319
2	−14	−76	80	70	90	91	3,1,2	471
3	15	−24	95	77	102	109	2,1,3	395
4	−45	−37	52	44	60	53	3,1,2	271
5	−3	68	85	88	70	80	2,1,3	319
6	27	35	50	44	51	54	1,2,3	275
7	−49	10	84	70	82	71	3,1,2	395

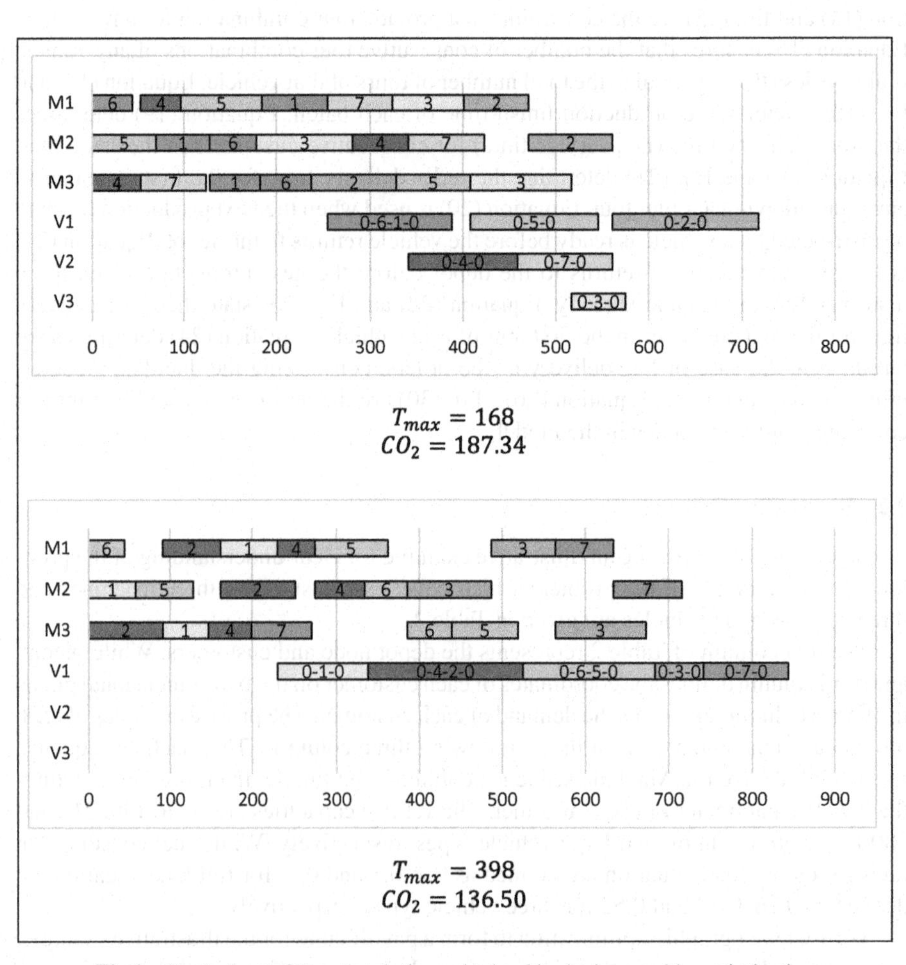

Fig.1. Gantt chart of Pareto solutions obtained by lexicographic optimization

the second objective function is optimized. In Fig. 1 we can show the Gantt chart of Pareto optimal solutions obtained by lexicographic optimization. According to the first solution, the available vehicles in the system are used simultaneously to minimize maximum tardiness. However, in this case, the total amount of emission is found to be larger than the second solution in which only a small vehicle is used. As seen from the non-dominated solutions, while one objective improves, the other worsens in multi objective optimization problems.

4 Augmented Epsilon Constraint Method

Epsilon constraint is one of the most common methods for multi-objective optimization problems due to its simplicity and easy applicability to many problems. It has been used effectively as the solution procedure in the field of supply chain management Talaei et al. [17], Toro et al. [18], as in other OR problems.

Mavrotas [19] states that some scaling problems may occur in the classical Epsilon Constraint method [20] as the only slack variable is used as an additional term in the objective function. So, a new version called as Augmented Epsilon Constraint Method is formed as seen in Eq. (31)–(33) where s_i is the slack variable of i. th objective function, r_i is the range of the i th objective function as calculated from the payoff table and δ is an sufficiently small number which is usually selected between 10^{-3} and 10^{-6} in the literature.

$$\min\left(f_1(x) - \delta\left(s_2/r_2 + s_3/r_3 + \cdots + s_p/r_p\right)\right) \tag{31}$$

$$f_i(x) + s_i = e_i i = 2, .., p \tag{32}$$

$$x \in S, s_i \in R^+ \tag{33}$$

In this section, we apply the AUGMECON method to illustrative example which is demonstrated in previous section. After generating the payoff table, we calculate the range of the first objective function as 107. Then, we divide this range into 10 equal sub-ranges from more relaxed to tighter as right-hand side and obtain 11 grid points as seen in the first column of Table 3. Thus, if an infeasible solution is obtained with the given right-hand side, the algorithm is stopped, and unnecessary runs are avoided.

CPLEX solver embedded to GAMS 24.2 is used as the MILP-solver. We easily adapt the sample code in [21] to our bi-objective model and we can obtain 9 Pareto optimal solutions with 11 grid points as seen in Fig. 2 in 624.07 s. The number of intervals can be increased so, a denser Pareto surface can be obtained, but this will also increase the computation time.

5 Computational Results

In order to evaluate the AUGMECON performance for bi-objective joint production and distribution scheduling problem, we generate test problems which covers three levels

Table 3. Grid points for AUGMECON

Grid points	f_1	f_2
≤398	398	136.50
≤375	367	139.88
≤352	303	141.33
≤329	303	141.33
≤306	303	141.33
≤283	280	144.17
≤260	252	146.55
≤237	211	158.25
≤214	211	158.25
≤191	182	167.59
≤168	168	187.34

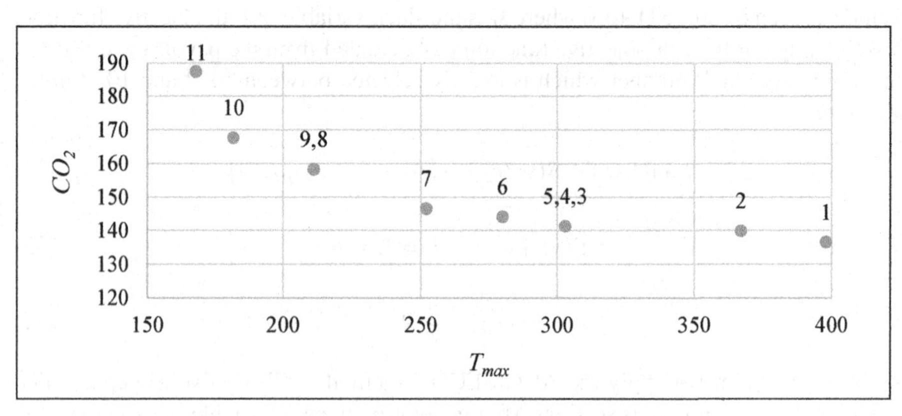

Fig. 2. Pareto optimal solutions of illustrative example

for number of customers (5,6 and 7 customers); two levels for number of machines (2 and 3 machines) and one level for number of vehicles (small, medium and large sized vehicles). In addition, while the due date parameter is selected three levels as wide, tight, and common due date (i.e., all orders have same due date) and the vehicle capacity is selected two levels to see the effect of vehicle capacity on bi-objective model. So, 36 test instances (3 × 2 × 1 × 3 × 2) are randomly generated based on the instance generation procedure in [8].

We examine the effects of the due date and vehicle capacity levels on the problem in terms of average objective functions, the number of non-dominated solutions and CPU times. When we examine the Table 4 the number of non-dominated solutions found by CPLEX increase from wide to tight and from tight to common due date. In addition to this we can say that as the number of non-dominant solutions found increases, the CPU time also increases. Especially when the size of the problem increases with the number of customers, the number of non-dominant solutions found has more impact on CPU time. Other remarkable result from the Table 4 is that average T_{max} values increase as the due date parameter gets tighter.

Table 4. CPLEX solutions based on due-date parameter

Due date	*N*	T_{max}	CO_2	*# Non-dominated solutions*	*CPU (s)*
Wide	5	36.7	128.7	4.3	9.9
	6	70.4	155.5	4.3	24.1
	7	118.1	144.7	3.5	268.2
Tight	5	150.8	132.1	5.5	12.1
	6	206.7	161.4	5.3	85.4
	7	257.4	147.9	5.3	284.5
Common	5	123.0	134.1	5.8	14.3
	6	162.9	171.3	6.0	103.6
	7	187.0	153.4	6.0	601.8

In table 5 we assess the effect of vehicle capacity on the problem. The average CPU times increase as the vehicle capacity increase due to increasing solution space. Another finding from Table 5 is that average T_{max} values decrease as the vehicle capacity gets larger because of the consolidation ability.

Table 5. CPLEX solutions based on capacity parameter

Capacity	*N*	T_{max}	CO_2	*# Non-dominated solutions*	*CPU (s)*
Small	5	103.6	132.1	5.0	11.5
	6	174.5	164.5	6.0	39.2
	7	188.1	149.3	4.8	296.1
Large	5	103.4	131.2	5.3	12.7
	6	118.9	160.9	4.3	102.9
	7	186.9	148.0	5.0	473.5

As seen from the results in this section we can say the problem complexity has exponentially increased by the number of customers. Therefore, for medium and large sized instance, bespoke multi-objective solution methods such as NSGA-2, PLS can be developed.

6 Conclusions

In this paper, we proposed a new mathematical model for a joint production and distribution problem which covers job shop scheduling and multiple tour heterogenous vehicle routing. The objectives of the model are structured as minimizing the maximum tardiness and the total amount of CO_2 emission. After we show that the objectives are conflicting through an illustrative example, we use AUGMECON method for solving the small sized randomly generated instances.

In further research, heuristic algorithms can be developed for the problem for more practical big sized instances and different performance metrics of multi-objective optimization such as hypervolume, spacing metric.

References

1. Chandra, P., Fisher, M.L.: Coordination of production and distribution planning. Eur. J. Oper. Res. **72**(3), 503–517 (1994)
2. Moons, S., Ramaekers, K., Caris, A., Arda, Y.: Integrating production scheduling and vehicle routing decisions at the operational decision level: a review and discussion. Comput. Ind. Eng. **104**, 224–245 (2017)
3. Scholz-Reiter, B., Makuschewitz, T., Novaes, A.G., Frazzon, E.M., Lima, O.F., Jr.: An approach for the sustainable integration of production and transportation scheduling. Int. J. Logist. Syst. Manag. **10**(2), 158–179 (2011)
4. Meinecke, C., Scholz-Reiter, B.: A heuristic for the integrated production and distribution scheduling problem. Int. Sci. Index **8**(2), 290–297 (2014)
5. Ramezanian, R., Mohammadi, S., Cheraghalikhani, A.: Toward an integrated modeling approach for production and delivery operations in flow shop system: trade-off between direct and routing delivery methods. J. Manuf. Syst. **44**, 79–92 (2017)
6. Wang, S., Wu, R., Chu, F., Yu, J.: Variable neighborhood search-based methods for integrated hybrid flow shop scheduling with distribution. Soft. Comput. **24**(12), 8917–8936 (2019). https://doi.org/10.1007/s00500-019-04420-6
7. Mohammadi, S., Al-e-Hashem, S.M., Rekik, Y.: An integrated production scheduling and delivery route planning with multi-purpose machines: a case study from a furniture manufacturing company. Int. J. Prod. Econ. **219**, 347–359 (2019)
8. Yağmur, E., Kesen, S.E.: A memetic algorithm for joint production and distribution scheduling with due dates. Comput. Ind. Eng. **142**, 106342 (2020)
9. Farahani, P., Grunow, M., Günther, H.-O.: Integrated production and distribution planning for perishable food products. Flex. Serv. Manuf. J. **24**(1), 28–51 (2012)
10. Jamili, N., Ranjbar, M., Salari, M.: A bi-objective model for integrated scheduling of production and distribution in a supply chain with order release date restrictions. J. Manuf. Syst. **40**, 105–118 (2016)
11. Ganji, M., Kazemipoor, H., Molana, S.M.H., Sajadi, S.M.: A green multi-objective integrated scheduling of production and distribution with heterogeneous fleet vehicle routing and time windows. J. Clean. Prod. **259**, 120824 (2020)
12. Xiao, Y., Zhao, Q., Kaku, I., Xu, Y.: Development of a fuel consumption optimization model for the capacitated vehicle routing problem. Comput. Oper. Res. **39**(7), 1419–1431 (2012)
13. Zhang, S., Lee, C., Choy, K., Ho, W., Ip, W.: Design and development of a hybrid artificial bee colony algorithm for the environmental vehicle routing problem. Transp. Res. Part D: Transp. Environ. **31**, 85–99 (2014)

14. Bektaş, T., Laporte, G.: The pollution-routing problem. Transp. Res. Part B: Methodol. **45**(8), 1232–1250 (2011)
15. Kirschstein, T., Meisel, F.: GHG-emission models for assessing the eco-friendliness of road and rail freight transports. Transp. Res. Part B: Methodol. **73**, 13–33 (2015)
16. Franceschetti, A., Demir, E., Honhon, D., Van Woensel, T., Laporte, G., Stobbe, M.: A metaheuristic for the time-dependent pollution-routing problem. Eur. J. Oper. Res. **259**(3), 972–991 (2017)
17. Talaei, M., Moghaddam, B.F., Pishvaee, M.S., Bozorgi-Amiri, A., Gholamnejad, S.: A robust fuzzy optimization model for carbon-efficient closed-loop supply chain network design problem: a numerical illustration in electronics industry. J. Clean. Prod. **113**, 662–673 (2016)
18. Toro, E.M., Franco, J.F., Echeverri, M.G., Guimarães, F.G.: A multi-objective model for the green capacitated location-routing problem considering environmental impact. Comput. Ind. Eng. **110**, 114–125 (2017)
19. Mavrotas, G.: Effective implementation of the ε-constraint method in multi-objective mathematical programming problems. Appl. Math. Comput. **213**(2), 455–465 (2009)
20. Haimes, Y.: On a bicriterion formulation of the problems of integrated system identification and system optimization. IEEE Trans. Syst. Man Cybern. **1**(3), 296–297 (1971)
21. Mavrotas, G.: Generation of efficient solutions in multiobjective mathematical programming problems using GAMS. Effective implementation of the ε-constraint method. Lecturer, Laboratory of Industrial and Energy Economics, School of Chemical Engineering. National Technical University of Athens (2007)

On the Effect of Product Demand Correlation on the Storage Space Allocation Problem in a Fast-Pick Area of a Warehouse

Felipe I. Gré Carafí[1], Alberto Ossa-Ortiz de Zevallos[1], Rosa G. González-Ramírez[1](✉), and Mario C. Velez-Gallego[2]

[1] Universidad de los Andes, 12455 Santiago, RM, Chile
{figre,aossa}@miuandes.cl, rgonzalez@uandes.cl
[2] Universidad EAFIT, Medellin, Colombia
marvelez@eafit.edu.co

Abstract. The storage location assignment problem (SLAP), also known as the slotting problem involves the decisions of how much and where should be stored each stock keeping unit (SKU) in the fast-pick area with the aim to minimize total order-picking and replenishment costs associated to the distance traveled by the picking operators. Motivated by this, we propose to analyze the impact of SKUs demand correlation on the slotting decisions. Based on an experimental design, the effects of SKUs with correlated demand are analyzed. Results show that the most significant factor with respect to the total distance traveled is the number of orders, followed by the capacity of the bins and the number of bins in each location. Results of an instance solved to optimality by a commercial solver and a greedy heuristic in which the latter does not consider the demand correlation illustrate the impact that demand correlation has on the solution obtained.

Keywords: Warehousing · Slotting problem · Order-picking · Demand Correlation · Fast-pick area

1 Introduction

Warehousing and its basic functions (receiving, storage, order picking and shipping) have a significant impact on the efficiency of the supply chains (Gu et al. 2007; Bartholdi and Hackman 2008; Boysen et al. 2019). A warehouse can be defined as a material handling station dedicated to receiving, storing, order-picking, accumulating, sorting and shipping goods (Van den Berg 1999). Among the different functions of warehousing, storage has a direct impact on direct variable costs, and it also improves delivery times and reliability. For this reason, efficient management systems are required (Gu et al. 2007; Revillot-Narváez et al. 2020).

Given that order-picking operations are one of the most labor-intensive activities, it is quite common that many warehouses concentrate the picking activities in a compact area to reduce the distance traveled by the pickers. This is known as the fast-pick area, and items at the fast-pick area are replenished from the reserve area of the warehouse

M. Mes et al. (Eds.): ICCL 2021, LNCS 13004, pp. 282–295, 2021.
https://doi.org/10.1007/978-3-030-87672-2_19

(Bartholdi and Hackman 2019). Order-picking can be improved in several ways: assigning appropriate storage locations to items; routing appropriately the picking tour; and picking orders in batches (Boysen et al. 2019; van Gils et al. 2019; Zhang et al. 2019).

The storage location assignment problem (SLAP), also known as the slotting problem involves the decisions of how much and where should be stored each stock keeping unit (SKU) in the fast-pick area so as to minimize the total order picking costs (Yingde and Smith 2012; Zhang et al. 2019). If this problem is solved for the fast-pick area, then it is necessary to include the replenishment costs as well. Several models and methodologies have been proposed in the literature to address this problem. However, not all the approaches have considered the demand correlation among the SKUs. Under a dynamic demand system setting, the SKU flow patterns change dynamically or periodically due to factors such as turnover rate, seasonality, life cycle, etc. Hence, the slotting of the SKUs should be adjusted to reflect this change over time. In these cases, demand correlation of SKUs in each order may have a significant impact on the slotting decisions (Yingde and Smith 2012).

Motivated by this, we propose to analyze the impact of demand correlation on the slotting decisions for a company in which the warehouse is divided into a reserve area and several fast-pick areas, each one dedicated to a given product category. Velez-Gallego and Smith (2018) proposed a mathematical model to address this problem and presented preliminary results based on a computational experiment that used randomly generated instances that were not generated with intrinsic demand correlation among the SKUs.

In this work we consider the model proposed by Velez-Gallego and Smith (2018) to evaluate the effects of SKUs with correlated demand on the solution obtained. The proposed model not only allows to support storage allocation but also the size of the layout. A factorial design is proposed, evaluating the impacts on the objective function that the different parameters of the model have on the total distance traveled. In order to evaluate the effect of correlation, we generated instances that present correlation among the SKUs. Results show that the most significant factor with respect to the total distance traveled is the number of orders, followed by the capacity of the bins and the number of bins in each location. Results of an instance solved to optimality by a commercial solver and by a greedy heuristic in which the latter does not consider demand correlation illustrate the impact that demand correlation of SKUs has on the solution obtained.

The remainder of this paper is organized as follows. Section 2 presents a literature review regarding the storage space allocation and order-picking problems in warehousing. Section 3 describes the mathematical model. Section 4 presents the experimental design and numerical results. Conclusions and recommendations for future research are given in Sect. 5.

2 Literature Review

We can find several contributions that deal with the design of a fast-pick area, and particularly the SLAP. Hackman et al. (1990) and Bartholdi and Hackman (2008) are some of the earlier contributions in the literature that proposed fluid models to address the problem of assigning storage space to SKUs. Subramanian (2013) introduces some heuristic algorithms to address this problem. As indicated by Velez-Gallego and Smith

(2018), the main drawback of such contributions is that they do not consider a discrete assignment of bins or storage positions to the SKUs and accordingly, they propose a mathematical model that considers a discrete assignment. Other related approaches are proposed by Walter et al. (2013) and Gu (2005) in which they also determine the optimal size of the fast-pick area that also consider a discrete assignment of bins or storage positions to the SKUs. However, they neglect traveling time within the fast-pick area.

In contrast, in a situation in which the size of the fast-pick area is not small and thus, the time that pickers spend preparing customer orders is hence, not negligible. These aspects are considered in (Thomas and Meller 2014; Velez-Gallego and Smith 2018; Wu et al. (2020). Thomas and Meller (2014) address the forward-reserve problem, in which traveling times for the replenishment from of the fast-pick area is considered. They present analytical models for put-away, order picking and replenishment operations for random storage and two class-based storage policies. In the same line, Velez-Gallego and Smith (2018) considered also the traveling times within the fast-pick area and in contrast to previous contribution, they propose a mathematical model in which total traveling distance is minimized. This is then, the main contribution in their proposed model. On the other hand, Wu et al. (2020) present an analysis to quantify the benefits of forward-reserve strategies. They conclude that in a Forward-Reserve storage system where forward and reserve stocks are stored in the same rack, this policy results beneficial if the ratio of average picks per replenishment is sufficiently larger than 1. The response time savings can go up to 50% when it is larger than 10.

None of the previous works take into consideration demand correlation among the SKUs, including the contribution of Velez-Gallego and Smith (2018). Demand correlation among SKUs have been considered in some contributions in the literature (Xiao and Zheng 2010; Chuang et al. 2012; Chiang et al. 2014), by a clustering-assigning approach. In this case, highly correlated items are grouped and then the storage space is assigned to those groups. To develop the storage strategies, a pairwise correlation scheme is considered. In contrast, Zhang et al. (2019) consider demand correlation patterns of SKUs and propose a non-linear and non-convex mathematical model to solve the SLAP. Due to the difficulty in solving the model, they propose two heuristics, one based on a simulated annealing framework. None of the previous research considered a fast-pick area of the warehouse and hence, they do not incorporate into their traveling distance computation, the traveling distance required for replenishment. A comprehensive review of the storage allocation problem is presented by Rojas Reyes et al. (2019).

Although the correlation of SKUs has been identified as an important factor to be considered when assigning storage locations, none of the previous research has studied the impacts that such correlations may have in the efficiency of the order-picking. Hence, this paper aims to fill this gap and propose a factorial design to evaluate it along with the illustration of some instance's solutions to contrast the effects of SKUs demand correlation.

3 Mathematical Formulation

We consider the mathematical formulation to solve the slotting problem or storage space allocation to SKUs in the fast-pick area in a warehouse proposed by Velez-Gallego and

Smith (2018). The warehouse is divided into two areas: the fast-pick area, where the processes of selection and collection of products are carried out manually by the pickers who processes a single order at a time, and the reserve area, which fulfills the function of restocking the fast-pick area. To complete each order, the pickers follow a U-shaped route through the warehouse aisle, as it is illustrated in Fig. 1. They go down the first side of the aisle in search of the products ordered until they reach the SKU stored in the furthest position from the starting point; then they cross to the other side of the aisle and take the same route in the opposite direction and continue with the search for the requested products until they reach point O. The distance between every position is 2 m and the aisle width is not considered as the picker always follows a U-shaped route. The size of the fast-pick area also requires to be defined.

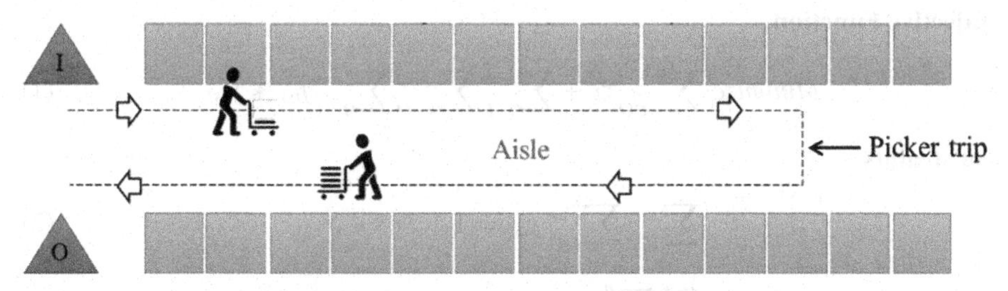

Fig. 1. Picking process. Source: Velez-Gallego and Smith (2018).

For the Velez-Gallego and Smith (2018) model it is important to take into consideration three key factors:

1. The smaller the fast-pick area, the shorter the distance each picker has to travel, however, they will have to go more times to the reserve area.
2. The warehouse used for the formulation of this model is already designed and operational, therefore, it is not possible to make structural changes.
3. Only the distances traveled by the picking operators are considered, neglecting the picking times at both the reserve and the fast-pick area.

3.1 Problem Formulation

The sets, parameters and variables are listed below, followed by the model formulation.

Sets

O: Customer orders.
S: SKUs.
Q_k: SKUs in customer order $k \in O$.
L: Storage locations.

Parameters

h_i: Distance to storage location $i \in L$.

r: Distance between the fast-pick and reserve areas.
q_s: Units of SKU $s \in S$ that can be stores in one bin.
d_s: Demand of SKU $s \in S$ in customer order O.
n_i: Number of bins at location $i \in L$.
f_{ist}: Distance traveled to replenish SKU $s \in S$ if assigned to t bins at location $i \in L$.

Decision Variables

y_{ist}: Binary variable. $y_{ist} = 1$ if SKU $s \in S$ is assigned to t bins at location $i \in L$, and. $y_{ist} = 0$ otherwise.
z_k: Distance traveled to prepare customer order $k \in O$.

Objective Function

$$Minimize \sum_{k \in O} z_k + \sum_{i \in L} \sum_{k \in O} \sum_{t=1} f_{ist} * y_{ist} \tag{1}$$

Constraints

$$\sum_{i \in L} \sum_{t=1}^{n_i} y_{ist} = 1 \quad \forall\, s \in S \tag{2}$$

$$\sum_{s \in S} \sum_{t=1}^{n_i} t \cdot y_{ist} \leq n_i \quad \forall\, i \in L \tag{3}$$

$$z_k \geq \sum_{t=1}^{n_i} 2 \cdot h_i \cdot y_{ist} \quad \forall\, i \in L,\ k \in O,\ s \in Q_k \tag{4}$$

$$y_{ist} \in \{0, 1\} \quad \forall\, s \in S,\ i \in L,\ 1 \leq t \leq n_i \tag{5}$$

The expression (1) is the objective function that aims to minimizing the distance traveled by the picker due to picking and replenishment activities. The first term is the distance traveled by the picker during the picking process and the second term is the distance traveled during replenishment activities. The possible values of the distance traveled to replenish each SKU must be computed based on the number of bins assigned to it, assuming that an operator replenishes one SKU at a time. The values of parameter f_{ist} are computed prior to solve the model as in expression (6). This parameter is defined assuming a constant distance r between the fast-pick area and the reserve area, which in turn implies that this distance is large enough to make the particular storage position of an SKU in the reserve is negligible.

$$f_{ist} = 2 \cdot r \cdot \left\lceil \frac{d_s}{t \cdot q_s} \right\rceil \quad \forall\, i \in L,\ s \in S,\ 1 \leq t \leq n_i \tag{6}$$

Constraint (2) ensures that a SKU partially or totally occupies a location and that it is not distributed in more than one location. Constraint (3) ensures that the number of storage bins assigned to a SKU does not exceed the number of bins available at that storage location. Constraint (4) defines the distance traveled by the picker to prepare the order k. Finally, constraint (5) determines the binary domain of the decision variable y_{ist}.

4 Instance Generation

4.1 Experimental Design

In order to measure the effect that some parameters have on the location chosen to store the products, an experiment design is proposed. For this, 5 factors are varied in 2 levels (low and high), resulting in a 2^5 factorial experiment, that is, 32 treatments. For each treatment, 5 replicates were created. The factors to vary are the following:

- *Orders*: Number of customer orders.
- *Nº Bins*: Number of storage bins in the location $i \in L$.
- *Bins Capacity*: Capacity of the storage bins, that is, how much of the SKU $s \in S$ can be stored in a bin. All bins are identical, this means they are the same in shape and volume and can only hold one SKU.
- *Nº Locations:* Number of storage locations in the fast-pick area.
- *Pairing*: Parameter that considers the percentage of appearance of the different SKUs and the combinations between them in the customer orders. The higher this value, the more SKUs are expected to be correlated in the instance generation.

Table 1 presents the values of the parameters for the low and high levels. The values were defined arbitrarily based on Velez-Gallego and Smith (2018).

Table 1. Low and high level of factors in the design of experiments.

Parameter	Level	
	Low (−1)	High (+1)
Orders	1,000	2,000
Nº Bins	5	10
Bins Capacity	20–50	70–100
Nº Locations	10	20
Pairing	1%	5%

4.2 Orders Generation Procedure

To generate orders in which the SKUs are correlated, we considered the instance generator proposed by Ansari et al. (2018) and Ansari et al. (2020) using the same *Fruithut* order database used by the authors to test their algorithm. This database is publicly available by (Mitchell 2016). This database is composed by the ticket number (order number), the name of the products requested in each order, the requested quantity of each product in each order and the category to which each product belongs to, and additional information that was not used for the instance generation procedure.

The procedure consists of two phases. As indicated by Ansari et al. (2018), the instance generator *"mimics the essence of real order data and generates a new set with an unlimited number of orders and SKUs"*. As the instance generator does not indicate the amount of each product that is requested in each order, which is something required in our model, we extended the proposed procedure to determine these values. So, it was necessary to incorporate a third phase to generate such values, extending the instance generator proposed by Ansari et al. (2018) and Ansari et al. (2020). A description of the three-phase procedure is presented in Fig. 2.

PHASE 1 (CORE GENERATION)
The parameters of the instances that resemble the characteristics of the original database are generated considering the correlation between SKUs. This is possible thanks to the association rule mining procedure that allows to identify the relationships between the SKUs in the database. In turn, this mining process generates loss of non-relevant SKUs (SKUs whose percentage of appearance is less than a predefined threshold value).

PHASE 2 (DETAILING)
The data lost during the first phase is compensated by adding SKUs to the orders, which generates a synthetic database with a similar average number of items per order and total number of SKUs than in the original database.

PHASE 3 (QUANTITY GENERATION)
The number of units ordered of each SKU in each customer order is defined, based on the probability distributions of the SKUs in the database. For this, the number of orders in which the SKU appears and the requested quantity of each of them is considered and the distribution that best fits the observed data of each category is obtained. Accordingly, the required number of SKUs per order is defined.

Fig. 2. Instance Generator. Source: Adapted from Ansari et al. (2018) and Ansari et al. (2020).

5 Numerical Results

During these tests, an MSI computer with an Intel Core i5-8250U 1.80 GHz CPU processor, with 8 GB of memory installed and Windows 10 with a 64-bit system is used. The algorithm is executed in Python version 3.9.1 using Pulp, Pandas, Math and Numpy libraries in conjunction with the Gurobi 9.1 solver.

For this paper, instead of using each product as an SKU, the category to which these products belong is used as the SKU. This is done to reduce the number of SKUs in the test instance and thus, generate an instance that can be solved to optimality within a reasonable amount of time. This is justified as the maximum number of orders generated for an instance is 2,000, and we need to be able to generate synthetic databases whose products can satisfy the characteristics of the original database.

The maximum computational time was set to 3,600 seconds and a maximum gap to 5%. Table 2 presents the general results for the distance traveled by the picker to complete the activities of picking and replenishment, as well as the computational times, indicating the minimum, maximum, average and median values of the five replicates of the instances that were solved.

Table 2. General results.

	General results			
	Minimum	Maximum	Average	Median
Distance (m)	13.380	78.668	34.773	34.204
Time (s)	3	3.600	1.645	1.022
Gap (%)	4,4%	9,1%	5,4%	5,0%

Figure 3 presents the box-plots of the total distance traveled by the picker, with respect to the levels of each factor that were evaluated. We can observe that when moving from a low level to a high level in the number of customers orders, the distance traveled by the picker increases considerably. On the other hand, it can be noted that by varying the level of pairing, number of bins and bins capacity the distance traveled by the picker to complete the picking and replenishment activities tends to decrease. Finally, when moving from a low level to a high level in the number of locations available in the warehouse the distance traveled by the picker does not present a significant variation.

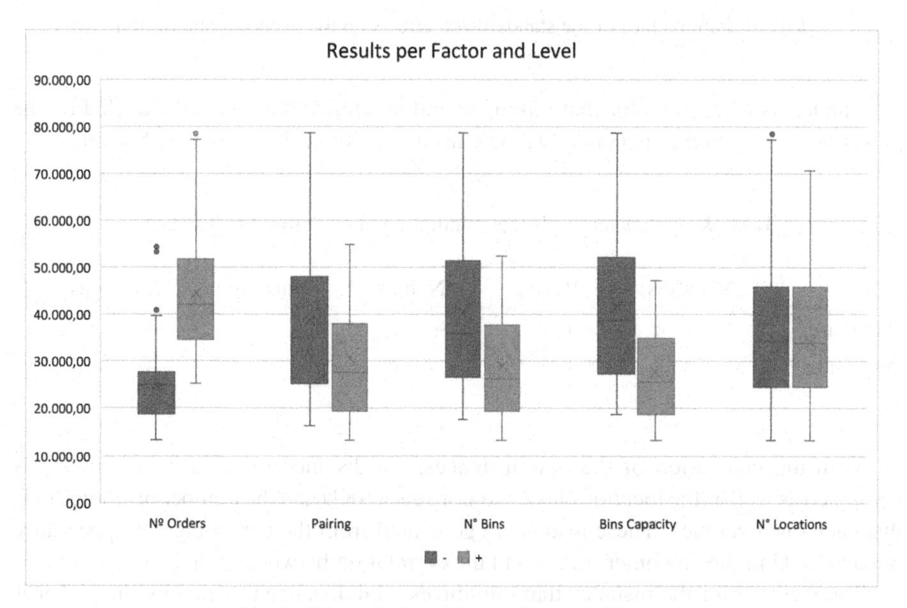

Fig. 3. Numerical results per factor and level.

We performed an analysis to determine the factors that are more relevant in the variation of the distance traveled by the picker. According to the results, the most relevant factors were the number of orders, the bins capacity, the number of available bins and the pairing (see Fig. 4). This means that a change in the level of any of the factors will have a significant effect on the distance traveled by the picker.

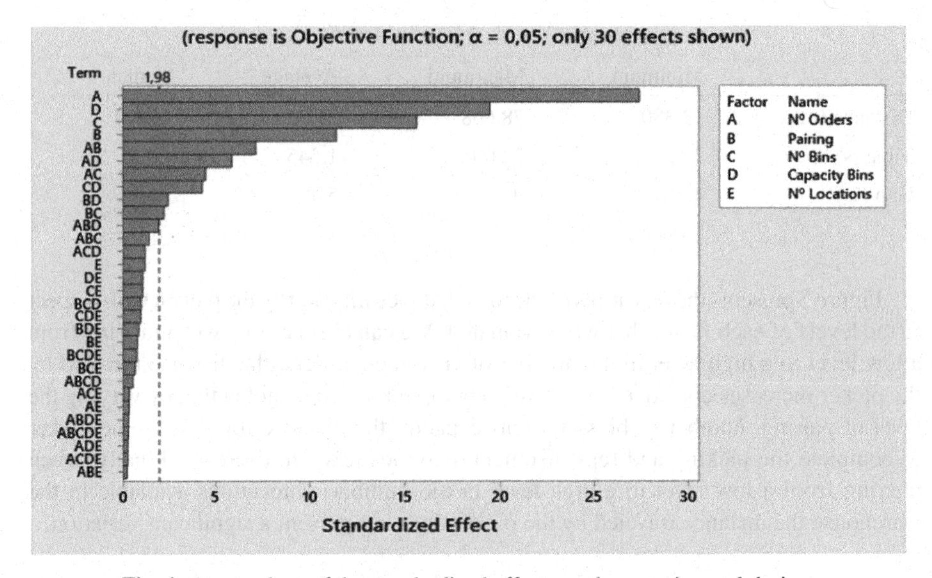

Fig. 4. Pareto chart of the standardized effects on the experimental design.

The levels of each factor that minimize and maximize the distance traveled by the picker are considered to generate two new instances. Table 3 presents such levels.

Table 3. Parameter levels that minimize and maximize the distance.

	N° orders	Pairing	N° bins	Bins cap	Locations
Minimize	–	+	+	+	+
Maximize	+	–	–	–	–

With the generation of the new instances, the distance traveled by the picker is computed as well as the location chosen to store each SKU and the number of storage bins that each one occupies. These results are generated from the percentage of appearance of each SKU in the customer orders and the correlation between them.

The solution for the instance that minimizes total distance traveled by the picker is illustrated in Fig. 5, that shows the position that each SKU occupies in the warehouse. This instance is characterized by being less restricted in terms of the number of locations available and the decision maker can also conclude that instead of 20 locations, it is necessary only 10 locations as part of the layout sizing decisions. So, the proposed mathematical model not only can be used to define the storage location assignments to the SKUs but also the size of the fast-pick area if required.

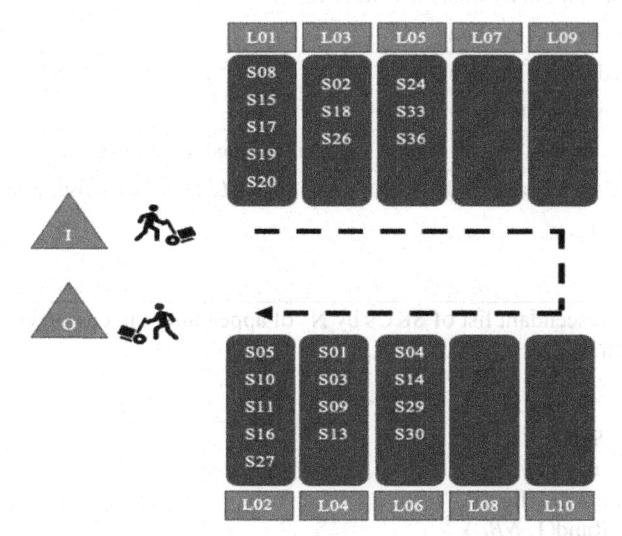

Fig. 5. Location of SKUs in the warehouse when minimize distance traveled. Source: Own elaboration.

From the design of experiments, the factor that has a greater effect on the variation of the distance traveled is the number of orders and accordingly, the SKUs that are ordered a higher number of times should be located at the most privileged positions (positions closest to the start and end-point of the tour) in the warehouse.

To contrast the effect of the SKUs correlation, the results obtained for an instance with SKUs that show demand correlation are analyzed. For this, an instance solved to optimality by a commercial solver and a greedy heuristic that assigns storage space considering only the frequency in which the SKUs appear in an order are compared. The greedy heuristic basically orders the SKUs in a descendant list based on the number of appearances in custom orders. Table 4 and Fig. 6 describes the notation and pseudocode of the heuristic.

In the algorithm presented above, as a first step, SKUs are sorted in descendent order with respect to the number of times that appear in the orders. Then, the SKU at the top of the list is selected and assign it to the first location which corresponds to the most preferred one (closest to the entry point). In this way, the better locations are assigned to SKUs with a higher number of times being ordered. Next, for each SKU the algorithm computes the minimum value for the parameter f, to determine the required number of bins to assign to the selected SKU.

To do this assignment, the algorithm analyzes the available number of bins with respect to the required bins by the SKU. The assignment of the bins is performed randomly verifying if the available number of bins is higher than those required. In this case the number of bins is determined based on a random number between 1 and the required

Table 4. Table of notation.

Q	Binary matrix with SKUs per order with size $O \times S$
$NB_{i,s}$	Necessary bins in location i to store SKU s
AB_i	Available bins in location i
OB_i	Occupied bins in location i
TB_i	Total bins in location i
$f_{i,s}$	Distance traveled to replenish SKU $s \in S$ if assigned to location $i \in L$
$y_{i,s,t}$	1 if SKU $s \in S$ is assigned to t bins at location $i \in L$, and 0 otherwise; where $t \leq (TB_i - OB_i)$

```
Algorithm: Descendant list of SKUs by N° of appearances in customer orders
1  sort Q.sum desc.
2  i=1
3  OB_i=0
4  for s=1.... S do
5        NB_i,s = min(f_i,s)
6        if AB_i > NB_i,s then
7            t = Rand(1, NB_i,s)
8        else
9            t = Rand(1, AB_i)
10        y_i,s,t = 1
11        OB_i+=t
12        if OB_i = TB_i then
13            i++
14 end for s
```

Fig. 6. Greedy heuristic pseudocode. Source: Own elaboration.

number of bins. Otherwise, this is determined randomly between 1 and the available number of bins. This procedure is repeated while there are available bins in the selected location considering the next SKU at the top of the ordered list. When no bins are left in the location, the algorithm selects the next location to continue assigning bins to the remaining SKUs. The procedure is repeated until all the SKUs have been assigned to a location with a determined number of bins.

Figure 7 presents the solutions obtained by the mathematical model and greedy heuristic. The solution to the left corresponds to the optimal solution whose objective function is equal to a total distance of 41,436 m. The solution to the right is obtained by a greedy heuristic with an objective function value of 65,680 m. As observed in the heat maps of each solution, the optimal solution not only considers the number of times that the SKU appears in the orders, but also its demand correlation with other SKUs in the instance. The greedy heuristic on the other hand, assigns storage space to SKUs based only on the number of times that an SKU appears in an order. The greedy solution distance value is 59% higher than the optimal, as it does not take into consideration the demand correlation among the SKUs.

Optimal Solution OF: 41,436				Greedy Solution OF: 65,680			
Location	bins	sku	y	Location	bins	sku	y
l01	1	s09	111	l01	1	s18	238
l01	1	s16	184	l01	4	s05	258
l01	2	s17	102	l02	1	s02	180
l01	1	s19	160	l02	2	s15	186
l02	1	s08	84	l02	2	s16	184
l02	1	s10	177	l03	1	s09	111
l02	1	s13	85	l03	1	s10	177
l02	1	s20	83	l03	1	s19	160
l02	1	s22	47	l03	2	s03	138
l03	2	s05	258	l04	2	s17	102
l03	3	s18	238	l04	3	s14	99
l04	1	s02	180	l05	1	s13	85
l04	2	s14	99	l05	1	s26	97
l04	2	s15	186	l05	1	s27	85
l05	1	s21	46	l05	2	s08	84
l05	1	s27	85	l06	1	s01	76
l05	1	s29	66	l06	1	s20	83
l05	1	s31	38	l06	1	s24	70
l05	1	s36	64	l06	2	s11	76
l06	1	s04	54	l07	1	s04	54
l06	1	s12	49	l07	1	s29	66
l06	1	s24	70	l07	1	s30	59
l06	1	s28	33	l07	1	s33	63
l06	1	s32	37	l07	1	s36	64
l07	2	s01	76	l08	1	s12	49
l07	2	s03	138	l08	1	s21	46
l07	1	s40	18	l08	1	s22	47
l08	2	s26	97	l08	1	s25	49
l08	2	s33	63	l08	1	s31	38
l08	1	s35	25	l09	1	s07	21
l09	2	s11	76	l09	1	s28	33
l09	1	s23	15	l09	1	s32	37
l09	1	s41	14	l09	1	s35	25
l09	1	s42	14	l09	1	s40	18
l10	1	s07	21	l10	1	s23	15
l10	3	s25	49	l10	1	s41	14
l10	1	s30	59	l10	1	s42	14

Fig. 7. Comparative Heat Map of two solutions: optimal (left) and greedy solution (right). Source: Own elaboration.

6 Conclusions

In this work, we address the allocation problem within a small warehouse for fruits and vegetables composed of a fast-pick area with a single aisle and a reserve area that only fulfills the function of replenishing the former. A mixed integer linear programming (MILP) model presented by Velez-Gallego and Smith (2018) is considered. The model aims to minimize the distance traveled for the picking and replenishment of products in the fast-pick area. We propose an experimental design to evaluate the factors that are more significant for the distance traveled by a picker, being the number of orders and the most important, followed by the capacity of the bins and the number of bins at each location. The instances were generated by the synthetic instance generator proposed by Ansari et al. (2018) and Ansari et al. (2020), incorporating a third phase to determine the quantity of each SKU that is requested in each order.

We also illustrate the differences among the solutions found by the mathematical model and a greedy heuristic that considers only the number of times that an SKU appears in an order, observing clear differences in the solutions as the mathematical model not only considers the number of times that an SKU appears in the orders, but also the demand correlation among SKUs.

For future research, the mathematical model can be adjusted, so that the distance traveled during the replenishment process may be modeled with more detail according to the storage strategy. In addition, we propose to consider different warehouse configurations and evaluate the impacts of SKU correlations accordingly. This may consider some adjustments in the traveling distance estimation. It is also possible to consider traveling times instead of distance, by assuming an average speed of the pickers. In this way we can incorporate in the workload computation the picking times in the reserve area for the replenishment operations.

Different slotting policies could be considered such as the possibility to assign multiple positions (scattered or dispersed storage strategy) proposed by Weidinger and Boysen (2018) and evaluate the impact of SKUs correlation under different slotting policies. The effects of assigning a unique storage location to SKUs versus multiple locations as it is allowed in this strategy can be assessed. Furthermore, it is possible to evaluate under which conditions one or the other strategy provides lower traveling times or distances.

Another research avenue can consider a heuristic solution approach, given that the problem is NP-Hard and it is possible to solve only small instances to optimality. As in practice, a warehouse can have thousands of SKUs, efficient solution methodologies are required to be able to solve real-size instances.

References

Ansari M., Rasoolian B., Smith J.: Synthetic order data generator for picking data. In: 15th IMHRC Proceedings, Savannah, Georgia, USA (2018)

Ansari M., Smith J., Rasoolian B.: Hybrid Synthetic Order Data Generator: A Two-Phase Process in Generating Correlated Order Picking Data. Working paper

Bartholdi, J.J., III., Hackman, S.T.: Allocating space in a forward pick area of a distribution center for small parts. IIE Trans. **40**(11), 1046–1053 (2008)

Bartholdi, J.J., Hackman, S.T.: Warehouse and distribution science. Version 0.98.1 (2019). https://www.warehouse-science.com/

Boysen, N., De Koster, R., Weidinger, F.: Warehousing in the e-commerce era: a survey. Eur. J. Oper. Res. **277**(2), 396–411 (2019)

Chuang, Y.F., Lee, H.T., Lai, Y.C.: Item-associated cluster assignment model on storage allocation problems. Comput. Ind. Eng. **63**(4), 1171–1177 (2012)

Gu, J.: The forward reserve warehouse sizing and dimensioning problem. Doctoral dissertation, Georgia Institute of Technology (2005)

Gu, J., Goetschalckx, M., McGinnis, L.F.: Research on warehouse operation: a comprehensive review. Eur. J. Oper. Res. **177**(1), 1–21 (2007)

Hackman, S.T., Rosenblatt, M.J., Olin, J.M.: Allocating items to an automated storage and retrieval system. IIE Trans. **22**(1), 7–14 (1990)

Ming-Huang Chiang, D., Lin, C.P., Chen, M.C.: Data mining based storage assignment heuristics for travel distance reduction. Exp. Syst. **31**(1), 81–90 (2014)

Mitchell, D.: Fruithut order data (2016). https://data.world/digitalbias/operationsanalysis/workspace/file?filename=fruithut_data_ordered_csv_file_1_1.csv

Reyes, J., Solano-Charris, E., Montoya-Torres, J.: The storage location assignment problem: a literature review. Int. J. Ind. Eng. Comput. **10**(2), 199–224 (2019)

Revillot-Narváez, D., Pérez-Galarce, F., Álvarez-Miranda, E.: Optimising the storage assignment and order-picking for the compact drive-in storage system. Int. J. Prod. Res. **58**(22), 6949–6969 (2020)

Subramanian, S.: Managing space in forward pick areas of warehouses for small parts. Doctoral dissertation, Georgia Institute of Technology (2013)

Thomas, L.M., Meller, R.D.: Analytical models for warehouse configuration. IIE Trans. **46**(9), 928–947 (2014)

Van Den Berg, J.P.: A literature survey on planning and control of warehousing systems. IIE Trans. **31**(8), 751–762 (1999)

van Gils, T., Caris, A., Ramaekers, K., Braekers, K., de Koster, R.B.: Designing efficient order picking systems: the effect of real-life features on the relationship among planning problems. Transp. Res. Part E Logist. Transp. Rev. **125**, 47–73 (2019)

Velez-Gallego, M.C., Smith, A.E.: Optimization of a fast-pick area in a cosmetics distribution center. In: 15th IMHRC Proceedings, Savannah, Georgia, USA (2018)

Walter, R., Boysen, N., Scholl, A.: The discrete forward–reserve problem–allocating space, selecting products, and area sizing in forward order picking. Eur. J. Oper. Res. **229**(3), 585–594 (2013)

Weidinger, F., Boysen, N.: Scattered storage: how to distribute stock keeping units all around a mixed-shelves warehouse. Transp. Sci. **52**(6), 1412–1427 (2018)

Wu, W., de Koster, R.B., Yu, Y.: Forward-reserve storage strategies with order picking: when do they pay off? IISE Trans. **52**(9), 961–976 (2020)

Xiao, J., Zheng, L.: A correlated storage location assignment problem in a single-block-multi-aisles warehouse considering BOM information. Int. J. Prod. Res. **48**(5), 1321–1338 (2010)

Yingde, L.I., Smith, J.S.: Dynamic slotting optimization based on SKUs correlations in a zone-based wave-picking system. In: 12th IMHRC Proceedings, Gardanne, France (2012)

Zhang, R.Q., Wang, M., Pan, X.: New model of the storage location assignment problem considering demand correlation pattern. Comput. Ind. Eng. **129**, 210–219 (2019)

Urban Transport and Collaborative Logistics

Real-Time Dispatching with Local Search Improvement for Dynamic Ride-Sharing

Martin Pouls[1(✉)], Anne Meyer[2], and Katharina Glock[1]

[1] FZI Research Center for Information Technology, 76131 Karlsruhe, Germany
{pouls,kglock}@fzi.de

[2] TU Dortmund University, 44221 Dortmund, Germany
anne.meyer@lfo.tu-dortmund.de

Abstract. Dynamic ride-sharing services such as UberPool or MOIA are becoming increasingly popular as they offer a cheap and flexible mode of transportation and reduce traffic compared to traditional taxi and ride-hailing services. One key optimization problem when operating ride-sharing services is the assignment of trip requests to vehicles to maximize the service rate while minimizing operational costs. In this work, we propose a real-time dispatching algorithm capable of quickly processing incoming trip requests. This dispatching algorithm is combined with a local search that aims to improve the current routing plan. Both algorithms are embedded into a planning and simulation framework for dynamic ride-sharing and evaluated through simulation studies on real-world datasets from Hamburg, New York City, and Chengdu. The results show that the local search improvement phase can improve the request acceptance rate as well as vehicle travel times. We achieve an average reduction of the request rejection rate by 1.62% points and a decrease in vehicle travel time per served request of 6.5%. We also study the influence of pre-booked rides and show that the local search yields even larger benefits when part of the trip requests are known in advance.

Keywords: Vehicle routing · Ride-sharing · Dial-a-ride-problem

1 Introduction

Ride-sharing services such as UberPool, Moia and GrabShare have recently emerged as suitable new modes of transportation for highly urbanized areas. They offer increased convenience and flexibility compared to public transportation as well as lower fares than classical taxi or ride-hailing services. At the same time, the increased usage of ride-sharing may serve as a tool to reduce traffic congestion as well as emissions.

Planning vehicle routes for such dynamic ride-sharing services has proven to be a difficult optimization problem that has attracted a significant amount of research attention (e.g. [1,10,12,15]). From a modelling perspective, the problem may be seen as a dynamic dial-a-ride problem with its standard constraints on

M. Mes et al. (Eds.): ICCL 2021, LNCS 13004, pp. 299–315, 2021.
https://doi.org/10.1007/978-3-030-87672-2_20

vehicle capacities, customer pickup time windows and customer ride times [5]. The particular challenges in this application setting arise mainly from the large number of trip requests (up to 20,000 per hour in this study) and the real-time requirements regarding computational times as customers expect a near instantaneous response to their requests.

In this paper, we propose a new algorithmic approach for tackling the vehicle routing problem in dynamic ride-sharing applications. So far, most existing approaches either sequentially process singular requests and therefore leave little room for optimization or are batch-based and therefore do not provide an immediate assignment of trip requests to a vehicle. In this work, we propose an approach that combines a sequential cheapest insertion heuristic (similar to [11,12]) with a local search. Our main contributions are:

- A cheapest insertion dispatching algorithm that processes incoming trip requests and facilitates fast response times for customers.
- A local search algorithm to improve the current plan via simple operators.
- A comprehensive simulation & planning framework to evaluate the performance of our algorithms on real-world data.

To cope with the running time requirements, we try to minimize shortest path calculations on the road network and use pre-calculated travel time estimations when possible. In contrast to many prior works in this field, we also allow for pre-booked trip requests. In this case customers do not desire immediate service but rather want to book a ride for some future point in time. We evaluate our approach through extensive simulation studies on four real-world datasets from Hamburg, New York City, Manhattan and Chengdu. Across all datasets, our results show that the local search yields improvements regarding the total vehicle travel times and thereby operational costs as well as the acceptance rate of trip requests. This effect is even more pronounced in scenarios with large portions of pre-booked trips.

The remainder of this work is structured as follows. In Sect. 2 we present an overview of related work regarding vehicle routing algorithms for dynamic ride-sharing. Section 3 introduces a formal problem description as well as our framework for evaluating dynamic ride-sharing services. Subsequently, Sect. 4 details our vehicle routing algorithms consisting of (1) the online dispatching algorithm and (2) the local search improvement phase. We present the results of our computational studies in Sect. 5. Finally, Sect. 6 summarizes our contributions and illustrates potential future research topics.

2 Related Work in Vehicle Routing for Dynamic Ride-Sharing

There is a large amount of extant research concerning dynamic dial-a-ride problems or more generally pickup-and-delivery vehicle routing problems summarized in several reviews (e.g. [3,13]). However, most classical applications differ significantly from the dynamic ride-sharing setting considered in this paper. In particular, they tend to consider significantly smaller instance sizes and have more lenient requirements concerning response times. Therefore, we focus our literature review on prior work concerning vehicle routing algorithms for dynamic ride-sharing. Existing approaches may be roughly divided into two groups (see also [10]): (1) sequential algorithms that process trip requests individually, and (2) batch-based algorithms that collect trip requests over a given time period and process them in batches.

An insertion-based sequential approach is presented in [11,12]. The authors first find suitable vehicles for an incoming trip request and subsequently determine the cheapest feasible insertion. Similar insertion-based approaches have been proposed by several authors. For instance, in [8], the authors present a combination of cheapest insertion with a kinetic tree data structure that maintains a set of potential routes for each vehicle. When inserting a new trip request they may evaluate the insertion into several feasible routes besides the one currently in execution. To cope with the complexity they propose a hotspot clustering algorithm that clusters similar nodes in their search tree. The authors of [4] also use this combination of cheapest insertion and kinetic trees but extend the approach in such a way that multiple non-dominated options are offered to customers that may differ in the promised pickup time or the proposed price.

In contrast to sequential algorithms, batch-based approaches promise a better solution quality at the expense of an increased computational complexity and the lack of instantaneous customer responses. One of the earliest batch-based approaches for dynamic ride-sharing was proposed in [1]. The authors utilize a graph-based approach that matches potential combinations of trip requests with vehicles. In a follow-up work [2], the authors propose a non-myopic algorithm including sampled future trip requests to ensure that the vehicle fleet is well-positioned to serve these anticipated trip requests. Several works have built upon the graph-based algorithm in [1]. In [9], the authors utilize a clustering algorithm to aggregate multiple pickup and drop-off locations. Subsequently, they build paths through these zones and match suitable vehicles to these so called zone paths. In an extended version of the algorithm [10], the same authors present a non-myopic variant of their algorithm that incorporates future anticipated trip requests. A neural-network based approximate dynamic programming (ADP) algorithm is proposed in [16]. The approach also builds on the graph-based approach in [1]. The authors utilize a trained neural network to score feasible trips and subsequently assign trips to vehicles in such a way that they maximize the scores. In [15] a column generation approach is presented. It utilizes an anytime algorithm for solving the pricing problem that is guaranteed to return a feasible solution even if interrupted by a time limit. In contrast to our approach the authors do not allow for the rejection of trip requests, but rather use an increasing penalty term that penalizes the delay of trip requests.

Our local search algorithm employs common operators for vehicle routing problems. There are countless works using local search methods on different VRP variants, for a thorough overview concerning the usage of local search techniques for vehicle routing we refer the reader to [7].

3 Dynamic Ride-Sharing: Problem Description and System Design

In this section, we first introduce the vehicle routing problem for dynamic ride-sharing along the necessary notation and subsequently give an overview of our overall system design for evaluating a dynamic ride-sharing service.

3.1 Problem Description and Notation

From a modelling perspective, the problem at hand may be formulated as a dynamic dial-a-ride-problem. Hence, our notation is largely consistent with the DARP formulation in [5]. Let r be an incoming trip request for a given number of passengers q_r, a pickup location p_r, and a drop-off location d_r. We denote the direct travel time between pickup and drop-off as tt_r^d. For the purpose of this work, we assume $q_r \in \{1, 2\}$, as larger groups are generally requested to resort to ride-hailing services offered by the same company (see e.g. UberPool [17]). Each request is associated with a creation time t_r that denotes the time at which it enters the system. In addition, it has a pickup time window given by an earliest pickup time e_r and a latest pickup time l_r. The size of this time window, i.e. $l_r - e_r$ is also referred to as the maximum waiting time w_r of a customer. Customers either want to be served immediately, i.e. $e_r = t_r$ or they submit a pre-booked request. In the latter case they want to reserve a ride for some future point in time. Thus, e_r corresponds to t_r plus a pre-booking time a_r. In this study, we focus on short pre-booking times. In general, a_r may range from a couple of minutes to several days. In addition to the time window on the pickup of a customer, there is also a temporal constraint on the customer's ride time given by the maximum ride time L_r. This is motivated by the fact that, although ride-sharing allows for detours compared to direct taxi services, these detours should be limited. In practice, the acceptable detour could either be specified by the customer or determined by the operator. In this work, the maximum ride is determined as $L_r = \max(tt_r^d \cdot \alpha, tt_r^d + L^{min})$. Here α corresponds to a detour factor and L^{min} denotes a minimum acceptable detour. The latter is required for trip requests covering very short distances as otherwise few feasible combinations with other trip requests would be possible. To serve the trip requests, we need a fleet of vehicles K. Each vehicle $k \in K$ is associated with a maximum capacity Q_k equivalent to the maximum number of passengers.

The hierarchical objective of our problem is to primarily maximize the number of served trip requests and secondarily minimize the total travel time of the vehicle fleet. These travel times serve as a proxy for the operational cost of the vehicles. In addition, the constraints concerning pickup time windows of requests $[e_r, l_r]$, maximum ride times L_r and vehicle capacities Q_k must be met.

3.2 A System Design for Dynamic Ride-Sharing

Our overall system for evaluating dynamic ride-sharing services is illustrated in Fig. 1. It is partitioned into two parts: (1) the planning service encompassing all relevant planning components, and (2) the discrete-event simulation that mimics the behavior of real-world vehicles and customers.

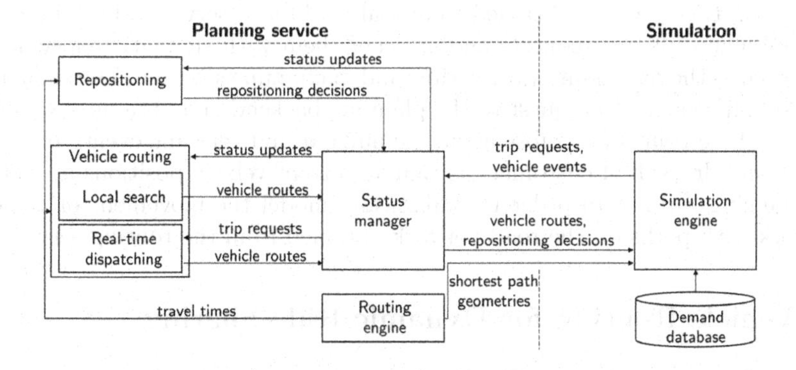

Fig. 1. Planning service, simulation and relevant communication.

Planning Service. The planning service maintains the current system state and is responsible for handling incoming trip requests and planning vehicle movements. All communication with the simulation takes place via the status manager. In addition, this component stores the current state of the vehicle fleet and all trip requests.

Vehicle routing is handled by two components. First, the dispatching algorithm handles incoming trip requests and assigns them to a suitable vehicle or rejects them if no feasible insertion into a route is possible. Second, the local search improves the current routing plan. In case of a centralized real-world vehicle operator, the local search could operate continuously whenever no trip request is being processed by the dispatching algorithm. However, in our simulation studies we perform the local search with a given time limit after a trip request is processed by the dispatching module. The vehicle routing algorithms are detailed in Sect. 4.

The repositioning component is responsible for repositioning idle vehicles. For details regarding this optimization problem, we refer the reader to our prior work [14]. For the remainder of this paper, we use a simple reactive repositioning algorithm inspired by [1]. Given a rejected trip request r, we greedily reposition the nearest idle vehicle to the pickup location p_r. The assumption behind this is that trip requests are highly spatially and temporally correlated. Hence, we may assume that additional trip requests may arise near a rejected request.

Lastly, the main purpose of the routing engine is to calculate realistic travel times on the road network which are used by the planning components. In this study we work with road networks based on OpenStreetMap (OSM) data and use a contraction-hierarchy based routing solver [6]. As these shortest path queries on the road network are expensive when performed at high volumes, we also use pre-computed travel times as an approximation. This is described in Sect. 4.

Simulation. Our discrete-event simulation operates on a demand database containing historic trip requests. Each request consists of the request time, pickup and destination coordinates, and the number of passengers. In this work, we operate on trip requests obtained from real-world taxi services. The simulation engine replays all trip requests and enriches them with simulation-specific settings such as the maximum waiting time and maximum ride-time. The simulation then submits each trip request to the planning backend and obtains new vehicle routes. These routes are subsequently simulated and relevant events regarding pickup and drop-off of customers as well as current vehicle positions are sent to the planning service. In order to realistically model the movement of vehicles, we work with paths on the road network obtained from the routing service.

4 Vehicle Routing for Dynamic Ride-Sharing

Our solution approach for vehicle routing for dynamic ride-sharing consists of two algorithms. First, a sequential real-time dispatching component that handles incoming trip requests and assigns them to a suitable vehicle if possible. Each new trip request is processed by the dispatching algorithm and either inserted into a route or rejected if no feasible insertion was found. Second, a local search that tries to improve the current routing plan via simple search operators. The local search is run for a given time period after each trip request was processed in order to exploit available computational time between requests. In the following, we present these two algorithms in detail.

4.1 Real-Time Dispatching

The aim of the real-time dispatching step is to quickly insert an incoming trip request into the current routing plan if possible and otherwise reject the request. For this purpose, we utilize a cheapest insertion heuristic similar to the one proposed in [11,12]. Our approach consists of three steps:

1. Find a set of candidate vehicles $K_r^c \subseteq K$ for a given trip request r.
2. Sort the candidate vehicles according to their estimated suitability for r.
3. Find the cheapest insertion.

Vehicle Selection. To select suitable vehicles for a given trip request r, we utilize a grid-based index data structure similar to the one proposed by [11,12]. We partition the area under study into a set of grid cells G as depicted in Fig. 2a. Each grid cell $g \in G$ covers an area of 750 m × 750 m and has a center c_g defined as the road node closest to the centroid of the cell. We pre-calculate and store a matrix of travel times $tt_{g,h}$ between the centers of each pair of cells g and h. This pre-calculated travel time matrix is used at several points in our algorithm to estimate travel times and save computational time. Furthermore, we store and continuously update a set of vehicles currently situated in a given cell g denoted as K_g.

Given a trip request r, we first determine the grid cell g_r in which the desired pickup location p_r is situated. Subsequently, we determine the neighborhood N_{g_r} of g_r as the set of grid cells from which g_r could be reached before the end of the pickup time window according to our pre-calculated travel times, i.e. $N_{g_r} = \{g \in G | tt_{g,g_r} \leq l_r - t_r\}$. We now select all vehicles situated within this neighborhood as our set of candidates K_r^c for inserting r, i.e. $K_r^c = \bigcup_{g \in N_{g_r}} K_g$. As we only utilize our pre-computed travel time matrix in this vehicle selection step, we avoid expensive shortest path queries.

Vehicle Sorting. After selecting the candidate vehicles K_r^c, we sort these vehicles according to their estimated suitability for the trip request r. Given a vehicle $k \in K_r^c$, we estimate the increase in total travel time incurred by inserting r into the current route of k. For this purpose, we calculate a representative pickup time tp_r^{avg} as the middle of the pickup time window $[e_r, l_r]$. Similarly we calculate a representative drop-off time td_r^{avg} as the middle of the feasible drop-off time window $[e_r + tt_r^d, l_r + L_r]$. As illustrated in Fig. 2b, we now determine a suitable index i in the current route of k for inserting the pickup location of r. We select i in such a way that tp_r^{avg} lies in the interval between the departure time t_{i-1}^d at the preceding stop and the arrival time t_i^a at the following stop. If no such position is found, the pickup location is inserted at the end of the route. The same procedure is performed for the drop-off location of r and index j.

The estimated detour (i.e. additional travel time) by inserting r into the route of k is denoted as $\hat{\Delta}_{r,k}^+$ and determined via our pre-calculated travel time matrix. Hence, in this step we approximate the actual travel time via the grid cells in which the tour stops are located and do not perform shortest path queries. The set of candidate vehicles K_r^c is now sorted according to $\hat{\Delta}_{r,k}^+$ and the resulting list is denoted as K_r^s.

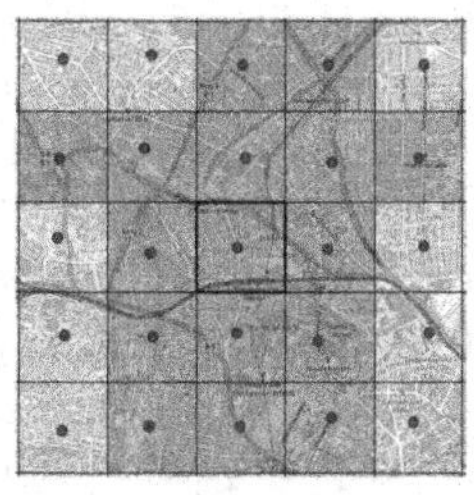

(a) Grid partitioning.

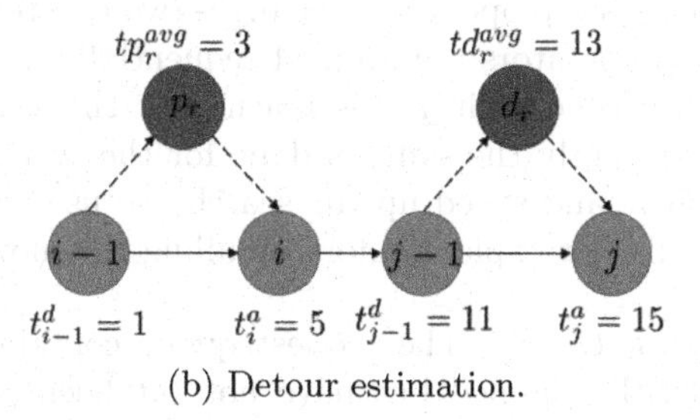

(b) Detour estimation.

Fig. 2. (a) Grid partitioning of an area with blue points representing the grid cell centers. The neighborhood N_g of the red grid cell is marked in green. (b) Detour estimation during vehicle sorting. (Color figure online)

Cheapest Insertion. In the third and final step of our dispatching algorithm we iterate over $k \in K_r^s$ and determine the cheapest feasible insertion position of p_r and d_r into the current route of k. For this purpose, we evaluate all potentially feasible combinations to insert p_r and d_r while leaving the order of the existing stops intact. If all constraints defined in Sect. 3 are satisfied, a feasible insertion was found. Feasible insertion are ranked by their increase in vehicle travel time $\Delta_{r,k}^+$. In contrast to the prior steps, we now use travel times from our routing solver to guarantee that time window and ride time constraints are met. The insertion with the minimal $\Delta_{r,k}^+$ among all evaluated vehicles is performed. After checking k^{max} vehicles, we abort the search if a feasible insertion has been found. Due to our prior sorting phase, we have checked the most promising vehicles at this point. Otherwise, we continue the search until a feasible insertion is found or all vehicles have been evaluated. If no feasible insertion is found, r is rejected. At this stage the customer is notified that the trip request is either accepted or rejected. Therefore, the local search cannot change this decision.

4.2 Local Search

The aim of our local search is to use available computational time and improve the current routing plan. Hence, it is triggered after a request was processed by the dispatching algorithm and consists of several separate phases:

1. Inter-route search: modifies two routes simultaneously.
   (a) Inter-route move: moves a single request from one vehicle to another.
   (b) Inter-route swap: swaps two requests between vehicles.
2. Intra-route search: improves the route of a single vehicle.

Inter-route Search. The inter-route phase utilizes two simple operators: the move operator moves a single trip request from its current route to another one, and the swap operator exchanges two trip requests between their current routes. These operators are applied sequentially, i.e. we first apply the move operator until a time limit T^m is reached or the search space has been exhausted and subsequently the same is done for the swap operator with a time limit T^s. To improve and speed up the search, we use two important data structures which we will first explain before detailing the move and swap operators themselves.

Request Queue. The request queue contains the set of trip requests that are currently planned but have not yet been picked up. The queue is ordered in decreasing order by the travel time contribution of each request Δ_{r,k_r}^-. This corresponds to the travel time that would be saved if the trip request r were removed from its current vehicle k_r and is determined based on the travel times from our routing engine. The request queue is updated every time a trip request is inserted into or removed from a route. The reasoning for this is that we want to prioritize requests with a large Δ_{r,k_r}^- during our search, as they offer the most potential for improvement and we may not be able to evaluate all requests due to running time restrictions. Our two inter-route search neighborhoods each utilize a separate request queue RQ^m (move) and RQ^s (swap) to determine the next request for evaluation. The usage of separate queues is necessary due to the tabu mechanism described in the next paragraph.

Tabu List. In addition to the request queue, our inter-tour operators each use a tabu list denoted as TS^m (move) and TS^s (swap). These contain requests that have recently been evaluated without finding an improving move or swap respectively. For a given tabu interval i^{tabu}, a request on the tabu list is not evaluated for a potential move or swap again.

Algorithm 1. Move Operator

1: Finds the first improving move and performs it. Repeats until the request queue is empty. Terminates at any point, if the time limit T^m was reached.
2: **for** $r_1 \in RQ^m$ **do**
3: $\quad K^t \leftarrow K \setminus \{k_1\}$
4: $\quad K^t \leftarrow sort(K^t, r_1)$ ▷ Sorts $k_2 \in K^t$ by $\hat{\Delta}^+_{r_1,k_2}$ (ascending)
5: $\quad K^t \leftarrow \{k_2 \in K^t | \hat{\Delta}^+_{r_1,k_2} \leq \beta \cdot \Delta^-_{r_1,k_1}\}$
6: $\quad$ **for** $k_2 \in K^t$ **do**
7: $\quad\quad m \leftarrow findBestMove(r_1, k_1, k_2)$
8: $\quad\quad$ **if** m is an improving move **then**
9: $\quad\quad\quad$ perform m, update RQ^m and go to line 2
10: $\quad\quad$ **end if**
11: $\quad$ **end for**
12: $\quad TS^m \leftarrow TS^m \cup r_1$ ▷ No improving move found, add r_1 to tabu list
13: **end for**

Move Operator. The move operator is outlined in Algorithm 1. It tries to move a single request from its currently assigned vehicle to another one and improve the overall vehicle travel time. We first select the next request r_1 from RQ^m (line 2). The corresponding vehicle to which r_1 is assigned is denoted as k_1. In lines 3–5 we select a set of potential target vehicles and subsequently sort and filter this set. Sorting is performed according to the estimated travel time increase $\hat{\Delta}^+_{r_1,k_2}$ for $k_2 \in K^t$ with the same procedure as in Sect. 4.1. We filter vehicles where the estimated detour by inserting r_1 exceeds the saved travel time from removing r_1 from the route of k_1 significantly. For this purpose, we define a factor $\beta > 1$. Hence, K^t only contains vehicles k_2 with $\hat{\Delta}^+_{r_1,k_2} \leq \beta \cdot \Delta^-_{r_1,k_1}$. Subsequently, in lines 6–10, we iterate over K^t and find the best feasible move for each vehicle with the same approach as in Sect. 4.1. We follow a first-improvement scheme. Hence, the first move that reduces the overall travel time is performed, i.e. $\Delta^+_{r_1,k_2} < \Delta^-_{r_1,k_1}$. If no improving move is found, r_1 is added to TS^m (line 12). The move operator is applied iteratively to all requests in RQ^m. The procedure is interrupted whenever the time limit T^m has been reached.

Algorithm 2. Swap Operator

1: Finds the first improving swap and performs it. Repeats until the request queue is empty. Terminates at any point, if the time limit T^s was reached.
2: **for** $r_1 \in RQ^s$ **do**
3: $K^t \leftarrow K \setminus \{k_1\}$
4: $K^t \leftarrow sort(K^t, r_1)$ ▷ Sorts $k_2 \in K^t$ by $\hat{\Delta}^+_{r_1,k_2}$ (ascending)
5: $K^t \leftarrow \{k_2 \in K^t | \hat{\Delta}^+_{r_1,k_2} \leq \beta \cdot \Delta^-_{r_1,k_1}\}$
6: **for** $k_2 \in K^t$ **do**
7: **for** $r_2 \in R_{k_2}$ **do**
8: $s \leftarrow findBestSwap(r_1, r_2, k_1, k_2)$
9: **if** s is an improving swap **then**
10: perform s, update RQ^s and go to line 2
11: **end if**
12: **end for**
13: **end for**
14: $TS^s \leftarrow TS^s \cup r_1$ ▷ No improving swap found, add r_1 to tabu list
15: **end for**

Swap Operator Subsequently, we apply the swap operator as outlined in Algorithm 2. This operator tries to exchange two requests between their current vehicles and improve the overall vehicle travel time. In the same manner as with the move operator, we select the next trip request r_1 from RQ^s and sort and filter the potential target vehicles K^t (lines 2–5). Let k_1 denote the vehicle to which r_1 is currently assigned. Proceeding with the next target vehicle $k_2 \in K^t$, let R_{k_2} denote all requests assigned to k_2 that have not been picked up yet. These are the potential swap partners. In lines 7–8, we iterate over $r_2 \in R_{k_2}$ and find the best swap for the pair of requests (r_1, r_2). The first swap that decreases the overall travel time is performed, i.e. $\Delta^+_{r_1,k_2} + \Delta^+_{r_2,k_1} < \Delta^-_{r_1,k_1} + \Delta^-_{r_2,k_2}$ (lines 9–11). If no improving swap is found for r_1, it is added to the tabu list TS^s (line 14). The procedure terminates when the request queue is empty or the time limit T^s was reached.

Intra-route Search. The intra-route search improves the order of stops within a single route. It is applied to all vehicles whose routes have been modified since the last run either by the inter-route search or the dispatching algorithm. We use two operators: (1) an intra-route stop move that re-inserts a single stop into the route, and (2) an intra-route request move that re-inserts all stops associated with a single trip request r into the route. Insertion is performed with the same approach as in Sect. 4.1. If an improvement over the old route is found, the insertion is performed. The intra-route search has no time limit and is applied exhaustively until no improvements are found. However, due to the small search space and the small number of vehicles that need to be evaluated in each iteration, the running time is negligible compared to the inter-route search and the dispatching algorithm.

5 Computational Results

We evaluate our algorithms on multiple real-world datasets. In the following, we first introduce these datasets and subsequently the specific scenarios and algorithms settings for this study. Finally, we present our results and findings.

5.1 Data and Setup

We perform simulation studies with real-world datasets from Hamburg (HH)[1], New York City (NYC)[2] and Chengdu (CH)[3]. In addition, we build a fourth dataset based on the one from NYC that only contains trips within Manhattan (MANH). The Manhattan dataset is considered separately to evaluate scenarios with high demand in a small area. In contrast, the other datasets also include suburban areas with relatively low demand. All these datasets contain the same basic information: trip requests with the desired pickup and drop-off coordinates and the pickup time. Moreover, the NYC dataset contains the number of passengers per trip request. As the HH and CH datasets are missing this information, we assume the same distribution as in the NYC data. We perform some basic data filtering and cleaning by removing records with missing information. Moreover, we only consider trip requests within the respective area of study and with at most two passengers as mentioned in Sect. 3.1. Our software components are implemented in C++. We use RoutingKit [6] as our routing engine and OSM data to derive the road network. All experiments are performed on the same computer with an Intel i7-6600U CPU and 20 GB of RAM.

5.2 Scenarios and Algorithm Settings

For each dataset we generate a set of scenarios by varying the following: weekday and pre-booking probability. We consider two weekdays, Wednesday and Sunday. The precise dates are given in Table 1. Note that our simulation uses a warm-up phase of 6 h in simulated time that immediately precedes the selected dates.

Our settings for maximum waiting times, maximum ride times, and pre-booking times are given in Table 2. For the pre-booking time we use a single value of $a_r = 20$ min. A given fraction of 0%, 25% or 50% of requests is pre-booked, all other requests desire immediate service. Unless noted otherwise, the default setting of 0% is used. We performed preliminary test to determine an adequate fleet size per dataset: 75 (HH), 1175 (NYC), 700 (MANH), and 1150 (CH). With this fleet size we should be able to serve roughly 90% of all requests.

The parameter settings for our dispatching and local search algorithms are summarized in Table 2. These were selected based on preliminary tests to work well with all datasets. Theoretically, these parameters could be tuned specifically for each dataset. All scenarios are run with two algorithm settings: (1) only with

[1] Provided by PTV Group, Haid-und-Neu-Str. 15, 76131 Karlsruhe, Germany.
[2] https://www1.nyc.gov/site/tlc/about/tlc-trip-record-data.page.
[3] https://outreach.didichuxing.com/appEn-vue/KDD_CUP_2020.

Table 1. Temporal scenario settings per dataset with dates and trip requests.

	Wednesday	Requests	Sunday	Requests
HH	20 Mar 2019	13,556	24 Mar 2019	10,669
NYC	16 Mar 2016	376,526	20 Mar 2016	368,508
MANH	16 Mar 2016	297,457	20 Mar 2016	269,346
CH	16 Nov 2016	239,037	20 Nov 2016	237,037

Table 2. Scenario and algorithm settings.

Fraction of pre-booked requests [%]	$[\mathbf{0}, 25, 50]$
Pre-booking time (a_r) [min]	20
Maximum waiting time (w_r) [s]	600
Detour factor (α)	1.5
Minimum allowed detour (L^{min}) [s]	300
Vehicle capacity (Q_k)	4
Vehicle limit (k^{max})	50
Time limit inter-route move (T^m) [ms]	80
Time limit inter-route swap (T^s) [ms]	120
Tabu interval (i^{tabu}) [min]	3
Grid cell size [m]	750×750
Vehicle filtering factor (β)	1.5

the dispatching algorithm from Sect. 4.1 ("DIS"), and (2) with the dispatching algorithm followed by the local search improvement phase described in Sect. 4.2 ("DIS+LS"). The time limit of 200 ms on the inter-route search ensures that even during peak demand times all trip requests may be processed in real time.

5.3 Performance Indicators

We use several performance measures to assess the impact of our local search. Firstly, the trip request rejection rate ("Rej") measures the fraction of trip requests that could not be served. Our main goal is to minimize this rejection rate. Secondly, to gauge the operational cost we consider the average travel time per vehicle ("TTv"). As this value is distorted by serving more requests, we also consider the average vehicle travel time per served trip request ("TTr"). Lastly, we introduce metrics to measure the customer satisfaction, in particular the average waiting time ("Wait") and average delay ("Del"). This delay corresponds to the increase compared to the direct travel time between pickup and drop-off of a trip request. In addition, we also report the total running time ("RT").

5.4 Results Overview

Table 3. Average results for all datasets and vehicle routing modes.

Data	Mode	Rej [%]	Wait [s]	Del [s]	TTv [min]	TTr [s]	RT [min]
HH	DIS	10.01	397.68	148.98	1012.97	419.42	2.38
	DIS+LS	**9.24**	**389.74**	**145.45**	**985.81**	**404.45**	2.94
NYC	DIS	8.09	**494.27**	170.04	1127.35	232.14	373.38
	DIS+LS	**7.41**	496.72	**163.63**	**1030.04**	**210.58**	1218.41
MANH	DIS	6.21	496.53	161.87	1061.27	168.05	187.58
	DIS+LS	**3.98**	**490.51**	**149.94**	**981.62**	**151.78**	453.81
CH	DIS	10.17	465.76	200.64	1030.70	351.66	128.78
	DIS+LS	**7.37**	**461.02**	**194.91**	**992.23**	**328.30**	303.33
ALL	DIS	8.62	463.56	170.38	1058.07	292.82	173.03
	DIS+LS	**7.00**	**459.50**	**163.48**	**997.43**	**273.78**	494.62

Table 3 summarizes the results for all datasets and vehicle routing modes with the default scenario settings. The values for each dataset correspond to the average of the two weekday scenarios. The rows denoted as "ALL" contain averages across all datasets. The results show that our local search manages to improve the results in several ways. Firstly, the rejection rates are reduced by an average of 1.62% points. At the same time vehicle routes become more efficient. Despite the increase in served trip requests, the average travel time per vehicle is reduced by 5.73% (60.64 s). The effect is even more pronounced when considering the travel time per served request with a reduction of 6.5% (19.04 s). Besides improving the vehicle fleet performance, the local search also provides minor benefits concerning the customer convenience by reducing the average waiting time and in-car travel delay. As expected, the local search increases the running times. However, these are still manageable and the algorithm may be used in real-time even with large-scale scenarios. The overall running time is still lower than the simulated time period of 30 h. In particular, even on the NYC dataset, the average running time for processing one trip request in the dispatching algorithm is 6.70 ms. Even during peak demand times with up to 20,000 trip requests per hour, we have roughly 200 ms to process a single request. Given the performance of our dispatching algorithm and the time limit on the local search, our approach is able to process requests in real-time while leaving enough computational time for the local search to improve the solution.

5.5 Impact of Pre-booking

One factor that has a large impact on the performance of DIS+LS compared to DIS is the percentage of pre-booked requests. Figure 3 shows the average trip request rejection rates by different percentages of pre-booked requests. Note that the values are again averages of the two weekdays. The performance with DIS deteriorates on some datasets (HH and CH) as the percentage of pre-booked requests increases. We assume that the dispatching algorithm takes sub-par decisions given the limited information when inserting a pre-booked request. In contrast, with DIS+LS an increased fraction of pre-booked requests leads to a decrease in rejection rate across all datasets. This is due to the fact that the local search has more trip requests to work with and therefore the search space is larger, providing more room for improvements.

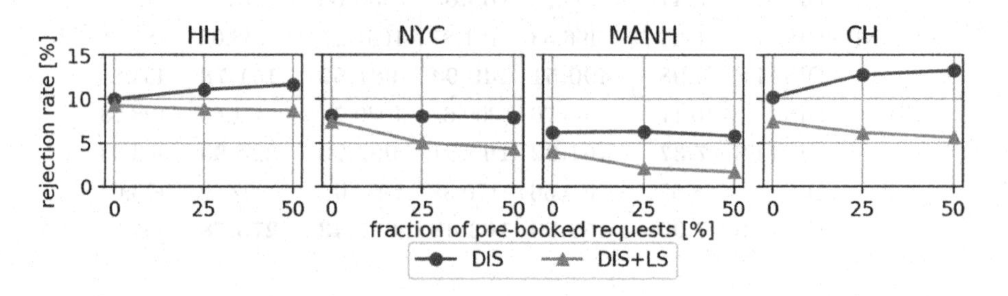

Fig. 3. Average rejection rate by pre-booking probability.

5.6 Vehicle Utilization

As a last analysis, we take a look at the utilization of the vehicle fleet throughout the day. Figure 4 shows the vehicle utilization throughout the day for a single scenario (MANH, 50% pre-booking) compared between DIS and DIS+LS. It illustrates the fraction of vehicles that are serving a route (active), repositioning or idle. In addition, we show the number of total and rejected requests over time.

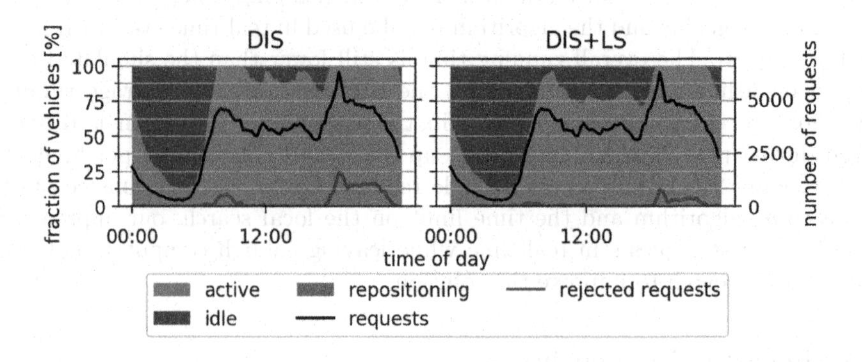

Fig. 4. Vehicle utilization over time.

In a similar manner, Fig. 5 shows the passenger occupation of the vehicle fleet, i.e. the fraction of vehicles that have a certain number of passengers aboard. The two figures clearly show that by using the local search we utilize the vehicle fleet more efficiently and the number of rejected requests is reduced. At the same time, the percentage of idle vehicles is increased and we have a larger fraction of vehicles with 3 or 4 customers aboard.

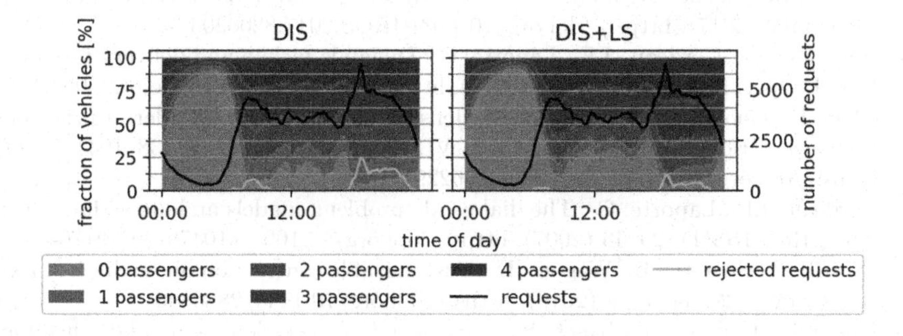

Fig. 5. Passengers per vehicle over time.

6 Conclusions and Outlook

In this work, we presented a real-time dispatching algorithm for dynamic ride-sharing enhanced by a local search phase. Our approach provides quick response times for incoming trip requests as well as an improved solution quality through the local search. We embed our algorithm into a simulation framework and perform extensive computational evaluations on several real-world datasets. The results show that we are able to reduce the rejection rates of trip requests by an average of 1.62% points and at the same time reduce the vehicle travel times and thereby operational costs. With a large fraction of pre-booked requests the improvement regarding the rejection rate increases up to almost 10% points in some datasets.

In the future, we would like to improve our algorithm by including more complex local search operators and search strategies such as simulated annealing. Moreover, we see potential for a more detailed study regarding the incorporation of pre-booked requests into dynamic ride-sharing services. In use cases where a significant portion of requests is pre-booked or we are working with longer pre-booking times of several days, we could solve a static routing vehicle problem in advance and use this solution as a starting point. We would also like to address additional real-world requirements such as the consideration of traffic congestion or the possibility of cancellations.

References

1. Alonso-Mora, J., Samaranayake, S., Wallar, A., Frazzoli, E., Rus, D.: On-demand high-capacity ride-sharing via dynamic trip-vehicle assignment. Proc. Natl. Acad. Sci. **114**(3), 462–467 (2017). https://doi.org/10.1073/pnas.1611675114
2. Alonso-Mora, J., Wallar, A., Rus, D.: Predictive routing for autonomous mobility-on-demand systems with ride-sharing. In: 2017 IEEE/RSJ International Conference on Intelligent Robots and Systems (IROS), pp. 3583–3590. IEEE, Vancouver (September 2017). https://doi.org/10.1109/IROS.2017.8206203
3. Berbeglia, G., Cordeau, J.F., Laporte, G.: Dynamic pickup and delivery problems. Eur. J. Oper. Res. **202**(1), 8–15 (2010). https://doi.org/10.1016/j.ejor.2009.04.024
4. Chen, L., Gao, Y., Liu, Z., Xiao, X., Jensen, C.S., Zhu, Y.: PTrider: a price-and-time-aware ridesharing system. Proc. VLDB Endow. **11**(12), 1938–1941 (2018). https://doi.org/10.14778/3229863.3236229
5. Cordeau, J.F., Laporte, G.: The dial-a-ride problem: models and algorithms. Ann. Oper. Res. **153**(1), 29–46 (2007). https://doi.org/10.1007/s10479-007-0170-8
6. Dibbelt, J., Strasser, B., Wagner, D.: Customizable contraction hierarchies. J. Exp. Algorithmics **21**(1), 1–49 (2016). https://doi.org/10.1145/2886843
7. Funke, B., Grünert, T., Irnich, S.: Local search for vehicle routing and scheduling problems: review and conceptual integration. J. Heuristic **11**(4), 267–306 (2005). https://doi.org/10.1007/s10732-005-1997-2
8. Huang, Y., Bastani, F., Jin, R., Wang, X.S.: Large scale real-time ridesharing with service guarantee on road networks. Proc. VLDB Endow. **7**(14), 2017–2028 (2014). https://doi.org/10.14778/2733085.2733106
9. Lowalekar, M., Varakantham, P., Jaillet, P.: ZAC: A zone path construction approach for effective real-time ridesharing. In: Proceedings of the International Conference on Automated Planning and Scheduling, vol. 29, no. 1, pp. 528–538 (2019)
10. Lowalekar, M., Varakantham, P., Jaillet, P.: Zone pAth Construction (ZAC) based approaches for effective real-time ridesharing. J. Artif. Intell. Res. **70**, 119–167 (2021). https://doi.org/10.1613/jair.1.11998
11. Ma, S., Zheng, Y., Wolfson, O.: T-share: a large-scale dynamic taxi ridesharing service. In: 2013 IEEE 29th International Conference on Data Engineering (ICDE), Brisbane, QLD, pp. 410–421. IEEE(April 2013). https://doi.org/10.1109/ICDE.2013.6544843
12. Ma, S., Zheng, Y., Wolfson, O.: Real-time city-scale taxi ridesharing. IEEE Trans. Knowl. Data Eng. **27**(7), 1782–1795 (2015). https://doi.org/10.1109/TKDE.2014.2334313
13. Molenbruch, Y., Braekers, K., Caris, A.: Typology and literature review for dial-a-ride problems. Ann. Oper. Res. **259**(1–2), 295–325 (2017). https://doi.org/10.1007/s10479-017-2525-0
14. Pouls, M., Meyer, A., Ahuja, N.: Idle vehicle repositioning for dynamic ride-sharing. In: Lalla-Ruiz, E., Mes, M., Voß, S. (eds.) ICCL 2020. LNCS, vol. 12433, pp. 507–521. Springer, Cham (2020). https://doi.org/10.1007/978-3-030-59747-4_33
15. Riley, C., Legrain, A., Van Hentenryck, P.: Column generation for real-time ride-sharing operations. In: Rousseau, L.-M., Stergiou, K. (eds.) CPAIOR 2019. LNCS, vol. 11494, pp. 472–487. Springer, Cham (2019). https://doi.org/10.1007/978-3-030-19212-9_31

16. Shah, S., Lowalekar, M., Varakantham, P.: Neural approximate dynamic programming for on-demand ride-pooling. Proc. AAAI Conf. Artif. Intell. **34**(01), 507–515 (2020). https://doi.org/10.1609/aaai.v34i01.5388
17. Uber: Uberpool (2021). https://www.uber.com/us/en/ride/uberpool. Accessed 08 Jan 2021

A Learning and Optimization Framework for Collaborative Urban Delivery Problems with Alliances

Jingfeng Yang(✉) and Hoong Chuin Lau

School of Computing and Information Systems, Singapore Management University, 80 Stamford Road, Singapore 178902, Singapore
jfyang.2018@phdcs.smu.edu.sg, hclau@smu.edu.sg

Abstract. The emergence of e-Commerce imposes a tremendous strain on urban logistics which in turn raises concerns on environmental sustainability if not performed efficiently. While large logistics service providers (LSPs) can perform fulfillment sustainably as they operate extensive logistic networks, last-mile logistics are typically performed by small LSPs who need to form alliances to reduce delivery costs and improve efficiency and compete with large players. In this paper, we consider a multi-alliance multi-depot pickup and delivery problem with time windows (MAD-PDPTW) and formulate it as a mixed-integer programming (MIP) model. To cope with large-scale problem instances, we propose a two-stage approach of deciding how LSP requests are distributed to alliances, followed by vehicle routing within each alliance. For the former, we propose machine learning models to learn the values of delivery costs from past delivery data, which serve as a surrogate for deciding how requests are assigned. For the latter, we propose a tabu search heuristic. Experimental results on a standard dataset show that our proposed learning-based optimization framework is efficient and effective in outperforming the direct use of tabu search in most instances. Using our approach, we demonstrate that substantial savings in costs and hence improvement in sustainability can be achieved when these LSPs form alliances and requests are optimally assigned to these alliances.

Keywords: Alliances · Collaboration · Machine learning · Pickup-and-delivery · Tabu search

1 Introduction

With rapid urbanization, urban delivery systems need to be optimized for capacity and efficiency. High delivery demands not only bring challenges to large LSPs such as Amazon and Cainiao, but also create more intense competition among small and medium-sized LSPs. Due to the high uncertainty of demands and locations in daily delivery, LSPs face operational issues from one end of the spectrum (idle capacity) to the other hand (vehicle and manpower shortage).

M. Mes et al. (Eds.): ICCL 2021, LNCS 13004, pp. 316–331, 2021.
https://doi.org/10.1007/978-3-030-87672-2_21

To overcome these issues, one approach is to establish collaboration with fellow logistics players. As described by Savelsbergh and Woensel [20], collaboration or cooperation is often regarded as a useful path to consolidating freight volumes, leading to a higher and efficient utilization of resources. An alliance by two or more companies offers opportunities for sharing of information and resources to jointly handle delivery tasks. Collaboration in city logistics systems has been widely studied during past few years.

In this paper, we study the pickup and delivery routing problem in a collaborative setting. In particular, we consider the problem that frequently occurs in urban delivery: LSPs perform their daily operations to pickup goods from one location and deliver to another location, and each request has a delivery time window. In an uncooperative setting, each LSP make route plans with their respective requests. For collaborative routing, we assume there exists multiple alliances in the market, and LSPs in the same alliance can share requests and execute the joint routing decision. For simplicity, we assume LSPs in a given alliance will share the same depot to locate their vehicles. Furthermore, an LSP may participate in more than one alliances (perhaps to service different types of goods). Note that this paper is not concerned with the coalition structure generation problem, which focuses on partitioning the set of agents into mutually disjoint coalitions so that the overall total reward is maximized in the long haul. Rather, we assume that the structure of the alliances (composition of LSPs in each alliance) is given as input parameters for our model, and deal with the operational problem of efficient deliveries in an environment where an LSP may belong to multiple alliances.

From the sustainability perspective, it would be ideal to consider the setting where the LSPs are co-operative, and the problem of how planning can be performed on an existing alliance structure that maximizes the system wide objective of total travel cost. We formulate this problem as a multi-alliance multi-depot pickup and delivery problem with time windows (or MAD-PDPTW).

The main contributions of this work are summarized as follows: (1) We propose a MIP model to formulate the MAD-PDPTW; (2) We develop a tabu search based heuristic method to solve the problem on large instances; (3) To solve the problem more efficiently, we decompose MAD-PDPTW to a two-stage problem, which first learns the delivery cost from data and then optimizes the request reassignment and vehicle routing; (4) We demonstrate the significance of the proposed learning and optimization framework (achieve lower delivery cost with less computational time) and obtain managerial insights for LSPs.

2 Related Work

This section provides a summary of existing studies which focus on collaboration in logistics and distance approximation in vehicle routing problems (VRP). Collaboration in logistics industry has been a prevalent topic in urban logistics studies which normally can be achieved in two ways: vertically and horizontally [20]. In this paper, we focus on the horizontal collaboration which involves logistic service providers (LSPs) at the same level in supply chains. A comprehensive description about the opportunities and impediments of horizontal collaborative logistic service was conducted by [3]. They did a survey include 1537 LSPs in Belgium, and the results shows that most of LSPs believe collaboration will increase their profits and improve service quality.

Horizontal Collaboration: Various studies for horizontal collaboration in logistic systems have been published in last decades. Readers can refer to [7], [22] for more details. Two main themes can be further summarized: (1) develop optimization models and mechanisms for collaborative network planning and design problems to help LSPs increase profits or decrease costs; and (2) propose cooperative and non-cooperative game theory methods for cost/gain sharing to establish and keep better collaborations. This study will focuses on the optimization models for collaborative multi-LSP delivery problem, the literature review is conducted accordingly. Berger et al. [1] proposed a decentralized control and auction exchange mechanisms to maximize total profits through collaboration among individuals carriers. Similar research has been conducted by Lai et al. [12], which focus on a centralized control with iterative auction to minimize empty traveling miles. Dahl and Derigs [4] studied a pickup and delivery vehicle routing problem with time windows (PDVRPTW) to minimize total delivery cost. Li et al. [14] also studied the pickup and delivery problem with requests exchange to maximize total profits. [19] solved a multi-depot vehicle routing problem to minimize the total distance traveled with a local search method. Unlike exchange requests or vehicle sharing, [6] introduced a new vehicle routing problem that customers can be served by more than one carrier. It aims to minimize overall operational cost by such collaboration. For more various vehicle routing problem in a collaborative setting, readers can refer to the survey investigated by Gansterer and Hartl [9].

Approximations of Routes: The VRP has been well studied from the last decade. Many exact and heuristic algorithms have been investigated to solve it optimally or in a short computational time. Different from the optimization algorithms which aims to get the optimal or good solutions, the continuum approximation (CA) models are used to approximate the travel distance of routes without solve the complex routing problem. Those CA models can provide faster and good approximation of route distance, which are developed and applied for many applications, such as terminal design problem [18], supply chain distribution network design [15] and collaboration mechanisms design [8]. While, in the face of large-scale complex problems, most CA approaches hold a low accuracy performance. Recently, few studies [16,17] use machine learning approaches to direct estimate the total travel distance of routes. In this paper, we develop a machine learning approach to estimate the delivery cost for pickup and delivery problem with time windows (PDPTW). And with the help of the learned cost, we can further integrate it in requests assignment procedure, and decide which alliance the order should be allocate to.

As discussed above, most studies have devoted to optimize collaborative planning and operation problems from a perspective of entire coalition, whereas all LSPs take part in a single coalition. Zhang et al. [23] investigated the less-than-truck collaboration decision making problem for the e-commerce logistic network, which objective is to maximize the total profit of the entire alliance. To our best of knowledge, Guajardo et al. [11] is the first work that studied the coalition configuration problem which allows company can collaborate in more than one coalition (we prefer to use the term 'alliance' in this paper) in collaborative transport. They developed optimization model to help finding the best coalition configuration. Hence, research gaps are identified from the review of extant literature. In our paper, alliances have been established as inputs in our model, and LSPs in one alliance can share requests and do centralized planning for urban delivery services. More specifically, we focus on optimizing collaborative urban delivery service, with some LSPs can collaborate in more than one alliance.

3 Problem Formulation

In this section, we present our collaborative urban logistic delivery problem in the context of multiple LSPs and multiple alliances. Since each LSP may specialize in fulfilling different types of goods (e.g. groceries and electronics) which may or may not be loaded in the same vehicle, and each may have their own trusted partners, it is plausible to have multiple alliances with overlapping participants.

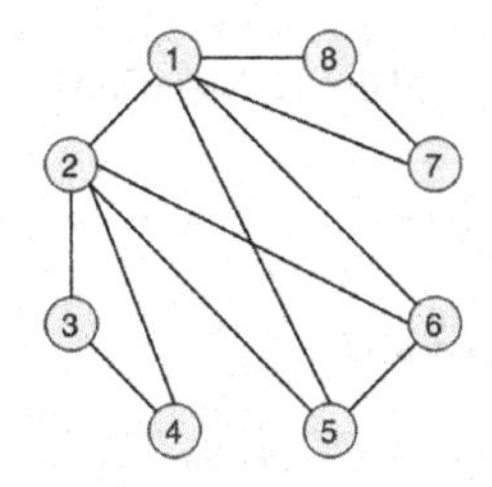

Fig. 1. Multiple alliances with overlapping LSPs

Figure 1 gives an example comprising 8 LSPs and 3 alliances. Each node is defined as a LSP, and if two nodes are connected with an edge, it represents those two LSPs that can share requests. So the alliance is defined as a complete sub-graph, in which a unique edge connects every pair of distinct vertices. Here, we have alliances $[2, 3, 4]$, $[1, 2, 5, 6]$ and $[1, 7, 8]$ in this example. This study aims to assess the potential benefits of collaborative routing among LSPs by sharing requests and joint planning, which means a centralized platform will decide the optimal assignment of requests among each alliance with the constraint that requests cannot be shared between different alliances.

Note that if this problem were to be treated as a whole, one will need to simultaneously decide how LSPs' own requests are distributed to different alliances, and how routing is performed on the assigned requests within each alliance. It is worth noting that even for a small-scale problem instance, a straightforward meta-heuristic approach such as Tabu Search may not be computationally efficient and may not provide an effective solution, as our experiment would show.

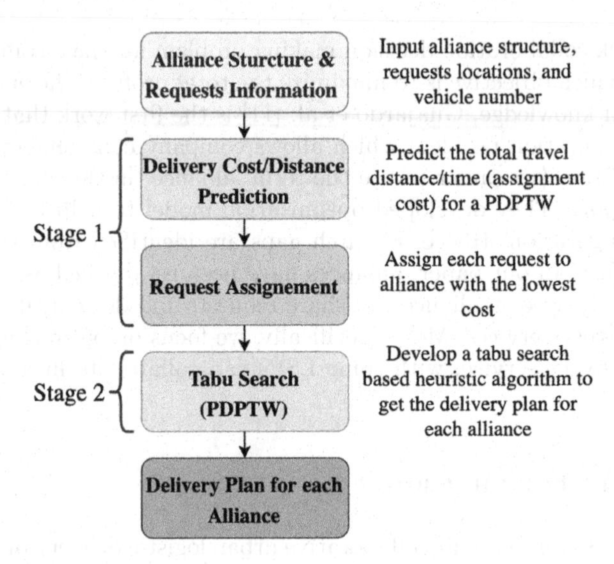

Fig. 2. A two-stage learning and optimization framework to solve the MAD-PDPTW

Before presenting our mathematical programming model, we first introduce notations in Table 1.

Table 1. Notations

Notation	Description
G	A complete direct graph
N	Set of all LSPs
A	Set of all alliances as well as depots
K	Set of vehicles
R	Set of all requests, each request r has a pickup node and delivery node
P	Set of pickup nodes
D	Set of delivery nodes
V	Set of all nodes in graph G
K_a	Set of vehicles only belong to alliance a
d_a	Depot node for alliance a
$[e_i, l_i]$	Time windows for node i, earliest pickup time and latest delivery time
s_i	Service time at location i
q_i	Weight of goods to pickup or delivery at node i
c_{ij}	Travel cost between node i and node j
t_{ik}	Time node i served by vehicle k
w_{ik}	Weight of vehicle k after visit node i
Q	Vehicle capacity
y_{ijk}	Binary variable, 1 if the vehicle k visited node j directly after visited node i, 0 otherwise

Given the above notations, we formulate the multi-alliance multi-depot vehicle routing problem with pickup and delivery (MAD-PDPTW) as follows:

$$\textbf{minimize} \quad \sum_{i\in V}\sum_{j\in V}\sum_{k\in K} c_{ij}y_{ijk} \tag{1}$$

$$\textbf{subject to} \quad \sum_{i\in V}\sum_{k\in K} y_{ijk} = 1 \quad \forall j \in P\cup D \tag{2}$$

$$\sum_{i\in V} y_{ijk} - \sum_{i\in V} y_{jik} = 0 \quad \forall j \in P\cup D, \forall k \in K \tag{3}$$

$$\sum_{i\in V} y_{id_ak} = \sum_{j\in P\cup D} y_{d_ajk} \leq 1 \quad \forall a \in A, \forall k \in K_a \tag{4}$$

$$\sum_{j\in P\cup D} y_{ijk} - \sum_{j\in P\cup D} y_{(i+|R|)jk} = 0 \quad \forall i \in R, \forall k \in K \tag{5}$$

$$t_{ik} + s_i + c_{ij} - M(1-y_{ijk}) \leq t_{jk} \quad \forall i,j \in P\cup D \tag{6}$$

$$e_i \leq t_{ik} \leq l_i \quad \forall i \in P\cup D, \forall k \in K \tag{7}$$

$$t_{ik} \leq t_{(i+|R|)k} \quad \forall i \in P \tag{8}$$

$$t_{ak} = 0 \quad \forall a \in A, \forall k \in K \tag{9}$$

$$w_{jk} \leq w_{ik} + q_j + M(1-y_{ijk}) \; \forall i,j \in V, \forall k \in K \tag{10}$$

$$w_{jk} \geq w_{ik} + q_j - M(1-y_{ijk}) \; \forall i,j \in V, \forall k \in K \tag{11}$$

$$w_{ik} \leq Q \quad \forall i \in V, \forall k \in K \tag{12}$$

$$y_{ijk} = 0 \quad \forall i \notin K_a, \forall j \notin K_a, \forall k \in K_a \tag{13}$$

We divide constraints into four groups. The first group of constraints deals with the in and out flow between each pickup and delivery node. Constraint (2) guarantees that each pickup or delivery node will be visited exactly once. Constraint (3) ensures that each pickup or delivery node, it must be served by the same vehicle k. Constraint (4) imposes constraints on each depot and ensures that each vehicle k belongs to depot K_l will start and back to depot d with at most once. Constraint (5) guarantees the pickup node i and delivery node $i+|R|$ belonging to one request will be served within the same tour.

The second group of constraints deals with visiting precedence of pickup nodes, delivery nodes and time windows. And constraint (6) is the Miller-Tucker-Zemlin (MTZ) sub-tour elimination constraint. If $y_{ijk} = 1$, then we have $t_{ik}+s_i+c_{ij} \leq t_{jk}$, otherwise we have a constraint with right hand side (RHS) is a enough big positive value. Constraint (7) is time windows constraints, which guarantee the delivery time for each request must in the time window. Constraint (8) is precedence constraint that ensure each request is serviced at its pick up node first before the delivery. Constraint (9) denotes the arriving time for each vehicle at the depots equals to 0.

The third group of constraints are the capacity constraints. Constraints (10) and (11) calculate the vehicle weight after visiting each node. In addition, we have $q_i = -q_{i+r}$ for $i \in R_p$. And constraint (12) means for each vehicle k after serve node i, the weight of it cannot exceed the capacity.

The final constraint is the request assignment constraint. It ensures that vehicles belonging to one alliance cannot deliver a request belonging to other alliance. In other words, each alliance is responsible for its own requests.

4 Two-Stage Learning and Optimization Framework

The above section introduces a MIP model to determine the optimal request assignment as well as routing of multiple alliances. In the MIP model, the decision variable y_{ijk} not only decide the delivery sequence from node i to node j, but also make decision for LSPs participating in multiple alliances on request assignment (choose the alliance to share requests). However, the underlying problem is NP-hard, which is computationally intractable to cope with larger instances. In this section, we propose a learning and optimization framework consisting of two stages from requests assignment to vehicle routing. Specifically, the first stage makes decisions for LSPs participating in multiple alliances, which alliance each request should be assigned to (Sect. 4.1). The second stage adopts a tabu search based heuristic algorithm to solve the PDPTW for each alliance with the assigned requests (Sect. 4.2). The whole framework is depicted in Fig. 2.

4.1 Delivery Cost Prediction and Request Assignment

In this subsection, we first discuss the prediction model for the delivery cost for each alliance. Second, we use the estimated delivery cost as input parameters for requests assignment.

Cost Prediction: Previous research has proposed approximate analytic formulas for TSP and VRP under various application scenarios as described in the literature review. However, those analytic based methods always have poor performance on larger problems or complex real-world constraints (e.g., capacity vehicle routing problem with time windows). Here, we use machine learning models to predict the delivery cost for PDPTW (in this paper, we take the delivery cost as the total travel distance). We first generate the promising features for the total travel distance. The general classes of predictors are based on the number of locations, visiting area, the distance between nodes, node dispersion, time windows, and the number of routes. Table 2 lists all the features included in our learning model, and there is a total of 19 features used in our prediction model.

After extracting the features of the PDPTW, the second step is to get the actual solutions to the problem instances as labeled data. Since PDPTW is an NP-hard problem, it would be computationally challenging to generate a large number of exact solutions to be used as label data. In this paper, we find a proxy for the best solution by applying our tabu search algorithm (presented in the next section) instead. Experimental results show that our algorithm comes within a 5% gap on average compared with the best known solutions, and this gives the assurance that the labeled data generated by this approach is accurate and precise. The next step is to select the appropriate machine learning model

Table 2. Features for total travel distance prediction

Features	Definitions
f_1	Number of locations need to be visited
f_2, f_3	Min/max distance between customers and depots
f_4, f_5	Min/max x distance between customers and depots
f_6, f_7	Min/max y distance between customers and depots
f_8	Average distance between customers and depots
f_9	Average x distance between customers and depots
f_{10}	Average y distance between customers and depots
f_{11}	Standard deviation of distance between customers (and depots)
f_{12}	Area of the smallest rectangle covering customer locations
f_{13}	Area of the smallest rectangle covering customer and depot locations
f_{14}	Sum of the length of time windows
f_{15}	Standard deviation of the length of time windows
f_{16}	Sum of the length of overlap time windows
f_{17}	Standard deviation of the length of overlap time windows
f_{18}	Total demand/vehicle capacity ratio
f_{19}	Vehicle capacity/average demand ratio

compatible with the request assignment optimization. We tried a wide range of machine learning regression models in this work, including linear models, such as ordinary least square, LASSO and ridge regression, and nonlinear models, e.g., decision trees and random forest. In summary, we want to identify prediction models that can achieve both good performance and interpretability. In the numerical experiments of Sect. 5, we show the selection details considering the above criteria.

Request Assignment: Note that in MAD-PDPTW, each request can only be assigned to one alliance. Since requests belong to LSPs that may participate in more than one alliances (e.g., LSP 1 in Fig. 1), we need to assign each request to an alliance. Assume there are $I = \{1, 2, \ldots, |I|\}$ requests to be assigned to the set of alliances $A = \{1, 2, \ldots, |A|\}$. Assume each request must be served and there is no limit on how many requests each alliance can have. The request assignment problem is to find a partition of the requests with the minimum total cost, which can be modelled as the set partitioning problem. Let the subset j of locations assigned to an alliance $a \in A$ be associated with an estimated delivery cost c_j, whose value can be predicted by our machine learning models. Since this predicted cost may contain errors, we handle the prediction uncertainly via adding an error term $\widetilde{e}$ which represents the prediction error. Similar to the method use in [16], we use the empirical distribution of $\widetilde{e}$ to generate scenarios for our sample average approximation (SAA) scheme.

$$\textbf{minimize} \quad \mathbb{E}_{\xi}\big[\sum_{j \in Z}(c_j + \widetilde{e}_j(\xi))v_j\big] \tag{14}$$

$$\textbf{subject to} \quad \sum_{j \in Z} \delta_{ij} v_j = 1 \quad \forall i \in I \tag{15}$$

$$\sum_{j \in J} v_j = |A| \tag{16}$$

$$v_j \in \{0, 1\} \tag{17}$$

Z is the set of all possible partition of requests. Decision variable v_j equals to 1 if subset j is selected. δ_{ij} equals to 1 if request i belongs to subset j, and 0 otherwise. Here, we can convert the objective function into a mean value formulation $\frac{1}{|\xi|} \cdot \sum_{\xi} \sum_{j \in Z}(c_j + \widetilde{e}_j(\xi))v_j$. As stated in [5], the expected value (solution of the mean value problem) can provide a robust solution to original stochastic problem. Constraints (15) ensure that every request is assigned to an alliance and constraint (16) ensures the number of selected subsets equal to the number of alliances $|A|$. The problem involves an exponential number of variables (columns) since the number of possible subsets grows exponentially in the number of requests waiting for assignment. And predict the cost c_j of all possible partition of request is also very time-consuming. Instead of enumerating all the possible partitions, we provide a simply greedy heuristic approach to solve the request assignment iteratively. We randomly rank requests sequence of unassigned requests, and assign one request to one alliance at each iteration. Here, c_{ia} denotes the cost for request i assigned to alliance a, which equals to the predict expected cost: $\widetilde{c}_j = \frac{1}{|\xi|} \cdot \sum_{\xi} \sum_{j \in Z}(c_j + \widetilde{e}_j(\xi))v_j$. Predicted by our machine learning models introduced in **cost prediction**. In this case, the problem becomes a simple facility location problem and we can simply assign each request i to the alliance a with lowest cost c_{ia}. Then the total cost equals to $\sum_{i \in I, j \in J} c_{ia}$.

4.2 Tabu Search Algorithm

In this subsection, we first develop an efficient tabu search algorithm to solve the PDPTW for each alliance. Furthermore, we make minor adjustments by including constraint (13) into the tabu search algorithm in Step 3 when we do insertion and removal operations. It makes sure that the candidate requests can only be inserted or removed to routes (denoted by k) that belong to the same alliances. We can use the adjusted tabu search algorithm to directly solve the MAD-PDPTW as a baseline method in our numerical experiments in Sect. 5.

Tabu search [10] is one of the well-known meta-heuristics. It takes a potential solution and search its neighborhood iteratively to find improved solutions. It has been applied successfully to various routing problems [2,21]. In what follows, we introduce the full framework of our algorithm, including initial solution construction and the tabu search algorithm. The procedure to construct an initial solution s_0 is described here. We construct the initial solution s_0 where not all the constraints defined in PDPTW need be satisfied. Given requests set R, pickup node set P, delivery node set D, and available vehicles K as inputs, for each vehicle $k \in K$, we iteratively select request c from the pickup set P, and check whether it satisfy the earliest pickup time constraint $e_i \leq e_c \leq e_{i+1}$. If yes, we add the both the pickup node and delivery node of requests c to vehicle k, otherwise, we put it in a new vehicle $k+1$. When there are no requests in set P, we end up with the initial solution s_0.

Algorithm 1. Tabu search algorithm

Input: s_0, best solution $s^* = s_0$, tabu list $\mathcal{L} = \emptyset$
Output: Best solution s^*
1: Let current solution $s_c = s_0$
2: **while** $i \leq I_{max}$ **do**
3: Do insertion and removal operation
4: Get the neighborhood solution N_s of s_c
5: **for** $s_i \in N_s$ **do**
6: Calculate fitness function $f(s_i)$
7: **if** $s_i \notin \mathcal{L}$ **and** $f(s_i) \leq f(s_c)$ **then**
8: $s_c = s_0$
9: **end if**
10: **end for**
11: **if** $f(s_c) \leq f(s^*)$ **then**
12: $s^* = s_c$
13: **end if**
14: **if** Size of $\mathcal{L} \geq L_{min}$ **then**
15: Update $\mathcal{L}$
16: **end if**
17: **end while**

Based on the initial solution found, the tabu search based heuristic algorithm is described in Algorithm 1. In our paper, the termination condition is that the maximum number of iterations I_{max} is reached. And the fitness function is described as $f(s) = C(s) + \alpha \cdot Q(s) + \beta \cdot T(s)$, where $C(s)$ is the value of objective function (1), $Q(s)$ denotes the total amount of weights that exceed the vehicle capacity and $T(s)$ represents the total unit of times that violate the time windows constraint. As can be seen, the fitness function consists of two parts: the original objective function and the penalty cost. Parameters α and β are both positive penalty terms that make the solution s become more likely to meet the capacity and time windows constraints, respectively. To achieve this, we introduce a new parameter θ with small value (e.g., 0.1) as step size to adjust the value of α and β. If either $Q(s)$ or $T(s)$ not equals to 0, we multiply it by $(1+\theta)$ in the next iteration. Another important part is the tabu list, which represents a set of solutions that have been visited in the recent past. In this paper, we define the maximum length of tabu list is L_{max} and use it to memorize the insertion operations when we insert the pickup node i and delivery node $i+r$ to route k.

In order to solve the MAD-PDPTW, we only need to add constraint (12) before doing insert and remove operation, to ensure that nodes i and $i + r$ can only be added to or removed from route $k \in K_a$ that belongs to the same alliance.

5 Numerical Experiments

This section presents experimental setup for problem instance generation, delivery cost prediction and compares our learning and optimization framework against tabu search in solving the MAD-PDPTW. Computational experiments are conducted to validate the developed framework's performance for multiple alliances under different kinds of settings. All computational experiments are conducted on a desktop computer with Intel Core i5 2.3 GHz with 16 GB RAM. The tabu search algorithm are implemented in Java, while the machine learning models are coded in Python 3.7.

5.1 Problem Instance Generation

The dataset proposed by [13] is a popular standard dataset in the study of PDPTW, and is used to generate sampled PDPTW instances in our paper. We need to construct two types of instances, which are synthesized from the PDPTW benchmark dataset, with the first one used as a training and testing dataset for delivery cost prediction, while the second one is prepared for running MAD-PDPTW. For the first type of instances, we randomly sample with the total number of requests of each instance are in the range of 100 to 200. The labeled data of each PDPTW instance is computed by the tabu search algorithm described in Sect. 4. We obtain 500 instances in total, of which 400 are randomly selected as the training set, and the remaining 100 serve as the test set. To set up the multiple alliance structures, we construct a second type instance by sampling from the original data and randomly reallocating the requests to LSPs and alliances. Compared to the first type of instances, the second type

Table 3. Instances generated from the PDPTW benchmark dataset

Notations	Description
x	The x coordinate of the pickup/delivery locations
y	The y coordinate of the pickup/delivery locations
q_i	Demand of node i
e_i	Earliest pickup/delivery time of node i
l_i	Latest pickup/delivery time of node i
s_i	Service time of node i
p_i	Pickup (index to sibling) of node i
d_i	Delivery (index to sibling) of node i
L_i	LSP index of node i

has one more column with request ownership information. Table 3 gives a brief description of the second type of sampled instances. Table 4 and Fig. 3 list the detailed parameters and shows the alliance structures for all second type test instances.

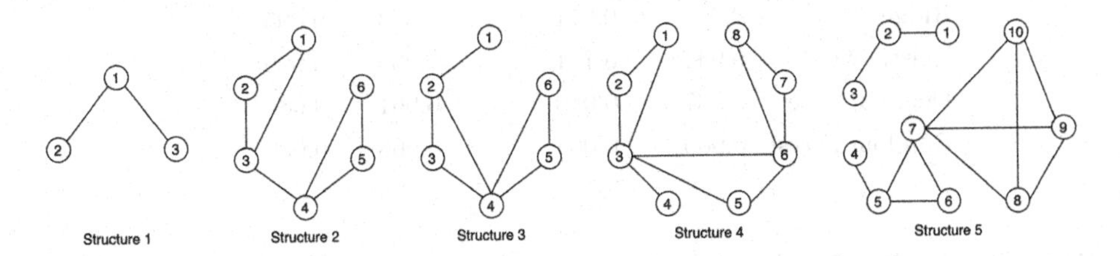

Fig. 3. Alliances structure setting for the case study

Table 4. Detail parameters setting for all test instances

No.	Structure	Alliance	LSPs	Requests	Size	Request configuration
1	1	2	3	9	Small	[3, 3, 3]
2	1	2	3	18	Small	[6, 6, 6]
3	1	2	3	20	Small	[7, 7, 6]
4	1	2	3	24	Small	[8, 8, 8]
5	3	3	6	30	Small	[5, 5, 5, 5, 5, 5]
6	1	2	3	65	Medium	[30, 10, 25]
7	1	2	3	65	Medium	[20, 15, 30]
8	1	2	3	65	Medium	[25, 5, 35]
9	2	2	6	60	Medium	[10, 10, 10, 10, 10, 10]
10	3	3	6	105	Large	[30, 10, 20, 10, 20, 15]
11	3	3	6	105	Large	[40, 5, 25, 10, 15, 10]
12	3	3	6	120	Large	[30, 15, 30, 20, 15, 10]
13	4	4	8	135	Large	[20, 10, 5, 30, 15, 10, 15, 30]
14	4	4	8	135	Large	[15, 15, 5, 20, 25, 15, 20, 20]
15	4	4	8	135	Large	[30, 5, 15, 25, 10, 20, 10, 20]
16	5	5	10	150	Large	[20, 10, 20, 5, 15, 10, 15, 15, 20, 20]
17	5	5	10	180	Large	[30, 15, 25, 20, 15, 30, 10, 10, 10, 15]
18	5	5	10	185	Large	[25, 5, 25, 30, 10, 20, 15, 20, 10, 25]

Table 5. Performance evaluation of the machine learning models

Model	5-CV R^2	5-CV MAPE	Test R^2	Test MAPE
LR	**0.969**	**0.067**	0.904	0.140
LASSO	0.966	0.072	**0.972**	**0.066**
Ridge	0.967	0.071	0.953	0.095
Elastic Net	0.947	0.101	0.939	0.099
Decision Tree	0.937	0.089	0.961	0.085
Random Forest	0.965	0.068	0.966	0.069

5.2 Prediction Model Selection

We test 5 different machine learning models: linear regression, LASSO regression, ridge regression, elastic net, decision trees, and random forest. To achieve the best performance, we implement 5-fold cross-validation (5-CV) to select the best hyper-parameters (e.g., coefficient value for the regulation term, maximum depth of the tree) for all models. All the training and validation procedures are implemented in Python 3.7. Table 5 summarizes the average cross-validation R^2 value and the mean absolute percentage value (MAPE). Let l_s denotes the best solutions we get by tabu search, $\hat{l}_s$ denotes the predicted delivery cost for a sample s in each fold S of the training set. The MAPE is defined as: $\frac{1}{|S|}\sum_{t \in S} \frac{|l_s - \hat{l}_s|}{l_s}$.

Based on the evaluation results, all the above machine learning models achieve reasonably good performance on delivery cost prediction. In particular, the LASSO regression model has the lowest test error and highest R^2 score. Besides, LASSO estimates sparse coefficients that reduce the number of features in the model and maintain good interpretability. Hence, we decide to use LASSO as the prediction model in our framework.

5.3 Performance Comparison

This subsection, we compare the results on delivery costs obtained by (1) self-routing by LSPs without collaboration, (2) collaborative routing with alliances solving by tabu search heuristic alone, (3) collaborative routing with alliances solving by proposed learning-based optimization framework and (4) collaborative routing with fully collaboration, which means each LSP can cooperate with each other and exchange requests from both the computational and management perspectives. And experiments results of all instances are shown in Table 6 (small instances) and Table 7 (medium and large instances).

We find our Tabu search method, learning method with and without error term estimation for small size instances can obtain solutions with small optimally gap compared with Gurobi. As we state previously in Sect. 4.1, errors will always exist in **cost prediction**. Here, to better evaluate the benefits of our learning-based approach, we also investigate the influence of error cascade. In Table 6, the last column shows the results of the learning method which incorporates the errors in request assignment. This indicates that the learning method considering the prediction error is more accurate.

Table 6. Experimental results for small size test instances

No.	Gurobi	Tabu	Learning	Learning + error
1	996	1026	996	996
2	1609	1609	1642	1621
3	1709	1715	1724	1715
4	1934	2035	2012	1987
5	2838	2972	2851	2851

For medium and large size instances, Gurobi fails to give feasible solutions in 3600 s, while both our tabu search and learning-based framework can find good solutions in less than 1 min. Columns $\mathcal{I}$, $\mathcal{F}$ and $\mathcal{L}$ denote the delivery costs obtained by self-routing without collaboration, collaborative routing with fully collaboration (represents the upper bound of the cost savings via collaboration) and collaborative routing with alliances solving by our learning-based approach, respectively. Columns $\mathcal{A}_{min}$, $\mathcal{A}_{max}$ denotes the minimal and maximal delivery cost obtained for collaborative routing with alliances after run the tabu search alone 5 times. Columns $\mathcal{S}_1$ is the cost savings in percentage achieved by collaborative routing with alliance compare to self-routing.

Table 7. Experimental results for medium and large size test instances

No.	$\mathcal{I}$	$\mathcal{F}$	$\mathcal{L}$	$\mathcal{A}_{min}$	$\mathcal{A}_{max}$	$\mathcal{S}_1$ (%)	$\mathcal{S}_{min}$ (%)	$\mathcal{S}_{max}$ (%)	$\mathcal{T}_1$ (s)	$\mathcal{T}_2$ (s)
6	5699	4165	4660	4915	5096	18.23	5.18	8.50	20	50
7	5263	3958	4706	4587	4784	10.58	−2.59	1.63	21	51
8	5067	3855	4804	4747	4887	5.19	−0.64	1.67	19	47
9	5659	3705	4550	4678	4785	18.45	2.81	5.16	18	49
10	8990	5685	7337	7805	7935	18.39	5.99	7.17	43	57
11	8499	5711	7342	7613	7763	13.61	3.79	5.65	40	62
12	12857	8955	11058	11286	11886	13.99	2.02	6.96	59	94
13	14978	9472	12856	13434	13628	14.17	4.30	5.66	65	93
14	14969	10187	12711	13205	13676	15.08	3.74	7.06	64	102
15	14416	10536	12302	13313	13700	14.66	**7.59**	10.20	55	104
16	17060	10572	14005	14986	15241	17.91	6.55	8.11	81	134
17	21029	11375	16802	17512	20297	**20.10**	4.05	**17.22**	106	226
18	19850	13645	16550	17166	17576	16.62	3.58	5.83	87	223

As shown in the table, we can find that both collaboration with alliance and fully collaboration always lead to fewer delivery costs compares to self-routing. Column $\mathcal{S}_{min}$ and $\mathcal{S}_{max}$ are the minimum and maximum savings that the learning framework can achieve compare to the direct use heuristic method (tabu search) alone. We find that for the small size of instances (No. 1 to No. 5), our learning and optimization framework can obtain solutions as good as tabu search. While for moderate or larger test instances with denser alliance structure graph (No. 6 to No. 18), our learning framework is about 2% to 10% better than use heuristic method (tabu search) alone, and it can achieve up to 17% cost savings. We also compare the running times of our proposed learning and optimization framework and directly using heuristic method (tabu search), as shown in column $\mathcal{T}_1$ and $\mathcal{T}_2$. It shows that the our new approach needs less computing resources compare to the heuristic method (tabu search), especially in large scale cases.

6 Conclusion

This paper attempted to address an emerging concept in a collaborative urban delivery problem involving multiple alliance structures. Compared to individuals performing optimal planning by LSPs themselves, our experiments show that centralized collaborative routing can potentially reduce the total operating cost by about 20%. Compared to centralized collaborative routing with the direct use of a heuristic algorithm, our experiments show that our learning-based optimization approach can reduce the total operating cost up to 17% with the less computational time required. Furthermore, the learning-based approach is a framework so methodologically, which means we can replace tabu search with any other heuristic methods to improve the results. We observe that (1) more LSPs joining alliances generally produces more cost savings; (2) the alliance structure has a significant impact: the denser the alliance structure is, the more substantial savings we can achieve, which suggests that overlapping alliance structure allows us to perform logistics more sustainably. This saving can be translated into profit-sharing schemes among participating LSPs, thereby incentivizing them to join such an alliance structure. Profit sharing mechanisms are another topic worthy of future works which fall outside the scope of this paper. In the future, we also aims to provide a robust optimization model to handle the errors for cost prediction in the first stage.

References

1. Berger, S., Bierwirth, C.: Solutions to the request reassignment problem in collaborative carrier networks. Transp. Res. Part E Logistics Transp. Rev. **46**(5), 627–638 (2010)
2. Cordeau, J.F., Laporte, G., Mercier, A.: A unified tabu search heuristic for vehicle routing problems with time windows. J. Oper. Res. Soc. **52**(8), 928–936 (2001)
3. Cruijssen, F., Cools, M., Dullaert, W.: Horizontal cooperation in logistics: opportunities and impediments. Transp. Res. Part E Logistics Transp. Rev. **43**(2), 129–142 (2007)

4. Dahl, S., Derigs, U.: Cooperative planning in express carrier networks–an empirical study on the effectiveness of a real-time decision support system. Decis. Support Syst. **51**(3), 620–626 (2011)
5. Delage, E., Arroyo, S., Ye, Y.: The value of stochastic modeling in two-stage stochastic programs with cost uncertainty. Oper. Res. **62**(6), 1377–1393 (2014)
6. Fernández, E., Roca-Riu, M., Speranza, M.G.: The shared customer collaboration vehicle routing problem. Eur. J. Oper. Res. **265**(3), 1078–1093 (2018)
7. Ferrell, W., Ellis, K., Kaminsky, P., Rainwater, C.: Horizontal collaboration: opportunities for improved logistics planning. Int. J. Prod. Res. **58**(14), 4267–4284 (2020)
8. Gansterer, M., Hartl, R.F.: Request evaluation strategies for carriers in auction-based collaborations. OR Spectrum **38**(1), 3–23 (2015). https://doi.org/10.1007/s00291-015-0411-1
9. Gansterer, M., Hartl, R.F.: Collaborative vehicle routing: a survey. Eur. J. Oper. Res. **268**(1), 1–12 (2018)
10. Glover, F., Laguna, M.: Tabu search. In: Handbook of Combinatorial Optimization, pp. 2093–2229. Springer, Heidelberg (1998). https://doi.org/10.1007/978-1-4613-0303-9_33
11. Guajardo, M., Rönnqvist, M., Flisberg, P., Frisk, M.: Collaborative transportation with overlapping coalitions. Eur. J. Oper. Res. **271**(1), 238–249 (2018)
12. Lai, M., Cai, X., Hu, Q.: An iterative auction for carrier collaboration in truckload pickup and delivery. Transp. Res. Part E Logistics Transp. Rev. **107**, 60–80 (2017)
13. Li, H., Lim, A.: A metaheuristic for the pickup and delivery problem with time windows. Int. J. Artif. Intell. Tools **12**(02), 173–186 (2003)
14. Li, J., Rong, G., Feng, Y.: Request selection and exchange approach for carrier collaboration based on auction of a single request. Transp. Res. Part E Logistics Transp. Rev. **84**, 23–39 (2015)
15. Lim, M.K., Mak, H.Y., Shen, Z.J.M.: Agility and proximity considerations in supply chain design. Manage. Sci. **63**(4), 1026–1041 (2017)
16. Liu, S., He, L., Max Shen, Z.J.: On-time last-mile delivery: order assignment with travel-time predictors. Manage. Sci. **67**, 3985–4642 (2020)
17. Nicola, D., Vetschera, R., Dragomir, A.: Total distance approximations for routing solutions. Comput. Oper. Res. **102**, 67–74 (2019)
18. Ouyang, Y., Daganzo, C.F.: Discretization and validation of the continuum approximation scheme for terminal system design. Transp. Sci. **40**(1), 89–98 (2006)
19. Pérez-Bernabeu, E., Juan, A.A., Faulin, J., Barrios, B.B.: Horizontal cooperation in road transportation: a case illustrating savings in distances and greenhouse gas emissions. Int. Trans. Oper. Res. **22**(3), 585–606 (2015)
20. Savelsbergh, M., Van Woensel, T.: 50th anniversary invited article–city logistics: challenges and opportunities. Transp. Sci. **50**(2), 579–590 (2016)
21. Taillard, É., Badeau, P., Gendreau, M., Guertin, F., Potvin, J.Y.: A tabu search heuristic for the vehicle routing problem with soft time windows. Transp. Sci. **31**(2), 170–186 (1997)
22. Verdonck, L., Caris, A., Ramaekers, K., Janssens, G.K.: Collaborative logistics from the perspective of road transportation companies. Transp. Rev. **33**(6), 700–719 (2013)
23. Zhang, M., Pratap, S., Huang, G.Q., Zhao, Z.: Optimal collaborative transportation service trading in b2b e-commerce logistics. Int. J. Prod. Res. **55**(18), 5485–5501 (2017)

Analysis of Schedules for Rural First and Last Mile Microtransit Services

Christian Truden[1(✉)], Mario Ruthmair[2], and Martin J. Kollingbaum[1]

[1] Lakeside Labs GmbH, Lakeside 04b, 9020 Klagenfurt, Austria
{truden,kollingbaum}@lakeside-labs.at

[2] Department of Mathematics, Alpen-Adria-Universität Klagenfurt, Universitätsstraße 65-67, 9020 Klagenfurt, Austria
mario.ruthmair@aau.at

Abstract. Low and infrequent demand in rural areas poses a problem for public transport providers to run cost-effective services and individual car use is usually the main means of transportation. We investigate how microtransit services can be integrated with existing public transport solutions (bus, train) as a flexible shared mobility alternative in rural areas and how to make them attractive alternatives to individual car use. We combine large neighborhood search with agent-based modeling and simulation to validate generated schedules for a microtransit service in terms of vulnerability to tardiness in passenger behavior or service provision. This includes the study of how disturbances, such as delays in service provision or late arrivals of passengers affect the stability of a transport schedule concerning a reliable timely delivery to transfer stops. We explore how simulation can be utilized as a means to fine-tune provider policies, e.g., how long vehicles may wait for late passengers before they depart.

Keywords: Mobility · Agent-based simulation · Ride-sharing

1 Introduction

Demand for transport in rural areas arises from the need to reach urban centers for work, schools, and the utilization of various services. Due to low population density, this demand usually peaks at particular times of a day. As a result, public transport provisions are concentrated around these times and otherwise operate with low frequency, and with transport services covering few select locations only. Individual car use is the main (and, often, only) means of transport available when ad-hoc demand arises. In particular, there is a lack of transport provisions for the first/last mile to/from public transport system corridors, where timetabled services are available at high frequency. There is a need to integrate microtransit services with existing (timetabled, high-volume) public transport systems to increase adoption of shared mobility solution in areas with low population density [6]. Improving access to and use of public transportation by refining the quality of the first/last mile connections is in the focus of

M. Mes et al. (Eds.): ICCL 2021, LNCS 13004, pp. 332–346, 2021.
https://doi.org/10.1007/978-3-030-87672-2_22

many transport authorities around the globe [9]. New demand-responsive forms of transport gain popularity, in particular in urban centers, where ride-hailing services are now widely used. Microtransit systems are a demand-responsive ride-sharing option that are flexible in their service provision and are deployed in regions where public transport is not (or scarcely) available.

We investigate sustainable and reliable forms of rural passenger mobility. Shared transport modes are regarded as one of the measures to reduce carbon emissions in daily commuter traffic [4,7,10]. We are, therefore, interested in how to provide shared transportation in rural areas that can compete with private car use in terms of availability and convenience. This poses a challenge as using a car is typically the fastest and most convenient mode of transportation in rural areas.

In the modeling of transport scenarios, two complementary perspectives, capturing the passenger's and the service provider's view, respectively, can be distinguished, *a)* the *usage* behavior of customers using a transport system, which is captured in the form of basic transport requests or more complex activities (transport request chains). The main concern of a customer is to be transported without delay and in a reliable fashion; *b)* the *service provision*, where transport services may vary in terms of modality, purpose, flexibility (on-demand, timetabled), etc. The concern of a service provider is to optimize transport provision so that demand can be met. For demand-responsive services, customers may choose to request transport well ahead of the actual journey start time (*pre-booked*), or make *ad-hoc* requests that may occur close to the actual required travel time.

In our study, we investigate how microtransit systems can be integrated with existing public transport operations to meet transport demand. Certain behaviors in a population, such as passengers being late at agreed pickup locations, may lead to delays that make such a transport service unreliable. A balance has to be found in terms of providing a convenient service (all transport requests, including delayed departures for late passengers, are serviced) and a reliable service provision (reaching destinations in time). Whereas passenger behavior is beyond the control of a service provider, a provider can make particular decisions about its own service provision, such as allowing a certain waiting policy at stops that may influence the number of successfully serviced transport requests. This waiting policy may have to be calibrated to balance convenience with reliable service provision. We use agent-oriented simulation as a means to calibrate demand-responsive services in terms of convenience and reliability. Microtransit solutions follow a trip-sharing model, where multiple passengers are transported together on-demand to particular destinations. These systems are either flexible in terms of pickup and delivery locations (door-to-door) or operate within a network of possible, albeit fixed, locations (stops), where passengers can board and leave vehicles. We assume in our investigation a rural area with low population density, where a microtransit solution is introduced as a shuttle service to transfer people from their homes to a public transport system (train). In our rural transport scenario, therefore, people will conduct journeys with multiple legs and transfers in their commute, and where on-demand service elements are combined with timetabled public transport systems. Important aspects of service provision are *customer satisfaction* – a service provider has the capacity to provide a service when ad-hoc demand arises, and *trust* – customers can reach a destination in time (avoid being late for transfer to other modes of transport, or being late to work or school). We consider two performance indicators for microtransit systems to capture these two customer-specific notions, *a)* the percentage of transport requests made for a particular time horizon that can actually be serviced (capacity-related issue), and *b)* how many of these serviced transport requests are fulfilled in time (the passengers arrive at their destination at the specified time). Additional considerations are whether a customer with delayed pickup can still be delivered to their destination in time, or how much delay of a transport customers may accept before they switch to alternative modes of transportation. The main concern in this study is how *lateness* of customers or transport services has an impact on service provision and customer acceptance of these new transport modes. Given these considerations regarding performance, we distinguish passengers either arriving at a pickup location in a timely manner, or them being late. We consider passenger populations with a mix of these two behaviors and analyze how our planning approach can cope with late arrivals. We modeled this scenario as a multi-objective variant of the *dial-a-ride-problem* (DARP) [2,3], and developed a planning system based on a large neighborhood search heuristic for creating transport schedules for microtransit systems. The problem formulation aims at finding routes for a fleet of vehicles that satisfy transport requests of passengers. These requests are defined by *a)* a pickup location where passengers may board a microtransit vehicle and an associated pick-up time, and *b)* a destination location with an associated arrival time window. In our scenario, the destination location is a public transport stop, therefore, the time tables of the public transport services frequenting this stop may influence the chosen size of the arrival time window. We use agent-based modeling [16] to develop a simulation of a rural commuter scenario with a microtransit shuttle service. With this *agent-based modeling and simulation* (ABMS) approach, lateness of passengers or road disturbances can be simulated to verify whether a transport schedule can cope with these kinds of problems. With such a microsimulation approach, we investigate how planning results perform in terms of sensitivity to disturbances and in terms of stability with respect to arriving in time for transfers between modalities at stops, or in terms of transport capacities made available.

2 Related Work

Mobility solutions that are demand-responsive, such as ride-sharing or car pooling, are promoted as new forms of transport in urban and wider metropolitan areas to meet transport demand [7]. Riley et al. [12] present a real-time dispatching solution for a ride-sharing service with a rolling horizon that utilizes a column-generation approach. A computational study shows that their approach scales very well in practice. However, their approach is tailored towards large-scale systems used for highly populated urban areas such as New York City. In contrast to our approach, where pickup and drop-off time windows are essential for scheduling trips, the approach presented in [12] can neglect time windows due to a large number of available vehicles and, typically, relatively short travel times in the urban environment. In our study, a flexible microtransit service shuttles passengers to a train or bus station where they may transfer to a timetabled public service. Therefore, the choice of arrival time may be influenced by the time tables of public transport services that passengers want to reach. However, this is taken into account in a pre-processing phase where a set of typical transport requests are generated (synthetic population data), and not a concern of the actual planning and optimization algorithm we developed (variants of DARP, such as IDARP [11], in contrast, include a mix of flexible (bookable) and fixed timetabled services in the model). We use agent-based modeling [16] and microsimulation to investigate how a demand-responsive transportation service can be delivered efficiently ([1] provide a review of agent-based transportation systems). Microsimulation allows the modeling of individual behavior of agents (passengers, vehicles, etc.) in a particular transport system. Ronald et al. [13] discuss agent-based simulations for studying demand-responsive transportation systems.

3 Rural Commuter Scenario

A rural commuter scenario forms the basis for the investigation presented in this paper. This scenario is situated in an assumed rural area where a central transport corridor, consisting of a major motorway and a rail line, connects two urban centers. Public transport is concentrated in this corridor, whereas outside in the wider rural region, no such services are available. In this scenario, inhabitants of a rural area commute to a workplace in an urban center. They either use a car, leading to congestion and pollution, or find a way to use the rail line. There is usually no service for the first/last mile of daily commutes.

3.1 Transportation Network

In the modeling of this scenario, we assume that a demand responsive mobility system is available for servicing the first/last mile travel of a rural population. A microtransit service will serve fixed locations where passengers may be picked up or transported to. We assume that there are stops specific to the microtransit service, but that also existing public transport stops (e.g. railway stations, bus stations) are frequented by such a service. This is necessary to allow a transition of passengers from a demand-responsive to a public transport system. We, therefore, distinguish two types of stops: *i)* Public Transport (PT) stops are provided by transit authorities; *ii)* Microtransit (MT) stops specific to such a mobility service. Microtransit stops are used on-demand – they are only frequented when passengers request transport from such a location. There are two main reasons that drive the creation of a network of such stops, *a)* it is demand-oriented – because of a certain population density, or through the initiative of local authorities, MT stops are established, or *b)* there are special points of interest that warrant good accessibility. In the considered region, there are 5 PT (train stations) and 97 MT stops. The road network of the rural region under consideration is represented by a travel matrix with distances and travel times (computed by OSRM [5] based on OpenStreetMap) between each pair of PT and MT stops with the default OSRM car profile.

3.2 Vehicle Fleet

The vehicle fleet of the microtransit service is comprised of small buses with limited seating. Vehicles are specified by the following parameters: *a)* number of passenger seats, and *b)* availability, i.e., earliest start time, latest end time, depot location. This microtransit fleet is deployed demand-responsive and, therefore, is not subject to fixed service times. However, we assume that the complementary public transport system is deploying services according to fixed timetables and, therefore, the scheduling of public transport resources cannot be changed.

3.3 Transport Demand

Transport demand arises through passengers (alone or in groups) issuing *transport requests*. A transport request is characterized through an OD-pair, describing a single journey from an origin to a destination location. In our current study, a passenger may issue multiple transport requests in one booking. Such a set of transport requests is then regarded as related and either all of them can be scheduled for transport at the requested times and from/to the requested locations, or all of them are rejected (no partial scheduling of a set of requests is allowed). Several such requests may form a *transport request chain*, if the destination of one request is the origin of the next and there is a timely correlation between arrivals and departures. However, they can also be unrelated in terms of timing and locations. In principle, transport requests are either pre-booked well before a defined planning horizon, or they are placed on short notice (ad-hoc transport requests). For now, we consider only a set of pre-booked requests. In our study, a pre-processing step is generating a set of such request chains (*synthetic population*) that represent the typical travel behavior of a particular rural population as close as possible. Such a pre-processing step is necessary as data about actual transport demand is not available. In the generation of such a synthetic population, we assume that passengers book a chain of requests, describing situations where passengers are taken from an MT stop to the closest PT stop (representing the commute from a rural microtransit stop to a selected public transport stop), from where the public transport system will get them to their work place (or any other destination), and the corresponding return request between these chosen stops. Passengers will start their journey at an MT stop, and, with a scheduled return request, also end their journey at the same MT stop.

For generating the first (*outgoing*) OD-pair in such a transport request chain (passengers commute to a train station), we assume that *a)* each origin stop in a generated OD-pair is a randomly chosen MT stop in the pilot region, *b)* each destination stop in a generated OD-pair is the PT stop closest to the MT stop (shortest Euclidean distance). In order to use these scenarios in the planning and simulation work, we limit what public transport is available to commuters. We assume that passengers start their commute in a time period between 05:00 and 09:00 in the morning from Monday to Friday. Timetable information at public transport stops is used in the calculation of the required arrival time windows for the microtransit services at such a PT stop. The arrival time window is currently set to 10 min and is correlated with timetabled departures of transport services at the PT stop. The pick-up time of the microtransit service is calculated from this arrival time window, using minimal duration of a transport request as well as its maximum allowed duration (ride time limit) between the two selected stops. For the generation of the corresponding *return* transport request, we assume that the origin and destination of the first request are used in reverse. Commuters returning from a train station to the original MT stop are assumed to do this in a time period from 15:00–19:00 in the evening from Monday to Friday.

The approach for generating a synthetic population presented here is currently limited to commuter trips, as no reliable data about other types of trips exist in the chosen rural area of study. Given that our work aims at showing the use of agent-based simulation for shaping policies for mobility service provision in general, we do not consider this focus on a particular type of trips a limitation. Clearly, in the long run, for a successful service also trips for shopping or recreational activities and trips leading to journeys local to the rural area (coordinated with the timetabled services) have to be included in policy-shaping procedures.

3.4 Constraints

The following constraints are considered in planning and execution of the transport schedules: *a)* the number of used seats in a vehicle cannot be exceeded, and *b)* the time windows and ride time limit defined by the passengers cannot be exceeded. Currently, multi-modality and, consequently, transfer times for changes between different modes of transportation are not considered by the optimization algorithm.

4 Approach

For the study of demand responsive transport systems, we use a combination of combinatorial optimization and microsimulation. In a first step (Fig. 1), an optimization algorithm takes a set of transport requests of customers and constructs a transport schedule for a fleet of vehicles. For each vehicle, a route is defined as a timed sequence of stops with additional information about performed transport requests. Currently, we only consider transport requests that are pre-booked in advance of the actual journey. In a second step, this transport schedule is executed in a simulation environment and tested under various conditions, introducing stochastic events such as late arrivals or disturbances.

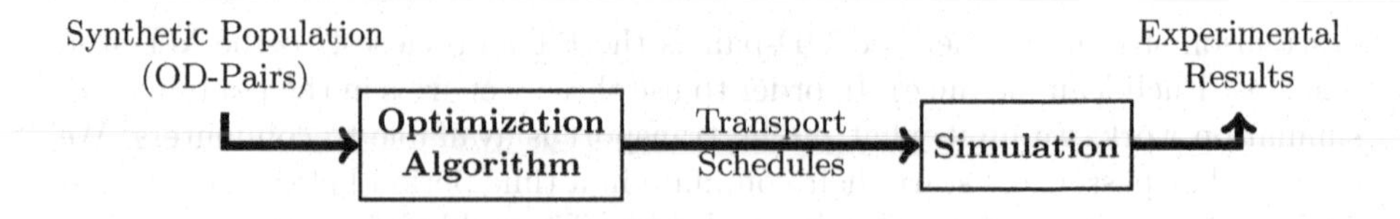

Fig. 1. Evaluation of transport schedules.

Agent-based modeling [16] is used to create a simulation of the rural area where passengers request transport at particular times and microtransit systems operate on a network of stops (pickup and delivery locations for passengers).

4.1 Optimization Algorithm

We generate feasible vehicle schedules by solving the considered DARP variant with a greedy heuristic to construct initial solutions, followed by a large neighborhood search (LNS) to improve them with respect to the following three objectives: *i)* maximizing the number of accepted requests, *ii)* minimizing the total distance driven by all vehicles, and *iii)* minimizing the total excess ride time of passengers exceeding their request's direct travel time. These goals have been selected to achieve both high customer satisfaction and carbon emission reduction in service provision. The latter is based on the assumption of a linear relationship between emissions and the total distance driven that is widely accepted in the literature [4,10]. Since we are dealing with three potentially conflicting objective functions, we need to adapt classical single-objective meta-heuristics to work with multiple objectives. We decided to use a large neighborhood search (LNS) similar to the one in [14], since it is considered to be one of the state-of-the-art heuristics for a wide class of vehicle routing problems. Additionally, we use some ideas from a bi-objective LNS in [8]. To preserve diversification throughout the search, we maintain a pool of non-dominated vehicle schedules that is continuously improved and updated. Initial schedules are constructed as follows: *i)* we sort all transport requests by ascending latest arrival times, *ii)* iteratively select the (initially empty) best schedule in our pool (based on the objective ordering above), *iii)* extend it with the current request in all feasible ways, and *iv)* add all obtained solutions to our pool. Deciding the feasibility of an insertion is non-trivial for the DARP and done by using the method described in [2]. The obtained solutions are then iteratively improved via LNS by *i)* randomly selecting one of the schedules in our pool, *ii)* removing random transport request chains from it, *iii)* trying to insert to it as many not yet served request chains as possible (by using greedy and regret insertion), and *iv)* feeding all intermediately obtained schedules back to the pool. These steps are repeated for 100 iterations.

4.2 Agent-Based Modeling

Microsimulation is used for the evaluation of the transport schedule generated by the optimization algorithm. In a process of agent-based modeling, we identify the stakeholders in the chosen rural transport scenario, such as passengers and vehicle fleets of service providers. In the execution of a transport schedule, concerns regarding delivery (passengers reach their connections in time) are investigated through simulation.

The following agent types are considered: *i) passengers* who require transport (and issue single transport requests or book whole transport chains), *ii) vehicles* that conduct these transports (following the transport schedule), *iii) stops (PT and MT)* that are agentified in this scenario, in order to model and control arrival, pickup and delivery procedures of passengers and vehicles at PT and MT-transit stops, respectively, and a *iv) disturbance agent* that is used to introduce randomness into the execution of the transport schedules.

Passenger and Vehicle Agents. Both passenger and vehicle agents execute information derived from the transport schedule. Vehicle agents receive a schedule comprised of a sequence of stops where passengers are either picked up or delivered at particular times. Assuming a microtransit system using a fleet of mini-buses, such a vehicle usually has a capacity of around eight seats. In addition, mini-buses require, in contrast to city buses with large seating and standing capacities, a one-to-one seat assignment. Passenger agents receive information about their chain of transport requests (scheduled according to the passenger's transport requests), they are required to arrive at a specified stop at a given time (within a time window) so that they can board a mini-bus. Vehicle agents, passenger agents and stop agents interact when arrival events occur at a particular stop. Vehicle agents perform the following actions: *a)* transfer between stops (starting from a depot), such a transfer ends with an arrival event at a stop, *b)* arrive/register at a stop, *c)* start waiting time at stop (wait for a period of time or until arrival events of registered passenger agents occur), *d)* drop off passenger agents according to transport schedule (seats become available, passenger agent is un-registered from the vehicle agent), *e)* pickup of all registered passenger agents (passenger agent is regarded as occupying a seat on the mini-bus and is registered by the vehicle agent), *f)* depart/un-register from stop when all pickup requests fulfilled or waiting time expires. Passenger agents perform the following actions at the pickup location: *a)* depart from home; *b)* arrive/register at stop; *c)* start waiting time at stop (record waiting time); *d)* wait for arrival event of vehicle that fulfills transport request (information received from stop agent); *e)* board vehicle (register with vehicle, occupy seat); *f)* un-register from stop; *g)* start recording transfer time. Passenger agents perform the following actions at the drop-off location: *a)* stop recording transfer time; *b)* arrive/register at stop (stop agent acknowledges arrival). At this point in the process, passenger agents either leave the stop (they un-register) to reach their final destination on foot, or they start the next leg of their request chain, where the current stop is the next pickup location.

Stop Agents. Stop agents act as arbitrators between passenger and vehicle agents and keep track of the registration and waiting of both agents at a particular stop of the transport network. Agents that are not registered with stop agents, are regarded "in transit". A stop agent keeps track of the passengers arriving, waiting, and departing at this stop.

4.3 Disturbance Events

A separate system agent, the so-called *Disturbance Agent*, generates disturbance events. In the first instance, two events are considered: *a)* longer (or shorter) travel times than expected, leading to delays or early arrival, and *b)* tardiness of passenger or vehicle agents when leaving from a location. Arrival delays are modeled implicitly. Each time a vehicle is in transit between two locations of the transport network, the disturbance agent adds a random delay (or reduction) to the travel times. Hence, a vehicle can arrive later (or earlier) than planned at its next destination. We assume that the travel times recorded in the travel matrix (representing the transport network) are not biased, i.e., they represent the expected value of the underlying (unknown) distribution of the travel times, where travel times for each vehicle and each trip are independent, which is a reasonable assumption in case that travel times depend on the condition of the vehicles, a driver's skills, or minor roadside obstacles (slower vehicles impeding the traffic, red lights) [15]. For now, extreme events such as traffic accidents, blocked roads, or mechanical failures of the vehicles are not considered.

Influences on Passenger Agents. We distinguish two types of passenger agents: *a) punctual*, showing little to no tardiness, *b) tardy*, being late most of the time. Assuming that a passenger has a scheduled pickup time α at a MT-stop that is β walking minutes from his/her home address, the arrival of the passenger is determined by two random processes. *a) Departure from home/work.* follows a normal distribution $X \sim \mathcal{N}(\mu, \sigma^2)$. Punctual (tardy) passengers leave at $\mu = \alpha - \beta - \gamma$ with a standard deviation of $\sigma^2 = 2$ min, where γ defines the "slack" of the passenger leaving earlier. Punctual passengers allow a slack of $\gamma = 5$ min, while tardy passengers only allow $\gamma = 3.5$ min. *b) Walking time to the stop* deviates from the expected walking time following a truncated normal distribution $X \sim \mathcal{N}(\mu, \sigma^2, a, b)$, where $a, b \in \mathbb{R}$ define upper/lower bounds. For punctual customers we assume the parametrization $\mu = 0$, $\sigma^2 = 10\%$, $a = -5\%$, $b = 5\%$ and for tardy customers we assume $\mu = 0$, $\sigma^2 = 15\%$, $a = -5\%$, $b = 5\%$. The actual arrival time at the pick-up location $\hat{\alpha}$ is determined by adding up both random values. In Fig. 2 we show the empirical density of the punctuality $(\hat{\alpha} - \alpha)$ of passengers at the pick-up locations. We notice that 0.65% of the *punctual* and 5.38% of the *tardy* passengers arrive later than the scheduled pick-up time α. Further, we assume that passengers wait up to 5 min after the scheduled pick-up time for the vehicle to arrive before they abort the request.

Influences on Vehicle Agents. Vehicle agents have the following behavior. The vehicles leave from a location towards the next location according to their schedule once *a)* all passengers that are scheduled for pick-up have arrived, or,

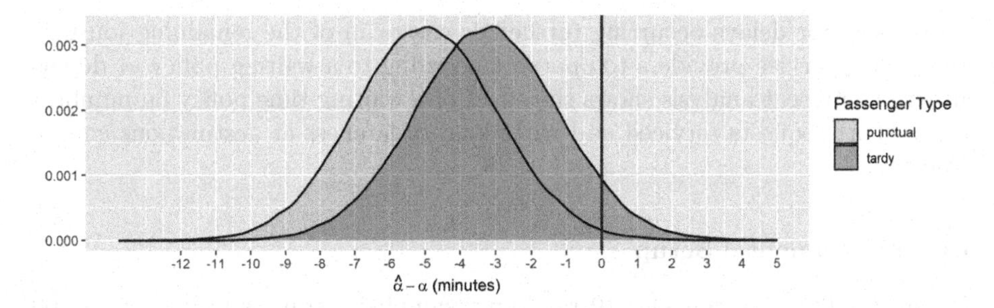

Fig. 2. Empirical density of punctuality for the two passenger types *punctual* and *tardy*, $n = 100\,000$ samples each. The punctuality is determined as the difference between the scheduled pick-up time α and the actual arrival time at the pick-up location $\hat{\alpha}$.

b) the scheduled boarding time (plus a waiting time $\omega \geq 0$) of the passenger(s) has passed and the passenger(s) have not arrived. However, their arrival times are subject to random influence through the disturbance agent. Clearly, travel times must always have positive values. In our experiments, we assume that they follow a *truncated Normal distribution.* $X \sim \mathcal{N}(\mu, \sigma^2, a, b$, where $a, b \in \mathbb{R}$ define upper/lower bounds. The bounds ensure that only "reasonable" values are sampled. The expected values mu (for each edge of the travel matrix) are obtained from the OSRM routing engine [5]. For now, we assume the following parameters $\mu = 0$, $\sigma^2 = 15\%$, $a = -10\%$, $b = 20\%$. Obviously the parameterization must be individually adjusted for other rural regions of study for which simulations would be done.

5 Analysis

We study how disturbances, such as tardiness of passengers and delayed services, affect the provision of a transport service with respect to transfers at stops and timely delivery at destinations. In the analysis performed, the focus is on the effect of passenger tardiness on service lateness at destinations and the number of transport requests that are aborted (rather than determining appropriate fleet sizes through simulation). Clearly, the punctuality of the passengers affects the efficiency and stability of the service. In particular, services should not arrive late at destinations. However, service providers may have a choice to wait for tardy passengers in order to maximize the number of transport requests that are serviced. Of interest is the sensitivity of a transport schedule for a microtransit service to a population of tardy passengers and delays in service provision. A proper management of a demand responsive mobility system, in particular timely delivery at transfer stops to other modes of transport, is important to ensure customer satisfaction and to establish trust in the reliability of the microtransit service. Demand-responsive services have a certain flexibility in terms of departure from stops as they are not bound to a fixed time table and may leave as soon as all booked passengers have boarded. Time savings like these may compensate for delays occurring further downstream of the remaining journey. This allows service providers to operate according to a waiting policy at departures. The present analysis shows the effect of a waiting time policy on numbers of transport requests serviced and what kind of lateness at destinations can be expected.

5.1 Experimental Setup

For our analysis, we consider 10 randomly sampled instances that contain 100 transport request chains each that are served by a fleet of 10 vehicles (minibuses), each with a capacity of 8 passenger seats. For each instance, we consider all those transport schedules from the pool of solutions for which all 100 chains have been accepted. Further, we perform 100 simulation runs for each transport schedule.

Passenger Agent Populations. The punctuality of the passengers affects the efficiency and stability of a transport service. To elaborate this point in more detail, we run our simulations with different "populations" of passenger, i.e., different mixes of *punctual* and *tardy* passengers. We compare the simulation results for the following three population types: *a)* 20% *punctual* passengers (and 80% *tardy* passengers), *b)* 50% *punctual* passengers, *c)* 80% *punctual* passengers. For each population, a total of around 1.34 million transport request chains have been generated in our experiments.

Lateness at the Destination. A late departure at pick-up locations, caused by vehicles being late or waiting for tardy passengers may lead to late arrivals at destinations. We measure this as the *lateness* ℓ, which is the difference between the actual (maybe late) arrival of the passenger at the destination and the end of the arrival time window (of 10 min length). We report this late arrival time (in minutes and seconds) at the destination for all requests (*outgoing* and *return*). In that sense, $\ell = -10$ min means arriving at the beginning of the arrival time window, while $\ell = 5$ min means arriving 5 min after the end of the time window (being late).

Results - Vehicles with Zero Waiting Time. At first, we assume that vehicles do not wait for the passengers beyond the scheduled pick-up time, i.e., waiting time $\omega = 0$ min. We summarize the aborted transport requests in Table 1 and report the lateness (passengers arriving late at their destinations) in Table 2. Additionally, we illustrate the lateness in Fig. 3. We notice that there are between 5.4% and 10% incomplete transport chains, depending on the population mix, while the q_{95} quantiles for the lateness range from 1 min 27 s to 1 min 40 s. The 25% quantile q_{95} is around −13 min meaning that these passengers arrive 3 min prior to the arrival time window. Also, there is no excessive lateness with the q_{95} being at around 4 min. Overall, the rate of aborted requests seems to be inversely proportional to the percentage of punctual passengers. Similarly, we notice that the passenger population mix also influences the lateness. Although we can clearly observe this effect in Fig. 3, it is limited as we can only report lateness for completed transport requests. Depending on how much slack for transferring to the next means of transportation at the destination was added when defining the arrival time windows, these results seem very promising due to the absence of excessive lateness or overly early arrivals. However, with the percentage of aborted requests being rather high the reliability of the service (even if this is induced by the tardiness of the passengers) is not guaranteed. In an effort to reduce the number of aborted requests, the service provider may introduce a waiting policy such that the drivers, who are represented by the vehicle agents in the ABM, may wait for the passenger to arrive past the schedule arrival time, i.e., $\omega > 0$ min. If such a policy is put in place, it is pertinent to provide additional date what a driver may allow in terms of lateness and still will be able to compensate for the passenger's lateness up to a certain amount of time.

Table 1. Percentage of aborted transport request chains reported for the three population types, $\omega = 0$ min. We report the percentages of transport request chains aborted at the *outgoing* or *return* transport request, and the completed transport chains. A chain being aborted at the *return* transport request prerequisites that the corresponding *outgoing* transport request was successful.

punctual (%)	Aborted *outgoing* (%)	Aborted *return* (%)	Both completed (%)
80	3.56	1.82	94.6
50	4.74	3.06	92.2
20	5.83	4.22	90.0

Table 2. Lateness ℓ reported for the three population types, $\omega = 0$ min. The percentage of (completed) transport requests with $\ell > 0$ is reported, and the 25%, 50%, 75%, 95%, 99% quantiles are reported as well. Given are combined numbers for all completed *outgoing* and *return* transport requests (if completed).

punctual (%)	$\ell > 0$ (%)	q_{25} (mm.ss)	q_{50} (mm.ss)	q_{75} (mm.ss)	q_{95} (mm.ss)	q_{99} (mm.ss)
80	10.80466	−13.53	−7.58	−2.45	1.27	4.07
50	12.11865	−13.20	−7.31	−2.15	1.34	4.10
20	13.26574	−12.52	−7.10	−1.51	1.40	4.14

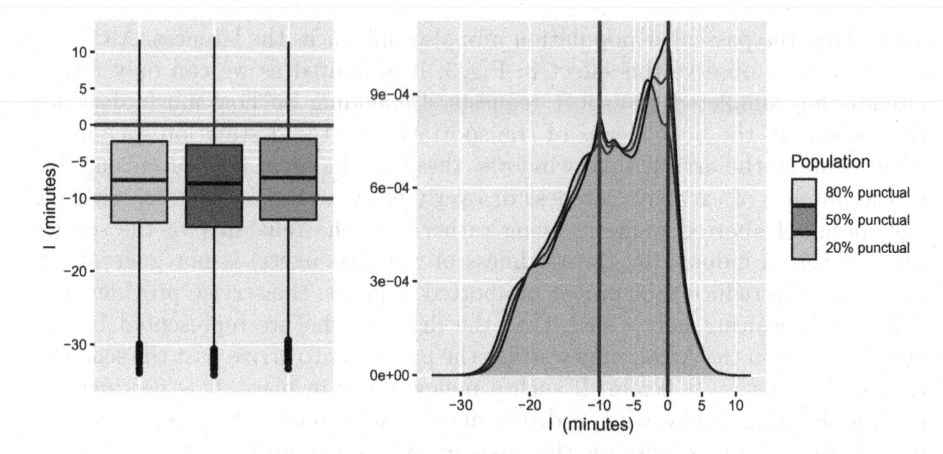

Fig. 3. Illustration of the lateness ℓ for the three population types (for the completed transport requests). The green (red) line marks the beginning (end) of the arrival time windows of the requests. All time windows are of 10 min length. (Color figure online)

Results - Vehicles with Waiting Times up to 10 min. We repeat our experiments with waiting times $\omega = \{1 \text{ min}, 2 \text{ min} \ldots, 10 \text{ min}\}$. The data show that the quantile values for ℓ increase for growing ω. However, this effect is rather modest as the values grow no more than a minute when ω is increased from 0 min to 10 min. In Fig. 4, we illustrate the percentages of aborted requests for changing ω. However, we notice a minimum for the *return* requests at $\omega = 2$ min, that is followed by an increase for $\omega > 2$ min. The *outgoing* requests show a similar behavior but the increase for $\omega > 3$ min is less strong. In that sense, the tardiness of passengers at pickup influences arrival times at destinations, depending on the chosen waiting time ω. All later arrivals lead to aborted requests (chains). Overall, we observe a consistent effect of the passenger population mix in terms of punctuality across all experiments, i.e., lateness at passenger destinations and percentage of aborted requests is always negatively affected by a lack of punctuality of the passengers. Passengers being tardy at their pickup location usually lead to aborted transport requests, which can be counteracted by introducing a vehicle waiting policy that increases the number of serviced requests. In summary, we see that introducing a waiting policy is beneficial to avoid aborted transport requests, while the effect on the lateness ℓ is rather small and therefore acceptable. The above results suggest that our approach can be a valuable decision-support tool for mobility providers that want to fine-tune their vehicle waiting policy in order to maximize the number of serviced transport requests.

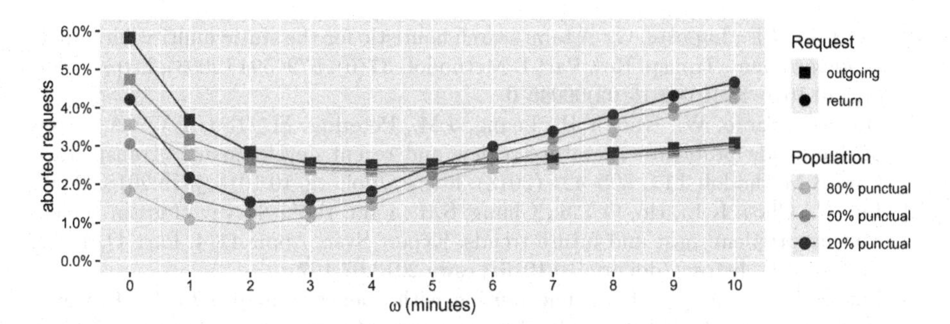

Fig. 4. Percentages of aborted transport request chains compared for different vehicle wait times $\omega = 0, 1, \ldots, 10$ min. We distinguish if a transport request chain is aborted at the *outgoing* or the *return* request.

6 Conclusion

We modeled a rural commuter scenario as a set of agents, where transport resources, passengers and elements of the transportation network (stops) are modeled as agents. At this stage of the project, generated requests represent passenger travel between dedicated stops of different mobility systems (on-demand transit, public transport). In a process of generating representative sets of passenger transport requests, these transport system stops are selected according to criteria such as whether they are the most plausible entry/exit points into a transport system that are closest to a person's start or to their vicinity to a person's intended destination, or their reachability from rural population centers, or whether they are public transport stops that allow a transfer from a microtransit system to a public transport system. We analyzed how late arrival of passengers and/or transport impacts on service quality (reaching transfer stops or final destinations in time). Our study shows that introducing a vehicle waiting policy is beneficial for service provision, resulting in less aborted trips. In future research, agent-based modeling and simulation (ABMS) will be used to investigate additional aspects, such as bottlenecks in service provision, how to optimize traffic and passenger flows, or how changes in procedures impact on the performance of an overall mobility system.

Acknowledgements. This work is supported by Lakeside Labs GmbH, Klagenfurt, Austria, and funding from the European Regional Development Fund and the Carinthian Economic Promotion Fund (KWF) under grant 20214/31942/45906.

References

1. Chen, B., Cheng, H.H.: A review of the applications of agent technology in traffic and transportation systems. IEEE Trans. Intell. Transp. Syst. **11**(2), 485–497 (2010). https://doi.org/10.1109/TITS.2010.2048313

2. Cordeau, J.F., Laporte, G.: A tabu search heuristic for the static multi-vehicle dial-a-ride problem. Transp. Res. Part B Methodol. **37**(6), 579–594 (2003). https://doi.org/10.1016/S0191-2615(02)00045-0
3. Ho, S.C., Szeto, W., Kuo, Y.H., Leung, J.M., Petering, M., Tou, T.W.: A survey of dial-a-ride problems: literature review and recent developments. Transp. Res. Part B Methodol. **111**, 395–421 (2018). https://doi.org/10.1016/j.trb.2018.02.001
4. Lin, C., Choy, K.L., Ho, G.T.S., Chung, S.H., Lam, H.Y.: Survey of green vehicle routing problem: past and future trends. Expert Syst. Appl. **41**(4, Part 1), 1118–1138 (2014). https://doi.org/10.1016/j.eswa.2013.07.107
5. Luxen, D., Vetter, C.: Real-time routing with openstreetmap data. In: Proceedings of the 19th ACM SIGSPATIAL International Conference on Advances in Geographic Information Systems, pp. 513–516. ACM (2011). https://doi.org/10.1145/2093973.2094062
6. Macfarlane, G.S., Hunter, C., Martinez, A., Smith, E.: Rider perceptions of an on-demand microtransit service in Salt Lake County, Utah. Smart Cities **4**(2), 717–727 (2021). https://doi.org/10.3390/smartcities4020036
7. Martins, L.d.C., de la Torre, R., Corlu, C.G., Juan, A.A., Masmoudi, M.A.: Optimizing ride-sharing operations in smart sustainable cities: challenges and the need for agile algorithms. Comput. Ind. Eng. **153**, 107080 (2021). https://doi.org/10.1016/j.cie.2020.107080
8. Matl, P., Nolz, P.C., Ritzinger, U., Ruthmair, M., Tricoire, F.: Bi-objective orienteering for personal activity scheduling. Comput. Oper. Res. **82**, 69–82 (2017). https://doi.org/10.1016/j.cor.2017.01.009
9. Mohiuddin, H.: Planning for the first and last mile: a review of practices at selected transit agencies in the United States. Sustainability **13**(4) (2021). https://doi.org/10.3390/su13042222
10. Nocera, S., Ruiz-Alarcón-Quintero, C., Cavallaro, F.: Assessing carbon emissions from road transport through traffic flow estimators. Transp. Res. Part C Emerg. Technol. **95**, 125–148 (2018). https://doi.org/10.1016/j.trc.2018.07.020
11. Posada, M., Andersson, H., Häll, C.H.: The integrated dial-a-ride problem with timetabled fixed route service. Public Transp. **9**(1), 217–241 (2016). https://doi.org/10.1007/s12469-016-0128-9
12. Riley, C., Legrain, A., Van Hentenryck, P.: Column generation for real-time ride-sharing operations. In: Rousseau, L.-M., Stergiou, K. (eds.) CPAIOR 2019. LNCS, vol. 11494, pp. 472–487. Springer, Cham (2019). https://doi.org/10.1007/978-3-030-19212-9_31
13. Ronald, N., Thompson, R., Winter, S.: Simulating demand-responsive transportation: a review of agent-based approaches. Transp. Rev. **35**(4), 404–421 (2015). https://doi.org/10.1080/01441647.2015.1017749
14. Ropke, S., Pisinger, D.: An adaptive large neighborhood search heuristic for the pickup and delivery problem with time windows. Transp. Sci. **40**(4), 455–472 (2006). https://doi.org/10.1287/trsc.1050.0135
15. Topaloglu, H.: A parallelizable dynamic fleet management model with random travel times. Eur. J. Oper. Res. **175**(2), 782–805 (2006). https://doi.org/10.1016/j.ejor.2005.06.024
16. Wilensky, U., Rand, W.: An Introduction to Agent-Based Modeling. MIT Press, Cambridge (2015)

The Share-A-Ride Problem with Integrated Routing and Design Decisions: The Case of Mixed-Purpose Shared Autonomous Vehicles

Max van der Tholen, Breno A. Beirigo, Jovana Jovanova, and Frederik Schulte(✉)

Delft University of Technology, Delft, The Netherlands
{m.p.vandertholen,b.alvesbeirigo,j.jovanova,f.schulte}@tudelft.nl

Abstract. The shared autonomous vehicle (SAV) is a new concept that meets the upcoming trends of autonomous driving and changing demands in urban transportation. SAVs can carry passengers and parcels simultaneously, making use of dedicated passenger and parcel modules on board. A fleet of SAVs could partly take over private transport, taxi, and last-mile delivery services. A reduced fleet size compared to conventional transportation modes would lead to less traffic congestion in urban centres. This paper presents a method to estimate the optimal capacity for the passenger and parcel compartments of SAVs. The problem is presented as a vehicle routing problem and is named variable capacity share-a-ride-problem (VCSARP). The model has a MILP formulation and is solved using a commercial solver. It seeks to create the optimal routing schedule between a randomly generated set of pick-up and drop-off requests of passengers and parcels. The objective function aims to minimize the total energy costs of each schedule, which is a trade-off between travelled distance and vehicle capacity. Different scenarios are composed by altering parameters, representing travel demand at different times of the day. The model results show the optimized cost of each simulation along with associated routes and vehicle capacities.

Keywords: Shared autonomous vehicles · Capacity optimization · Vehicle routing problem

1 Introduction

Urbanization is a phenomenon that is becoming ever more apparent across the world. Already in 2018, over 55% of the world's population was located in urban areas with prospects of an increase to almost 70% mid-21st century [22]. The ongoing demographic changes within cities give cause for new developments in the transportation of people and goods. Other trends too will have an impact on transport within urban centres. E-commerce is growing fast, with a massive demand for business-to-customer movements. On top of that comes a desire for

M. Mes et al. (Eds.): ICCL 2021, LNCS 13004, pp. 347–361, 2021.
https://doi.org/10.1007/978-3-030-87672-2_23

quick deliveries, sometimes even as fast as a couple of hours. Another trend is that of the sharing economy, in which customers and businesses share resources, potentially reducing freight movements and fleet sizes. Finally, climate change awareness and sustainability play an ever-increasing role in reducing emissions and improve quality of life in heavily congested areas [20]. With these aspects in mind, new approaches to vehicle design are taken.

The development of autonomous vehicles is expected to bring significant changes to the mobility patterns of vehicle users. Connecting vehicles through an internet of autonomous vehicles enables services such as intelligent transportation and ridesharing [10]. The concept of ridesharing promises to improve the efficiency of individual, on-demand transportation in densely populated areas. Combining the benefits of autonomous driving and ridesharing allows for the introduction of autonomous mobility-on-demand (AMoD). This approach consists of fully autonomous driving vehicles that can combine multiple traveling requests into one journey. AMoD has the potential to reduce traffic congestion and parking problems while offering fast, on-demand mobility, relieving passengers from the task of driving, and improving safety [24].

Naturally, not only the transport of passengers in urban areas is growing. Transportation of goods through parcel delivery is increasing and is required to become faster and cheaper. Short trips through cities, such as last-mile delivery services, could potentially be done by purpose-built autonomous vehicles (AV) [4].

While AMoD can already be more efficient and sustainable than a conventional approach, the results heavily depend on traveller demand. Passenger request numbers are typically much higher at day time than at night and peak during morning and afternoon rush hours [3]. As a result, large portions of the fleet of vehicles will be idle or inefficiently occupied during low-demand periods. Unifying the separate vehicle fleets for passenger and parcel transport into one fleet with mixed-purpose vehicles can be a solution to improve occupancy levels and further reduce the number of vehicles on the roads. This share-a-ride approach was introduced by Li et al. [12], where taxis can combine single parcel and rider requests, and later expanded by Beirigo et al. [1] in the context of the share-a-ride with parcel lockers problem (SARPLP).

The SARPLP considers autonomous vehicles with passenger compartments and parcel lockers, such that both commodity types can be transported simultaneously on single journeys. The effectiveness of this concept has been proven, and 92% of simulated scenarios result in higher profit with one mixed-purpose fleet rather than two single-purpose fleets [1]. However, the model relies on a fixed vehicle capacity. While proving effective, this approach hardly links the logistic challenges of shared autonomous driving to the design of shared autonomous vehicles (SAV). Varying the capacity of SAVs is still an unexplored area.

This study seeks to create a model for a people and freight integrated transportation system (PFIT) in an AMoD setting with variable vehicle capacity. We aim to find the optimal capacity for mixed purpose SAVs whose internal space can be outfitted to simultaneously transport passengers and parcels. This optimal capacity can then be used to constrain the design and give an early approximation of the dimensions and features of SAVs according to the demand patterns of the service area. The outcome shows whether the SAV will look like an ordinary passenger car, a large bus, or anything in between. From this point on, the problem will be referred to as the variable capacity share-a-ride problem (VCSARP).

This paper will continue by mentioning some of the relevant literature that is related to the subject. After that, the problem definition and model formulation are explained. Next, the experimentation section will picture the scenarios and give the results of the simulations. The final section will summarize the findings, conclude the current research and give recommendations for further research.

2 Related Work

The vehicle routing problem (VRP) is a classical optimization problem that aims to determine the optimal set of routes to be taken by a fleet of vehicles to serve a set of customers [21]. The first mathematical programming formulation and algorithmic approach for the VRP was the truck dispatching problem by Dantzig and Ramser [7] from 1959. Ever since, efforts have been made to extend VRPs and make them more realistic. The VCSARP is based on a combination of previous VRPs. The Dial-a-Ride Problem (DARP) is a well-known VRP variation that consists of designing vehicle routes for on-demand pick-up and delivery requests. The DARP is built up from a combination of existing VRPs, including the Pick-up and Delivery Vehicle Routing Problem and the Vehicle Routing Problem with Time Windows [6]. With the popularization of app-based transportation services, the DARP has been the basis for passenger ridesharing services (see, e.g., [13,19]). More recently, the advantages of autonomous vehicles for ridesharing have been explored, for instance, considering service quality improvements when platforms activate idle/ parked vehicles [2].

However, short-haul integration of passenger and good flows is hardly observed both in practice [20] and in the literature [15], especially in a ridesharing context. Among the few models for PFIT systems is the SARP, an extension of the DARP introduced by Li et al. [12] that allows taxis to transport one passenger and one parcel simultaneously. This problem has been further covered by Nguyen et al. [16] and Do et al. [8], where a taxi is able to carry one passenger and multiple parcels. A more flexible version of the SARP is the SARPLP [1], in which the vehicles consist of passenger compartments and parcel lockers, thus being able to serve several customers at once. This study shows that a shared, mixed-purpose fleet proves more profitable to the transport company. More work on passenger and parcel ridesharing was done by Ronald et al. [18]. Their model considers passengers requesting transport between homes and activity locations and parcels that are transported from shops to homes. Ultimately, they find that ridesharing resources in vans and taxis results in shorter waiting and travel times. Finally, to the best of our knowledge, the only noteworthy contribution considering variable capacity and VRPs is by Louveaux and Salazar-González [14]. Their model, however, does not consider any ridesharing.

3 Problem Description

The VCSARP aims to create the most cost-effective routing schedule for SAVs across a city with a known set of transport requests. The requests consist of pick-up and delivery points and need to be satisfied within a certain time window. The SAVs start and end their routes at a centrally located depot. The output of the model gives the total cost of the routing and the optimal vehicle capacity that is needed to achieve this. The MILP model is explained in the remainder of this section.

3.1 Model Formulation

The virtual city in which the SAVs operate is expressed as a 20 by 20 grid structure with intervals of 100 m between each node. Within this grid structure, a total of n transport requests is generated, split up into n_h passenger and n_p parcel requests. V represents the complete set of nodes, including requests and start/end depot, while A represents the arcs connecting all those nodes. The distances and travel times between all nodes are captured by $d_{i,j}$ and $t_{i,j}$ respectively.

The requests are characterised by their quantity q_i^r, where resource r is 0 for a passenger request or 1 for a parcel request. Pick-up quantities q_i^r are generated as positive amounts and drop-off quantities q_{i+n}^r are of the same magnitude but negative, representing a loss of load. Each request has a pickup time window $[e_i,\, l_i]$, which is T^r wide, and a maximum travel time delay δ^r. Both must be satisfied for a feasible solution. All requests are generated within a time horizon H. Each pick-up and delivery stop has a service time delay s.

The set of vehicles is K. Each vehicle has a capacity Q^r, which is variable but constrained by an upper and lower bound $[Q^r_{min},\, Q^r_{max}]$ for realism and computational speed. The velocity of the vehicles is assumed to be constant and the same across each arc in A.

The model seeks to minimize the total costs of a routing schedule. The total costs are calculated as the product of travelled distance and cost per unit of distance that varies with vehicle type). The vehicle capacity impacts the travel cost per kilometer because larger-capacity vehicles are heavier and bulkier, thus consuming more energy. This relationship between vehicle capacity and operating costs is visualized in Fig. 1. A complete overview of variables, parameters, sets, and indices needed for the formulation of the VCSARP can be seen in Table 1.

Figure 2 shows some examples of how the VCSARP model works. Each scenario has two passenger and two parcel requests. When requests are positioned close together and very few detours have to be made, the model will most likely choose a larger capacity vehicle to serve multiple requests simultaneously. Figure 2a shows this case. Assuming that each request quantity is equal to 1, the minimum vehicle capacity for the vehicle in this scenario must be 2 for passengers and 2 for parcels. Figure 2b shows a case where the requests are not all located favourably. Here one large vehicle is not able to fulfill all the requests within their time windows. Using smaller vehicles covering slightly larger distances might even be more efficient. The vehicle capacity in this scenario is 1 for passengers and 2 for parcels.

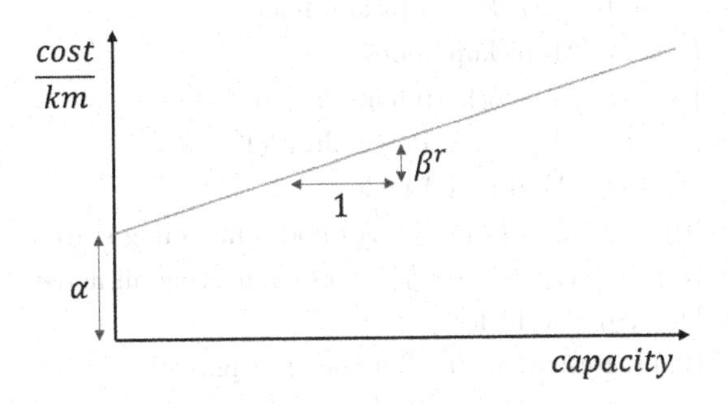

Fig. 1. Linear relation between vehicle capacity and operating cost per unit of distance. Constants α and β^r are explained in Sect. 4.

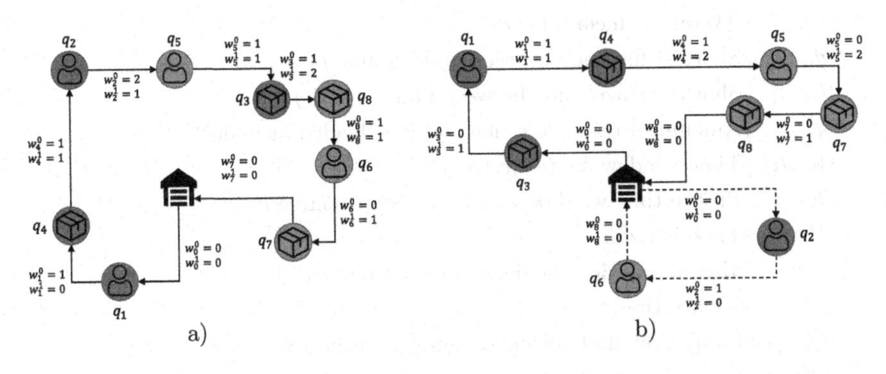

Fig. 2. Two examples of the VCSARP. Blue nodes are pick-ups and green nodes are drop-offs. Example a) shows a scenario in which only one vehicle (solid path) is deployed from the depot. Example b) shows a situation where one vehicle is not sufficient to meet all the constraints, such that two vehicles are deployed (solid and dashed paths). Vehicle loads along each path are displayed for each resource, where q_i^r is the amount of resources that must be loaded at node i and w_i^k is the load of resources on vehicle k after node i. (Color figure online)

Table 1. Sets, indices, parameters and variables of the VCSARP.

Sets and indices	
P_h	$= \{1, ..., n_h\}$. Human pickup nodes
P_p	$= \{n_h + 1, ..., n\}$. Parcel pickup nodes
P	$= P_h \cup P_p$. All pickup nodes
D_h	$= \{n + 1, ..., n + n_h\}$. Human drop-off nodes
D_p	$= \{n + n_h + 1, ..., 2n\}$. Parcel drop-off nodes
D	$= D_h \cup D_p$. All drop-off nodes
V	$= \{0\} \cup P \cup D \cup \{2n + 1\}$. All nodes including start/end depots
A	$= \{i, j : i \in V, j \in V, i \neq j\}$. Arcs connecting all nodes
K	$= \{1, ..., n_k\}$. Vehicles
R	$= \{0, 1\}$. Resources (0 = human, 1 = parcel)
Parameters	
n_h	Number of passenger requests
n_p	Number of parcel requests
n	$= n_h + n_p$. Total number of requests
n_k	Number of vehicles
v_{avg}	Average vehicle velocity
$d_{i,j}$	Shortest distance between nodes i and j
$t_{i,j}$	Shortest travel time between nodes i and j
q_i^r	Amount of resource r that must be loaded at node i
$\{e_i, l_i\}$	Time window for request i
T^r	Pickup time window width for each resource r
H	Time horizon
δ^r	Maximum ride time delay for each resource r
s	Service time
Q_{min}^r	Lower bound of vehicle capacity for each resource r
Q_{max}^r	Upper bound of vehicle capacity for each resource r
α	Constant of the objective function
β^r	Slope of the objective function for each resource r
pp_h	Passenger pickup probability [residential, industrial, campus]
dp_h	Passenger delivery probability [residential, industrial, campus]
pp_p	Parcel pickup probability [residential, industrial, campus]
dp_p	Parcel delivery probability [residential, industrial, campus]
Variables	
$x_{i,j}^k$	Traveled arcs (i, j) of each vehicle k
τ_i^k	Arrival time of vehicle k at node i
$w_i^{r,k}$	Load of each resource r on vehicle k after node i
tr_i^k	Ride time of pickup i on vehicle k
Q^r	Vehicle capacity for each resource r

3.2 Model Formulation

The MILP formulation of the model is as follows:

Minimize:

$$C = \sum_{i,j \in A} \sum_{r \in R} \sum_{k \in K} (\alpha + \beta^r Q^r) d_{i,j} x_{i,j}^k \tag{1}$$

Subject to:

$$\sum_{\substack{j \in V \\ j \neq i}} \sum_{k \in K} x_{i,j}^k = 1 \qquad \forall i \in P \tag{2}$$

$$\sum_{\substack{j \in V \\ j \neq 0}} x_{0,j}^k = \sum_{\substack{i \in V \\ i \neq 2n+1}} x_{i,2n+1}^k = 1 \qquad \forall k \in K \tag{3}$$

$$\sum_{\substack{j \in V \\ j \neq i}} x_{i,j}^k = \sum_{\substack{j \in V \\ j \neq i+n}} x_{i+n,j}^k \qquad \forall i \in P, \forall k \in K \tag{4}$$

$$\sum_{\substack{j \in V \\ j \neq i}} x_{j,i}^k = \sum_{\substack{j \in V \\ j \neq i}} x_{i,j}^k \qquad \forall i \in N, \forall k \in K \tag{5}$$

$$\tau_j^k = (\tau_i^k + t_{i,j} + s) x_{i,j}^k \qquad \forall i,j \in A, \forall k \in K \tag{6}$$

$$\tau_{i+n}^k \geq \tau_i^k \qquad \forall i \in P, \forall k \in K \tag{7}$$

$$e_i \leq \tau_i^k \leq l_i \qquad \forall i \in P, \forall k \in K \tag{8}$$

$$w_j^{r,k} \geq (w_i^{r,k} + q_j^r) x_{i,j}^k \qquad \forall i,j \in A, \forall r \in R, \forall k \in K \tag{9}$$

$$\max(0, q_i^r) \leq w_i^{r,k} \leq Q^r \qquad \forall i \in V, \forall r \in R, \forall k \in K \tag{10}$$

$$Q_{min}^r \leq Q^r \leq Q_{max}^r \qquad \forall r \in R \tag{11}$$

$$tr_i^k = \tau_{i+n}^k - \tau_i^k \qquad \forall i \in P, \forall k \in K \tag{12}$$

$$t_{i,i+n} \leq tr_i^k \leq t_{i,i+n} + \delta^r \qquad \forall i \in P, \forall r \in R, \forall k \in K \tag{13}$$

$$x_{i,j}^k \in \{0,1\} \qquad \forall i,j \in A, \forall k \in K \tag{14}$$

$$\tau_i^k \in \mathbb{N} \qquad \forall i \in V, \forall k \in K \tag{15}$$

$$w_i^{r,k} \in \mathbb{Z} \qquad \forall i \in V, \forall r \in R, \forall k \in K \tag{16}$$

$$tr_i^k \in \mathbb{N} \qquad \forall i \in V, \forall k \in K \tag{17}$$

$$Q^r \in \mathbb{Z} \qquad \forall r \in R \tag{18}$$

The objective function (1) aims to minimize the total cost, which is calculated as (cost/km)*(travelled distance). Based on [11] and [23], we consider the energy consumption and running costs (denoted by cost/km) increase linearly with vehicle capacity. Parameters α and β^r are the intercept and slope of the linear function and are quantified in Sect. 4. Equation (2) guarantees that each request is served once. Each vehicle must start and end its route at the depot, which is controlled by (3), while (4) ensures that pick-up and delivery of one request are done by the same vehicle. The final routing constraint (5) guarantees conservation of flow, meaning that a vehicle entering a node must also leave that node again. The definition of the arrival time of SAVs at nodes is given by (6). Vehicles must first complete the pick-up of a request before the drop-off, which is guaranteed by (7). Equation (8) ensures that arrival at the pick-up nodes is on time and within the required time window. The vehicle load or weight after each node is defined by Eq. (9). This load must never become negative, be always larger than the previous request quantity, and never exceed the maximum loading capacity, of which (10) makes sure. The vehicle capacity has a lower and upper bound, which are imposed by (11). Each request has a total time spent on the vehicle. The minimum ride time is given by (12). The actual ride time cannot exceed this by more than the maximum ride time delay, which is guaranteed by (13). The model's five decision variables are traveled arcs, arrival times, compartment loads, ride times, and vehicle capacities. These are defined respectively by (14), (15), (16), (17), and (18).

4 Experimental Study

Once traffic flows and transportation demand fluctuate throughout the day, we carry out an experimental study to obtain insights into the ideal capacity of an SAV considering different demand scenarios. Ideally, an SAV system should function efficiently at any time, consistently featuring high occupancy rates and low idleness. First, to simulate the various times of day in which an SAV operates, the following scenarios have been considered:

- **Morning rush hour:** During morning hours, there is a high density of passenger requests from homes to workplaces. This causes traffic to flow from residential areas to industrial and commercial areas. The amount of parcel movements is significantly smaller.
- **Afternoon:** A well-mixed blend of parcel and passenger movements. Parcels tend to move from industrial areas towards residential areas, while passenger travel patterns are more evenly distributed across the city.
- **Late-afternoon rush hour:** Similar to the morning hours, passenger demand is higher than parcel demand. However, passenger traffic flow is reversed from industrial/commercial areas to residential areas.
- **Evening:** The time of day at which most people are at home creates great opportunities for parcel deliveries. This scenario is dominated by parcel transport requests from industrial to residential areas, but some scattered passenger transport occurs too.

Considering that previously the city map was formulated as a rather abstract grid structure without any information on the function of each node, neighbourhoods are added to the grid. These are described as four rectangular sections on the map. Two sections are representing residential areas, one section is an industrial area, and the final section is a campus with a university and offices. These sections or neighbourhoods can now be used to create scenarios. Request locations are generated using a probability distribution that can shape the traffic flow patterns of the scenario. For example, in the morning scenario, there will be a high probability that passenger pick-ups will occur on any of the nodes within the residential areas and a much lower probability of them occurring at industrial or campus nodes. Likewise, passenger destinations will more likely occur in the industrial or campus neighbourhood, rather than in one of the residential areas. A visual representation of the map with the different neighbourhoods is shown in Fig. 3.

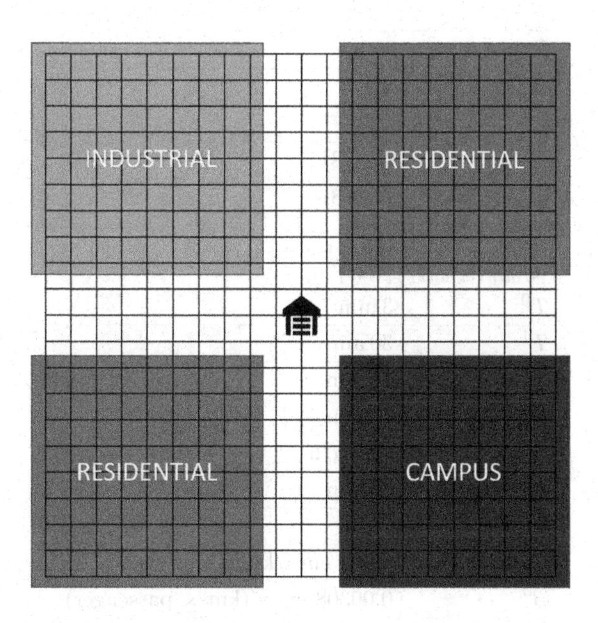

Fig. 3. City grid divided into neighbourhoods with different functions. The depot is located in the middle.

The general parameters of the VCSARP are constant across all scenarios. The total number of requests and available vehicles are set to 8 and 4, respectively. These relatively low numbers are needed to limit computation times, which can become very large due to the complexity of the model. To make up for the low number of requests, request quantities are chosen randomly from a $U(4, 8)$ distribution. The case of a few large requests (quantities between 4 and 8) can be considered analogous to a larger number of smaller requests with similar origin and destination, essentially creating a scenario that serves much more customers. The parcel and passenger capacity lower and upper bounds are both set to 4 and 24, respectively. Assuming that one passenger seat takes up the space of about 4 large parcels, that results is a total "passenger size" capacity between 4 and 30. Passenger time windows and delivery delays are set quite tight once most people would not want to experience much delay during their trip. Conversely, parcel time constraints are much less strict, allowing for a maximum delay of 1 h at delivery. The average speed is set at 20 km/h, which is realistic in cities with short stopping intervals [17]. A short 10-min total time horizon is chosen because of the small number of requests. To simulate the (un)loading of resources, a 1-min service time at each node is added. The objective function parameters α and β^r are retrieved from real-life electricity costs and consumption data of electric vehicles (see [9] and [5]) (Tables 2 and 3).

Table 2. General parameter values.

Parameter	Value
n	8
n_k	4
q_i^0	$U(4,8)$
q_i^1	$U(4,8)$
$[Q_{min}^0, Q_{max}^0]$	[4,24]
$[Q_{min}^1, Q_{max}^1]$	[4,24]
T^0	3 min
T^1	30 min
δ^0	10 min
δ^0	10 min
v_{avg}	20 km/h
H	10 min
s	1 min
α	0.022 euro/km
β^0	0.00308 euro/(km × passenger)
β^1	0.00077 euro/(km × parcel)

Table 3. Scenario specific parameter values.

	Scenario			
Parameter	Morning	Afternoon	Late-afternoon	Evening
Number of requests				
n_h	6	4	6	2
n_p	2	4	2	6
Pickup and delivery probabilities [res, ind, cam]				
pp_h	[1, 0, 0]	[1/3, 1/3, 1/3]	[0, 1/2, 1/2]	[1/3, 1/3, 1/3]
dp_h	[0, 1/2, 1/2]	[1/3, 1/3, 1/3]	[1, 0, 0]	[1/3, 1/3, 1/3]
pp_p	[0, 1, 0]	[1/5, 4/5, 0]	[0, 1, 0]	[1/5, 4/5, 0]
dp_p	[1, 0, 0]	[4/5, 1/5, 0]	[1, 0, 0]	[4/5, 1/5, 0]

5 Results

The computations were performed by an Intel Core i7 @ 2.20 GHz processor, 16 GB RAM computer. The programming was done in Python, and the MILP model was solved using Gurobi Optimizer 9.0.2.

Table 4 shows for each scenario the average optimal vehicle capacity for both resources and the total vehicle capacity (i.e., the combination of passenger and parcel capacities). One passenger space is considered the same size as 4 parcel spaces. A vehicle with a passenger capacity of 4 and a parcel capacity of 8 would thus have a total capacity of 6. This value determines the overall interior volume of the vehicle, which can be of use for the design of the vehicle, and, ultimately, the number of vehicles used. Simulations that did not converge to a 0% optimality gap, thus not reaching the most optimal solution, within a 30-min time limit were discarded. In total, at least 30 simulations with optimal solutions were generated for each of the four scenarios. Table 5 shows the average total costs, distance travelled, and cost/km across scenarios and Fig. 4 illustrates the outcome routes of a single simulation for each scenario.

Table 4. The results of each scenario that are related to vehicle capacity and design. These are average values from 30 instances on each scenario.

Scenario	Passenger cap.	Parcel cap.	Total cap.	# of vehicles
Morning	12.8	8.23	14.9	3.60
Afternoon	8.50	11.9	11.5	3.31
Late-afternoon	12.9	8.93	15.2	3.50
Evening	6.75	17.6	11.1	2.05

Table 5. Average costs, distance travelled, and cost/km for each scenario across 30 different instances.

Scenario	Costs [e]	Distance travelled [km]	Cost/km [e /km]
Morning	1.84	20.6	0.0898
Afternoon	1.46	18.3	0.0793
Late-afternoon	1.67	18.5	0.0907
Evening	1.14	15.1	0.0783

Table 4 shows that the optimal vehicle composition and fleet size differ markedly for each scenario. As input parameters heavily influence the model's outcome, engineers can take multiple approaches to design suitable SAVs that ultimately match the operational requirements. The most straightforward approach would be designing vehicles using the maximum capacity for each resource across all scenarios, resulting in a vehicle with around 15 passenger seats and 18 parcel spaces. Naturally, these vehicles would end up under-occupied most of the time, and costs would be higher than calculated, but most requests could be satisfied easily. Another approach consists of repeating the simulations for the afternoon and evening scenarios with adjusted capacity constraints. Considering that vehicles tend to be idler in these scenarios, using a smaller parcel capacity and deploying more vehicles would also satisfy the conditions. This results in higher overall costs but more efficient vehicle occupation.

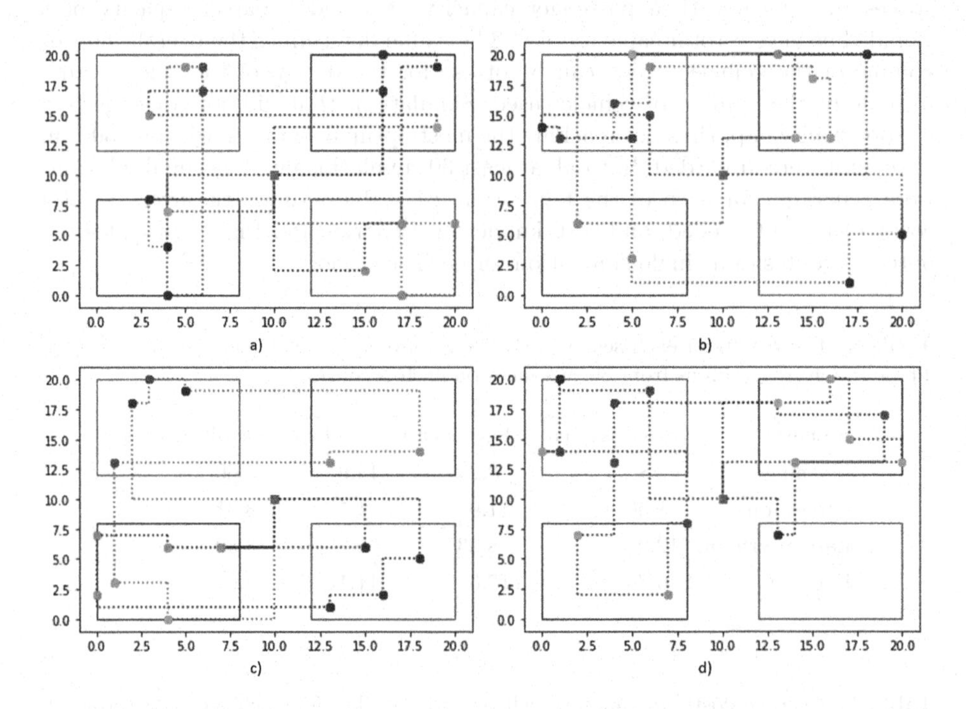

Fig. 4. The results of 4 simulations, where a) is a morning scenario, b) is an afternoon scenario, c) is a late-afternoon scenario, and d) is an evening scenario. The red node is the depot, the blue nodes are human pick-ups, the cyan nodes are human drop-offs, the green nodes are parcel pick-ups and the yellow nodes are parcel drop-offs. Different colour dotted lines represent different vehicles in operation. The neighbourhoods are also displayed, as rectangles (see Fig. 3). (Color figure online)

6 Conclusions

This research paper presented a MILP formulation for the variable capacity share-a-ride problem. The SARP was reformulated to set vehicle capacity as a decision variable and allow for shared autonomous vehicles. The objective was to find the optimal capacity of SAVs to give insights into the design of such vehicles given several operational scenarios, featuring various parcel and people demand patterns.

Overall, the model provides a basis for estimating the optimal capacity of SAVs in a static scenario. The parameter inputs are flexible and allow for a wide variety of scenarios. A point of critique is the limited number of requests that the model is able to solve within a reasonable time. This was countered by using larger quantity requests, each representing multiple single requests that have similar origins and destinations.

The results show that the optimal capacity is highly dependent on the scenario parameters. Scenarios with high passenger transport demand call for vehicles with large passenger capacity and smaller parcel capacity. Scenarios with more parcel movements have an opposite effect on both capacities. While this is to be expected, it makes it hard to find an optimum capacity that will satisfy all scenarios and be cost-efficient at the same time. The concept of SAV in this paper uses separate compartments for passenger and parcel transport. Empty passenger spaces cannot be used for parcels and vice versa. One could rethink this concept and go for a more flexible utilization of interior space. Possible solutions include foldable seats for additional parcel storage, under-seat storage of parcels, or a simple flat floor with standing/leaning space for passengers and freely usable space for parcels. These concepts could easily be implemented into the current model with simple reformulations. Alternatively, a more flexible solution could be considered, where AVs are dynamically outfitted at service points, having their people and parcel compartment capacity adjusted to match the demand changes. Future work will focus on modeling such a flexible setting considering additional operational characteristics such as different revenues for passengers and parcels, penalties for extended ride time, premiums for private travel requests, vehicle and compartment purchasing costs.

References

1. Beirigo, B.A., Schulte, F., Negenborn, R.R.: Integrating people and freight transportation using shared autonomous vehicles with compartments. IFAC-PapersOnLine **51**(9), 392–397 (2018). https://doi.org/10.1016/j.ifacol.2018.07.064
2. Beirigo, B.A., Schulte, F., Negenborn, R.R.: A learning-based optimization approach for autonomous ridesharing platforms with service level contracts and on-demand hiring of idle vehicles. Transp. Sci. (2021, in press). https://doi.org/10.1287/trsc.2021.1069
3. Boesch, P.M., Ciari, F., Axhausen, K.W.: Autonomous vehicle fleet sizes required to serve different levels of demand. Transp. Res. Rec. **2542**, 111–119 (2016). https://doi.org/10.3141/2542-13
4. Buchegger, A., Lassnig, K., Loigge, S., Mühlbacher, C., Steinbauer, G.: An autonomous vehicle for parcel delivery in urban areas. In: Proceedings of the IEEE Conference on Intelligent Transportation Systems, pp. 2961–2967 (2018). https://doi.org/10.1109/ITSC.2018.8569339
5. Černý, J.: Testing of five different types of electric buses. In: Proceedings of the CIVITAS Forum Conference 2015, Ljubljana, Slovenia (2015)
6. Cordeau, J.F., Laporte, G.: The dial-a-ride problem (DARP): variants, modeling issues and algorithms. 4OR **1**(2), 89–101 (2003). https://doi.org/10.1007/s10288-002-0009-8

7. Dantzig, G.B., Ramser, J.H.: The truck dispatching problem. Manage. Sci. **6**(1), 80–91 (1959)
8. Do, P.T., Nghiem, N.V.D., Nguyen, N.Q., Nguyen, D.N.: A practical dynamic share-a-ride problem with speed windows for Tokyo city. In: Proceedings of the 8th International Conference on Knowledge and Systems Engineering, pp. 55–60. IEEE (2016). https://doi.org/10.1109/KSE.2016.7758029
9. Electric Vehicle Database: Energy consumption of full electric vehicles cheatsheet - EV Database (2020). https://ev-database.uk/cheatsheet/energy-consumption-electric-car, https://ev-database.uk/cheatsheet/energy-consumption-electric-car. Accessed 15 Jan 2021
10. Jameel, F., Chang, Z., Huang, J., Ristaniemi, T.: Internet of autonomous vehicles: architecture, features, and socio-technological challenges. IEEE Wirel. Commun. **26**(4), 21–29 (2019)
11. Jung, H., Silva, R., Han, M.: Scaling trends of electric vehicle performance: driving range, fuel economy, peak power output, and temperature effect. World Electr. Veh. J. **9**(4), 1–14 (2018). https://doi.org/10.3390/wevj9040046
12. Li, B., Krushinsky, D., Reijers, H.A., Van Woensel, T.: The share-a-ride problem: people and parcels sharing taxis. Eur. J. Oper. Res. **238**(1), 31–40 (2014). https://doi.org/10.1016/j.ejor.2014.03.003
13. Lin, Y., Li, W., Qiu, F., Xu, H.: Research on optimization of vehicle routing problem for ride-sharing taxi. Procedia Soc. Behav. Sci. **43**, 494–502 (2012). https://doi.org/10.1016/j.sbspro.2012.04.122
14. Louveaux, F.V., Salazar-González, J.J.: Solving the single vehicle routing problem with variable capacity. Transp. Sci. **50**(2), 708–719 (2016). https://doi.org/10.1287/trsc.2014.0556
15. Mourad, A., Puchinger, J., Chu, C.: A survey of models and algorithms for optimizing shared mobility. Transp. Res. Part B Methodol. **123**, 323–346 (2019)
16. Nguyen, N.Q., Tuan, K.L., Nghiem, N.V.D., Nguyen, M.S., Thuan, P.D., Mukai, N.: People and parcels sharing a taxi for Tokyo city. In: Proceedings of the Sixth International Symposium on Information and Communication Technology, pp. 90–97 (2015). https://doi.org/10.1145/2833258.2833309
17. Oskarbski, J., Birr, K., Miszewski, M., Zarski, K.: Estimating the average speed of public transport vehicles based on traffic control system data. In: Proceedings of the 2015 International Conference on Models and Technologies for Intelligent Transportation Systems, pp. 287–293 (2015). https://doi.org/10.1109/MTITS.2015.7223269
18. Ronald, N., Yang, J., Thompson, R.G.: Exploring co-modality using on-demand transport systems. Transp. Res. Procedia **12**, 203–212 (2016). https://doi.org/10.1016/j.trpro.2016.02.059
19. Santos, D.O., Xavier, E.C.: Taxi and ride sharing: a dynamic dial-a-ride problem with money as an incentive. Expert Syst. Appl. **42**(19), 6728–6737 (2015). https://doi.org/10.1016/j.eswa.2015.04.060
20. Savelsbergh, M., Van Woensel, T.: 50th anniversary invited article-city logistics: challenges and opportunities. Transp. Sci. **50**(2), 579–590 (2016). https://doi.org/10.1287/trsc.2016.0675
21. Toth, P., Vigo, D.: The vehicle routing problem. SIAM (2002)
22. United Nations; Department of Economic and Social Affairs; Population Division: World Urbanization Prospects: The 2018 Revision. United Nations, New York (2019)

23. Weiss, M., Cloos, K.C., Helmers, E.: Energy efficiency trade-offs in small to large electric vehicles. Environ. Sci. Eur. **32**(1), 1–17 (2020). https://doi.org/10.1186/s12302-020-00307-8
24. Zhang, R., Spieser, K., Frazzoli, E., Pavone, M.: Models, algorithms, and evaluation for autonomous mobility-on-demand systems. In: Proceedings of the 2015 American Control Conference, pp. 2573–2587. American Automatic Control Council (2015). https://doi.org/10.1109/ACC.2015.7171122

Algorithms for the Design of Round-Trip Carsharing Systems with a Heterogeneous Fleet

Pieter Smet[1(✉)], Emmanouil Thanos[1], Federico Mosquera[1], and Toni I. Wickert[1,2]

[1] Department of Computer Science, KU Leuven, Leuven, Belgium
pieter.smet@kuleuven.be
[2] Institute of Informatics, Federal University of Rio Grande do Sul, Porto Alegre, Brazil

Abstract. Carsharing has become a viable mode of transport which not only contributes to improving the environment and traffic congestion, but is often also cheaper for its users. It is a challenging task for carsharing providers to design an effective system which meets user demand while at the same time limiting expenses. This paper introduces an integer programming formulation and a simulated annealing metaheuristic to optimize the location of vehicles for round-trip systems with a heterogeneous fleet. An extensive computational study is carried out to understand the impact of fleet heterogeneity, request generality and the number of possible vehicle locations on the algorithms' performance. Problem instances derived from a case study are shown to be edge cases in terms of fleet heterogeneity and request generality, for which the proposed integer programming formulation performs exceptionally well. Finally, solutions of the case study are analyzed to demonstrate the effect of spatial flexibility on the system's costs.

Keywords: Carsharing · Vehicle location · Integer programming · Simulated annealing

1 Introduction

Carsharing systems have existed since the 1970s and have become increasingly popular in recent years as a consequence of increased congestion and environmental concerns [15]. Carsharing is based on shared usage of a fleet of vehicles by users who pay a monthly or yearly subscription fee to a carsharing provider. In return, users may make use of the available vehicles for a cost which typically depends on trip duration or distance. Many studies have shown that carsharing users tend to drive less, thereby directly contributing to improving some of the pressing mobility issues currently faced by society [18].

M. Mes et al. (Eds.): ICCL 2021, LNCS 13004, pp. 362–376, 2021.
https://doi.org/10.1007/978-3-030-87672-2_24

Carsharing systems are categorized as either round-trip or one-way[1]. In round-trip systems, users should return the vehicle to the same station where it was picked up. One-way systems offer more flexibility to their users by allowing vehicles to be picked up and dropped off at different locations. A further distinction is made between station-based and free-floating one-way systems. The former uses fixed locations for its vehicles while in free-floating systems, vehicles are bound only by a fixed geographic area. Despite the increased flexibility offered by one-way systems, round-trip systems are still predominantly used in practice. A recent survey reported that globally, 69% of carsharing users make use of round-trip systems, while only 31% use one-way systems [19].

Optimizing the design of round-trip systems, more specifically deciding where to place vehicles, is crucial to ensure a high quality of service. The problem studied in this paper addresses this strategic optimization problem for a round-trip carsharing systems operating with a fixed heterogeneous fleet. The quality of a solution is determined by how well user demand is matched. This is quantified by solving a subproblem in which the expected user demand is considered. User demand is determined by the users' requests for a specific type of vehicle, for example a small city car or an SUV, during a fixed period of time. These requests should be representative of future demand, and are typically generated based on historical data. The objective is to maximize the number of serviced requests while taking into account proximity of users to vehicles. Costs associated with opening stations for the vehicles are also minimized in the objective function.

This paper investigates the impact of three problem characteristics on the performance of two solution approaches: an integer programming (IP) formulation and a simulated annealing metaheuristic. A computational study is carried out using a dataset consisting of 415 problem instances, which have been made publicly available. The proposed solution approaches are also used to solve problem instances from practice which have been provided by a carsharing company. These real-world problem instances take extreme values for two of the three studied problem characteristics. Because of this, the IP formulation is able quickly to find optimal solutions for these instances. Finally, the optimal solutions to two problem instances from practice are analyzed in detail to demonstrate how user spatial flexibility can decrease operational costs for the carsharing provider.

The remainder of this paper is organized as follows. Section 2 provides an overview of related work. Section 3 defines the problem which is solved in this paper. Section 4 proposes two solution approaches for the considered problem: an IP formulation and a simulated annealing metaheuristic. Section 5 analyzes the performance of the different solution approaches and investigates the impact of various problem characteristics. This section also analyzes the impact of user spatial flexibility on operational costs. Finally, Sect. 6 concludes the paper and outlines directions for future research.

[1] A third, less common type of carsharing is peer-to-peer carsharing in which users rent out their private vehicle to other users.

2 Related Work

The design of carsharing systems has been the subject of many studies in the academic literature [3]. Three categories of solution approaches can be distinguished: exact methods using integer programming, (meta)heuristic algorithms and simulation. The following discussion on related work is limited to strategic optimization problems related to station location and vehicle allocation. For a recent literature survey on models and algorithms for operational problems related to carsharing, we refer to [9].

The problem addressed in [22] considers both spatial and temporal flexibility when determining which vehicle to use to serve a user. It is shown that these two types of flexibility are highly complementary and lead to cost reduction of almost 20%. By limiting the degree of flexibility and considering a scheduling period of one week, the problem could be solved as an IP problem using state of the art solvers. Integer programming is also used in [8] to determine the number of vehicle depots, their location and fleet size for a one-way system. Three IP formulations are introduced which model different request rejection policies. [8] extended these models by allowing spatial flexibility for the system's users. A bi-objective integer programming problem is solved in [2] to optimize the design of a one-way system with electric vehicles. [5] use Benders decomposition to determine the location of recharging stations for a one-way electric car sharing system.

The time-dependent IP formulation proposed in [4] is used to optimize the location of charging stations in electric carsharing systems. While state of the art IP solvers are able to solve small and medium-sized instances, a heuristic method is required to address larger problem instances. Heuristics and metaheuristics are also commonly used when the objective function involves complex calculations. [6] use simulated annealing to determine fleet size and its allocation to stations. The goal is to minimize user waiting time which is computed using simulation. Other examples of complex objective functions are discussed in [14] and [7]. In these papers, regression models are used to estimate trip demand and user membership, which are then optimized using heuristic algorithms.

Simulation has also been used as a stand-alone tool in the literature to evaluate design decisions, thereby allowing some problem characteristics to be modeled as stochastic variables. In [10], solutions from a deterministic IP model are used to analyze the impact of vehicle relocation operations in between users' requests, while considering demand variability. The methodology proposed in [1] uses simulation to evaluate the performance of two approaches which aim to increase the utilization of vehicles.

3 Problem Description

This section formally defines the optimization problem addressed in the remainder of this paper. The set of vehicles in the system is denoted by V. It is common for carsharing companies to have multiple vehicles of the same type. Let T be the set of types by which the individual vehicles may be grouped. Let $V_t \subseteq V$ be the set of vehicles of type t, with $q_t = |V_t|$ the total number of vehicles of type t. The geographic area within which the vehicles must be placed is discretized into a set of non-overlapping zones Z. Two zones z and z' are considered *adjacent* if a user is able to walk from z to z' without having to pass through another intermediary zone.

Let R be the set of requests made during a planning horizon consisting of days D. A request r is defined by its start time s_r, end time f_r, duration $u_r = f_r - s_r$ and day d_r on which the request begins. Two requests are said to be overlapping in time if the intersection of their time intervals is non-empty. The home zone of request r is the zone in which the request is made and is denoted by h_r. Let $Z_r \subseteq Z$ be a subset of zones consisting of the home zone of request r and all zones adjacent to h_r. The subset of vehicle types to which request r may be assigned is denoted by $T_r \subseteq T$. The subset of individual vehicles suitable for request r is denoted as the set $V_r \subseteq V$. We refer to V_r as the set of feasible vehicles for request r.

The primary objective is to find an assignment of vehicles to zones such that as many requests as possible are assigned to a feasible vehicle. A request may be feasibly assigned to a vehicle if it is located in the request's home zone or in an adjacent zone. Requests which overlap in time cannot be assigned to the same vehicle. If request r is not assigned to any vehicle, a cost $\mathcal{C}_r^1$ is incurred. If request r is assigned to a vehicle in a zone adjacent to h_r, a cost $\mathcal{C}_r^2$ is added to the total solution cost. Finally, there is also a cost $\mathcal{C}_{vz}^3$ associated with placing vehicle v in zone z. This third cost component corresponds to expenses for the carsharing company incurred by, for example, commissions to the city administration for reserving parking spaces. The objective function to be minimized is the weighted sum of these three costs.

4 Solution Approaches

This section introduces two solution approaches for the considered problem. First, an IP formulation of the problem is presented which is solved using a general purpose IP solver. Second, a problem-specific simulated annealing metaheuristic is discussed.

4.1 Integer Programming Formulation

The IP formulation uses two main sets of decision variables. For each $t \in T$ and $z \in Z$, an integer variable x_{tz} equals the number of vehicles of type t assigned to zone z. For each $r \in R$, $t \in T_r$ and $z \in Z_r$, a binary variable y_{rtz} equals one if request r is assigned to a vehicle of type t in zone z, and zero otherwise. Additionally, for each $r \in R$, an auxiliary binary variable p_r equals one if request r is unassigned, and zero otherwise.

Request overlap is modeled using sets $C_t = \{K_1, \dots, K_n\}$, $\forall t \in T$. Each set $K_i \in C_t$ consists of requests for which $t \in T_r$ and which pairwise overlap in time.

Moreover, each K_i is maximal in the sense that there does not exist any request in $R \backslash K_i$ which overlaps with all requests in K_i. Note that the same request r for which $t \in T_r$ may appear in multiple sets in C_t. The complete set C_t can be constructed by finding all maximal cliques in a conflict graph where nodes correspond to requests for which $t \in T_r$, and where two nodes are connected by an edge if their corresponding requests overlap in time. Given that this conflict graph is an interval graph, all maximal cliques can be found in polynomial time by the algorithm described in [12]. The formulation also uses the set $\tilde{Z}_r \subseteq Z$ consisting of only those zones adjacent to h_r, that is, $Z_r \backslash \{h_r\}$. The IP model may now be formulated as follows:

$$min \sum_{r \in R} C_r^1 p_r + \sum_{r \in R} \sum_{t \in T_r} \sum_{z \in \tilde{Z}_r} C_r^2 y_{rtz} + \sum_{t \in T} \sum_{z \in Z} C_{tz}^3 x_{tz} \tag{1}$$

$$s.t. \sum_{z \in Z} x_{tz} = q_t \qquad \forall t \in T \tag{2}$$

$$y_{rtz} \leq x_{tz} \qquad \forall r \in R, t \in T_r, z \in Z_r \tag{3}$$

$$\sum_{t \in T_r} \sum_{z \in Z_r} y_{rtz} + p_r = 1 \qquad \forall r \in R \tag{4}$$

$$\sum_{r \in K} y_{rtz} \leq x_{tz} \qquad \forall t \in T, z \in Z, K \in C_t \tag{5}$$

$$x_{tz} \in \{0, 1, ..., q_t\} \qquad \forall t \in T, z \in Z \tag{6}$$

$$y_{rtz} \in \{0, 1\} \qquad \forall r \in R, t \in T_r, z \in Z_r \tag{7}$$

$$p_r \in \{0, 1\} \qquad \forall r \in R \tag{8}$$

Objective function (1) minimizes the three aforementioned objective terms in a weighted sum. Constraints (2) ensure that the assigned number of vehicles of each type matches the available number of vehicles. Constraints (3) require that requests are only assigned to a vehicle type in a zone if at least one vehicle of that type is available in that zone. Constraints (4) make sure each request is assigned to at most one vehicle and sets the binary penalty variable p_r accordingly. Constraints (5) ensure overlapping requests are assigned to different vehicles. Constraints (6)–(8) enforce bounds on the decision variables.

Given that model (1)–(8) aggregates individual vehicles into types, a post-processing procedure is required to obtain an assignment of individual vehicles to zones and requests. Our proposed methodology for doing so consists of two steps. First, the vehicle-to-zone assignments are determined. For each zone $z \in Z$ and type $t \in T$, exactly x_{tz} vehicles of type t are assigned to zone z. Vehicles are randomly selected in such a way that each vehicle is assigned to exactly one zone. Second, the request-to-vehicle assignments are derived from the formulation's solution. Request r is assigned to a vehicle of type t in zone z based on the value of decision variable y_{rtz} and the assignments made during the first step. This corresponds to solving a fixed interval scheduling problem [13]. If the sets T_r are disjoint, this problem may be solved in polynomial time by first sorting all requests based on their start time and then assigning them to the first available vehicle. However, if the sets T_r are non-disjoint, something which may occur if, for example, vehicle types are defined in a hierarchical structure, then deriving the request-to-vehicle assignments becomes an NP-hard problem. Typically, V_r will be relatively small due to the vehicle-to-zone assignments and this subproblem may thus be solved in acceptable computation time using IP or problem-specific algorithms from the literature [17,20].

4.2 Simulated Annealing

Simulated annealing is a single solution, iterative metaheuristic in which one neighboring solution is sampled per iteration. Algorithm 1 outlines the main steps of the algorithm which takes as input parameters an initial solution S_0, a starting temperature $\mathcal{T}_0$, a minimum temperature $\mathcal{T}^s$, a cooling rate α and the number of equilibrium iterations $\mathcal{I}$. In the function `accept`(S'), the probability of accepting a neighboring solution S' depends on the difference in solution cost Δ between the current and neighboring solution, and the current temperature $\mathcal{T}$, calculated as $\exp(-\Delta/\mathcal{T})$ [11]. A fixed number of iterations $\mathcal{I}$ is performed at each temperature level before decreasing the temperature using an exponential cooling schedule $\alpha\mathcal{T}$, with $0 < \alpha < 1$. When the minimum temperature level $\mathcal{T}^s$ is reached the algorithm stops and the best solution S^* is returned.

Algorithm 1: Simulated annealing

```
   Data: S_0, T_0, T^s, α, I
   Result: S*
 1 S* ← S_0;
 2 T ← T_0;
 3 while T ≥ T^s do
 4     i ← 0;
 5     while i ≤ I do
 6         S' ← N(S);
 7         if accept(S') then
 8             S ← S';
 9             if S' < S* then
10                 S* ← S';
11         i ← i + 1;
12     T ← αT;
13 return S*;
```

A direct solution representation is used based on two data structures corresponding to the two main assignment decisions. The first is a set of vehicle-to-zone assignments $\mathcal{V}$ which consists of $|V|$ tuples of the form (v, z), indicating that vehicle $v \in V$ is assigned to zone $z \in Z$. The second is a set of request-to-vehicle assignments $\mathcal{R}$ which consists of $|R|$ tuples of the form (r, v) or $(r, \emptyset)$, indicating that request $r \in R$ is assigned to vehicle $v \in V_r$ or remains unassigned, respectively.

The initial solution S_0 is constructed by first assigning each vehicle to a random zone. Then, each request is assigned to a randomly selected feasible vehicle, that is, a vehicle with no overlapping requests and which is assigned to the request's home zone or an adjacent zone. If no such vehicle exists, the request remains unassigned.

In each iteration of the algorithm, one of two possible neighborhoods is randomly sampled for a new solution. The first neighborhood modifies assignments in both $\mathcal{V}$ and $\mathcal{R}$ while the second only modifies the set $\mathcal{R}$. The neighborhoods are defined by the following two operators:

- $move(v, z)$: move vehicle v to a new zone z. Each request r previously assigned to v for which $z \notin Z_r$ initially becomes unassigned. These requests are then either immediately re-assigned to a different feasible vehicle without causing overlap conflicts and with minimal cost increase, or they remain unassigned if no such assignment exists. Additionally, each request r' for which $z \in Z_{r'}$ is assigned to vehicle v if this leads to a lower cost without incurring any conflicts.
- $assign(r, v)$: assign request r to a new vehicle $v \in V_r$ which is currently assigned to a zone in Z_r. All requests which were previously assigned to v and which overlap with r become unassigned and are re-assigned to a different feasible vehicle which leads to the minimal cost increase, or remain unassigned if no such feasible vehicle exists.

5 Computational Study

This section presents the results of a computational study on the performance of the two proposed solution approaches. Section 5.1 provides details regarding the used data sets and computational environment. Section 5.2 analyzes the impact of three problem characteristics on algorithm performance. Section 5.3 compares and evaluates the performance of the two approaches on real-world cases.

5.1 Data Generation and Experimental Setup

A set of problem instances was generated based on properties identified in real-world data provided by an industry partner which described the available fleet and users' requests during the first six months of 2017[2]. Request start day, start time and duration were sampled from distributions fitted to the probability density functions observed in the real-world data. Further analysis of the data revealed a recurring pattern concerning where exactly requests are made. There are typically a few zones with a high number of requests which are located close to each other. The number of requests per zone then decreases as one moves away from this high-density cluster. This pattern was replicated in the generated data set. Figure 1 shows the number of requests per zone in six example problem instances.

[2] The generated instances can be downloaded at https://people.cs.kuleuven.be/~pieter.smet/carsharing.

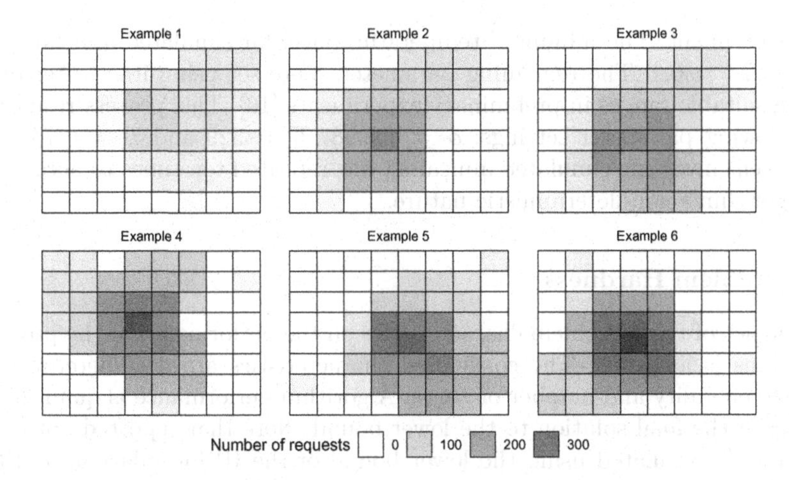

Fig. 1. Number of requests per zone in randomly generated instances with an 8×8 grid zone layout.

Two parameters are introduced to quantify fleet heterogeneity and variation in the requested vehicle types. First, let $\sigma^s = 1 - |T|/|V|$ be a metric for the degree of fleet heterogeneity. Low values for σ^s indicate that the fleet is strongly heterogeneous whereas high values indicate that the fleet is strongly homogeneous and mainly consists of vehicles of the same type. Second, let $\phi_r = \sum_{r \in R}(|T_r|/|T|)/|R|$ be a metric for the level of request generality. Specifically, it represents how many of the different vehicles types are considered feasible for a request, averaged over all requests. If $\phi_r = 1$, then requests can be assigned to all vehicle types. By contrast, low values for ϕ_r indicate that requests can only be assigned to a limited number of vehicle types.

All terms in the objective function are expressed in the same unit (Euro), which allows them to be summed directly without having to include any additional multipliers. The first cost $\mathcal{C}_r^1$, associated with leaving request r unassigned, is determined by multiplying the duration of request r by the price per minute of use. This value was determined based on a survey of carsharing providers in Belgium and was set to €0.30/min. The second cost $\mathcal{C}_r^2$ corresponds to the incentive given by the provider to users whose requests are assigned to vehicles in adjacent zones, thus requiring spatial flexibility. This cost is calculated as $0.5 \times \mathcal{C}_r^1$. The third cost $\mathcal{C}_{tz}^3$ is associated with placing vehicle of type t in zone z and is set to €2/day. Note that, in practice, this cost may vary depending on the zone and type of vehicle. For example, a large parking space in the city center will cost more than a small space located at the edge of the city. For the purpose of the present computational study, a random increase or decrease of at most 10% was applied to the average to simulate such variations.

All experiments were conducted on a Dell Poweredge T620, 2x Intel Xeon E5-2670 with 128 GB RAM. Gurobi 8.1 was used as IP solver with default settings and configured to use one thread. A time limit of five hours was imposed for each experiment. The minimum temperature level for simulated annealing was fixed to $\mathcal{T}^s = 0.1$. The remaining parameters were set using irace after determining suitable ranges in preliminary experiments [16]. This process resulted in the following parameter settings: $\alpha = 0.9998$, $\mathcal{T}_0 = 329$ and $\mathcal{I} = 1775$. Each experiment involving simulated annealing was repeated ten times to account for the algorithm's non-deterministic nature.

5.2 Problem Hardness

The impact of three problem characteristics on the performance of the proposed algorithms is analyzed. The considered characteristics are fleet heterogeneity, request generality and number of zones. Algorithm performance is quantified as the gap of the final solution to the lower bound. Note that reported optimality gaps are all calculated using the lower bound of the IP formulation, and thus depend on the solver's performance.

For each experiment, a set of instances was generated with a planning horizon of 28 days. The number of requests was varied between 700 and 4200 in increments of 700, while the number of vehicles was calculated as $|V| = \alpha(|R|/|D|)$ where $\alpha \in \{0.8, 1.3\}$ in order to vary vehicle occupancy rates.

Fleet Heterogeneity. The level of fleet heterogeneity σ^s was varied between 0 and 1 in increments of 0.1, considering 64 zones organized in an 8×8 grid and a request generality of $\phi_r = 0.3$. Table 1 shows the average optimality gap for the two solution approaches. The performance of the IP formulation is inversely correlated with σ^s. For instances with a strongly homogeneous fleet, the IP formulation is able to quickly find optimal solutions. Less fleet heterogeneity results in fewer vehicle types and thus smaller IP problems. However, for instances with a more heterogeneous fleet, the IP formulation is unable to find good solutions, with average optimality gaps of up to 48% for $\sigma^s \leq 0.1$. Simulated annealing follows a similar pattern for the optimality gaps, however, its gaps are, on average, smaller. Overall, simulated annealing proves to be more suited for addressing instances with varying levels of fleet heterogeneity, independent of their size.

Table 1. Optimality gaps for varying levels of fleet heterogeneity.

	Fleet heterogeneity σ^s											Average
	0.0	0.1	0.2	0.3	0.4	0.5	0.6	0.7	0.8	0.9	1.0	
Integer programming	48%	48%	44%	42%	41%	40%	37%	31%	24%	6%	0%	33%
Simulated annealing	31%	31%	25%	24%	24%	22%	21%	17%	14%	7%	0%	20%

Request Generality. The level of request generality ϕ_r was varied between 0.1 and 1 in increments of 0.1, considering 64 zones and a fleet heterogeneity of $\sigma^s = 0.6$. Table 2 again reports the average optimality gap for the two approaches.

The IP formulation is clearly affected by varying request generality. Problem instances with high values for ϕ_r are generally easy to solve for the IP solver. The most computationally challenging instances are those where $\phi_r \in [0.2, 0.5]$. When only a few vehicle types are considered feasible, the instances again become easier to solve. The IP formulation is able to find all optimal solutions when ϕ_r is close to one. However, on average, simulated annealing finds solutions which are closer to the optimum.

Table 2. Optimality gaps for varying levels of request generality.

	Request generality ϕ_r										Average
	0.1	0.2	0.3	0.4	0.5	0.6	0.7	0.8	0.9	1.0	
Integer programming	14%	25%	36%	34%	31%	15%	5%	1%	0%	0%	16%
Simulated annealing	9%	17%	20%	19%	16%	11%	8%	6%	5%	0%	11%

Zone Size. From the user's perspective, a zone's size has a major impact on how convenient they experience the carsharing system. If zones are too large, users will have to walk far to access their vehicle, even if it is located in their home zone. However, small zones increase the size of the problem, making it more difficult for algorithms to find good solutions. This section investigates the impact of the number of zones on the performance of the proposed approaches.

The number of zones was varied from 64 to 900, with $\phi_r = 0.3$ and $\sigma^s = 0.6$. All instances considered zones that are organized in a square grid. Table 3 shows the average optimality gap for the two solution approaches. For both approaches, the number of zones does not significantly affect the optimality gaps. Nevertheless, there is a clear difference in performance overall between the approaches, which follows the previously identified trends. Simulated annealing finds the best solutions overall, with a gap of 21% to the best lower bounds obtained by the IP formulation.

Table 3. Optimality gaps for varying number of zones.

	Number of zones											Average
	64	81	100	121	144	169	196	225	400	625	900	
Integer programming	33%	37%	35%	35%	37%	34%	37%	36%	37%	36%	36%	36%
Simulated annealing	19%	21%	19%	20%	21%	20%	21%	22%	21%	21%	23%	21%

5.3 Real-World Instances

A set of problem instances was also derived from real-world data concerning two cities, denoted here by A and B. These two cities cover an area of 3.9 km × 4.9 km and 9.6 km × 10.5 km, respectively. Both areas were divided into zones by overlaying a grid of ten by ten rectangles resulting in 100 zones of approximately

0.2 km^2 and 1 km^2 each, respectively. By varying the length of the planning horizon, nine instances were constructed for each city, labeled A1-A9 and B1-B9.

The fleet available in city A is considerably smaller compared to city B, both in terms of number of vehicles and vehicle types. However, in both cases, the fleet is rather homogeneous, with $\sigma^s = 0.77$ and $\sigma^s = 0.93$ for A and B, respectively. In terms of request generality, the two cities are identical: each request can only be serviced by one type of vehicle, that is, $|T_r| = 1$ for all requests $r \in R$. In city A, $|T| = 3$ and $\phi_r = 0.33$ while in city B $|T| = 13$ and $\phi_r = 0.08$. As demonstrated by the results in Sect. 5.2, model (1)–(8) is very effective for solving instances with these characteristics.

Table 4 compares the performance of the IP formulation and simulated annealing. For both approaches, the final solution cost (*Cost*), optimality gap (*Gap*) and computation time in seconds (*Time*) are shown. For simulated annealing, the reported solution cost is the average of ten repeated runs with different seed values for the algorithm's random number generator.

The IP formulation found optimal solutions for all instances in very limited computation time. Even for the largest instance with 5306 requests and 193 vehicles, the solver found an optimal solution in under two seconds. In general, computation time increased with problem size (number of requests and vehicles). For simulated annealing, the average optimality gap was 0.8%, with a small relative standard deviation which varied between 0.02% and 0.31%. However, the required computation time was significantly larger compared to the IP formulation.

Table 5 shows details of the optimal solutions for two of the real-world instances, A9 and B9. In addition to the results for the proposed problem, values are also shown for a scenario without spatial user flexibility, that is, in which requests can only be assigned to vehicles located in their home zones. First, data related to the number of assigned requests is shown. Second, details on the solutions' objective values are given. Finally, the average vehicle utilization rate is reported which is defined here as the average of the total time a vehicle is used divided by the total time the vehicle may potentially be used. This value is calculated as shown in Eq. (9), with $\mathcal{A}_v \subset R$ the subset of requests assigned to vehicle v and δ the duration of one day in the planning horizon expressed in the same unit as request duration u_r.

$$\text{Avg. vehicle utilization rate} = \frac{1}{|V|} \sum_{v \in V} \frac{\sum_{r \in \mathcal{A}_v} u_r}{|D| \times \delta} \quad (9)$$

When allowing spatial flexibility, most of the requests can be assigned. Moreover, the majority of these requests are assigned to vehicles in their home zones. In the larger city (B9), relatively more requests can be assigned and home zones can be better matched compared to the smaller city (A9). Analyzing the objective values, the largest component comes from the incentives paid by the carsharing provider to users who are serviced in an adjacent zone. The reported utilization rates are in line with those found in the literature [21]. Without spatial flexibility, a considerable drop in solution quality is observed. This is

Table 4. Characteristics and algorithm performance for IP and simulated annealing. All computation times are reported in seconds.

Instance							IP formulation			Simulated annealing		
Id	$\|D\|$	$\|R\|$	$\|T\|$	$\|V\|$	σ^s	ρ_r	Cost	Gap	Time	Cost	Gap	Time
A1	84	470	3	13	0.77	0.3	20718	0.0%	0.4	20832.0	0.5%	365.8
A2	84	474	3	13	0.77	0.3	20487	0.0%	0.4	20653.0	0.8%	379.8
A3	84	473	3	13	0.77	0.3	18161	0.0%	0.4	18494.8	1.8%	386.1
A4	126	709	3	13	0.77	0.3	31166	0.0%	0.5	31397.0	0.7%	590.1
A5	126	699	3	13	0.77	0.3	31555	0.0%	0.6	31789.0	0.7%	587.0
A6	126	695	3	13	0.77	0.3	30584	0.0%	0.4	30865.0	0.9%	579.5
A7	168	943	3	13	0.77	0.3	43728	0.0%	0.4	44129.6	0.9%	777.5
A8	168	945	3	13	0.77	0.3	43402	0.0%	0.4	43836.0	1.0%	779.9
A9	168	937	3	13	0.77	0.3	42067	0.0%	0.4	42298.0	0.5%	761.7
B1	21	3081	13	193	0.93	0.1	26753	0.0%	1.0	26962.6	0.8%	3945.1
B2	21	3169	13	193	0.93	0.1	29149	0.0%	0.9	29431.4	1.0%	3966.2
B3	21	2763	13	193	0.93	0.1	17544	0.0%	0.8	17612.0	0.4%	3360.3
B4	28	3934	13	193	0.93	0.1	37440	0.0%	1.4	37643.8	0.5%	5069.8
B5	28	4061	13	193	0.93	0.1	43075	0.0%	1.2	43656.6	1.3%	5130.4
B6	28	4095	13	193	0.93	0.1	50300	0.0%	1.3	50563.0	0.5%	5358.0
B7	35	5218	13	193	0.93	0.1	51498	0.0%	1.7	52001.0	1.0%	6952.6
B8	35	5306	13	193	0.93	0.1	74174	0.0%	2.0	74576.4	0.5%	7062.0
B9	35	5181	13	193	0.93	0.1	51669	0.0%	1.8	52164.4	0.9%	6900.5

Table 5. Solution details for two instances.

	Instance A9		Instance B9	
	Spatial flexibility	Only home zones	Spatial flexibility	Only home zones
Total number of requests	937	937	5181	5181
Requests assigned	93.4%	62.2%	98.4%	88.4%
Assigned in home zone	58.2%	62.2%	87.4%	88.4%
Assigned in adjacent zone	35.2%	0.0%	11.0%	0.0%
Requests unassigned	6.6%	37.8%	1.6%	11.6%
Total objective value	€42,067	€105,046	€51,669	€149,084
Unassigned requests cost	€18,060	€102,400	€15,510	€143,220
Adjacent-assigned requests cost	€21,570	€0	€30,585	€0
Vehicle location cost	€2,437	€2,646	€5,574	€5,864
Avg. vehicle utilization rate	10.9%	8.2%	17.6%	16.3%

especially noticeable in the smaller city (A09) as there are fewer vehicles available for the same number of zones. As fewer requests are assigned in total, the vehicle utilization rates also decrease.

6 Conclusions and Future Work

Increased traffic congestion and environmental concerns have motivated the use of a variety of shared mobility systems. Carsharing is one such system in which different users make use of a shared fleet of vehicles. Different types of carsharing systems may be distinguished depending on whether they operate using stations or free-floating zones and whether they allow one-way trips or require users to return their vehicle to the pickup location.

For round-trip systems, the location of vehicles is one of the primary design decisions which impacts service quality. The quality of a solution is determined by how well users' requests can be fulfilled. This paper models this as a scheduling subproblem in which individual requests are assigned to allocated vehicles. To solve the resulting optimization problem, two solution approaches were presented: an IP formulation and a simulated annealing metaheuristic.

A computational study on a new publicly available dataset investigated how the algorithms' performance is affected by three problem characteristics: fleet heterogeneity, request generality and the number of possible vehicle locations. First, it was demonstrated that the IP formulation is well-suited for problems with a strongly homogeneous fleet or when requests can be fulfilled by only a limited number of vehicle types. Second, simulated annealing outperformed the IP solver when considering a broad range of instances. Finally, increasing the number of possible vehicle locations in a geographic area did not affect the approaches' performance. Problem instances derived from real-world data were shown to be edge cases in terms of fleet heterogeneity and request generality. As a result, these instances could be solved to optimality in very limited time using IP. Simulated annealing, while requiring more computation time, did find near-optimal solutions.

Future research may challenge some of the assumptions in the proposed problem in order to arrive at a more realistic model. Possible extensions include stochastic demand, dynamic arrival of requests, choice-based optimization or different types of time discretization for request duration. These extensions would prove useful when it comes to understanding the trade-offs which exist when implementing carsharing systems in practice. From a computational perspective, an analysis of our algorithms' performance revealed that many of the generated instances could not be solved to optimality. The performance of simulated annealing may be further improved by using an alternative solution representation which avoids possible symmetries in a manner similar to the proposed IP formulation. Furthermore, previous research has demonstrated that decomposition approaches such as Benders decomposition are well-suited for solving challenging large-scale problems and may therefore prove useful in addressing these open instances.

Acknowledgements. This research was supported by the Strategic Basic Research project 'Data-driven logistics' (S007318N), funded by the Research Foundation Flanders (FWO). Editorial consultation provided by Luke Connolly (KU Leuven).

References

1. Balac, M., Ciari, F.: Enhancement of the carsharing fleet utilization. In: 15th Swiss Transport Research Conference, Ascona, Switzerland (2015)
2. Boyacı, B., Zografos, K.G., Geroliminis, N.: An optimization framework for the development of efficient one-way car-sharing systems. Eur. J. Oper. Res. **240**(3), 718–733 (2015)
3. Brandstätter, G., et al.: Overview of optimization problems in electric car-sharing system design and management. In: Dawid, H., Doerner, K.F., Feichtinger, G., Kort, P.M., Seidl, A. (eds.) Dynamic Perspectives on Managerial Decision Making. DMEEF, vol. 22, pp. 441–471. Springer, Cham (2016). https://doi.org/10.1007/978-3-319-39120-5_24
4. Brandstätter, G., Kahr, M., Leitner, M.: Determining optimal locations for charging stations of electric car-sharing systems under stochastic demand. Transp. Res. Part B Methodol. **104**, 17–35 (2017)
5. Çalık, H., Fortz, B.: A benders decomposition method for locating stations in a one-way electric car sharing system under demand uncertainty. Transp. Res. Part B Methodol. **125**, 121–150 (2019)
6. Cepolina, E.M., Farina, A.: A new shared vehicle system for urban areas. Transp. Res. Part C Emerg. Technol. **21**(1), 230–243 (2012)
7. Ciari, F., Weis, C., Balac, M.: Evaluating the influence of carsharing stations' location on potential membership: a Swiss case study. EURO J. Transp. Logist. **5**(3), 345–369 (2016)
8. Correia, G., Jorge, D.R., Antunes, D.M.: The added value of accounting for users' flexibility and information on the potential of a station-based one-way car-sharing system: an application in Lisbon, Portugal. J. Intell. Transp. Syst. **18**(3), 299–308 (2014)
9. Illgen, S., Höck, M.: Literature review of the vehicle relocation problem in one-way car sharing networks. Transp. Res. Part B Methodol. **120**, 193–204 (2019)
10. Jorge, D., Correia, G., Barnhart, C.: Testing the validity of the MIP approach for locating carsharing stations in one-way systems. Procedia Soc. Behav. Sci. **54**, 138–148 (2012)
11. Kirkpatrick, S., Gelatt, C.D., Vecchi, M.P.: Optimization by simulated annealing. Science **220**(4598), 671–680 (1983)
12. Krishnamoorthy, M., Ernst, A., Baatar, D.: Algorithms for large scale shift minimisation personnel task scheduling problems. Eur. J. Oper. Res. **219**, 34–48 (2012)
13. Kroon, L.G., Salomon, M., Van Wassenhove, L.N.: Exact and approximation algorithms for the tactical fixed interval scheduling problem. Oper. Res. **45**(4), 624–638 (1997)
14. Kumar, P., Bierlaire, M.: Optimizing locations for a vehicle sharing system. In: Proceedings of the Swiss Transport Research Conference, pp. 1–30 (2012)
15. Laporte, G., Meunier, F., Wolfler Calvo, R.: Shared mobility systems: an updated survey. Ann. Oper. Res. **271**(1), 105–126 (2018). https://doi.org/10.1007/s10479-018-3076-8
16. López-Ibánez, M., Dubois-Lacoste, J., Cáceres, L.P., Birattari, M., Stützle, T.: The irace package: iterated racing for automatic algorithm configuration. Oper. Res. Perspect. **3**, 43–58 (2016)

17. Niraj Ramesh, D., Krishnamoorthy, M., Ernst, A.T.: Efficient models, formulations and algorithms for some variants of fixed interval scheduling problems. In: Sarker, R., Abbass, H.A., Dunstall, S., Kilby, P., Davis, R., Young, L. (eds.) Data and Decision Sciences in Action. LNMIE, pp. 43–69. Springer, Cham (2018). https://doi.org/10.1007/978-3-319-55914-8_4
18. Shaheen, S., Cohen, A.: Carsharing and personal vehicle services: worldwide market developments and emerging trends. Int. J. Sustain. Transp. **7**(1), 5–34 (2013)
19. Shaheen, S., Cohen, A., Jaffee, M.: Innovative mobility: Carsharing outlook. UC Berkeley: Transportation Sustainability Research Center (2018). https://doi.org/10.7922/G2CC0XVW
20. Smet, P., Wauters, T., Mihaylov, M., Vanden Berghe, G.: The shift minimisation personnel task scheduling problem: a new hybrid approach and computational insights. Omega **46**, 64–73 (2014)
21. Sprei, F., Habibi, S., Englund, C., Pettersson, S., Voronov, A., Wedlin, J.: Free-floating car-sharing electrification and mode displacement: travel time and usage patterns from 12 cities in Europe and the United States. Transp. Res. Part D Transp. Environ. **71**, 127–140 (2019)
22. Ströhle, P., Flath, C.M., Gärttner, J.: Leveraging customer flexibility for car-sharing fleet optimization. Transp. Sci. **53**(1), 42–61 (2019)

Exact Separation Algorithms for the Parallel Drone Scheduling Traveling Salesman Problem

Tobias Klein(✉) and Peter Becker

Hochschule Bonn-Rhein-Sieg (H-BRS) - University of Applied Sciences, Sankt Augustin, Germany
tobias.klein@smail.inf.h-brs.de, peter.becker@h-brs.de

Abstract. The joint delivery of parcels by trucks and drones is a futuristic scenario that already started. As a result of this development, new optimization problems are defined and studied. In this paper, exact separation algorithms for the Parallel Drone Scheduling Traveling Salesman Problem are presented. Known separation algorithms for subtour elimination constraints and 2-matching inequalities are modified and applied to the new context. In addition, a new valid inequality for an invalid drone-truck subtour is given. For this inequality, a simple separation algorithm with a runtime of $O(n^2)$ for n nodes (customers) is presented. It is shown that this problem specific separation algorithm reduces the total runtime of problem instances more effectively compared to modified TSP approaches, especially for instances with a large number of customers that can be served by a drone. These separation algorithms are used to solve instances with up to 127 customers by a branch and cut algorithm.

Keywords: Parallel drone scheduling traveling salesman problem · PDSTSP · Integer programming · Branch and cut · Separation algorithm

1 Introduction

In the last few decades, a truck has been the vehicle to cover the route from a depot to the final customer. But in the last years, the development of drone technology opened an additional option for parcel delivery on the "last mile". The german company DHL created the Parcelcopter and tested it around the Lake Victoria for medicine shipment over a period of six month [7]. The company WING, a subsidiary of Alphabet Inc., uses autonomous drones already in Helsinki, Canberra, Logan, and Christiansburg to deliver parcels and food to customers[1].

[1] https://wing.com.

M. Mes et al. (Eds.): ICCL 2021, LNCS 13004, pp. 377–392, 2021.
https://doi.org/10.1007/978-3-030-87672-2_25

In urban areas or in areas without a road network as in a coastal environment, drones have a navigational advantage over trucks due to their ability to fly. They are able to avoid heavy traffic and natural obstacles like lakes, rivers, and parks. However, trucks have a higher capacity and can therefore carry more and heavier parcels.

The development of drones for parcel delivery leads to the creation of new optimization problems for drones and trucks. Murray and Chu [20] were one of the first to define such optimization problems like the Flying Sidekick Traveling Salesman Problem (FSTSP) and the Parallel Drone Scheduling Traveling Salesman Problem (PDSTSP), where a fixed set of customers must be supplied by either a drone or a truck. Following their paper, other authors adapted and modified the FSTSP or PDSTSP to new problems like the Traveling Salesman Problem with Drone (TSP-D) [1] and solved them mostly with heuristic approaches. However, these optimization problems are not deeply studied, as the focus is mainly on the creation of new problems and not on further algorithmic investigations.

This paper aims to provide exact separation algorithms for the PDSTSP on basis of exact approaches for the Traveling Salesman Problem (TSP) in combination with the branch and cut algorithm. First, an overview over the current research for optimization problems with drones and trucks is given in Sect. 2. In Sect. 3, the PDSTSP by Murray and Chu is presented, modified, and a relaxation of this problem is described. Section 4 covers the creation of valid inequalities. Separation algorithms for these inequalities are given in Sect. 5 and they are applied in a computational experiment in Sect. 6. This paper summarizes the key findings of the unpublished german master's thesis by one of the authors [16].

2 Related Literature

Murray and Chu [20] defined the PDSTSP in addition to the Flying Sidekick Traveling Salesman Problem (FSTSP) and created mixed integer linear program (MILP) models for both of them. Moreover, the authors developed a two-phase heuristic algorithm to solve the PDSTSP. Initially, they assign all customers to drone delivery if they can be supplied by one, the other customers are supplied by the truck. In the first phase, they solve the truck route as a TSP and the routes of the drone as a Parallel Machine Scheduling Problem (PMS). In the second phase, they improve the actual solution iteratively by swapping the delivery type of customers, if that reduces the overall delivery time. Through the reduction of the PDSTSP to a TSP and a PMS they reduce the complexity and use well known algorithms for both of these problems.

Based on this work, Saleua et al. [19] created a similar approach to solve the PDSTSP through a two-phase heuristic. In contrast to Murray and Chu, they use dynamic-programming in the first phase to partition the customers between the truck and the drones. With this algorithm, they are able to improve the runtime for problem instances that Murray and Chu used in their experiment.

A more detailed view on heuristic algorithms for the PDSTSP is provided by Dell'Amico, Montemanni, and Novellani [6]. The authors developed four different heuristics to analyze the trade-off between runtime and the gap to an optimal solution. In contrast to the work of Saleua et al., they change the MILP formulation of the PDSTSP for problem instances with more than 20 customers. They add additional inequalities to the MILP that increase the number of iterations for solving a problem. However, the added inequalities reduce the overall runtime of the heuristic. They report more exact solutions in 28 out of 90 problem instances that were used by Saleu et al. in the analyzation.

The master's thesis by Klohn [17] uses exact separation algorithms to solve the PDSTSP and the FSTSP in addition to heuristics. The author reuses the known subtour elimination constraints (SEC) from the TSP and applies them to a simplified version of the MILP formulation from Murray and Chu. The author showed that the use of SECs in addition to heuristic algorithms reduces the total runtime of both optimization problems.

Similar to the thesis from Klohn, the master's thesis by van Dijck [22] uses also exact separation algorithms for SECs and comb inequalities to solve the Traveling Salesman Problem with Drone (TSP-D) through a branch and cut approach. A heuristic algorithm to separate the comb inequalties by Padberg and Rinaldi is used by the author in addition to an exact approach through the solution of a mixed integer program (MIP). The separation algorithm was applied in different combinations to analyse the optimal configuration for solving the TSP-D exactly.

A variant of the PDSTSP was developed by Ham [11]. The author defines the PDSTSP with "drop and pickup" (PDSTSP+DP). In the PDSTSP+DP the customers themselves can send parcels to the depot via drones as a return. Further, customers can order multiple parcels Constraint programming (CP) was used in order to solve the PDSTSP+DP exactly. Ham concludes that more CP should be used for such optimization problems instead of MILPs.

In the master's thesis of Klein [16] exact separation approaches for the PDSTSP were created. In addition to modified separation approaches from the TSP for SECs and 2-matching inequalities, a new valid inequality and a corresponding exact separation algorithm is given. This paper summarizes the key findings of this master's thesis.

3 The Parallel Drone Scheduling TSP

The first description and MILP model of the PDSTSP was given by Murray and Chu [20]. With the PDSTSP, a fixed set of known customers must be supplied with a parcel. A truck and a fleet of homogeneous drones are available for delivery. The goal is to minimize the total delivery time, that is either determined by the truck or the drones, whichever is longer. Related problems that use trucks as mobile stations as the FSTSP or the Hybrid Vehicle-Drone Routing Problem (HVDRP) [13] minimize the total operational cost, similar to the Ring Star Problem (RSP) [18] and the Capacitated m-Ring-Star Problem (CmRSP) [4].

As the drones fly directly between a customer and the depot, the total operational cost can be simplified to the total delivery time as it is done by Hà, Vu and Vu [10].

A drone can only carry one parcel at a time. Therefore, a drone always has to return to the depot before delivering a new parcel. Also, not all customers can be served by a drone due to flight range limitations or other restrictions like no-fly zones. The truck, on the other hand, can carry an unlimited amount of parcels and can access every customer. The route of the truck starts and ends at the depot.

In this paper, only the distances between the customers and the speed of the truck and the drones have influence on the total delivery time. The conditions for the truck are identical to the conditions for a TSP. Therefore, the truck tour can be seen as a TSP tour. Furthermore, the speed of the truck can differ from that of the drones.

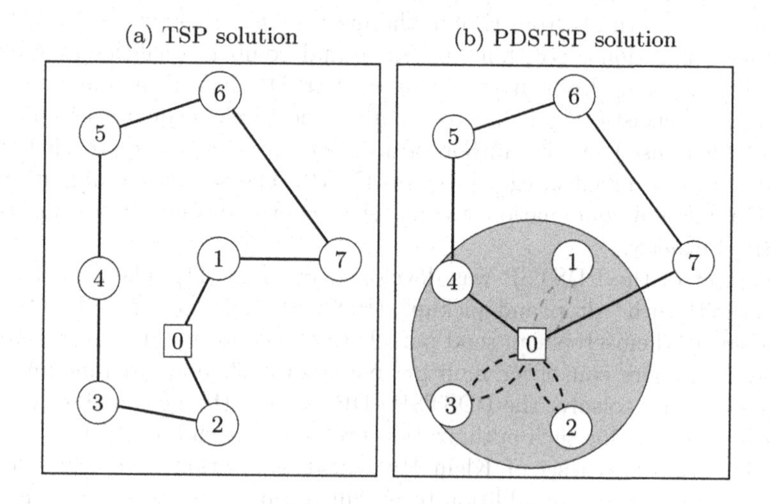

Fig. 1. Comparison of a TSP and a PDSTSP solution (Color figure online)

Figure 1(b) shows a feasible solution to a PDSTSP with two drones. The depot where the drones and the truck start is marked by the node 0. The green area around the depot symbolizes the flight range of the drones. Therefore, the customers 1 to 4 can be supplied by a drone. But in the solution, only the customers 1 to 3 are supplied by a drone, whereas the truck delivers parcels to the customers 4 to 7. Customer 1 is served by drone 1, customers 2 and 3 are served by drone 2.

Compared to the PDSTSP solution, the total delivery time of the TSP tour in Fig. 1(a) is longer. In the TSP solution the truck delivers parcels to the customers 1 to 7, in the PDSTSP only to the customers 4 to 7. If the total delivery time could be reduced in the PDSTSP solution by changing one of the drone customers 1 to 3 to a truck customer, it would have already happened, since this is the optimal solution with respect to the total delivery time.

MILP Formulation

We use a modified version of the MILP model created by Murray and Chu [20]. Our model only uses symmetrical distances, similar to the modification done in [17]. The set D is the set of all drones, C the set of all customers. The set $C'' \subseteq C$ represents the customers that can be served by a drone and are in flight range.

In our model the flight range of a drone must be halved initially, as a drone route consists of the flight from the depot to a customer and a flight back to the depot. The depot is modelled by node 0. As the depot is not a customer but part of the truck and drone tour, the set $V = \{0\} \cup C$ is the set of all nodes. Let E be the set of edges of the complete graph with V as node set. For a proper node subset $S \subset V$ with $S \neq \emptyset$ the set

$$\delta(S) = \{\{i,j\} \in E | i \in S, j \notin S\} \subset E$$

is the cut-set that is induced by S, i. e. the set of edges that connect a node $i \in S$ with a node $j \in V \setminus S$. We simply write $\delta(i)$ instead of $\delta(\{i\})$, if S consists of a single node i. So the MILP model is:

$$\min \; z \tag{1}$$

$$\text{subject to } z \geq \sum_{e \in E} \tau_e x_e \tag{2}$$

$$z \geq 2 \sum_{i \in C''} \tau_i' y_{i,d} \quad \forall d \in D \tag{3}$$

$$\sum_{e \in \delta(i)} x_e + 2 \sum_{d \in D} y_{i,d} = 2 \quad \forall i \in C'' \tag{4}$$

$$\sum_{e \in \delta(i)} x_e = 2 \quad \forall i \in V \setminus C'' \tag{5}$$

$$\sum_{e \in \delta(S)} x_e \geq 2 \quad \forall S \subset C, 2 \leq |S| \leq |C| - 1 \tag{6}$$

$$x_e \in \{0,1\} \quad \forall e \in E \tag{7}$$

$$y_{i,d} \in \{0,1\} \quad \forall i \in C'', d \in D \tag{8}$$

The MILP model contains to different sets of decision variables. The decision variables x_e, with $e = \{i,j\} \in E$, represent, whether the route between node i and node j is part of the truck tour. If $x_e = 1$, the edge e, i. e. the route from i to j, is part of the truck tour, otherwise not. For a graph-theoretical point of view, we call the edge e of a decision variable x_e a *truck edge.*

Setting the decision variable $y_{i,d}$ to 1 represents, that customer $i \in C''$ is served by drone $d \in D$. Otherwise, customer i is served by the truck. We call the edge $\{0,i\}$ of a decision variable $y_{i,d}$ a *drone edge.*

The parameter τ_e is the travel time of a truck edge e. In contrast, τ_i' is the flight time of a drone edge $\{0,i\}$.

We want to minimize the total delivery time z (1) which is lower bounded by the delivery time of the truck (2) and the flight time of each individual drone

(3). Any customer node that can be served by a drone must be incident with two truck edges or one drone edge (4). Customers who cannot be supplied by a drone must be incident with two truck edges (5). This also applies to the depot. Finally, we have to eliminate truck subtours, which is done by subtour elimination constraints (6).

In contrast to the MILP model by Murray and Chu, we simplified the time bound of the truck tour in (2) and the drone flight time in (3), as we use only symmetric distances. We also reduced the condition, that a truck must leave every customer it visits, to the condition, that the truck must start and end at the depot (5). Based on our symmetric assumption in combination with (4), (5), and (6) the truck tour is fully bounded. Further, we use in (6) a formulation for subtour elimination, that matches with the separation algorithms in this paper. This formulation states that the minimum cut between two nodes in the graph that represents the truck tour must be at least 2. Our relaxation of this MILP excludes the constraint (6), (7), and (8).

4 Valid Inequalities for the PDSTSP

The TSP can be interpreted as a special PDSTSP where only the truck is allowed to deliver parcels to customers. Therefore, known TSP separation algorithms can be modified and used for the truck route in a PDSTSP. To analyze invalid structures in a relaxed solution of the PDSTSP, we modified the TSP instance `d493` from the TSPLIB [21] to a PDSTSP instance. We moved the depot node 0 to a more centered location and added parameters for the drone count, drone speed, truck speed, flight range, and a set of customers that can be accessed by a drone. We set the speed of the drones to double the speed of the truck, that is $\tau'_i = \tau_{0,i}/2$ for $i \in C''$. Moreover, all drones have an unlimited flight range. Finally, we created a number of problems with varying amounts of customers that can be served by a drone. For example, in the PDSTSP instance `d493_10` only the first 10% of the customers (1 to 49) can be served by a drone[2]. The Gurobi solver version 9.1.1 was used to calculate relaxed solutions.

The solution of the relaxed `d493_10` problem instance contained subtours in the truck route that are addressed in the TSP through SECs and 2-matching inequalities. As `d493_10` is close to the corresponding TSP instance `d493`, this result met our expectations. Consequently, SECs and 2-matching inequalities can be used for the separation of relaxed PDSTSP instances.

But with an increasing number of customers that can be served by a drone, another invalid structure occurred in the relaxation solutions. As far as we know, this structure is not described in other papers. This structure can be interpreted as a subtour consisting of truck and drone routes. Therefore, we call this structure *Drone-Truck SEC (DT-SEC)* in relation to the known SEC. The graph in Fig. 2 shows a simplified DT-SEC.

[2] All problem instances used in this paper can be found at https://github.com/to-klein/PDSTSP_Instances.

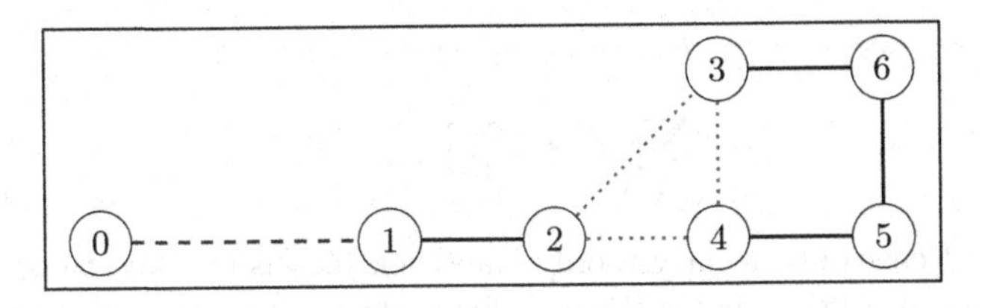

Fig. 2. DT-SEC with SECs and 2-matching-inequalities (Color figure online)

In this figure, the blue dashed line represents a drone edge (variable $y_{1,d}$ for some $d \in D$) with a value of 0.5, the black lines represent truck edges (variables x_e) with a value of 1 and the red dotted lines represent truck edges with a value of 0.5. Although no relaxed inequality constraint of the MILP model is violated, this structure is invalid for a feasible solution as customer 1 cannot be served half by a drone and half by a truck. Additionally, the nodes 3 to 6 form a subtour and the nodes 1 to 6 violate a 2-matching inequality.

Definition 1. *Let G be the graph that consist of all truck edges e with $x_e > 0$ and all drone edges $\{0, i\}$ with $y_{i,d} > 0$ for some $d \in D$. A* DT-SEC *is then a connected component of nodes where at least one node is incident with at least one truck and one drone edge.*

A quadratic node based inequality as in Theorem 1 can be used to detect such invalid structures, as all customer nodes in a feasible solution fulfill this inequality.

Theorem 1. *All nodes $i \in C''$ of a feasible solution fulfill the following inequality:*

$$\left(\sum_{e \in \delta(i)} x_e - 2 \sum_{d \in D} y_{i,d} \right)^2 \geq 4.$$

Proof. The validity of this inequality for optimal solutions can be proven by using the MILP formulation. Every customer $i \in C''$ is either served by a drone or the truck, as the edges must have a value of either 0 or 1 as described in (7) and (8).

Due to constraint (4) we get: If customer i is served by the truck the minuend equals 2 and the subtrahend is 0. Therefore, the left side of the inequality equals 4. If a customer is served by a drone the minuend equals 0 and the subtrahend equals -2, so the left side is also 4.

Corollary 1. *The quadratic DT-SEC inequality can be split into two disjoint linear cases.*

Case 1. Only one truck edge $e = \{i, j\}$is incident to node i:

$$x_e + \sum_{d \in D} y_{i,d} \leq 1.$$

Case 2. More than one truck edge is incident to node i:

$$\sum_{e \in \delta(i)} x_e + 2 \sum_{d \in D} y_{i,d} \leq 2.$$

In the first case of Corollary 1, only one truck edge is incident to node i. In the given example in Fig. 2, this is the case for node 1. This case can be interpreted that either the truck edge or the drone edge is part of a feasible solution, but not both. The second case can be seen as a more compact node degree condition from the MILP (4). Here, $\delta(i)$ contains only the truck edges e that have a value $x_e > 0$, while in (4) all truck edges of the complete graph that are incident with node i are part of the constraint. This inequality can never be violated as this would also violate (4).

DT-SECs can also include SECs and 2-matching inequalities as shown in Fig. 2. Instead of adding all these inequalities during the branch and cut algorithm, the DT-SEC can be used to forbid the complete structure. The DT-SEC specifies that either the drone edge $\{0, 1\}$ or the truck edge $\{1, 2\}$ can be part of a feasible solution. If one of these edges is removed, the rest of the structure can not occur in a solution of a relaxation, because then one of the nodes would violate (4).

Theorem 2. *Drones can be arranged to reduce the set of feasible solutions through the following inequality:*

$$\sum_{j \in C''} y_{j,d} \geq \sum_{j \in C''} y_{j,d+1} \quad \forall d \in \{1, \ldots, |D| - 1\}. \tag{9}$$

Proof. This theorem is proven by the problem description itself. Since the fleet of drones is homogenous, the time to deliver a set of customers only depends on their location, but not on which drone is used for this purpose. Two different situations can occur in a solution: All drones deliver the same count of parcels or the number varies. If the number varies, any drone can be selected to deliver parcels to the largest customer set. By this theorem, always the first drone will delivery the most parcels.

Arranging the drones as in Theorem 2 can drastically reduce the count of feasible solutions. Assume two sets A and B of customers, with $|A| > |B|$, are waiting for their parcels and two identical drones can be used. The drones can be assigned to either set A or set B. Therefore, two different solutions exist. As the assignment has no impact on the objective function value (total delivery time), both assignments lead to the same solution. By stating that drone 1 always delivers the most parcels, the number of solutions is halved in this example. For instances with more customers and more drones, the reduction can be even greater.

5 Separation Algorithms

A known strategy for the separation of SECs is to calculate all minimum cuts in a graph. A cutset of a cut with a capacity less than 2 violates a SEC inequality.

We use a Gomory-Hu tree [9] to reduce the total number of all cut calculations from $\frac{n(n-1)}{2}$ to $n-1$. Before calculating the Gomory-Hu tree, we apply the graph reduction of Padberg and Crowder [5] to reduce the size of the graph. To implement this algorithm, we use an union-find data structure with path compression.

Two different algorithms were implemented to compute the minimum cuts from the Gomory-Hu tree. Since in [15] the cut computation was determined to be a bottleneck, we would like to analyze the impact of the minimum cut calculation. The first algorithm we use is the exact preflow-push algorithm with the highest-label strategy (HLPP) [8]. This algorithm has a runtime of $O(n^2\sqrt{m})$ for a graph with n nodes and m edges [2]. Newer preflow-push algorithms like the variant from Henzinger, Rao, and Wang [12] with a theoretical runtime of $O(m\log(n)\log(\log(n)))$ are faster than the highest-label preflow-push algorithm, but these runtimes are in practice hard to achieve. In addition to this exact minimum cut algorithm, we choose the Karger-Stein algorithm as a heuristic algorithm for comparison [14]. This algorithm can be seen as a recursive variant of the graph reduction algorithm from Padberg and Crowder, as the graph is heuristically reduced to only two nodes, where the edges between them represent the minimum cut.

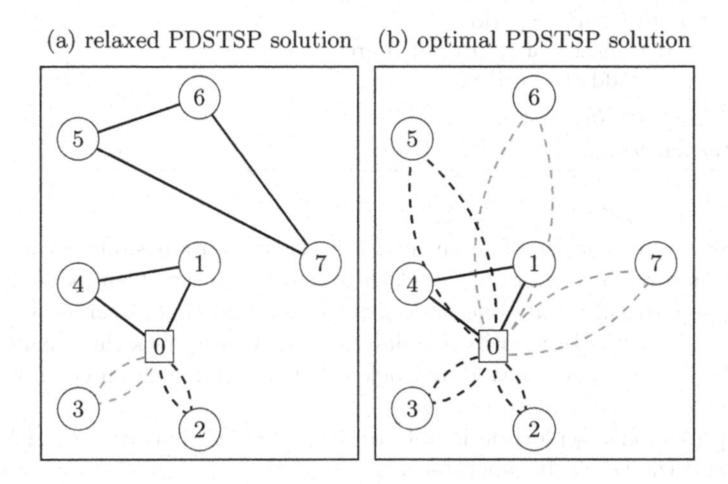

Fig. 3. SECs in a relaxed PDSTSP solution compared with the optimal solution (Color figure online)

In contrast to a TSP instance, not all cuts with a capacity less than 2 represent an invalid subtour. In Fig. 3(a) a solution of a relaxed PDSTSP instance is shown. A black edge represents a truck edge, a dotted arc is a drone edge, where each drone has its own colour. As the minimum cut between node 1 and 5 equals 0 in this example, the node sets $\{0, 1, 4\}$ and $\{5, 6, 7\}$ form a subtour. But as shown in Fig. 3(b), the truck route between the nodes $\{0, 1, 4\}$ is part of the optimal solution. Therefore, the route between the nodes $\{0, 1, 4\}$ cannot be considered invalid. Consequently, in order to find only invalid subtours, all DT-SEC structures and drone customers must be removed, before reducing the graph with the algorithm by Padberg and Crowder. All cutsets of a minimum cut from the calculated Gomory-Hu tree with a capacity less than 2 are also removed, if they include the depot. The remaining subsets form invalid subtours.

For the separation of 2-matching inequalities, we implemented the separation algorithm by Aráoz, Fernandez, and Meza [3]. This algorithm with a runtime of $O(|V|^4)$ is one of the fastest algorithms for the exact separation of 2-matching inequalities. In contrast to SECs, no modification of this algorithm is needed to be used on truck routes for the PDSTSP.

Pseudocode 1.1. Separation of DT-SECs

```
 1: function SEPARATE(Set of DroneEdges E_D(V, D), Set of TruckEdges E_T(V, V))
 2:     Let S_dtsec be an empty set
 3:     for all i ∈ V, i ≠ 0 do
 4:         Let S_d be an empty set of drone edges and S_t an empty set of truck edges
 5:         for all Drones d ∈ D do
 6:             if Capacity of e_{i,d} < 1 and e_{i,d} > 0 then
 7:                 Add e(i, d) to S_d
 8:         for all k ∈ V, k ≠ i do
 9:             if Capacity of e(i, k) > 0 then
10:                 Add e(i, k) to S_t
11:         Add (S_d, S_t) to S_dtsec
12:     Return S_dtsec
```

DT-SECs can consist of many nodes and can contain subtours and other invalid structures. However, to detect them and to construct an inequality that forbids this structure, only one node has to be found that is served by both a drone and the truck. To identify a violated DT-SEC, we check the capacity of all drone edges for a value above 0 and below 1. If we find such an edge, we check the node for any incident truck edge.

Our presented pseudocode is able to detect nodes that are served by both a drone and the truck. It generates a set consisting of pairs of drone and truck edges that are needed to formulate a DT-SEC inequality. This pseudocode has a runtime of $O(n^2)$ as it can be understood as a simple iteration over all drone and truck edges in the graph. It usually results in a sparse graph if we internally store only the truck and drone edges that have a positive value. This will allow us to further improve the runtime.

6 Computational Results

The presented separation approaches were analyzed on the Platform for Scientific Computing at Bonn-Rhein-Sieg University[3]. As hardware a Gigabyte R182-Z92

[3] https://wr0.wr.inf.h-brs.de/wr/.

server was used with a limit to 200 GB RAM and 32 threads with 2 GHz. The separation algorithms were programmed in Java in combination with the Gurobi solver version 9.1.1. The time limit for all calculations was set to 5 h.

The impact of each separation algorithm was analyzed by combining them in different variants. These different variants present the impact of each approach on the total runtime. A total of eight variants were constructed:

1. Separation of all SECs (HLPP) in integer solutions
2. Separation of every SEC (HLPP) and DT-SEC for integer and non-integer solutions, one violated 2-matching inequality is added if no violated SEC is found, the MILP is expanded with inequality (9) to arrange the drones
3. Separation of every SEC (Karger-Stein) and DT-SEC for integer and non-integer solutions, one violated 2-matching inequality is added if no violated SEC is found, the MILP is expanded with the inequality (9) to arrange the drones
4. Identical with variant 3 but without the inequality (9) to arrange the drones
5. Separation of every SEC (Karger-Stein) in integer solutions, separation of every DT-SEC, inequality (9) to arrange the drones
6. Identical with variant 3 but without the separation of DT-SECs
7. Separation of every SEC (Karger-Stein) for integer and non-integer solutions
8. Identical with variant 3 but without 2-matching inequalities

The first variant can be seen as the worst-case variant, as only SECs are used in integer solutions. That is the minimum of separation needed to find an optimal solution for the presented MILP.

For an analyzation, we modified seven different TSP instances to PDSTSP instances. For each TSP instances, six PDSTSP instances were created, three of them with 4 drones and the other with 8 drones. Each of the three instances have a different percentage of customers that can be served by a drone. For example, in `rd100_50` the customers 1 to 49 can be served by a drone.

A first overview of the impact of the implemented algorithms is given in Table 1. In this table, variant 1 is compared with the fastest variant for each problem instance. We see that the instances where 50% of the customers can be served by drones are the most complex, as no solution could be calculated with variant 1 in 5 h. Further, in all cases expect `hk48_25` with 8 drones variant 1 is the slowest variant. The instance `bier127_35` with 8 drones was solved with variant 2 in less than 1% of the time of variant 1.

Additionally, it can be stated that the PDSTSP instances become more complex with a higher number of customers that can be served by drones. It also becomes clear that there is no one variant that is best for all problem instances. The analysis of all runtimes showed that variants 2 and 3, although not always the fastest, were always close to the fastest for each problem instance. Thus, we can consider these variants as the ones with the best performance.

Table 1. Excerpt of runtime comparisons of different separation variants for the PDSTSP

Instance	4 drones		8 drones	
	Variant 1	Fastest variant	Variant 1	Fastest variant
hk48_25	19.62 s	1.93 s (Var. 4)	3.27 s	3.90 s (Var. 2)
hk48_35	2.39 s	0.69 s (Var. 3)	1.25 s	0.56 s (Var. 4)
hk48_50	>5 h	446.36 s (Var. 2)	>5 h	1181.02 s (Var. 2)
pr76_25	1210.45 s	29.76 s (Var. 5)	2399.78 s	32.81 s (Var. 4)
pr76_35	210.35 s	58.76 s (Var. 5)	319.16 s	25.49 s (Var. 5)
pr76_50	>5 h	916.66 s (Var. 2)	>5 h	1492.77 s (Var. 8)
gr96_25	7.06 s	2.04 s (Var. 4)	4.05 s	3.37 s (Var. 6)
gr96_35	2.95 s	1.97 s (Var. 2)	6.47 s	2.67 s (Var. 6)
gr96_50	>5 h	>5 h	>5 h	6740.66 s (Var. 2)
eil101_25	0.76 s	0.46 s (Var. 3)	1.04 s	1.04 s (Var. 1)
eil101_35	18.00 s	3.66 s (Var. 4)	53.24 s	12.80 s (Var. 3)
eil101_50	>5 h	560.33 s (Var. 8)	>5 h	648.50 s (Var. 4)
bier127_25	450.05 s	23.65 s (Var. 7)	98.29 s	35.65 s (Var. 3)
bier127_35	448.34 s	17.81 s (Var. 8)	2922.57 s	22.04 s (Var. 2)
bier127_50	>5 h	145.08 s (Var. 3)	>5 h	118.80 s (Var. 8)

The highest positive impact on the total runtime is achieved by the use of DT-SEC separation as shown in Table 2. As the number of DT-SECs increase with an increasing amount of customers that can be served by a drone, the separation of DT-SECs has a higher impact. With variant 6 the instance rd100_50 cannot be solved in under 5 h, whereas variant 3 needs less than one minute to find an optimal solution. For most other analyzed problem instances a similar reduction in runtime was observed, especially for instances with a larger amount of drone customers. However, in some cases with just a few seconds of runtime variant 6 was the fastest. This can be argued by the low amount of DT-SECs in non-integer solutions for these instances. This table further shows, by comparing variants 3 and 5, that the overall impact of 2-matching inequalities is neither positive nor negative.

The separation of SECs without the separation of DT-SECs is not an advisable strategy for PDSTSP instances with a high percentage of drone customers. In most cases, the runtime of variant 7 is a multiple of the runtime of variant 3. In 13 out of 42 instances, variant 7 fails to find a solution within the 5-hour time limit. In contrast, with variant 3 only one instance cannot be solved within the time limit. The difference in the runtime can be explained by the effectiveness of the DT-SECs. As mentioned in Sect. 4, a DT-SEC structure can include multiple SECs and 2-matching inequalities. Instead of adding all these inequalities to the MILP, only one short inequality consisting of only one node and its incident

Table 2. Impact of DT-SEC separation on the total runtime

Instance	Variant 3	Variant 5	Variant 6	Variant 7
hk48_25 (8 drones)	4.48 s	8.20 s	537.71 s	17.87 s
hk48_35 (8 drones)	0.91 s	1.91 s	1.34 s	1.08 s
hk48_50 (8 drones)	3066.65 s	3374.23 s	>5 h	>5 h
pr76_25 (4 drones)	55.05 s	29.76 s	3481.91 s	1260.61 s
pr76_35 (4 drones)	71.34 s	58.76 s	11500.34 s	14420.18 s
pr76_50 (4 drones)	1012.01 s	4496.30 s	>5 h	>5 h
gr96_25 (8 drones)	4.44 s	5.01 s	3.37 s	3.79 s
gr96_35 (8 drones)	2.86 s	2.78 s	2.67 s	6.33 s
gr96_50 (8 drones)	10773.79 s	6645.40 s	>5 h	>5 h
eil101_25 (8 drones)	1.15 s	1.46 s	1.46 s	1.63 s
eil101_35 (8 drones)	12.80 s	29.70 s	6904.91 s	2.72 s
eil101_50 (8 drones)	1331.42 s	1897.05 s	>5 h	>5 h
rd100_25 (4 drones)	72.36 s	7.46 s	11.70 s	15.08 s
rd100_35 (4 drones)	28.65 s	23.93 s	13.26 s	14.38 s
rd100_50 (4 drones)	53.80 s	188.77 s	>5 h	>5 h
bier127_25 (4 drones)	37.30 s	49.54 s	164.30 s	23.65 s
bier127_35 (4 drones)	32.59 s	85.53 s	41.56 s	55.11 s
bier127_50 (4 drones)	145.08 s	184.60 s	>5 h	>5 h

edges needs to be added to the MILP. For instances like hk48_35 and gr96_25, which only have a runtime of a few seconds, the difference in runtimes between the variants is also small. The number of DT-SEC structures during the branch and cut is small here and therefore has only a minor impact on the total runtime.

A comparison between variant 2 and variant 3 reveals the impact of the choosen cut algorithm, as in variant 2 the HLPP and in variant 3 the Karger-Stein algorithm is applied. The difference between these two algorithm is, that the Karger-Stein algorithm is capable of finding SEC structures faster than the HLPP, although the HLPP is supported by Gomory-Hu trees and by the graph reduction by Crowder and Padberg. However, the SECs found by the Karger-Stein algorithm contain in most cases more nodes than the SECs found by the HLPP. As the runtime of a branch and cut algorithm is influenced by the size of added inequalities, both algorithms have their advantages and disadvantages. In the analyzed instances, variant 3 with the Karger-Stein algorithm calculates the optimal solution faster than variant 4 for a majority of instances. Nevertheless, with variant 2 some instances can be solved in a fraction of the time needed with variant 3.

The inequalities (9) for arranging the drones have a positive impact on average over all analyzed instances. Although these inequalities enlarge the MILP model, these restrictions reduce the total runtime for the majority of analyzed instances.

Some of the PDSTSP instances like pr76_25 with 8 drones can be split up into a TSP and a PMS instance and solved separately. With the described separation of SECs and 2-matching inequalities the TSP for pr76_25 is solved in 8.29 s. The PMS is solved in 1.88 s by the Gurobi-Solver. In total the splitted pr76_25 instance is solved in around 10 s, whereas the PDSTSP instances has a runtime of around 82 s.

7 Conclusion

In this paper, the DT-SEC was presented as a new inequality for the PDSTSP. Previously published papers with the aim of solving this problem exactly by branch and cut applied only separation approaches from the TSP. DT-SECs are a special type of subtour where at least one node is served by both the truck and a drone. With an increasing number of customers that can be served by a drone, the total number of DT-SECs in a relaxed solution rises. DT-SECs can include multiple truck subtours and 2-matching inequalities. As the DT-SEC inequality only consists of incident edges from a node that is served by both the truck and a drone, large invalid structures can be forbidden efficiently.

The computational results show that only applying TSP separation approaches take a multiple of runtime compared to the addition of DT-SECs. An instance with 127 nodes, where 50% customers can be served by a drone, could not be solved by the Gurobi-Solver using only TSP separation approaches within 5 h. With the additional use of DT-SECs, the instance is solved in under 3 min. This shows that this problem specific separation approach is far more effective compared to modified TSP separation approaches.

Inspired by Dell'Amico, Montemanni, and Novellani [6] the MILP formulation from Murray and Chu for the PDSTSP was extended with an additional inequality for arranging the homogenous drones. The results show that the total runtime was reduced by around 30% on average by adding the inequality to the MILP.

Further research on additional problem specific inequalities for the PDSTSP or related problems is recommended. Other problems like the FSTSP could include similar invalid structures that can be handled by DT-SECs. Although the TSP can be interpreted as a PDSTSP where no drone can deliver a customer, TSP separation approaches should not be used exclusively for the PDSTSP as they only consider invalid structures in the truck-route without any drones.

References

1. Agatz, N., Bouman, P., Schmidt, M.: Optimization approaches for the traveling salesman problem with drone. Transp. Sci. **52**(4), 965–981 (2018). https://doi.org/10.1287/trsc.2017.0791

2. Ahuja, R., Magnanti, T., Orlin, J.: Network Flows: Theory, Algorithms, and Applications. Prentice-Hall, Englewood Cliffs (1993)
3. Aráoz, J., Fernández, E., Meza, O.: A simple exact separation algorithm for 2-matching inequalities (2007). http://www.optimization-online.org/DB_FILE/2007/11/1827.pdf
4. Baldacci, R., Dell'Amico, M., González, J.S.: The capacitated m -ring-star problem. Oper. Res. **55**(6), 1147–1162 (2007). https://doi.org/10.1287/opre.1070.0432
5. Crowder, H., Padberg, M.: Solving large-scale symmetric travelling salesman problems to optimality. Manage. Sci. **26**(5), 495–509 (1980). https://doi.org/10.1287/mnsc.26.5.495
6. Dell'Amico, M., Montemanni, R., Novellani, S.: Matheuristic algorithms for the parallel drone scheduling traveling salesman problem. Ann. Oper. Res. **289**(2), 211–226 (2020). https://doi.org/10.1007/s10479-020-03562-3
7. Deutsche Post DHL Group: Schnelle Hilfe aus der Luft: Medikamentenversorgung mit Paketdrohne in Ostafrika erfolgreich erprobt, 04 Oct 2018. https://www.dpdhl.com/content/dam/dpdhl/de/media-relations/press-releases/2018/pm-dhl-paketkopter-tansania-20181004.pdf mit Paketdrohne in Ostafrika erfolgreich erprobt, 04 Oct 2018. https://www.dpdhl.com/content/dam/dpdhl/de/media-relations/press-releases/2018/pm-dhl-paketkopter-tansania-20181004.pdf
8. Goldberg, A., Tarjan, R.: A new approach to the maximum-flow problem. J. ACM **35**(4), 921–940 (1988). https://doi.org/10.1145/48014.61051
9. Gomory, R.E., Hu, T.C.: Multi-terminal network flows. J. Soc. Ind. Appl. Math. **9**(4), 551–570 (1961). https://doi.org/10.1137/0109047
10. Hà, M., Vu, L., Vu, D.: The two-echelon routing problem with truck and drones (2020)
11. Ham, A.: Integrated scheduling of m-truck, m-drone, and m-depot constrained by time-window, drop-pickup, and m-visit using constraint programming. Transp. Res. Part C Emerg. Technol. **91**, 1–14 (2018). https://doi.org/10.1016/j.trc.2018.03.025
12. Henzinger, M., Rao, S., Wang, D.: Local flow partitioning for faster edge connectivity (2017). https://doi.org/10.1137/1.9781611974782.125
13. Karak, A., Abdelghany, K.: The hybrid vehicle-drone routing problem for pick-up and delivery services. Transp. Res. Part C Emerg. Technol. **102**, 427–449 (2019). https://doi.org/10.1016/j.trc.2019.03.021
14. Karger, D., Stein, C.: A new approach to the minimum cut problem. J. ACM **43**(4), 601–640 (1996). https://doi.org/10.1145/234533.234534
15. Klein, T.: Konzeption und Realisierung von Separationsalgorithmen für das Traveling Salesman Problem. Master's project, Hochschule Bonn-Rhein-Sieg (2019)
16. Klein, T.: Konzeption und Implementierung von Separationsverfahren für das Parallel-Drone-Scheduling-Problem. Master's thesis, Hochschule Bonn-Rhein-Sieg (2021)
17. Klohn, H.: Optimierung von TSP-Varianten mit Drohnen durch Branch and Cut Verfahren. Master's thesis, Hochschule Bonn-Rhein-Sieg (2019)
18. Labbé, M., Laporte, G., Martín, I., González, J.: The ring star problem: polyhedral analysis and exact algorithm. Networks **43**(3), 177–189 (2004). https://doi.org/10.1002/net.10114
19. Mbiadou Saleu, R., Deroussi, L., Feillet, D., Grangeon, N., Quilliot, A.: An iterative two-step heuristic for the parallel drone scheduling traveling salesman problem. Networks **72**(4), 459–474 (2018). https://doi.org/10.1002/net.21846

20. Murray, C., Chu, A.: The flying sidekick traveling salesman problem: optimization of drone-assisted parcel delivery. Transp. Res. Part C Emerg. Technol. **54**, 86–109 (2015). https://doi.org/10.1016/j.trc.2015.03.005
21. Reinelt, G.: Tsplib (2021). http://comopt.ifi.uni-heidelberg.de/software/TSPLIB95/
22. van Dijck, E.: A branch-and-cut algorithm for the traveling salesman problem with drone. Master's thesis, Erasmus University Rotterdam (2018). https://thesis.eur.nl/pub/44107/Dijck-van.pdf

A Multi-start VNS Algorithm for the TSP-D with Energy Constraints

Giovanni Campuzano([✉]), Eduardo Lalla-Ruiz, and Martijn Mes

University of Twente, Drienerlolaan 5, 7500 AE Enschede, The Netherlands
{g.f.campuzanoarroyo,e.a.lalla,m.r.k.mes}@utwente.nl

Abstract. The Traveling Salesman Problem (TSP) is a well-known optimization problem with a wide range of extensions and applications in delivery systems. In this paper, we consider a recent extension of the TSP where a truck in collaboration with a single drone should visit a set of customers while minimizing the transportation times. We propose a Variable Neighbourhood Search (VNS) and a Multi-Start VNS (MS-VNS) algorithm, develop new neighbourhood structures, and compare the solutions against an existing mixed-integer linear programming (MILP) formulation. We take a set of instances based on existing benchmarks from the related literature. Results point out that the new neighbourhood structures substantially improve the performance of the VNS algorithms. Furthermore, results also show that the exact method is only able to find competitive solutions for small sets of instances, whereas our MS-VNS approach reaches better solution quality for large instances.

Keywords: Traveling salesman problem · Drones · UAV · Last-mile delivery · Multi start · Variable neighbourhood search

1 Introduction

Logistics plays an important role in today's economy. It controls the forward and reverse flows of goods from producers to consumers. In this context, the logistic sector is currently facing an era of unprecedented change with developments in digitization, autonomous vehicles, urbanization, increasing customers demands, and the rise of e-commerce. These changes have led companies to search for more efficient and sustainable management of urban freight distribution, in order to improve their service's quality, diminish greenhouse gas emissions, and reduce operational costs. Specifically in last-mile delivery, companies are seeing many opportunities to improve their business by incorporating autonomous vehicles into their logistic operations, giving rise to a wide range of new optimization problems in parcel distribution.

In this context, the Traveling Salesman Problem with Drone (TSP-D) is a recent extension of the TSP, in which a truck should work in collaboration with a single drone to serve a predefined group of customers while minimizing the transportation times or makespan. The TSP-D has shown distinct advantages

M. Mes et al. (Eds.): ICCL 2021, LNCS 13004, pp. 393–409, 2021.
https://doi.org/10.1007/978-3-030-87672-2_26

in last-mile operations, due to the drone's ability to transport parcels, food, medicine, and several other goods [8]. These autonomous vehicles have gained the attention of international companies, being involved in promising projects such as Zookal's project, Wing's project, and Parcelcopter, among others [20]. Additionally, drones have also shown strong applicability in other fields, such as energy, agriculture, forestry, environmental protection, and emergency [4].

The research community has also paid close attention to truck-and-drone problems. For example, [20] introduce the Flying Sidekick Traveling Salesman Problem (FSTSP) and the Parallel Drone Scheduling TSP (PDSTSP). They develop mathematical formulations and a two-phased heuristic approach to solve these problems. They show that the mathematical formulations are able to find efficient solutions for small-size instances, while the heuristic provides better solutions for large-size instances. [1] introduce the TSP-D and face this problem by developing a mathematical formulation and a heuristic approach. They also provide insights into the savings from the incorporation of a drone into the delivery system, showing the advantages of the truck-and-drone approach increase with faster drone speeds. [25] develop a compact formulation and several extensions for the TSP-D. They solve these problems by implementing a Branch-and-Price scheme, which is able to find solutions for instances of up to 39 customers. Branch-and-Cut procedures have been proposed in [6,9,26], and [2]. A Bender's Decomposition algorithm is developed in [28], reaching solutions for instances of up to 20 customers. In [21], authors study an extension of the FSTSP called the Multiple Flying Sidekicks Traveling Salesman Problem (mFSTSP). They develop a mathematical formulation and a three-phased heuristic, demonstrating that by incorporating more drones into the delivery system, it is possible to reduce the makespan for instances involving a large number of customers. The same authors study another extension, namely the mFSTSP with Variable Drone Speeds (FSTSP-VDS; [24]) in which the drone is able to fly at slower speed levels. They solve this problem by adapting the heuristic from [21]. They demonstrate that by allowing the drone to fly at slower speed levels, it is able to visit more customers due to a reduction in energy consumption. To provide a realistic representation of the delivery system, most of the works on truck-and-drone problems limit the drone flying range by using an energy consumption function, a maximum flying time, or a maximum flying distance ([6,11,12,16,19,27]). The reader is referred to [29], for a comparison of several energy consumption models used in truck-and-drone related problems.

In this paper, we propose two meta-heuristic approaches: a Variable Neighbourhood Search (VNS) and a Multi-Start Variable Neighbourhood Search (MS-VNS), which use 11 neighbourhood structures to explore the solution space. Those structures are a combination of neighbourhoods from the literature with some of our own. We take existing benchmarks from the literature to conduct the experiments and compare the results with the MILP formulation proposed in [25]. Furthermore, we provide insights into the heuristic performance improvements when incorporating the two new neighbourhood structures. Accordingly, we adapt to our algorithms the neighbourhood structures of the hybrid general VNS algorithm proposed in [13] and analyse the results. In addition, we study these algorithms for the case where the drone flying range is infinite and when the drone has a limited energy endurance. Consequently, we show that the effectiveness of the MS-VNS is not only suitable for a theoretical version of the TSP-D, but also for a more realistic representation of the problem.

The remainder of this paper is organized as follows. Section 2 provides a detailed description of the TSP-D. In Sect. 3, the VNS and MS-VNS heuristics are presented. The experiments are conducted and analyzed in Sect. 4. Finally, Sect. 5 provides the main conclusions and suggests several directions for further research.

2 The Traveling Salesman Problem with Drone

The TSP-D is a problem that has gained increasing attention in last-mile delivery, due to the advantages that the incorporation of a drone provides. In this sense, the drone flies faster than the truck is able to drive and it avoids traffic congestion, however, its flying range is particularly restricted due to the battery endurance. Especially in urban areas, last-mile delivery faces crowded places with a high customers density, stimulating the collaboration between a truck and a drone. Figure 1 provides an illustrative example of a delivery schedule carried out by a truck and a drone. Here, solid arcs represent the truck arcs, dashed arcs represent the drone arcs, and double-solid arcs represent those that are traversed by the truck with the drone on board. Hence, in the example, there are customers that are served only by the truck (e.g., $\{8, 1\}$), customers that are visited by the drone (e.g., $\{4, 6\}$), and customers that are served by the truck with the drone on board (e.g., $\{3, 5, 7, 2\}$). We refer to those customers (or nodes) as truck nodes, drone nodes, and combined nodes, respectively.

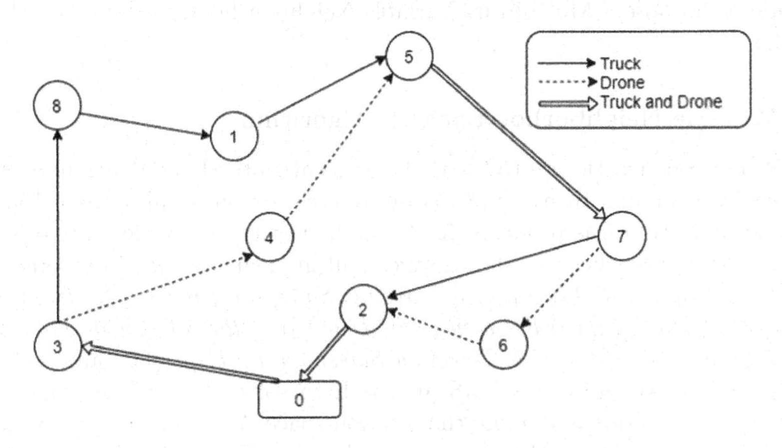

Fig. 1. Description of a feasible TSP-D route.

In real implementation scenarios, the TSP-D is a problem characterized by the restricted drone flying range. Therefore, when the drone is serving a given customer, the energy consumed by the trip cannot exceed the energy capacity constraints. For this reason, it is important to properly coordinate the launching and landing times of the drone. Thus, we assume that every time the drone meets the truck in a combined node, the drone swaps the battery and loads a new package on board. Furthermore, to avoid feasibility issues, the vehicle that arrives first at a combined node should wait for the other vehicle before leaving to serve a new customer. To face the TSP-D described in [25], the following assumptions are made:

1. The truck capacity is large enough to carry all the packages and the drone batteries for the route.
2. The drone can only carry one package on board. For this reason, every time the drone is launched to serve a given customer, the drone should fly back after delivering the package, to meet the truck at a combined node.
3. The drone cannot take off and land in the same location. Therefore, once the drone is launched from a combined node, the truck cannot wait for the drone in the same place.
4. The drone has a fixed battery capacity and the drone's trips cannot exceed its battery's endurance. Furthermore, every time the drone meets the truck, it is assumed that the drone either is kept on the truck or swaps the battery and loads a new package on board.

3 Metaheuristic Algorithms

This section introduces two metaheuristic approaches to solve the TSP-D as presented in [25]. We first present two new neighbourhood structures and develop a Variable Neighbourhood Search algorithm in Sect. 3.1. Then, we extend the VNS heuristic to propose a Multi-Start Variable Neighbourhood Search algorithm in Sect. 3.2.

3.1 Variable Neighborhood Search Algorithm

VNS is a metaheuristic approach that systematically changes neighbourhood structures within a local search procedure to escape from local optima [15]. The first VNS algorithm was proposed in [14] and, ever since, a wide body of literature on VNS approaches has shown successful implementations in several routing problems, e.g., *the Asymmetric Traveling Salesman Problem* [3], *the Vehicle Routing Problem* [5], *the VRP with Multi-Depot* [17], *the VRP with Time Windows* [7], *the Flying Sidekick Traveling Salesman Problem* [11], and so forth. In this section, we present a VNS approach to solve the TSP-D, which uses eleven neighbourhoods to explore the solution space. We combine several neighbourhood structures from the literature with some of our own. The neighbourhoods *Make flying, Push left and Push right* are proposed in [1]. The neighbourhoods *Two optimality, Exchange 1.1, Exchange 2.1, Exchange 2.2, Reinsertion and Relocate customer* are proposed in [11]. The structures *Exchange 3.1 and Exchange 3.2* are developed in this research and are defined as follows:

- *Exchange 3.1*: This neighbourhood exchanges one node that is visited by the drone with another node that is visited by the truck.
- *Exchange 3.2*: This operator swaps two nodes that are visited by the drone.

A description for the VNS scheme developed in this research is provided in Algorithm 1. The input data of Algorithm 1 are the number of *iterations*, the number of shaking applications (*stopping*), the drone time matrix t^D, and the truck time matrix t^T. The output is the best TSP-D solution found in the search space S^* (incumbent). The objective function value of each solution is given by $f(\cdot)$, S_i stores the solution found when exploring the search space, and $S_i^{'}$ is used as a transition variable that stores the solution from the previous iteration. The index i represents the current iteration, l identifies the neighbourhood $\mathcal{N}_l(\cdot)$, p provides the number of shaking procedures applied during the search, and l_m gives the number of neighbourhoods.

Algorithm 1 consists of an *initialization phase* in which an initial TSP-D solution is built, and an *iterative phase* in which the solution space of the TSP-D is explored. An initial TSP solution is built in $SolveTSP(\cdot)$ by using the truck's time matrix t^T and applying a nearest neighbour search that starts from the depot (line 2). The TSP route is stored in S. Next, $MakeFlying(\cdot)$ builds a TSP-D solution by using the drone time matrix t^D and TSP route (line 3). This operation selects three consecutive nodes and builds a TSP-D solution in such a way that the truck visits the first and last node of the operation and the remaining node is visited by the drone. Therefore, given three selected nodes, the $MakeFlying$ movement creates a TSP-D operation leaving as a drone node the one that minimizes the costs the most. On the other hand, if the TSP-D operation does not reduce the transportation costs, the three selected nodes remain as a TSP section of the route within the TSP-D solution, i.e., visited by the truck and drone at the same time.

The iterative phase begins once an initial TSP-D solution has been constructed (line 6). This phase is repeated until either i or p reaches the maximum number of iterations or the maximum number of shaking procedures, respectively (line 7). The set of neighbourhoods is explored within the second while-loop until the current solution S_i is improved or all the neighbourhoods in $\mathcal{N}_l$ are checked (lines 9–11). To explore the solution space, the VNS algorithm applies the set of neighbourhoods $\mathcal{N}_l$ starting from the structure with the smallest search space to the neighbourhood with the largest one. Further, a given neighbourhood $\mathcal{N}_l$ is explored until its best solution is found. Once the algorithm finishes exploring the set of neighbourhoods, if S_i is better than $S_i^{'}$ and in addition the current solution is better than the incumbent ($S_i < S^*$), the incumbent is updated and p is set to zero (lines 12–15). Additionally, every time S_i is better than the previous solution $S_i^{'}$, l is set to 0 (lines 16). Otherwise, if the current solution S_i is worse than the previous solution $S_i^{'}$, it means the algorithm has found a local optimum, so the shaking procedure is applied and the solution is stored in S_i (lines 17–19). Finally, S_{i+1} and $S_{i+1}^{'}$ store the solution from the current iteration S_i (line 20).

Algorithm 1: Variable Neighbourhood Search

```
   Data: (iterations, stopping, t^D, t^T)
   Result: (S*)
1  Initialization phase:
2  S ← SolveTSP(t^T)
3  S_i, S'_i, S* ← MakeFlying(S, t^D)
4  f(S*) ← +∞
5  i, p ← 0
6  Iterative phase:
7  while i < iterations || p < stopping do
8      l ← 0
9      while f(S_i) ≥ f(S'_i) and l ≤ l_m do
10         S_i ← N_l(S_i)
11         l ← l + 1
12     if f(S_i) < f(S'_i) then
13         if f(S_i) < f(S*) then
14             S* ← S_i
15             p ← 0
16         l ← 0
17     else
18         S_i ← Shaking(S_i)
19         p ← p + 1
20         if p then
21             =
22         =
23     S'_{i+1}, S_{i+1} ← S_i
24     i ← i + 1
```

In a TSP-D solution, every time the drone visits a customer, the part of the route between the combined take-off and landing nodes is referred to as an *Operation.* In this context, the shaking procedure consists of firstly taking apart two operations to let the truck visit the customers and then applying three nested neighbourhoods, which are selected randomly. When applying a given random neighbourhood, the solution selected from this neighbourhood is chosen randomly without considering the objective function value.

3.2 Multi-Start Variable Neighborhood Search Algorithm

The VNS approaches have been broadly accepted by the research community and, over the years, several VNS schemes with additional mechanisms have been developed to improve their performances, for instance, *Descent Search*, *Reduced VNS*, *Basic VNS*, *General VNS*, *Skewed VNS*, *Decomposition Search*, *Parallel*, and *Primal-Dual*, among others [15]. In addition, multi-start mechanisms are applied to add diversification to the heuristic optimization process by re-starting the search once a certain criterion is met [23]. Consequently, with the purpose of exploring deeply promising areas of the search space, we incorporate a multi-start mechanism into the VNS heuristic of Sect. 3.1. As such, we develop an effective multi-start VNS which, after a certain number of iterations without improving the solution, re-starts the search from those solutions that were chosen as *incumbents* throughout the optimization process. As a result, we re-start the exploration from previous local optima, which are adjusted by a *Shaking* procedure, to explore new regions of promising areas already identified. A description of the MS-VNS heuristic is provided in Algorithm 2. The input data of Algorithm 2 is the same as Algorithm 1 plus the number of consecutive shakings until a re-starting procedure is applied (*Restarting*). The output is the best TSP-D solution found in the search space S^*. The counter i represents the iteration number, p shows the number of re-starting procedures applied, and h displays the number of shakings.

As mentioned, Algorithm 2 is an adaptation of Algorithm 1, which incorporates a multi-start method. Note that different from the VNS, the MS-VNS randomizes the order in which the neighbourhoods $\mathcal{N}_l$ are explored in every iteration i (line 9). Therefore, in order to re-start the search, the MS-VNS uses *list()*, which is a list that stores each new incumbent found in the iterative phase (line 16). Every time the incumbent is updated, the number of shaking procedures h is set at zero (line 17). On the other hand, if after exploring the set of neighbourhoods $\mathcal{N}_l$ the solution from the previous iteration is not improved, one shaking procedure is applied and the counter h is increased by one (lines 19–21). Additionally, when h reaches the number of consecutive iterations without improving S_i (*Restarting*), the re-starting mechanism is activated (line 22). Consequently, S_i randomly takes one of the solutions stored in $list()$ and a shaking procedure is applied to this new solution in S_i (lines 23–24). Then, the counter of shakings h is set at zero and the counter of re-starting procedures p is increased by one (lines 25–26). Furthermore, it is worth mentioning that the MS-VNS heuristic also adjusts the *Shaking* procedure. The shaking procedure consists of firstly taking apart seven operations, if possible, and then three nested neighbourhoods, which are selected randomly, are applied. After that, the shaking uses the $MakeFlying()$ neighbourhood structure to create a new operation. Finally, similarly as done in the VNS heuristic, the solutions selected from the neighbourhoods in the shaking procedure are randomly chosen.

4 Numerical Experiments

In this section, we provide the computational experiments on the TSP-D. A detailed description of the computational settings, instances, heuristic tuning, and MILP model parameters are presented in Sect. 4.1. Computational results are presented in Sect. 4.2.

Algorithm 2: Multi-Start Variable Neighbourhood Search

```
   Data: (iterations, stopping, t^D, t^T)
   Result: (S*)
 1 Initialization phase:
 2 S ← SolveTSP(t^T)
 3 S_i, S'_i, S* ← MakeFlying(S, t^D)
 4 f(S*) ← +∞
 5 i, p ← 0
 6 Iterative phase:
 7 while i < iterations || p < stopping do
 8     l ← 0
 9     N_l() ← randomize()
10     while f(S_i) ≥ f(S'_i) and l ≤ l_m do
11         S_i ← N_l(S_i)
12         l ← l + 1
13     if f(S_i) < f(S'_i) then
14         if f(S_i) < f(S*) then
15             S* ← S_i
16             list(S*)
17             h ← 0
18         l ← 0
19     else
20         S_i ← Shaking(S_i)
21         h ← h + 1
22         if h == Restarting then
23             S_i ← randomize(list())
24             S_i ← Shaking(S_i)
25             h = 0
26             p ← p + 1
27     S'_{i+1}, S_{i+1} ← S_i
28     i ← i + 1
```

4.1 Experimental Settings

This section describes the parameters used by the heuristics from Sect. 3 as well as the computational settings. To conduct the experiments, we take the *single center* set of benchmark instances from [1], and execute the test on a computer equipped with a 2.29 GHz Intel(R) Xeon(R) Gold 5218 CPU with 64 GB of RAM. The algorithm was coded in C++ and solved using CPLEX 12.10. The truck and drone speeds considered are 7 and 15 [m/s], respectively.

Given two locations i and j, and their corresponding $x-y$ coordinates (x_i, y_i), the time it takes the drone to go from i to j is computed as $t^D_{ij} = t^{tk}_i + t^h_{ij} + t^l_j$, where t^h_{ij} is the horizontal flying time, t^{tk}_i the takeoff time, and t^l_{ij} the landing time. t^{tk}_i and t^l_i are constant parameter of values 60 an 30 [s], respectively. The horizontal flying time is computed as $t_{ij}^h = \frac{\sqrt{(x_j-x_i)^2+(y_j-y_i)^2}}{v_{ij}^D} \forall(i,j) \in A$ (i.e., considering Euclidean distance). Likewise, when the truck is traversing the arc $(i,j) \in A$, at a given speed v_{ij}^T, the time it takes the truck to go from i to j is computed as $t_{ij}^T = \frac{\lfloor |x_j-x_i|+|y_j-y_i| \rfloor}{v_{ij}^T} \forall(i,j) \in A$ (i.e., considering Manhattan distance). We expand the $x-y$ coordinates by 100 to compute the transportation times.

We base the calculations of the energy consumption of the drone on the consumption model of [18], which depends on the selected drone speed, taking off speed, landing speed, and payload. The consumption model is described by Eqs. (1)–(3). Further, [18] propose an energy model $f(\cdot) = p^{tl} + p^c + p^h$, where p^{tl} is the induced power of the vertical flow (either take off or landing), p^c represents the profile power consumption during the horizontal cruise, and p^h is the energy consumed during hover. In this regard, when the drone is traversing the arc $(i,j) \in A$, transporting a load equal to weight w_j of customer $j \in N$, at a vertical drone speed (v_{ve}), and at a selected drone speed level v_{ij}^D, the time t_{ij}^D to traverse the arc (i,j) is given by: $t_{ij}^D = t_i^{tk} + t_{ij}^h + t_j^l$. Consequently, by making use of these different flying times of the drone, the battery consumption is set by Eqs. (1)–(2) as: $b_{ij}(w_j, v_{ij}^D) = t_i^{tk} \cdot p_i^{tl}(w_j, v_{ve}) + t_{ij}^h \cdot p_{ij}^c(w_j, v_{ij}^D) + t_j^l \cdot p_j^{tl}(w_j, v_{ve})$. Note, when traveling to meet the truck without a parcel on board, the same Eqs. (1)–(3) are used to compute b_{ij} with the payload w_j equal to 0.

$$p_{ij}^{tl} = k_1(W + w_j)g \left[\frac{v_{ve}}{2} \sqrt{\left(\frac{v_{ve}}{2}\right)^2 + \frac{(W + w_j)g}{k_2^2}} \right] + c_2\left((W + w_j)g\right)^{3/2} \qquad (i,j) \in A \quad (1)$$

$$p_{ij}^c = (c_1 + c_2)\left(\left((W + w_j)g - c_5(v_{ij}^D \cos\theta)^2\right)^2 + \left(c_4(v_{ij}^D)^2\right)^2\right)^{3/4} + c_4(v_{ij}^D)^3 \quad (i,j) \in A \quad (2)$$

$$p_{ij}^h = (c_1 + c_2)\left((W + w_j)g\right)^{3/2} \qquad (i,j) \in A \quad (3)$$

In Eqs. (1)–(3), c_1, c_2, c_4 c_5, k_1, k_2, W, θ, g, and v_{ve} are model coefficients, whose values are derived from [18] and provided in Table 1. This way, c_1, c_2, c_4 c_5, k_1 and k_2 are model constant, W represents the drone frame weight, θ is the angle of attack, i.e., the vertical angle with which the drone faces the wind, and g is the gravitational constant. Note that the model of [18] assumes a fixed angle of attack for a certain range of payloads. Depending on the type of drone considered, other models could be applicable here.

Table 1. Coefficient values for the energy consumption models presented in [24] and [18].

Coefficient:	k_1	k_2	c_1	c_2	c_4	c_5	W	g	θ	$V_{ve}(takeoff)$	$V_{ve}(landing)$
Value:	0.8554	0.3051	2.8037	0.3177	0.0296	0.0279	1.5	9.8	10	5	10
Units:	$[unitless]$	$\sqrt{kg/m}$	$\sqrt{m/kg}$	$\sqrt{m/kg}$	kg/m	Ns/m	kg	m/s^2	Degrees	m/s	m/s

To properly select the parameter values of the algorithms presented in Sect. 3, the Friedman non-parametric statistical test is applied over the performance of the heuristic approaches [22]. Table 2 shows the parameters assessed in the Friedman's test to state the most suitable settings, where *iterations* represents the number of iterations, *stopping* establish the maximum number of shaking procedures allowed for the VNS and the maximum number of re-starting procedures allowed for the MS-VNS, and *Restarting* is the maximum number of shaking procedures before the re-starting mechanism is activated for the MS-VNS. In this regard, the multiple parameters Friedman's test is performed in those cases in which the null hypothesis is rejected. A total of 5 representative instances for the combination of the parameters from Table 2 are solved by the VNS and MS-VNS algorithms. A significance level of $\alpha_{friedman} = 0.05$ is used for the objective functions to indicate a significant difference among the parameter assessed. After applying the Friedman non-parametric statistical test, we decided to use the parameter settings $Iterations = 9000$ and $Stopping = 70$ for the VNS and $Iterations = 21000$, $Stopping = 200$, and $Restarting = 90$ for the MS-VNS.

Table 2. Parameter values used to configure heuristics VNS and MS-VNS.

Parameter	VNS	MS-VNS
Iterations	{11000, 9000, 7000}	{42000, 21000, 7000}
Stopping	{90, 70, 50}	{4000, 300, 200}
Restarting	–	{150, 90, 30}

4.2 Computational Results

In this section, we conduct three sets of experiments with the purpose of measuring the effectiveness of our algorithms. The first set of experiments provides a comparison of the heuristic approaches with infinite energy capacity in the drone's battery. More precisely, we compare our VNS and MS-VNS algorithms using both the original 9 neighbourhoods (as used in the hybrid general VNS proposed in [10]) and the 11 neighbourhoods that include the 2 neighbourhoods proposed in this work. We denote the neighbourhood structures using 9 or 11 as subscript. The second set of experiments presents the results of the exact formulation and the metaheuristic from the first set of experiments with the best performance, for the TSP-D with infinite energy capacity. The third set shows the results of the TSP-D where the drone flying range is restricted by a limited energy capacity, using the same methods as used in the second set of experiments.

The tables presented below report the results of the optimization model and the metaheuristics VNS and MS-VNS, for instances of up to 75 nodes. The tables show the best solution found of 10 runs for each heuristic. The experiments are carried out with a time limit of 3600 (s) for the MILP model. The columns show the objective function value (Z), initial constructive TSP-D solution that the metaheuristic algorithms use to explore the solution space (Z_{init}), Gap (%), average computation time in seconds of the experiments for each instance, a percentage difference of the heuristic objective function Z compared to the MILP formulation as: $\Delta Z(\%) = 100 \cdot \frac{Z_{MILP}-Z}{Z_{MILP}}$, and a percentage difference of the heuristic objective function compared to the initial constructive TSP-D solution as: $\Delta Z_{TSP-D}(\%) = 100 \cdot \frac{Z_{TSP-D}-Z}{Z_{TSP-D}}$.

Assessment of the Neighbourhood Structures for VNS Algorithms. Table 3 reports the results of the two new neighbourhood structures when studying the performance of the VNS and MS-VNS algorithms for the TSP-D with unlimited battery capacity. Best values are bold-faced. In addition, Fig. 2 provides a summary of the results, where for every heuristic algorithm the set of instances is averaged by size n.

Table 3. Comparison of the VNS and MS-VNS algorithms for the TSP-D.

Instance		VNS_9		VNS_{11}		MS-VNS_9		MS-VNS_{11}	
Name	n	Z	Time (s)	Z	Time (s)	Z	Time (s)	Z	Time (s)
51	10	4695	0.02	4118	0.03	3793	1.95	**3625**	2.12
52		6480	0.03	**5280**	0.02	**5280**	2.02	**5280**	1.94
53		4388	0.04	5629	0.03	8022	3.23	**3608**	1.73
54		6485	0.03	3883	0.03	4759	1.74	**3833**	1.54
55		6429	0.03	5023	0.02	4897	2.23	**4787**	2.09
61	20	7586	0.07	6345	0.12	6232	8.16	**4884**	7.57
62		7840	0.12	5087	0.07	5859	7.20	**4500**	6.98
63		11832	0.06	7603	0.08	6072	6.99	**5818**	7.09
64		7079	0.05	5112	0.07	5859	6.87	**4343**	6.24
65		7767	0.07	5592	0.06	4715	6.35	**4707**	6.41
71	50	9828	0.86	8344	1.66	7660	83.50	**7513**	88.25
72		12810	1.40	9799	1.12	8569	76.67	**8402**	83.21
73		8777	1.20	7271	1.87	6566	82.81	**6521**	87.66
74		13794	0.68	10502	1.15	9699	79.04	**9183**	83.41
75		14331	1.38	10430	1.43	12263	78.90	**9349**	82.95
81	75	19308	4.06	14226	12.32	14318	315.94	**12704**	323.11
82		16282	5.67	11776	10.87	10660	310.28	**10134**	337.81
83		16449	5.95	13050	7.20	12878	334.56	**11646**	335.94
84		18089	3.67	14068	6.89	12666	309.05	**12434**	333.00
85		15070	3.92	12413	12.05	12022	325.37	**10682**	336.90
Avg.		10765.95	1.47	8277.55	2.85	8139.45	102.14	7197.65	106.80

Results highlight the effectiveness of the multi-start mechanism in the performance of heuristic MS-VNS_{11}, although obviously at the expense of increasing computational time. As such, we can see that MS-VNS_{11} is the algorithm that finds the best solution for all the instances studied, where only for instance 52 the heuristics VNS_{11} and MS-VNS_9 were able to find the same result as MS-VNS_{11}.

Moreover, results also point out the effect of the two new neighbourhoods, which help the heuristics algorithms VNS and MS-VNS perform a better exploration of the solution space.

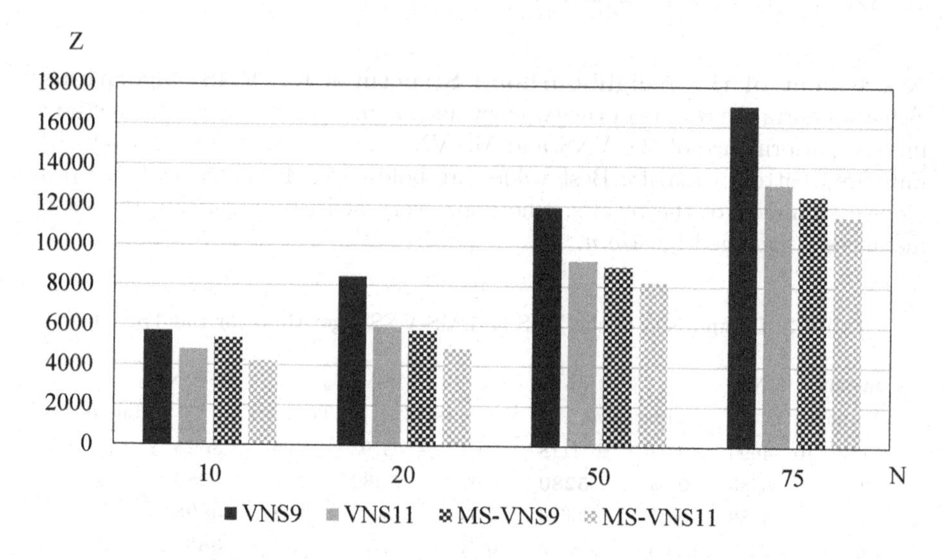

Fig. 2. Comparison of VNS_9, VNS_{11}, MS-VNS_9, and MS-VNS_{11} for unlimited drone flying range.

Assessment of Optimization Approaches for Unlimited Drone Flying Range. Table 4 reports the results of the MILP formulation and the MS-VNS_{11} algorithm for the TSP-D when considering unlimited battery capacity. Best values are bold-faced.

Results point out that the TSP-D formulation is able to find good quality solutions for the small instance set (10–20 nodes) in computational times under 50 s. The formulation reaches optimality for all the instances of 10 nodes and gaps under 17% for the instances of 20 nodes. With respect to the MS-VNS_{11} algorithm, the MS-VNS_{11} is able to reach optimal solutions for all the instances of 10 nodes and improve the performance of the MILP formulation for instances 61 and 64 by 2.3% and 0.84%, respectively. Further, the MILP formulation present better solutions than the MS-VNS_{11} for instances 62, 63, and 65 by 1.99%, 0.14%, and 0.84%, respectively. Additionally, the MS-VNS_{11} improves the initial constructive TSP-D solution that the algorithm uses to start the iterative search by at least 21%.

When conducting the experiments on the large set of instances, we see that the optimization model finds feasible solutions with gaps over 74% for all the instances. Nevertheless, MS-VNS_{11} presents better results by at least 64% for all the instances of the large set and improves the constructive initial solution by at least 35%. Furthermore, the initial constructive TSP-D solution shows better solutions than the final value of the MILP model for all the instances. This

Table 4. Comparison of the MILP formulation and the MS-VNS$_{11}$ algorithm for the TSP-D with unrestricted drone flying range.

Instance		MILP			Constructive	MS-VNS$_{11}$			
Name	n	Z	Gap (%)	Time (s)	Z_{TSP-D}	Z	Time (s)	ΔZ_{MILP} (%)	ΔZ_{TSP-D} (%)
51	10	**3625**	0.00	25.45	8653	**3625**	2.12	0.00	58.11
52		**5280**	0.00	17.80	6754	**5280**	1.94	0.00	21.82
53		**3608**	0.00	20.38	5901	**3608**	1.73	0.00	38.86
54		**3833**	0.00	18.19	6988	**3833**	1.54	0.00	45.15
55		**4787**	0.00	49.00	6913	**4787**	2.09	0.00	30.75
61	20	4999	17.12	3620.92	8504	**4884**	7.57	2.30	42.57
62		**4412**	3.38	3600.59	9603	4500	6.98	−1.99	53.14
63		**5810**	8.58	3632.08	9479	5818	7.09	−0.14	38.62
64		4380	9.34	3631.00	7819	**4343**	6.24	0.84	44.46
65		**4668**	4.20	3616.73	8913	4707	6.41	−0.84	47.19
71	50	33597	84.24	3613.98	12370	**7513**	88.25	77.64	39.26
72		57116	89.64	3656.48	17595	**8402**	83.21	85.29	52.25
73		18608	74.83	3670.56	12994	**6521**	87.66	64.96	49.82
74		52287	88.18	3615.36	14444	**9183**	83.41	82.44	36.42
75		45991	85.72	3618.97	20191	**9349**	82.95	79.67	53.70
81	75	87452	90.12	3601.52	22590	**12704**	323.11	85.47	43.76
82		76357	90.63	3600.64	17403	**10134**	337.81	86.73	41.77
83		75689	90.09	3601.08	20393	**11646**	335.94	84.61	42.89
84		84827	90.74	3600.59	19173	**12434**	333.00	85.34	35.15
85		70425	89.79	3602.06	16787	**10682**	336.90	84.83	36.37
Avg.		32387.55	45.83	2720.67	12673.35	7197.65	106.80	40.86	42.60

demonstrates that, for the large set, the initialization phase of the algorithm generates solutions that are more effective than the optimization model. According to this, we conclude that the MS-VNS$_{11}$ presents competitive behaviour for the small and large set of instances, where for the majority of the instances the VM-VNS$_{11}$ reaches the same or better solutions than the MILP formulation in smaller computational times.

Assessment of the Optimization Approaches for Limited Drone Flying Range. Table 5 reports the results of the MILP formulation and the MS-VNS$_{11}$, considering a battery capacity of 600 [KJ]. Best values are bold-faced.

When studying the TSP-D with a restricted drone energy capacity, results show that the MILP model is able to reach optimality for all the instances of 10 nodes under 49 s, and present gaps under 17% for instances of 20 nodes. In this regard, when studying the heuristic, we see that the MS-VNS$_{11}$ is able to reach the optimal solution for all instances of the small set and it finds better solutions than the MILP model for instances 62 and 64 by 3.97% and 5.21%, respectively. The MILP formulation reaches better solutions than the MS-VNS$_{11}$ for instances 61 and 65 by 0.48% and 0.45%. In addition, the MS-VNS$_{11}$ improves the initial TSP-D solution by at least 21% for the small set.

Table 5. Comparison of the MILP formulation and the MS-VNS algorithm for the TSP-D with restricted drone flying range.

Instance		MILP			Constructive	MS-VNS$_{11}$			
Name	n	Z	Gap (%)	Time (s)	Z_{TSP-D}	Z	Time (s)	ΔZ_{MILP} (%)	ΔZ_{TSP-D} (%)
51	10	**3793**	0.00	24.16	8653	**3793**	1.95	0.00	56.17
52		**5280**	0.00	10.02	6754	**5280**	2.02	0.00	21.82
53		**8022**	0.00	10.19	8760	**8022**	3.23	0.00	8.42
54		**4759**	0.00	12.53	9028	**4759**	1.74	0.00	47.29
55		**4897**	0.00	5.52	7144	**4897**	2.23	0.00	31.45
61	20	**6202**	21.94	3600.83	10556	6232	8.16	−0.48	40.96
62		6101	14.21	3682.98	9027	**5859**	7.20	3.97	35.09
63		**6072**	8.05	3607.05	10921	**6072**	6.99	0.00	44.40
64		6181	27.86	3600.63	7819	**5859**	6.87	5.21	25.07
65		**4694**	5.03	3600.30	10529	4715	6.35	−0.45	55.22
71	50	46679	88.53	3615.20	12939	**7660**	83.50	83.59	40.80
72		54923	89.37	3607.88	17905	**8569**	76.67	84.40	52.14
73		27755	83.11	3648.28	12994	**6566**	82.81	76.34	49.47
74		40209	84.46	3609.17	19496	**9699**	79.04	75.88	50.25
75		64628	88.13	3606.97	20191	**12263**	78.90	81.03	39.27
81	75	85509	89.15	3600.78	24700	**14318**	315.94	83.26	42.03
82		73198	90.25	3601.31	17403	**10660**	310.28	85.44	38.75
83		285228	97.19	3601.47	22404	**12878**	334.56	95.49	42.52
84		89872	91.18	3600.95	19173	**12666**	309.05	85.91	33.94
85		56462	86.94	3601.67	22083	**12022**	325.37	78.71	45.56
Avg.		44023.20	48.27	2712.39	13923.95	8139.45	102.14	41.91	40.03

On the other hand, we see that the MILP model provides feasible solutions with gaps over 83% for all the instances of the large set. In this respect, the MS-VNS$_{11}$ algorithms report better solutions than the MILP model, improving the initial TSP-D solution by at least 33%. Besides, results show that the initial TSP-D solutions are better than the solutions provided by the optimization model. Consequently, this demonstrates that the initialization phase of the VNS scheme reaches better solutions than the MILP model for these instances when studying the TSP-D with a restricted drone flying range. Similarly as for the previous set of experiments, we conclude that for the small and large set of instances the MS-VNS$_{11}$ presents competitive solutions for the TSP-D with a restricted drone flying range and small computational times. This shows that the effectiveness of the multi-start mechanism does not only apply to the unrestricted drone flying range, but also to the case in which limited energy capacity is considered. Consequently, the multi-start mechanism allows the MS-VNS to effectively explore the reduced solution space given the energy constraints, which provides a more realistic representation of the last-mile delivery operations.

5 Conclusions and Future Work

In this paper, we studied the Traveling Salesman Problem with Drone (TSP-D), with the objective of minimizing the makespan. Because of the NP-Hardness of the TSP-D, we proposed two heuristic approaches, a VNS and a MS-VNS, and compared their performance with a formulation from the literature [25]. Our heuristic schemes consist of two phases, where up to 11 neighbourhoods can be explored in every iteration of the search. We proposed two new neighbourhood structures and studied their effectiveness by comparing with the neighbourhoods from [13]. We showed that the MS-VNS algorithm presents competitive solutions for a small and large set of instances when comparing the solutions with the MILP formulation, with smaller computational times. Moreover, we demonstrate that the two neighbourhood structures substantially improve the performance of the heuristic approaches, finding better solutions in all the considered instances. In addition, we also demonstrate that the multi-start approach improves the performance of the VNS algorithm, finding better solutions for all the considered instances, but at the expense of higher computational times. Consequently, we conclude that the effectiveness of the two new neighbourhoods and the multi-start approach holds for the theoretical version of the TSP-D as well as for the more realistic representation of the problem including energy constraints.

With regards to future research, more emphasis should be placed on developing more efficient algorithms to reach better results in smaller computational times. Hence, algorithms that include learning, such as look ahead, arrangement of neighbourhoods based on their performance, and selection of a subgroup of neighbourhoods in different stages of the optimization process, seem particularly interesting as an object of study.

References

1. Agatz, N., Bouman, P., Schmidt, M.: Optimization approaches for the traveling salesman problem with drone. Transp. Sci. **52**(4), 965–981 (2018)
2. Boccia, M., Masone, A., Sforza, A., Sterle, C.: A column-and-row generation approach for the flying sidekick travelling salesman problem. Transp. Res. Part C Emerg. Technol. **124**, 102913 (2021)
3. Burke, E.K., Cowling, P.I., Keuthen, R.: Effective local and guided variable neighbourhood search methods for the asymmetric travelling salesman problem. In: Boers, E.J.W. (ed.) EvoWorkshops 2001. LNCS, vol. 2037, pp. 203–212. Springer, Heidelberg (2001). https://doi.org/10.1007/3-540-45365-2_21
4. Carlsson, J.G., Song, S.: Coordinated logistics with a truck and a drone. Manage. Sci. **64**(9), 4052–4069 (2018)
5. Chen, P., Huang, H., Dong, X.: Variable neighborhood search algorithm for fleet size and mixed vehicle routing problem. J. Syst. Simul. **23**(9), 1945–1950 (2011)
6. Dell'Amico, M., Montemanni, R., Novellani, S.: Drone-assisted deliveries: new formulations for the flying sidekick traveling salesman problem. Optim. Lett. **15**(5), 1617–1648 (2019). https://doi.org/10.1007/s11590-019-01492-z
7. Dhahri, A., Mjirda, A., Zidi, K., Ghedira, K.: A VNS-based heuristic for solving the vehicle routing problem with time windows and vehicle preventive maintenance constraints. Procedia Comput. Sci. **80**, 1212–1222 (2016)

8. Dorling, K., Heinrichs, J., Messier, G.G., Magierowski, S.: Vehicle routing problems for drone delivery. IEEE Trans. Syst. Man Cybern. Syst. **47**(1), 70–85 (2016)
9. El-Adle, A.M., Ghoniem, A., Haouari, M.: Parcel delivery by vehicle and drone. J. Oper. Res. Soc. 1–19 (2019)
10. de Freitas, J.C., Penna, P.H.V.: A randomized variable neighborhood descent heuristic to solve the flying sidekick traveling salesman problem. Electron. Notes Discrete Math. **66**, 95–102 (2018)
11. de Freitas, J.C., Penna, P.H.V.: A variable neighborhood search for flying sidekick traveling salesman problem. Int. Trans. Oper. Res. **27**(1), 267–290 (2020)
12. Gonzalez-R, P.L., Canca, D., Andrade-Pineda, J.L., Calle, M., Leon-Blanco, J.M.: Truck-drone team logistics: a heuristic approach to multi-drop route planning. Transp. Res. Part C Emerg. Technol. **114**, 657–680 (2020)
13. Ha, Q.M., Deville, Y., Pham, Q.D., Hà, M.H.: On the min-cost traveling salesman problem with drone. Transp. Res. Part C Emerg. Technol. **86**, 597–621 (2018)
14. Hansen, P., Mladenović, N.: An introduction to variable neighborhood search. In: Voß S., Martello, S., Osman, I.H., Roucairol, C. (eds.) Meta-heuristics, pp. 433–458. Springer, Boston (1999). https://doi.org/10.1007/978-1-4615-5775-3_30
15. Hansen, P., Mladenović, N., Pérez, J.A.M.: Variable neighbourhood search: methods and applications. Ann. Oper. Res. **175**(1), 367–407 (2010)
16. Jeong, H.Y., Song, B.D., Lee, S.: Truck-drone hybrid delivery routing: Payload-energy dependency and no-fly zones. Int. J. Prod. Econ. **214**, 220–233 (2019)
17. Kocatürk, F., Tütüncü, G.Y., Salhi, S.: The multi-depot heterogeneous VRP with backhauls: formulation and a hybrid VNS with GRAMPS meta-heuristic approach. Ann. Oper. Res. 1–26 (2021). https://doi.org/10.1007/s10479-021-04137-6
18. Liu, Z., Sengupta, R., Kurzhanskiy, A.: A power consumption model for multi-rotor small unmanned aircraft systems. In: 2017 International Conference on Unmanned Aircraft Systems (ICUAS), pp. 310–315. IEEE (2017)
19. Marinelli, M., Caggiani, L., Ottomanelli, M., Dell'Orco, M.: En route truck-drone parcel delivery for optimal vehicle routing strategies. IET Intell. Transport Syst. **12**(4), 253–261 (2017)
20. Murray, C.C., Chu, A.G.: The flying sidekick traveling salesman problem: optimization of drone-assisted parcel delivery. Transp. Res. Part C Emerg. Technol. **54**, 86–109 (2015)
21. Murray, C.C., Raj, R.: The multiple flying sidekicks traveling salesman problem: parcel delivery with multiple drones. Transp. Res. Part C Emerg. Technol. **110**, 368–398 (2020)
22. Oda, T., Liu, Y., Sakamoto, S., Elmazi, D., Barolli, L., Xhafa, F.: Analysis of mesh router placement in wireless mesh networks using Friedman test considering different meta-heuristics. Int. J. Commun. Netw. Distrib. Syst. **15**(1), 84–106 (2015)
23. Qi, X., Fu, Z., Xiong, J., Zha, W.: Multi-start heuristic approaches for one-to-one pickup and delivery problems with shortest-path transport along real-life paths. PloS One **15**(2), e0227702 (2020)
24. Raj, R., Murray, C.: The multiple flying sidekicks traveling salesman problem with variable drone speeds. Transp. Res. Part C Emerg. Technol. **120**, 102813 (2020)
25. Roberti, R., Ruthmair, M.: Exact methods for the traveling salesman problem with drone. Transp. Sci. **55**(2), 315–335 (2021)
26. Schermer, D., Moeini, M., Wendt, O.: A branch-and-cut approach and alternative formulations for the traveling salesman problem with drone. Networks **76**(2), 164–186 (2020)

27. Tu, P.A., Dat, N.T., Dung, P.Q.: Traveling salesman problem with multiple drones. In: Proceedings of the Ninth International Symposium on Information and Communication Technology, pp. 46–53 (2018)
28. Vásquez, S.A., Angulo, G., Klapp, M.A.: An exact solution method for the TSP with drone based on decomposition. Comput. Oper. Res. **127**, 105127 (2020)
29. Zhang, J., Campbell, J.F., Sweeney, D.C., II., Hupman, A.C.: Energy consumption models for delivery drones: a comparison and assessment. Transp. Res. Part D Transp. Environ. **90**, 102668 (2021)

Formal Methods to Verify and Ensure Self-coordination Abilities in the Internet of Vehicles

Vahid Yazdanpanah(✉), Enrico H. Gerding, and Sebastian Stein

Agents, Interaction and Complexity Research Group, University of Southampton, Southampton, UK
v.yazdanpanah@soton.ac.uk, {eg,ss2}@ecs.soton.ac.uk

Abstract. The emerging Internet of Vehicles (IoV) is a distributed multiagent network that utilises the potentials for *collaboration of vehicles* with the aim to improve the reliability and safety of transportation and logistic systems. IoV systems require operational methods to reason about the capacity of the involved (human and artificial) agents to form strategically *capable* coalitions as a means to ensure safety. In this work, we (1) develop a logic-based machinery to represent and reason about strategic abilities in IoV systems, (2) provide a process to verify whether a given IoV system is capable to safely self-coordinate, and (3) introduce a mechanism to ensure such an ability in a *temporal*, *strategic*, and *normative* setting.

Keywords: Multiagent systems · Computational logic · Smart logistics · Self-coordination · Internet of Vehicles · Formal reasoning

1 Introduction

With the increasing need for safe transportation, the rapid development of communication technologies, and the ongoing race for developing autonomous vehicles, the Internet of Vehicles (IoV) is emerging as one of the hottest topics in the AI research community with high relevance and potential impact in the transportation industry. The IoV is a distributed multiagent network that utilises the potentials for *collaboration of vehicles* (via coalition forming, information sharing, or collaborative task execution) with the overarching aim to improve the reliability and safety of transportation and logistic systems [9,12]. Realising such a form of collaboration can be reduced to classic optimisation problems with standard solution concepts when we deal with obedient non-autonomous

M. Mes et al. (Eds.): ICCL 2021, LNCS 13004, pp. 410–425, 2021.
https://doi.org/10.1007/978-3-030-87672-2_27

vehicles (as tools) that merely follow instructions. However, in IoV systems[1] that involve autonomous (human, vehicular, and infrastructure) agents, it leads to new forms of coordination and control problems. In principle, giving more autonomy raises the challenging problem of *whether* and *to what extent* such autonomous systems are capable of collaboration and self-coordination towards collective-level safety concerns. For instance, in an intersection, it is crucial to realise whether the set of involved IoV agents can collaboratively self-coordinate towards ensuring the safety of the intersection.

To that end, IoV systems require operational methods to reason about the capacity of the involved (human and artificial) agents to form *feasible* and *capable* coalitions as a means to ensure safety. Here, feasibility refers to the absence of potential incompatibilities of involved agents (i.e., their preference to get involved or avoid collaborating with one another) and capability refers to being able to strategically ensure safety (i.e., having a chain of actions—a strategy—to guarantee safety regardless of what agents outside the coalition do). We argue that such methods should satisfy the following desiderata:

- *Expressivity:* to capture the temporal, strategic, and normative aspects of IoV systems in coalitional settings. The behaviour of the IoV evolves over time (temporal dimension) and is a result of the interaction of various agents where the combination of their actions (strategic dimension) may result in desirable outcomes such as safety or undesirable ones such as a collision (normative dimension). A reasonable analysis of whether an IoV system can maintain such desirable outcomes requires modelling what different groups of agents, and not merely individuals, are able to ensure collectively (coalitional dimension). E.g., the collective-level safety is a property that in most multiagent IoV cases is achievable not by an individual but is a result of group-level coalitional strategies.
- *Coalitional Feasibility:* it is not realistic to assume that any subset of involved (human, vehicle, and infrastructure) agents in an IoV system can and is willing to collaborate. So, even if a subgroup is theoretically capable of ensuring a desirable outcome, it is crucial to verify whether it is a feasible coalitional collaboration[2]. For instance, two vehicles may be close enough to communicate, share information, and as a coalition find a way to coordinate who goes

[1] We say *"IoV system"* to refer to a particular instance, e.g., a multiagent intersection scenario, in the IoV as a whole (with millions of nodes). In the following, whenever it is clear from the context that we are focused on a specific IoV system, we may simply refer to it as the *"IoV (case/scenario)"*. Indeed, one may approach the IoV in a granular level, with all the nodes and sensors, but this work approaches the topic on a system level and focuses on reasoning about the self-coordination abilities of coalitions in IoV systems.

[2] Coalitional *feasibility* is different from stability in cooperative games and coalition structure generation (see e.g., [19,21]). There, the focus is mainly on post-conditions and what agents gain as a result of coalition formation while what we call coalitional feasibility is about pre-conditions on inherent incompatibilities of agents and verifying whether the formation of a collaborative coalition is feasible.

first in an intersection but not willing to do so due to privacy concerns avoiding them to share information with vehicles from other manufactures. In this case, we say the coalition is not feasible to form. We argue that contextual requirements such as privacy, and in turn the feasibility of coalitions, need to be integrated into smart IoV coordination frameworks.
- *Generality:* to be generic enough for reasoning about various forms of normative outcomes and not hard-wired to a given safety concern. E.g., a safe behaviour in a T-shape intersection may be unsafe in crossroads.

Reviewing the literature on IoV coordination and control, no method capable of capturing all these principles currently exists [8,14,16,22,24]. In [22], the authors use a graph theoretical representation that is temporally expressive. However, the results are specific to platoon formation cases in a static topology with no room for expressing scenarios in which the set of available actions to each agent varies over time and with respect to the state of the IoV system. The presented approach in [8] removes the need for traffic lights and solves the coordination problem in crossroads but under the strong assumption that the involved vehicles are fully collaborative and that all the potential two-member coalitions are feasible to form in order to avoid pairwise collisions. Finally, in [14,16], the safety can be guaranteed under the assumption that vehicles share information regardless of their potential incompatibilities, e.g., caused by being designed by different manufacturers with justifiable conflicts of interests.

For the first time in this work, we present a formally evaluated methodology to verify the self-coordination capacity of IoV systems using formal strategic reasoning and develop a mechanism to ensure it. We employ the semantic machinery of Alternating-time Temporal Logic (ATL) [1] as the basis for reasoning about strategic abilities in the IoV and employ its coalitional extension [4] to model endogenous and exogenous constraints in terms of argumentative notions of *incompatibility* and *priority*. In practice, this framework can be embedded in the IoV infrastructure as a safety-ensuring coordination service. In IoV scenarios, e.g., in an intersection, it is common that vehicles communicate with a trusted Infrastructure Agent (IA) a short time before reaching the intersection. Using our presented methods, an IA can evaluate the self-coordination ability of the agents and if needed apply the mechanism to ensure it.

2 Conceptual Analysis and Game Structures

In the IoV, we encounter situations in which the traditional forms of coordination may be ineffective. For instance, think of an intersection with a fixed interval traffic light. When two vehicles—both travelling from east to west—arrive at the intersection, they have to wait for the green light even if there is no harm if they pass. In the IoV, given the possibility of communication among the involved agents, such inefficiencies can be handled using smart and dynamic forms of self-coordination techniques. Thus, this paper focuses on developing reasoning methods to verify whether self-coordination is feasible in an IoV system and establishing a framework to coordinate IoV scenarios to ensure safety. We later elaborate on safety as a principle to guide the process of designing coordination mechanisms for IoV systems. If one aims to replace static external coordination techniques (e.g., traditional traffic lights) with a dynamic self-coordination mechanism, safety is the main concern. Imagine the simple two-agent scenario in Fig. 1(a). Having no external signal, how can they decide to go or to stop? They can stop at the intersection forever or may cause a crash.

Even with full awareness about the environment and observability over the surrounding traffic, ensuring the safety of the intersection requires verifying if the group of vehicles (in this case *blue* and *red*) can come up with a plan to pass the intersection safely. Here, no single one of them is able to ensure that both a crash and a deadlock is avoided. The only safe way is if one goes before the other—which requires them acting together as a group, i.e., taking a coalitional action. Then it is crucial to consider that, e.g., for privacy concerns, they may avoid sharing information with one another, hence cannot form a coalition and execute a collective plan. This necessitates capturing what we introduced as the *coalitional feasibility* condition while verifying self-coordination abilities in the IoV.

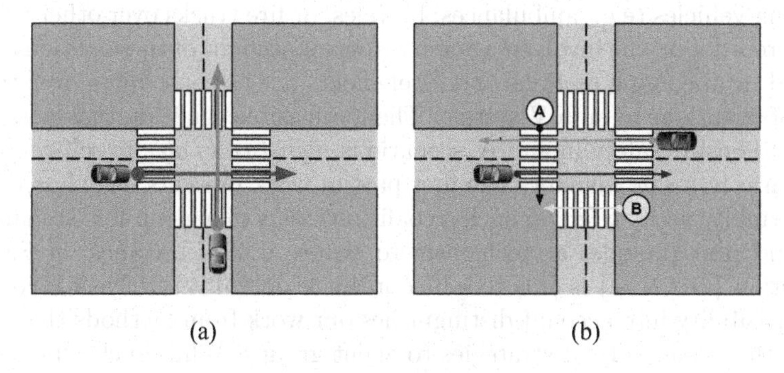

Fig. 1. The Intersection Scenario: Vehicles (*red*, *blue*, *green*) and pedestrians (A and B) can form coalitions in order to self-coordinate and ensure safety. (Color figure online)

Note that many intersections and traffic junctions currently operate without any traffic light and merely rely on the self-coordination capacity of human-driven vehicles (e.g., in roundabouts). To keep them operational and safe for a mixed group of human-driven and autonomous vehicles, we argue that the IoV infrastructure needs to be enriched with computational methods to (1) *verify* whether—in a given scenario—involved agents are able to safely self-coordinate and (2) when they are not able to do so, to *ensure* safety using verifiably reliable coordination techniques.

Using multiagent strategic reasoning semantics (e.g., [1]), one can evaluate whether an agent or agent group is able to avoid the crash or to ensure that no deadlock takes place. For instance, we can verify whether the coalition of agents $\{red, blue\}$ is capable of ensuring that neither a crash nor a deadlock takes place. However, such a realisation (that such a group with a successful strategy exists) does not necessarily imply that such a group will form and accordingly will execute their collective strategy to avoid undesirable situations. In the context of the IoV—with vehicles from different manufacturers and with different priority levels—it is not a reasonable assumption that all agents can communicate, form such coalitions, and successfully execute actions in a particular order. There might be incompatibilities among agents. These are mostly caused by inherent characteristics of the involved agents (endogenous constraints). E.g., differences in the technologies or manufacturers may prevent agents from forming a collaborative coalition with each other[3]. Moreover, in real-life and physically bounded multiagent systems such as the IoV, there exist external conditions that may require them to do specific actions or to form coalitions (exogenous constraints). For instance, consider regulatory rules and road priorities that expect agents to give way if an ambulance arrives at the intersection.

To model the external as well as internal constraints of coalition formation in real-life applications of the IoV, we capture inherent incompatibilities of agents (e.g., being produced by competing manufacturers) using the argumentative notion of "attack" relations [17]. We follow [4] and consider potential priorities among vehicles (e.g., ambulances, bicycles, or fire trucks over other vehicles) using a preorder on the involved agents. A formal account of these notions will be presented in upcoming sections. These elements act as constraints over the feasibility of coalitions in an IoV system. Then, a feasible coalition that possesses a strategy to ensure safety in an IoV scenario is *realistically able* to self-coordinate the scenario in a safe manner. The first part of work presents logic-based methods to formally verify whether such a coalition exists in a given IoV scenario and the second part provides a mechanism to ensure it. For instance, in Fig. 1(a), the coalition $\{red, blue\}$ is able to self-coordinate only if it is a feasible coalition. Taking feasibility into account distinguishes our work from methods that merely focus on the availability of strategies to agent groups. While such a perspective is applicable for closed worlds with benevolent and cooperative agents (e.g., software agent systems in [18]), physically bounded IoV systems demand methods capable of addressing feasibility constraints. Our concern to capture feasibility of coalitions in the IoV relates to the notion of natural abilities in [11], where the authors formalise natural strategies as those with a feasible degree of complexity. Then it is a question whether a feasible coalition to ensure safety necessarily exists. Do we fail to self-coordinate if *red* and *blue* cannot form a feasible coalition? In real-life scenarios, e.g., traffic handling in intersections with no traffic light, priority rules act as predesigned mechanisms for coordination. Inspired

[3] We would like to emphasise that ensuring the safety and reliability of dynamic systems such as the IoV needs to balance the trade-offs between strict, design-time, offline standardisation (e.g., see [20]) and flexible run-time online coordination (e.g., see the interdisciplinary study in [15]). While standardisation is effective for parts of the IoV with less complexity and more predictability, for other parts in which standardisation is out of reach, the IoV requires flexible techniques able to capture the characteristics of specific scenarios and able to be exploited at the run time. This work focuses on such instances where incompatibilities are probable and provides a base for verifying and ensuring self-coordination in the run-time.

by this, we introduce a model-updating mechanism to ensure the existence of a minimal coalition, feasible to be formed and capable of ensuring the safety of an IoV system.

To reason about the abilities of agents and agent coalitions in a multiagent system such as the IoV, we employ Concurrent Game Structures (CGS). CGS, as the semantics machinery of Alternating-time Temporal Logic (ATL) [1], enables modelling the behaviour of IoV systems over time (capturing temporality) and is expressive for representing the ability of individual agents as well as coalitions to ensure/avoid a given situation (strategic abilities). Moreover, CGS-based notions can be implemented using established model-checking tools [5,13]. Using CGS, we reason about coalitional aspects of the IoV and model the temporal and strategic aspects of IoV scenarios.

Concurrent Game Structures: Formally, a *Concurrent Game Structure* (CGS) is a tuple $\mathcal{M} = \langle \Sigma, Q, \Pi, \pi, Act, d, o \rangle$ where: $\Sigma = \{a_1, \ldots, a_n\}$ is a finite, non-empty set of *agents*; Q is a finite, non-empty set of *states*; Π is a set of atomic propositions; $\pi : Q \mapsto \mathcal{P}(\Pi)$ is a valuation of propositions; Act is a finite set of atomic *actions*; function $d : \Sigma \times Q \mapsto \mathcal{P}(Act)$ specifies the sets of actions available to agents at each state; and o is a transition function[4] that assigns the outcome state $q' = o(q, \alpha_1, \ldots, \alpha_n)$ to state q and a tuple of actions $\alpha_i \in d(a_i, q)$ that can be executed by Σ in q. To represent *strategies* and *outcomes* we make use of the following auxiliary notions.

Successors and Computations: For two states q and q', we say q' is a *successor* of q if there exist actions $\alpha_i \in d(a_i, q)$ for $a_i \in \{1, \ldots, n\}$ in q such that $q' = o(q, \alpha_1, \ldots, \alpha_n)$, i.e., agents in Σ can collectively guarantee in q that q' will be the next system state. A *computation* of a CGS $\mathcal{M}$ is an infinite sequence of states $\lambda = q_0, q_1, \ldots$ such that, for all $i > 0$, we have that q_i is a successor of q_{i-1}. We refer to a computation that starts in q as a *q-computation*. For $k \in \{0, 1, \ldots\}$, we denote the k'th state in λ by $\lambda[k]$.

Strategies and Outcomes: A *memoryless strategy*[5] for an agent $a \in \Sigma$ is a function $\zeta_a : Q \mapsto Act$ such that, for all $q \in Q$, $\zeta_a(q) \in d(a, q)$. For a coalition of agents $C \subseteq \Sigma$, a *collective strategy* $Z_C = \{\zeta_a \mid a \in C\}$ is an indexed set of strategies, one for every $a \in C$. Then, $out(q, Z_C)$ is defined as the set of potential q-computations that agents in C can enforce by following their corresponding strategies in Z_C.

CGS is expressive to capture complex scenarios where we are interested in a combination of temporal, coalitional, and strategic properties. Note that reasoning about normative *(un)desirable* properties in IoV scenarios, e.g., to verify

[4] CGS can be extended to probabilistic forms by adding probabilities to transitions. For the purpose of this work, our approach can be applied after a standard determinisation procedure [23]. If one focuses on finite traces (as argued for in [2,7]), a probabilistic automaton can be reduced to a deterministic one.

[5] We focus on memoryless strategies to avoid the strong assumption that agents necessarily recall the evolution of the IoV system.

whether a crash is possible, avoidable, or inevitable, requires going beyond single-shot properties in contrast to merely analysing whether a crash happens in the immediate next state of the system. For instance, our running example with two vehicles can be modelled as a 3-state CGS (Fig. 2).

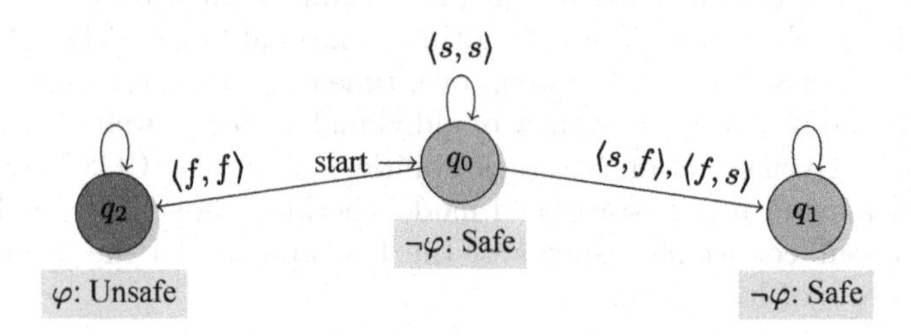

Fig. 2. We model the scenario presented in Fig. 1(a) as $\mathcal{M} = \langle \Sigma, Q, \Pi, \pi, Act, d, o \rangle$ where: $\Sigma = \{red, blue\}$; $Q = \{q_0, q_1, q_2\}$; $\Pi = \{\varphi\}$ (φ represents the breach of safety, i.e., the crash, in the intersection); $Act = \{f, s\}$ to represent going forward "f" and stopping "s"; $d(a_i, q) = Act$ for all $a_i \in \Sigma$ and $q \in Q$; and π and o are represented in the automaton.(Color figure online)

Being concerned about the occurrence of a crash or passing safely, we use the unique proposition φ to represent a crash (in general, we can represent a finite set of propositions, relevant to our scenario, in Π). Then, we are interested in verifying if individual agents or coalitions of agents are capable of avoiding such a crash using their available actions or sequences of actions. To reason about agents' potentials, we have that in q_0, *red* and *blue* may each stop or go forward. With no communication, hence no coordination, the two vehicles may remain in a deadlock situation in q_0 or cause a crash in q_2. However, coalition $\{red, blue\}$ has a strategy to avoid this (e.g., *red* goes first and *blue* second). But the question is whether such a coalition is a reasonable one to form in the context of IoV systems and given potential incompatibilities among the involved agents. In real-life IoV scenarios, we cannot assume benevolent agents but need to restrict theoretically formable coalitions to a subset of feasible ones.

3 Coalitional IoV Systems

Inspired by the presented approach in [4], we build on the CGS machinery, adopt elements from the literature on formal argumentation theory [17], and model coalitional IoV systems in Definition 1. The first element, adopted from argumentation theory, is a relation to represent potential incompatibilities between agents (known in the formal argumentation community as the *attack* relation). Furthermore, the second adopted element is a preorder to represent priorities among agents in the IoV context, e.g., an ambulance has priority over a truck. Note that notions of computation and strategy are syntactic elements of CGS and do not capture the feasibility of coalitions. In a CGS, a group C may have a collective strategy to enforce a particular form of computation. This merely means that there exists a chain of collective actions to do so and disregards that some agents in C may hesitate to (form a collaborative coalition and) execute the collective strategy. By adding new elements to CGS, we can distinguish groups with a strategy from feasible groups with a strategy. This allows capturing, not only the existence of a collective strategy but also, the feasibility of forming a coalition to execute the strategy.

Definition 1 (Coalitional IoV). *A coalitional IoV is given by a tuple* $\mathcal{I} = \langle \Sigma, Q, \Pi, \pi, Act, d, o, \Im, \prec \rangle$ *where* $\langle \Sigma, Q, \Pi, \pi, Act, d, o \rangle$ *is a CGS modelling the behaviour of the IoV,* $\Im \subseteq \Sigma \times \Sigma$ *is an antireflexive relation representing the potential incompatibilities between agents, and* $\prec$ *is a preorder on* Σ *representing priorities between agents.*

Here, relation $\Im$ reflects potential incompatibilities not as misalignment in goals but in terms of inherent characteristics, e.g., competing manufacturers of vehicles. Making incompatibilities explicit in the modelling relaxes the common assumption that IoV systems are operating either in a cooperative or fully non-cooperative setting and enables modelling realistic scenarios, where some agent groups (but not necessarily all) may be able to form coalitions. Furthermore, the ordering $\prec$ is a contextual element in the IoV system that models priorities. In real-life cases—e.g., in applying our method to traffic coordination scenarios—priorities act as regulatory norms and override potential incompatibilities. For instance, regardless of the brand of an ambulance, other vehicles collaborate with it as it has a higher priority. Using CGS for specifying IoV systems enables capturing the temporal, strategic, and normative aspects. In addition, it is implementable using computationally affordable model-checking tools, e.g., [5,13].

Definition 2 (Feasibility). *Let* $\mathcal{I}$ *be a coalitional IoV and* $C \subseteq \Sigma$ *an agent group in* $\mathcal{I}$*. We say* C *is a feasible coalition if and only if for all* $a_i, a_j \in C$ *we have that* $(a_i, a_j) \in \Im \Rightarrow a_i \prec a_j$*. The set of all feasible coalitions is denoted with* $\mathfrak{F}$*.*

Notice that the feasibility of a coalition is defined in terms of incompatibilities that are not overruled by priorities. Moreover, note that we do not assume symmetry on the incompatibility relation $\Im$. This reflects the reality of the IoV context. For instance, vehicles from a manufacturer A may avoid forming coalitions with others from B while vehicles from B are not necessarily concerned with being in a coalition with vehicles from A. This is also the case when we consider vehicle-human or vehicle-infrastructure relations. For instance, imagine extending our two-vehicle scenario with $\Im = \{(red, blue)\}$ and $red \prec blue$. Then we have an incompatibility but it is "overruled" by the priority preorder. Thus, the grand coalition is feasible and able to safely self-coordinate towards q_1. By the corollary to Definition 2, we have:

Corollary 1. *Let* $\mathcal{I}$ *be a coalitional IoV. Then we have: 1. (conflict-freeness) Any* $C \subseteq \Sigma$ *is a feasible coalition (i.e., a member of* $\mathfrak{F}$*) if* $\Im = \varnothing$*; and 2. (maximality) If* C *is a feasible coalition (*$C \in \mathfrak{F}$*), then any* $C' \subseteq C$ *is a feasible coalition.*

The first point shows that in a conflict-free IoV—i.e., with no incompatibility among the agents—all potential coalitions are feasible. Then the second point shows that being feasible is a maximal quality for a coalition. The latter is crucial to presenting the first main result of our work on the ability of an IoV to safely self-coordinate (Theorem 1). The former point relates to what we later discuss as *weak homogeneity* in IoV systems. Note that the feasibility of coalitions does *not* reflect the post-conditions in the IoV (i.e., what they can ensure via their actions) but is defined with respect to their inherent characteristics (e.g., potential manufacturer differences and type of the vehicle) and the context of the IoV (i.e., what priorities and codes of conduct are in place). Next, we move to our focal question on "whether agents in an IoV system can safely self-coordinate". This is to verify if agents can *feasibly* ensure a given S that represent safe states and *maintain* it over time. In a CGS, the capacity of a set of agents C to ensure a subset of states S from a state q requires C to have a strategy Z_C such that all $q-$computations in $out(q, Z_C)$ include *a* state in S. In other words, all such computations should meet a state in S at least once. Then, maintaining S is a specific form of ensuring as it requires that all such computations consist of states in S and only states in S. Note that it is not necessary that computations C enforces (using Z_C) consist of identical states but any state in S. In other words, even if $q \in S$, group C does not need to enforce the system to remain in q but can enforce computations that go through other states in S. In the following, we use the term *"to ensure and maintain"* to convey the point that maintaining S (even from a $q \in S$) can be achieved by ensuring other states in S and not necessarily by remaining in q.

Theorem 1. *Let $\mathcal{I}$ be a coalitional IoV and $S \subseteq Q$ a safety state of affairs represented by a set of states in the IoV. From a safe state $q \in S$, the IoV $\mathcal{I}$ can safely self-coordinate w.r.t. S if (1) there exists a $C \subseteq \Sigma$ with a collective strategy Z_C to ensure and maintain S, (2) C is a feasible coalition, and (3) no $C' \subset C$ satisfies both (1) and (2).*

Proof. To guarantee that $\mathcal{I}$ can ensure and maintain S, agents in Σ should have the capacity to do so. A minimal (part 3) and feasible (part 2) sup-group of agents $C \subseteq \Sigma$ with a collective strategy Z_C (part 1) is capable of forming C (thanks to its feasibility) and ensuring that all the q-computations λ in $out(q, Z_C)$ are merely composed of states that are in S. This is thanks to its capacity to ensure and in addition maintain (i.e., remaining in) S. As $\mathcal{I}$ is safe in the first place (in q) and any potential q-computation remains in S if C executes Z_C, we have the coordination capacity within $\mathcal{I}$. $\square$

From an implementability point of view, note that the problem to verify whether an IoV is capable of self-coordination in a safe manner reduces to a model-checking problem, that takes the set of feasible coalitions as its input, and hence is implementable using standard tools (e.g., using [5,13]). First, any state of affairs S translates to a valid proposition in $\varphi \in \mathcal{L}_{ATL}$ (see [25] for a similar approach). Basically, S consists of those states in Q that satisfy φ. Then we have a straightforward process to verify if in q, any feasible coalition has a strategy to minimally ensure S. As shown in Corollary 1, subsets of a feasible coalition preserve the feasibility. Hence, instead of model-checking on possible coalitions in Σ, we only consider feasible coalitions.

Based on Theorem 1, the presented 2-vehicle IoV can safely self-coordinate as the grand coalition is feasible and is the only minimal coalition capable of ensuring and maintaining q_1. In the scenario displayed in Fig. 1-b, we have vehicles *blue* and *green*; pedestrians Alice (A) and Bob (B); and the following action space, incompatibilities, and priorities: $d(a, q) = \{s, f\}$ meaning that any agent a can either stop (s) or go forward (f) in any state q; $\Im = \{(green, blue)\}$ meaning that *green* is incompatible with *blue*; $green \prec blue, green \prec A, blue \prec A$ meaning that *blue* has priority over *green* and Alice has priority over both the vehicles. The partial CGS in Fig. 3 models this scenario. Here, the three-member coalition of Alice, *green*, and *blue* satisfies the conditions of Theorem 1. Their coalition is feasible thanks to the priority of *green* over *blue* and minimal as no member can be excluded.

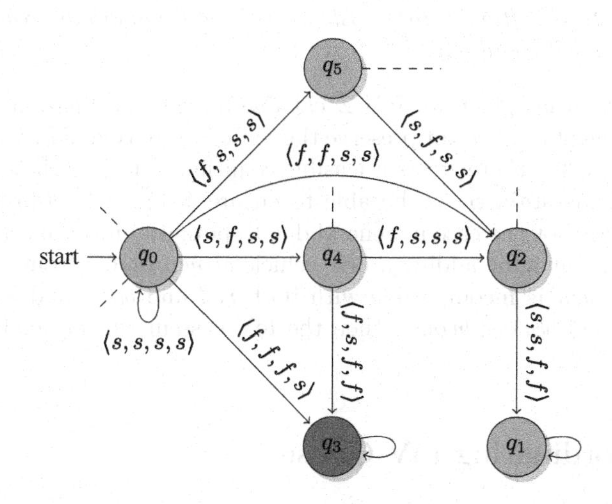

Fig. 3. This CGS represents various paths which the coalition composed by A, *green*, and *blue* can ensure and maintain safety in q_1. E.g., $q_0, q_4, q_2, q_1, \ldots$. Action profiles on each arrow are actions for A, B, *blue*, and *green*, respectively. Dashed lines represent other paths that may result from unrepresented actions. (Color figure online)

Building feasible coalitions (i.e., expecting to collaborate with others) is not necessarily required in all the IoV cases. Our framework can capture such cases too. For instance, imagine Fig. 1-(b) with only the two vehicles *green* and *blue*–with neither Alice nor Bob. Then all the potential outcomes are safe, hence the empty set is the feasible and minimal coalition capable of ensuring a safe state. In general:

Proposition 1. *For any incompatibility relation* $\Im$ *and priority* $\prec$, $\mathcal{I}$ *can safely self-coordinate w.r.t.* S *if* $S = Q$.

Proof. If all states are in S, any successor of any $q \in Q$ is a safe state. Formally, for any strategy $Z_{\varnothing}$, all the states in $out(q, Z_{\varnothing})$ are in S. Hence the empty set is the unique minimal coalition capable of ensuring safety from any q. □

While we use a state-based formalisation of safety in the IoV, our approach is not limited to state-based normativity and safety on the state level but has behavioural connotations. Basically, a state of affairs in Q represents all the states that satisfy a $\varphi \in \mathcal{L}_{ATL}$ (e.g., see Fig. 2). Using the state-based notation is more intuitive and enables colour-based visual labelling. To see the behavioural aspect, notice that our notion of strategy is a multi-step temporal notion as it involves chains of actions to reach a $q \in S$. Moreover, verifying if such a strategy is available to a group requires model-checking the potential evolution/behaviour of the MAS using the semantic notion of path/computation (in contrast with the purely syntactic notions of action and state). As observed, adding agents to an IoV does not preserve the self-coordination ability.

Proposition 2. *Coalitional IoV systems are non-monotonic regarding their ability to safely self-coordinate.*

Proof. We show that given an IoV $\mathcal{I}$ capable of self-coordination w.r.t. to S, adding a new agent a_i does not preserve the self-coordination ability of the newly formed $\mathcal{I}'$. Basically, there exists a feasible coalition C in $\mathcal{I}$ with a strategy Z_C to ensure and maintain S. To be able to ensure S in $\mathcal{I}'$, $C \cup \{a_i\}$ should be feasible, which is not the case in general due to potential incompatibilities. As a counterexample, imagine adding a new vehicle *brown* (going from east to west) to Fig. 1. If *brown* is incompatible with both *red* and *blue*, and we have that *red* $\prec$ *brown* and *blue* $\prec$ *brown*, then the IoV system will be unable to safely self-coordinate. □

4 Self-coordinating IoV Classes

The IoV has the potential to be deployed in various contexts with domain-specific characteristics. For instance, in the context of rail freight transport, one may face a homogeneous set of autonomous train vehicles operating with no incompatibility. This form of context-dependent homogeneity arises when we deal with collaborative agents with negligible incompatibilities. E.g., when we are dealing with a set of vehicles from the same manufacturer. This translates into having $\Im = \varnothing$ in our formalisation. In this section, we focus on the ability of such sub-classes to safely self-coordinate. This has direct implications for deployability of real-life IoV systems.

Proposition 3. *Let $\mathcal{I}$ be an IoV with $\Im = \varnothing$, $S \subseteq Q$ be a non-empty safety state of affairs, and q be a state in S. $\mathcal{I}$ can safely self-coordinate from q if there is a safe path in S that contains a loop.*

Proof. The antecedent is identical to saying that S is reachable from q and maintainable; formally, that for a q-computation λ, there exists a finite prefix

$\lambda[j] = q_0, \ldots, q_i, \ldots, q_j$ s.t. all the states in the prefix are in S and $q_i = q_j$. To prove, we show how such an IoV satisfies the three conditions of Theorem 1 on: (1) S-ensuring and maintenance ability, (2) coalitional feasibility, and (3) minimality. For part one, the existence of safe finite prefix implies that Σ can ensure q_i in the non-empty S. Given that S has a loop, Σ can maintain S by ensuring that $\mathcal{I}$ loops back to q_i. Then, for part two, relying on the conflict-freeness of $\mathcal{I}$ (see Corollary 1), we have that Σ is feasible. Finally, for part three, either Σ is minimal or by excluding excess agents, we reach to a minimal subset C where we have that C necessarily preserve the feasibility condition (maximality term in Corollary 1). □

Having no incompatibility relaxes the feasibility check, but observe that the ability to ensure and to maintain S are independent. For instance, in Fig. 3, $\{Bob\}$ can ensure S but has no strategy to maintain it. Moreover, a group may be capable of maintaining S in a state q' but not in a different state q, e.g., the empty set can preserve S in q_1 but has no strategy to ensure it from q_0. Our notion of self-coordination in the IoV is essentially a *local* notion as it is about the ability of the IoV to remain safe from a given safe state. In the context of the IoV, we deem that safety is a requirement that is necessary to be respected in all the possible outcomes. In IoVs, one cannot be satisfied with a path of states that reaches a safe state via unsafe states. For instance, in the intersection scenario, given that q_0 is a safe state itself, we are interested to see if agents can self-coordinate towards q_1 via a safe path. One may argue that *Alice* can stop and individually avoid going into a red state. Note that she can only ensure that the next state after q_0 is either q_0 or q_4 but cannot maintain remaining in a safe state. We need to consider that the capacity to self-coordinate (Theorem 1) is in terms of and bounded to a given set of states, without imposing any requirements on these states, and here we are focused on q_1. Next, we focus on how the global self-coordination capacity (in contrast to a state-bounded local capacity) can be verified.

Theorem 2. *IoV $\mathcal{I}$ can safely self-coordinate w.r.t. $S \subseteq Q$ if in all $q \in S$, there is a coalition $C \in \mathfrak{F}$ with a strategy Z_C such that in any $\lambda \in out(q, Z_C)$, we have that $\lambda[1]$ is in S.*

Proof. To prove, we show how such an IoV satisfies the local conditions of Theorem 1 in any $q \in S$. In any q, if a feasible coalition C with such a S-ensuring strategy exists, then the set of possible next immediate states that C can ensure consists of either only q itself or has other states $q' \neq q$ in S. In the former case, C can make a loop in S that satisfies the maintaining condition. Such a C is either minimal or by excluding excess members we reach to a minimal subset. In the later case, when we may have $q' \neq q$ in the set of possible next immediate states that C can ensure, we have a coalition C' to ensure S by looping on q' or by reaching another state in S. Note that we are not extending C to $C \cup C'$ hence coalitional feasibility is not an issue. Intuitively, each $C, C', \cdots \in \mathfrak{F}$ is responsible just to ensure that the next state is in S. The antecedent indicates that in all $q \in S$, such a coalition exists. This ensures the maintenance condition. □

In this class of IoVs, $\mathcal{I}$ is capable of self-coordination globally only if in every safe state, there exists at least a feasible coalition capable of ensuring at least an immediate safe state. For instance, imagine the scenario in Fig. 1-b but without $green \prec blue$. Then, *blue* has no priority over *green* and their incompatibility results in the infeasibility of their coalition with *Alice*. Such practical concerns in the context of the IoV make self-coordination in coalitional IoVs distinguishable from strategic reasoning in software systems, with fully-collaborative agents, and motivates using mechanisms to update priority rules in order to ensure safety properties.

5 A Safety-Ensuring Mechanism

In cases where the conditions of Theorem 1 are not satisfied, the IoV is incapable of self-coordination in a safe manner, thus requires intervening mechanisms to ensure safety. Next, we show that in our formalisation of IoV systems, this can be ensured using a mechanism that minimally introduces priority rules. In this context, minimality refers to the introduction of l new priority rules such that the IoV is incapable of safe self-coordination under any $k < l$ added rules. Mechanism $\mathfrak{M}^S$ (Algorithm 1) is designed to ensure safety concern S as a function that takes $\mathcal{I}$ and updates it to $\mathcal{I}'$ capable of self-coordination with respect to S.

Theorem 3. *Let $\mathcal{I}$ be an IoV and non-empty $S \subseteq Q$ a safety state of affairs. If S is reachable and maintainable from $q \in Q$ (in the sense of Proposition 3), mechanism $\mathfrak{M}^S$ ensures the self-coordinating capacity of the updated IoV $\mathcal{I}'$ to ensure S via minimal priority updates.*

Proof. To prove, we show the correctness of Algorithm 1 using the following lemmas. In Lemma 1 and 2, we respectively establish the non-emptiness of $\hat{C}_q^S$ and that $\mathcal{I}'$ satisfies the self-coordinating conditions of Theorem 1. □

Lemma 1. *If S is reachable and maintainable from $q \in Q$ (in the sense of Proposition 3), then $\hat{C}_q^S \neq \varnothing$.*

Proof. Building on Proposition 3, we have that for such a S, the grand coalition Σ is a member of $\hat{C}_q^S$ as it is necessarily capable of ensuring and maintaining S. Note that in this stage, the mechanism is merely focused on the strategic ability of groups regardless of their coalitional feasibility. □

By focusing on the class of reachable and maintainable S from q, we are making the assumption that it is not reasonable to specify an inevitable situation with S. An IoV neither can nor is expected to avoid inevitable situations. Moreover, for generating $\hat{C}_q^S$, using standard model-checking tools, we highlight that the notion of state of affairs in terms of a subset of Q is translatable to an ATL-based verifiable proposition. φ corresponds to a fixed set of states $S \subseteq Q$. From a computational complexity point of view, verifying whether a group is able to ensure a φ is P-complete w.r.t. the size of $\mathcal{I}$ and φ [3].

Lemma 2. *From q, IoV $\mathcal{I}'$ generated by $\mathfrak{M}^S$, can safely self-coordinate w.r.t. S.*

Algorithm 1: Safety Ensuring Mechanism $\mathfrak{M}^S$

Input: IoV $\mathcal{I}$; state $q \in Q$; state of affairs $S \subseteq Q$.
Result: $\mathcal{I}'$ capable of safe self-coordination w.r.t. S.

1 **Initialisation:** $\varphi \leftarrow$ ATL formula (from $\mathcal{L}_{ATL}$) corresponding to S; $\hat{C}_q^S \leftarrow$ indexed set of groups C_k able to ensure and maintain φ from q (standard ATL model-checking [3]); $index \leftarrow 0$; $u \leftarrow |\Im|$;
2 **forall the** $C_k \in \hat{C}_q^S$ **do**
3 $\quad u_k \leftarrow 0$; $U_k \leftarrow \varnothing$;
4 $\quad$ **forall the** $(i,j) \in \Im$ **do**
5 $\quad\quad$ **if** $i \in C_k$ *and* $j \in C_k$ *and* $j \not\prec i$ **then**
6 $\quad\quad\quad u_k \leftarrow u_k + 1$;
7 $\quad\quad\quad U_k \leftarrow U_k \cup \{(i,j)\}$;
8 $\quad$ **if** $u_k = 0$ **then**
9 $\quad\quad$ **return** $\mathcal{I}$;
10 $\quad$ **else**
11 $\quad\quad$ **if** $u_k \leq u$ **then**
12 $\quad\quad\quad u \leftarrow u_k$;
13 $\quad\quad\quad index \leftarrow k$;
14 $\prec' \leftarrow \prec$;
15 **forall the** $(i,j) \in U_{index}$ **do**
16 $\quad$ **if** $j \prec i \notin \prec'$ **then**
17 $\quad\quad \prec' \leftarrow \prec' \cup \{k \prec j : k \prec i \in \prec'\}$;
18 **return** $\mathcal{I}' = \langle \Sigma, Q, \Pi, \pi, Act, d, o, \Im, \prec' \rangle$;

Proof. We show that $\mathcal{I}'$ satisfies the conditions of Theorem 1. The first condition is fulfilled because $\hat{C}_q^S$ is non-empty (Lemma 1). For the second condition, Algorithm 1 finds $C_{index} \in \hat{C}_q^S$ with the least number of unresolved intra-group incompatibilities and introduces the required priorities. For the third condition, as coalitional feasibility is a maximal notion (Corollary 1), excluding excess members leads to the minimal group without harming the feasibility. □

As presented, this approach aims at intervening as little as possible as we focus on the coalition in $\hat{C}_q^S$ with the smallest number of incompatibilities. If there exists a feasible group to form a coalition (that is capable of ensuring safety), we say the IoV system is capable of safe self-coordination without any external intervention. This is a crucial background for reasoning about (1) whether intervention, and applying the presented mechanism, is necessary (i.e., by introducing external rules) and (2) as a secondary outcome, whether we can see agents responsible for an unsafe behaviour, i.e., we can justifiably see them responsible if they could ensure safety but failed to do so [26].

6 Conclusions

We proposed a coalitional representation of IoV systems rooted in a logic for strategic reasoning in multiagent systems. In our approach, integrating elements from formal argumentation theory resulted in a contextual representation and reasoning framework that captures key aspects of the IoV. For the first time, this work enables automated reasoning about the temporal, strategic, and normative aspects of the IoV using logic-based methods. Specifically, it enables verifying whether a given IoV system is capable of self-coordination with respect to a safety concern. We designed an algorithmic mechanism that guarantees this ability in IoV systems and formally evaluated our results. In this work, we focused on the availability of strategies to coalitions. An interesting extension is to rank groups based on the hardness of their strategy. E.g., in some traffic contexts, a strategy is useful only if it can be executed fast. We aim to integrate the notion of *"natural strategy"* [11] to address this in terms of the number of state transitions. Another direction is to formulate agents' *responsibility* for an undesirable situation, e.g., a crash. We aim to link our approach to epistemic strategic logics [10] and integrate it with multiagent responsibility reasoning tools [6,25].

Ethical Statement. For an effective deployment of trustworthy IoV systems, verifiable computational models to capture and preserve social values, safety concerns, and ethical norms play a key role. The results of this study can be embedded in the IoV infrastructure as a safety-ensuring coordination service and are adaptable for preserving social values and ethical norms in human-centred AI systems.

Acknowledgements. This work is supported by the UK Engineering and Physical Sciences Research Council (EPSRC) through the Trustworthy Autonomous Systems Hub (EP/V00784X/1), the platform grant entitled "AutoTrust: Designing a Human-Centred Trusted, Secure, Intelligent and Usable Internet of Vehicles" (EP/R029563/1), and the Turing AI Fellowship on Citizen-Centric AI Systems (EP/V022067/1).

References

1. Alur, R., Henzinger, T.A., Kupferman, O.: Alternating-time temporal logic. J. ACM **49**(5), 672–713 (2002)
2. Brafman, R.I., Giacomo, G.D.: Planning for ltlf /ldlf goals in non-markovian fully observable nondeterministic domains. Proc. IJCAI **2019**, 1602–1608 (2019)
3. Bulling, N., Dix, J., Jamroga, W.: Model checking logics of strategic ability: complexity, pp. 125–159. Springer, Heidelberg (2010). https://doi.org/10.1007/978-1-4419-6984-2_5
4. Bulling, N., Dix, J., Chesñevar, C.I.: Modelling coalitions: ATL + argumentation. In: Proceedings of AAMAS'08, pp. 681–688 (2008)
5. Cermák, P., Lomuscio, A., Mogavero, F., Murano, A.: MCMAS-SLK: a model checker for the verification of strategy logic specifications. In: Proceedings of the 26th International Conference on Computer Aided Verification, pp. 525–532 (2014)
6. Chockler, H., Halpern, J.Y.: Responsibility and blame: a structural-model approach. J. Artif. Intell. Res. **22**, 93–115 (2004)
7. Fogarty, S., Kupferman, O., Vardi, M.Y., Wilke, T.: Profile trees for büchi word automata, with application to determinization. Inf. Comput. **245**, 136–151 (2015)
8. Ghaffarian, H., Fathy, M., Soryani, M.: Vehicular ad hoc networks enabled traffic controller for removing traffic lights in isolated intersections based on integer linear programming. IET Intell. Transp. Syst **6**(2), 115–123 (2012)

9. Hammoud, A., Sami, H., Mourad, A., Otrok, H., Mizouni, R., Bentahar, J.: Ai, blockchain, and vehicular edge computing for smart and secure iov: Challenges and directions. IEEE Internet Things Mag. **3**(2), 68–73 (2020)
10. van der Hoek, W., Wooldridge, M.J.: Cooperation, knowledge, and time: alternating-time temporal epistemic logic and its applications. Studia Logica **75**(1), 125–157 (2003)
11. Jamroga, W., Malvone, V., Murano, A.: Natural strategic ability. Artif. Intell. **277** (2019)
12. Kaiwartya, O., et al.: Internet of vehicles: motivation, layered architecture, network model, challenges, and future aspects. IEEE Access **4**, 5356–5373 (2016)
13. Kurpiewski, D., Jamroga, W., Knapik, M.: STV: model checking for strategies under imperfect information. In: Proceedings of AAMAS'19, pp. 2372–2374 (2019)
14. Lee, J., Park, B.: Development and evaluation of a cooperative vehicle intersection control algorithm under the connected vehicles environment. IEEE Trans. Intell. Transp. Syst **13**(1), 81–90 (2012)
15. Malone, T.W., Crowston, K.: The interdisciplinary study of coordination. ACM Comput. Surv. (CSUR) **26**(1), 87–119 (1994)
16. Milanés, V., Alonso, J., Bouraoui, L., Ploeg, J.: Cooperative maneuvering in close environments among cybercars and dual-mode cars. IEEE Trans. Intell. Transp. Syst **12**(1), 15–24 (2011)
17. Modgil, S., Prakken, H.: A general account of argumentation with preferences. Artif. Intell **195**, 361–397 (2013)
18. Mohamed, A.M., Huhns, M.N.: Benevolent agents in multiagent systems. In: Proceedings Fourth International Conference on MultiAgent Systems, pp. 419–420. IEEE (2000)
19. Perez-Diaz, A., Gerding, E., McGroarty, F.: Coordination of electric vehicle aggregators: a coalitional approach. In: Proceedings of AAMAS'18, pp. 676–684 (2018)
20. Priyan, M., Devi, G.U.: A survey on internet of vehicles: applications, technologies, challenges and opportunities. Int. J. Adv. Intell. Paradigms **12**(1–2), 98–119 (2019)
21. Rahwan, T., Michalak, T.P., Wooldridge, M., Jennings, N.R.: Coalition structure generation: a survey. Artif. Intell **229**, 139–174 (2015)
22. Santini, S., Salvi, A., Valente, A.S., Pescapè, A., Segata, M., Cigno, R.L.: A consensus-based approach for platooning with inter-vehicular communications. In: 2015 IEEE Conference on Computer Communications, pp. 1158–1166 (2015)
23. Vardi, M.Y.: Automatic verification of probabilistic concurrent finite-state programs. In: 26th Annual Symposium on Foundations of Computer Science, Portland, Oregon, USA, 21–23 October 1985, pp. 327–338 (1985)
24. Yang, F., Li, J., Lei, T., Wang, S.: Architecture and key technologies for internet of vehicles: a survey. J. Communi. Inform. Netw. **2**(2), 1–17 (2017)
25. Yazdanpanah, V., Dastani, M., Jamroga, W., Alechina, N., Logan, B.: Strategic responsibility under imperfect information. In: Proceedings of AAMAS '19, pp. 592–600 (2019)
26. Yazdanpanah, V., Gerding, E.H., Stein, S., Dastani, M., Jonker, C.M., Norman, T.J.: Responsibility research for trustworthy autonomous systems. In: Proceedings of AAMAS'21, pp. 57–62 (2021)

Routing, Dispatching, and Scheduling

Equipment Dispatching Problem for Underground Mine Under Stochastic Working Times

Nour El Houda Hammami, Amel Jaoua, and Safa Bhar Layeb(✉)

LR-OASIS, National Engineering School of Tunis, University of Tunis El Manar, Tunis, Tunisia
nourelhouda.hammami@etudiant-enit.utm.tn, safa.layeb@enit.utm.tn

Abstract. This work investigates an underground mine equipment dispatching problem under equipment stochastic working times. First, a mathematical model is developed for solving the Equipment Dispatching problem while considering the machines working times as deterministic parameters. Then, Monte Carlo simulation is implemented in order to assess the reliability of the deterministic dispatching under stochastic environment, i.e. stochastic working times that include travel times between stopes, settlement times, and breakdown times. For this challenging problem, an illustrative variability effect analysis is proposed. Promising preliminary results highlight the importance of considering machines working times as stochastic parameters in the case of medium and high variability levels.

Keywords: Underground mine equipment dispatching · Stochastic working times · Monte Carlo simulation

1 Introduction

With a worldwide high consumption of mineral products, and with a huge resurgence to "Open pit to underground mining transition" (King et al. 2017) in mining industry, underground mining projects are considered among the most significantly rewarding businesses (Campeau and Gamache, 2020). Consequently, it is crucial for underground mining companies to optimize their processes, mainly, their equipment dispatching process in cited rigid environments (Yu et al. 2017). A real fact leading to controlling several uncertain parameters related to machines performance, that can be tracked and recorded easily by IoT sensors in the context of mining 4.0 era.

In a real-world context, the uncertainty of underground mining equipment parameters, particularly machine working times, have a significant impact on the dispatching process and the mining activities short-term planning. However, such stochastic processing times have not been clearly investigated in the existing literature on underground mine dispatching problems, as highlighted in the recent study of Hou et al. (2020). It is worthy to note that machines working times could include effective processing times, machines settlement times and mobile equipment traveling times between stopes etc. (Samatemba et al. 2020). In fact, the work of Hou et al. (2020) is among the very few

M. Mes et al. (Eds.): ICCL 2021, LNCS 13004, pp. 429–441, 2021.
https://doi.org/10.1007/978-3-030-87672-2_28

works that have considered uncertainty and in particular only breakdowns are taken as uncertain, and not the other parameters affecting the mining dispatching problem.

Underground mine mobile equipment dispatching is defined as the process of allocating available equipment of the mine to the different developed stopes, for the execution of the production operations, and scheduling working times of the allocated machines for every phase of the same production sequence (Hou et al. 2020). In this context, our study aims at quantifying the effect of stochastic processing times on an underground mine mobile equipment dispatching problem, in order to highlight the risk of not considering stochastic equipment working times. For that purpose, we first develop a mathematical model to solve the equipment dispatching problem while considering the machines working times as deterministic. Then, Monte Carlo simulation (MCS) is implemented to assess the reliability of the deterministic dispatching under stochastic environment. This relevant resolution scheme has been recently used in transportation context (Elgesem et al. 2018; Guimarans et al. 2018; Layeb et al. 2018) to measure the risk of ignoring the real stochastic nature of the environment.

The rest of the paper is structured as follows. Section 2 reviews the most related work on machines dispatching issues in underground mines, and highlights the importance of considering stochastic working times for such problems. Section 3 presents the main characteristics of the considered equipment dispatching problem. Section 4 reports the deployed mathematical model as well as the MCS-based approach. Section 5 discusses our preliminary experimental results. Finally, Sect. 6 draws conclusions and avenues for future research.

2 Related Work

Mobile equipment dispatching is defined as the process of allocating available equipment of the mine to the different developed stopes, mining production sites, for the execution of the production operations, and scheduling working times of the allocated machines for every phase of the same production sequence (Hou et al. 2020). During the last decades, some studies were conducted in order to treat dispatching issues in the mining industry, especially in the context of mining short-term scheduling. Beaulieu and Gamache (2006) implemented an enumeration tree based algorithm for solving a fleet management problem in underground mines. The sequence of the tree's states presents the shortest routes for the vehicles. The proposed dynamic algorithm showed efficiency in controlling underground mines' fleet system at the short-term level. Paduraru and Dimitrakopoulos (2019) worked differently on real time dispatching and scheduling issues in mining complex. Precisely, the authors introduce the reinforcement learning to respond to the insertion of new information that can be related to geological characteristics, availability of material transportation; i.e. mobile equipment availability, and processing characteristics. Geostatistical simulations were used to model these features in order to determine adequate destinations policy of extracted material, in real time. After determining several causes of uncertainty related to mines' processing activities (material extraction and transportation), Paduraru and Dimitrakopoulos (2019) state that this uncertainty is triggered by the interactions between several field's activities such as relations of cause effect between equipment queuing times and cycle times, extraction rates etc. Recently, Manríquez et al. (2020) introduced a Simulation-Optimization framework for the consideration of operational uncertainty in the generation of the mine's short-term schedule based on equipment KPIs: Availability and Utilization as introduced by Mohammadi et al. (2015). The yielding underground mine short-term schedule includes mobile machines allocation to stopes while considering uncertainty affecting their performance indicators.

Hou et al. (2020) investigated the problem of simultaneously dispatching the mobile equipment and sequencing stopes. The authors established a link between production process and dynamic resource scheduling. A multi-objective optimization model was proposed to minimize the gap time between two consecutive phases of mining production as well as the makespan of the stopes' production cycle. For the problem resolution, Hou et al. (2020) used a stope production phase algorithm that allows having possible scenarios of stopes schedule. The algorithm helps finding a coherent equipment assignment responding to the problem dynamic properties. After preparing different scenarios of stopes sequencing with equipment assignment, a genetic algorithm is introduced to reach the best solution meeting the two objectives of the main model, with the initial individuals being the established scenarios of the equipment assignment algorithm.

Differing from (Hou et al. 2020), our work considers clearly stochastic working times for the dynamic mobile equipment dispatching problem in underground mines and assesses the effect of their variability on the generated deterministic solution.

In the context of transportation, Jaoua et al. (2020) considered urban vehicles travel times as stochastic variables, fitted with the skewed Lognormal distribution, to investigate the effect of the flow pattern on the reliability of routes planning. They measured the risk of missing predefined time windows for vehicles when stochastic travel times are not considered. They found that at high variability levels, the deterministic solution is no longer efficient, thereby the need to consider stochastic work times for route planning.

Analogically to the transportation context, underground mining equipment working times are herein fitted with the unimodal Lognormal distribution. Their impact on the reliability of the deterministic dispatching is evaluated using Monte Carlo simulation.

3 Problem Description

3.1 Production Sequences in Underground Mines

Let's begin by presenting underground mines production main characteristics. Although several methods of ore extraction are used within underground mines, the production cycle remains the same and is based on six-unit operations: Rock drilling, charging, blasting, ventilation support, stope supporting, ore extraction, loading and transporting (Åstrand et al. 2018a,2018b; Hou et al. 2020; Song et al. 2015).

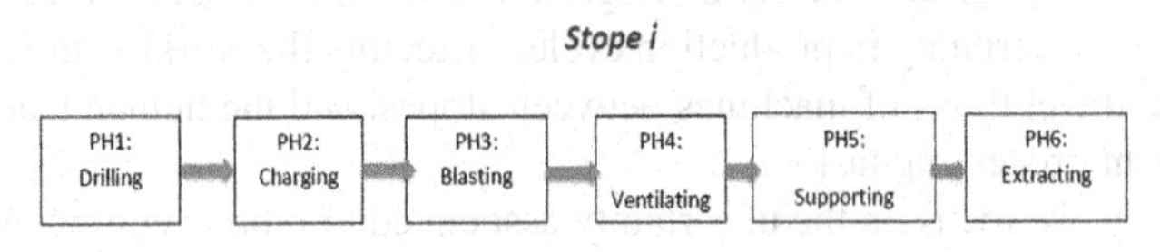

Fig. 1. A stope's production sequence

Figure 1 presents the six phases of production sequences. In an underground mine, every existing stope is in one of these six phases. Moreover, underground mining production sequence defines stopes' states or phases. The equipment dispatching realized in the present study is based on these phases.

3.2 Hypotheses Consideration

This study aims at resolving the problem of mobile equipment dispatching in an underground mine which consists of assigning machines executing the different phases of a stope production cycle to the different stopes being operated or planned to beoperated. This is a process evoked by the short-term planning of the mine under consideration.

Precisely, to solve the yielding equipment dispatching problem, the following assumptions are considered:

- The underground mine is composed of N stopes planned to be-operated for ore extraction. Every stope is operated according to the above mentioned six-phase production sequence.
- For each phase of the production cycle, a specific fleet of mobile equipment is dedicated. For each of the first five phases, a particular fleet of Jumbos is associated and for the extracting phase, a set of Load Haul Dumps (LHDs) is dedicated.
- For every phase, the machines are not identical. In fact, every single machine of a certain phase is characterized by its capacity, expressed in mass unit per time unit. It is worth to mention that capacity differs from one equipment to another which leads to variations in machines working times. For the computational experimentation, we used the average working times of the machines per stope and per phase.
- Each machine can break down during working time. breakdowns are possible at any time during the effective work in the mine and are part of the factors that create uncertainty in the working times of the machines.
- Based on the ore reserve and average machine working times, a predetermined set of machine types required for each stope is established.
- It is assumed that the work of the machines in each stope cannot be interrupted by moving to other stopes.
- Each machine cannot be operated in two or more stopes in parallel, and when a machine is inactive, it is considered to be available for each stope that requires it.
- It is possible to have many stopes of a same phase being operated at the same time.

3.3 Equipment Stochastic Working Times

Underground mines have always been known for their rigid environment due mainly to natural factors. This fact makes underground mining a difficult process, since its rigidity leads to uncertainty in production cycles, affecting the working times of mobile machines, the travel times of machines between stopes, and the human teams carrying out the different processing tasks.

In our study, we focus on the uncertainty associated to mobile equipment in underground mines. According to Mohammadi et al. (2015) and Samatemba et al. (2020), for underground mining equipment, the time to complete its assigned task, which refers to the machine's working time, is equal to the sum of the effective processing time, the machine's settlement time, and delays. Precisely, the equipment effective processing time is the time of executing the real task for which the machine is constructed; for example, for a drilling jumbo, it is the real time spent in drilling stopes. Here, the notion of effective processing time is considered as presented by Manríquez et al. (2020). Then, the settlement time is the time spent setting up the machine to perform a specific task. It can be considered in the case of the same mobile machine ensuring two or more different tasks. Finally, delays present the time spent waiting for the start time of the processing. For our case, delays can include machines failure or breakdowns or travel times between the different stopes of the underground mine. All of these parameters related to the machine's working time are considered as uncertain. It is assumed that working times are stochastic input parameters to our problem.

At this stage, it is worthy to note that when fitting an equipment working times, for a specific production phase at a specific stope, with the Lognormal distribution at different variability levels (as defined later in Sect. 4.2), the higher the coefficient of variation *(CV)*, the greater the variability in equipment working time. Illustrative histograms are reported in Fig. 2.

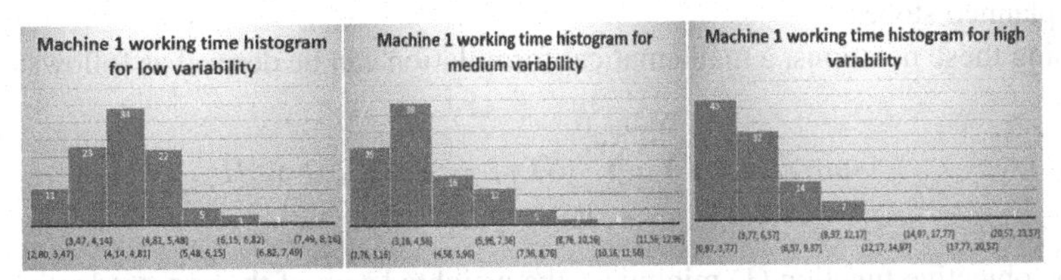

Fig. 2. Histogram of the working time of machine 1 of phase 1 at stope 1, respectively from the left: CV = 20%, CV = 40%, and CV = 70%

4 Problem Formulation

4.1 Mathematical Model

Now, let's turn our attention to proposing a mathematical model for this challenging equipment dispatching problem.

Sets

I = {1..., N}:Set of N stopes,

P = {1..., L}:Set of L possible phases,

K_p = {1...,E}:Set of available equipment for phase p.

Indices

i refers to stope *i*.

p refers to phase *p*.

p' refers to the phase following the phase *p*. For example, if phase *p* is "charging", *p'* is "blasting".

k refers to machine *k* of the set K_p of phase *p*.

Parameters
PR_{ipk}: Working time of machine k of phase p at stope i,
N_{ip}: Number of required machines for the execution of phase p at the stope i,
B: Big positive number.
Decision Variables
S_{ipk}: continuous non-negative variable that reflects the start time of the execution of phase p at stope i by equipment k,
E_{ipk}: continuous non-negative variable that reflects the end time of the execution of phase p at stope i by equipment k,
W_{ipk}: continuous non-negative variable that reflects the start time of phase p at stope i of assigned equipment k,
V_{ipk}: continuous non-negative variable that reflects the end time of phase p at stope i of assigned equipment k,
ST_{ip}: continuous non-negative variable that reflects the start time of phase p at stope i,
ET_{ip}: non-negative variable that reflects the end time of phase p at stope i,
X_{ipk}: binary decision variable that takes value 1 if equipment k is assigned to phase p at stope i, 0 otherwise,
H: continuous non-negative variable that reflects the completion time of the execution of the planned stopes.

Using these notations, a mathematical formulation can be derived as follows:

$$\min Z = w_1 \sum_{i=1}^{N} \sum_{p=1}^{L-1} (ST_{i(p+1)} - ET_{ip}) + w_2 H \tag{1}$$

The objective function (1) minimizes the weighted sum of the non-productive time between two consecutive phases and the end time of stopes processing completion H. Expression (1) should be minimized subject to the following constraints:

$$E_{ipk} = S_{ipk} + PR_{ipk} \quad \forall i \in I, p \in P, k \in K_p \tag{2}$$

Constraints (2) express the end working time of each machine k of phase p at stope i.

$$\text{Either } (\mathrm{E}_{i'pk} \leq S_{ipk}) \text{ Or } (\mathrm{E}_{ipk} \leq S_{i'pk}) \quad \forall i, i' \in I / i' \neq i, p \in P, k \in K_p \tag{3}$$

To express the prohibition of a same machine parallel work in two different stopes or more, an "either-or" relationship is defined in (3). It avoids overlapping both intervals of working times of the same machine k at two different stopes.

$$\sum_{k=1}^{N_p} X_{ipk} = N_{ip} \quad \forall i \in I, p \in P \tag{4}$$

Constraints (4) ensure that the sum of machines operating at phase p for every stope is equal to the defined required number of machines per specific phase, for the same stope.

$$\text{If } (X_{ipk} = 1) \text{ Then } (W_{ipk} = S_{ipk}) \quad \forall i \in I, p \in P, k \in K_p \tag{5}$$

In order to define working start times of assigned machines, “If, then” relationships are introduced in (5).

$$\text{If } (X_{ipk} = 0) \text{ Then } (W_{ipk} = B) \quad \forall i \in I, p \in P, k \in K_p \tag{6}$$

Relationships (6) allow the start time of equipment that is not assigned to a specific stope at a phase p to be ignored when calculating the start time of that phase at that stope.

$$\text{If } (X_{ipk} = 1) \text{ Then } (V_{ipk} = E_{ipk}) \quad \forall i \in I, p \in P, k \in K_p \tag{7}$$

Relationships (7) express the working end times of the assigned equipment.

$$\text{If } (X_{ipk} = 0) \text{ Then } (V_{ipk} = 0) \quad \forall i \in I, p \in P, k \in K_p \tag{8}$$

Relationships (8) enable the end time of equipment that is not assigned to a specific stope at a phase p to be ignored when calculating the end time of that phase at that stope.

$$ST_{ip} \leq W_{ipk} \quad \forall i \in I, p \in P, k \in K_p \tag{9}$$

Constraints (9) define the start time of each phase per stope.

$$ET_{ip} \geq V_{ipk} \quad \forall i \in I, p \in P, k \in K_p \tag{10}$$

Constraints (10) define the end time of each phase per stope.

$$H \geq ET_{ip} \quad \forall i \in I, p \in P, k \in K_p \tag{11}$$

Constraints (11) establish the end time of stopes process completion time as it is the maximum of different phases end times of all planned stopes.

$$ST_{ip'} \geq ET_{ip} \quad \forall i \in I, p \in P, p' \in \{2...L-1\} \tag{12}$$

Constraints (12) ensure that the start time of the next phase is necessarily greater than or equal to the end time of the current phase.

$$S_{ipk}, W_{ipk}, ST_{ip} \geq 0 \quad \forall i \in I, p \in P, k \in K_p \tag{13}$$

$$X_{ipk} \in \{0, 1\} \quad \forall i \in I, p \in P, k \in K_p \tag{14}$$

Constraints (13)-(14) define the nature of the decision variables.

To conclude, Model (1)-(14) is a valid formulation for the considered mobile equipment dispatching Problem in underground mine.

4.2 Monte Carlo Simulation-Based Sampling Approach

To analyze the effect of equipment stochastic working times on the underground mine machines dispatching problem, a MCS-based sampling approach is used. It consists of introducing stochastic input parameters, i.e. stochastic machines working times, to Model (1)–(14), for generating objective functions values of different stochastic instances. More precisely, M independent scenarios of equipment working times per instance are considered. A scenario is composed of machines sets of working times per phase and stope. Let $f_{1l}, f_{2l}..., f_{Ml}$ denote stochastic objective functions of the problem, for instance l. For scenarios generation, the Lognormal distribution is adopted with three different coefficients of variation. After obtaining stochastic objective functions values of the problem, the absolute gap between both, the deterministic objective function value Z_l and the stochastic one, is calculated for every scenario realization. The absolute gap and the mean absolute gap are defined as follows:

$$Gap_l^j = \left|Z_l - f_l^j\right| \quad \forall j \in \{1...M\} \tag{15}$$

$$MeanGap_l = \frac{1}{M}\sum_{j=1}^{M} Gap_l^j \tag{16}$$

Moreover, let **T** be the vector of random working times variables of machine per phase and per stope. Actually, each machine working time per phase and per stope is a random variable following the Lognormal distribution. Thus, **T** is defined as follows:

$T = (T_1,...,T_s)$ where $s = \sum_{p=1}^{L} K_p N$.

Then, the location and the scale parameters of the lognormal distribution are calculated as follows:

$$\mu_t = \ln(E[T_t]) - 0.5\ln(1 + \frac{Var[T_t]}{(E[T_t])^2}) \quad \forall t \in \{1...s\} \tag{17}$$

$$\sigma_t^2 = \ln(1 + \frac{Var[T_t]}{(E[T_t])^2}) \quad \forall t \in \{1...s\} \tag{18}$$

where E[T*t*] is the arithmetic mean of T*t*, and Var[T*t*] is its variance.

5 Results and Discussion

The mathematical model (1)–(14) was implemented in OPL language and solved using the commercial state-of-the-art solver CPLEX, version 12.6 with its default settings, on a computer processor intel Core i5, 7th generation running at up to 3.1GHz. In order to understand the utility of the mathematical formulation, an illustrative example is first introduced. Then, the results of six different instances of equipment dispatching problem for an underground mine are presented. Results for the stochastic case are reported to extract some useful insights about its effect on the proposed dispatching solution.

5.1 The Deterministic Case

Let's begin by treating an illustrative example that presents the case of three to-be-operated phases for three different stopes. We assume having as number of available machines $K1 = 3$, $K2 = 3$, $K3 = 4$ (and as weights in the objective function (1) $w1 = w2 = 0.5$).

For the first phase's execution, 3 machines are required for the different stopes. For the second phase, 3 machines are required at the first stope, and 2 machines are needed at stopes 2 & 3. And finally, for the third phase's execution, 2 machines are required at the different stopes. Mean equipment working times are extracted from (Hou et al. 2020).

The optimal assignment for this illustrative example is deployed in Fig. 3.

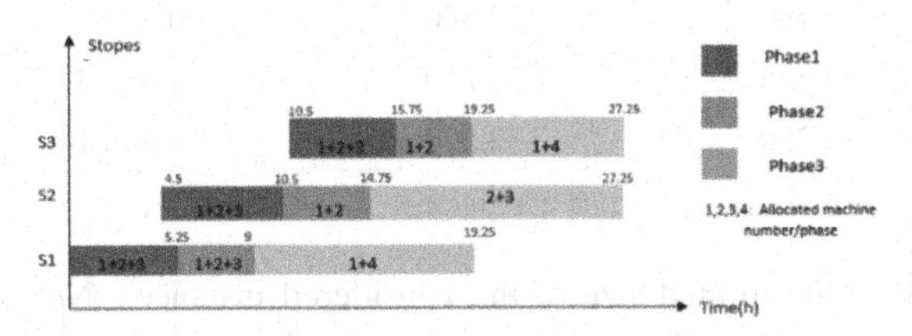

Fig. 3. Diagram of stopes production phases for the illustrative example of 3 phases & 3 stopes

Considering the execution of Phase 2, it is noticed that Stope 2 and Stope 3 use both the same equipment, namely Machine 2. In fact, Machine 2 starts operating Phase 2 at Stope 2 (the priority is for the second stope for this case), at time $h_1 = 10.5$ h. It finishes its job at time $h_2 = 14.75$ h, and becomes then "Available". At $h_3 = 15.75$ h, Machine 2 starts operating Phase 2 at Stope 3, and finishes its work at $h_4 = 19$ h. It is also the same case for Machines 1, 2, and 3 of Phase 1. The priority is first considered for the first stope, after that the second stope is processed, and the three assigned machines finish their work at the third stope. We do not have overlapping working intervals for a same machine proceeding at two different stopes. It is the case for all of the shared machines of different phases. The end completion time of stopes processing H is equal to 27.25 h.

Depending on the start and end times of the phases per stope, the production cycles of the phases differ from one stope to another. This can be due to the difference between the working times of the machines for the same phase and the different stops, and to the dispatching generated. For example, the first phase's execution lasts 5.25 h for both Stopes 1 and 3, whereas it lasts 6h for the second stope. The second phase lasts 3.75 h, 4.25 h and 3.5 h for stopes 1, 2, and 3, respectively. For the first and second phase, the gap between phases' production cycles for many stopes is not large when compared with the gap between stopes phase 3 execution times. In fact, a gap of 4.5 h is found between the execution times of Phase 3 for Stope 3 and Stope 2.

Now, based on this illustrative example and the case of "Sanshandao" gold mine in china treated by Hou et al. (2020), six realistic instances are derived by varying the number of stopes and phases. Deterministic results are as reported in Table 1. Precisely, Model (1)–(14) was solved to optimality while taking the mean values as the machines working times, for each instance. The column headings of Table 1 are as follows: Inst. = name of the instance; Z^* = value of the optimal solution in hours, H^* = end time of the stopes processing in hours, CPU Time = CPU time required to compute Z^*.

Recall that the objective function presents the equally weighted sum of the total production end time and between-phases gaps. Table 1 shows that Model (1)–(14) were successful in achieving equipment dispatching with zero inter-phase deviations. This

Table 1. Results of the deterministic case

Inst	Z^*	H^*	CPU time
1	7.000	14.000	1.27 s
2	8.625	17.250	4.27 s
3	10.375	20.750	70.53 s
4	10.750	21.500	0,52 s
5	13.625	27.250	4 min 30,78 s
6	16.250	32.500	5 min 14,8 s

fact may be due to the limited size of the considered instances. Not surprisingly, by increasing the constraints and variables of the model (i.e., increasing the number of stops and phases), its CPU time increases.

5.2 The Stochastic Case

In order to measure the variability effect of stochastic machines working times on the underground mine equipment dispatching problem, we use the MCS Sampling. This method is tested on instances Inst.1- Inst.6. 100 stochastic objective functions are generated, for every variability level, and for every instance, basing on the absolute gap between the stochastic and the deterministic optimal solutions, and the mean absolute gap are calculated. Numerical results showed that the higher the variability level is, the higher the Mean Absolute Gap (*MeanGap*) is, for all of the instances. The highest *MeanGap* is noticed for Inst.4, for an arithmetic coefficient of variability $CV = 70\%$; $MeanGap = 5.329$ h. It presents 50% of the instance's deterministic objective function. For low variability level, *MeanGap* is of low values indicating that for low working times variability rate, the objective function of the dispatching problem lightly varies. For medium variability levels ($CV = 40\%$), *MeanGap* values present higher values when compared to *MeanGap* of low variability. The impact of variability is clearer at the high level.

For a detailed study of the absolute gap between stochastic and deterministic solutions, Box & Whisker plots representation are used, as shown in Fig. 4.

For a low variability level ($CV = 20\%$), maximum values of absolute gap (*Gap*) do not exceed 1.8 h for Inst.1, Inst.2 and Inst.3 instances. For a medium variability level ($CV = 40\%$), 50% of *Gap* values do not exceed 1.6 h, and 75% of the Gaps data do not exceed the value of 2 h for the three first instances. However, for a high variability level ($CV = 70\%$), 75% of the first instance *Gap* values are below 4 h (below 46% of the deterministic objective function of Inst.2), and maximum *Gap* values can reach 8.3 h for the first instance and 5.5 h for the Inst.3 instance. As for the last three instances, for low variability level, 75% of absolute gap is below 2 h. For high variability, maximum *Gap* can reach 44% of the objective function for Inst.5, and 49% of the deterministic objective function of Inst.6. For medium variability level, maximum *Gap* values can reach 27% of the deterministic solution. So as a first conclusion, for the medium variability (CV

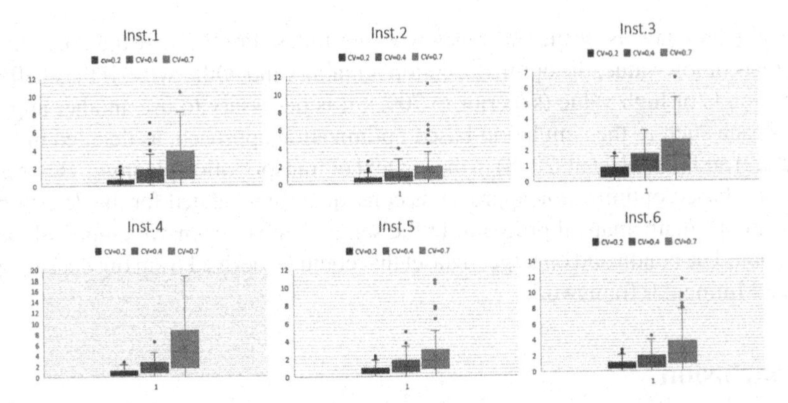

Fig. 4. Box & Whisker Plots

= *40%)* and high variability *(CV = 70%)*, it may be necessary to consider stochastic equipment working times for the dispatching problem: The problem's deterministic solution is no longer optimal, so there is a need to use other approaches for solving the problem such as the simulation-based optimization approach (Jaoua et al. 2020).

Beside their effect on the variation of the objective function, equipment stochastic working times affect the assignment of machines within the underground mine. This effect can be illustrated in the case of the illustrative example by comparing one of its stochastic solutions to the optimal deterministic solution. The example of the 76th stochastic generation at a high variability level *(CV = 70%)* is displayed in Fig. 5. The gap between stochastic and deterministic solutions is equal to 7.94 h. Ac: for this specific generation, the end time of the total production cycle has increased up to 43.12 h and the assignment has changed too for specific stopes at phase 2 and phase3. With the change of assignment, clearly considering this stochasticity level induces a complete change of priorities as well as phases start and end times per stope.

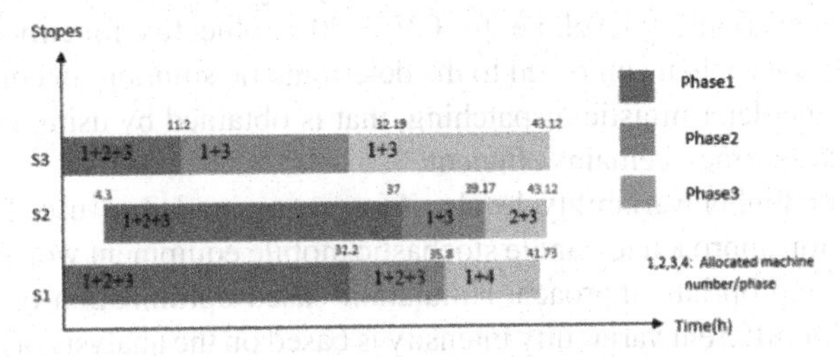

Fig. 5. Diagram of stopes production phases for the illustrative example at the 76th stochastic working times generation

These preliminary computational results illustrate the validity of the implemented model and confirm the intuitive outcomes of the proposed approach. It is clearly seen, that at low variability levels, the solution of the mathematical program is robust, which is not the case for higher variability levels. For enhancing the effectiveness of the planning system optimization, it may be necessary for underground mines planners to refer to the historical data analysis of the machines working times. Precisely, if the coefficient of variation is of low value, the optimal solution remains valid. Otherwise, if the coefficient of variation is of high value (superior to 70%), it is necessary to use another appropriate approach such as the simulation-based optimization approach, as detailed in (Jaoua et al. 2020) and (Layeb et al. 2018) in the context of transportation planning. Besides, the simulation-based optimization approach is consequently validated for the deterministic case, with our mathematical program. In the same context, extensive empirical experiments should be conducted on larger dataset that could be derived from big data analytics within a Mining 4.0 framework.

6 Conclusion

This work investigated the variability effect of the equipment working times on the equipment dispatching in the context of underground mines. It underlines the disadvantages of not taking into account the stochastic working times of the mobile equipment on the quality of the equipment dispatching solution, for the short-term planning of an underground mine.

The proposed approach consists of first, generating the fleet dispatching by solving a mathematical model that considers deterministic mean values of the equipment working times. Then, MCS is used to assess the reliability of the deterministic dispatching under stochastic environment, i.e. under stochastic working times. In the present work, mobile equipment working times of six realistic instances are fitted with the skewed unimodal Lognormal distribution. Preliminary computational results illustrative that for high variability level of the equipment working times, the mean absolute gap between the stochastic solution and the deterministic one, can exceed the value of the deterministic objective function, by 50% of its value. Whereas for low variability level, the deterministic objective function of the dispatching problem lightly varies. Furthermore, results reveal that for high variability level, absolute gap per scenario can reach 49% and 27% of the deterministic objective function, for respectively high and medium variability levels. In the case of low variability level, i.e. for CV = 20%, objective functions of generated scenarios lightly vary when compared to the deterministic solution. In consequence, for low variability, the deterministic dispatching, that is obtained by using mean values of equipment working times, remains efficient.

However, for higher variability levels, it is recommended to use the simulation-based optimization approach to handle stochastic mobile equipment working times. The decision of the appropriate approach, simulation-based optimization or mathematical programming, for different variability intensity is based on the analysis of the equipment historical data.

As a future work, we aim to conduct an extensive computational experimentation on larger dataset to assess the proposed approach. Such larger sets of data could be obtained using big data analytics in the new mining era: Mining 4.0. In the same vein, the stochastic waiting and queuing times of mobile machines, which are included in stochastic working times, could be obtained from real-time records of connected IoT sensors in the mining environment.

References

Åstrand, M., Johansson, M., Greberg, J.: Underground mine scheduling modelled as a flow shop: a review of relevant work and future challenges. J. South Afr. Inst. Min. Metall. **118**(12), 1265–1276 (2018a)

Åstrand, M., Johansson, M., Zanarini, A.: Fleet scheduling in underground mines using constraint programming. In: van Hoeve, W.-J. (ed.) CPAIOR 2018. LNCS, vol. 10848, pp. 605–613. Springer, Cham (2018b). https://doi.org/10.1007/978-3-319-93031-2_44

Beaulieu, M., Gamache, M.: An enumeration algorithm for solving the fleet management problem in underground mines. Comput. Oper. Res. **33**(6), 1606–1624 (2006)

Campeau, L.P., Gamache, M.: Short-term planning optimization model for underground mines. Comput. Oper. Res. **115**, 104642 (2020)

Elgesem, A.S., Skogen, E.S., Wang, X., Fagerholt, K.: A traveling salesman problem with pickups and deliveries and stochastic travel times: An application from chemical shipping. Eur. J. Oper. Res. **269**(3), 844–859 (2018)

Guimarans, D., Dominguez, O., Panadero, J., Juan, A.A.: A simheuristic approach for the two-dimensional vehicle routing problem with stochastic travel times. Simul. Modell. Pract. Theory **89**, 1–14 (2018)

Hou, J., Li, G., Wang, H., Hu, N.: Genetic algorithm to simultaneously optimise stope sequencing and equipment dispatching in underground short-term mine planning under time uncertainty. Int. J. Min. Reclam. Environ. **34**(5), 307–325 (2020)

Jaoua, A., Layeb, S.B., Rekik, A., Chaouachi, J.: The shared customer collaboration with stochastic travel times for urban last-mile delivery. In: Sustainable City Logistics Planning: Methods and Applications, vol. 1, chapter III, pp. 63–96. Nova science publishers (2020)

King, B., Goycoolea, M., Newman, A.: Optimizing the open pit-to-underground mining transition. Eur. J. Oper. Res. **257**(1), 297–309 (2017)

Layeb, S.B., Jaoua, A., Jbira, A., Makhlouf, Y.: A simulation-optimization approach for scheduling in stochastic freight transportation. Comput. Ind. Eng. **126**, 99–110 (2018)

Manríquez, F., Pérez, J., Morales, N.: A simulation–optimization framework for short-term underground mine production scheduling. Optim. Eng. **21**(3), 939–971 (2020). https://doi.org/10.1007/s11081-020-09496-w

Mohammadi, M., Rai, P., Gupta, S.: Performance measurement of mining equipment. Int. J. Emerg. Technol. Adv. Eng. **5**(7), 240–248 (2015)

Paduraru, C., Dimitrakopoulos, R.: Responding to new information in a mining complex: fast mechanisms using machine learning. Min. Technol. **128**(3), 129–142 (2019)

Samatemba, B., Zhang, L., Besa, B. Evaluating and optimizing the effectiveness of mining equipment; the case of Chibuluma South underground mine. J. Cleaner Prod. **252**, 119697 (2020)

Song, Z., Schunnesson, H., Rinne, M., Sturgul, J.: Intelligent scheduling for underground mobile mining equipment. PLoS One, **10**(6), e0131003 (2015)

Yu, X., Zhou, L., Zhang, F.: Self-localization algorithm for deep mine wireless sensor networks based on MDS and rigid subset. In: 2017 IEEE 2nd International Conference on Opto-Electronic Information Processing (ICOIP), pp. 6–10. IEEE, July 2017

Vertical Stability Constraints in Combined Vehicle Routing and 3D Container Loading Problems

Corinna Krebs[1](✉) and Jan Fabian Ehmke[2]

[1] Department of Management Science, Otto von Guericke University Magdeburg, Universitätsplatz 2, 39106 Magdeburg, Germany
Corinna.Krebs@ovgu.de

[2] Department of Business Decisions and Analytics, University of Vienna, Kolingasse 14-16, 1090 Vienna, Austria
Jan.Ehmke@univie.ac.at

Abstract. The vertical stability of the cargo is one of the most important loading constraints, since it ensures parcels from falling on the ground. However, frequently considered constraints either lead to unstable positions, are too restrictive or have high complexity. This paper focuses on the evaluation of different vertical stability constraints, analyses corner cases and introduces a new improved constraint. For the first time, constraints based on the science of statics are considered in the context of combined Capacitated Vehicle Routing Problem with Time Windows and 3D Loading (3L-VRPTW). All constraints are embedded in an established hybrid heuristic approach, where an outer Adaptive Large Neighbourhood Search tackles the routing problem and an inner Deepest-Bottom-Left-Fill algorithm solves the packing problem. For the computational tests, we use a well-known instance set enabling a comparison w.r.t. the number of customers, the number of items and the number of item types. Based on the impact on the objective values and on the performance, we give recommendations for future work.

Keywords: Vehicle Routing Problem · 3D loading · Vertical stability

1 Introduction

Fueled by the Corona pandemic, the number of shipped parcels worldwide has increased significantly. In 2019, the worldwide parcel volume exceeded the mark of 100 billion parcels for the first time. However, it is estimated to be even between 115 and 132 billion parcels worldwide in 2021. High growth rates are also forecasted for the coming years. Within the next six years, the parcel volume could double (220 to 262 billion) according to PitneyBowes [2020]. Along with this, the daily challenges of packing parcels into cargo spaces are intensifying. As a result, the number of conciliation applications to the Bundesnetzagentur (Federal Network Agency for Electricity, Gas, Telecommunications, Post and

M. Mes et al. (Eds.): ICCL 2021, LNCS 13004, pp. 442–455, 2021.
https://doi.org/10.1007/978-3-030-87672-2_29

Railway) in Germany increased by 28% to 1.861 applications in the year 2020 (see BNetzA [2020]). Thereby, 25.3% of applications accounted for damaged consignments. Consequently, the stable loading of parcels is still a major challenge in the planning process and will continue to be an important constraint in the future.

This paper evaluates vertical stability constraints in the field of the 3L-VRPTW, which is a combination of the Vehicle Routing with Time Windows and the 3D Container Loading Problem. In practice, high stability requirements are met by forbidding the parcels to overhang (Full Base Support). Due to this restrictive constraint, other solutions that also achieve stable positions and incur less costs are excluded. For ensuring vertical stability, most approaches require the support of a certain ratio of the base area by directly underlying items. As stated in Ceschia et al. [2013] and in Krebs et al. [2021], this requirement can lead to unstable stacks so that two new vertical stability constraints are regarded. Hence, this paper compares five vertical stability approaches from the literature, which are based on the support of the base area and/or on the science of statics. The latter is evaluated for the first time in the context of the 3L-VRPTW. By indicating common weaknesses and corner cases, we introduce a new vertical stability constraint which enlarges the solution space and is based on the support ratio of the base area and on the science of statics.

We use an established hybrid heuristic for tackling the Vehicle Routing and the Container Loading Problem. The routing heuristic is based on the Adaptive Large Neighbourhood Search (ALNS) by Koch et al. [2018] calling for each route a modified packing heuristic based on the Deepest-Bottom-Left-Fill (DBLF) algorithm proposed by Krebs et al. [2021]. For the computational tests, we use a well-known instance set from the literature, where the number of items, item types and customers varies systematically, and evaluate the received results in terms of performance and solution quality in comparison to the established Full Base Support constraint.

The related literature is reviewed in Sect. 2. The considered problem (3L-VRPTW) is formulated in Sect. 3. In Sect. 4, the vertical stability constraints are described in detail. The hybrid heuristic is briefly explained in Sect. 5. Section 6 presents computational results analysing the impact of the vertical stability approaches on the 3L-VRPTW. Finally, conclusions are drawn in Sect. 7.

2 Literature Review

In this section, we provide a brief literature overview over different constraints dealing with vertical stability. First, we show the relevant literature in the context of the 3D Container Loading Problem (3D CLP). Then, we present the literature for its combination with the Vehicle Routing Problem (3L-CVRP).

2.1 3D Container Loading Problem

Vertical stability constraints prevent items from falling down on the ground, on top of other items or at the operator while (un-)loading. Various approaches have been formulated in the 3D Container Loading Problem, where a set of items has to be arranged within a container. Most approaches consider the support of the base area, which must be supported either with a defined ratio (Partial Base Support) or completely (Full Base Support) by directly underlying items. This support ratio ranges between 0.55 (Mack et al. [2004]) and 1.0 (Full Base Support, e.g. Ngoi et al. [1994], Fanslau and Bortfeldt [2010], Ceschia and Schaerf [2010]). Since these constraints can lead to either unstable stacks (Partial Base Support) or are too restrictive (Full Base Support), other constraints based on the science of statics have been introduced, where in general the center of gravity of an item or a stack must be supported (e.g. in De Castro Silva et al. [2003], Lin et al. [2006] and Ramos [2015]). Mack et al. [2004] use a combination of the support of the center of gravity and of the item base area w.r.t. items laying at the ground. However, the support of the center of gravity does not guarantee highly stable item stacks as will be demonstrated in Sect. 4.

2.2 3L-CVRP and Extensions

When introducing the combined Vehicle Routing and the 3D Container Loading Problem (namely "3L-CVRP"), Gendreau et al. [2006] use the Minimal Supporting Area (aka Partial Base Support) constraint to ensure stable item positions with a support ratio of 0.75. Therefore, this constraint is the most established one in the 3L-CVRP and its extensions. However, Ceschia et al. [2013] show that unstable item stacks can occur when using the Minimal Supporting Area constraint and propose therefore the Multiple Overhanging constraint. In Krebs et al.[2021], several vertical stability constraints, such as the Minimal Supporting Area and the Multiple Overhanging, are examined, and the Top Overhanging constraint is introduced. So far, vertical stability constraints based on the science of statics have not been considered yet in this problem field. Hence, this paper is intended to fill this area.

3 Problem Formulation

Following the formulation as used in Krebs et al. [2021], the 3L-VRPTW is described as follows: Let $G = (N, E)$ be a complete, directed graph, where N is the set of $n+1$ nodes including the depot (node 0) and n customers to be served (node 1 to n), and E is the edge set connecting each pair of nodes. Each edge $e_{i,j} \in E$ $(i \neq j, i, j = 0, ..., n)$ has an associated routing distance $d_{i,j}$ $(d_{i,j} > 0)$. The demand of customer $i \in N \setminus \{0\}$ consists of c_i cuboid items. Let m be the total number of all demanded items. Moreover, time windows are considered by assigning three times to each node i: the ready time RT_i, which is the earliest possible start time of service, the due date DD_i, the latest possible start time, and the service time ST_i, which specifies the needed time to (un-)load all c_i items of a customer i.

Each item $I_{i,k}$ $(k = 1, ..., c_i)$ is defined by mass $m_{i,k}$, length $l_{i,k}$, width $w_{i,k}$ and height $h_{i,k}$. Each item has a fragility flag $fg_{i,k}$, grouping items into fragile items ($fg_{i,k} = 1$) or not fragile items. The items are delivered by at most v_{max} available, homogenous vehicles. Each vehicle has a maximum load capacity D and a cuboid loading space defined by length L, width W and height H.

The point of origin of a Cartesian coordinate system is assumed to be located in the deepest, bottom, leftmost point of the loading space. The driver's cab is located behind it accordingly. The length, width and height of the loading space are parallel to the x-, y- and z-axes. The placement of an item $I_{i,k}$ is defined by $(x_{i,k}, y_{i,k}, z_{i,k})$ of the corner which is closest to the point of origin.

It is assumed that each vehicle has a constant speed of 1 distance unit per time unit. If a vehicle arrives at a node before its ready time, it has to wait until the ready time is reached.

Let v_{used} be the number of used vehicles in a solution. A solution is a set of v_{used} pairs of routes R_v and packing plans PP_v, whereby the route R_v ($v = 1, ..., v_{used}$) is an ordered sequence of at least one customer and PP_v is a packing plan containing the position within the loading space for each item included in the route. The total number of items in a route R_v is described by t_v.
A solution is feasible if

(S1) All routes R_v and packing plans PP_v are feasible (see below);
(S2) Each customer is visited exactly once;
(S3) The number of used vehicles v_{used} does not exceed the number of available vehicles v_{max};
(S4) Each packing plan PP_v contains all t_v items.

A route R_v must meet the following routing constraints:

(R1) Each route starts and terminates at the depot and visits at least one customer;
(R1) The vehicle does not arrive after the due date DD_i of any location i.

Each packing plan must the following loading constraints.

(C1) *Geometry*: The items must be packed within the vehicle without overlapping;
(C2) *Orthogonality*: The items can only be placed orthogonally inside a vehicle;
(C3) *Rotation*: The items can be rotated 90° only on the width-length plane;
(C4) *Load Capacity*: The sum of masses of all included items t_v of a vehicle does not exceed the maximum load capacity D;
(C5) *LIFO*: No item is placed above or in front of item $I_{i,k}$, which belongs to a customer served after customer i;
(C6) *Vertical Stability*: Each item is placed stable either on the vehicle ground or on top of other items;
(C7) *Fragility*: No non-fragile items are placed on top of fragile items.

The 3L-VRPTW aims at determining a feasible solution minimizing the objective values, e.g. number of used vehicles v_{used} and the total travel distance ttd, and meeting all constraints.

4 Vertical Stability Constraints

In this section, we first explain the calculation of the center of gravity and then, we summarize six constraints for the consideration of vertical stability and present the new Static Stability constraint. Each constraint has been implemented in our solution validator, which can check the feasibility of solutions[1]. The implementation for Ramos [2015] is available here[2].

4.1 Calculation of Center of Gravity

For the approaches based on the science of statics, the center of gravity of an item ($CG_{i,k}$) must be calculated. It is supposed that the mass of an item is homogeneously distributed so that the center of gravity corresponds to the center of volume. Therefore:

$$CG_{i,k} = (x_{i,k} + \frac{l_{i,k}}{2} \quad , \quad y_{i,k} + \frac{w_{i,k}}{2} \quad , \quad z_{i,k} + \frac{h_{i,k}}{2}). \tag{1}$$

Moreover, the center of gravity of a group of items can be calculated. Let G be a set of items belonging to a group. The center of gravity of the group of items is calculated by weighting the center of gravity of each item r ($r \in G$). The equation is as follows:

$$CG_{group} = (\frac{\sum_{r \in G}(x_{CG_r} \cdot m_r)}{\sum_{r \in G}(m_r)} \quad , \quad \frac{\sum_{r \in G}(y_{CG_r} \cdot m_r)}{\sum_{r \in G}(m_r)} \quad , \quad \frac{\sum_{r \in G}(z_{CG_r} \cdot m_r)}{\sum_{r \in G}(m_r)}). \tag{2}$$

4.2 Minimal Supporting Area

The Minimal Supporting Area (aka Partial Base Support) constraint is one of the most considered vertical stability constraints in the field of the combined Vehicle Routing and 3D Container Loading Problem (3L-CVRP), since it is included in the original problem formulation by Gendreau et al. [2006]. For the Minimal Supporting Area constraint, a certain ratio α of the base area of an item must be supported by the upper surface of the directly underlying items. The parameter α is set to 0.75 in the field of the 3L-CVRP.

However, corner cases can occur leading to theoretically feasible, but actually unstable item arrangements, since it is assumed that all items have the same density. From one level to another, the item's length (or width) is enlarged by $\frac{1}{\alpha}$. This may lead to an item stack as visualized in Fig. 1. Since the support is calculated based on the directly underlying items, the stack is feasible. However, when calculating the center of gravity of the stack (see Eq. 2), the dimensions lay outside of the first item. Therefore, the stack would topple.

[1] see https://github.com/CorinnaKrebs/SolutionValidator.

[2] see https://github.com/CorinnaKrebs/StaticStabilityRamos.

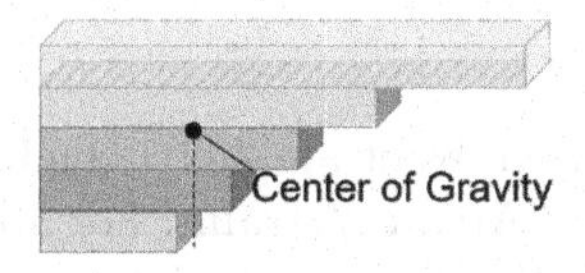

Fig. 1. Feasible, but unstable stack

4.3 Full Base Support

As the name suggests, the base area of each item must be fully supported in this constraint. Consequently, overhanging items are not allowed. The approach is the same as for the Minimal Supporting Area though the support parameter α must be one ($\alpha = 1.0$). Apparently, this constraint is the most restrictive one.

4.4 Multiple Overhanging

This constraint has been introduced in Ceschia et al. [2013] and is described in more detail in Krebs et al. [2021]. Hereby, all items of a stack are allowed to overhang. However, the Minimal Supporting Area must be obeyed at each level of the stack.

Let $I_{i,k}$ be an item for which the constraint should be checked. First, it is necessary to determine all items which directly or indirectly support item $I_{i,k}$ and store them in set U. An item I_u supports $I_{i,k}$ directly if the upper surface of item I_u has direct contact with the base area of item $I_{i,k}$. An item I_u supports $I_{i,k}$ indirectly if I_u directly supports any placed item which directly supports $I_{i,k}$. Based on the set U, the levels are created. Each upper surface of an item I_a ($I_a \in U$) defines a level (see Fig. 2b). Another item I_b ($I_b \in U, I_a \neq I_b$) counts to the level if the upper surface of I_b is at the same level as of the level or if the upper surface of I_b is above the level and the base area of I_b is below the level (see Fig. 2c). The supporting area units of each level are summed to calculate the total support for item $I_{i,k}$ for each level. Then, it can be checked if each level obeys the Minimal Supporting Area constraint.

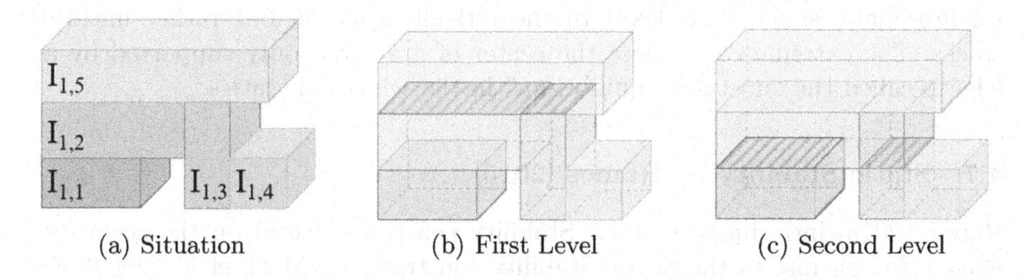

(a) Situation (b) First Level (c) Second Level

Fig. 2. Creation of Levels for Support Area Calculation, exemplary for $I_{1,5}$

4.5 Top Overhanging

This constraint proposed in Krebs et al. [2021] combines the Full Base Support and the Minimal Supporting Area constraint: For all items except the topmost item of a stack, the Full Base Support constraint is applied, while for the topmost item, the Minimal Supporting Area constraint must be obeyed. Consequently, only the topmost item of a stack is allowed to overhang.

The implementation is rather simple: when placing an item $I_{i,k}$ on top of another one, the distance between the upper surface of the last placed item and the vehicle ceiling is calculated. Then, the item with the smallest height of all items not packed yet is determined. If the distance to the ceiling is smaller than the smallest height, it is not possible to pack another item on top of that stack. Therefore, the item $I_{i,k}$ is allowed to overhang obeying the Minimal Supporting Area constraint. Otherwise, another item could be stacked on top of item $I_{i,k}$ and therefore, $I_{i,k}$ must fulfil the Full Base Support constraint.

4.6 Static Stability by Mack et al. [2004]

In the Static Stability constraint introduced by Mack et al. [2004], the center of gravity of each item is calculated according to Eq. 1 and must be supported by the directly underlying items. Moreover, it is required that a certain ratio of the base area of an item is supported indirectly by the items laying on the ground.

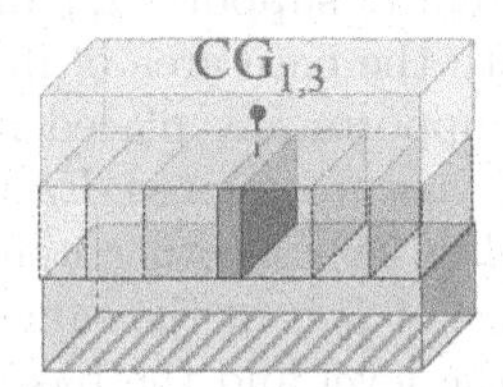

Fig. 3. Example for a feasible stack applying Mack et al. [2004]

Figure 3 indicates that the item supporting the center of gravity could be infinitesimally small. This leads to theoretically feasible, but rather unstable stacks. The extreme case where the center of gravity is only supported by an edge is called the "unstable equilibrium" in the science of statics.

4.7 Static Stability by Ramos [2015]

Ramos [2015] introduces a Static Stability constraint based on the science of statics. In contrast to the Static Stability constraint by Mack et al. [2004], the mass of all items placed above the current placed item $I_{i,k}$ is considered, which can shift the center of gravity. This new acting point can be calculated as shown in Eq. 2.

Then, two cases can occur: If the acting point is supported by directly underlying items, then the position of item $I_{i,k}$ is stable. If the acting point is not supported, then all supporting edges between the item $I_{i,k}$ and its directly underlying items are determined and stored in set T. Based on set T, a convex hull representing the support polygon is calculated. This enables a point-in-polygon to check if the application point lies within the support polygon indicating a stable position for item $I_{i,k}$. If item $I_{i,k}$ is stable according to one of the two cases, the check of the Static Stability constraint is recursively repeated starting from all items directly supporting item $I_{i,k}$ to the items placed on the ground (Fig. 4).

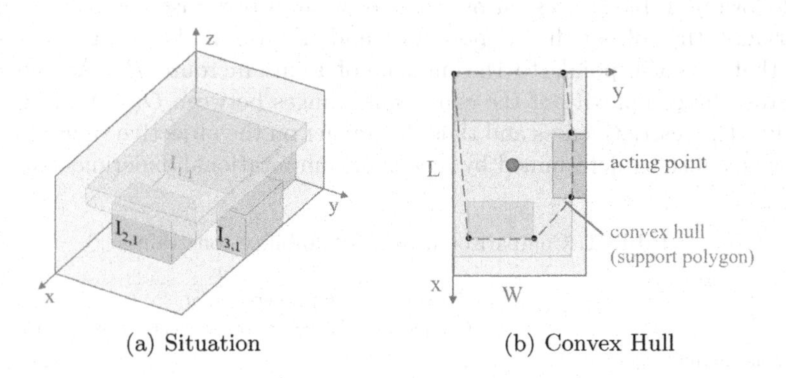

(a) Situation (b) Convex Hull

Fig. 4. Example for the convex hull determination

Similarly to the constraint by Mack et al. [2004], relatively small items lead to rather unstable stacks.

4.8 New Static Stability

In this paper, we want to introduce a new Static Stability constraint. It is inspired by encountered corner cases in the Minimal Supporting Area constraint, science of statics based approaches, and by the Multiple Overhanging constraint. The main idea is that the center of gravity of an item must be supported at each level of the stack. A level is created in the same way as described in Sect. 4.4. For each level, the minimum and maximum edges of the level are determined. In particular, the rightest edge, the leftest edge, the foremost edge and the rearmost edge of all items belonging to the level are searched. Then, it is checked whether the center of gravity is within the frame spanned by the edges (see Fig. 5). Therefore, it is not required that the center of gravity is directly supported as shown in Fig. 5c, Level 1. Due to the support of the center of gravity at each level, item stacks as in Fig. 1 are prevented.

As shown before, the sole support of the center of gravity does not ensure highly stable stacks. Therefore, the Minimal Supporting Area constraint with $\alpha = 0.75$ must also be obeyed. This prevents arrangements as shown in Fig. 3.

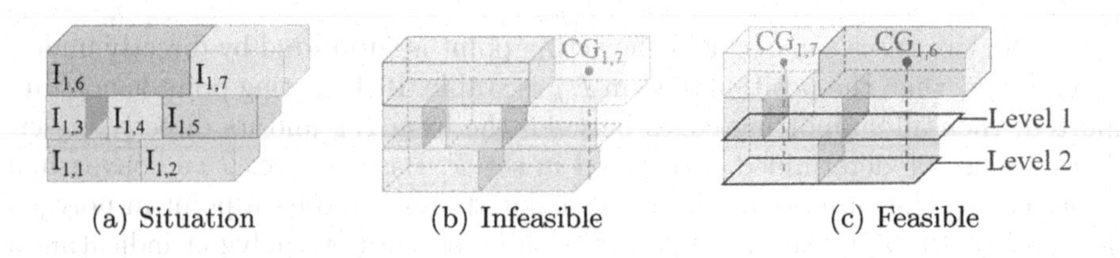

(a) Situation (b) Infeasible (c) Feasible

Fig. 5. Example for Consideration of Static Stability

4.9 Summary

The following Table 1 gives an overview over the worst-case complexity of the constraints, the role of the support area and the use of the science of statics. Note that t_v corresponds to the number of items in route R_v. As the table indicates, the complexity of the constraints ranges between $O({t_v}^2)$ and $O({t_v}^3)$. However, the restrictiveness and thus the impact on the objective values for each constraint must be determined by means of computational experiments.

Table 1. Comparison of vertical stability constraints

	Worst Case Complexity	Consideration of		Corner Cases
		Support area	Science of statics	
Minimal supporting area	$O({t_v}^2)$	✓		✓
Full base support	$O({t_v}^2)$	✓		
Multiple overhanging	$O({t_v}^3)$	✓		
Top overhanging	$O({t_v}^2)$	✓		
Static stability by Ramos [2015]	$O({t_v}^2 \log(t_v))$		✓	✓
Static stability by Mack et al. [2004]	$O({t_v}^2)$	✓	✓	✓
New static stability	$O({t_v}^2)$	✓	✓	

t_v = total number of items in R_v

5 Hybrid Algorithm

We implement the previously described vertical stability constraints in a hybrid heuristic approach, which is described in detail in Krebs et al. [2021] and in Koch et al. [2018]. In the following, we briefly describe the algorithm and focus on the extensions required.

Since the 3L-VRPTW is a combination of the Capacitated Vehicle Routing Problem with Time Windows (VRPTW) and the 3D Container Loading Problem, the problem is decomposed and separate algorithms are used to solve each subproblem. The routing part is tackled with an Adaptive Large Neighbourhood Search as proposed in Krebs et al. [2021] and in Koch et al. [2018], which is a modification of the heuristic by Ropke and Pisinger [2006a] and by Ropke and Pisinger [2006b]. The algorithm is shown in Algorithm 1.

The initial solution is constructed by using the Savings Heuristic proposed by Clarke and Wright [1964]. In every iteration, it is tried to improve the current

Algorithm 1. Hybrid Heuristic Algorithm

Input: Instance Data, parameters
Output: best feasible solution s_{best}

```
 1: construct s_init by Savings Heuristic
 2: s_best := s_curr := s_init
 3: do
 4:     select number of customers to be removed n_rem
 5:     select destroy operator dest and repair operator rep
 6:     determine next feasible³ solution s_next := rep(dest(s_curr, n_rem))
 7:     for each route R_v in s_next do
 8:         feasible := true
 9:         if Deepest-Bottom-Left-Fill(R_v) not feasible then
10:             feasible := false
11:             break
12:         end if
13:     end for
14:     if feasible AND Simulated Annealing (s_next) then
15:         s_curr := s_next
16:         update s_best
17:     end if
18:     update selection probabilities of operators after defined number of iterations
19: while no Stopping Criterion reached
20: return s_best
```

solution. Hereby, a set of v_{used} routes is created. The set of routes is feasible if the routing constraints as described in Sect. 3 are obeyed. Each feasible set of routes is evaluated concerning the objective values, whereby the minimization of the number of used vehicles has the highest priority, the total travel distance the second highest priority. A Simulated Annealing approach as proposed by Kirkpatrick et al. [1983] enables the acceptance of inferior feasible solutions in order to enlarge the search.

For each feasible route, the packing algorithm (Deepest-Bottom-Left-Fill algorithm) is called, which is shown in Algorithm 2. As the name suggests, the items are placed in the deepest, bottommost, leftmost position. Hereby, a list of possible placement spaces is sorted according to the DBL policy. Then, for every item of the route, it is tried to find a feasible position by checking every space of the list until a feasible position is found. A position is feasible if all loading constraints are obeyed. If no feasible position can be found, the Adaptive Large Neighbourhood Search must find a new set of routes in the next iteration. If feasible positions for every item of the set of routes can be found, the feasible solution is stored. The entire algorithm stops either when reaching a defined run time limit, after conducting a defined number of total iterations, or after a total number of iterations without improvement (Stopping Criteria). In the end, the overall best solution is presented.

[3] According to Routing Constraints, see Sect. 3

Algorithm 2. Deepest-Bottom-Left-Fill with Spaces

Input: Instance data
Output: Feasibility, Packing Plan PP_v

```
1: initialize sorted sequence of items IS
2: initialize set of unique available spaces S
3: for each item I_c ∈ IS do
4:     for each space sp ∈ S do
5:         for each permitted orientation do
6:             if item I_c fits in space sp AND placement is feasible then
7:                 save placement for I_c
8:                 create new spaces
9:                 sort spaces based on DBL
10:                erase space sp and too small spaces
11:                continue with next item
12:            end if
13:        end for
14:    end for
15:    if no feasible position found then return false
16: end for
17: return true
```

6 Computational Studies

In this section, we evaluate the impact of the vertical stability constraints and show their performance in comparison. Hereby, we use our instance set[4] enabling the evaluation w.r.t. the number of customers, items and item types. The instances have either 20, 60 or 100 customers, which demand either 200 or 400 items in total. These items differ in their homogeneity: Either there are only three item types (very homogeneous), 10 item types or 100 different item types (very heterogeneous). Each instance is tested five times and we present the average results. All results along with detailed packing plans are available via Github[5]. The hybrid algorithm is implemented in C++ as single-core, x64-application and is compiled using the GCC version 4.8.3 compiler. The experiments were executed on a High Performance Cluster, Haswell-16-Core with 2.6 GHz. In terms of the routing parameters, the same are used as described in Koch et al. [2018]. The loading parameters are set according to the values used in the literature.

As shown before, the Full Base Support constraint is the most restrictive one but ensures stable arrangements. The target of this paper is to find a vertical stability constraint that guarantees highly stable positions of items and is less restrictive. Therefore, we exclude approaches where the unstable equilibrium can occur. Consequently, the impact of the constraints Full Base Support, Multiple Overhanging, Top Overhanging and the new Static Stability is investigated. The following Table 2 shows the impact of these constraints w.r.t. to the

[4] see https://doi.org/10.24352/UB.OVGU-2020-139.
[5] see https://github.com/CorinnaKrebs/Results.

average number of used vehicles (v_{used}), total travel distance *ttd* and run time in comparison to the Full Base Support constraint. Nevertheless, we tested all approaches described in in Sect. 4 and provide the results online.

Table 2. Average results per vertical stability constraint

		n			*m*		*Item types*			**Total**
		20	60	100	200	400	3	10	100	
Full base support	Sum v_{used}	549.20	4,392.40	4,713.60	3,449.40	6,205.80	2,677.40	3,186.40	3,791.40	**9,655.20**
	Sum *ttd*	54,956.51	340,558.21	406,434.31	326,656.33	475,292.70	233,987.06	267,122.94	300,839.03	**801,949.03**
	Avg. *time* [s]	1,809.61	1,435.49	2,559.26	1,658.60	2,261.04	1,680.67	2,103.83	2,094.97	**1,959.82**
Multiple Overhanging	Diff. v_{used}	–6.34%	–11.86%	–1.95%	0.61%	–10.78%	1.53%	–4.80%	–14.13%	**–6.71%**
	Diff. *ttd*	–2.28%	–6.90%	–1.71%	0.55%	–7.05%	1.99%	–3.31%	–9.14%	**–3.95%**
	Diff. *time*	59.51%	147.96%	40.67%	101.79%	56.35%	102.47%	61.14%	68.51%	**75.58%**
Top Over-hanging	Diff. v_{used}	–0.95%	–8.36%	1.35%	3.50%	–6.93%	2.14%	–2.24%	–7.78%	**–3.20%**
	Diff. *ttd*	0.00%	–4.49%	0.62%	2.29%	–4.26%	2.30%	–1.25%	–4.92%	**–1.59%**
	Diff. *time*	61.94%	147.84%	40.55%	102.74%	56.27%	102.64%	61.08%	69.43%	**75.93%**
Static stability	Diff. v_{used}	–5.79%	–9.05%	–7.29%	–4.91%	–9.73%	–0.63%	–6.20%	–14.73%	**–8.01%**
	Diff. *ttd*	–2.07%	–6.31%	–5.31%	–3.60%	–6.83%	–0.48%	–4.65%	–10.20%	**–5.51%**
	Diff. *time*	24.81%	25.06%	12.71%	24.08%	14.52%	31.94%	14.74%	11.67%	**18.56%**

As expected, the enlargement of the solution space leads to better objective values (lower number of used vehicles and the total travel distance) for all constraints. Regarding the total level of restrictiveness (impact on the objective values), the following descending order occurs: Full Base Support, Top Overhanging, Multiple Overhanging and Static Stability. Hereby, the Static Stability constraint creates route plans with 8% fewer vehicles and a shorter total travel distance by 5.5% on average. However, compared to the Full Base Support constraint, the Static Stability constraint causes an increase of the run time by 18.56% on average.

Regarding the number of customers or items, the correlations between these instance features and the impact on the objective values are not evident. However, the Static Stability constraint shows significantly smaller fluctuations than the Multiple Overhanging and the Top Overhanging constraint.

Concerning the number of item types, the Top Overhanging and the Multiple Overhanging constraints lead to an increase in the number of used vehicles and total travel distance by around 2% for instances with three item types. This is due to the fact that the constraints enable overhanging and therefore prevent homogeneous item stacks at the same time. Therefore, gaps between items can occur. Consequently, more vehicles are needed, which also results in a longer total travel distance. As the Static Stability constraint is less restrictive than the Top Overhanging and the Multiple Overhanging constraint, gaps are more likely to be filled. However, as the number of item types increases, the reduction of the objective values is achieved by all constraints. In terms of the Multiple Overhanging constraint, 14.13% fewer vehicles are used, the Top Overhanging shows a decline of 7.78%, the Static Stability of even 14.73%.

All constraints have in common that the average run time increases significantly compared to the Full Base Support constraint. On average, the Multiple Overhanging and the Top Overhanging constraints lead to an increase of the run time of around 75%; for the Static Stability constraint, the increase is only 18.5%. In general, the higher the number of customers or items, the higher the run time. However, according to the results, this is not the case (see $n = 100$ or $m = 400$). The reason is that at the same time, the difference to the maximum run time gets smaller or is exploited.

Based on the described effects, we generally recommend using the Static Stability constraint. However, for time critical or highest stability requirements, the Full Base Support constraint should be used.

7 Conclusion

In this paper, we compared six vertical stability constraints and introduced a new one (Static Stability) in the context of combined Vehicle Routing Problem with Time Windows and 3D Container Loading (3L-VRPTW). The constraints from the literature are based on a defined support ratio of the base area and/or on the support of the center of gravity of each item. We showed that most approaches can lead to unstable item stacks for specific corner cases. Therefore, we introduced a new approach in this paper, which covers common corner cases. This constraint is based on the science of statics and on the support ratio of the base area of an item. As the computational experiments show, the new Static Stability constraint is less restrictive than most of the other approaches and therefore achieves a reduction of the number of used vehicles by 8% and the total travel distance by 5.5% on average, compared to the most restrictive constraint – the Full Base Support. However, the run time increases by almost 19%. Therefore, we recommend the new Static Stability constraint if the reduction of the objective values is of first priority. If the run time or a high stability of items is more important, the Full Base Support constraint should be used.

References

BNetzA. Tätigkeitsbericht schlichtungsstelle post 2020. Technical report, Bundesnetzagentur für Elektrizität, Gas, Telekommunikation, Post und Eisenbahnen, Tulpenfeld 4, 53113 Bonn (2020)

Ceschia, S., Schaerf, A.: Local search for a multi-drop multi-container loading problem. J. Heurist. **19**, 01 (2010). https://doi.org/10.1007/s10732-011-9162-6

Ceschia, S., Schaerf, A., Stützle, T.: Local search techniques for a routing-packing problem. Comput. Ind. Eng. **66**(4), 1138–1149 (2013). https://doi.org/10.1016/j.cie.2013.07.025. ISSN 0360-8352

Clarke, G., Wright, J.W.: Scheduling of vehicles from a central depot to a number of delivery points. Oper. Res. **12**(4), 568–581 (1964). http://www.jstor.org/stable/167703. ISSN 0030364X, 15265463

De Castro Silva, J.L., Soma, N.Y., Maculan, N.: A greedy search for the three-dimensional bin packing problem: the packing static stability case. Int. Trans. Oper. Res. **10**(2), 141–153 (2003). https://doi.org/10.1111/1475-3995.00400

Fanslau, T., Bortfeldt, A.: A tree search algorithm for solving the container loading problem. INFORMS J. Comput. **22**, 222–235 (2010). https://doi.org/10.1287/ijoc.1090.0338

Gendreau, M., Iori, M., Laporte, G., Martello, S.: A tabu search algorithm for a routing and container loading problem. Transp. Sci. **40**(3), 342–350 (2006). https://doi.org/10.1287/trsc.1050.0145. ISSN 0041–1655

Kirkpatrick, S., Gelatt, C.D., Vecchi, M.P.: Optimization by simulated annealing. Science **220**(4598), 671–680 (1983)

Koch, H., Bortfeldt, A., Wäscher, G.: A hybrid algorithm for the vehicle routing problem with backhauls, time windows and three-dimensional loading constraints. OR Spectr. **40**(4), 1029–1075 (2018). https://doi.org/10.1007/s00291-018-0506-6

Krebs, C., Ehmke, J.F., Koch, H.: Advanced loading constraints for 3D vehicle routing problems. OR Spectr. **4**, 1–41 (2021). https://doi.org/10.1007/s00291-021-00645-w

Lin, J.-L., Chang, C.-H., Yang, J.-Y.: A study of optimal system for multiple-constraint multiple-container packing problems. In: Ali, M., Dapoigny, R. (eds.) IEA/AIE 2006. LNCS (LNAI), vol. 4031, pp. 1200–1210. Springer, Heidelberg (2006). https://doi.org/10.1007/11779568_127

Mack, D., Bortfeldt, A., Gehring, H.: A parallel hybrid local search algorithm for the container loading problem. Int. Trans. Oper. Res. **11**(5), 511–533 (2004). https://doi.org/10.1111/j.1475-3995.2004.00474.x

Ngoi, B.K.A., Tay, M.L., Chua, E.S.: Applying spatial representation techniques to the container packing problem. Int. J. Prod. Res. **32**(1), 111–123 (1994). https://doi.org/10.1080/00207549408956919

PitneyBowes. Pitney bowes parcel shipping index reports continued growth as global parcel volume exceeds 100 billion for first time ever. Pitney Bowes Parcel Shipping Index (2020). https://www.pitneybowes.com/au/newsroom/press-releases/pitney-bowes-parcel-shipping-index-reports-continued-growth-as-global-parcel.html

Ramos, A.G.: Analysis of cargo stability in container transportation (2015)

Ropke, S., Pisinger, D.: A unified heuristic for a large class of vehicle routing problems with backhauls. Eur. J. Oper. Res. **171**(3), 750–775 (2006). https://doi.org/10.1016/j.ejor.2004.09.004. ISSN 0377–2217

Ropke, S., Pisinger, D.: An adaptive large neighborhood search heuristic for the pickup and delivery problem with time windows. Transp. Sci. **40**(4), 455–472 (2006). https://doi.org/10.1287/trsc.1050.0135

Automated Tour Planning for Driving Service of Children with Disabilities: A Web-Based Platform and a Case Study

Mahdi Moeini(✉) and Lukas Mees

Chair of Business Information Systems and Operations Research (BISOR), Technische Universität Kaiserslautern, 67663 Kaiserslautern, Germany
mahdi.moeini@wiwi.uni-kl.de, lmees@rhrk.uni-kl.de

Abstract. In this paper, we focus on a real-world problem called *Kindergarten Tour Planning Problem* (KTPP), which corresponds to a case study. In the KTPP, the objective consists in running a driving service for a group of children with disabilities to a central kindergarten. We formulate this problem as a *Mixed-Integer Linear Program* (MILP), which can be solved by any standard MILP solver. However, for practical use, we designed a simple yet effective heuristic to find good-quality solutions in short computation time. We conducted computational experiments, on randomly generated instances, to verify effectiveness of our heuristic by benchmarking it versus the state-of-the-art solver Gurobi Optimizer. Moreover, we introduce and present a publicly-available web-based platform that we have developed for practical use.

Keywords: Transportation on Demand · Open vehicle routing problem · School bus routing problem · Health care services · Heuristic

1 Introduction

Transportation systems are essential and non-negligible parts of our daily life. In this context, *Transportation on Demand* (ToD) is defined based on the request of users for transporting passengers and/or goods from some specified points to destinations [13]. A well-known example of ToD is *dial-a-ride* services which can be provided for a set of handicapped or elderly people [8,12]. In fact, in such a dial-a-ride service, based on the requests of passengers, each of them is picked up from a point to be dropped off at another location.

In a similar context, a regional branch of the *Deutsches Rotes Kreuz* (DRK), the German Red Cross, located in the city of Landstuhl (Rhineland Palatinate, Germany), offers a driving service to multiple facilities that take care of people with disabilities. More precisely, the driving service picks up people in an area of about 25 km radius covering these facilities, and then drives them to their respective destinations. One of these facilities is an integrative kindergarten in the city of Landstuhl, where about 70 children are currently served by the driving

M. Mes et al. (Eds.): ICCL 2021, LNCS 13004, pp. 456–470, 2021.
https://doi.org/10.1007/978-3-030-87672-2_30

service of the DRK. Indeed, there are currently 13 equipped vans such that each van can provide enough space and equipment for up to six passengers in addition to a driver and a co-driver or supervisor. Even though the number of vans is relatively flexible, the amount of passengers per vehicle may change due to different restrictions, e.g., respecting social distance during a pandemic. The main task is twofold: determining optimal allocations of passengers to the vans and optimizing driving routes from pickup positions to the kindergarten. In addition, the drivers are volunteers who keep the vans at home and start and end each tour at their respective private address. Finally, because of the limited patience of disabled children, the travel time should be as short as possible. Consequently, it is important to take this fact into consideration to generate short driving tours.

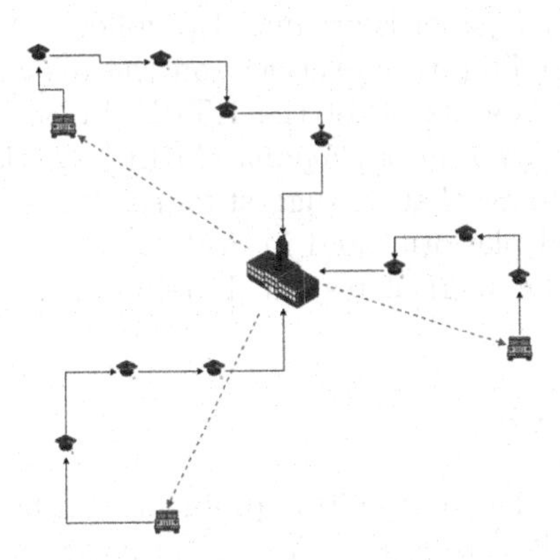

Fig. 1. A visualization of the Kindergarten Tour Planning Problem.

Figure 1 visualizes the general setup of the problem, where the central school-like building mimics the referred kindergarten, the little graduation hats symbolize the locations of children to be picked up, and the yellow school buses represent the location of the available drivers. While the black arrows illustrate the routes on which each child is picked up and brought to the facility, the dashed arrows represent the empty way back for the driver after dropping off the children.

Currently, all the planning of the tours is done manually, which is quite time consuming, and without a distinctive approach to minimize operational costs. Even though the planning might be suitable for a time horizon of one year, any fluctuation, e.g., absence or sickness of a driver, the planning requires a maintenance, which make a challenge for the responsible employees because the process requires considerable additional time and energy. Moreover, the recent Corona pandemic, with its direct contact restrictions, showed that new circumstances can impose a full rescheduling of all tours in a short amount of time. Such a situation can not only affect the group of children that need to be transported or the available drivers, but also the permitted capacity of each van. Consequently, for the cooperating branch of the DRK, it is quite advantageous to have an assistant software/tool that allows for fast and efficient tour planning of the driving service. The current paper is the result of our cooperation with the DRK through which we solve their tour planning problem that we call *Kindergarten Tour Planning Problem* (KTPP). For this purpose, we have designed and implemented a heuristic solution method as well as a web-based platform that has problem solving and visualization features to address the practical challenges and the context of the KTPP. Moreover, with the objective of verifying efficiency of the heuristic, we have conducted computational experiments on randomly-generated instances. The results of the heuristic are compared with those of the state-of-the-art solver Gurobi Optimizer [17]. The numerical results show the effectiveness of the heuristic.

The remainder of this paper is organized as follows. After this introduction, Sect. 2 presents a short literature overview on the relevant existing literature, and we give a formal description of the KTPP. In Sect. 3, we formulate the KTPP an a *Mixed-Integer Linear Program* (MILP). Section 4 is devoted to the presentation of the heuristic that we suggest for solving the KTPP. In Sect. 5, we describe our web-based platform, and in Sect. 6, we report our computational experiments and their numerical results. Finally, in Sect. 7, we present some concluding remarks.

2 Literature Review

A detailed literature review on the ToD problems can be found in [13]. In this section, we present a short overview on some problems which are quite similar to the KTPP, and we provide its formal description.

The *Dial-a-Ride Problem* concerns a typical ToD service, where a set of users (customers) has pickup and delivery requests for two distinct points, e.g., from home to a hospital. The requests are fulfilled by a set of vehicles with the aim of minimizing the total cost [8,12].

Another ToD problem is the *School Bus Routing Problems* (SBRP), where the general purpose consists in picking up students to bring them to their respective schools [15,16,18–20]. In this context, a school represent a predetermined location, which can be any other central institution, e.g., a kindergarten.

Thereby, the SBRP can be divided into several sub-problems, e.g., the bus stop selection, bus route generation, or bus route scheduling problems [16]. In fact, the bus stop selection problem copes with selecting the right locations for picking up students, based on the number of students nearby. The bus route generation problem deals with finding the optimal routes to pick up the respective students in the best way, while the bus route scheduling problem considers ensuring specific time windows [16]. Further, different kinds of SBRPs can be classified along various attributes. For example. Ellegood et al. (2019) classify such problems into seven different categories, ranging from the actual type of sub-problem over the number of schools, service environment, load type, fleet mix, different objectives, and up to various kinds of constraints [16].

Indeed, the core of the SBRP might be the route planning, which can be seen as a *Vehicle Routing Problem* (VRP) [14], where a set of vehicles start their journey from a depot to serve a given set of customers, and to return to the depot by the end of mission. The objective of a VRP can be serving all customers in shortest possible time or just by the smallest number of vehicles. In this context, among all variantes of the VRPs, the routing part of the SBRP can actually be modelled by a specific form of the VRP, i.e., the *Open Vehicle Routing Problem* (OVRP) [9].

The OVRP considers a VRP where the vehicles do not need to return to their depot, but all tours end at the last stop, which transforms the tours into trips or more precisely into paths and open tours. To illustrate the similarity of the OVRP to the SBRP, consider an empty bus that starts from an origin point, e.g., the bus depot, picks up students by visiting bus stops, and ends at a destination point, where all students are dropped off at the school [16,21].

In this paper, with a motivation from a real-world context, we study a problem that we name the *Kindergarten Tour Planning Problem* (KTPP). To define the KTPP, assume that a set of children (kids) with disabilities, a central kindergarten, and a set of volunteer drivers as well as their equipped vans are given. The objective of the KTPP consists in finding the shortest possible tours to pick up all registered children and to bring them to the kindergarten. More precisely, in each tour, a driver starts at his/her private address (home), picks up a subset of disabled children, and drives them to the kindergarten. Once the children are dropped off, the driver returns to his/her private address. As soon as the kindergarten time is over, the same driver picks up the same group of children as in the morning to drive them back to their home, after which the driver returns to his/her private address. Once we have the tours and their length (in terms of time), the scheduling is done easily because there is no specific time windows and the children are available at the requested time.

Finally, in the KTPP, we make the following additional assumptions:

- We assume that each child is picked up at his/her respective home individually.
- The vans, which are used by the drivers, are standardized and all vans have a same limited capacity.
- Even though there might be more vans and drivers than necessary to transport the children, there is no need to use all of the vans.
- The drivers start their tours according to the assigned schedule and the children are ready to be picked up when they are told to be.

In the next section, we provide a mathematical formulation for route generation in the KTPP.

3 Mathematical Formulation of the KTPP

To present the mathematical formulation for the route generation of the KTPP, let us start with the notation that we require. Assume that a graph $G = (V, A)$

is given where V is the set of nodes and A is the set of arcs. Moreover, the symmetric matrix $D := [d_{ij}]$ contains the distance values d_{ij} between each pair of nodes $i, j \in V$. In addition, the set of nodes is partitioned into $V = \{0\} \cup I$, where node 0 being the central depot, or in the case of KTPP the kindergarten, and I is the set of all child locations. We assign to each node $i \in I$ a positive integer number q_i, which indicates the number of children that must be picked up at node i. In case of the KTPP, we assume that $q_i = 1$ for all $i \in I$. Nevertheless, it could also happen that multiple children from the same address want go to a same kindergarten, e.g., this can happen in the case of brothers or sisters with close age, twins or multi-apartment buildings.

Furthermore, we consider a set of k vans, each with identical capacity of Q, to pick up the children to bring them to the kindergarten. A route solution to the KTPP is a planar graph with star-shaped topology composed of k paths (open tours) that originate from the central depot (kindergarten). In fact, this graph corresponds to an OVRP solution.

Moreover, to simplify the modeling, we introduce an auxiliary "dummy" node d into the graph G [9]. The new graph G' is defined by the set of nodes $V' = V \cup \{d\}$ and the set of arcs A', where each node in I is connected to the dummy node d by using the following arc weights:

$$d'_{ij} = \begin{cases} 0 & : \quad \text{if } i \in I \text{ and } j \equiv d, \\ M & : \quad \text{if } i \equiv 0 \text{ and } j \equiv d, \\ d_{ij} & : \quad \text{otherwise,} \end{cases} \tag{1}$$

where M is a sufficiently large positive number.

To complete all required notation for the mathematical model, we define the decision variables u_i indicating the total number of children picked up by a van once leaving node i, and the binary decision variables x_{ij} as follows:

$$x_{ij} = \begin{cases} 1 & : \quad \text{if the arc } (i,j) \text{ is part of the route,} \\ 0 & : \quad \text{otherwise.} \end{cases} \tag{2}$$

Using the presented notation, similar to the case of the SBRP [9], the following *Mixed-Integer Linear Program* (MILP) serves at finding routes to pick up the children.

$$\min \left(\sum_{i \in V'} \sum_{j \in V'} c_{ij} x_{ij} + f \cdot k \right) \tag{3}$$

$$\text{subject to} \quad \sum_{i \in I} x_{0i} \leq k, \tag{4}$$

$$\sum_{i \in I} x_{id} \leq k, \tag{5}$$

$$\sum_{j \in I \cup \{d\}} x_{ij} = 1 \; : \; \forall i \in I, \tag{6}$$

$$\sum_{i \in I \cup \{0\}} x_{ij} = 1 \; : \; \forall j \in I, \tag{7}$$

$$u_i - u_j + Qx_{ij} + (Q - q_i - q_j)x_{ji} \leq Q - q_j \ : \forall i, j \in I \text{ where } i \neq j, \quad (8)$$

$$u_i \geq q_i \ : \forall i \in I, \quad (9)$$

$$u_i - q_i x_{0i} + Qx_{0i} \leq Q \ : \forall i \in I, \quad (10)$$

$$x_{ij} \in \{0, 1\} \ : \forall i \in I, \quad (11)$$

$$u_i \in \mathcal{Z}^{\geq 0} \ : \forall i \in I. \quad (12)$$

In this MILP formulation, objective function (3) defines the total traveling costs and fixed costs for all vans in use. Here, c_{ij} represents the costs for traversing arc (i, j) and is defined by $c_{ij} := \alpha \cdot d_{ij}$, where α is the traveling cost per distance unit. In the objective function, the second term represents the fixed costs for having used the k vans, where f is the actual fixed costs per van. Constraints (4) and (5) guarantee that at most k vans depart from the depot node 0 and end their tour at the "dummy" node d. The flow-conservation conditions, described by Constraints (6) and (7), assure that each intermediate node is traversed exactly once. Finally, Constraints (8)–(10) deal with the capacity of the vans and, in particular, certify that the capacity restrictions of vans are not exceeded (for more details, refer to [9] and references therein).

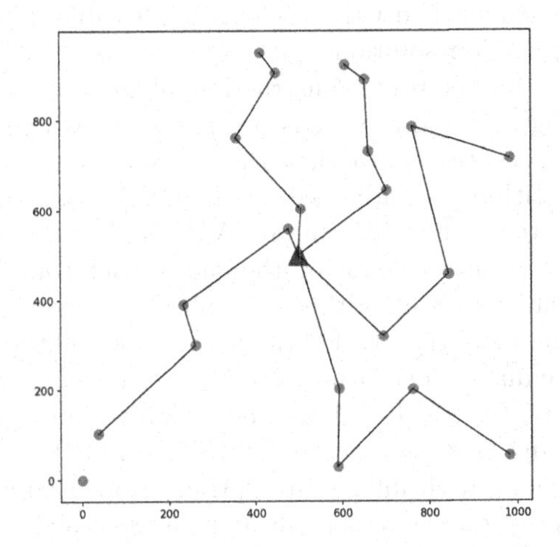

Fig. 2. The MILP solution for an instance of size 20 nodes, solved by the solver Gurobi Optimizer (with time limit of 1 h).

The MILP formulation (3)–(12), which allows to generate routes for the KTPP, can be solved by any standard MLIP solver, e.g., Gurobi Optimizer [17]. Figure 2 shows a sample solution that is obtained by solving MILP model (3)–(12). On this figure, we solve a randomly-generated instance composed of 20 children. The depicted set of open routes needs still assignment of drivers to become a complete solution to the KTPP. In Fig. 2, each red dot represents a pickup location of children, and the black links between the red dots are the selected paths for the vans to traverse. Furthermore, the kindergarten (central depot) is visible at $(500, 500)$ to which all open tours are connected. The aforementioned "dummy" node is located at $(0, 0)$, which, as intended, none of the tours are actually connected to.

As a variant of the VRP and the OVRP, the KTPP is an NP-hard problem. Consequently, solving KTPP is only possible for small-sized instances. To solve larger instances in reasonable computation time, we need to utilize heuristics and metaheuristics. Additionally, for the case study of the DRK, we cannot employ commercial solvers. Hence, for the practical use, we have designed a simple and fast heuristic to solve the KTPP, and we present the heuristic in the next section.

4 A Heuristic for Solving the KTPP

The algorithm that we suggest for solving the KTPP is composed of two main phases: *route generation* and *driver assignment.*

Due to large similarity of the KTPP to the bus route generation problem and the OVRP, a natural heuristic choice for creating KTPP routes consists in using similar heuristics that are used for solving the VRPs [14,16]. In this context, the VRP heuristics are classified as *construction* and *improvement* heuristics [14]. In fact, while the goal of construction heuristics is to build up initial feasible solutions, improvement heuristics use existing feasible solutions and refining techniques to obtain better solutions.

As stated in [16], for the route construction, algorithms based on the Clarke & Wright [11] savings method as well as two-phase algorithms perform well and they are commonly accepted approaches. Hence, we use a variant of the Clarke & Wright savings algorithm [11], adjusted for the OVRP [10], to construct KTPP routes. Indeed, due to the fact that in an OVRP and in the KTPP, there is no need to return to the central depot by the end of each tour, an adjustment to the Clarke & Wright savings algorithm is required to generate open tours [10].

According to [16], more than half of the SBRP-related literature use improvement heuristics to enhance algorithms. For this purpose, intra-route; in particular, k-opt operators, for $2 \leq k \leq 4$, as well as inter-route operations are commonly used. Following this approach, to solve the KTPP, we use 2-opt heuristic for intra-route along with inter-route optimization techniques. Algorithm 1 summarizes all steps of the route generation phase for solving the KTPP. More precisely, once a set of k routes is generated, where k is the number of available vans/drivers, the heuristic improves the routes by a sequence of intra-route and inter-route operations.

Once open tours, i.e., OVRP routes, for the KTPP are generated, we need to assign adequate drivers to the vans for which we use a simple heuristic. More precisely, once the open tours are constructed, we compute a distance matrix of all endpoints of these tours and the positions of the available drivers. Afterwards, for each open tour, we pick up the closest available driver in regard to the endpoint of the corresponding tour, and assign it to the tour. Once a driver gets assigned, he/she is then removed from the list of available drivers. This procedure is repeated until all tours receive a driver. The pseudocode of the heuristic is given in Algorithm 2.

Algorithm 1. Route Generation Heuristic (RGH)

```
1: procedure GENERATE AND IMPROVE(Routes)
2:     Inputs: set of nodes (V), vans' number (k), vans' capacity (Q), Routes = {}
3:     Output: Routes
4:     Routes ← use the adjusted savings algorithm of Clarke & Wright to create k
   OVRP tours
5:     improvedRoutes = {}
6:     for each tour in Routes do
7:         new = swapTwoNodes()
8:         if distance(new) < distance(tour) then
9:             improvedRoutes.append(new)
10:        end if
11:    end for
12:    Routes ← imporvedRoutes
13:    for each tour in Routes do
14:        for each nextTour in Routes do
15:            newRoutes = swapNodesBetweenTours(tour, nextTour)
16:            if distance(newRoutes) < distance(Routes) then
17:                improvedRoutes = newRoutes
18:            end if
19:        end for
20:    end for
21:    return imporvedRoutes
22: end procedure
```

Algorithm 2. Driver Assignment Heuristic (DAH)

```
1: procedure ASSIGN(tourEndpoints, driversSet)
2:     assignments = {}
3:     for each endpoint in tourEndpoints do
4:         for each driver in driversSet do
5:             if distance(endpoint,driver) < best then
6:                 best = driver
7:             end if
8:         end for
9:         assignments.update({endpoint:best})
10:        drivers.remove(best)
11:    end for
12:    return assignments
13: end procedure
```

5 A Web-Based Platform

For the case study, which motivates this contribution, we designed and implemented a web-based platform for which we provide a concise description in this section[1].

5.1 Architecture of the Platform

The implemented web-based platform permits an installation on a central web-server. Hence, even though all users will have access to the provided information, all possible future changes can only to be deployed to a single server. In other words, if multiple people use the application, possible changes in the application data do not need to be synchronized among many clients as all changes should happen on a single central database. However, this functionality might seem as a disadvantage because the current implementation does not provide any automated session management. Consequently, while different employees may work with the platform, no simultaneous change should be permitted; otherwise, there might be incorrect data in the database. To avoid such an issue, for each database, the tool should be installed on a server and get assigned to an administrator as the session manager. These conditions are fulfilled in our case study at the DRK.

We implemented the platform in Python through the web-framework *Flask* [2]. Further, the platform uses a MySQL database to store various information. As usual for web applications, the graphical user interface (frontend) is written in HTML, CSS, and JavaScript. Further, the CSS-framework *Bootstrap* [1] was used in implementing the frontend design, and through JavaScript, the library jQuery [4] is used to handle user interactions.

Another important technology used on the frontend is a Python library called *Folium* [3] that uses the JavaScript library *Leaflet* [5], which helps creating interactive maps, which are displayed to the user on the frontend. The last technology that we have used, and should be mentioned is the *Openrouteservice*, which is a geo-coding API [6]. It is used for retrieving important geo-data, which provides the longitude and latitude indices for locations of children and drivers.

For the end-user application, there are only two kinds of information that need to be stored in the database: children (or passengers to be transported) and drivers, where both groups share most of their respective information types. More precisely, next to a unique ID, personal data, e.g., name, address, and availability are stored for each object. Additionally, for both entities, the Openrouteservice API is used to automatically retrieve the geo-data in form of latitude and longitude indices for the given address to be stored in the database for the respective objects.

The actual core functionality provided by the platform is to apply the previously presented algorithms to the stored data and to determine which drivers

[1] The full package of the platform, including the source codes, is publicly available on: https://github.com/moeini-mahdi/AutomatedTourPlanning.git.

are supposed to drive which group of children and in which order. In addition to the stored active drivers and children, the only input that is given for each calculation is the amount of passengers that each van can transport. Once the algorithms are executed, their results are then fed to the Openrouteservice API. Afterwards, a route on real-world roads is displayed on an interactive map. Moreover, additional information about the length of the respective tour in kilometers as well as an estimated driving time are provided.

5.2 Design of the Platform

With the goal of developing an easy-to-use application with a clear design structure, we have selected a simple two-column design with a menu bar on the left-hand-side to switch between the different views and the actual content, which is on the right-hand-side. In terms of color scheme, the platform aligns with the official corporate design of the DRK [7].

In the following, we describe briefly different features of the platform. In particular, as it is shown in Fig. 3, the information of the children as well as the available drivers can be inserted into, edited, or deleted from the database in a simple way.

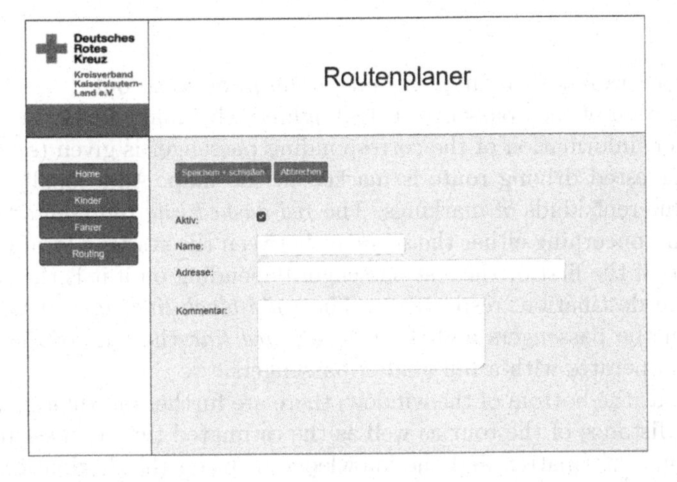

Fig. 3. A view of database object to insert the corresponding information. This view looks exactly the same for both, the children as well as the drivers, because they share the same set of attributes.

Figure 4 illustrates a sample output of a routing plan, which is created by the platform. The user can select the desired vehicle capacity and then start the calculation. Once the calculation is done (by the heuristic), the resulting tours are displayed on the same window. Figure 5 shows, in more detail, how an output tour is visualized and what information about the tour are given. The visualized map combines multiple functional features and information:

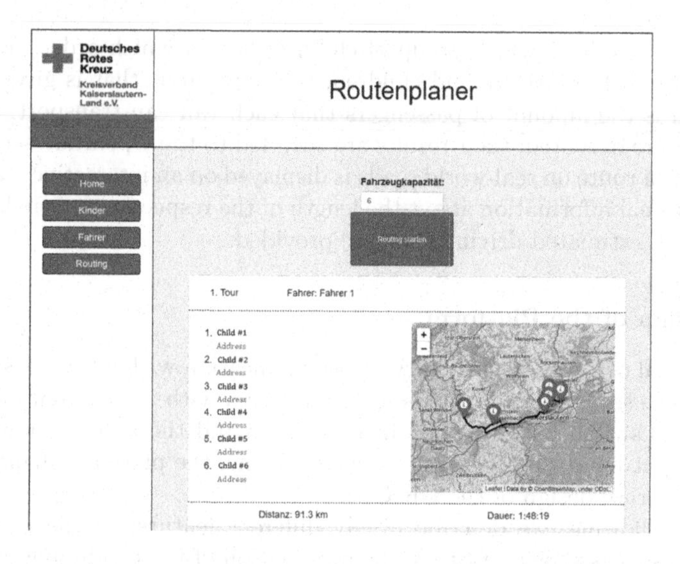

Fig. 4. A sample output view, where the name of the driver (Fahrer 1), the list of the registered children (to be picked up) together with their respective home address, a detailed map of the route, and its length (Distanz) as well as its duration (Daurer) are reported.

- The user can move the map around, zoom in, or zoom out.
- The position of each passenger is highlighted with markers. By hovering over a marker, information of the corresponding passenger is given (see Fig. 5).
- The suggested driving route is marked in the map. Additionally, there are three different kinds of markings. The *red-dashed line* represents the part of the tour concerning either the segment between the starting location and the position of the first or the last passenger depending on if it is the route to or from the destination, respectively. The *solid-black line* represents the route between the passengers and the *black-dashed line* the part of the tour that the van operates with a full load of passengers.
- Finally, at the bottom of the window, there are further information about the overall distance of the tour as well as the estimated time it takes to drive it. From this information and the knowledge of when the destination needs to be reached as well as how long it takes to pick up a passenger, the starting time is calculated easily.

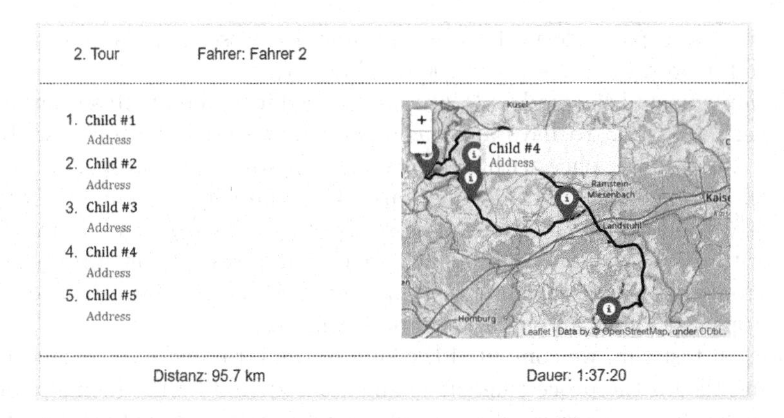

Fig. 5. A sample zoomed output, where by hovering over a marker, we observe the address of the corresponding child as well as his/her address.

6 Computational Experiments

From mathematical point of view, we have verified the model and the results by visualizing them and investigating the solutions numerically, in particular, by comparing the results of heuristic versus those of the MILP solver. From practical perspective, we had several discussion sessions with the DRK to incorporate exactly their assumptions in the model and to design the web-based tool to cover entirely the objectives of the DRK. The model is checked with real-world data, provided by the DRK; however, for the sake of privacy protection, their detailed results cannot be presented in this paper. Nevertheless, Figs. 4 and 5 illustrate some results after removing private data of the children.

Since we had the plan to provide the web-based platform at the service of the German Red Cross, we could not use a commercial solver for solving the MILP model. Hence, we designed and implemented a simple two-phase heuristic to solve the KTPP. Even though a heuristic does not guarantee global optimality of the solution, it is important to verify the quality of the solutions provided by the heuristic. For this purpose, we conducted computational experiments on randomly-generated instances, and compared the results of the heuristic versus those of the standard solver Gurobi Optimizer, which solves the MILP model (3)–(12). Hence, in our computational experiments, the driver-assignment phase is not taken into consideration because it is identical for both approaches.

6.1 Test Setting

We generated random instances with five different sizes, i.e., 10, 25, 50, 75, and 100 passengers (children). Each child is a node of the graph instance, generated by the Python's random class in an area of 1000×1000 distance units, where the kindergarten (depot) is located at the center of the area. In addition, we considered two capacity levels for the vans: 4 and 8 passengers. Moreover, in the objective function (3), we set $f = 1000$ and $\alpha = 1$.

We implemented the model (3)–(12) and solved it by the MILP solver Gurobi Optimizer 9.0.1 [17]. To have a fair comparison, we used the solver with two different time limits: once with only 5 s, i.e., treating the solver as a kind of heuristic, and once more with a time limit of 10 min with the aim of obtaining sufficiently good-quality solutions. The results of the heuristic are obtained by the algorithm of Clarke & Wright, which are then improved by intra-route as well as inter-route operations. Despite Gurobi, there is no time limit on the heuristic, i.e., it stops as soon as all operations are accomplished.

For each test run, we computed the sum of the total distance covered by all tours, and record the processing time, which is required by the heuristic or the solver to reach the corresponding results. Furthermore, we note different gap values for each experiment. More precisely, for the solver results, the Gap_s is the MIP gap provided by the solver. However, regarding the heuristic, we compute the gap, in percent and denoted by Gap_h, as follows:

$$Gap_h = \frac{\text{result of the heuristic} - \text{result of the solver}}{\text{result of the heuristic}} \ (\%). \tag{13}$$

All experiments have been conducted on a laptop under Windows 10 with an Intel Core i5-5200U CPU @ 2,20 GHz and 8 GB of RAM.

6.2 Numerical Results

Tables 1 and 2 show the results of the experiments for vehicle capacities of 4 and 8 children, respectively, where for each van capacity $Q \in \{4, 8\}$, we have five different instance sizes $|I| \in \{10, 25, 50, 75, 100\}$. These tables provide, for each instance, the required distance L to be traveled (according to the solutions of the solver and the heuristic), the MIP gap of the solver Gap_s (in percent), the relative gap of the heuristic Gap_h (in percent), and the runtime t (in seconds) required to each approach to obtain the corresponding solutions.

Comments on the Numerical Results:

- Due to the NP-hardness of the KTPP, it is not surprising that solving the problem is a challenging task. In fact, as we observe in Tables 1 and 2, the optimality gap of the solver remains high even for medium-sized instances.
- By comparing the results of the solver with 5 s runtime versus the presented heuristic, we see that our heuristic outperforms the internal heuristic of the solver in most of the instances.
- By benchmarking the heuristic, the results show that the relative gap of the solutions found by the heuristic are in average smaller than 7% and 14% for KTPP with van capacity of $Q = 4$ and $Q = 8$, respectively.
- In terms of computation time, as it is expected, the heuristic solves the instances almost instantly.

Table 1. Results of the experiments with the van capacity $Q = 4$.

	Gurobi (5 s)			Gurobi (10 min)			Heuristic		
$\|I\|$	L	$Gap_s(\%)$	t(s)	L	$Gap_s(\%)$	t(s)	L	$Gap_h(\%)$	t(s)
10	2231.86	30	5.04	2220.18	0	17.34	2574.45	14	0.04
25	5091.17	63	5.05	4719.56	49	600.59	5344.50	12	0.01
50	9137.79	75	5.15	8187.42	71	600.36	9061.72	10	0.05
75	13816.29	81	5.46	11990.60	75	600.34	12008.86	0	0.13
100	16683.10	84	5.47	14838.79	82	600.51	14901.19	0	0.29
Average		67			55			7	

Table 2. Results of the experiments with the van capacity $Q = 8$.

	Gurobi (5 s)			Gurobi (10 min)			Heuristic		
$\|I\|$	L	$Gap_s(\%)$	t(s)	L	$Gap_s(\%)$	t(s)	L	$Gap_h(\%)$	t(s)
10	2422.26	21	5.03	2422.26	0	22.07	2866.62	16	0.008
25	4122.26	51	5.06	3722.92	43	600.76	4769.62	22	0.03
50	10184.83	67	5.26	6417.14	57	600.53	7865.61	18	0.14
75	11226.29	68	5.32	9726.21	65	600.37	10564.12	8	0.20
100	12858.80	71	5.46	11582.36	69	600.51	12159.60	5	0.33
Average		56			47			14	

7 Conclusion

In this paper, we introduced the *Kindergarten Tour Planning Problem* (KTPP), which is motivated by a practical case study in cooperation with the regional branch of the *Deutsches Rotes Kreuz* (DRK), the German Red Cross, located in the city of Landstuhl. A valid KTPP solution is composed of two parts: a set of routes (to pick up passengers) and driver assignment to the routes. After a short literature review on similar problems, e.g., the OVRP and the SBRP, we formulated route planing part of the KTPP as a MILP and presented a two-phase heuristic to solve the KTPP. In addition, we presented the results of our computational experiments on randomly generated instances to verify the quality of the solutions obtained by the heuristic. The preliminary results show that the heuristic can find promising solutions, but they can still be improved.

Moreover, we have designed and implemented an open-access web-based platform to solve the KTPP, and to provide an assistance service to the DRK (the branch that is located at Landstuhl). We dedicated a full section to describe main features of the platform.

We have several research plans for the future; in particular, an enhanced heuristic or metaheuristic should be designed, implemented, and integrated into the platform to solve not only the current instances but also the ones with asymmetric distance matrix D. In addition, it would be interesting and useful to implement or integrate an exact algorithm (or free exact solver) into the platform. To give an international dimension to the platform, we need to equip it with several languages. The research and work in these directions are in progress, and the results will be reported in the future.

References

1. Bootstrap. https://getbootstrap.com/. Accessed 15 Apr 2021
2. Flask. https://flask.palletsprojects.com/en/1.1.x/. Accessed 15 Apr 2021
3. Folium. python-visualization.github.io/folium/. Accessed 15 Apr 2021
4. jquery. https://jquery.com/. Accessed 15 Apr 2021
5. Leaflet. https://leafletjs.com/. Accessed 15 Apr 2021
6. Openrouteservice. https://openrouteservice.org/. Accessed 15 Apr 2021
7. Styleguide Deutsches Rotes Kreuz. https://styleguide.drk.de/deutsches-rotes-kreuz/basiselemente/farben. Accessed 15 Apr 2021
8. Baugh, J.W., Jr., Kakivaya, G.K.R., Stone, J.R.: Intractability of the dial-a-ride problem and a multiobjective solution using simulated annealing. Eng. Optim. **30**(2), 91–123 (1998)
9. Bektas, T., Elmastas, S.: Solving school bus routing problems through integer programming. J. Oper. Res. Soc. **58**, 1599–1604 (2007)
10. Bodin, L., Golden, B., Assad, A., Ball, M.: Routing and scheduling of vehicles and crews: the state of the art. Comput. Oper. Res. **10**(2), 63–211 (1983)
11. Clarke, G., Wright, J.: Scheduling of vehicles from a central depot to a number of delivery points. Oper. Res. **12**(4), 568–581 (1964)
12. Cordeau, J.F., Laporte, G.: The dial-a-ride problem: models and algorithms. Ann. Oper. Res. **153**, 29–47 (2007)
13. Cordeau, J.F., Laporte, G., Potvin, J.Y., Savelsbergh, M.W.: Transportation on Demand, vol. 14. Elsevier (2007)
14. Cordeau, J.F., Laporte, G., Savelsbergh, M.W., Vigo, D.: Vehicle Routing, vol. 14. Elsevier (2007)
15. Dulac, G., Ferland, J., Forgues, P.: School bus routes generator in urban surroundings. Comput. Oper. Res. **6**(3), 199–213 (1980)
16. Ellegood, W., Solomon, S., North, J., Campbell, J.: School bus routing problem: Contemporary trends and research directions. Omega 95 (2020)
17. Gurobi Optimization: Gurobi Optimizer Reference Manual (2019)
18. Jaradat, A., Shatnawi, M.: Solving school bus routing problem by intelligent water drops algorithm. J. Comput. Sci. **16**(1), 25–34 (2020)
19. Lekburapa, A., Boonperm, A., Sintunavarat, W.: A new integer programming model for solving a school bus routing problem with the student assignment. In: Vasant, P., Zelinka, I., Weber, G.-W. (eds.) ICO 2020. AISC, vol. 1324, pp. 287–296. Springer, Cham (2021). https://doi.org/10.1007/978-3-030-68154-8_28
20. Ozmen, M., Sahin, H.: Real-time optimization of school bus routing problem in smart cities using genetic algorithm. In: 6th International Conference on Inventive Computation Technologies (ICICT), pp. 1152–1158 (2021)
21. Sariklis, D., Powell, D.: A heuristic method for the open vehicle routing problem. J. Oper. Res. Soc. **51**, 564–573 (2000)

A Multi-objective Biased Random-Key Genetic Algorithm for Service Technician Routing and Scheduling Problem

Ricardo de Brito Damm(✉) and Débora P. Ronconi

University of São Paulo, Polytechnic School, São Paulo, Brazil
dronconi@usp.br

Abstract. Every day many service companies need to plan the tasks that will be carried out by its field staff. Maintenance service technicians have to perform a set of jobs at different locations in a city or state. This problem can be defined as the Service Technician Routing and Scheduling Problem in which tasks have different priorities and time windows, and technicians have different skills and working hours. Scheduling must account for technicians' lunch breaks, which must be respected. Each task is performed by only one technician. To ensure quality customer service and consumer rights are upheld, a novel approach is proposed: to address the problem in a multi-objective context aiming to execute the priority tasks and, simultaneously, to serve the customers at the beginning of their time windows. A Multi-objective Biased Random-Key Genetic Algorithm (BRKGA) was customized to tackle this NP-hard optimization problem and then compared with the Non-dominated Sorting Genetic Algorithm II (NSGA-II). The analyzed methods showed similar performance for small instances, but for medium- and large-sized instances the proposed method presented superior performance and more robust results.

Keywords: Multiple objective programming · Routing and scheduling technicians · Time windows · Biased Random-Key Genetic Algorithm · NSGA-II

1 Introduction and Background

This paper proposes a multi-objective integer programming (MOIP) model and heuristic methods for a version of the Service Technician Routing and Scheduling Problem (STRSP), which involves daily planning of technicians' activities (such as installing a device or providing equipment maintenance to houses or businesses in a medium-sized or big city). In this version of the problem, each task is assigned a priority level, representing the importance of the customer, how urgent

This research has been partially supported by FAPESP (Grants 13/07375-0 and 16/01860-1) and CNPq (grant 311536/2020-4).

M. Mes et al. (Eds.): ICCL 2021, LNCS 13004, pp. 471–486, 2021.
https://doi.org/10.1007/978-3-030-87672-2_31

the job is or whether it has been postponed for some time. Time windows are also assigned to represent customer availability in a range of hours. Deterministic times are stipulated for service and travel times. Technicians have a set of skills that allow them to perform each task; technicians' working hours and lunch breaks must be respected. Although technicians obviously need lunch breaks or rest breaks, it is not usual to find studies that take this parameter into account. According to Kovacs et al. [1], Xu and Chiu [2] and Pillac et al. [3], the STRSP is a generalization of the Vehicle Routing Problem with Time Windows, which is NP-hard. Considering the complexity of the problem and its practice relevance, in this paper, heuristics are developed. The problem was modeled to prioritize one customer service quality aspect, punctuality, which is highly relevant both to companies and customers in a large city.

Tsang and Voudouris [4] and Xu and Chiu [2] were among the first authors to study the STRSP. In 2007, the French Operational Research Society and France Telecom launched this problem as a challenge to researchers and Cordeau et al. [5] and Hashimoto et al. [6] won this contest. Other important publications are worth mentioning: Kovacs et al. [1], who studied an extension of the two previously mentioned research studies; Pillac et al. [3], who analyzed the similarity between the STRSP and the Vehicle Routing Problem with Time Windows.; and Souyris et al. [7] and Fikar et al. [8] who analyzed the problem in a stochastic context. In the literature, several real-life business problems have been studied, involving companies such as British Telecom [4], United Technologies Corporation [9], military maintenance scheduling [10], repair services for printers and copiers [7,11], and, as mentioned earlier, France Telecom [5,6].

Many different objective functions were found in the literature for the STRSP: to maximize the number of performed tasks [2,10,12,13]; to minimize the weighted sum of the completion time of task sets [5,6] and to minimize the completion time of the last task [14]; travel costs, overtime, and outsourcing [1,4]; to optimize the number of instances where time windows are violated, the number of customers served and the total travel time [11]. As it can be observed, most authors studied the problem by grouping different objectives into a single function. This occurs because, for this problem, several objectives are important in the decision-making process so they should be considered simultaneously. For this reason, we propose the use of multi-objective optimization techniques to study the STRSP. As far as we know, these techniques have never been applied to this problem.

Two objectives are proposed for the STRSP: to maximize the sum of priority values associated with the tasks performed and to serve priority customers as soon as possible. The first objective is based on Damm et al. [12] and the second one was inspired by the works of Cordeau et al. [5], Hashimoto et al. [6], Cortés et al. [11], and Xu et al. [15].

In the literature, heuristics and metaheuristics are the main methods adopted to solve the STRSP. Xu and Chiu [2], and Hashimoto et al. [6] applied the Greedy Randomized Adaptive Search Procedure (GRASP) metaheuristic; Cordeau et al. [5], Kovacs et al. [1], and Pillac et al. [3] developed constructive heuristics and customized the Adaptive Large Neighborhood Search (ALNS); Tsang and Voudouris [4] developed two constructive heuristics for the problem; Tang et al. [9] developed a tabu search metaheuristic; Dohn et al. [13] and Cortés et al. [11] used a branch-and-price approach; Xu et al. [15] a Genetic Algorithm and Damm et al. [12] the Biased Random-Key Genetic Algorithm (BRKGA).

In this work, a multi-objective version for the BRKGA was developed, trying to properly explore the characteristics of the problem and its search space. We chose the BRKGA because it can solve several combinatorial optimization problems successfully, e.g. divisible load scheduling [16], constrained clustering problem [17,18], multiproduct capacitated facility location problem [19], and tool switches problem [20]. In particular, BRKGA was successfully applied to routing-related problems such as the STRSP [12], the problem of routing and wavelength assignment in optical networks [21], the family traveling salesman problem [22], and traffic congestion in road networks [23].

The main contributions of this paper are described below. First, a MOIP model for the STRSP is proposed. The objective is to perform the priority tasks in a day and also to serve customers as soon as possible, within their time windows. In addition, the model considers technicians' lunch breaks and respects the time windows assigned for this purpose. Note that, as far as we know, this feature has never been considered in an integer programming model of the SRTSP with time windows. Taking into account the good results obtained by the BRKGA in mono-objective combinatorial problems, a multi-objective version was developed for this GA. Although it is a natural candidate to be applied to multi-objective problems while working with a population of solutions, there is only one research that proposes a multi-objective version for the BRKGA (see Tangpattanakul et al. [24,25]). This is a promising area of research that remains largely unexplored. We intend to customize the main components of BRKGA to propose a method that can effectively tackle the multi-objective SRTSP.

This paper is organized as follows. Section 2 presents the mathematical model of the problem. Section 3 introduces the Biased Random-Key Genetic Algorithm BRKGA and its components. Section 4 describes the computational experiments performed and the analysis of their results. The last section summarizes the main results and presents possible future works.

2 Problem Description and a Multi-objective Integer Programming Model

The STRSP problem involves a set $S = \{1, \ldots, n\}$ of n tasks or services and a set $K = \{1, \ldots, m\}$ of m technicians. A dummy task 0 represents the origin or the home base of the company. Each pair of tasks i and j is associated with a travel time c^p_{ij}, which is equal to the Euclidean distance between the coordinates of points i and j. It is worth mentioning that this parameter can assume a value associated with a distance derived from a real road network without any drawback for the model. Each task $i \in S$ has a positive integer value w_i representing its priority (indicating the importance or urgency of the task), a processing time p_i, and a time window $[e_i, \ell_i]$ within which the task must be executed. A technician is allowed to reach the location of a task i before the beginning of its time window, but he/she should wait for time e_i to start processing the task. Each technician $k \in K$ has a daily work schedule defined by $[a_k, b_k]$ and must leave the base after the start and return before the end of the working hours. The skill of a technician k to perform a task i is given by s_{ik}, a binary parameter, where 1 means "being capable to perform the task" and 0 means "unable to perform it". The lunch break of each technician $k \in K$ is represented by a dummy task $n+k$ with duration or processing time p_{n+k}, time window $[e_{n+k}, \ell_{n+k}]$, and such that $s_{n+k,k} = 1$ and $s_{n+k,\nu} = 0$ for all $\nu \neq k$. We define $J = S \cup \{n+1, \ldots, n+m\}$. It should be noted that the flexibility of the time windows for this activity can be established in the definition of e_{n+k} and ℓ_{n+k}. In addition, we assume that since the technician must perform the selected tasks, this activity will likely occur near some of these tasks.

Seven types of decision variables are adopted: y_{ik} is a binary variable that equals 1 if task i is assigned to technician k and 0, otherwise; x_{ijk} is also a binary variable which takes value 1 when task i precedes task j in the route of technician k and 0, otherwise; u_k is a binary variable that is equal to 1 when technician k has their lunch break at the location of the task that precedes task $n+k$ and it is equal to 0 when the lunch break happens at the location of the task that succeeds task $n+k$; t_i is the start time of the execution of task i; c_{ij} is the travel time between two tasks $i, j \in J$ (that coincides with c^p_{ij} when $i, j \in S \cup \{0\}$); and c^b_k and c^a_k are the travel times between task $n+k$ (lunch break of technician k) and its predecessor and successor tasks, respectively. The model is given by:

$$\max f_1 = \frac{1}{W} \sum_{i \in S} \sum_{k \in K} w_i y_{ik} - \frac{1}{C} \sum_{k \in K} \left(\sum_{i \in V} \sum_{j \in V} c^p_{ij} x_{ijk} + c^b_k + c^a_k \right) \tag{1}$$

$$\max f_2 = \frac{1}{W} \sum_{i \in S} w_i \gamma_i \tag{2}$$

Subject to:

$$y_{ik} \leq s_{ik} \;\; k \in K, i \in J \tag{3}$$

$$\sum_{k \in K} y_{ik} \leq 1 \;\; i \in J \tag{4}$$

$$y_{n+k,k} = 1 \;\; k \in K \tag{5}$$

$$\sum_{j \in J \cup \{0\} \setminus i} x_{ijk} = \sum_{j \in J \cup \{0\} \setminus i} x_{jik} = y_{ik} \;\; k \in K, i \in J \tag{6}$$

$$\sum_{i \in J} x_{0ik} \leq 1 \;\; k \in K \tag{7}$$

$$e_i \leq t_i \leq l_i - p_i + M(1 - \sum_{k \in K} y_{ik}) \;\; i \in J \tag{8}$$

$$t_i \geq l_i - p_i - M \sum_{k \in K} y_{ik} + \varepsilon \;\; i \in J \tag{9}$$

$$t_i \leq l_i - p_i + M \sum_{k \in K} y_{ik} + \varepsilon \;\; i \in J \tag{10}$$

$$t_i + p_i + c_{ij} \leq t_j + M(1 - \sum_{k \in K} x_{ijk}) \;\; i \neq j \in J \tag{11}$$

$$\sum_{k \in K} x_{0jk} a_k + c_{0j} \leq t_j + M(1 - \sum_{k \in K} x_{0jk}) \;\; j \in J \tag{12}$$

$$t_i + p_i + c_{i0} \leq \sum_{k \in K} x_{i0k} b_k + M(1 - \sum_{k \in K} x_{i0k}) \;\; i \in J \tag{13}$$

$$c_{ij} = c^p_{ij} \;\; i, j \in S \cup \{0\} \tag{14}$$

$$c_{j,n+k} \geq \sum_{i \in J \cup \{0\} \setminus j} x_{n+k,i,k}\, c^p_{ji} - M u_k \;\; j \in S, k \in K \tag{15}$$

$$c_{n+k,i} \geq \sum_{j \in J \cup \{0\} \setminus i} x_{j,n+k,k}\, c^p_{ji} - M(1 - u_k) \;\; i \in S, k \in K \tag{16}$$

$$c^b_k \geq c_{i,n+k} - M(1 - x_{i,n+k,k}) \;\; i \in S, k \in K \tag{17}$$

$$c^a_k \geq c_{n+k,i} - M(1 - x_{n+k,i,k}) \;\; i \in S, k \in K \tag{18}$$

$$c_{j,n+k} \geq 0 \;\; j \in S, k \in K \tag{19}$$

$$c_{n+k,i} \geq 0 \;\; i \in S, k \in K \tag{20}$$

$$c^b_k \geq 0 \;\; k \in K \tag{21}$$

$$c^a_k \geq 0 \;\; k \in K \tag{22}$$

$$\gamma_i = \frac{(l_i - t_i - p_i + \varepsilon)}{(l_i - e_i - p_i + \varepsilon)} \;\; i \in S \tag{23}$$

$$x_{ijk}, y_{ik}, u_k \in \{0, 1\} \;\; i, j \in J, k \in K \tag{24}$$

$$t_i \in \mathbf{R}_+ \;\; i \in J, k \in K \tag{25}$$

where:

$$\varepsilon \text{ is a small number}$$
$$W = \min_{i \in J} w_i$$
$$C = \sum_{k \in K} (b_k - a_k)$$
$$M = \max\left(\max_{i \in J} l_i, \max_{k \in K} a_k\right) + \max_{i,j \in J \cup \{0\}} c^p_{ij}$$
$$J = \{1, ..., n, n+1, ..., n+m\}$$
$$S = \{1, ..., n\}$$

The first objective (1) aims to maximize the sum of priority values associated with the tasks performed each day and to minimize the total travel time. The purpose of the denominator is two-fold: to obtain dimensionless values of the sum and ensure that it is always better to perform a task and increase the total travel time, but never vice versa. Note that, when a task is associated with a technician ($y_{ik} = 1$), the expression $\frac{w_i y_{ik}}{W}$ will always be greater than or equal to one, whereas the term C guarantees that the second term of f_1 will be always strictly less than one.

The second objective aims to prioritize the most important customers by placing them at the beginning of their time windows. Thus, a variable *gap* (γ) is defined between 0 and 1, representing the fraction of idle time between the end of a task and the end of its time window, indicating how early the task is performed. For example, when task i is performed exactly at the beginning of its time window ($t_i = e_i$), then its *gap* reaches its maximum value ($\gamma_i = 1$); when it has finished exactly at the end of the time window ($t_i = l_i - p_i$), the *gap* is equal to a very small number; the *gap* is null ($\gamma_i = 0$) only when a task is not performed. Thus, the variable *gap* is defined as $\gamma_i = \frac{(l_i - t_i - p_i + \varepsilon)}{(l_i - e_i - p_i + \varepsilon)}$, where ε is a very small number. Note that parameter ε also ensures that the denominator of γ is never equal to zero, which can occur when the time window of a task is equal to its processing time ($p_i = l_i - e_i$). Thus, the second objective weighs the priorities of the tasks (w_i) with their respective *gaps* (γ_i), so that higher priority tasks have larger values associated with their *gaps*. Denominator W was included so that the second objective has the same order of magnitude of the first one (1).

Constraints (3), (4) and (5) ensure that each task can only be performed by a skilled technician, assigned at most to one technician, and that the task that represents the lunch break of each technician is scheduled. In (6), when a task i is assigned to a technician k, there is only one predecessor and one successor of i. Each technician must leave the origin at most once, as indicated by (7). Constraints (8) guarantee that each task is performed within its time window while the constraints (9–10) guarantee that, when a task is not executed, the gap variable γ is null. Constraints (11–13) define the relationship between the beginning of the processing time of a task and the execution of its predecessor or the beginning of the working day of a technician; these restrictions avoid the *subtour*. The travel time between two tasks is determined by constraint (14) and, between the dummy task (lunch break) and its predecessor (c_k^b) and successor (c_k^a), by constraints (15–22). Note that when $u_k = 1$, technician k has a lunch break right after performing task j and then goes to the next task i (then $c_k^a = 0$ and $c_k^b > 0$); otherwise, after performing task j, the technician goes to the site where task i will be performed, but has a lunch break before starting it (then $c_k^a > 0$ and $c_k^b = 0$). It should be noted that the positioning of the lunch break (before or after a task) in some situations may influence the objective values of the function (e.g., allowing a priority customer to be served earlier or taking advantage of idle time before a task). The variable γ is defined by (23), while the domain of the remaining variables is defined by (24) and (25).

3 Multi-objective Biased Random Key Genetic Algorithm

For the proposed problem, a GA was applied that uses random keys and prevents the generation of new infeasible solutions. The Random-Key Genetic Algorithm (RKGA) was proposed by Bean [26] to improve the performance of the traditional GA in combinatorial problems. The RKGA exploits the feasible region indirectly through a search in the space of the random keys (problem-independent) and uses a decoder (problem-dependent) to solve the problem [22]. It should be emphasized that an essential part of the RKGA is the decoder, which should be developed to avoid the generation of unfeasible solutions to the optimization problem. The latest version of this algorithm was called Biased Random-Key Genetic Algorithm (BRKGA), which differs from the previous version particularly in the generation of new solutions: one of the best (elite) solutions always participates in the crossover and this elite solution is more likely to transmit its genes. Comparisons between standard GA, RKGA, and BRKGA for different optimization problems can be found in [27,28]; the authors concluded that the BRKGA is more effective than the RKGA.

Here is a brief overview of a mono-objective BRKGA. The initial population is composed of vectors (λ) of random real numbers (random keys) in the interval (0; 1], which will be transformed by an algorithm (decoder) into feasible solutions of the problem. Constructive heuristic solutions can be coded and included in the initial population. The current population is divided into two groups: a first smaller group with the best solutions (elite) and a second larger group with the other solutions (non-elite). The next generation is formed by all elite solutions of the previous generation, by some new solutions (mutants) randomly generated (thus increasing population diversity), and by solutions created by the parameterized uniform crossover. In this kind of crossover, one offspring is generated by two chromosomes (one selected from the elite and other, from the non-elite group). To decide if each gene of the new chromosome comes from the elite or non-elite solution, a real number between 0 and 1 is randomly generated and the probability of the elite solution transmitting its gene is greater than that of the non-elite solution; in other words, the random keys of the elite solution tend to predominate in the new offspring generated. Once a new population is formed, all operations are repeated until a stopping criterion is reached. A detailed description of the BRKGA can be found Morán-Mirabal et al. [22] and in Damm et al. [12].

3.1 Elite Set

The main question that arises when adapting the mono-objective BRKGA for a multi-objective BRKGA is how to determine an order for solutions of each population to select the elite group. As far as we know, only Tangpattanakul et al. [24,25] developed a multi-objective BRKGA, which was used to select and schedule observations for an agile Earth-observing satellite. In [25], the authors chose to include only the non-dominated solutions in the elite set and, therefore, to adopt a variable size for the elite set (dependent on the cardinality of the set of non-dominated solutions). While in another paper [24], these authors adopted traditional multi-objective GA methods to construct the elite set of the BRKGA: the crowding distance (Non-dominated Sorting Genetic Algorithm II – NSGA-II), the metric selection (Evolutionary Multi-objective Optimization Algorithm – EMOA), and the indicator based on the hypervolume concept (Indicator-Based Evolutionary Algorithm – IBEA). The authors concluded that the three methods performed similarly and have not identified any superior method. Therefore, in our research, one strategy of each previous paper (the non-dominated solutions [25] and crowding distance [24]) was adopted. Furthermore, two other methods (fitness sharing and cell-based density) were analyzed. These methods are described below.

Deb et al. [29] proposed the Crowding distance, which is an approach for the Non-dominated Sorting Genetic Algorithm (NSGA-II) where chromosomes are classified by the non-dominated fronts and solutions in the same front are ordered by the distance of the nearest solutions.

Fitness sharing was proposed by Goldberg and Richardson [30]. This strategy penalizes solutions located in a region of high concentration of solutions so that the most isolated solutions (i.e., areas less exploited by the search) are favored, thus increasing diversity in the Pareto set throughout the GA generations.

In Cell-based density the objective function space is divided into cells or hypercubes: each axis of the objective function is divided into equal sizes. Solutions of the first fronts and in lower densities cells are favored.

3.2 Decoder

This section presents the decoder applied in the proposed BRKGA. The decoder has $n+1$ random keys. Each of the n tasks has one random key and the last one ($\lambda[n+1]$) is a weighted factor between the objectives, which allows to decide if a new task can or cannot be included in the route of a skilled technician, based on the weighted sum of the objectives. The main details of the decoder are listed below.

1. *Task Ordering:* Sort the tasks in decreasing order of random keys. Select the first task i. For each technician k, do $f^k = 0$.
2. *Evaluation of candidate technicians:* for each technician k capable of performing task i (i.e. $s_{ik} = 1$), evaluate all the possible positions of this task in their route. Choose the position that provides a feasible solution with the greatest value of f_w^k (if $f_w^k \geq f^k$), where f_w^k is given by:

$$f_w^k = \lambda[n+1] \cdot f_1^k + (\lambda[n+1] - 1) \cdot f_2^k \tag{26}$$

   where f_1^k and f_2^k are the objective function values associated with technician k.
3. *Selection of a technician:* if Step 2 provides no solution, go to the next step. Otherwise, select the technician who can perform task i with the greatest value of $f_w^k - f^k$. If more than one technician has the same greatest value, the second and the third decision criteria are lowest total travel time (from the beginning until return to the home base) and the identification number of the technician, respectively. Add task i to the route of technician k and do $f^k = f_w^k$.
4. *Stop condition:* if task i is the last task sorted in Step 1, stop. Otherwise, select the next task i classified in the first step and go back to Step 2.

3.3 Initial Population

The initial population of the BRKGA for the STRSP includes random chromosomes and a number of solutions found by a constructive heuristic (ncs) similar to the decoder (Sect. 3.2), except for two issues.

In the first step, tasks are ordered by:

$$\rho_i = w_i - \frac{NST_i}{m+1} \tag{27}$$

where NST_i is the number of skilled technicians for task i. In Step 2, instead of $\lambda[n+1]$ in formula (26), the constructive heuristic uses values between 0 and 1 to generate different solutions. To include the constructive heuristic solutions in the initial population, do $\lambda[i] = \frac{\rho_i}{\rho_{max}}$, for $i = 1, ..., n$ where $\rho_{max} = \max\{\rho_i | i = 1, ..., n\}$.

4 Computational Results

Initially, this section describes the instances and the performance measures used to compare heuristic methods. Then, the results obtained by the proposed BRKGA and a comparison with traditional GA approach (NSGA-II) are presented. Codes were written in C programming language and tests were conducted on an Intel (R) Core (TM) i7-5500U CPU 2.40 GHz with 8 GB of RAM memory. The MOIP model was solved on a 2.93 GHz Intel processor with 16 GB of RAM memory.

4.1 Instances

A total of 90 instances were generated in the experiments, with considerably diverse parameters that can impact on the behavior of the heuristics: geographical distribution of customers, length of time windows of tasks (short, long, or random), processing time, task priority, number of tasks and technicians. Three different geographical distributions were used for tasks: random uniform distribution (R), clustered distribution (C), and semi-clustered distribution (RC). These distributions were suggested by Solomon [31]. For each geographical distribution, 3 instances were generated for 10 different numbers (or cases) of tasks and technicians. Table 1 shows the number of tasks and technicians for each case.

Table 1. Dimension of the instances generated

Case	#tasks (n)	#technicians (m)
1	16	2
2	26	2
3	30	3
4	39	3
5	45	7
6	64	5
7	80	13
8	100	10
9	150	15
10	200	33

Instance parameters were generated in the following range of discrete uniform distribution:

- Priority of tasks (w_i): 1 (low), $2, \ldots, 10$ (high).
- Processing time of each task (p_i): $30, 35, \ldots, 120$ min.
- Beginning of time windows of tasks (e_i): $7, \ldots, 19$ h.
- Length of time windows of tasks: short ($1.5, \ldots, 3.5$ h.), long ($6.5, \ldots, 9.0$ h.), and random ($1.5, \ldots, 9.0$ h.).
- End of time windows of tasks (l_i): e_i plus the length of the window.
- Beginning of time windows of technicians (a_k): $7, 7.5, 8, \ldots, 12$ h.
- End of time windows of technicians (b_k): $a_k + 9$ h.
- Lunch break of each technician: duration (60 min.), beginning ($e_i = a_k + 3$ h.) and end of time window ($l_i = b_k - 3$ h.).
- Skill level of technician k to perform task i (s_{ik}): 0 (not allowed) or 1 (can execute).

Taillard's random number generator and seeds were used to generate random numbers for the discrete uniform distribution of the parameters described above (see [32]).

4.2 Performance Measures

To evaluate the analyzed methods, three comparative measures were adopted: Proportion of Pareto-optimal objective vectors found (only for small instances), Hypervolume Indicator, and Epsilon Multiplicative Indicator (for all instances) [33].

The indices obtained by each method were compared with the indices of the reference set (the set of the best results among all methods compared) for the medium and large cases and the indices of the Pareto-optimal set for small instances. To find the Pareto-optimal set in cases 1 to 4, the MOIP model presented in Sect. 2 was solved by the ε-constraint method, using the ILOG CPLEX software, version 12.6, with a limited processing time of forty hours. To obtain as many Pareto-optimal sets as possible, complete enumeration was also applied. In total, the Pareto-optimal sets of 29 instances were found. Table 2 gives some details of the obtained results. The first column identifies the case. The following columns show (for instances with short, long, and random time windows of tasks) the number of instances in which the Pareto-optimal set was found, their average cardinality (Avg. Card.), and the average CPU time to find the Pareto-optimal set by CPLEX (the "-" character indicates that optimal solutions were not found within the time limit and the Pareto-optimal set was obtained by complete enumeration).

Table 2. Pareto-optimal set known for small instances

Case	Short time windows			Long time windows			Random time windows		
	#instances	Avg. card	CPU (s)	#instances	Avg. card	CPU (s)	#instances	Avg. card	CPU (s)
1	3	5.3	3.3	3	19.0	–	3	7.7	2200.3
2	3	8.3	35.3	3	20.3	–	3	14.3	–
3	3	8.7	169.9	2	25.0	-	3	17.3	–
4	3	11.7	1105.9	–	–	–	–	–	–
Average		8.5			21.4			13.1	

Avg. Card. is the average cardinality Pareto-optimal set

4.3 BRKGA

Performance of different elite strategies. The four methods of solution classification and construction of the elite set were applied to the BRKGA, for cases 1 to 9, for 5 runs of each instance. Table 3 show the performance measures (the numbers shown in bold in this table and in the following tables indicate the best result of each row). For small instances, the performance of the classification methods is similar. For medium and large instances, the performance measures indicate that the classification of solutions and elite strategies affect results differently and the average indices show that the best strategy is fitness sharing. The method that includes only non-dominated solutions ($F1$) in the elite set, adopted by Tangpattanakul et al. [25], had a below-average performance when compared to all other strategies. This result supports the hypothesis that, by

Table 3. Average (Avg.) and Standard Deviation (σ) of the performance of the **BRKGA** for different elite strategies

(a) For small instances (cases 1 to 4) in relation to the Pareto-optimal set.								
	F1		Crowding distance		Fitness sharing		Cell-based density	
	Avg	σ	Avg	σ	Avg	σ	Avg	σ
I_{PF} (%)	89.2	15.0	**95.0**	8.2	94.7	8.3	91.4	16.2
I_{PH} (%)	99.6	1.1	**100.0**	0.1	**100.0**	0.1	99.8	0.7
I_{ϵ}	0.995	0.008	**0.999**	0.004	0.998	0.004	0.998	0.004
(b) For medium and large instances (cases 5 to 9) in relation to the Reference set.								
	F1		Crowding distance		Fitness sharing		Cell-based density	
	Avg	σ	Avg	σ	Avg	σ	Avg	σ
I_{PH} (%)	84.7	16.9	94.5	4.9	**94.7**	6.6	94.0	8.5
I_{ϵ}	0.977	0.014	0.988	0.008	**0.989**	0.007	**0.989**	0.007

including some selected dominated solutions in the elite set, the performance of the BRKGA is likely to improve. In the following sections, the BRKGA will use the best classification method.

4.4 Comparison with Multi-objective GAs of the Literature

Genetic Algorithms are popular among the metaheuristics used in multi-objective optimization. Since 1985, many multi-objective GAs have been proposed. In this paper, the BRKGA is compared with the NSGA-II, which is widely used and particularly well-suited for many combinatorial problems. The same representation of the chromosomes was used to compare only the differences between the BRKGA and other GA. Therefore, the NSGA-II algorithm was also customized with chromosomes of $n+1$ random numbers between 0 and 1, which are transformed into solutions of the STRSP using the decoder.

Table 4. Average (Avg.) and Standard Deviation (σ) of the performance measures of the **BRKGA** and **NSGA-II**

(a) For small instances (cases 1 to 4) in relation to the Pareto-optimal set.				
	BRKGA		NSGA-II	
	Avg	σ	Avg	σ
I_{PF} (%)	94.0	12.3	**95.0**	11.6
I_{PH} (%)	**99.8**	1.2	**99.8**	1.0
I_{ϵ}	0.998	0.005	**0.999**	0.005
(b) For medium and large instances (cases 5 to 10) in relation to the Reference set.				
	BRKGA		NSGA-II	
	Avg	σ	Avg	σ
I_{PH} (%)	**92.3**	6.4	90.4	8.5
I_{ϵ}	**0.988**	0.007	0.986	0.009

Table 4 presents the average performance indices of the BRKGA and NSGA-II, with 20 runs of each instance. For small instances, these methods show similar performance. Note that, if the average cardinality of the Pareto-optimal set is 14.3 (see Table 2), then the BRKGA can find 13.5 (94.0%) of these solutions, on average. For medium and large instances, the BRKGA obtained an average performance superior to the NSGA-II.

Statistical analysis of the performance indices confirmed these results, which allows us to conclude that the BRKGA shows superior performance indeed. In order to perform this analysis, the Kolmogorov-Smirnov test (with a significance level $\alpha = 5\%$) was initially applied to accept or reject the hypothesis of normality of the distribution of the Hypervolume and Epsilon indices (and Proportion of Pareto-optimal Objective vectors Found for small instances). As a result, we used the statistical technique Analysis of Variance (ANOVA) or the nonparametric Wilcoxon rank-sum test, both methods with a significance level $\alpha = 5\%$.

For small instances, the hypothesis of normality was rejected for all performance measures and, according to the Wilcoxon test, the null hypothesis is accepted. Therefore, it was concluded that there is no difference between the averages presented in Table 4a.

For medium and large instances, the hypothesis of normality for performance measures was rejected too and, in the comparison of the BRKGA with other methods, the null hypothesis for the Wilcoxon test was rejected for the two performance indices and, therefore, it was concluded that, on average, the BRKGA indices are superior.

Aiming to identify if these conclusions can be applied to all instances with more than 45 tasks, a detailed analysis was carried out. The average values of Hypervolume for cases 5 and 6 are similar for all methods, and in four larger cases, the BRKGA has significantly higher average results.

5 Conclusions and Future Works

This paper addressed the multi-objective Service Technician Routing and Scheduling Problem (STRSP). In addition to the usual characteristics of the problem at hand, technicians' lunch breaks were included in the schedule. As far as we know, lunch breaks have never been considered in an integer programming model of SRTSP with time windows. A multi-objective Biased Random-Key Genetic Algorithm (BRKGA) was customized to tackle this problem and a comparison with the latest developments in Non-dominated Sorting Genetic Algorithm II (NSGA-II) was presented.

In a comparison made with the optimal Pareto for small problems (up to 39 tasks and 3 technicians), on average, the BRKGA found 94% of the Pareto-optimal solutions, achieving 99.8% of the optimal Hypervolume and 0.998 of the Epsilon Multiplicative Indicator.

In a comparison with the NSGA-II algorithm, for problems up to 64 tasks and 5 technicians, the performance of the BRKGA was similar. For medium- and large-sized instances (more than 80 tasks and 10 technicians), statistical tests indicated that, on average, BRKGA outperformed NSGA-II.

Suggestions for future work include developing new decoders and applying the multi-objective BRKGA proposed in this paper to other combinatorial problems. Furthermore, considering the practical relevance of the considered problem, different strategies for solving this bi-objective problem can be investigated, such as the one proposed in [34].

References

1. Kovacs, A.A., Parragh, S.N., Doerner, K.F., Hartl, R.F.: Adaptive large neighborhood search for service technician routing and scheduling problems. J. Sched. **15**(5), 579–600 (2012)
2. Xu, J., Chiu, S.Y.: Effective heuristic procedures for a field technician scheduling problem. J. Heurist. **7**(5), 495–509 (2001)
3. Pillac, V., Guéret, C., Medaglia, A.L.: A parallel matheuristic for the technician routing and scheduling problem. Optim. Lett. **7**(7), 1525–1535 (2012). https://doi.org/10.1007/s11590-012-0567-4
4. Tsang, E., Voudouris, C.: Fast local search and guided local search and their application to British Telecom's workforce scheduling problem. Oper. Res. Lett. **20**(3), 119–127 (1997)
5. Cordeau, J.F., Laporte, G., Pasin, F., Ropke, S.: Scheduling technicians and tasks in a telecommunications company. J. Sched. **13**(4), 393–409 (2010)
6. Hashimoto, H., Boussier, S., Vasquez, M., Wilbaut, C.: A GRASP-based approach for technicians and interventions scheduling for telecommunications. Ann. Oper. Res. **183**(1), 143–161 (2011)
7. Souyris, S., Cortés, C.E., Ordóñez, F., Weintraub, A.: A robust optimization approach to dispatching technicians under stochastic service times. Optim. Lett. **7**(7), 1549–1568 (2012). https://doi.org/10.1007/s11590-012-0557-6
8. Fikar, C., Juan, A.A., Martinez, E., Hirsch, P.: A discrete-event driven metaheuristic for dynamic home service routing with synchronised trip sharing. Eur. J. Ind. Eng. **10**(3), 323–340 (2016)
9. Tang, H., Miller-Hooks, E., Tomastik, R.: Scheduling technicians for planned maintenance of geographically distributed equipment. Transp. Res. Part E Logist. Transp. Rev. **43**(5), 591–609 (2007)
10. Overholts, D.L., II., Bell, J.E., Arostegui, M.A.: A location analysis approach for military maintenance scheduling with geographically dispersed service areas. Omega **37**(4), 838–852 (2009)
11. Cortés, C.E., Gendreau, M., Rousseau, L.M., Souyris, S., Weintraub, A.: Branch-and-price and constraint programming for solving a real-life technician dispatching problem. Eur. J. Oper. Res. **238**(1), 300–312 (2014)
12. Damm, R.B., Resende, M.G., Ronconi, D.P.: A biased random key genetic algorithm for the field technician scheduling problem. Comput. Oper. Res. **75**, 49–63 (2016)
13. Dohn, A., Kolind, E., Clausen, J.: The manpower allocation problem with time windows and job-teaming constraints: A branch-and-price approach. Comput. Oper. Res. **36**(4), 1145–1157 (2009)
14. Chen, X., Thomas, B.W., Hewitt, M.: The technician routing problem with experience-based service times. Omega **61**, 49–61 (2016)
15. Xu, Z., Ming, X.G., Zheng, M., Li, M., He, L., Song, W.: Cross-trained workers scheduling for field service using improved NSGA-II. Int. J. Prod. Res. **53**(4), 1255–1272 (2015)

16. Brandão, J.S., Noronha, T.F., Resende, M.G., Ribeiro, C.C.: A biased random-key genetic algorithm for scheduling heterogeneous multi-round systems. Int. Trans. Oper. Res. **24**(5), 1061–1077 (2017)
17. de Oliveira, R.M., Chaves, A.A., Lorena, L.A.N.: A comparison of two hybrid methods for constrained clustering problems. Appl. Soft Comput. **54**, 256–266 (2017)
18. Chaves, A.A., Goncalves, J.F., Lorena, L.A.N.: Adaptive biased random-key genetic algorithm with local search for the capacitated centered clustering problem. Comput. Ind. Eng. **124**, 331–346 (2018)
19. Mauri, G.R., Biajoli, F.L., Rabello, R.L., Chaves, A.A., Ribeiro, G.M., Lorena, L.A.N.: Hybrid metaheuristics to solve a multiproduct two-stage capacitated facility location problem. Int. Trans. Oper. Res. (2021)
20. Chaves, A.A., Lorena, L.A.N., Senne, E.L.F., Resende, M.G.: Hybrid method with CS and BRKGA applied to the minimization of tool switches problem. Comput. Oper. Res. **67**, 174–183 (2016)
21. Noronha, T.F., Resende, M.G., Ribeiro, C.C.: A biased random-key genetic algorithm for routing and wavelength assignment. J. Glob. Optim. **50**(3), 503–518 (2011)
22. Morán-Mirabal, L.F., González-Velarde, J.L., Resende, M.G.: Randomized heuristics for the family traveling salesperson problem. Int. Trans. Oper. Res. **21**(1), 41–57 (2014)
23. Stefanello, F., Buriol, L.S., Hirsch, M.J., Pardalos, P.M., Querido, T., Resende, M.G.C., Ritt, M.: On the minimization of traffic congestion in road networks with tolls. Ann. Oper. Res. **249**(1–2), 119–139 (2015). https://doi.org/10.1007/s10479-015-1800-1
24. Tangpattanakul, P., Jozefowiez, N., Lopez, P.: Biased random key genetic algorithm for multi-user earth observation scheduling. In: Fidanova, S. (ed.) Recent Advances in Computational Optimization. SCI, vol. 580, pp. 143–160. Springer, Cham (2015). https://doi.org/10.1007/978-3-319-12631-9_9
25. Tangpattanakul, P., Jozefowiez, N., Lopez, P.: A multi-objective local search heuristic for scheduling Earth observations taken by an agile satellite. Eur. J. Oper. Res. **245**(2), 542–554 (2015)
26. Bean, J.C.: Genetic algorithms and random keys for sequencing and optimization. ORSA J. Comput. **6**(2), 154–160 (1994)
27. Gonçalves, J.F., Resende, M.G., Toso, R.F.: An experimental comparison of biased and unbiased random-key genetic algorithms. Pesquisa Operacional **34**(2), 143–164 (2014)
28. Gonçalves, J.F., Resende, M.G.: Biased random-key genetic algorithms for combinatorial optimization. J. Heurist. **17**(5), 487–525 (2011)
29. Deb, K., Pratap, A., Agarwal, S., Meyarivan, T.A.M.T.: A fast and elitist multiobjective genetic algorithm: NSGA-II. IEEE Trans. Evol. Comput. **6**(2), 182–197 (2002)
30. Goldberg, D. E., Richardson, J.: Genetic algorithms with sharing for multimodal function optimization. In Genetic Algorithms and their Applications: Proceedings of the Second International Conference on Genetic Algorithms, pp. 41–49. Lawrence Erlbaum, Hillsdale (1987)
31. Solomon, M.M.: Algorithms for the vehicle routing and scheduling problems with time window constraints. Oper. Res. **35**(2), 254–265 (1987)
32. Taillard, E.: Benchmarks for basic scheduling problems. Eur. J. Oper. Res. **64**(2), 278–285 (1993)

33. Zitzler, E., Knowles, J., Thiele, L.: Quality assessment of pareto set approximations. In: Branke, J., Deb, K., Miettinen, K., Słowiński, R. (eds.) Multiobjective Optimization. LNCS, vol. 5252, pp. 373–404. Springer, Heidelberg (2008). https://doi.org/10.1007/978-3-540-88908-3_14
34. Matl, P., Hartl, R.F., Vidal, T.: Leveraging single-objective heuristics to solve bi-objective problems: heuristic box splitting and its application to vehicle routing. Networks **73**, 382–400 (2019)

Optimization of Green Pickup and Delivery Operations in Multi-depot Distribution Problems

Alejandro Fernández Gil[1(✉)], Eduardo Lalla-Ruiz[2], Martijn Mes[2], and Carlos Castro[1]

[1] Departamento de Informática, Universidad Técnica Federico Santa María, Valparaíso, Chile
affernan@jp.inf.utfsm.cl, carlos.castro@inf.utfsm.cl

[2] Department of Industrial Engineering and Business Information Systems, University of Twente, Enschede, The Netherlands
{e.a.lalla,m.r.k.mes}@utwente.nl

Abstract. In this work, the Multi-Depot Green VRP with Pickups and Deliveries (MDGVRP-PD) is studied. It is a routing optimization problem in which the objective is to construct a set of vehicle routes considering multiple depots and one-to-one pickup and delivery operations that minimize emissions through fuel consumption, which depends on weight and travel distance. In one-to-one problems, goods must be transported between a single origin and its single associated destination. Practical considerations imply addressing the pickup and delivery of customers from multiple depots, where a logistics service company can efficiently combine its resources, thus reducing environmental pollution. To tackle this problem, we develop a mathematical programming formulation and matheuristic approach based on the POPMUSIC (Partial Optimization Metaheuristic under Special Intensification Conditions) framework. The results show that if the weight carried on the routes as part of the fitness measure is considered, our matheuristic approach provide an average percentage improvement in emissions of 30.79%, compared to a fitness measure that only takes into account the distances of the routes.

Keywords: Multi-depot · Green VRP · PDVRP · Matheuristic

1 Introduction

Road freight transportation is crucial for societies' economic and industrial development [13]. However, the distribution of goods negatively affects local air quality, generates noise and vibration, causes accidents, and contributes to global warming [25,26]. Thus, reducing emissions in the road transport sector has been a central topic in international agreements on climate change since greenhouse gases (GHGs) are considerably associated with environmental pollution. Moreover, the main source of energy used by the global transport sector has

M. Mes et al. (Eds.): ICCL 2021, LNCS 13004, pp. 487–501, 2021.
https://doi.org/10.1007/978-3-030-87672-2_32

been petroleum products (e.g., gasoline, diesel, etc.), whose demand is expected to increase by 30% and 82% between 2010 and 2050. The CO_2 emissions are expected to increase from 16% to 79% [6]. As a result, logistics and freight transportation companies are investigating sustainability concepts around several dimensions such as financial, environmental, and social [1].

Several researchers mainly focused on complex planning and routing problems that address the challenges mentioned above for road freight transportation, which is one of the main sources of CO_2 emissions [37]. More specifically, in the supply chain design scope, the green distribution network planning leads to the Green Vehicle Routing Problems (GVRPs), which is a variant of the well-known operational problem in transportation, namely the Vehicle Routing Problem (VRP, [10]). The GVRP considers environmental issues in routing problems [14,24,27]. The routing process involves designing routes for a fleet of vehicles and customers, subject to given constraints. The customers must be served according to particular features (e.g., delivery time, priority, low delivery costs, etc.). In GVRPs, besides optimizing cost or profit, environmental costs are also explicitly considered (e.g., CO_2 emission, fuel consumption, energy minimization, etc.). The emissions of CO_2 are directly proportional to the amount of fuel consumed by a vehicle and this amount depends on several factors such as the environment, traffic congestion, roadway gradient, curb-weight, and payload [12].

One way to reduce the carbon footprint of vehicles is using better operational strategies and establishing sustainable supply chains in the logistics industry. Therefore, it is essential to achieve efficient vehicle routing plans that properly considers sustainability factors. In the related literature, several variants of the VRP have been considered in the field of green logistics. The cumulative vehicle routing problem (CumVRP) introduced by [18,19] is a widely studied optimization problem that involves a weighted load function (load multiplied by distance). Another variant is the pickup and delivery vehicle routing problem (PDVRP) [33], which involves satisfying a set of pickup and delivery requests between location pairs. According to the type of demand and route structure, the PDVRP can be done in the following forms [5]: *many-to-many* (multiple products are transported between multiple origins and destinations), *one-to-many-to-one* (multiple products are transported from one depot to many clients and vice-versa), and *one-to-one* (each product is transported from a single origin to a single destination). An interesting aspect of the PDVRP is the influence of the load on fuel consumption when the delivery trip is made to a certain location. In addition, the multi-depot vehicle routing problem (MDVRP) has received increasing attention [28], because is it essential for companies with a wide range of business fields and having multiple depots, because the solution of the MDVRP could support these companies decrease their transport costs and improve their economic fulfillment. In the MDVRP, vehicles serve customers from several depots and return to the same depot. There are a few studies that consider MDVRPs combined with environmental factors (see [15,17,23,37]). However, to the best of our knowledge, the multi-depot green vehicle routing problem with pickups and deliveries (MDGVRP-PD) has not yet been investigated in the literature.

In this work, we study the MDGVRP-PD that consists of a one-to-one variant of the PDVRP in a multi-depot context considering the load on each arc, with the objective to minimize fuel consumption. For this one-to-one variant, we only work with the restriction that goods must first be picked up before they can be delivered, but these processes do not have to be carried out consecutively. To provide feasible solutions for the MDGVRP-PD, this study proposes a POPMUSIC matheuristic approach [21]. POPMUSIC is capable of addressing large scenarios by decomposing them into subsets of parts. Subsets of parts are bundled and used to create sub-problems, which are then solved by means of a mathematical programming approach.

The remainder of the paper is organized as follows. Section 2 reviews related works. Section 3 describes the mathematical formulation of the MDGVRP-PD. The POPMUSIC approach is presented in Sect. 4. Computational experiments and results are given in Sect. 5, and finally, we present the conclusions and future work in Sect. 6.

2 Related Works

Freight transportation is a significant component in logistic distribution activities and inevitably the largest consumer of fuel compared to other forms of transportation [6]. Therefore, the importance of achieving optimal routing plans that include sustainability factors is increasing as societies need transportation services that are economical, societal, but also environmentally sustainable. Furthermore, the green routing problems are characterized by achieving a sustainable supply chain network design and have attracted the scientific community's interest, especially in Operations Research and Artificial Intelligence fields (e.g., [6,14,24,27]).

Several authors have proposed diverse optimization models and solution approaches for green vehicle routing problems (GVRPs), considering the effects of vehicle's load on fuel consumption to reduce environmental pollution. The studies [18] and [19] present the energy minimizing vehicle routing problem and cumulative vehicle routing problem (CumVRP), which are the first vehicle routing studies proposing a cost function as a sum of the product between vehicles' load and distance for each arc in the vehicles' routes. The authors of [8] present a two-phase constructive heuristic approach to solve the CumVRP with limited duration restrictions and minimizing the fuel consumption. In this work, we use the fuel consumption parameters of vehicle categories proposed in [20]. In [7], the authors present the pollution routing problem, which seeks to minimize both operational and environmental costs by taking into account customers' time-windows constraints. The total travel distance, the amount of load carried per distance unit, the vehicle speeds, and the duration of the routes are the main costs.

While the GVRPs are suitable for solving single depot problems, supply chain networks primarily consist of multi-depots and multiple delivery points, which require more practical approaches such as the multi-depot vehicle routing problems (MDVRP) [9]. Furthermore, there exists a requirement for transportation and logistics businesses to minimize their environmental footprints. In comparison to the MDVRP, the MDGVRP considers more factors that might affect emissions (e.g., load, speed, traffic congestion, etc.), which, at the same time, increases the complexity of the problem and the difficulty of solving it. In [23], the authors study the MDGVRP maximizing the income (multiplication between the total demand for a product and the price) and minimizing costs, time, and emissions using an improved ant colony optimization algorithm. The authors of [37] propose a bi-objective model for the MDGVRP to minimize total carbon emissions and operational cost by considering the sharing of transportation resources within the same depot and among multiple depots. Similarly, to solve the MDGVRP efficiently, [17] presents a hybrid approach based on ant colony optimization and variable neighborhood search approaches to minimize cost and emission.

MDGVRP variants are considered in several studies, e.g., [15,30,38], but they do not take into account the influence of the vehicle's weight during the course of its route on fuel consumption. In this work, we estimate the fuel consumption using the approach defined by [32], where the weight of the vehicle is considered in the objective function. According to [20] and [34], the effect of vehicle payload on fuel consumption can be characterized by a linear function dependent on the payload, the consumption per unit of distance for the empty vehicle, and the consumption of moving the unit weight of goods per unit distance.

Besides the aforementioned works, some authors investigate PDVRPs, in which a set of pickup and delivery requests between customer couples are satisfied. In [5], the authors classify three different forms of pickup and delivery: one-to-one, one-to-many-to-one, and many-to-many. The PDVRP is classified as a $\mathcal{NP}$-hard problem due to complexity and the excessive consumption of computational time in its resolution. There are a few PDVRP studies where the emissions are taking into account [4,33,36]. However, as far as we know, the green PDVRP variant has not yet been investigated in a multi-depot scenario. In this study, we present a green variant of the multi-depot one-to-one pickup and delivery problem, where the objective is to design a set of optimal routes starting and ending at different depots to satisfy pickup and delivery requests under minimal emissions.

Other than in previous works, we develop a matheuristic that relies on the Partial Optimization Metaheuristic under Special Intensification Conditions (POPMUSIC, [35]) matheuristic proposed in [21] to solve the MDGVRP-PD. Matheuristics have been used successfully in routing problems [3]. In this context, POPMUSIC has been successfully utilized in various studies associated with the VRP (see [2,22,29]) for address large instances by decomposing them into a set of parts.

3 Problem Definition

The Multi-Depot Green VRP with Pickups and Deliveries (MDGVRP-PD) can be defined as follows. Let $G = (V, A)$ be a complete directed graph, where

$V = \{N \cup M\}$ is the node set that contains all customer and depot nodes, and $A = \{(i, j) : i, j \in V, i \neq j\}$ is the arc set. Node set $N = \{P \cup D\}$ with $P = \{1, \ldots, n\}$ represents the set of pick-up nodes and $D = \{n+1, \ldots, 2n\}$ the sets of delivery nodes, whereas node set $M = \{1, 2, \ldots, m\}$ represents the set of m uncapacitated depots. Furthermore, we have a homogeneous set of vehicles $K = \{1, 2, \ldots, k\}$, each with capacity Q. The following parameters and decision variables are defined in the problem:

Parameters:

- d_{ij}, the travel distance between each *arc* $(i, j) \in A$,
- q_i, the demand of each node,
- Q, the maximum load weight for each vehicle,
- tl, the time limit duration of the each subtour,
- α, the cost of moving an empty vehicle per unit of distance,
- β, the cost of moving the unit weight of goods per unit distance, respectively.

Decision variables:

- x_{ij}^k, 1 if a vehicle k travels from node $i \in V$ to node $j \in V$, and 0 otherwise,
- w_{ij}^k, the total load transported from node i to node j by vehicle k for $i, j \in V$.

We estimate the fuel consumption following the approach of [32]. Also, we consider the total vehicles' weight, represented by a carried load w_{ij}^k on arc (i, j), and it is a measure that helps reducing fuel consumption and environmental pollution. The objective function (1) of MDGVRP-PD minimizes the fuel consumption based on traveled distance d_{ij}, the load w_{ij}^k carried by the vehicle k on each arc $i, j \in V$, and fuel consumption parameters (see [20]). The parameters α and β represent the cost of moving an empty vehicle per unit of distance and the cost of moving the unit weight of goods per unit distance, respectively.

Objective function:

$$\text{Minimize} \quad \sum_{i=0}^{|V|} \sum_{j=0}^{|V|} \sum_{k=0}^{|K|} d_{ij}(\alpha x_{ij}^k + \beta w_{ij}^k) \tag{1}$$

In the MDGVRP-PD, the following constraints are considered:

(a) Each node has to be served exactly by one vehicle.
(b) Each vehicle visits a delivery location once it has visited the corresponding pickup location.
(c) Each vehicle starts and finishes at a depot.
(d) For each tour, the flow on the arcs accumulates as much as preceding node's supply in the case of pickup or diminish as much as preceding node's demand in the case of delivery.
(e) For each pickup or delivery node, the required demand must be satisfied.
(f) For each node, the in-degree must the equal to the out-degree.
(g) Total demand must not exceed vehicle capacity.
(h) For each vehicle, the maximum tour duration is not exceeded.

4 Proposed Algorithm

The POPMUSIC approach is a decomposition-based method, originally proposed by [35] as a metaheuristic and revised as a matheuristic in [21]. This approach divides a problem into smaller subproblems. Some or all of these subproblems are solved through metaheuristics or mathematical programming methods to optimality or suboptimality (i.e., matheuristic version).

To be precise, POPMUSIC works on an initial generated solution of the problem $\mathcal{S}$, which then will be decomposed into t parts $\{s_1, \ldots, s_t\}$. Each part corresponds to a subtour of $\mathcal{S}$. Next, some of these parts will be joined to build a subproblem SP using a proximity measure between parts. Subproblems are built by first selecting one of the t parts (called *seed-part*) and taking into account the r nearest parts ($SP = \{s_{seed}, s_1, s_2, \ldots, s_r\}$) according to the lexicographic strategy. The parameter r delimits the size of the subproblems. A mathematical programming method is used to optimize SP, and if there is an improvement over SP, then this improvement contributes to the total solution $\mathcal{S}$.

Algorithm 1 shows our matheuristic approach for solving the MDGVRP-PD. Initially, a feasible starting solution is constructed by a greedy strategy consisting of a set of parts or subtours (line 1). After the initial solution has been generated, the solution $\mathcal{S}$ is divided into t parts creating the set $H = \{s_1, \ldots, s_t\}$ (line 2). A set U is created to control the set of parts that have not been used as *seed-part* for building a subproblem (lines 3 and 4). Then, a *seed-part* is selected randomly (line 5). A subproblem SP is constructed by considering its r closest parts and is locally optimized by an exact method (lines 6 and 7). If the solution has been improved, then the solution $\mathcal{S}$ is updated (lines 8-10). Once U contains all the parts of the complete solution (line 4), the process ends as all sub-problems have been explored without improved results.

Algorithm 1: POPMUSIC pseudocode

```
1  Generate an initial solution S;
2  Decompose S into t parts, H = {s_1, ..., s_t};
3  Set U = ∅;
4  while (U ≠ {s_1, ..., s_t}) do
5      Select a seed-part, s_seed ∈ H, at random and s_seed ∉ U;
6      Build a sub-problem SP composed of the r parts of S which are the closest
       to s_seed;
7      Optimize SP by using a mathematical programming approach;
8      if (SP improved) then
9          Update solution S with SP;
10         U = ∅;
11     else
12         Insert s_seed in U;
13     end
14 end
15 return S;
```

Solution Representation. To represent a solution, we use a solution structure based on a two-dimensional vector of parts s_i, where each item represents a set of routes belonging to the same depot (see Fig. 1). Furthermore, each route must comply with the pickup and delivery constraints. That is, the delivery nodes must be visited after visiting the corresponding pickup nodes.

We consider three types of nodes in our problem. Each route is represented by a depot node (o), pickup node (p), and delivery node (d) (see Fig. 1). The sequencing of visiting is related to the one-to-one variant of the PDVRP; for example, the visiting order can be done by firstly visiting a pickup node then going directly to deliver the goods to their corresponding delivery node and then performing another pickup, or first all pickups and then all deliveries; any other form of mixed pickup and deliveries satisfying the sequencing restrictions.

$$\mathcal{S} = \begin{array}{c} s_1 \\ s_2 \\ \vdots \\ s_t \end{array} \begin{bmatrix} [o_1, p_1, d_{11}], & [o_1, p_2, d_{12}, p_5, d_{15}] & \cdots & [o_1, p_i, d_{|P|+i}, \cdots] \\ [o_2, p_7, p_{11}, d_{17}, d_{21}], & [o_2, p_6, d_{16}] & \cdots & [o_2, p_i, d_{|P|+i}, \cdots] \\ \vdots & \vdots & \ddots & \vdots \\ [o_m, p_{13}, d_{23}], & [o_m, p_{16}, d_{26}] & \cdots & [o_m, p_{|P|}, d_{|D|}, \cdots] \end{bmatrix}$$

Fig. 1. Solution structure composed of parts.

Initial Solution Strategy. For the operation of POPMUSIC, a key point is the generation of an initial solution. To do this, we have developed and tested a greedy construction method that considers the *haversine distance* of each node to each depot.

Algorithm 2 shows the pseudocode of the initial greedy solution. Initially, for each node, the closest depot is determined (lines 1–3). Next, a pickup node and its corresponding delivery node, as well as its closest depot, are obtained (lines 7–11). With this, a check is performed whether the trip's time duration and the vehicle's capacity restriction are satisfied (lines 12 and 13). If the nodes can be assigned to the vehicle, then a tour is built until it is part of the problem's solution (lines 14 and 15); otherwise, we proceed with another vehicle and construct a new tour (lines 17–20). Finally, the assignment of the nodes to the tour is performed by fulfilling the sequencing restrictions of pickup and delivery nodes.

Subproblem Generation Strategy. The *lexicographic* strategy presented in [22] is used to group the parts of a subproblem. This strategy consists of randomly selecting a seed-part s_{seed} and r parts of increasing index concerning the index θ of the seed-part. For example, if the initial solution is divided into 4 parts, and we consider $r = 2$, then we can have the following subproblems: $SP = \{s_1, s_2, s_3\}$, $SP = \{s_2, s_3, s_4\}$, and $SP = \{s_3, s_4, s_1\}$. The previous strategy can be grouped as a disjoint set and can be generalized by $\bigcup_{p=\theta}^{\theta+r}$.

Algorithm 2: Greedy algorithm pseudocode

```
1  for i ∈ N do
2      for j ∈ M do
3          minD[i] = argmin(d_ij);
4      end
5  end
6  current_vehicle = 1;
7  for (i ∈ M) do
8      for (j ∈ N/2) do
9          p_node = N[j];
10         d_node = N[|P| + j];
11         if (minD[j]==i) then
12             time = calculate travel time;
13             if (time ≤ tl and p_node.q_j ≤ Q) then
14                 Q = Q − p_node.q_j;
15                 p_node, d_node addeed to route S[current_vehicle]
16             else
17                 current_vehicle = current_vehicle + 1;
18                 Q = max_payload;
19                 time = 0;
20                 i = i − 1;
21             end
22         end
23     end
24 end
```

5 Computational Results

This section is devoted to analyzing the performance of the POPMUSIC approach for solving the MDGVRP-PD variant. All implementations were done in C++11 using Visual Studio v15.9.2 IDE and IBM ILOG CPLEX v12.9.0 API on Windows 10 OS. The tests were performed on an Intel(R) Xeon(R) E3-1220L (Sandy Bridge) CPU 2.20 GHz with 16GB RAM memory. The matheuristic approach was tested for 10 executions for each instance. The POPMUSIC approach was run in single-thread mode.

5.1 Instances

To test our matheuristic approach for the MDGVRP-PD, we modified subsets of the $n100$ and $n200$ groups of instances proposed in [31], where the authors consider real urban locations, and where a set of routes can be performed in a single labor day (eight hours). An example of these modified instances can be seen in Fig. 2.

The modified instances have $n+m$ locations. There are n customer locations and m depots. The n locations are paired to form a total of n requests (pickup and delivery couples). The n locations for pickup (P) and other n locations for delivery (D) are paired, where $|P| = |D| = n$, in a one-to-one way. The instance set consists of three different groups, ranging from 10 to 200 customers and are classified in three different complexity levels: small-scale with the first 10 or 50 customers from the original n100 group from [31]; medium-scale with the first 70 or 100 customers from $\{n100; n200\}$, and the last group for the large-scale with the first 150 or 200 customers. For all instances considered in this section, the time limit of the tour duration is 240 min, and we add four depot locations from the remaining locations that were not used in the generated instances. The customers' demands and the capacity of the vehicles are considered in kilogram. Also, the vehicle parameters are based on light-duty type with a curb-weight of 3500 kg and a maximum payload of 4000 kg, known as the gross vehicle weight rating.

5.2 Parameter Setting

A parameter tuning process was performed by executing them on all problem set instances. The only tuned parameter for POPMUSIC is r, with $r \in \{1, 2\}$. We have run a Friedman-k Related Samples test [16] to show the importance of our results. The test indicates no significant differences for both samples, with mean rank $r = 1$ (1.44) and $r = 2$ (1.56), showing both parameters have a small difference. Due to this, an analysis of the results will be carried out, considering the case of $r = 1$ (POPMUSIC^{r1}) and $r = 2$ (POPMUSIC^{r2}).

5.3 Results

This section compares the performance of the POPMUSIC^{r1} and POPMUSIC^{r2}. In doing so, we assess the performance of these variants on all problem instances in terms of objective function values of all iterations performed. Table 1 shows the results provided by the POPMUSIC^{r1} and POPMUSIC^{r2}.

In this table, the first column reports the instance studied, and columns Z^{r1}_{min}, Z^{r2}_{min} and Z^{r1}_{max}, Z^{r2}_{max} provide the minimum and maximum objective function values found, respectively. Columns Z^{r1}_{start} and Z^{r2}_{start} show the objective values of the initial solution provided by the greedy algorithm. Columns λ^{r1} and λ^{r2} represent the improvement between the best objective value and the initial solution value for each instance. The values of λ^{r1} and λ^{r2} are calculated using $100{\times}(Z^{r}_{min}\text{-}Z^{r}_{start})/Z^{r}_{start}$, with $r = r1 \vee r2$. Columns t(s)^{r1} and t(s)^{r2} represent the computational times. The last column shows the relative difference Gap(%) between POPMUSIC^{r1} and POPMUSIC^{r2}. It is calculated according to $100 \times (Z^{r1}_{min} - Z^{r2}_{min})/Z^{r2}_{min}$, where POPMUSIC^{r1} represents the best values for fuel consumption provided by our matheuristic approach. A negative value in columns λ^{r1}, λ^{r2}, and Gap % shows improvements.

The results show that POPMUSIC^{r1} generally obtains better objective function values (Z^{r1}_{min}) than POPMUSIC^{r2} (Z^{r2}_{min}), having 9 best values obtained from a total of 14 instances with different complexities. The difference between the minimum and maximum values (Z^{r1}_{max}, Z^{r2}_{max}) obtained shows that the

(a) *n10-1* with 10 locations.

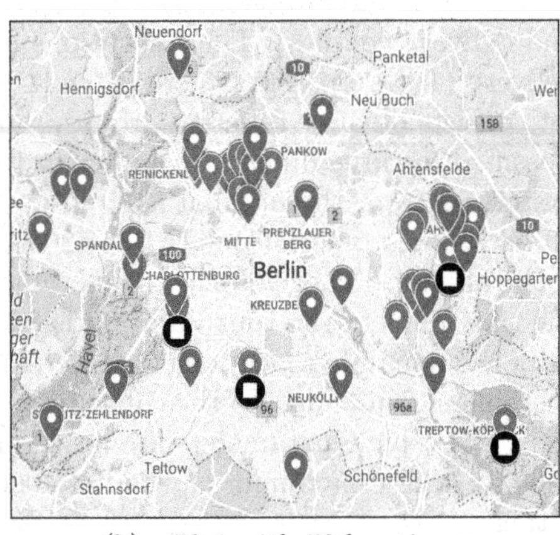

(b) *n50-1* with 50 locations.

Fig. 2. Example of two modified instances. Blue and red circles are pickups and delivery locations, respectively. The black circles are the depots. (Color figure online)

Table 1. Computational results for the POPMUSIC variants with $r = 1$ and $r = 2$ on modified instances. The best values are given in bold face.

Instance	POPMUSIC^{r1}					POPMUSIC^{r2}					
	Z^{r1}_{min}	Z^{r1}_{max}	Z^{r1}_{start}	λ^{r1}(%)	t(s)^{r1}	Z^{r2}_{min}	Z^{r2}_{max}	Z^{r2}_{start}	λ^{r2}(%)	t(s)^{r2}	Gap (%)
n10–1	**6.44**	7.00	11.53	−44.15	127.60	11.53	11.53	11.53	0.00	213.84	−44.15
n10–2	6.54	6.70	7.83	−16.48	276.16	**4.89**	7.07	7.83	−37.55	204.65	33.74
n30–1	**42.12**	56.37	51.48	−18.18	212.16	45.70	61.51	51.48	−11.23	208.8	−7.83
n30–2	**41.66**	50.26	53.08	−21.51	201.08	41.92	53.08	53.08	−21.02	206.25	−0.62
n50–1	**151.19**	175.05	166.58	−9.24	202.41	152.48	178.05	166.58	−8.46	207.32	−0.85
n50–2	120.66	133.51	133.42	−9.56	230.14	**114.45**	139.15	133.42	−14.22	211.56	5.43
n70–1	**99.11**	137.89	121.55	−18.46	204.68	105.05	132.46	121.55	−13.57	206.21	−5.65
n70–2	138.73	189.44	159.74	−13.15	249.90	**137.8**	161.18	159.74	−13.72	217.80	0.65
n100–1	108.17	111.82	110.71	−2.29	206.64	**103.42**	113.19	110.71	−6.58	289.79	4.59
n100–2	**112.87**	120.27	112.87	0.00	417.72	**112.87**	112.87	112.87	0.00	339.57	0.00
n150–1	214.44	234.04	219.40	−2.26	205.02	**214.07**	238.38	219.40	−2.43	205.53	0.17
n150–2	**288.96**	301.30	299.82	−3.62	232.79	297.19	304.79	299.82	−0.88	201.81	−2.77
n200–1	**271.66**	324.21	287.94	−5.65	201.93	272.03	289.09	287.94	−5.53	205.36	−0.14
n200–2	**418.66**	429.19	432.08	−3.11	202.58	422.33	439.84	432.16	−2.27	202.46	−0.87
Avg.	144.37	162.65	154.86	−11.98	226.49	145.41	160.16	154.87	−9.82	222.93	−1.31

matheuristic generally maintains a stable behavior. The initial solution values (Z^{r1}_{start}, Z^{r2}_{start}) concerning the best values obtained (Z^{r1}_{min}, Z^{r2}_{min}) are always improved during the POPMUSIC process (see λ^{r1} and λ^{r2}). Only for instance n10–2, the gap value of 33.74% is significant. Furthermore, these results also indicate, as discussed in [21], the suitability of the matheuristic POPMUSIC for using and exploiting the exact optimization method for subproblems that allow solving them to optimality within reasonable computational times.

5.4 Effects of Loading on Fuel Consumption

To analyze the effect of the load on fuel consumption and emissions and bearing in mind the previous results, we made a trade-off between the objective function (see Eq. (1)) and one of the most used objective functions in the literature for VRPs, i.e., the minimization of travel distances.

First, the set of values Z^{r1}_{min} and Z^{r2}_{min} for all instances with the estimation of the amount of fuel consumption using the objective function (1), in which the load carried over an arc w_{ij} was showed in Table 1. Second, let $Z^{'r1}_{min}$ and $Z^{'r2}_{min}$ denote the set of values for all instances with the estimation of the amount of fuel consumption considering a modified objective function considering only the distance d_{ij} traveled by the vehicle as $\sum_{i=0}^{V}\sum_{j=0}^{V}\sum_{k=0}^{K} d_{ij}x^k_{ij}$ and keeping the flow restrictions.

Table 2 shows a comparison of the experiments between fuel consumption and emission values. Columns Z^{r1}_{min} and Z^{r2}_{min} are the same as used for Table 1. The amount of fuel consumption without considering the weight component in the objective function is represented in columns $Z^{'r1}_{min}$ and $Z^{'r2}_{min}$. Also, we calculate the emissions of CO_2 for all fuel consumption values using the emission factor defined as 2.72 kg/L of fuel consumption [11], see columns CO^{r1}_2, $CO^{'r1}_2$, CO^{r2}_2, and $CO^{'r2}_2$. Furthermore, columns $Imp.^{r1}$ and $Imp.^{r2}$ show the environmental percentage of improvement between (CO^{r1}_2 and $CO^{'r1}_2$), and (CO^{r2}_2 and $CO^{'r2}_2$) in terms of emissions. The average percentage of improvement for POPMUSIC^{r1} and POPMUSIC^{r2} is 30.79% and 25.50%, respectively.

Table 2. Computational results of the POPMUSIC approaches for the MDGVRP-PD considering fuel consumption with and without the weight component in the objective function. The environmental improvement average % values are given in boldface.

Instance	POPMUSIC^{r1}					POPMUSIC^{r2}				
	Z^{r1}_{min}	CO^{r1}_2	$Z^{'r1}_{min}$	$CO^{'r1}_2$	$Imp.^{r1}$	Z^{r2}_{min}	CO^{r2}_2	$Z^{'r2}_{min}$	$CO^{'r2}_2$	$Imp.^{r2}$
n10-1	6.44	17.52	20.73	56.39	**−68.93**	11.53	31.36	20.73	56.39	**−44.38**
n10-2	6.54	17.79	31.80	86.50	**−79.43**	4.89	19.31	19.31	52.52	**−63.24**
n30-1	42.12	114.57	91.22	248.12	**−53.83**	45.70	124.30	91.22	248.12	**−49.90**
n30-2	41.66	113.32	77.76	211.51	**−46.42**	41.92	114.02	77.76	211.51	**−46.09**
n50-1	151.19	411.24	170.58	463.98	**−11.37**	152.48	414.75	170.58	463.98	**−10.61**
n50-2	120.66	328.20	180.65	491.37	**−33.21**	114.45	311.30	136.79	372.07	**−16.33**
n70-1	99.11	269.58	143.81	391.16	**−31.08**	105.05	285.74	143.81	391.16	**−26.95**
n70-2	138.73	377.35	197.30	536.66	**−29.69**	137.80	374.82	213.29	580.15	**−35.39**
n100-1	108.17	294.22	150.26	408.71	**−28.01**	103.42	281.30	122.17	332.30	**−15.35**
n100-2	112.87	307.01	145.95	396.98	**−22.67**	112.87	307.01	145.95	396.98	**−22.67**
n150-1	214.44	583.28	226.13	615.07	**−5.17**	214.07	582.27	225.66	613.80	**−5.14**
n150-2	288.96	785.97	293.39	798.02	**−1.51**	297.19	808.36	293.39	798.02	**1.30**
n200-1	271.66	738.92	319.23	868.31	**−14.90**	272.03	739.92	319.23	868.31	**−14.79**
n200-2	418.66	1138.76	439.86	1196.42	**−4.82**	422.33	1148.74	439.86	1196.42	**−3.99**
Avg.	144.37	392.69	177.76	483.51	**−30.79**	145.41	395.94	172.84	470.12	**−25.25**

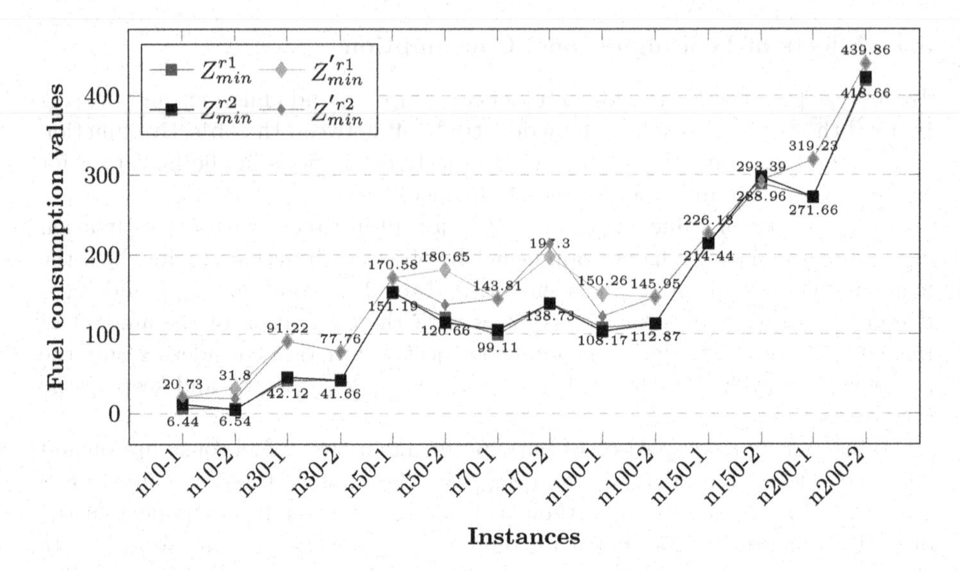

Fig. 3. Comparison between fuel consumption values for Z^{r1}_{min}, Z'^{r1}_{min}, and Z^{r2}_{min}, Z'^{r2}_{min} for all instances.

Figure 3 shows the values of Z^{r1}_{min}, Z'^{r1}_{min}, Z^{r2}_{min}, and Z'^{r2}_{min}, for all instances. The blue and red line represents the values for Z^{r1}_{min} and Z'^{r1}_{min}, which show a decrease in the fuel consumption values concerning Z'^{r1}_{min}. The black and gray line represents the values for Z^{r2}_{min} and Z'^{r2}_{min}, which shows similar behavior as the previous one. This illustration remarks the significance of considering the fuel consumption depending on weight and travel distance as it is proportional to the reduction in emissions.

6 Conclusion and Future Research

This work addresses the Multi-Depot Green Vehicle Routing Problem with Pick-ups and Deliveries (MDGVRP-PD). To solve this problem, we have designed a POPMUSIC matheuristic approach to take advantage of the efficiency of exact solutions to solve simple subproblems. We designed a set of instances for (MDGVRP-PD) based on real urban locations to test our approach. A parameters setting was conducted for POPMUSIC, showing minor differences in terms of the fitness function.

The results show that POPMUSIC can provide feasible solutions for all instances in reasonable computational times. POPMUSIC can use and exploit the exact optimization method for subproblems that allow solving them to optimality, showing that the decomposition methods can offer adequate and robust solutions for routing problems with several depots.

Furthermore, an analysis of the effect of the weight on fuel consumption which is proportional to the emissions showed a trade-off when it is considered that the weight carried by the routes with the fitness measure can produce significant reductions in the emissions as compared to the case when only distances are taken into account. As a result, our approach can provide an average percentage of improvement in the emission of almost 30.79%.

Finally, as future work, we want to investigate the heterogeneous fleet and speed variation in multi-depot green routing problems. Another issue concerns the consideration of clustering algorithms to improve the initial solution procedures of POPMUSIC.

Acknowledgments. This work has been partially supported by ANID-PFCHA/ Doctorado Nacional/2020-21200871, and in part by Proyectos de Línea de Investigación Regular (PI_LIR_2020_67, UTFSM) and Programa de Incentivo a la Iniciación Científica (PIIC, UTFSM).

References

1. Abdullahi, H., Reyes-Rubiano, L., Ouelhadj, D., Faulin, J., Juan, A.A.: Modelling and multi-criteria analysis of the sustainability dimensions for the green vehicle routing problem. Eur. J. Oper. Res. **292**(1), 143–154 (2021)
2. Alvim, A.C.F., Taillard, É.D.: Popmusic for the world location-routing problem. EURO J. Transp. Logist. **2**(3), 231–254 (2013)
3. Archetti, C., Speranza, M.G.: A survey on matheuristics for routing problems. EURO J. Comput. Optim. **2**(4), 223–246 (2014). https://doi.org/10.1007/s13675-014-0030-7
4. Asghari, M., Mirzapour Al-e-hashem, S.M.J.: A green delivery-pickup problem for home hemodialysis machines; sharing economy in distributing scarce resources. Transp. Res. Part E Logist. Transp. Rev. **134**, 101815 (2020)
5. Battarra, M., Cordeau, J.-F., Iori, M.: Chapter 6: pickup-and-delivery problems for goods transportation, chapter 6, pp. 161–191. SIAM (2014)
6. Bektaş, T., Ehmke, J.F., Psaraftis, H.N., Puchinger, J.: The role of operational research in green freight transportation. Eur. J. Oper. Res. **274**(3), 807–823 (2019)
7. Bektaş, T., Laporte, G.: The pollution-routing problem. Transp. Res. Part B Methodol **45**(8), 1232–1250 (2011)
8. Cinar, D., Gakis, K., Pardalos, P.M.: A 2-phase constructive algorithm for cumulative vehicle routing problems with limited duration. Expert Syst. Appl. **56**, 48–58 (2016)
9. Cordeau, J.-F., Gendreau, M., Laporte, G.: A tabu search heuristic for periodic and multi-depot vehicle routing problems. Networks **30**(2), 105–119 (1997)
10. Dantzig, G.B., Ramser, J.H.: The truck dispatching problem. Manag. Sci. **6**(1), 80–91 (1959)
11. U. DBEIS. Greenhouse gas reporting: Conversion factors 2018. Technical report, UK Department for Business, Energy and Industrial Strategy (2018). Accessed 01 June 2021
12. Demir, E., Bektaş, T., Laporte, G.: The bi-objective pollution-routing problem. Eur. J. Oper. Res. **232**(3), 464–478 (2014)
13. Demir, E., Bektaş, T., Laporte, G.: A review of recent research on green road freight transportation. Eur. J. Oper. Res. **237**(3), 775–793 (2014)
14. Erdelić, T., Carić, T.: A survey on the electric vehicle routing problem: Variants and solution approaches. J. Adv. Transp. **2019**, 5075671 (2019)

15. Fan, H., Zhang, Y., Tian, P., Lv, Y., Fan, H.: Time-dependent multi-depot green vehicle routing problem with time windows considering temporal-spatial distance. Comput. Oper. Res. **129**, 105211 (2021)
16. Friedman, M.: A comparison of alternative tests of significance for the problem of m rankings. Ann. Math. Stat. **11**, 86–92 (1940)
17. Jabir, E., Panicker, V.V., Sridharan, R.: Design and development of a hybrid ant colony-variable neighbourhood search algorithm for a multi-depot green vehicle routing problem. Transp. Res. Part D Transp. Environ. **57**, 422–457 (2017)
18. Kara, İ, Kara, B.Y., Yetis, M.K.: Energy minimizing vehicle routing problem. In: Dress, A., Xu, Y., Zhu, B. (eds.) COCOA 2007. LNCS, vol. 4616, pp. 62–71. Springer, Heidelberg (2007). https://doi.org/10.1007/978-3-540-73556-4_9
19. İ. Kara, B. Y. Kara, and M. K. Yetiş. Cumulative vehicle routing problems. In: Vehicle Routing Problem, pp. 85–98. IntechOpen (2008)
20. Kopfer, H.W., Schönberger, J., Kopfer, H.: Reducing greenhouse gas emissions of a heterogeneous vehicle fleet. Flex. Serv. Manuf. J. **26**(1), 221–248 (2014)
21. Lalla-Ruiz, E., Voß, S.: Popmusic as a matheuristic for the berth allocation problem. Ann. Math. Artif. Intell **76**(1), 173–189 (2016)
22. Lalla-Ruiz, E., Voß, S.: A popmusic approach for the multi-depot cumulative capacitated vehicle routing problem. Optim. Lett. **14**, 1–21 (2019)
23. Li, Y., Soleimani, H., Zohal, M.: An improved ant colony optimization algorithm for the multi-depot green vehicle routing problem with multiple objectives. J. Cleaner Prod. **227**, 1161–1172 (2019)
24. Lin, C., Choy, K.L., Ho, G.T., Chung, S.H., Lam, H.: Survey of green vehicle routing problem: past and future trends. Expert Syst. Appl. **41**(4), 1118–1138 (2014)
25. Macrina, G., Laporte, G., Guerriero, F., Di Puglia Pugliese, L.: An energy-efficient green-vehicle routing problem with mixed vehicle fleet, partial battery recharging and time windows. Eur. J. Oper. Res. **276**(3), 971–982 (2019)
26. McKinnon, A.: Environmental sustainability: a new priority for logistics managers. Kogan (2015)
27. Moghdani, R., Salimifard, K., Demir, E., Benyettou, A.: The green vehicle routing problem: a systematic literature review. J. Cleaner Prod. **279**, 123691 (2021)
28. Montoya-Torres, J.R., López Franco, J., Nieto Isaza, S., Felizzola Jiménez, H., Herazo-Padilla, N.: A literature review on the vehicle routing problem with multiple depots. Comput. Ind. Eng. **79**, 115–129 (2015)
29. Ostertag, A., Doerner, K.F., Hartl, R.F., Taillard, E.D., Waelti, P.: Popmusic for a real-world large-scale vehicle routing problem with time windows. J. Oper. Res. Soc. **60**(7), 934–943 (2009)
30. Sadati, M.E.H., Çatay, B.: A hybrid variable neighborhood search approach for the multi-depot green vehicle routing problem. Transp. Res. Part E Logist. Transp. Rev. **149**, 102293 (2021)
31. Sartori, C.S., Buriol, L.S.: A study on the pickup and delivery problem with time windows: Matheuristics and new instances. Comput. Oper. Res. **124**, 105065 (2020)
32. Singh, R.R., Gaur, D.R.: Cumulative VRP: a simplified model of green vehicle routing. In: Cinar, D., Gakis, K., Pardalos, P.M. (eds.) Sustainable Logistics and Transportation. SOIA, vol. 129, pp. 39–55. Springer, Cham (2017). https://doi.org/10.1007/978-3-319-69215-9_3
33. Soysal, M., Çimen, M., Demir, E.: On the mathematical modeling of green one-to-one pickup and delivery problem with road segmentation. J. Cleaner Prod. **174**, 1664–1678 (2018)

34. Suzuki, Y.: A new truck-routing approach for reducing fuel consumption and pollutants emission. Transp. Res. Part D Transp. Environ. **16**(1), 73–77 (2011)
35. Taillard, É.D., Voß, S.: Popmusic – partial optimization metaheuristic under special intensification conditions, pp. 613–629. Springer, Boston (2002). https://doi.org/10.1007/978-1-4615-1507-4_27
36. Wang, J., Yu, Y., Tang, J.: Compensation and profit distribution for cooperative green pickup and delivery problem. Transp. Res. Part B Methodol. **113**, 54–69 (2018)
37. Wang, Y., Assogba, K., Fan, J., Xu, M., Liu, Y., Wang, H.: Multi-depot green vehicle routing problem with shared transportation resource: integration of time-dependent speed and piecewise penalty cost. J. Cleaner Prod. **232**, 12–29 (2019)
38. Zhang, W., Gajpal, Y., Appadoo, S.S., Wei, Q.: Multi-depot green vehicle routing problem to minimize carbon emissions. Sustainability **12**(8), 3500 (2020)

Solving the Shipment Rerouting Problem with Quantum Optimization Techniques

Sheir Yarkoni[1,2(✉)], Andreas Huck[3], Hanno Schülldorf[3], Benjamin Speitkamp[3], Marc Shakory Tabrizi[3], Martin Leib[1], Thomas Bäck[2], and Florian Neukart[1,2]

[1] Volkswagen Data:Lab, Munich, Germany
sheir.yarkoni@volkswagen.de
[2] LIACS, Leiden University, Leiden, The Netherlands
[3] Deutsche Bahn AG, Berlin, Germany

Abstract. In this work we develop methods to optimize an industrially-relevant logistics problem using quantum computing. We consider the scenario of partially filled trucks transporting shipments between a network of hubs. By selecting alternative routes for some shipment paths, we optimize the trade-off between merging partially filled trucks using fewer trucks in total and the increase in distance associated with shipment rerouting. The goal of the optimization is thus to minimize the total distance travelled for all trucks transporting shipments. The problem instances and techniques used to model the optimization are drawn from real-world data describing an existing shipment network in Europe. We show how to construct this optimization problem as a quadratic unconstrained binary optimization (QUBO) problem. We then solve these QUBOs using classical and hybrid quantum-classical algorithms, and explore the viability of these algorithms for this logistics problem.

1 Introduction

Quantum computing has garnered increased interest in recent years in both research and industrial settings. This novel technology holds the promise of solving computationally intractable problems asymptotically faster than their classical counterparts in a variety of application areas [1–3]. The public availability of quantum devices from commercial entities such as D-Wave Systems, Google, and IBM have produced a variety of results showcasing novel algorithms and potential use-cases for quantum computing in fields such as quantum machine learning [4,5], logistics/scheduling [6,7], quantum chemistry [8,9], and more [10]. Of particular interest is the potential of quantum processing units (QPUs) to affect the field of optimization, making the technology attractive to both research and industry experts. Currently, variational quantum optimization techniques are the main targets of promising research, and hold the highest potential of gaining advantage using quantum processors. For more details about the different paradigms of quantum computing and their mathematical backgrounds we refer the reader to [11–13].

M. Mes et al. (Eds.): ICCL 2021, LNCS 13004, pp. 502–517, 2021.
https://doi.org/10.1007/978-3-030-87672-2_33

Optimization problems for quantum algorithms and other similar heuristics are typically formulated as either Ising Hamiltonians (posed in a $\{-1, 1\}$ basis) or quadratic unconstrained binary optimization (QUBO) problems (in a $\{0, 1\}$ basis). Finding the minimum of an Ising Hamiltonian, or its equivalent QUBO, is known to be an NP-hard problem in the worst case [14], meaning many difficult and well-known optimization problems have such representations [15]. For our work, we focus on the QUBO formulation of optimization problems:

$$\text{Obj}(Q, \mathbf{b}) = \mathbf{b} \cdot Q \cdot \mathbf{b}^T. \tag{1}$$

Here, Q is an $N \times N$ matrix representing interaction terms between variables in the binary vector $\mathbf{b}$ with N variables. Therefore, the first step in using quantum optimization algorithms is finding a valid QUBO representation of the problem to be solved. In this paper we focus on designing a QUBO representation for an industrially motivated logistics optimization problem, and attempt to optimize the QUBOs using both hybrid quantum-classical and purely classical QUBO solvers. Additionally, we consider a representation by a mixed integer program (MIP) which we optimize by the standard solver Gurobi.

We investigate a problem motivated by an application in logistics: the less-than-truckload network service design. Less-than-truckload (LTL) denotes shipments not exceeding a maximum weight significantly below a full truck load. The transport of a single shipment follows the sequence of (a) a collecting truck run, followed by (b) one or several linehaul truck runs (including handling of the shipment) and ending with (c) a distributing truck run to the shipment's final destination. Our work focuses on the design of the linehaul network, step (b). The linehaul network for LTL is made up by the set of terminals and timetable based truck runs, connecting the terminals and thereby producing the long-hauls of all the shipments entering the network. Taking limitations on transport times into account, the forwarding of the shipments shall be as cost efficient as possible. One key factor for cost efficiency is the consolidation of multiple shipments in jointly utilized trucks, at least regarding parts of their individual linehaul paths through the network. This measure targets an increase of truck utilization. However, the consolidation of multiple shipments with different origins and destinations in jointly utilized trucks requires detours of shipments. As detours come at a cost, the network design searches for an optimal trade-off between detour costs and the benefit of increased truck utilization. We focus on this central trade-off decision and call the reduced problem the *shipment rerouting problem* (SRP). We provide an illustrative example with two shipments in Fig. 1.

The input to the SRP includes a set of terminals, their distances to each other, and the numbers of available trucks connecting the terminals. Moreover, we have a set of shipments, each with a set of possible routes of intermediate terminals. These possible routes already comply with constraints like maximum transport time or maximum detour factor. They include the direct route from the origin to the destination of the shipment, which are the default for all shipments. Other candidate routes for rerouting are constructed in a pre-processing step based on the graphical structure of the terminals and the distances between them.

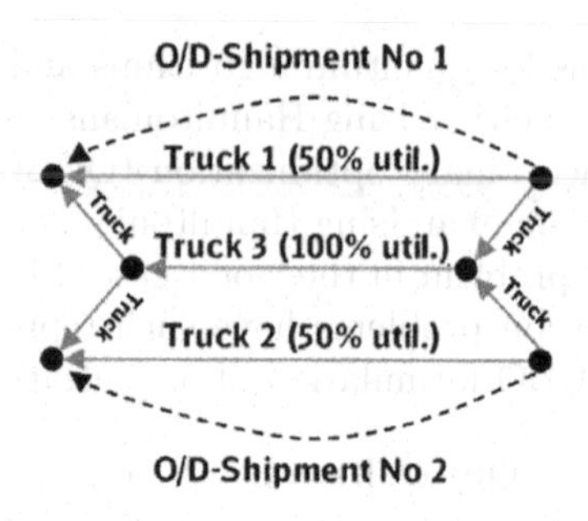

Fig. 1. An example of the SRP with two shipments. The default routing (Trucks 1 and 2 carrying their respective shipments at 50% capacity each) is optimized by replacing Trucks 1 and 2 with a single truck (Truck 3) which can be fully utilized. The cost of rerouting each shipment to the route serviced by Truck 3 is offset by the removal of Trucks 1 and 2, thereby reducing the overall distance travelled to deliver the shipments.

Thus, a subset of shipments may be rerouted through alternate routes in order to reduce the overall distance all trucks travel to deliver the shipments. Each shipment has a size (volume, weight, etc.), and each truck has a corresponding capacity, i.e. an upper bound for the total shipment size that can be loaded. For our purposes, we denote the shipment sizes and truck capacities with respect to volume, and refer to them as such throughout the rest of this work. We note that our mathematical formulations equally admit other quantities.

A shipment cannot be split across different routes. However, for transporting a shipment between two terminals, we may split it to distribute it on multiple trucks (this is necessary especially for shipments with a large volume). Given the input, the task is to decide on a route for each of the given shipments, which may include overlaps between shipments. Consequently, the result includes the number of required trucks in the network and which terminals are connected by truck runs in which frequency.

The rest of this paper is organized as follows: Sect. 2 discusses previous literature relevant to this analysis, and Sect. 3 motivates the QUBO construction for this problem based on a MIP representation and details the specifics of the QUBO construction used throughout this work. Section 4 outlines the input data, solvers, and experimental design of our analysis. Section 5 presents the results from those experiments, and Sect. 6 summarizes the conclusions derived from our work.

2 Previous Works

In [16], Ding et al. solve a network design problem with a quantum annealing approach. However, this problem is different as it searches for the best terminal locations while the arc costs are linear (a hub location problem). In [6] it is shown how to form QUBO representation of a simple traffic flow combinatorial optimization problem. In that work, individual vehicles are given multiple candidate routes whose intersection needs to be minimized. This route-generation procedure is used in our work, but with opposite intent, our objective is to consolidate as many routes as possible.

Other examples of logistics and scheduling applications in quantum computing include flight-path conflict resolution [17] and railway train rescheduling [18]. Both examples use elements from a generic job-shop scheduling formulation for quantum annealing [7]. While these applications are qualitatively similar to some aspects of our work, the shipment rerouting problem is unique for quantum annealing as not only the selection of routes for each shipment is variable, but also the number of trucks used on each edge along the path is selected by the optimization. Typically, for example in job-shop scheduling, the number of machines and jobs are inputs to the QUBO construction. Our formulation thus incorporates elements from both scheduling (route selection in [6] and [18]) and packing problems (canonical problems in NP [15]).

Because of its cost structure, the SRP problem is closely related to the fixed-charge multi-commodity network design (FCMND) problem which has not been investigated in the field of quantum computing. However, it has been studied extensively in the past. Exact algorithms are usually based on branch-and-cut approaches and Bender's decomposition– see [19] for an overview. A particular problem for exact algorithms is that the lower bound is hard to improve. There are also heuristic approaches to solve this problem, using evolutionary algorithms [20] and simulated annealing [21], among others.

3 Constructing MIP and QUBO Representations

The MIP representation of the SRP is straightforward to formulate, as we can use multiple kinds of variables (binary, integer, real) and constrain the solutions explicitly. Therefore, we start with a MIP representation which we then transform to a QUBO.

3.1 Constructing the MIP Representation

We assume that the connectivity of the terminals can be represented as a weighted directed graph G where the vertices V are the terminals and the edges E between them represent the ability to transport shipments from any single terminal to another; in other words, we have an edge $e \in E$ from a terminal a to a terminal b if there are trucks available driving from a to b. These trucks are called the *trucks on* e and their number is denoted by $t_{\max}(e)$. The weight of e is the distance from a to b and is denoted by $d(e)$. For each shipment s, $v(s)$ denotes its volume and $R(s)$ denotes the set of all routes that can be used to transport s (*candidate routes* of s). For each edge e, $R(e)$ denotes the set of all candidate routes containing e. A shipment s is *scheduled* on some edge e if s is transported using an associated candidate route r containing e, i.e. $r \in R(s) \cap R(e)$.

In our scenarios, all trucks have the same volume capacity, denoted by c_{vol}. Moreover, all shipments have different origin-destination pairs so that no two different shipments have common candidate routes (however, their candidate routes may overlap). Therefore, for each candidate route r, we have a unique shipment $s(r)$ that can be transported using r.

We want to transport each shipment on an associated candidate route such that the total distance of all used trucks in the network is minimized. To represent this problem by a MIP, we introduce a binary decision variable y_r for each candidate route r that is 1 if r is used to transport $s(r)$, and 0 otherwise. For each edge e, we introduce a non-negative integer variable t_e with maximal value $t_{\max}(e)$ representing the number of used trucks on e. We represent the problem by the following MIP:

Objective: Minimize the total truck distance

$$\sum_{e \in E} d(e) \cdot t_e \tag{2}$$

with respect to the following constraints:
Route-shipment constraints: For each shipment s, exactly one associated candidate route is used, i.e.

$$\sum_{r \in R(s)} y_r = 1. \tag{3}$$

Capacity constraints: For each edge e, the total volume of all shipments scheduled on e does not exceed the total volume capacity of the used trucks on e, i.e.

$$\sum_{r \in R(e)} v(s(r)) \cdot y_r \leq c_{\text{vol}} \cdot t_e. \tag{4}$$

The capacity constraints ensure that on each edge e, enough trucks are used to transport all shipments scheduled on e because we can split shipments to optimally exploit the truck capacities. Note that in an optimal solution, each truck number t_e is as small as possible, namely $\left\lceil \sum_{r \in R(e)} v(s(r)) \cdot y_r / c_{\text{vol}} \right\rceil$. In that case, for each edge e, we can completely fill all used trucks on e except possibly one truck that is partially filled.

3.2 Constructing the QUBO Representation

Contrary to a MIP, a QUBO only contains binary variables and an objective function to be minimized without explicit constraints. However, the quadratic summands arising from Eq. (1) allow us to include penalty terms that emulate the MIP constraints.

Our QUBO formulation uses the binary variables y_r for the candidate routes r. In replacement of the integer variables t_e for the edges e, we use modified binary representations of their values in the QUBO based on a concept in [15]: for each edge e, we define $T(e)$ to be the set of all powers of two less than or equal $t_{\max}(e)$, and for each $n \in T(e)$, we introduce a binary variable $t_{e,n}$ in order to represent the number of used trucks on e by $\sum_{n \in T(e)} n \cdot t_{e,n}$. In this way, we can represent at least each number up to $t_{\max}(e)$, i.e. each allowed truck number. However, the maximal representable number is $2n_{\max} - 1$ where $n_{\max}$ is the maximal value in $T(e)$. Therefore, to avoid representations of numbers greater than $t_{\max}(e)$, we reduce the coefficient $n_{\max}$ in $\sum_{n \in T(e)} n \cdot t_{e,n}$ by the surplus $s := 2n_{\max} - 1 - t_{\max}(e)$. The new expression is denoted by

$$\sum_{n \in T(e)} \overline{n} \cdot t_{e,n}, \tag{5}$$

i.e. we have $\overline{n_{\max}} = n_{\max} - s = 1 + t_{\max}(e) - n_{\max}$ and $\overline{n} = n$ for each $n \neq n_{\max}$. Now we can still represent each number up to $t_{\max}(e)$ but no other numbers. In our QUBO, we reformulate the total truck distance (2) as

$$\sum_{e \in E} d(e) \cdot \sum_{n \in T(e)} \overline{n} \cdot t_{e,n}. \tag{6}$$

To encode the route-shipment constraints (3), note that they are linear equalities of the form $A = B$ where only binary variables occur. Each such constraint is implemented in our QUBO by adding the summand $M \cdot (A - B)^2$ where M is a large penalty factor ensuring that the constraint is fulfilled at least in all optimal solutions of our QUBO. We will later discuss how to define a suitable penalty factor.

The capacity constraints (4) cannot be implemented in that way (after reformulation using the representations (5)) because they are inequalities of the form $A \leq B$. However, such a constraint can be transformed into an equality $A + \ell = B$ by using a non-negative slack variable ℓ. Note that the slack of the capacity constraint for each edge e is the wasted volume in the used trucks on e (*volume capacity slack* on e). Unfortunately, contrary to the numbers of used trucks, these slacks might be large or fractional values so that their representations might require many binary variables, making our QUBO intractable.

To overcome this problem, we discretize the shipment volumes into *bins*: We virtually divide the loading area of each truck into the same number c_{bin} of equally sized bins. c_{bin} is called the *bin capacity* of the trucks. Each bin can only be used for transporting one shipment and has the volume capacity $c_{\text{vol}}/c_{\text{bin}}$. Hence, for each shipment s, the number $b(s)$ of bins needed to transport s is given by $b(s) = \lceil v(s) \cdot c_{\text{bin}}/c_{\text{vol}} \rceil$. Instead of the volume capacity slacks, we now have to represent the *bin capacity slack* on each edge e, i.e. the number of unused bins in the used trucks on e. These slacks are more tractable because they are integers that can be assumed to be less than c_{bin} (we will see this later).

On the other hand, if c_{bin} is too small, then the bin volume capacity $c_{\text{vol}}/c_{\text{bin}}$ is large so that we may obtain several partially filled bins in the trucks, especially if shipments exist that are smaller than the bin volume capacity (recall that we cannot use a bin for transporting more than one shipment). Hence, we may not optimally exploit the truck capacities any more which may increase the number of used trucks. We can improve the situation by multiplying the bin capacity, i.e. by subdividing each bin into the same number of smaller bins.[1] Therefore, c_{bin} is a crucial parameter for the QUBO construction: more bins may lead to a better exploitation of the truck capacities, but at the cost of larger bin capacity slacks to be represented. In our experiments, we used the bin capacity 10, which was an empirically-determined compromise.

For each edge e, we introduce a non-negative integer variable ℓ_e representing the bin capacity slack on e. Then we obtain a new *discretized* MIP by modifying the capacity constraints (4) as follows:

Capacity Constraints: For each edge e, we have

$$\sum_{r \in R(e)} b(s(r)) \cdot y_r + \ell_e = c_{\text{bin}} \cdot t_e. \tag{7}$$

These constraints imply the former ones (because $v(s) \leq b(s) \cdot c_{\text{vol}}/c_{\text{bin}}$ for each shipment s) and may even be stronger (due to a worse exploitation of the truck capacities). Similar to the former MIP, in an optimal solution of the new MIP, each truck number t_e is as small as possible, namely $\left\lceil \sum_{r \in R(e)} b(s(r)) \cdot y_r / c_{\text{bin}} \right\rceil$. Therefore, each bin capacity slack ℓ_e is less than c_{bin} so that we can represent these values in the QUBO as follows: we define L to be the set of all powers of two less than c_{bin}, and for each edge e and for each $m \in L$, we introduce a binary variable $\ell_{e,m}$ such that the bin capacity slack of e is

$$\sum_{m \in L} m \cdot \ell_{e,m}. \tag{8}$$

In this way, we can represent at least each number less than c_{bin}, i.e. each relevant bin capacity slack (it doesn't matter if we can represent further numbers).

Using the representations (5) and (8), we can reformulate the capacity constraints (7) as follows:

Capacity Constraints: For each edge e, we have

$$\sum_{r \in R(e)} b(s(r)) \cdot y_r + \sum_{m \in L} m \cdot \ell_{e,m} = c_{\text{bin}} \cdot \sum_{n \in T(e)} \overline{n} \cdot t_{e,n}. \tag{9}$$

Similar to the route-shipment constraints, these capacity constraints are implemented in the standard way by summands of the form $M \cdot (A - B)^2$. Putting all components together, we obtain the following formulation of the QUBO:

[1] Simply increasing the bin capacity may worsen the situation. For instance, suppose that $v(s) = c_{\text{vol}}/2$ for each shipment s so that $b(s) = \lceil c_{\text{bin}}/2 \rceil$. If $c_{\text{bin}} = 2$, then $b(s) = 1$ so that we can put two shipments into a truck. But if $c_{\text{bin}} = 3$, then $b(s) = 2$ so that we can put only one shipment into a truck.

$$\sum_{e\in E} d(e) \cdot \sum_{n\in T(e)} \overline{n} \cdot t_{e,n} + M \cdot \sum_{s\in S} \left(\sum_{r\in R(s)} y_r - 1 \right)^2$$

$$+M \cdot \sum_{e\in E} \left(\sum_{r\in R(e)} b(s(r)) \cdot y_r + \sum_{m\in L} m \cdot \ell_{e,m} - c_{\text{bin}} \cdot \sum_{n\in T(e)} \overline{n} \cdot t_{e,n} \right)^2 . \quad (10)$$

Here, all variables are as before, and S is the set of all shipments in the problem.

We now choose the penalty factor M to ensure that only feasible solutions are present in the global optimum of the QUBO objective, so that it is never energetically favorable to violate one of the constraints in favor of minimizing the total track distance. In general, we may choose any M greater than the total truck distance *d(feas)* of any known feasible solution *feas* (for instance, the solution transporting each shipment on its direct route). To see the correctness of M, consider an optimal solution *opt* and suppose that *opt* violates a constraint. Then the *opt*-value of the QUBO objective is at least M and thus greater than *d(feas)*. But since *feas* is feasible, *d(feas)* is also the *feas*-value of the QUBO objective, contradicting the optimality of *opt*.

The QUBO requires many more variables than the MIP in Sect. 3.1. For each truck number variable t_e in the MIP, we have $|T(e)| = \lceil \log_2(t_{\max}(e) + 1) \rceil$ variables $t_{e,n}$ to represent its values. Additionally, we have $|L| \cdot |E| = \lceil \log_2 c_{\text{bin}} \rceil \cdot |E|$ variables $\ell_{e,m}$ to represent the bin capacity slacks.

3.3 Improvements to the QUBO

To make our QUBO more tractable to a QUBO solver, we now construct improvements which do not necessarily reduce the QUBO size but the range of the coefficients for implementing certain capacity constraints. We define the *potential shipments* on an edge e to be the shipments with a candidate route containing e (i.e. these shipments can be scheduled on e) and we denote by $S(e)$ the set of all these shipments. For each shipment set S, we define $t_{\text{vol}}(S)$ to be the minimal number of trucks that are sufficient to transport all shipments in S referring to the volume capacity, i.e. $t_{\text{vol}}(S) = \lceil \sum_{s\in S} v(s) \,/\, c_{\text{vol}} \rceil$.

Note that we must use at least one truck on an edge e if at least one shipment is scheduled on e. Moreover, if $t_{vol}(S(e)) = 1$, then that truck is sufficient to transport *all* that shipments. Therefore, we can replace the corresponding capacity constraints in (9) by simpler ones which do not require the bin discretization:

Simple Capacity Constraints : For each edge e with $t_{\text{vol}}(S(e)) = 1$ and for each $r \in R(e)$, we have

$$y_r \leq t_{e,1}. \quad (11)$$

This constraint can be implemented by adding the summand $M \cdot (y_r - y_r \cdot t_{e,1})$ to our QUBO. Note that $y_r - y_r \cdot t_{e,1} = 0$ if $y_r \leq t_{e,1}$, and that $y_r - y_r \cdot t_{e,1} = 1$ otherwise. Therefore, it is unfavorable to violate the constraint.

This concept can be generalized to a larger class of edges. To do that, we define the *shipment sets* of an edge e to be the subsets of $S(e)$, and we call a shipment set S *n-minimal* for some positive integer n if $t_{\text{vol}}(S) \geq n$ and $t_{\text{vol}}(S') < n$ for each proper subset S' of S. Moreover, S is called *minimal* if it is n-minimal for some n (n may be not unique if all shipments in S exceed the volume capacity). We generalize the concept of (11) to all edges that only have a few minimal shipment sets. These edges we define *good* and all other edges *bad*.

Note that the only 1-minimal shipment sets are the singletons $\{s\}$ with some shipment s. Therefore, if e is an edge with $t_{\text{vol}}(S(e)) = 1$ (as in (11)), then each minimal shipment set of e is 1-minimal and thus a singleton. Hence we may definitely define each such edge to be good. For our scenarios, we obtained the best results when defining an edge e to be good if $t_{\text{vol}}(S(e)) \leq 2$ or if there are at most 8 potential shipments on e. Then most edges in our scenarios are good, and for all other edges, the number of minimal shipment sets is usually such huge that applying our approach does not improve our QUBO.

We let E_{good} and E_{bad} denote the set of all good and bad edges, respectively. For each good edge e, we introduce new binary variables: the trucks on e are numbered by $1, ..., t_{\max}(e)$ and for each $n = 1, ..., t_{\max}(e)$, we have a binary decision variable $x_{e,n}$ that is 1 iff truck n on e is used. For technical reasons, we additionally define $x_{e,t_{\max}(e)+1}$ to be the constant 0. The total truck distance (6) is then reformulated as

$$\sum_{e \in E_{\text{bad}}} d(e) \cdot \sum_{n \in T(e)} \overline{n} \cdot t_{e,n} + \sum_{e \in E_{\text{good}}} d(e) \cdot \sum_{n=1}^{t_{\max}(e)} x_{e,n}. \tag{12}$$

As before, our QUBO contains implementations of all route-shipment constraints (3) and of the capacity constraints (9) for all bad edges. For the good edges, we have the following constraints which generalize the concept of (11):

Good Capacity Constraints : Let e be a good edge. Then for each n-minimal shipment set S of e with $n \leq t_{\max}(e) + 1$, we have

$$\prod_{s \in S} \sum_{r \in R(s,e)} y_r \leq x_{e,n} \tag{13}$$

where $R(s, e)$ denotes the set of all candidate routes of s containing e, i.e. $R(s, e) = R(s) \cap R(e)$. Particularly, if $n = t_{\max}(e)+1$, we have $\prod_{s \in S} \sum_{r \in R(s,e)} y_r = 0$.

Before showing how to implement these constraints, we first verify their correctness. We show that in a feasible solution of our QUBO, we have enough used trucks on each good edge e to transport all shipments scheduled on e. Let U denote the set of these shipments. For each $n = 1, 2, ..., t_{\text{vol}}(U)$, we choose a smallest possible subset S_n of U with $t_{\text{vol}}(S_n) \geq n$. Then each S_n is n-minimal and $\prod_{s \in S_n} \sum_{r \in R(s,e)} y_r$ is always 1 because each shipment s in S_n is scheduled on e so that $\sum_{r \in R(s,e)} y_r = 1$. Hence if $n \leq t_{\max}(e) + 1$, then by (13) applied to S_n, we obtain $x_{e,n} = 1$. Now since $x_{e,t_{\max}(e)+1} = 0$, we see that $t_{\mathrm{vol}}(U) \leq t_{\max}(e)$ and that truck n on e is used for each $n = 1, 2, ..., t_{\mathrm{vol}}(U)$. These are enough trucks to transport all shipments in U. On the other hand, the constraints (13) do not force $x_{e,n} = 1$ for some $n > t_{\mathrm{vol}}(U)$. To see that, consider an n-minimal shipment set S of e. Then since $t_{\mathrm{vol}}(S) > t_{\mathrm{vol}}(U)$, S contains a shipment s that is not in U, i.e. s is not scheduled on e. Hence $\sum_{r \in R(s,e)} y_r = 0$ and thus $\prod_{s \in S} \sum_{r \in R(s,e)} y_r = 0$.

To implement the constraints (13), we consider an n-minimal shipment set S of some good edge e with $n \leq t_{\max}(e) + 1$. Let $s_1, s_2, ..., s_k$ denote the shipments of S, and for each $i = 1, 2, ..., k$, let Σ_i denote the expression $\sum_{r \in R(s_i,e)} y_r$. Hence we have to implement the constraint $\Sigma_1 \cdot \Sigma_2 \cdot ... \cdot \Sigma_k \leq x_{e,n}$. We consider a solution fulfilling all route-shipment constraints (3) so that $\Sigma_i \leq 1$ for each $i = 1, 2, ..., k$. If $k = 1$, then similarly to (11), we implement the constraint $\Sigma_1 \leq x_{e,n}$ by the summand $M \cdot (\Sigma_1 - \Sigma_1 \cdot x_{e,n})$. Note that since $\Sigma_1 \leq 1$, we obtain $\Sigma_1 - \Sigma_1 \cdot x_{e,n} = 0$ if $\Sigma_1 \leq x_{e,n}$, and $\Sigma_1 - \Sigma_1 \cdot x_{e,n} = 1$ otherwise. Now assume $k \geq 2$. To emulate the product $\Sigma_1 \cdot \Sigma_2 \cdot ... \cdot \Sigma_k$, we introduce auxiliary binary variables $z_2, z_3, ..., z_k$. First, we implement the constraint $z_2 = \Sigma_1 \cdot \Sigma_2$ by the summand $\Omega_2 := M \cdot \left(\Sigma_1 \cdot \Sigma_2 + (3 - 2 \cdot \Sigma_1 - 2 \cdot \Sigma_2) \cdot z_2\right)$. Note that since $\Sigma_1 \leq 1$ and $\Sigma_2 \leq 1$, we obtain $\Omega_2 = 0$ if $z_2 = \Sigma_1 \cdot \Sigma_2$, and $\Omega_2 \in \{M, 3M\}$ otherwise. Now analogously, for each $i = 3, 4, ..., k$, we implement the constraint $z_i = z_{i-1} \cdot \Sigma_i$ by the summand $\Omega_i := M \cdot \left(z_{i-1} \cdot \Sigma_i + (3 - 2 \cdot z_{i-1} - 2 \cdot \Sigma_i) \cdot z_i\right)$. This forces $z_i = \Sigma_1 \cdot \Sigma_2 \cdot ... \cdot \Sigma_i$ for each $i = 2, 3, ..., k$. Finally, we implement the constraint $z_k \leq x_{e,n}$ by the summand $M \cdot (z_k - z_k \cdot x_{e,n})$.

Note that the summands Ω_i may be negative in a solution where $\Sigma_i \geq 2$ for some indices i. Therefore, it may be favorable (or at least not unfavorable) to violate some route-shipment constraints (3). To avoid such violations, we add the auxiliary summand $\Delta_i := \frac{M}{2} \cdot \Sigma_i \cdot (\Sigma_i - 1)$ to our QUBO for each $i = 1, 2, ..., k$ which is 0 if all route-shipment constraints are fulfilled. It is straightforward to verify that in all solutions, $\Omega_2 + \Delta_1 + \Delta_2 \geq 0$ and $\Omega_i + \Delta_i \geq 0$ for each $i = 3, 4, ..., k$ so that it is unfavorable now to violate any constraint.

4 Experiments and Data

The inputs used in this work were generated from a real-world network of hubs in Europe belonging to DB Schenker. The specific locations and distances between hubs have been abstracted to comply with data protection laws, but are representative of the real-world network. Connections between hubs correspond to serviced routes between hubs. We use one graphical model to represent the entire hub network, and generate multiple inputs based on different numbers of shipments: 30, 50, 80, and 100 shipments. In all inputs, every shipment s_{ij} travels from one hub (v_i) to another (v_j). The direct route, $v_i \rightarrow v_j$ along e_{ij}, is always the first candidate route for s_{ij}. The other candidate routes are generated by a staggered *k-shortest path* approach: shipments are categorized by their OD-distance, and for each category the k shortest paths are calculated where k

increases with respect to the OD-distance of the category. For example, shipments up to 200 km have one alternative route while shipments over 1000 km have up to 10 routes. The volume of the shipments is randomly generated using an adapted exponential distribution, resulting in many smaller shipments and few larger shipments.

In this study we use multiple solvers for our SRP instances and gauge the viability of QUBOs as representations of the problem. We provide a brief introduction and motivation for each solver.

Direct Shipments. We consider the "direct shipment" solution to the SRP as a simple baseline. The direct solution is computed by routing every shipment (s_{ij}) along its most direct path (e_{ij}). Since every shipment origin/destination is unique in our instances, this equates to using one truck per edge for every shipment.

Simulated Annealing. Simulated annealing is a well-known heuristic optimization algorithm for combinatorial optimization [22]. The algorithm involves probabilistically flipping individual variables' states proportionally to the objective value change such a flip would induce and the current "temperature" of the system. Candidate solutions are initialized at random, and the temperature parameter is initialized to infinity; solutions are slowly "cooled" and the temperature is lowered until the solutions settle in local optima and variable flips no longer occur. Simulated annealing has been used extensively in benchmarking studies related to quantum computing [23,24]. The specific implementation of simulated annealing in this analysis was from the Python package dimod [25].

Tabu Search. This algorithm is another metaheuristic for combinatorial optimization, operating on the principle that searching already-discovered solutions should be actively discouraged (a "tabu list"). Individual variables' states are flipped based on their likelihood of importance in the global optimum [26]. Solutions which worsen the objective function value may be explored by the search if no other variable flip is possible, which allows for both global and local refinement of solutions. The Python package used for Tabu can be found here [25].

Gurobi. Optimal solutions and optimality bounds were produced by solving the MIP in Sect. 3.1 using Gurobi, an exact branch-and-bound solver. The benefit of using Gurobi is that a bound on the optimality of the solutions is provided. Given that the objective function units are the same for all solvers, this optimality gap can be used for all solvers in this analysis. The run-time allocated to Gurobi was 24 h per input to obtain good bounds for each instance.

D-Wave Hybrid Solver. The smallest instance in our test set required 787 QUBO variables. While small for the application, this is larger than could be solved on D-Wave QPUs at the time of experiments. Instead, a proprietary hybrid classical-quantum algorithm offered by D-Wave Systems was used, called the Hybrid Solver Service (HSS), which has been used in previous applications [27], and admits QUBOs with up to 10k binary variables. The HSS uses a QPU to optimize clusters of variables, allowing one to leverage the use of a quantum processor without the overhead of embedding. However, this hybrid algorithm does not allow direct access to control the QPU in its inner loop. Therefore, we consider the HSS as a black-box optimizer, and measure the performance as a function of the timeout parameter, similar to Gurobi and other proprietary solvers.

5 Results

The consolidated results appear in Fig. 2. While the total run-time of Gurobi was 24 h to obtain good lower bounds, good solutions with an optimality gap of less than 10 percent were already found after a few minutes for all instances. For the 30 and 50 shipment instances, we also obtained provably optimal solutions within the first few minutes of optimization. The solutions from Gurobi were significantly better than those obtained by solving the QUBO formulation. However, this is possibly due to both the fact that Gurobi is an exact solver and the way in which the MIP is discretized, as explained in Sect. 3.2. Tabu search was able to find a near-optimal solution for the 30 shipment instance, but was unable to find even feasible solutions for any of the other instances. Simulated annealing was able to find feasible solutions, but only in the largest case of 100 shipments was the solution better than the direct shipment approach. The D-Wave HSS was able to find better-than-direct solutions for the 30, 50, and 80 shipment instances. To attempt a fair comparison, each QUBO solver was given roughly the same amount of time per test instance. However, the specific parameter choices corresponding to such times were found and set by hand. We include the parameter settings chosen in Appendix A.

Table 1. Number of variables and terms needed to describe the problem instances using binary encoding and good capacity constraints.

Shipments	Routes	QUBO variables	QUBO terms
30	223	787	4856
50	428	1526	16315
80	752	2305	40594
100	925	3318	59014

Throughout our initial experiments, we found that increasing the number of possible routes for each shipment does not directly correlate with improved solutions to the original problem (lower total truck km). This is due to the fact

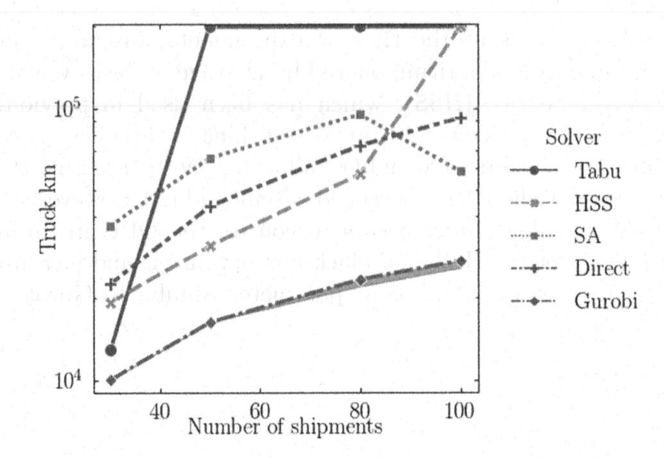

Fig. 2. Performance of all solvers used in the experiments. We display the results in units of truck kilometers for ease of comparison. Simulated annealing (SA), Tabu, and the D-Wave HSS are QUBO solvers, Gurobi is a MIP solver, and the direct solutions are the simple baseline of one truck per shipment (explained in Sect. 4).

that each additional route creates more minima and a more rugged landscape. It is important to note that given the way we construct the QUBO– no trucks along an edge is a valid solution– increasing the number of possible routes can only create additional minima, not remove minima that have already been created. Given this insight, it became even more important to consider the number of QUBO terms (shown in Table 1) and to improve the QUBO as outlined in Sect. 3.3.

6 Conclusions

In this work we motivated a logistics optimization problem based on a real-world use-case, the shipment rerouting problem. This problem models the distance minimization of a simple objective function– total number of truck kilometers used to send shipments between nodes in a graph. We presented methods to translate this problem to a QUBO form using both simple minimization objectives (truck kilometers as weights on decision variables), and hard constraints (knapsack-like constraints on edges in the graph) to test both quantum and classical optimization algorithms. We further presented methods to optimize the QUBO representation in attempts to improve the performance of the algorithms used in our experiments. We found that there was significant amount of work in finding such valid QUBO representations. Despite the relatively straightforward description of the problem, to correctly model the solution landscape was more subtle, and required multiple iterations of derivations, as explained throughout the text. Nonetheless, we found it an informative exercise, as the lessons learned can be applied to future work.

Of the algorithms tested, Gurobi performed the best despite being an exact branch-and-bound algorithm. Of the heuristics, we found that the D-Wave HSS was able to find better than greedy solutions for the smaller problem sizes tested. We stress that given our small test bed we cannot conclude any one solver being the best relative to the others, nor was this the intention. Furthermore, we find that the bar we define as "acceptable" (finding solutions that are better than direct shipments) was surprisingly difficult for the heuristics to beat. This is important to note since simulated annealing was able to find valid solutions for all the problem sizes, but better-than-direct for only the largest problem. From this, together with the long run-times required to find these valid solutions, we conclude that this type of logistics optimization problem may not benefit from transformation to a QUBO for the purpose of being solved with heuristics. The growth in the number of variables required to solve such relatively small problems was a bottleneck that could not be compensated for. However, with the advent of error-corrected quantum processors in the future, it is possible that this bottleneck can be overcome. Until then, our future research will be dedicated to finding real-world optimization problems that are better-suited for current quantum technologies.

A Solver Parameters

Here we present the time allocated to each solver in Table 2, and the corresponding parameters in Table 3. For the D-Wave HSS, we limit the 30 and 50 shipment instances to only 5 minutes of run-time. We note that these 5 minutes were sufficient for the problems tested. Because we could not control the usage of the QPU in the D-Wave HSS, we report the QPU run-time in the timing results rather than a parameter. All software solvers were executed using single-threaded programs.

Table 2. Table of run-time allocated to each solver in the experimental setup.

Instance	Simulated annealing	Tabu	HSS
30	1 h	1 h	5 min (QPU: 3.0 s)
50	1 h	1 h	5 min (QPU: 1.4 s)
80	1 h	1 h	1 h (QPU: 3.61 s)
100	1 h	1 h	1 h (QPU: 4.34 s)

Table 3. Parameter sets used for each solver. Parameters not mentioned were set to default values.

Instance	Simulated annealing	Tabu	HSS
30	2500 samples, 50000 sweeps	1 h timeout	5 min timeout, $use\_qpu = True$
50	1600 samples, 50000 sweeps	1 h timeout	5 min timeout, $use\_qpu = True$
80	1000 samples, 50000 sweeps	1 h timeout	1 hr timeout, $use\_qpu = True$
100	500 samples, 50000 sweeps	1 h timeout	1 hr timeout, $use\_qpu = True$

References

1. Shor, P.W.: Algorithms for quantum computation: discrete logarithms and factoring. In: Proceedings 35th Annual Symposium on Foundations of Computer Science, pp. 124–134 (1994)
2. Deutsch, D., Jozsa, R.: Rapid solution of problems by quantum computation. Proc. Roy. Soc. Lond. Ser. A Math. Phys. Sci. **439**(1907), 553–558 (1992)
3. Grover, L.K.: A fast quantum mechanical algorithm for database search. In: Proceedings of the Twenty-Eighth Annual ACM Symposium on Theory of Computing, STOC '96, pp. 212–219. Association for Computing Machinery, New York (1996)
4. Amin, M.H., Andriyash, E., Rolfe, J., Kulchytskyy, B., Melko, R.: Quantum Boltzmann machine. Phys. Rev. X **8**(2), 021050 (2018)
5. Alexander, C., Shi, L., Akhmametyeva, S.: Using quantum mechanics to cluster time series. arXiv:1805.01711 (2018)
6. Neukart, F., Compostella, G., Seidel, C., von Dollen, D., Yarkoni, S., Parney, B.: Traffic flow optimization using a quantum annealer. Front. ICT **4**, 29 (2017)
7. Venturelli, D., JJ Marchand, D., Rojo, G.: Quantum annealing implementation of job-shop scheduling. arXiv:1506.08479 (2015)
8. Streif, M., Neukart, F., Leib, M.: Solving quantum chemistry problems with a D-wave quantum annealer. In: Feld, S., Linnhoff-Popien, C. (eds.) QTOP 2019. LNCS, vol. 11413, pp. 111–122. Springer, Cham (2019). https://doi.org/10.1007/978-3-030-14082-3_10
9. Quantum, G.A.I.: Hartree-fock on a superconducting qubit quantum computer. Science **369**(6507), 1084–1089 (2020)
10. Venturelli, D., Kondratyev, A.: Reverse quantum annealing approach to portfolio optimization problems. Quant. Mach. Intell **1**(1), 17–30 (2019). https://doi.org/10.1007/s42484-019-00001-w
11. Johnson, M.W., et al.: Quantum annealing with manufactured spins. Nature **473**(7346), 194–198 (2011)
12. Farhi, E., Goldstone, J., Gutmann, S.: A quantum approximate optimization algorithm. arXiv:1411.4028 (2014)
13. Aharonov, D., Van Dam, W., Kempe, J., Landau, Z., Lloyd, S., Regev, O.: Adiabatic quantum computation is equivalent to standard quantum computation. SIAM Rev. **50**(4), 755–787 (2008)
14. Barahona, F.: On the computational complexity of ising spin glass models. J. Phys. A Math. Gener. **15**(10), 3241 (1982)
15. Lucas, A.: Ising formulations of many np problems. Front. Phys. **2**, 5 (2014)
16. Ding, Y., Chen, X., Lamata, L., Solano, E., Sanz, M.: Implementation of a hybrid classical-quantum annealing algorithm for logistic network design. SN Comput. Sci. **2**(2), 68 (2021)

17. Stollenwerk, T., et al.: Quantum annealing applied to de-conflicting optimal trajectories for air traffic management. IEEE Trans. Intell. Transp. Syst. **21**(1), 285–297 (2020)
18. Domino, K., Koniorczyk, M., Krawiec, K.,Jałowiecki, K., Gardas, B.: Quantum computing approach to railway dispatching and conflict management optimization on single-track railway lines. arXiv:2010.08227 (2021)
19. Costa, A.M.: A survey on benders decomposition applied to fixed-charge network design problems. Comput. Oper. Res. **32**(6), 1429–1450 (2005)
20. Paraskevopoulos, D.C., Bektaş, T., Crainic, T.G., Potts, C.N.: A cycle-based evolutionary algorithm for the fixed-charge capacitated multi-commodity network design problem. Eur. J. Oper. Res. **253**(2), 265–279 (2016)
21. Yaghini, M., Momeni, M., Sarmadi, M.: A simplex-based simulated annealing algorithm for node-arc capacitated multicommodity network design. Appl. Soft Comput. **12**(9), 2997–3003 (2012)
22. Kirkpatrick, S., Gelatt, C.D., Vecchi, M.P.: Optimization by simulated annealing. Science **220**(4598), 671–680 (1983)
23. Rønnow, T.F., Wang, Z., Job, J., Boixo, S., Isakov, S.V., Wecker, D., Martinis, J.M., Lidar, D.A., Troyer, M.: Defining and detecting quantum speedup. Science **345**(6195), 420–424 (2014)
24. King, J., et al.: Quantum annealing amid local ruggedness and global frustration. J. Phys. Soc. Jpn. **88**(6), 061007 (2019)
25. D-Wave Systems has produced an open-source library in Python (dimod) for solvers that optimize QUBOs and Ising Hamiltonians. More information can be found here. https://docs.ocean.dwavesys.com/en/stable/docs_dimod/
26. Glover, F.: Tabu search–part I. ORSA J. Comput. **1**(3), 190–206 (1989)
27. Yarkoni, S., et al.: Quantum shuttle: traffic navigation with quantum computing, pp. 22–30. Association for Computing Machinery, New York (2020)

Improving the Location of Roadside Assistance Resources Through Incident Forecasting

Roman Buil[1,2](✉), Santiago Garcia[2], Jesica de Armas[3], and Daniel Riera[1]

[1] Universitat Oberta de Catalunya, Rambla Poble Nou, 18, 08018 Barcelona, Spain
rbuilg@uoc.edu

[2] Accenture S.L., Passeig de Sant Gervasi, 51, 08022 Barcelona, Spain

[3] Department of Economics and Business, Universitat Pompeu Fabra, Ramon Trias Fargas, 25-27, 08005 Barcelona, Spain

Abstract. This paper presents a solution for a real world roadside assistance problem. Roadside assistance companies must allocate their specialised resources to minimize the operating cost associated with servicing when incidents occur. In this process, the location of these resources plays an important role. Therefore, this work proposes a study on the forecasting of incidents and their impact on the location of resources and operating costs. To do this, we have built a machine learning model competition enriched with new features drawn from traditional time series methods and external data such as weather, holidays, and client portfolios. The results show a significant reduction in operating costs thanks to the forecasting of incidents.

Keywords: Road incident forecasting · Model competition · Location of resources · Machine learning

1 Introduction

Roadside Assistance is a service for car, motorcycle and bicycle drivers whose vehicles have suffered a mechanical failure (or accident) leaving them stranded. These incidents can involve starting a car, diagnosing and repairing, towing a vehicle, changing a flat tire, removing a stuck vehicle, or helping people who cannot get into their car, among others. A company-specific (or external) resource will be dispatched if the incident can be repaired on-site. Otherwise, a tow truck will remove the vehicle.

One of the purposes of roadside assistance companies is to reduce operating costs, including those associated with their own resources, those of external suppliers, fuel and penalties for not complying with the Service Level Agreement (SLA), which is time maximum elapsed since the incident is reported until the arrival of the resource at the scene of the incident.

M. Mes et al. (Eds.): ICCL 2021, LNCS 13004, pp. 518–531, 2021.
https://doi.org/10.1007/978-3-030-87672-2_34

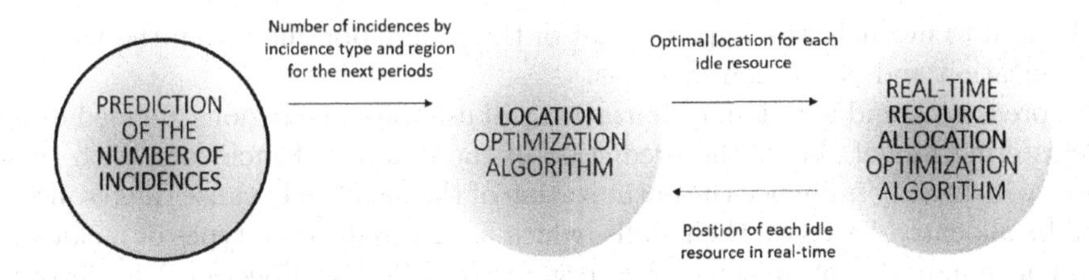

Fig. 1. Relationship between the three modules (encircled, main scope of this paper)

Resources are limited and specialized. Therefore, they need certain skills to be candidates for a certain task. Different resources incur different fixed and/or variable costs. Incidents have some specific requirements or characteristics that could also affect the assignment. Finally, regulations regarding lunch breaks and breaks, as well as contract requirements, will determine the feasibility of a solution.

`Real Automòbil Club de Catalunya` (`RACC`[1]) is a road-side assistance company that operates in Spain. It covers an average of 1600 incidents per day, reaching maximum values of 2000 incidents per day, manages approximately 135 resources, and uses autonomous cranes as well as external providers to cover those incidents that cannot be covered with its own resources. RACC is currently forecasting the number of incidents at a very high level. The accuracy of this aggregated forecast is below 65%; however, with this estimation they are able to make mid to long term decisions such as fleet sizing or determining the number of resources needed to work on each shift and vacations scheduling. Operational and tactical decisions require a lower granularity and fall out of the scope of this paper. Nowadays, all resources start their shifts in some fixed base locations.

RACC believes that current resources utilization rates and SLA times are below the potential level that could be reached using modern Data Science techniques. A solution has been designed to improve day-to-day operations, minimise the use of third-party services, reduce operating costs and meet customer SLAs. It integrates the following three modules (Fig. 1):

- Prediction of the number of incidents: Forecasting Machine Learning (ML) models the competition to obtain the number of incidents by predefined region, time period and type of incident.
- Location Optimization Algorithm: Dynamic decision making of the optimal location of each resource based on data history, clusters and current status.
- Real-time resource allocation optimization algorithm: automatic allocation of resources to incidents in real time using mathematical optimization models.

This document focuses on the first module, but also includes an initial basic solution for the second. The third has also been developed and is currently in production, but we consider that it is outside the scope of this work. It is

[1] https://www.racc.cat.

only used to evaluate the improvement of the results obtained from the location optimization and prediction modules.

Forecasting and location optimization will use the same regions, defined using a clustering model. Thus, the prediction of the number of incidents, with their type, will directly influence the optimization of the location because the resources will be allocated based on their skills, which are related to the types of incidents.

The remainder of this paper is organized as follows. Section 2 is devoted describe the problem. Section 3 illustrates the problem scope. Section 4 presents the solution approach. Section 5 discusses the results of the forecast at different geographical and time horizon and granularity levels, comparing with RACCs forecast when possible. Finally, Sect. 6 concludes this work and proposes possible future research lines.

2 Problem Statement

The prediction of the demand for vehicle-related services has been a focus of interest in recent years, since this forecasting allows improving the performance of companies in the sector. Recent works in this area apply different methods to predict the demand of services, such as deep learning [11,14,16,21], non-parametric density estimation [15], Petri nets [5], long short-term memory networks [22], back propagation neural network [23], or learning-based optimization [3]. The most common services that required this prediction are car-sharing networks or taxi networks, due to the advantages that relocation of vehicles involves in these dynamic environments [9].

In general, the articles mentioned above talk about predicting the general demand in a previously determined location. We not only look for a forecast of incidents, we first identify the optimal number of strategic locations and then predict the number of incidents that will occur at each location for different types of incidents and time periods. This must be done for different time horizons.

Generating a forecast that is sufficiently accurate is mandatory to have a solution that contains the best location of resources, thus reducing the arrival time to incidents. Therefore, RACC also requires a first version of that solution that will decide the location of the resources. Finally, these new locations should be tested using the real-time resource allocation optimization algorithm and compared to the benchmark results without using location optimization.

The proposed innovative approach to incident forecasting is to combine traditional time series-based forecasting methods with ML models. New features are generated from traditional time series-based methods and incorporated into ML models to enrich them. We are not just using a ML model, but doing ML models competition [20] in order to use the best model in each situation depending on the time horizon.

3 Problem Scope

The problem presented in this work consists on the prediction of number of incidents depending on the time horizon and granularity (hourly, daily), the location or region (higher disaggregation than the current one), and type of incident, with the aim of relocating the resources in the right positions to better serve the demand, adjust resources schedules and size the fleet.

The time horizons to consider are determined by business needs, since requirements are different when making strategic, tactical or operational decisions. The length of the time horizon determines the way to group the incidents (by hour, by set of hours, by day, etc.) In this work, three time horizons have been developed for each specific business need:

- Short-term: hourly forecast for a 7-day time window (one week ahead) to adjust resources schedules
- Mid-term: daily forecast for the following month to optimize resources location
- Long-term: daily forecast for the following 6–12 months to size fleet needs

Volume and density of incidents make Barcelona Metropolitan Area a region that can be used to compare results between current forecast used by RACC, and our new forecast. A definition of multiple dynamic clusters, instead of few static zones, has been performed by means of statistical methods based on mass centers taking into account historical incidents data. These clusters could be dynamically redefined according to the month, paying special attention to specific areas where more detail is needed, e.g., cities with higher activity, where a higher level of granularity can be of use.

Being able to compute a different forecast for each main resource category can help define the most appropriate resources to use in an incident, since different types of incidents require different kinds of resources. For this reason, based on incidents volumes and relevance, the type of incidents have been defined as:

- Non-repairable (NR)
- Repairable incidents related to a wheel problem (R_wheel)
- Repairable incidents with a battery problem (R_battery)
- Other repairable incidents (R_other)

In order to obtain appropriate forecasts, we have used different kind of data about the problem: historical data of incidents for around 3 years, clients portfolio during this period of 3 years, national and regional holidays of the period, weather related data for the period.

4 Solution Approach

4.1 Data Preparation

ML models are based on historical data, which usually needs to be cleaned before use. Particularly, all the data required for this work is related to incidents. In the available (real) historical data, an incident can be related to several services and covered by multiple resources. However, the forecast must be done based on the initial known information about the incident, and hence, on the first service recorded. For this reason, a cleansing on the historical data has been made in order to consider the first record for each incident before running the forecast. Once the data has been cleaned and there is one row per incident containing the timestamp, geographical coordinates and the type of incident, they will be used as input of the models.

Since one of the reasons to forecast incidents is trying to relocate resources to better serve demands, it is important to aggregate them in small regions, ensuring that all the area inside the scope of the project is covered. This is done by applying the well-known K-means algorithm [10] on the historical incidents.

An initial analysis has been performed on the data variance to find the optimal number of clusters using the elbow method [12]. However, this number, which is the best statistically speaking, did not match with RACC's strategy, and the final number has been selected by a compromise between RACC's strategy and the mathematical solution. This happens because although clustering algorithms provide the theoretical optimal location, at the end of the day, workers need to be sent to those locations from one of RACC's bases. This includes an extra cost to the problem: they spend part of their working shift travelling to their starting point (while they may have been already working on another incident).

For each cluster, data is split into training, testing and validation data sets. Due to its nature – time series with cycles in it – data is not split randomly, but maintaining order. For example, if a forecast of the next week is the objective, the testing and validation data also covers a week following the normal sequence of days. The training set would then be the rest of the data, maintaining the same sequential order. Additionally, to improve the model generalization, K-Folds Cross Validation [19] can also be performed, taking several time horizons of the same length from the data to use as validation sets. This can be especially interesting if trying to find a model that can perform well during the whole year, as using validation sets from different months could help the model to be more generalized to any event, although they can also be trained more specifically for a certain period, such as summer.

4.2 Additional Features

In general, once the data has been aggregated, classical methods such as ARIMA [8, Chapter 8], or Exponential Smoothing models [8, Chapter 7] are used to predict the outcome of the next period based on linear regressions or series that have trend and seasonal components. In this case, ML models are used to try to predict what will happen in the next period. This means that new features added to the data model could have a positive impact, since they provide useful information that can improve the accuracy of the forecast results. This information could be split between 2 categories: a) external, such as the weather, client portfolios or holidays; or b) calculated, information that is obtained from data statistics. Some details about this information are:

- *Weather*: Weather data can influence the behaviour of drivers at the wheel and the possibility of having an accident, so it is an external valuable information.
- *Client portfolios*: The total amount of clients that RACC has each month. This is an important variable, especially for the long-term forecast, as given the number of clients, you have an upper limit for the simultaneous incidents.
- *Holidays*: Holidays are an important factor for the model as there is a clear evidence of mobility changes around events such as bank holidays.
- *Season*: The season of the year may have an effect on where the incidents may occur. For instance, there is a greater mobility towards the beach in summer, whereas the Pyrenees might have more drivers during winter.
- *Data statistics*: Historical statistics can also provide useful information to the forecasting model. After the data exploration phase, it can be concluded that there is a cyclical effect on the occurrence of incidents. Therefore the model also receives as inputs maximums, minimums, averages, medians and standard deviations of the number of incidents calculated by combinations of time granularities, such as month-day, weekday-hour, season-day of week.
- *Lags*: Providing input data about a "similar" day is important. It can also be useful to provide the data of what happened the previous day, a week before, or if it is going to be a bank holiday the following Monday, which will affect Friday's traffic.
- *Cyclic transformations*: To let the ML model know that a feature is cyclical, e.g. time, sine and cosine transforms are applied to the year, month, weekday and day of the incident. By doing this, the model is provided with the meaning that although Monday is the 1st day of the week and Sunday is the 7th, they are actually only one day apart.

4.3 Model Competition

The final step to obtain the prediction results is to launch a ML model competition. The set of models tried are the following: Decision Tree Regressor (dt) [17], Extra Trees Regressor (extratr) [1], Gradient Boosting Regressor (gboost) [4], K-Neighbors Regressor (knn) [2], Multilayer Perceptron (MLP) network (neural) [13]. Random Forest Regressor (randf) [18], SVR (svm) [6], and XGB Regressor (xgboost) [7].

The model competition is first launched intra-model. This means each model is tested with different parameters to fine tune individually. This process is followed by comparing the best models for each type of incident. The comparison is performed by means of the root-mean-square error (RMSE), with the objective of minimizing the average of all the validation sets that are used during the training phase. The best model overall is then selected to compute the final prediction.

5 Results

The result of the forecast at region/cluster level, by type of incident and time is required to decide the best location for the resources. However, it is not

Algorithm 1. Main Algorithm

```
1: procedure RUNPREDICTION(parameters)
2:     workorders ← readAndCleanHistoricalWorkorders()            ▷ Data Cleansing
3:     aggregated_workorders ← aggregateWorkorders()             ▷ Data Preparation
4:     aggregated_workorders ← addAdditionalFeatures()         ▷ Additional Features
5:     X_train, Y_train, X_val, Y_val, X_test ← featureEngineering()    ▷ Standardize
   data, split into training, testing and validation sets
6:     best_model ← empty solution
7:     for all model in models do training_result ← trainModel(X_train, Y_train,
   X_val, Y_val)                  ▷ Intra-model competition with different parameters
8:         best_model ← compareModels(best_model, training_result)
9:     end for
10:    final_forecast ← predictModel(best_model, X_test)          ▷ Compute the Final
   forecast with the best model
11:    return final_forecast
12: end procedure
```

comparable with current RACC's forecast, that just predicts the number of incidents, at high level regions, and without distinguish between types neither timestamp. Therefore, first, we generate forecast at the same level of aggregation, so we can compare our accuracy; second, we compute the clusters and we generate forecast by cluster, incidents' type and timestamp; third, we use this disaggregated forecast to test the impact of locating resources in the different clusters at the beginning of the day and use the resource allocation module.

The input data set used to train the algorithms is made up from 3 years of data, aggregated by 30 min. Data from the Barcelona region has been used to compare against RACC's forecast. Data from Catalonia has been used to compute clusters and test the forecast together with the resource location and the resource allocation modules.

5.1 Comparison at RACCs Level of Aggregation

The scope of this experiment is only "repairable incidents in Barcelona"; otherwise we cannot compare against RACC's forecast. The comparison is made at 3 different time horizons:

- Short-term: Two weeks of November 2018 aggregated by hour.
- Mid-term: November and December 2018 aggregated by day.
- Long-term: From January to December 2018 aggregated by day.

Table 1 shows that the increment on forecast accuracy at any time horizon is more than 18 points, achieving more than 33 points for one of the long-term horizons.

Figure 2 shows the graphics for the mid-term horizon case. Lines correspond to RACC's prediction (usually over forecasting), historical data, and the Extra Trees Regressor model. The latter is very close to the historical data, being the model with higher accuracy for this case.

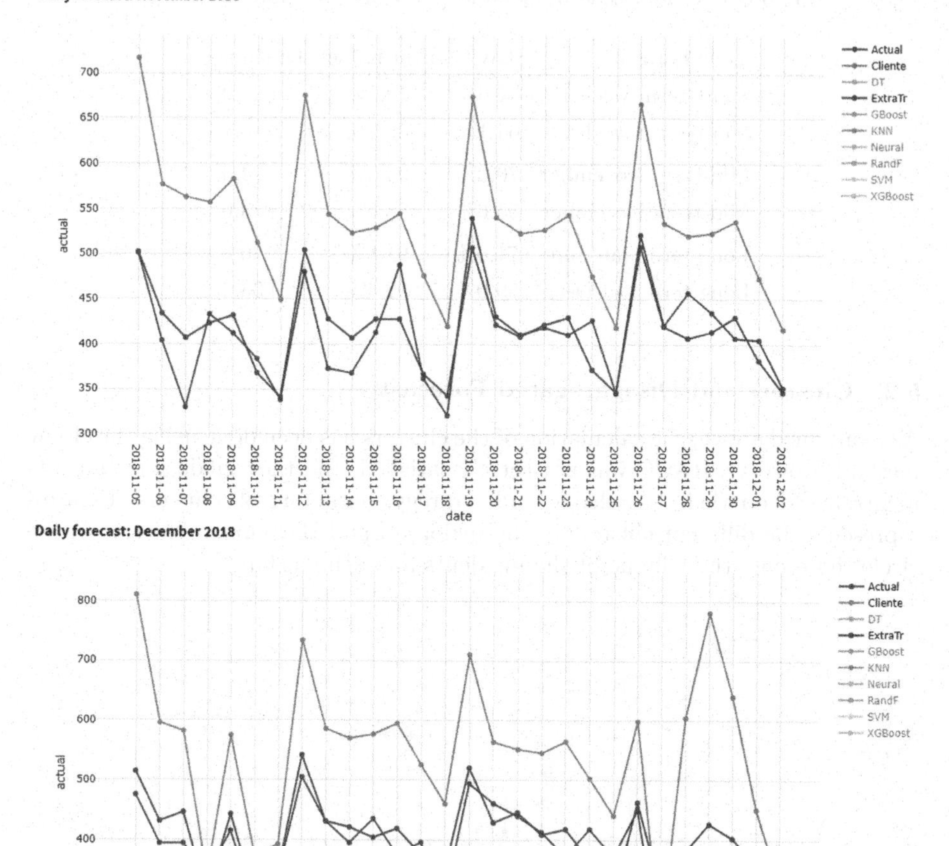

Fig. 2. Mid-term forecast comparison

Table 1. Comparison between RACC's and our work accuracy

Time horizon	RACC	Best model	Difference
Short-term week 1	57.6%	78.5%	+20.9
Short-term week 2	64.1%	82.2%	+18.1
Mid-term November	70.2%	94.5%	+24.3
Mid-term December	55.6%	88.6%	+33.0
Long-term Jan-Jun	71.0%	91.8%	+20.8
Long-term Jul-Dec	59.0%	92.4%	+33.4

5.2 Clusters and Disaggregated Forecast

As mentioned above, the calculation of the clusters has been done using the elbow method. However, RACC was not totally satisfied, and they made some adjustments based on their knowledge and strategy, obtaining 39 clusters. Figure 3 represents the different clusters in the region around Barcelona. The size of the circle represents the volume of the incidents for each cluster.

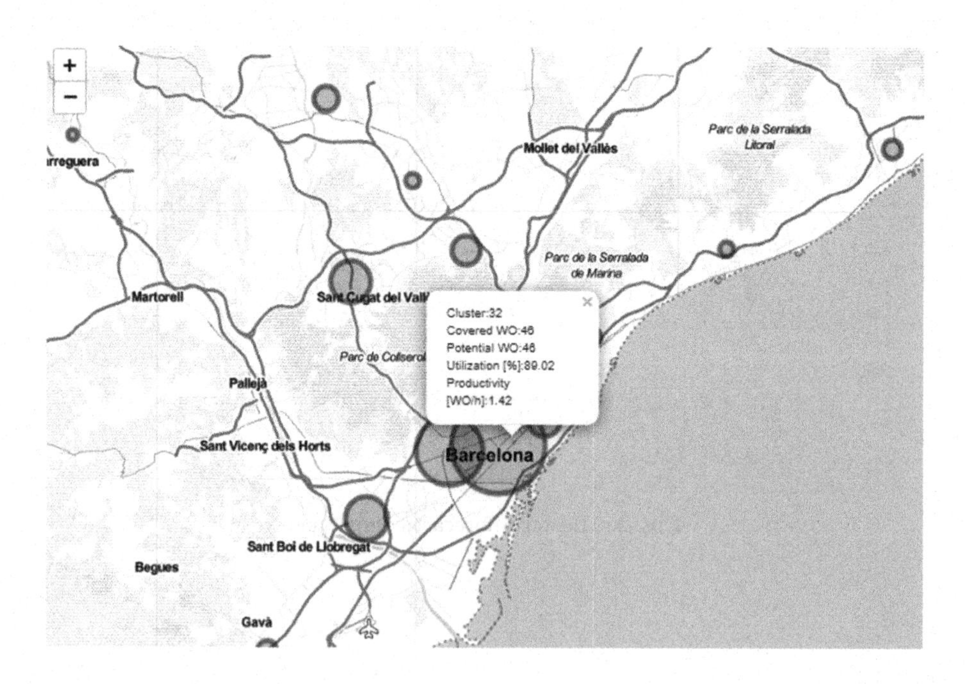

Fig. 3. Generated Clusters around Barcelona

Once the clusters are defined, incidents are grouped by cluster, type of incident and timestamp. Results are compared against the historical data for the following time periods:

- Short-term: Third week of September 2019 aggregated by hour.
- Mid-term: From January to February 2019 aggregated by day.

Figures 4 and 5 present the results of the disaggregated forecast for short-term (by hour) and mid-term (by day) and for one cluster.

Good forecast accuracy on short-term where incidents are grouped by hour is difficult to achieve. Only clusters with enough volume have more than 60% accuracy. In particular, one of the clusters in Barcelona (Fig. 4) has 80% accuracy using a Random Forest model. The mid-term forecast, where incidents are grouped by day, also present some cluster with low accuracy. However, in this case, there is a global accuracy of 63.7%, 81.2% in the best case (Fig. 5) obtained with the XGBoost model.

The accuracy of the results when disaggregating the data by clusters and incident types is clearly lower than when calculating only by one incident type and region. This is especially noticeable when aggregating by hour instead of day. The reason behind this is the increment of complexity of the problem by multiplying the number of features by the number of clusters times the number of incident types. It also means that the number of incidents per instant of time is also reduced, adding more bias to the data due to getting more zeroes in the time series, especially in clusters with lower activity.

5.3 Resource Location and Resources Allocation

Any forecast, no matter with which accuracy, is not useful if you do not use it with a certain objective. For example, using forecasting of sales to better plan the inventory management. In this case, we have calculated the forecast by cluster, incident type and timestamp to locate resources closer to the region where incidents will happen. Thus, we can reduce the time to arrive to incidents. Therefore, we decided to do a small test to see if locating the resources in different locations would improve the allocation of resources during the day to day operations. Since this is a preliminary test, there is no algorithm locating the resources by skills (meaning matching incidents types). The location is a distribution of the resources depending on the volume of incidents per cluster. The selection of the type of resource is random.

The day of activity has been selected randomly and corresponds to 17th September 2018, from 06:00 to 15:00 local time, in the area of Catalonia (Barcelona, Girona, Lleida, Tarragona). A set of 567 incidents has been extracted and used to test the algorithm, with 147 own resources (each of which has its own working schedule), 46 tow trucks and 427 external providers. These are the real numbers of incidents that happened and resources that were available during that day.

Table 2 presents the real allocation done by RACC, and the results of the resources allocation with an initial location of the resources based on the forecast of incidents.

Basically, results show the following:

- Increment of the number of own resources allocated by 15%

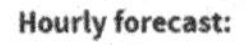

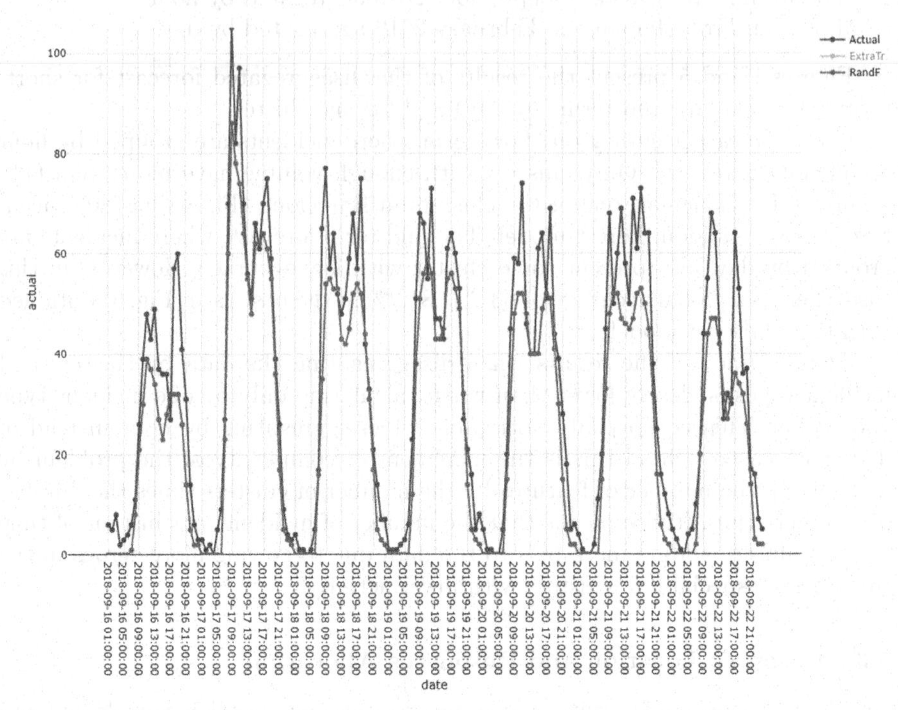

Fig. 4. Short-term forecast for the cluster with higher accuracy (80%)

Table 2. RACC real allocation vs new resource allocation with resource location based on incidents forecast

	Own Resources	Freelance Tow Trucks	External Providers	Own resources Occupation rate	SLA compliance	Cost reduction
RACC	256	96	215	60%	45%	
Our Solution	295	168	104	76%	85%	20%

- Increment of the number of freelance tow trucks allocated by 75%
- Decrease number of external providers allocated by 51%
- Increase own resources occupation rate by 16%
- Increase SLA Compliance by 40%
- Decrease operational costs by 20%

Therefore, we can state that even though the type of incidents is still not considered to locate the resources, the improvements on the main indicators are very promising.

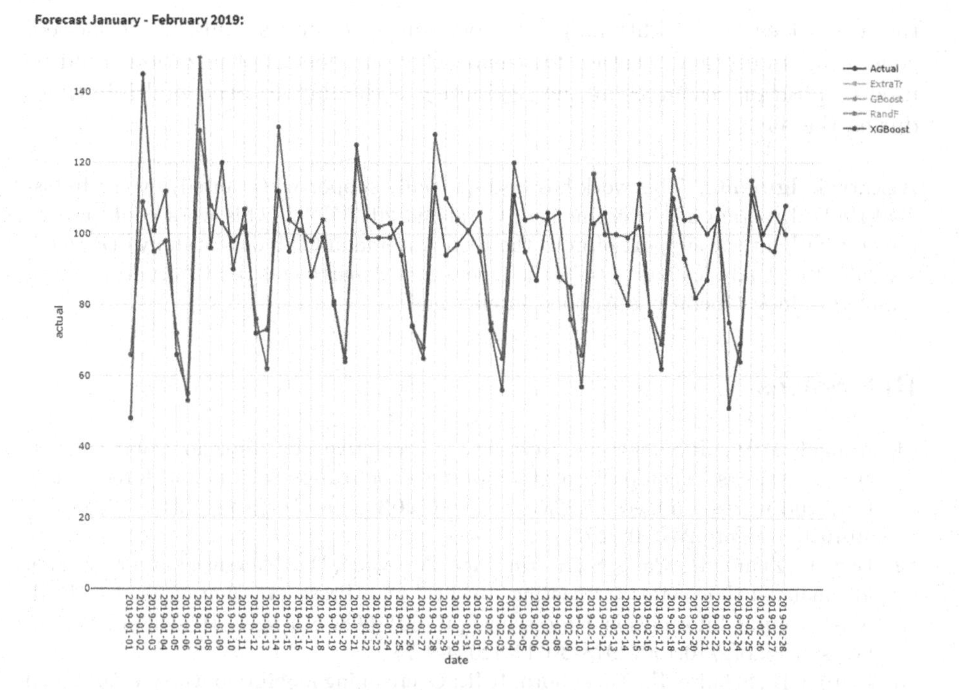

Fig. 5. Mid-term forecast for the cluster with higher accuracy (82.1%)

6 Conclusions

In this paper, we propose an innovative approach to forecast incidents using a ML model competition enriched with new features created based on traditional time series methods, mentioned in Sect. 4. Additionally, this approach uses external data such as weather, holidays, and customer portfolios. The ML model competition is used to select the best performing model for all clusters. The clusters have been generated using a mathematical calculation as a basis and then using the customer's information and knowledge to fit them. The best ML models for disaggregated forecasting are: Random Forest for the short term and XGBoost for the medium term. When generating a forecast to compare with the RACC, based on time horizon and time period, the best models are XGBoost, Multilayer Perceptron Network (MLP), and Extra Trees.

The test, which assesses both the placement of the proposed resources and the subsequent allocation of services, has demonstrated the potential of this solution. The results obtained show a 20% reduction in operating costs, while the use of own resources increased, the number of autonomous cranes also increased and the number of external suppliers decreased.

This work can be further developed with forecast enhancements by adding additional features to give more weight to the last few weeks of historical data. Therefore, unexpected variations of incidents could be automatically reflected in the forecast earlier. Additionally, the location of resources could be developed, generating an optimal location for them. This placement optimization could be used to position resources at the beginning of the day, or even multiple times during the day.

Acknowledgments. This work has been partially supported (granted) by the Industrial Ph.D. Program of Government of Catalonia 2017DI092. This work could not be possible without the support of both the Real Automóbil Club de Catalunya (RACC), specially the Analytics and Assistance Operations departments, and Accenture team, Supply Chain & Operations Applied Intelligence.

References

1. Ahmad, M.W., Reynolds, J., Rezgui, Y.: Predictive modelling for solar thermal energy systems: a comparison of support vector regression, random forest, extra trees and regression trees. J. Cleaner Prod. **203**, 810–821 (2018). https://doi.org/10.1016/j.jclepro.2018.08.207
2. Ban, T., Zhang, R., Pang, S., Sarrafzadeh, A., Inoue, D.: Referential knn regression for financial time series forecasting. In: Lee, M., Hirose, A., Hou, Z.G., Kil, R.M. (eds.) Neural Information Processing, pp. 601–608. Springer, Heidelberg (2013). https://doi.org/10.1007/978-3-642-42054-2_75
3. Beirigo, B., Schulte, F., Negenborn, R.R.: Overcoming mobility poverty with shared autonomous vehicles: a learning-based optimization approach for rotterdam zuid. In: Lalla-Ruiz, E., Mes, M., Voß, S. (eds.) ICCL 2020. LNCS, vol. 12433, pp. 492–506. Springer, Cham (2020). https://doi.org/10.1007/978-3-030-59747-4_32
4. Ben Taieb, S., Hyndman, R.J.: A gradient boosting approach to the kaggle load forecasting competition. Int. J. Forecast. **30**, 382–394 (2014). https://doi.org/10.1016/j.ijforecast.2013.07.005
5. Clemente, M., Fanti, M.P., Mangini, A.M., Ukovich, W.: The vehicle relocation problem in car sharing systems: modeling and simulation in a petri net framework. In: Colom, J.-M., Desel, J. (eds.) PETRI NETS 2013. LNCS, vol. 7927, pp. 250–269. Springer, Heidelberg (2013). https://doi.org/10.1007/978-3-642-38697-8_14
6. Fan, G.F., Peng, L.L., Hong, W.C., Sun, F.: Electric load forecasting by the svr model with differential empirical mode decomposition and auto regression. Neurocomputing **173**, 958–970 (2016). https://doi.org/10.1016/j.neucom.2015.08.051
7. Gregory, B.: Predicting customer churn: Extreme gradient boosting with temporal data (2018)
8. Hyndman, R., Athanasopoulos, G.: Forecasting: Principles and Practice, 2nd edn. OTexts, Melbourne (2018)
9. Illgen, S., Höck, M.: Literature review of the vehicle relocation problem in one-way car sharing networks. Transp. Res. Part B Methodol. **120**, 193–204 (2019). https://doi.org/10.1016/j.trb.2018.12.006
10. Jin, X., Han, J.: K-means clustering, pp. 563–564. Springer, Boston (2010). https://doi.org/10.1007/978-0-387-30164-8_425
11. Ke, J., Zheng, H., Yang, H., Chen, X.M.: Short-term forecasting of passenger demand under on-demand ride services: a spatio-temporal deep learning approach. Transp. Res. Part C Emerg. Technol. **85**, 591–608 (2017). https://doi.org/10.1016/j.trc.2017.10.016

12. Kodinariya, T., Makwana, P.: Review on determining of cluster in k-means clustering, vol. 1, pp. 90–95 (2013)
13. Koskela, T., Lehtokangas, M., Saarinen, J., Kaski, K.: Time series prediction with multilayer perception, fir and elman neural networks (1996)
14. Lei, Z., Qian, X., Ukkusuri, S.V.: Efficient proactive vehicle relocation for on-demand mobility service with recurrent neural networks, vol. 117 (2020). https://doi.org/10.1016/j.trc.2020.102678
15. Li, X., Wang, C., Huang, X.: Reducing car-sharing relocation cost through non-parametric density estimation and stochastic programming. In: 2020 IEEE 23rd International Conference on Intelligent Transportation Systems (ITSC), pp. 1–6 (2020). https://doi.org/10.1109/ITSC45102.2020.9294599
16. Liao, S., Zhou, L., Di, X., Yuan, B., Xiong, J.: Large-scale short-term urban taxi demand forecasting using deep learning. In: 2018 23rd Asia and South Pacific Design Automation Conference (ASP-DAC), pp. 428–433 (2018). https://doi.org/10.1109/ASPDAC.2018.8297361
17. Meek, C., Chickering, D., Heckerman, D.: Autoregressive tree models for time-series analysis, pp. 229–244. https://doi.org/10.1137/1.9781611972726.14
18. Mei, J., He, D., Harley, R., Habetler, T., Qu, G.: A random forest method for real-time price forecasting in new york electricity market, vol. 2014, pp. 1–5 (2014). https://doi.org/10.1109/PESGM.2014.6939932
19. Rodriguez, J.D., Perez, A., Lozano, J.A.: Sensitivity analysis of k-fold cross validation in prediction error estimation. IEEE Trans. Pattern Anal. Mach. Intell. **32**, 569–575 (2010). https://doi.org/10.1109/TPAMI.2009.187
20. Thakur, A., Krohn-Grimberghe, A.: Autocompete: a framework for machine learning competition (2015)
21. Vateekul, P., Sri-iesaranusorn, P., Aiemvaravutigul, P., Chanakitkarnchok, A., Rojviboonchai, K.: Recurrent neural-based vehicle demand forecasting and relocation optimization for car-sharing system: a real use case in Thailand, vol. 2021 (2021). https://doi.org/10.1155/2021/8885671
22. Wang, N., Guo, J., Liu, X., Fang, T.: A service demand forecasting model for one-way electric car-sharing systems combining long short-term memory networks with granger causality test, vol. 244 (2020). https://doi.org/10.1016/j.jclepro.2019.118812
23. Wang, N., Jia, S., Liu, Q.: A user-based relocation model for one-way electric carsharing system based on micro demand prediction and multi-objective optimization, vol. 296 (2021). https://doi.org/10.1016/j.jclepro.2021.126485

Solving a Multi-objective Vehicle Routing Problem with Synchronization Constraints

Briseida Sarasola[1(✉)] and Karl F. Doerner[1,2]

[1] Institut für Business Decisions and Analytics, Oskar-Morgenstern-Platz 1, Wien, Austria
briseida.sarasola@univie.ac.at
[2] Data Science @ Uni Vienna, Vienna, Austria
karl.doerner@univie.ac.at

Abstract. In this paper, we solve a multi-objective vehicle routing problem with synchronization constraints at the delivery location. Our work is motivated by the delivery of parcels and consumer goods in urban areas, where customers may await deliveries from more than one service provider on the same day. In addition to minimizing travel costs, we also consider a second objective to address customer preferences for a compact schedule at the delivery location, so that all deliveries to a customer happen within a non-predefined time interval. To determine the Pareto fronts, three metaheuristic methods based on large neighborhood search are developed. The results on small instances are compared with an ϵ-constraint method using an exact solver. Results for large real-world instances are also presented.

Keywords: Vehicle routing problem · Synchronization · Multi-objective optimization

1 Introduction

The vehicle routing problem with synchronization constraints at the delivery location (VRPSCDL) is motivated by a current situation in urban transportation, where a single recipient often expects several orders from more than one service provider on the same day [14]. Service providers aim at minimizing the costs, while recipients are customers that wish to receive all orders approximately at the same time. Therefore the decision involves several stakeholders that need to find a compromise between transportation costs and compact schedules at the delivery location.

A real-life application of the VRPSCDL arises in housing and decoration logistics [16,20], where several furniture suppliers offer their products on a shared online platform. A customer may buy products from different suppliers, that are later shipped with different logistics companies. However, customers wish to receive all products within a small time interval. As a result, furniture suppliers

M. Mes et al. (Eds.): ICCL 2021, LNCS 13004, pp. 532–546, 2021.
https://doi.org/10.1007/978-3-030-87672-2_35

need to find a way to collaborate with each other to improve customer service. Other possible applications include supermarkets, construction sites, and in general, any business that expects deliveries from several suppliers.

The VRPSCDL has been solved in the literature by allowing a maximum amount of idle time at the delivery location, where idle time is defined as non service time between the first and the last delivery to a given location. Existing results show that it is possible to substantially reduce the idle time by more than 40% without sacrificing the travel cost (3% longer travel times) [14]. However, the problem was addressed by first determining a fix idle time/service time ratio and then solving the single-objective optimization problem, but it is in general not easy to find appropriate values for the ratios. We aim at supporting decision makers by providing several solutions, so that a suitable trade-off solution can be chosen. To tackle this problem we model the multi-objective problem as an extension of the single-objetive problem. We formulate two objectives to minimize travel cost and idle time at the delivery location.

Many of the initial approaches to tackle multi-objective routing problems in the literature use highly specialized population-based metaheuristics [5,9], such as evolutionary algorithms [6,13,17]. It has been recently shown that combining several single-objective methods [3,8,18] as well as using a single-objective algorithm within a solution framework that treats all other objectives as contraints [1,12] provide excellent results in solving multi-objective vehicle routing problems. In this work, we focus on solution techniques that use powerful single-objective neighborhood search methods to build the Pareto set.

The remainder of this paper is structured as follows. Section 2 introduces the problem. Our solution methods are described in Sect. 3. Results are presented in Sect. 4. Finally, Sect. 5 gives some conclusions and lines of future work.

2 Problem Statement

The multi-objective VRPSCDL (MO-VRPSCDL) is based on the VRPSCDL, which is defined using a multi-commodity flow formulation [14]. We introduce here the necessary notation to later define the multi-objective problem. In the single-objective formulation, the problem consists of minimizing the total travel time to serve n deliveries to m delivery locations from p depots. A solution to the VRPSCDL is a set of routes that serves all deliveries once. Each route starts at a departure depot and ends at the corresponding arrival depot, and only serves deliveries associated to that depot. A delivery location receives one or more deliveries, which must be fulfilled in a compact schedule. The notation for the sets, data, and variables of the VRPSCDL are summarized next.

Sets:

- D is the set of deliveries.
- P_1 is the set of departure depots.
- P_2 is the set of arrival depots.
- $N = P_1 \cup D \cup P_2$ is the total set of nodes.
- U is the set of delivery locations.

- D'_l is the set of all deliveries at location $l \in U$.
- D''_i is the set of all deliveries to be fulfilled by depot $i \in P_1$.
- V is the set of all vehicles.
- V_i is the set of all vehicles associated to depot $i \in P_1$.

Problem data:

- c_{ij} is the travel time from node i to j, $i, j \in N$.
- d_i is the service time at node $i \in N$.
- w_l is the maximum idle time at delivery location $l \in U$.
- Q is the vehicle capacity.
- T is the return time at the depot.

Variables:

- x_{ijk} is equal to 1 if vehicle k travels from node i to node j, 0 otherwise.
- s_{ik} is the time at which service starts at delivery i by vehicle k.
- $start_l$ is the time at which the first delivery at location l starts.
- $last_l$ is the time at which the last delivery at location l ends.
- z_{ij} is 1 if, given two deliveries i, j to the same location, i is scheduled before j, 0 otherwise.

In previous work, the maximum idle time w_l at the delivery location has been defined as the maximum amount of non service time between the first and the last delivery at location l. The value of w_l depends on the total service time of all deliveries to location l and a parameter $\alpha \geq 0$, so that $w_l = \alpha \cdot \sum_{i \in D'_l} d_i$, where D'_l is the set of all deliveries to l. The value of α controls the percentage of allowed idle time at each delivery location. The maximum idle time is thus modeled as Constraint (1) in the single-objective problem.

$$last_l - start_l - \sum_{i \in D'_l} d_i \leq \alpha \cdot \sum_{i \in D'_l} d_i \qquad \forall l \in U \tag{1}$$

The model imposes that the service of two or more deliveries at a given location cannot overlap, so that a vehicle must wait if another vehicle is currently serving the customer at the delivery location. Moreover, a vehicle might also wait if it arrives to the location before service can start according to the synchronization constraints. Following previous work, we assume that vehicles leave each node as early as possible, so that they departure from the depot at time 0 and leave the delivery location right after serving the delivery.

The MO-VRPSCDL can be defined as a minimization problem of the form:

$$\min f(\boldsymbol{x}, \boldsymbol{\alpha}) = (f_1(\boldsymbol{x}), f_2(\boldsymbol{\alpha})) \tag{2}$$

s.t.

$$g_i(\boldsymbol{x}) \geq 0 \quad \forall i \tag{3}$$

$$h_j(\boldsymbol{x}) = 0 \quad \forall j \tag{4}$$

The first objective $f_1(\boldsymbol{x})$ is the total travel time of the VRPSCDL.

$$f_1(\boldsymbol{x}) = \sum_{i \in N} \sum_{j \in N} \sum_{k \in V} x_{ijk} \cdot c_{ij} \tag{5}$$

The second objective $f_2(\boldsymbol{\alpha})$ aims at minimizing the maximum α_l for every delivery location l in the problem.

$$f_2(\boldsymbol{\alpha}) = \max_{l \in D'_l} \left\{ \alpha_l \;\middle|\; \alpha_l = \frac{last_l - start_l}{\sum_{i \in D'_l} d_i} - 1 \right\} \tag{6}$$

The problem constraints of the MO-VRPSCDL are the same as in the VRP-SCDL except Constraint (1), which is removed and treated as the second objective.

3 Solution Techniques

This section provides a description of our solution methods based on neighborhood search. All described methods provide an approximation of the Pareto front in the MO-VRPSCDL and use an archive to maintain the set of non-dominated solutions. A solution s_i is non-dominated if there are no other solutions s_j, $i \neq j$, with $f_1(s_j) < f_1(s_i)$ and $f_2(s_j) \leq f_2(s_i)$, or $f_1(s_j) \leq f_1(s_i)$ and $f_2(s_j) < f_2(s_i)$. Our local search algorithms try to improve one objective at each step, while the other objective values might deteriorate, so that each step follows a "pure local search" scheme [11].

We use the Solomon C1 heuristic to build initial solutions that are feasible with respect to some value of α. The algorithm selects depots in a random order and builds a set of routes considering all deliveries of the selected depot. To ensure that the solution is feasible, we use "self-imposed time windows", which provide an earliest and latest start time for all deliveries to the same location [14]. These time windows are not predefined, so all locations have an initial self-imposed time window equal to the work day duration $[0, T]$. When a delivery is inserted in the solution, its time window is updated. We denote this algorithm $Solomon\_C1(\alpha)$. If $\alpha = \infty$, self-imposed time windows are not used.

3.1 Multi-Directional Local Search

We solve the MO-VRPSCDL by integrating it in Multi-Directional Local Search (MDLS) [18]. This framework requires the definition of a local search method for each objective. Therefore, since the MO-VRPSCDL is a bi-objective problem, we need to define two local search methods. In our case, each execution of a local search is a single iteration of an Adaptive Large Neighborhood Search (ALNS). We denote this method MDLS-ALNS (see pseudocode in Algorithm 1).

For the first objective, we use $ALNS_1$, which is based on the existing ALNS for the VRPSCDL [14]. Its destroy and remove operators are described in detail in previous work. Originally, it handles the maximum idle time constraint by setting self-imposed time windows on the delivery locations. However, MDLS considers the maximum idle time as a second objective, so self-imposed time windows are not used, i.e. an insertion is always feasible as long as other problem constraints are not violated (flow and timing constraints, vehicle capacity, return time at the depot, and deliveries served sequentially at the delivery location).

For the second objective, we define $ALNS_2$ as an ALNS with the following features. The destroy operators are the random removal and the worst removal operator. The first one selects ξ deliveries using a uniform distribution and removes them from the solution, whereas the second one removes ξ deliveries that correspond to delivery locations with high values of α. This latter uses a randomization factor to avoid always removing the same deliveries. The repair operators are the greedy insertion and the 2-regret insertion operator. Both of them consider the cost of inserting a delivery as the maximum value of α in the solution after inserting the delivery.

In our pseudocode, $ALNS_1$ and $ALNS_2$ are called with two parameters, where the first one is the initial solution of the ALNS and the second one is the maximum allowed value of α. In particular, MDLS relies on $ALNS_2$ to find solutions that are good with respect to the second objective, so it does not impose a constraint on the maximum allowed idle time and the second parameter is $\alpha = \infty$.

The initial solution s_0 of MDLS is generated by solving the VRPSCDL with $\alpha = \infty$ (line 1) and it is used to initialize the archive A (line 2). Then, MDLS iteratively selects one random solution z from the archive (line 5) at iteration i and applies $ALNS_1$ and $ALNS_2$ to obtain two new solutions, $s_{i,1}$ and $s_{i,2}$ (lines 6–7), that are used to update the archive (line 8).

Algorithm 1. MDLS-ALNS

1: $s_0 \leftarrow Solomon\_C1(\infty)$
2: $A \leftarrow \{s_0\}$
3: **while** stopping condition not met **do**
4: $\quad i \leftarrow 1$
5: $\quad$ Select a random solution z from A
6: $\quad s_{i,1} \leftarrow ALNS_1(z, \infty)$
7: $\quad s_{i,2} \leftarrow ALNS_2(z, \infty)$
8: $\quad$ Update archive A with $s_{i,1}$ and $s_{i,2}$
9: **end while**

3.2 ϵ-Constraint Method

The ϵ-constraint method (ECM) for multi-objective problems consists of optimizing one single objective, while formulating the second objective as a constraint [4]. Although it has been mainly used in association with exact algorithms, the ECM and some of its variants has been shown to provide good results in combination with heuristics to approximate the Pareto front [1,7,12].

We embed $ALNS_1$ in the ECM framework and denote the resulting method ECM-ALNS (see Algorithm 2). The method proceeds as follows. We first solve the single-objective sub-problem π_0 with $\alpha_0 = 0$, so that no idle time is allowed at any delivery location (line 1). To obtain this first solution s_0, the solver is allowed to run until I_0 iterations without improvement are reached. Next, we solve the single-objective sub-problem π_∞ with $\alpha_\infty = \infty$ to get a minimum reference value for the first objective (line 2). The initial archive A contains thus s_0 and s_∞ (line 3). Then, the value of the maximum allowed α is set back to 0 (line 4) and iteratively increased by ϵ, so that the single-objective sub-problem π_i with $\alpha_i = \alpha_{i-1} + \epsilon$ is solved, $i > 0$ (lines 6-8). After I_n iterations without improvement, $ALNS_1$ stops and returns solution s_i, which is used to update the archive (line 9). Following previous work [12] and our own preliminary experiments, we allocate comparatively longer execution times to obtain the initial solution s_0 by setting $I_0 \gg I_n$.

The ECM can proceed with both increasing and decreasing constraint values. We choose to increase the values of α because the solutions of a sub-problem with α_i are also feasible solutions of sub-problems with α_j if $\alpha_j > \alpha_i$, but the opposite is in general not true. Preliminary experiments show that, instead of generating a new solution from scratch for each problem π_i, $i > 0$, better results can be obtained by using the solution s_{i-1} of π_{i-1} found in the previous iteration as the initial solution of the $ALNS_1$ to solve π_i. This can be achieved without repairing the solution or using penalties if the ECM operates for increasing values of α.

Algorithm 2. ECM-ALNS

```
1: s_0 ← ALNS_1(Solomon_C1(0), 0)
2: s_∞ ← ALNS_1(Solomon_C1(∞), ∞)
3: A ← {s_0, s_∞}
4: α_0 = 0
5: while stopping condition not met do
6:     i ← 1
7:     α_i ← α_{i-1} + ε
8:     s_i ← ALNS_1(s_{i-1}, α_i)
9:     Update archive A with s_i
10: end while
```

3.3 Heuristic Box Splitting

Heuristic Box Splitting (HBS) addresses some of the problems arising in the ECM [12]. Instead of iteratively solving a single-objective sub-problem with increasing (or decreasing) constraint values, HBS forms a rectangle determined by the minimum and maximum values of each objective. This rectangle is iteratively split in halves, so that a single-objective sub-problem with a constraint determined by the splitting value is solved.

We embed $ALNS_1$ in HBS (HBS-ALNS) as follows (see Algorithm 3). Similar to the ECM, the single-objective sub-problem π_0 with $\alpha_0 = 0$ is solved to obtain s_0 after I_0 iterations without improvement (line 1). Then, we solve the single-objective problem with $\alpha_\infty = \infty$ and obtain thus s_∞ with the same termination criterion as before (line 2). The initial archive contains solutions s_0 and s_∞ (line 3). Points $(f_1(s_\infty), f_2(s_\infty))$ and $(f_1(s_0), f_2(s_0))$ form a rectangle that determines the area where HBS searches for new solutions (lines 4–5). The initial rectangle is added to the rectangle set S (line 6) and HBS runs until no more rectangles are available as follows. It selects the rectangle $R(y^1, y^2)$ with the larger area (line 9). The rectangle is split in two halves, so that the line that halves the rectangle determines the value of the constraint α_i, where $i > 0$ is the current iteration. For example, in the first iteration after creating the initial rectangle, $\alpha_1 = 0.5 \cdot f_2(s_\infty)$, and in general, $\alpha_i = 0.5 \cdot (y_2^1 + y_2^2)$ (line 10). The solver runs using this constraint until I_n iterations without improvement are reached (line 12), and the found solution s_i is used to update the archive (lines 13–16). The solution also allows the algorithm to discard areas that are dominated by the found solutions and to create new rectangles that are added to S (line 17).

Algorithm 3. HBS-ALNS

1: $s_0 \leftarrow ALNS_1(Solomon\_C1(0), 0)$
2: $s_\infty \leftarrow ALNS_1(Solomon\_C1(\infty), \infty)$
3: $A \leftarrow \{s_0, s_\infty\}$
4: $z^1 \leftarrow (f_1(s_\infty), f_2(s_\infty))$
5: $z^2 \leftarrow (f_1(s_0), f_2(s_0))$
6: $S \leftarrow \{R(z^1, z^2)\}$
7: **while** stopping condition not met **do**
8: $\quad i \leftarrow 1$
9: $\quad$ Select $R(y^1, y^2) \in S$ with the largest area
10: $\quad \alpha_i \leftarrow 0.5 \cdot (y_2^1 + y_2^2)$
11: $\quad$ Select $z \in A$ with $f_1(z) \leq f_1(z_j)$ such that $z_j \in A$ and $f_2(z_j) \leq \alpha_i$
12: $\quad s_i \leftarrow ALNS_1(z, \alpha_i)$
13: $\quad$ **if** s_i is dominated **then**
14: $\quad\quad y^2 \leftarrow (y_1^2, \alpha_i)$
15: $\quad$ **else**
16: $\quad\quad$ Update archive A with s_i
17: $\quad\quad$ Update S according to HBS rules
18: $\quad$ **end if**
19: $\quad$ **if** $S = \emptyset$ **then**
20: $\quad\quad S \leftarrow \{R(z^1, z^2)\}$
21: $\quad$ **end if**
22: **end while**

Similar to the ECM, preliminary experiments show that better results can be obtained by using solutions found with α_i as the initial solution for solving problems with $\alpha_j > \alpha_i$ (line 11). In the ECM this step is straightforward, since α increases monotonically. However, HBS needs to select a solution $z \in A$ that is feasible with respect to the current α_i. This is done by choosing the solution z with the best value of the first objective f_1 among those those solutions $z_j \in A$ that are feasible with respect to α_i (line 11).

The original HBS terminates when the rectangle set S is empty. We modify it to run until the maximum runtime is reached. If S is empty, the initial rectangle is added to S (lines 18–20).

4 Experiments and Results

This section presents the results obtained by our algorithms. We first compare the performance of our neighborhood search based methods with solutions obtained by an exact solver embedded in the ECM (see Sect. 4.1). Then, we evaluate our algorithms by solving large instances obtained from real-world data in Sect. 4.2. We use the set of instances for the VRPSCDL without instances 0, 1, 20, and 21, because those instances define a single depot and there is thus no synchronization involved.

Results report the normalized hypervolume (HV) as a percentage of the reference HV [21]. To calculate the reference HV for each instance, we build reference sets with all non-dominated solutions found in all our experiments. The nadir point is estimated to be 10% larger than the worst found values for each objective [10,12], so that extreme solutions also contribute to the HV. Although the HV is currently considered the most relevant performance indicator in multi-objective optimization [2], we also include some results concerning the overall non-dominated vector generation ($ONVG$) [19] and spacing (SP) [15] to obtain further information about the performance of our algorithms.

The exact solver is an implementation of the model of the VRPSCDL using CPLEX. We embed this solver in the ECM framework as described in Sect. 3.2, and denote this algorithm ECM-CPLEX. It runs for 86,400 s and dedicates a maximum of 3,600 s to solve each sub-problem. We use $\epsilon = 0.01$ in all experiments. Each call to CPLEX runs until it finds the optimum solution or the maximum runtime is reached. The exact solver is run only once for each instance.

We allow each combination of multi-objective framework and neighborhood search algorithm to run for a maximum time of 3,600 s. Following previous settings in the literature, each iteration of MDLS consists of one iteration of each ALNS. Same as described above, ECM-ALNS uses $\epsilon = 0.01$. ECM-ALNS and HBS-ALNS first obtain their reference points using $I_0 = 5{,}000$, and then solve each sub-problem with $I_n = 500$. For each instance and metaheuristic, we report the average and the best of 5 independent runs.

All algorithms were implemented in Java 1.7. The exact solver requires CPLEX 12.6.2. Each experiment runs on a Xeon core at 2.50 GHz with 64 GB shared RAM and deactivated hyperthreading.

4.1 Small Instances

First we compare our results on a small dataset with instances that contain $p \in [2, 3]$ depots, $n \in [10, 40]$ deliveries, and $m \in [6, 23]$ delivery locations.

Table 1 reports the HV obtained by the ECM-CPLEX as well as the average and best HV obtained by ECM-ALNS, HBS-ALNS, and MDLS-ALNS. Best results for each instance are highlighted in bold. Our results show that ECM-CPLEX is not always able to find the best-known Pareto front due to runtime restrictions. It obtains the best HV in 9 of 18 instances, but its results are weaker for the larger instances 8–9 and 16–19, each with 30 to 40 deliveries. In addition, the exact solver for the VRPSCDL only optimizes the travel cost, while its schedules are just determined to be feasible by CPLEX. For this reason, it often obtains worse Pareto fronts than the heuristic algorithms, which try to schedule deliveries in a compact manner. The best overall results are provided by HBS-ALNS, both regarding the average quality of the Pareto fronts as well as the best of 5 runs. In particular, it obtains the best average result in 11 of 18 instances with an average $HV = 98.5\%$. MDLS-ALNS provides competitive results and finds the best result for 8 of 18 instances. ECM-ALNS is outperformed by all algorithms.

Table 1. HV (%) obtained by each algorithm solving small instances.

Instance	ECM-CPLEX	ECM-ALNS		HBS-ALNS		MDLS	
		Avg	Best	Avg	Best	Avg	Best
2	99.8	97.9	99.8	**99.9**	**99.9**	97.0	98.9
3	**100.0**	98.8	**100.0**	**100.0**	**100.0**	**100.0**	**100.0**
4	**98.8**	97.1	98.0	98.5	98.7	98.1	98.2
5	95.9	94.9	95.4	**96.0**	**96.0**	89.9	91.8
6	**100.0**	97.5	98.7	**100.0**	**100.0**	95.7	97.7
7	**100.0**	97.6	**100.0**	**100.0**	**100.0**	**100.0**	**100.0**
8	97.1	99.1	99.9	**100.0**	**100.0**	94.3	94.9
9	90.2	97.4	99.2	**100.0**	**100.0**	95.0	98.1
10	**100.0**	97.7	**100.0**	**100.0**	**100.0**	**100.0**	**100.0**
11	97.7	95.2	96.8	**98.0**	**98.1**	95.0	95.2
12	**100.0**	94.4	99.5	96.3	**100.0**	**100.0**	**100.0**
13	**100.0**	97.9	99.1	99.6	99.9	**100.0**	**100.0**
14	**96.7**	93.1	95.7	96.0	**97.1**	95.8	96.3
15	**100.0**	79.9	86.7	97.3	98.5	**100.0**	**100.0**
16	97.8	96.7	97.9	**98.2**	98.3	98.0	**98.6**
17	95.8	95.6	96.6	97.5	**98.5**	**98.1**	98.4
18	91.1	94.7	96.4	**98.3**	**98.9**	95.3	96.0
19	83.6	95.2	96.6	97.6	98.9	**99.9**	**99.9**
Avg	96.9	95.6	97.6	**98.5**	**99.0**	97.3	98.0

4.2 Large Instances

Table 2 show the results obtained using the set of large instances with $p \in [2, 6]$ depots, $n \in [100, 300]$ deliveries, and $m \in [65, 100]$ delivery locations. For each algorithm and instance the average and the best HV (%) are reported. Our results show that HBS-ALNS outperforms the other solution methods in 15 of 18 instances regarding the average results, and it is able to find the best approximation of the Pareto front in 14 of 18 instances. In the remaining instances, ECM-ALNS finds the best results and always outperforms MDLS-ALNS.

Table 2. HV (%) obtained by each algorithm solving large instances.

Instance	ECM-ALNS		HBS-ALNS		MDLS-ALNS	
	Avg	Best	Avg	Best	Avg	Best
22	97.0	97.3	**98.4**	**98.5**	96.8	97.7
23	97.2	**98.0**	**97.9**	**98.0**	96.3	97.0
24	97.5	97.7	**98.2**	**98.4**	96.0	97.1
25	**95.9**	**96.6**	95.7	96.4	91.3	92.2
26	95.8	**96.5**	**96.3**	**96.5**	91.8	92.6
27	96.9	97.1	**97.8**	**98.2**	92.7	93.4
28	95.3	95.6	**96.3**	**96.9**	87.3	88.7
29	**95.1**	**95.6**	94.8	95.5	85.1	85.3
30	95.2	95.6	**96.6**	**97.0**	89.5	90.0
31	95.0	95.8	**96.4**	**96.6**	88.9	89.5
32	95.8	96.6	**96.8**	**97.1**	87.6	88.5
33	94.9	**96.3**	**96.0**	96.2	85.1	85.6
34	**95.9**	**96.9**	95.7	95.9	83.7	84.1
35	95.4	96.0	**96.4**	**96.9**	86.1	88.0
36	93.0	94.4	**95.6**	**95.9**	83.9	84.5
37	94.5	95.3	**95.1**	**95.9**	82.6	83.3
38	94.8	95.9	**95.6**	**96.0**	83.0	83.7
39	94.3	94.6	**95.6**	**96.3**	84.9	85.5
Avg	95.5	96.2	**96.4**	**96.8**	88.5	89.2

Table 3 shows average and best results of the $ONVG$ indicator obtained by each algorithm on each instance. This measure is the number of solutions in the Pareto front approximation and larger values are considered to be better. Although it poses some problems, as bad approximations with many non-dominated points are preferred over better approximations with only a few points, it can be used together with the HV to provide additional insights about the algorithms. Our results show that ECM-ALNS consistently finds mores solutions than the other algorithms, both in the average and in the best case. It also finds more solutions for larger instances (80–100) than for medium-sized ones (40–80), which seems a priori logical because the solution space is likely larger. HBS-ALNS finds more solutions than MDLS-ALNS, but it should be noted that the number of solutions for both algorithms does not depend on the instance size. The difference between ECM-ALNS and HBS-ALNS can be explained by how they explore the solution space. While ECM-ALNS increases α by very small values in each iteration and is therefore able to find many similar (but different) solutions, HBS-ALNS tries to explore different regions of the solution space and therefore finds less (but more diverse) solutions.

Table 3. *ONVG* obtained by each algorithm solving large instances.

Instance	ECM-ALNS		HBS-ALNS		MDLS-ALNS	
	Avg	Best	Avg	Best	Avg	Best
22	**45.2**	**57.0**	42.0	48.0	28.4	42.0
23	**42.4**	**51.0**	32.6	41.0	23.2	28.0
24	**57.6**	**71.0**	47.2	59.0	32.2	45.0
25	**69.8**	**80.0**	52.6	73.0	39.6	47.0
26	**63.0**	**75.0**	53.2	61.0	35.4	40.0
27	**65.6**	**78.0**	49.0	56.0	42.2	49.0
28	**76.9**	**86.0**	45.6	54.0	36.4	40.0
29	**85.4**	**96.0**	47.8	64.0	38.6	44.0
30	**79.1**	**92.0**	39.8	56.0	37.6	43.0
31	**78.7**	**90.0**	39.0	54.0	39.6	45.0
32	**81.3**	**88.0**	42.8	49.0	42.2	50.0
33	**91.1**	**100.0**	47.6	60.0	36.2	39.0
34	**84.3**	**94.0**	38.8	46.0	31.8	41.0
35	**89.8**	**98.0**	56.4	70.0	30.4	42.0
36	**86.1**	**93.0**	40.0	51.0	35.8	39.0
37	**96.1**	**112.0**	41.0	55.0	32.6	34.0
38	**90.5**	**100.0**	38.6	54.0	26.4	34.0
39	**95.2**	**106.0**	45.8	54.0	40.6	50.0
Avg	76.6	87.1	44.4	55.8	35.0	41.8

The performance of MDLS is hindered by the poor performance of $ALNS_2$ when the instance size increases. Although it obtains reasonably good HV values for middle-sized instances such as 22–24 (100 deliveries), its results quickly degradate for larger instances. Figure 1 shows the median Pareto front obtained by each method based on neighborhood search for two instances. The median Pareto front is the front that corresponds to the median HV of 5 independent runs. Figures 1a and 1b show the results obtained for instances 19 and 39 (40 and

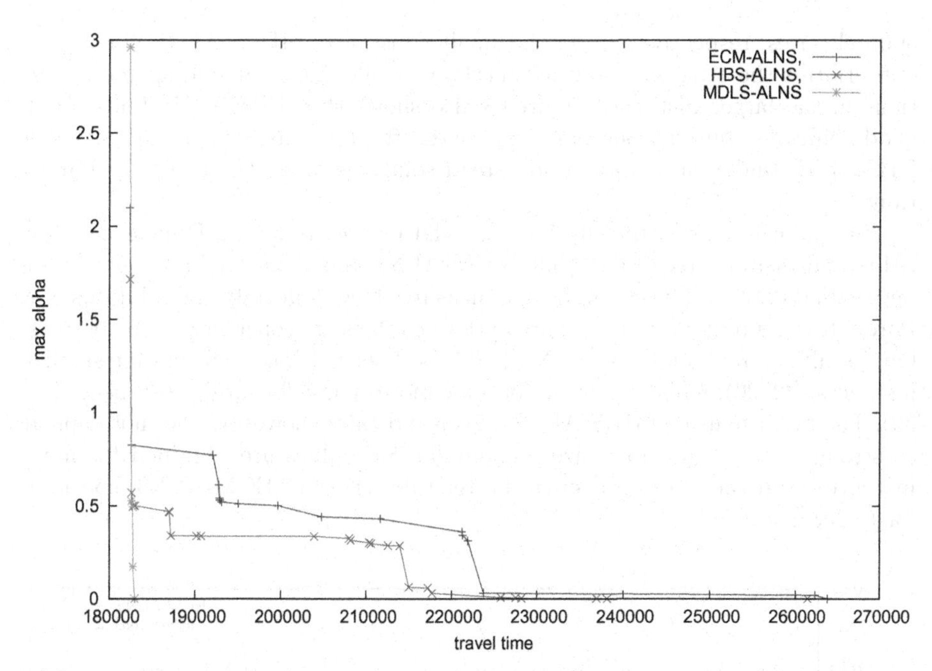

(a) Instance 19 with 3 depots, 40 deliveries, and 17 locations.

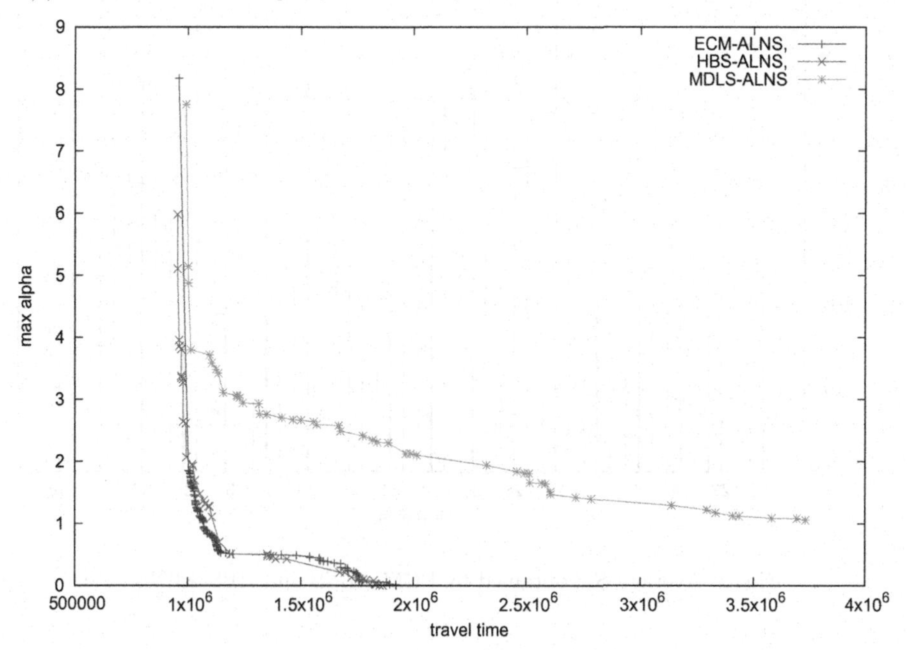

(b) Instance 39 with 6 depots, 300 deliveries, and 100 locations.

Fig. 1. Median Pareto front obtained by each solution method based on neighborhood search on (a) a small instance and (b) a large instance.

300 deliveries, respectively). On the smaller instance, MDLS-ALNS is fast and outperforms the other two solution methods, while it fails at finding good solutions in the larger instance. Figure 1b also shows that EPS-ALNS finds many good solutions, but it does not explore the front for high values of α, while HBS-ALNS finds reasonably good spread solutions along the complete Pareto front.

We quantify this latter effect by calculating the average SP for each algorithm and instance (see Fig. 2, only ECM-ALNS and HBS-ALNS are shown for better readability). The SP indicator measures how uniformly the solutions of a Pareto front approximation are spread (lower values are considered to be better). Our results show that ECM-ALNS finds good spread Pareto fronts for smaller instances (22–26), while the fronts become more sparse for larger instances (27–39). The performance of HBS-ALNS on this indicator, however, does not depend so strongly on the instance size, i.e. similar SP values are obtained for most instances, and they are on average better than those of ECM-ALNS, specially for larger instances.

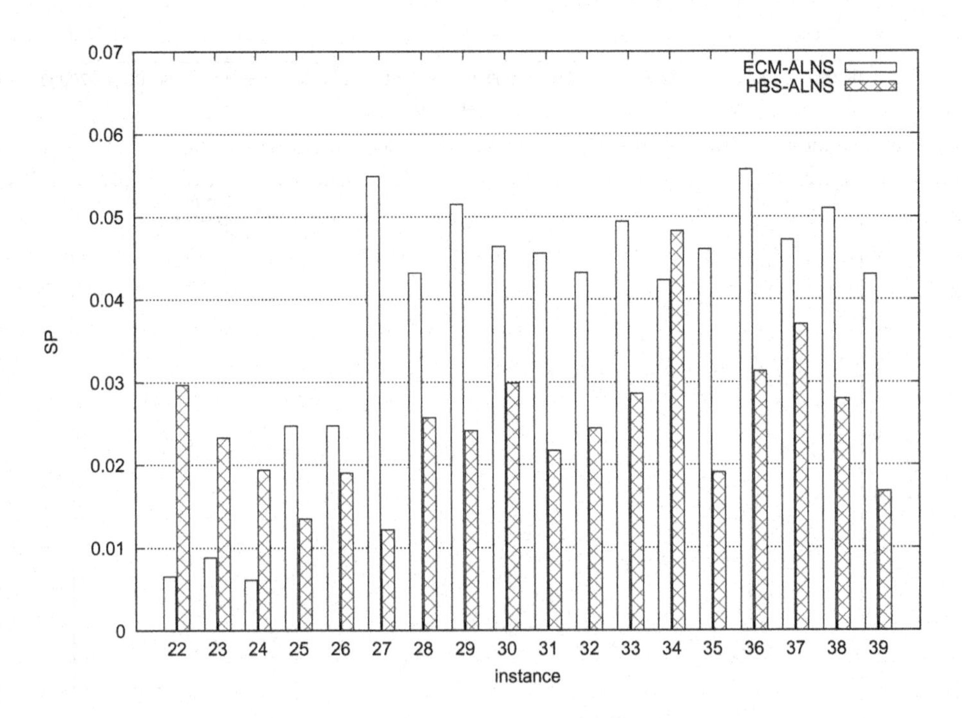

Fig. 2. Average SP obtained by ECM-ALNS and HBS-ALNS.

5 Conclusions

In this paper, we defined and solved a multi-objective vehicle routing problem with synchronization constraints at the delivery location. We proposed three methods based on single-objective neighborhood search that are embedded within more general optimization frameworks for multi-objective optimization. In particular, we used ECM, HBS, and MDLS together with two ALNS algorithms. Experiments showed that HBS-ALNS provide the best Pareto fronts regarding the HV, also outperforming an exact solver on small instances. Future research should consider other quality indicators for multi-objective optimization problems, additional analysis on how fast the methods are at finding new solutions, tune the ALNS for the second objective in MDLS, and implement some restart criteria to better explore the Pareto front in the ECM.

References

1. Anderluh, A., Nolz, P.C., Hemmelmayr, V.C., Crainic, T.G.: Multi-objective optimization of a two-echelon vehicle routing problem with vehicle synchronization and 'grey zone' customers arising in urban logistics. Eur. J. Oper. Res. **289**(3), 940–958 (2021). https://doi.org/10.1016/j.ejor.2019.07.049
2. Audet, C., Bigeon, J., Cartier, D., Digabel, S.L., Salomon, L.: Performance indicators in multiobjective optimization. Eur. J. Oper. Res. **292**(2), 397–422 (2021). https://doi.org/10.1016/j.ejor.2020.11.016
3. Eskandarpour, M., Ouelhadj, D., Hatami, S., Juan, A.A., Khosravi, B.: Enhanced multi-directional local search for the bi-objective heterogeneous vehicle routing problem with multiple driving ranges. Eur. J. Oper. Res. **277**(2), 479–491 (2019). https://doi.org/10.1016/j.ejor.2019.02.048
4. Haimes, Y.Y., Lasdon, L.S., Wismer, D.A.: On a bicriterion formation of the problems of integrated system identification and system optimization. IEEE Trans. Syst. Man Cybern., 296–297 (1971). https://doi.org/10.1109/TSMC.1971.4308298
5. Jozefowiez, N., Semet, F., Talbi, E.: Multi-objective vehicle routing problems. Eur. J. Oper. Res. **189**(2), 293–309 (2008). https://doi.org/10.1016/j.ejor.2007.05.055
6. Lacomme, P., Prins, C., Sevaux, M.: A genetic algorithm for a bi-objective capacitated arc routing problem. Comput. Oper. Res. **33**(12), 3473–3493 (2006). https://doi.org/10.1016/j.cor.2005.02.017. Part Special Issue: Recent Algorithmic Advances for Arc Routing Problems
7. Laumanns, M., Thiele, L., Zitzler, E.: An efficient, adaptive parameter variation scheme for metaheuristics based on the epsilon-constraint method. Eur. J. Oper. Res. **169**(3), 932–942 (2006). https://doi.org/10.1016/j.ejor.2004.08.029
8. Lian, K., Milburn, A.B., Rardin, R.L.: An improved multi-directional local search algorithm for the multi-objective consistent vehicle routing problem. IIE Trans. **48**(10), 975–992 (2016). https://doi.org/10.1080/0740817X.2016.1167288
9. Liu, Q., Li, X., Liu, H., Guo, Z.: Multi-objective metaheuristics for discrete optimization problems: a review of the state-of-the-art. Appl. Soft Comput. **93**, 106382 (2020). https://doi.org/10.1016/j.asoc.2020.106382
10. Maltese, J., Ombuki-Berman, B.M., Engelbrecht, A.P.: A scalability study of many-objective optimization algorithms. IEEE Trans. Evol. Comput **22**(1), 79–96 (2018). https://doi.org/10.1109/TEVC.2016.2639360
11. Martí, R., Campos, V., Resende, M.G., Duarte, A.: Multiobjective GRASP with path relinking. Eur. J. Oper. Res. **240**(1), 54–71 (2015). https://doi.org/10.1016/j.ejor.2014.06.042

12. Matl, P., Hartl, R.F., Vidal, T.: Leveraging single-objective heuristics to solve bi-objective problems: Heuristic box splitting and its application to vehicle routing. Networks **73**(4), 382–400 (2019). https://doi.org/10.1002/net.21876
13. Ombuki, B.M., Ross, B., Hanshar, F.: Multi-objective genetic algorithms for vehicle routing problem with time windows. Appl. Intell. **24**(1), 17–30 (2006). https://doi.org/10.1007/s10489-006-6926-z
14. Sarasola, B., Doerner, K.F.: Adaptive large neighborhood search for the vehicle routing problem with synchronization constraints at the delivery location. Networks **75**(1), 64–85 (2020). https://doi.org/10.1002/net.21905
15. Schott, J.R.: Fault Tolerant Design Using Single and Multicriteria Genetic Algorithm Optimization. Master's thesis, Department of Aeronautics and Astronautics, MIT (1995)
16. Shao, S., Xu, G., Li, M., Huang, G.Q.: Synchronizing e-commerce city logistics with sliding time windows. Transp. Res. Part E Logist. Transp. Rev. **123**, 17–28 (2019). https://doi.org/10.1016/j.tre.2019.01.007
17. Tan, K.C., Cheong, C.Y., Goh, C.K.: Solving multiobjective vehicle routing problem with stochastic demand via evolutionary computation. Eur. J. Oper. Res. **177**(2), 813–839 (2007). https://doi.org/10.1016/j.ejor.2005.12.029
18. Tricoire, F.: Multi-directional local search. Comput. Oper. Res. **39**(12), 3089–3101 (2012). https://doi.org/10.1016/j.cor.2012.03.010
19. Veldhuizen, D.A.V.: Multiobjective Evolutionary Algorithms: Classifications, Analyses, and New Innovations. Ph.D. thesis, Graduate School of Engineering, Air Force Institute of Technology (1999)
20. Xu, S.X., Shao, S., Qu, T., Chen, J., Huang, G.Q.: Auction-based city logistics synchronization. IISE Trans. **50**(9), 837–851 (2018). https://doi.org/10.1080/24725854.2018.1450541
21. Zitzler, E.: Evolutionary Algorithms for Multiobjective Optimization: Methods and Applications. Ph.D. thesis, ETH Zurich, Switzerland (1999)

Air Logistics and Multi-modal Transport

Analysis of the Impact of Physical Internet on the Container Loading Problem

Ana Rita Ferreira[1], António G. Ramos[1(✉)], and Elsa Silva[2]

[1] INESC TEC and School of Engineering, Polytechnic of Porto, Porto, Portugal
agr@isep.ipp.pt
[2] INESC TEC, Porto, Portugal

Abstract. In the Physical Internet supply chain paradigm, modular boxes are one of the main drivers. The dimension of the modular boxes has already been subject to some studies. However, the usage of a modular approach on the container loading problem has not been accessed. In this work, we aim to assess the impact of modular boxes in the context of the Physical Internet on the optimization of loading solutions. A mathematical model for the CLP problem is used, and extensive computational experiments were performed in a set of problem instances generated considering the Physical Internet concept. From this study, it was possible to conclude for the used instances that modular boxes contribute to a higher volume usage and lower computational times.

Keywords: Physical internet · Modular boxes · Container loading problem

1 Introduction

"Official statistics report that in the USA, trailers are approximately 60% full when travelling loaded" (Montreuil 2011), numbers that reflect the unsustainability in current logistics, mainly because the number of journeys is more than necessary, which leads to higher C02 emissions and higher costs to companies sustain. Thus, was born the Physical Internet to solve the problems of traditional logistics and, in particular, the problem of: "we are shipping air" (Montreuil 2011).

The Container Loading Problem (CLP) is a combinatorial optimization problem. The objective is to pack a set of boxes into a container to maximize the occupied volume without overlapping (Ramos et al. 2018).

The modularity of PI boxes combined with CLP models is the key to minimize the previously mentioned problem. Landschützer et al. (2014) defines Physical Internet boxes as a "modular set of load units forming a building block of smaller units that can be combined manifold while respecting current sizing requirements for efficient handling and space usage". This type of box has great relevance because there is a significant diversity of dimensions in the packaging (products are from different brands, and each has distinct measures) directly associated with the inefficient use of space (Landschützer et al. 2014).

M. Mes et al. (Eds.): ICCL 2021, LNCS 13004, pp. 549–561, 2021.
https://doi.org/10.1007/978-3-030-87672-2_36

This work aims to understand how the Physical Internet can positively impact solving the CLP, realizing the effect of modular boxes on the content occupation rate.

This work differs from the others found in the literature because it shows the advantages of Physical Internet (the modular boxes) to obtain better solutions in the CLP.

The Sect. 2 of the paper is dedicated to the literature review. Section 3 develops the methodology and highlights the 3 main steps: the choice of the model and dimensions for the modular boxes and the development of an instance generator to randomly develop the instances used in the tests. In Sect. 4 the results obtained are presented, and in the last Sect. 5, the conclusions are discussed.

2 Literature Review

Physical Internet (π or PI) is inspired by Digital Internet and was defined by Montreuil (2011) and Ballot et al. (2014) as a logistical system that operates globally, combining physical, digital and operational elements by "encapsulation", interfaces and protocols and by Tran-Dang et al. (2020) as having been "developed to be a global logistics system that aims to move, handle, store, and transport logistics products in a sustainable and efficient way".

This paradigm shift requires several changes and has a major impact on the way products are designed, handled, stored, manufactured and distributed, redefining business models and value creation (Montreuil et al. 2012; Ballot et al. 2014).

Thus, Physical Internet comes to solve the 13 problems of traditional logistics defined by Montreuil (2011): (1) We are shipping air and packing, (2) Empty travel is the norm rather than the exception, (3) Truckers have become the modern cowboys, (4) Products mostly sit idle, stored where unneeded, yet so often unavailable fast where needed, (5) Production and storage facilities are poorly used, (6) So many products are never sold, never used, (7) Products do not reach those who need them the most, (8) Fast and reliable intermodal transport is still a dream or a joke, (9) Getting products in through, and out of cities is a nightmare, (10) Products unnecessarily move, crisscrossing the world, (11) Networks are neither secure nor robust, (12) Smart automation and technology are hard to justify and (13) Innovation is strangled.

The PI logistics system is open, global and allows standard containers to easily move on various means of transport (Montreuil et al. 2012). Furthermore, since these networks are collaborative, there is the aggregation of containers from different origins, optimizing the loading and increasing the number of trips origin-destination (Montreuil et al. 2012; Montreuil et al. 2010; Crainic & Montreuil 2016). These standardized, modular and intelligent containers are called π-containers. They are monitored using a PI identifier during this path, just as with data packets on the internet.

In the literature, the problem presented in this paper is the CLP and addresses the disposal of rectangular boxes in the container minimizing the total wasted space, ensuring that the boxes fit in the container and do not overlap, besides ensuring other loading constraints (Meller et al. 2012).

The interested reader can consult important existing literature review in the CLP field. In Zhao et al. (2016), an exhaustive comparative review of the CLP algorithms is presented, and in Bortfeldt & Wäscher (2013), a state-of-the-art review is presented for the constraints in CLP. More recently, in Silva et al. (2019), a comparative study of the exact methods for CLP is also presented.

For greater efficiency in optimization, it is necessary to reflect on the dimensions of modular boxes, such as the Modulushca Project- an initiative that represents the first genuine contribution to the development of interconnected logistics at the European level and provides a basis for an interconnected logistics system until 2030. Regarding the dimensions of modular boxes, the Modulushca project developed, based on the dimensions traditionally used, such as euro, a modular platform that allows up to 440 different types of boxes. From that and to obtain a more accessible and effective solution, Landschützer et al. (2014) suggested 5 dimensions for each of the PI boxes, present in Fig. 1.

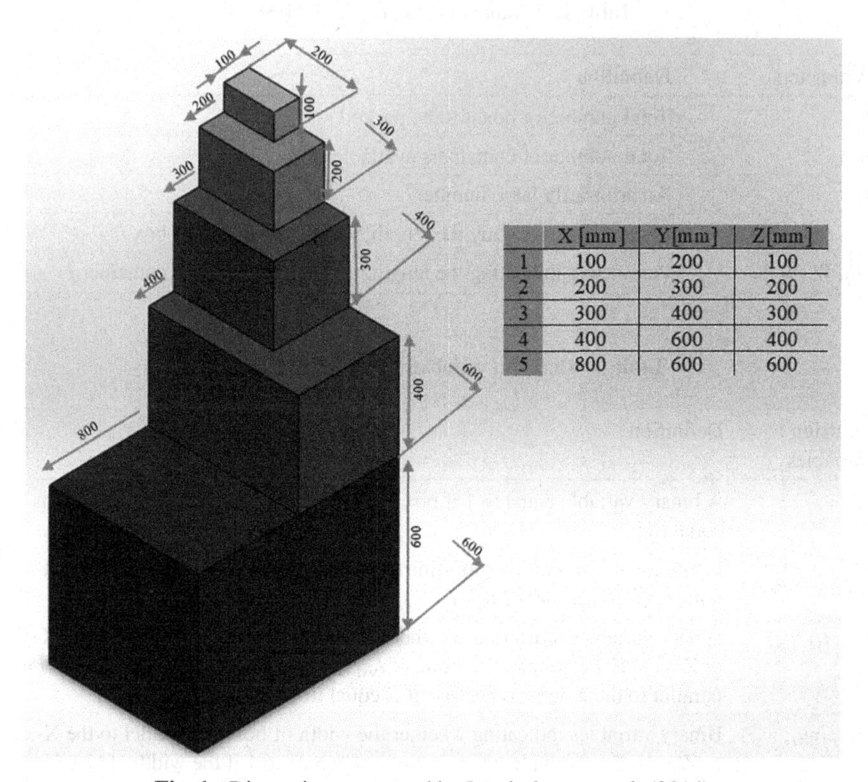

	X [mm]	Y[mm]	Z[mm]
1	100	200	100
2	200	300	200
3	300	400	300
4	400	600	400
5	800	600	600

Fig. 1. Dimensions proposed by Landschützer et al. (2014)

3 Methodology

3.1 Model for the CLP

Several authors have studied the CLP since the 60's and currently, there are multiple algorithms, which can be classified according to their objective function and existing restrictions (Pisinger, 2002). The chosen model was developed by Chen et al. (1995) and fell under the category of Multi-container loading model (Pisinger, 2002). This model is distinguished from the others because it considers having multiple containers of different dimensions.

In this work, the model of Chen et al. (1995) was adapted to respond to the objective. Instead of having multiple containers, there is only one (j = 1), and the goal is to maximize the volume of boxes in it. Despite using only one container in the computational experiments, the model is generic and can be used for multiple containers. Consequently, the objective function has changed. From the model adaptation, there is the objective function present in Eq. (1) and the constraints from (2) to (20). Parameters are defined in Table 1 and decision variables are defined in Table 2.

Table 1. Parameters of Chen et al. (1995).

Parameters	Definition
I	Total number of boxes to be packed
J	Total number of containers available
M	An arbitrarily large number
(p_i, q_i, r_i)	Parameters indicating the length, width, and height of box i
(L_j, W_j, H_j)	Parameters indicating the length, width, and height of container j

Table 2. Decision variables of Chen et al. (1995).

Decision Variables	Definition
s_{ij}	A binary variable equal to 1 if box i is placed in container j; otherwise it is equal to 0
(x_i, y_i, z_i)	Continuous variables (for location) indicating the coordinates of the front-left bottom (FLB) corner of box i
(l_{xi}, l_{yi}, l_{zi})	Binary variables indicating whether the length of box i is parallel to the X-, Y-, or Z-axis. For example, the value of l_{yxi} is equal to 1 if the length of box i is parallel to the X-axis; otherwise it is equal to 0
(w_{xi}, w_{yi}, w_{zi})	Binary variables indicating whether the width of box i is parallel to the X-, Y-, or Z-axis. For example, the value of w_{xi} equals to 1 if the width of box i is parallel to the X-axis; otherwise it is equal to 0
(h_{xi}, h_{yi}, h_{zi})	Binary variable indicating whether the height of box i is parallel to the X-, Y-, or Z-axis. For example, the value of h_{xi} is equal to 1 if the height of box i is parallel to the X-axis; otherwise it is equal to 0
a_{ik}	Binary variable indicating if box i is to the left of box k
b_{ik}	Binary variable indicating if box i is to the right of box k

(continued)

Table 2. (*continued*)

Decision Variables	Definition
c_{ik}	Binary variable indicating if box i is in front of box k
d_{ik}	Binary variable indicating if box i is behind box k
e_{ik}	Binary variable indicating if box i is on top of box k
f_{ik}	Binary variable indicating if box i is under box k

$$O.F.Maximize \quad \sum\nolimits_{j=1}^{J} \sum\nolimits_{i=1}^{I} p_i \cdot q_i \cdot r_i \cdot s_{ij} \tag{1}$$

$$x_i + p_i \cdot l_{xi} + q_i \cdot w_{xi} + r_i \cdot h_{xi} \le x_k + (1 - a_{ik}) \cdot M \quad \forall i, k, i < k \tag{2}$$

$$x_k + p_k \cdot l_{xk} + q_k \cdot w_{xk} + r_k v \cdot h_{xk} \le x_i + (1 - b_{ik}) \cdot M \quad \forall i, k, i < k \tag{3}$$

$$y_i + p_i \cdot l_{yi} + q_i \cdot w_{yi} + r_i \cdot h_{yi} \le y_k + (1 - c_{ik}) \cdot M \quad \forall i, k, i < k \tag{4}$$

$$y_k + p_k \cdot l_{yk} + q_k \cdot w_{yk} + r_k \cdot h_{yk} \le y_i + (1 - d_{ik}) \cdot M \quad \forall i, k, i < k \tag{5}$$

$$z_i + p_i \cdot l_{zi} + q_i \cdot w_{zi} + r_i \cdot h_{zi} \le z_k + (1 - e_{ik}) \cdot M \quad \forall i, k, i < k \tag{6}$$

$$z_k + p_k \cdot l_{zk} + q_k \cdot w_{zk} + r_k \cdot h_{zk} \le z_i + (1 - f_{ik}) \cdot M \quad \forall i, k, i < k \tag{7}$$

$$a_{ik} + b_{ik} + c_{ik} + d_{ik} + e_{ik} + f_{ik} \ge s_{ij} + s_{kj} - 1 \quad \forall i, k, j, i < k \tag{8}$$

$$\sum\nolimits_{j=1}^{J} s_{ij} \le 1 \quad \forall i \tag{9}$$

$$x_i + p_i \cdot l_{xi} + q_i \cdot w_{xi} + r_i \cdot h_{xi} \le L_j + \left(1 - s_{ij}\right) \cdot M \quad \forall i, j \tag{10}$$

$$y_i + p_i \cdot l_{yi} + q_i \cdot w_{yi} + r_i \cdot h_{yi} \le W_j + \left(1 - s_{ij}\right) \cdot M \quad \forall i, j \tag{11}$$

$$z_i + p_i \cdot l_{zi} + q_i \cdot w_{zi} + r_i \cdot h_{zi} \le H_j + \left(1 - s_{ij}\right) \cdot M \quad \forall i, j \tag{12}$$

$$l_{xi} + l_{yi} + l_{zi} = 1 \tag{13}$$

$$w_{xi} + w_{yi} + w_{zi} = 1 \tag{14}$$

$$h_{xi} + h_{yi} + h_{zi} = 1 \quad (15)$$

$$l_{xi} + w_{xi} + h_{xi} = 1 \quad (16)$$

$$l_{yi} + w_{yi} + h_{yi} = 1 \quad (17)$$

$$l_{zi} + w_{zi} + h_{zi} = 1 \quad (18)$$

$$l_{xi}, l_{yi}, l_{zi}, w_{xi}, w_{yi}, w_{zi}, h_{xi}, h_{yi}, h_{zi}, a_{ik}, b_{ik}, c_{ik}, d_{ik}, e_{ik}, f_{ik}, s_{ij}, n_j = 0 \, ou \, 1 \quad (19)$$

$$x_i, \; y_i, \; z_i \geq 0 \quad (20)$$

The objective function (1) aims to maximize the occupied volume of the container. Constraints (2)–(8) ensure that there is no overlapping. These constraints are helpful when comparing boxes from the same container. Constraints (9) guarantees that a box can only be placed in one container. Constraints (10)–(12) ensure that the boxes placed in a container fit within the physical dimensions of the container. Finally, the legitimacy of the box rotations is ensured by the conditions from (13)–(18). The decision variables domain is presented in constraints (19) and (20).

3.2 Boxes Dimensions Definition

Initially, measures were defined for 10 boxes so that tests could be carried out at a later stage. It was intended that the dimensions selected were not inconsistent with the current reality - for easier adaptation in the future - and, for this reason, the existing systems were used to define the dimensions of the modular boxes.

Table 3 shows the measurements for each of the systems used as a reference: Galia (GALIA), Modulushca (Landschützer et al. 2014) and EURO (Mokhlesi & Andersson 2009), as well as the similarities between them. In Table 3, it is possible to see the dimensions selected. The criterion for their selection is the greater similarity between the 3 systems mentioned (Table 4).

Table 3. Galia, Moduluscha and EURO dimensions in millimetres.

Galia			Modulushca		EURO	
L	W	H	L	W	L	W
1200	500	500	1200	800	1200	1000
1200	500	300	800	600	1200	800
1000	600	500	600	400	800	600

(continued)

Table 3. (*continued*)

Galia			Modulushca		EURO	
L	W	H	L	W	L	W
1000	600	300	400	300	600	400
1000	400	500	300	200	400	300
1000	400	300	200	100	300	200
600	400	300				
600	400	250				
600	400	200				
400	300	300				
400	300	200				
400	300	150				
300	200	200				
300	200	125				
300	200	90				

Table 4. Selected dimensions in millimetres.

L	W	H
1200	800	400
600	400	300
600	400	250
600	400	200
400	300	300
400	300	200
400	300	150
300	200	200
300	200	150
300	200	100

3.3 Instance Generation

The instance generator developed returned a distribution suggestion of the occupation of the container according to the volume of the container and the desired variability of boxes (the greater the variability of boxes, the greater the heterogeneity and vice versa). The flowchart of the instance generator is shown in Fig. 2.

It is important to note that the instance for modular and non-modular boxes is generated simultaneously because the data necessary is the same (variability and container size).

A total of 72 instances (36 modular and 36 non-modular) were generated, the variability of boxes is in the range of 2 to 9. As for the number of boxes, there is a smaller instance containing 30 boxes and the largest containing 138.

Although the instance generator is prepared for any container measurement, in this study was considered a container of size $1.2\,m \times 0.8\,m \times 2.4\,m$. Note that the dimensions of a Euro pallet were used as container measurements because they are widely used in practice and allow us to take advantage of a standard dimension.

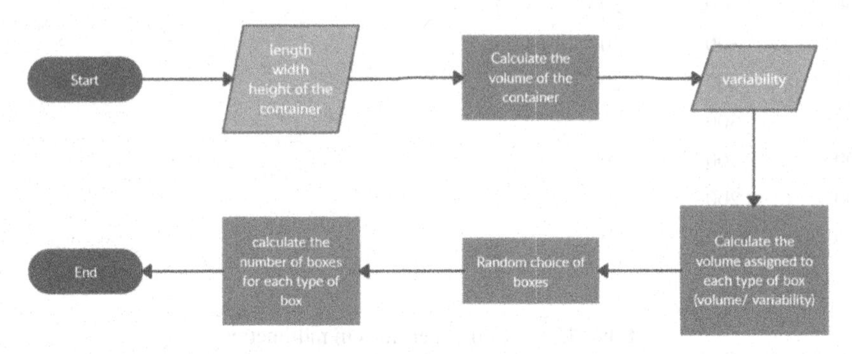

Fig. 2. Instance generator

4 Results and Discussion

The computational experiments were run on "IBM ILOG CPLEX Optimization Studio 12.10.0", using Intel Xeon Gold 6148 CPU2.40 GHz, with 96.0 GB of RAM with one hour of time limit. The computational experiments were performed considering two types of instances: modular boxes and non-modular boxes. An analysis of the optimization times, the occupied volume and the number of allocated boxes is conducted.

4.1 Comparing Modularity and Non-modularity

Each instance of modular boxes corresponds to an instance of non-modular boxes. The only differences between both are the dimensions (length, width and height) since the total volume, the volume of the box itself, the number of boxes and the variability are exactly the same.

Table 5 reports the results of occupied volume and GAP (reported by IBM ILOG CPLEX) for both scenarios.

All instances ran for 1 h (the limit) except modular instances 1 and 2, which ran for 126 s and 485 s, respectively.

Table 5. Computational results

Instance			Volume occupied [%]		GAP [%]	
N°	Total # boxes	# Types of boxes	Modular	Non-modular	Modular	Non-modular
1	**30**	**4**	**90,10**	**74,43**	**0,00**	**20,95**
2	**35**	**2**	**100,00**	**80,11**	**0,00**	**20,53**
3	44	3	86,46	78,07	13,25	25,33
4	51	2	87,50	70,76	14,29	17,64
5	58	5	81,25	67,90	14,42	36,77
6	61	5	73,44	66,02	25,53	39,36
7	61	4	64,32	57,74	53,04	70,23
8	65	3	83,33	69,18	18,75	42,82
9	65	3	63,54	53,58	53,28	81,53
10	67	2	61,46	58,27	61,86	70,53
11	74	5	65,10	54,60	43,20	70,34
12	82	3	78,13	63,39	28,00	57,28
13	82	6	53,13	43,38	81,86	122,22
14	83	2	59,38	56,43	67,54	76,00
15	89	4	52,61	50,47	87,13	94,61
16	90	5	63,28	40,01	51,03	138,22
17	93	4	57,81	40,52	70,27	142,27
18	94	6	66,15	52,03	46,46	85,93
19	96	6	45,05	44,34	116,19	119,25
20	101	3	45,31	44,63	120,69	122,10
21	102	6	61,46	52,54	57,63	84,14
22	104	4	60,42	31,92	65,52	212,42
23	105	6	58,07	40,52	69,96	142,94
24	106	8	51,56	34,66	89,90	181,77
25	109	7	50,00	46,46	87,50	101,25
26	109	7	46,09	42,12	103,39	122,22
27	110	5	56,25	47,83	67,59	96,77
28	113	7	48,05	45,42	97,29	108,11
29*	120	2	45,31	49,27	120,69	102,07
30*	122	8	41,15	43,64	137,98	123,84
31	123	7	45,83	36,09	105,68	160,57

(continued)

Table 5. (*continued*)

Instance			Volume occupied [%]		GAP [%]	
Nº	Total # boxes	# Types of boxes	Modular	Non-modular	Modular	Non-modular
32*	130	7	28,13	34,53	238,89	175,43
33*	130	8	11,33	33,27	764,37	193,68
34	134	9	48,18	35,95	97,84	164,41
35	138	8	44,01	40,38	122,49	141,88
36	142	8	35,68	8,31	174,45	1075,02

The results show that 89% of the instances obtained better results with modular boxes, with the opposite being only confirmed 4 times (instances highlighted with *). Also noteworthy is the fact that a GAP of 0 was reached twice.

This study demonstrates that transport will be more sustainable as it is possible to carry more volume with modular boxes.

In the particular case of 35 boxes with variability 2, when using modular boxes, it becomes clear that there is a better use of space, allowing to transport 3 more boxes than the instance of non-modular boxes. Figure 3.a. and Fig. 3.b. represent the organization of modular and non-modular boxes, respectively.

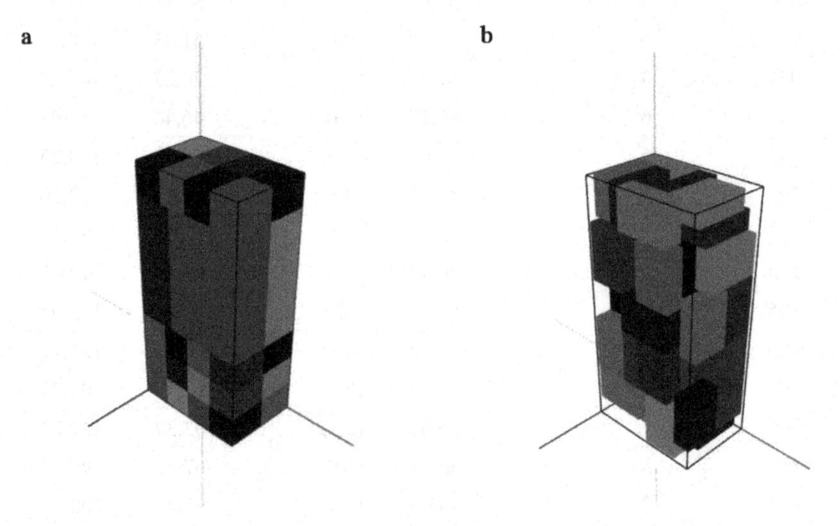

Fig. 3. Represent the organization of modular (a) and non-modular (b)

4.2 The Influence of the Number of Boxes

With this study, it was also possible to better understand the relation between the number of boxes, the volume occupied and the GAP. Figure 4 and Fig. 5 are graphics on which it is possible to observe, respectively, the relation boxes quantity – volume results and boxes quantity – GAP. Note that in the GAP graphic (Fig. 5), the outliers were removed: instance 33 of the modular boxes and instance 36 of the non-modular boxes.

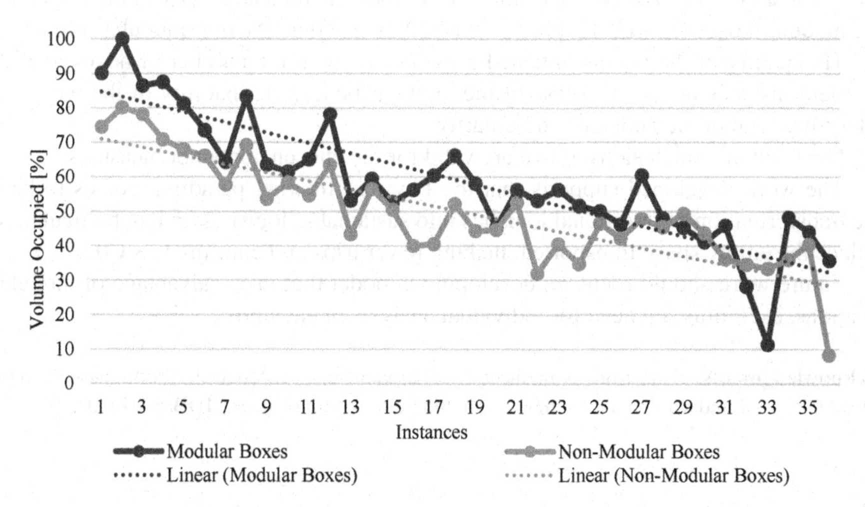

Fig. 4. Relation between boxes quantity and obtained volume results

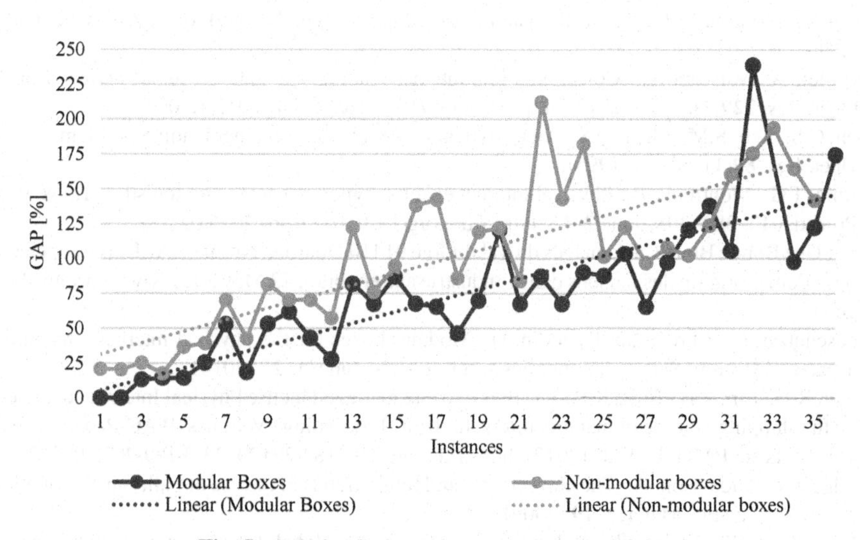

Fig. 5. Relation between boxes quantity and GAP

The previous figures show that the greater the number of boxes in the instance, the more difficult it becomes for the model to obtain an optimal solution (GAP = 0%). Consequently, the volume occupied also decreases with the increase in the number of boxes.

5 Conclusion

The developed study made it possible to understand the relation between Physical Internet modular boxes and the volume efficiency of loading models. In 89% of the instances, the modular boxes showed a higher volume efficiency than the non-modular boxes.

The quality of the results obtained is connected with the number of boxes in each problem instance and, in the case of the modular boxes, the inability of the model to take advantage of the geometric modularity.

Note that all conclusions drawn are valid for the chosen container measures.

The work developed supports that the Physical Internet paradigm comes to help the transformation of traditional logistics into sustainable logistics and, in particular, to reduce the empty space transported, making fewer trips and emitting less C02.

Future work should focus on developing a model that takes advantage of modular relations, excluding the need for individual analysis of each box.

Acknowledgments. This work is financed by National Funds through the Portuguese funding agency, FCT - Fundação para a Ciência e a Tecnologia, within project UIDB/50014/2020.

References

Ballot, E., Montreuil, B., Meller, R.: The Physical Internet: The Network of Logistics Networks (2014)

Bortfeldt, A., Wäscher, G.: Constraints in container loading – a state-of-the-art review. Eur. J. Oper. Res. **229**(1), 1–20 (2013). https://doi.org/10.1016/J.EJOR.2012.12.006

Chen, C.S., Lee, S.M., Shen, Q.S.: An analytical model for the container loading problem. Eur. J. Oper. Res. **80**(1), 68–76 (1995)

Crainic, T.G., Montreuil, B.: Physical internet enabled hyperconnected city logistics. Trans. Res. Procedia **12**, 383–398, June 2015. https://doi.org/10.1016/j.trpro.2016.02.074

Galia: Une Bonne Boite Cartonnages du Val d'Orge 91410 Dourdan Norme GALIA (Groupement pour l'amélioration des liaisons dans l'industrie automobile), 13 May 2021. www.cartonvaldorge.fr

Landschützer, C., Ehrentraut, F., Jodin, D.: Modular boxes for the physical internet – technical aspects. Literature Series - Economics and Logistics, June, 1–27 (2014)

Meller, R.D., Lin, Y.H., Ellis, K.P.: The impact of standardized metric Physical Internet containers on the shipping volume of manufacturers. In: IFAC Proceedings Volumes, IFAC-PapersOnline, vol. 14, Issue PART 1, IFAC (2012). https://doi.org/10.3182/20120523-3-RO-2023.00282

Mokhlesi, J., Andersson, A.: The current state and future trends in the use of pallets in distribution systems. Logistics Manage. **19** (2009)

Montreuil, B.: Toward a physical internet: meeting the global logistics sustainability grand challenge. Logist. Res. **3**(2–3), 71–87 (2011). https://doi.org/10.1007/s12159-011-0045-x

Montreuil, B., Meller, R.D., Ballot, E.: Towards a physical internet: the impact on logistics facilities and material handling systems design and innovation. Prog. Mater. Handling Res., 305–327 (2010)

Montreuil, B., Rougès, J.-F., Cimon, Y., Poulin, D.: The physical internet and business model innovation. Technol. Innov. Manag. Rev. **2**(6), 32–37 (2012)

Pisinger, D.: Heuristics for the container loading problem. Eur. J. Oper. Res. **141**(2), 382–392 (2002). https://doi.org/10.1016/S0377-2217(02)00132-7

Ramos, A.G., Silva, E., Oliveira, J.F.: A new load balance methodology for container loading problem in road transportation. Eur. J. Oper. Res. **266**(3), 1140–1152 (2018). https://doi.org/10.1016/j.ejor.2017.10.050

Silva, E.F., Toffolo, T.A.M., Wauters, T.: Exact methods for three-dimensional cutting and packing: a comparative study concerning single container problems. Comput. Oper. Res. **109**, 12–27 (2019). https://doi.org/10.1016/J.COR.2019.04.020

Tran-Dang, H., Krommenacker, N., Charpentier, P., Kim, D.S.: Toward the internet of things for physical internet: perspectives and challenges. IEEE Internet Things J. **7**(6), 4711–4736 (2020)

Zhao, X., Bennell, J.A., Bektaş, T., Dowsland, K.: A comparative review of 3D container loading algorithms. Int. Trans. Oper. Res. **23**(1–2), 287–320 (2016). https://doi.org/10.1111/ITOR.12094

Applying Constraint Programming to the Multi-mode Scheduling Problem in Harvest Logistics

Till Bender, David Wittwer(✉), and Thorsten Schmidt

TU Dresden, Professur für Technische Logistik, 01062 Dresden, Germany
david.wittwer@tu-dresden.de

Abstract. In this paper, we present a Constraint Programming (CP) based model for scheduling forage harvesters and transport vehicles during corn harvest. The key aspects are the synchronization of the two resource types and a forage harvester utilization depending on the number of transport vehicles supporting the harvester. The process is modelled as a pre-emptive multi-mode resource-constraint project scheduling problem with fast-tracking, sequence-dependent time lags and synchronization. We use the specialized scheduling features of CP Optimizer for modelling and solving the harvest logistics problem. The results show the suitability of the CP-based approach for modelling the problem in terms of representability of its characteristics. In computational experiments, a solution is found for any of the test instances. Proving optimality, however, is found to be difficult, especially for larger instances. Further variants of the model without pre-emption and fast-tracking and with fewer modes per activity are introduced and tested, showing improvements in computation time and the number of optimal solutions found for the prior variant.

Keywords: Harvest logistics · Constraint programming · Resource-constraint project scheduling

1 Introduction

This paper covers the application of constraint programming-based modelling to a multi-mode resource constraint scheduling problem (MRCPSP) stemming from the field of harvest logistics. It is inspired by the dispatching of forage harvesters and transport vehicles from the view of a contractor. Varying customer demands with high peaks, short harvest periods and strong weather dependency require efficient use of high-priced agricultural machinery.

In the harvest process, the grain is cropped by the forage harvester and transferred directly to the transport vehicle, driving in the field right next to the harvester. Once the capacity of the transport vehicle is reached, it transports the corn to a storage facility nearby. The number of transport vehicles required for a harvester to work at full utilization depends on the capacity of the transport

M. Mes et al. (Eds.): ICCL 2021, LNCS 13004, pp. 562–577, 2021.
https://doi.org/10.1007/978-3-030-87672-2_37

vehicles, the throughput of the harvester and the distance to the storage facility. If the number of transporters required for maximum utilization of the forage harvester is not reached, waiting times occur, which prolong the service time (time to harvest a single field). In this paper, instead of modelling the movement of transport vehicles between harvester/field and storage facility explicitly, we consider that different numbers of transport vehicles assigned to a field result in different service times.

To meet the practical requirements of the problem, focusing on forage harvester utilization, we model it as a pre-emptive MRCPSP with fast-tracking, sequence-dependent setup times and synchronization of two resource types. The MRCPSP is time-based and consists of activities performed by limited resources [15]. The duration of an activity depends on the mode in which it is performed. The mode of an activity is determined by the resources used. An activity represents either harvesting a field or supporting a harvester. In the presented problem, the mode corresponds to the number of transport vehicles supporting a forage harvester. The aim is to find a schedule for these activities such that the harvest is completed in the shortest time possible (makespan) with a given set of resources (forage harvesters and transport vehicles) and tasks (fields). Considering pre-emptive activities with fast-tracking refers to a relaxation of the classic resource-constraint project scheduling problem where activities are split into subactivities that can be executed independently (parallel, iteratively or overlapping) [16]. We use this variant to model several harvesters working independently from each other at the same field. In this model a harvester cannot return to a field visited before. Thus, in case of pre-emption, a different harvester will finish the harvest of that field. Since the harvest process requires both forage harvesters and transport vehicles, vehicles are synchronized to service a field simultaneously. The number of transport vehicles supporting a forage harvester determines the mode and thus the service time required to harvest a field. In addition, to account for transfer times between fields, we implement sequence-dependent time lags where the length of the time lags corresponds to the travel times between fields.

We formulate the problem described in this article as a constraint programming (CP) problem as the application of CP to scheduling problems has shown promising results in recent years. [18] and [12] applied CP and mixed integer programming (MIP) to scheduling problems of an operating theatre and a semiconductor fabrication plant, respectively. When minimizing the makespan, CP outperforms MIP in both studies regarding the objective function value, but fails to prove optimality in most cases. Laborie [9] proved the superiority of CP to various optimization techniques for a resource allocation problem. The literature presented shows that CP can be advantageous over MIP, but it does not demonstrate a general superiority of this method. Nevertheless, it encourages us to solve the problem at hand with CP. In this study we use the IBM ILOG CP Optimizer to address the scheduling problem of forage harvesters and transport vehicles.

Various aspects of agricultural machinery management and biomass logistics have already been studied. For an overview of these topics, we refer to Bochtis et al. [4] and Zhai et al. [20]. We briefly describe a few relevant examples of this vast field in the following. Basnet et al. [2] developed an integer programming model to schedule the harvest of rape seed at various farms. To solve greater instances a heuristic approach was developed. Guan et al. [7] use a two-phase meta-heuristic to calculate a schedule with a high resource utilization ratio in sugarcane production. El Hachemi et al. [6] apply CP to a synchronized log-truck scheduling problem, where the operations of both vehicles types are synchronized regarding temporal and spacial aspects. Bochtis et al. [3] formulate the problem of handling several sequential biomass operations on geographically dispersed fields as a flow shop problem with the aim of optimizing the makespan while incorporating predicted task times. Their approach was extended by Orfanou et al. [13], allowing multiple machinery types per operation type. Aguayo et al. [1] present a mixed integer programming model for scheduling the corn stover collection by two types of harvesters. A static and a dynamic variant, where demands are added over time, were developed. They greatly improve the objective value and decrease the optimality gap by passing a solution calculated by a simplified model. Lin et al. [11] present a mixed-integer linear programming problem for scheduling the distribution of perishable goods, considering real-time quality information. He et al. [8] present a MIP-Model and a heuristic approach, to schedule combine harvesters taking into account a measurement of fairness between combine harvester owners. Transport vehicles though are not considered.

Although a vast range of problems were introduced and various methods have been applied in the fields of agricultural machinery management and biomass logistics, CP has not been applied to solve the MRCPSP to dispatch forage harvesters and transport vehicles in the harvest process. We make the following contributions to address this gap:

1. We apply CP to solve a MRCPSP to minimize the harvest makespan of forage harvesters and transport vehicles with focus on the utilization rate of the forest harvesters.
2. We show the applicability of the model to problem sizes of practical relevance in computational experiments.
3. We present model variants and show the beneficial influence for greater instances on computation time and makespan of the variant without fast-tracking and pre-emption compared to the standard model and the variant with only two available modes.

The remainder of this paper is structured as follows. We present a CP model of the harvest process by forage harvesters and transport vehicles as MRCPSP in Sect. 2. Section 3 contains computational experiments and discussions regarding the model introduced in this paper and its variants. Finally, we summarize the presented work and give an outlook on future work in Sect. 4.

2 Problem

In this section, we first present the assumptions made in modeling the harvest logistics problem. Then a CP model is proposed and its constraints are explained in detail.

2.1 Assumptions

The harvest logistics described in Sect. 1 is modelled as a pre-emptive MRCPSP with fast-tracking and synchronization. The dimension used is time in minutes. To account for travel times between fields, we introduce sequence-dependent time lags. We make the following assumptions on the operations of forage harvesters and transport vehicles:

- Each vehicle starts and ends at the depot.
- Harvesting a field requires one harvester and at least one transport vehicle.
- The harvesters are heterogeneous in terms of throughput of corn harvested per time unit while the fleet of transport vehicles is homogeneous.
- The amount of transport vehicles supporting a forage harvester corresponds to the mode a field is serviced in. Additional transport vehicles decrease the service time at a field due to a higher utilization of the forage harvester, until a harvester- and field-specific maximum of transport vehicles is reached. The number of transport vehicles at a field and its service time are inversely proportional.
- Multiple forage harvesters can service a field independently (simultaneously, iteratively or overlapping).
- A harvester can visit a field not more than once. Thereby we reduce the number of variables and exclude solutions that are very unlikely to be optimal.
- The minimum service time at a field is 10 min to meet the requirements of a practical application.

Figure 1 illustrates an optimal solution to the model for an examplary instance. It shows that minimizing the makespan requires the harvest of field 1 to be split into two parts. While harvester k_2 harvests field 1 at the beginning of the schedule and leaves before it is finished, harvester k_1 finishes harvesting the field at the end of the schedule. Furthermore, transport vehicle l_1 switches from harvester k_2 to harvester k_1 when traversing from field 3 to field 4. Gaining an additional transport vehicle enables harvester k_1 to harvest fields 4 and 1 in a higher mode and thus in less time.

2.2 Model

To model the problem, the scheduling concepts of CP Optimizer are used, including interval and sequence variables and associated constraints.[1]

[1] For further information about the concepts available in CP Optimizer see [10].

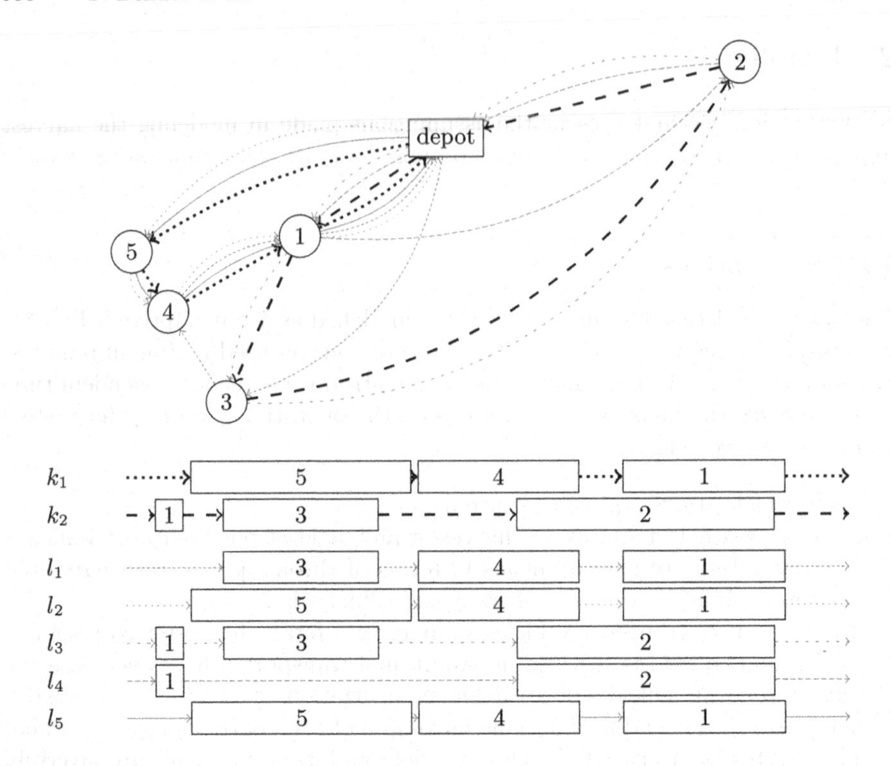

Fig. 1. Representation of an optimal solution of an instance with five fields, two harvesters (k_1, k_2) and five transport vehicles (l_1, l_2, l_3, l_4, l_5) as graph and Gantt-Chart

Activities are defined as a set $N := \{1, \ldots, f\}$ with f being the number of customer fields. The set $V = \{0\} \cup N \cup \{f + 1\}$ represents the customer fields including the start and end of each vehicle's trip, the depot. Splitting the depot into a start and end point facilitates modelling and is a common technique when modelling scheduling problems. K is the set of forage harvesters, L is the set of transport vehicles.

Activities are modelled as optional interval variables. An interval variable in CP Optimizer has a start time value (*start*), as well as an end time value (*end*) with a $length = end - start$. An interval variable can be optional to model a task or an activity which is not necessarily executed [10]. This is useful because a vehicle might only visit some of the fields. The decision variables used in the model can be found in Table 1. A minimal duration of 10 min for each interval except for the ones representing the start and end of the schedule is defined to prevent short impracticable stops at a field.

The number of transport vehicles necessary for harvester k to work at its optimal utilization on field i is given by m_{ik}^{opt}. In this paper we assume that the relation between service time and the number of transport vehicles supporting a harvester is inversely proportional.

Table 1. Decision variables used in the CP model

Variable	Variable type	Explanation
m_{ik}	Integer	Mode that harvester k works in at field i (this corresponds to the number of transport vehicles supporting a harvester)
ha_{ik}	Interval	Duration for which harvester k services field i. The intervals for the depot are dummy variables with a duration of 0 defined as $ha_{0,k}$ and $ha_{f+1,k}$ $\forall k \in K$
tva_{ikl}	Interval	Duration for which transport vehicle l supports harvester k at field i. The intervals for the depot are dummy variables with a duration of 0 and $k = 0$ defined as $tva_{0,0,l}$ and $tva_{if+1,0,l}$ $\forall l \in L$
hr_k	Sequence	Contains the interval variables of harvester k. It contains the order of the interval variables i.e. the route of a harvester
tvr_l	Sequence	Contains the interval variables of transport vehicle l. It contains the order of the interval variables i.e. the route of a transport vehicle

The interval variables for each vehicle are grouped as a sequence variable. A sequence variable contains interval variables. The solution to a sequence variable contains the order of the associated interval variables. It therefore represents the route that a vehicle takes during the schedule. A transition matrix can be assigned to sequence variables, defining time lags between each interval in the sequence. Transition matrix T^h contains the transition times $t^h_{i,j}$ defining the time that harvesters need to travel from field i to field j (or depot). The transition matrix T^{tv} for transport vehicles is composed similarly.

The objective of the model is to minimize the duration of the harvest process. This results in the following objective function:

$$minimize\big(\max(endOf(ha_{f+1,k})\ \forall k \in K)\big) \tag{1}$$

The objective function (1) minimizes the maximum of the end of the interval variables $ha_{f+1,k}$ for every harvester. The interval variable $ha_{f+1,k}$ represents the last activity of a schedule, the return to the depot. Thus, the arrival time at the depot of the latest forage harvester, i.e. the makespan, is minimized. The minimization of the makespan ensures an even distribution of workload between the harvesters.

Subject to the assumptions stated above, the constraints are as follows:

$$noOverlap(hr_k, T^h) \quad \forall k \in K \tag{2}$$

$$noOverlap(tvr_l, T^{tv}) \quad \forall l \in L \tag{3}$$

$$first(hr_k, ha_{0,k}) \quad \forall k \in K \tag{4}$$

$$first(tvr_l, tva_{0,0,l}) \quad \forall l \in L \tag{5}$$

$$last(hr_k, ha_{f+1,k}) \quad \forall k \in K \tag{6}$$

$$last(tvr_l, tva_{f+1,0,l}) \quad \forall l \in L \tag{7}$$

$$endOf(tva_{f+1,0,l}) \leq \max(endOf(ha_{f+1,k,0}) \, \forall k \in K) \quad \forall l \in L \tag{8}$$

$$alternative(ha_{ik}, (tva_{ikl} \, \forall l \in L), m_{ik}) \quad \forall i \in V, \forall k \in K \tag{9}$$

$$\sum_{k \in K} size(ha_{ik}) \cdot tp_k \cdot \frac{m_{ik}}{m_{ik}^{opt}} \geq yield_i \quad \forall i \in N \tag{10}$$

$$\sum_{k \in K} presenceOf(ha_{ik}) \geq 1 \quad \forall i \in N \tag{11}$$

Constraints (2) and (3) use the *noOverlap()* constraint featured in CP Optimizer. It ensures that the activities of a route of a vehicle cannot overlap. The constraints take the transition matrices T^h and T^{tv} as an argument, declaring that the time between two intervals is greater than or equals time given by the respective time in the transition matrix.

The depot is always the start of a schedule. The constraints (4) and (5) enforce the intervals $ha_{0,k}$ and $tva_{0,0,l}$ to be ordered as the first interval in the respective sequences hr_k and tvr_l. While the constraints do not affect the start or end time of the intervals, they ensure that no interval can be sequenced before the start intervals. The end of a sequence at the depot is modeled similarly by constraints (6) and (7).

In the objective function, the end of the harvesters' sequences is minimized. The end of the transport vehicles' sequences is not restricted. To limit the search space, constraints (8) are introduced. They define that a transport vehicle's schedule ends before or with the maximum of the end of the harvesters' sequences. This constraint is modelled under the consideration that the driving speed of the harvesters is less than or equal to the driving speed of the transport vehicles. In practice, this is usually the case.

Synchronization of the two vehicle types is one of the key aspects of the model. The constraints (9) ensure that for every present harvester activity ha_{ik} exactly m_{ik} transporter activities from the tuple $(tva_{ikl} \, \forall l \in L)$ are present. The constraints also ensure that the activities of the harvester and the transport vehicles start and end at the same time. The number of transport vehicles to be synchronized corresponds to the mode that the harvester is working in.

The amount of biomass harvested from a field i is called $yield_i$ and mainly depends on the field size. To schedule the time of an activity at a field depending on the yield and to guarantee that all fields are harvested, the constraints (10) and (11) are introduced. Constraints (10) ensure that the harvested amount of corn of each field corresponds to the yield of the field. The constant tp_k represents the maximum throughput of a harvester when harvesting a field. Depending on the mode that the harvester is working in, the throughput is reduced by factor $\frac{m_{ik}}{m_{ik}^{opt}}$. The product of the interval's size and throughput can be larger than the yield of the field, due to both factors being discrete. This might lead to slightly longer durations of activities than necessary but can be neglected due to the length of the intervals being modeled in minutes. Constraints (11) ensure that each field is harvested by at least one harvester. Changing it to an equality disables pre-emption and fast-tracking of activities.

3 Computational Experiments

In this section, the general setting of the studies is presented and major findings are outlined. The complete results of the computational experiments are summarized in Appendix A.

3.1 Setting and Test Instances

The model is tested on instances[2] defined by the number of fields, the number of harvesters and the number of transport vehicles. We generate instances for 4, 6, 8, 10, 12, 14 and 16 fields. Instances with greater field numbers include the same fields as smaller instances. For each number of fields we vary the number of harvesters and transport vehicles as presented in Table 2.

Table 2. Transport vehicle number per harvester

Harvesters	2				3				4			
Transp. Veh.	4	5	6	7	6	7	8	9	8	9	10	11

The field locations are randomly distributed on a 50 $\times$ 50 km plane. The transition times of harvesters and transport vehicles between the fields correspond to the euclidean distances between the fields and a constant velocity for both vehicle types of 30 km h^{-1}. Field sizes range from 1 to 8 ha according to field sizes common in Bavaria, Germany [19]. The yield of a field is calculated using its size and an average yield of 38.44 t ha^{-1} [14]. The throughput of the harvesters ranges between $0.5 \cdot tp$ and $1.5 \cdot tp$ with $tp = 81\,\text{t h}^{-1}$ [5].

All instances are solved with a time limit of 1200 s. The time limit of 20 min is chosen to represent a practical case where results are needed within a short time due to the necessity to adapt to changes at short notice. If the time limit is reached, the gap as relative difference between upper bound (best solution found) and lower bound (computed e.g. by linear relaxation of the model [17]) is returned. The tests are conducted on a system with an AMD Ryzen 7 4750u processor with 8 Cores, a base clock of 1.7 GHz and 16 GB of RAM. We use IBM ILOG CP Optimizer Version 12.10 via its Python API and its default configuration. The number of threads used in CP Optimizer is set to 8.

[2] The instances are available for download at https://tud.link/47mz.

3.2 Influence of the Number of Forage Harvesters and Transport Vehicles on the Makespan

The runtime to optimality, the gap and the resulting makespan were monitored during the tests. A solution was found to any of the 84 test instances of the standard variant, while 14 of the solutions were proved to be optimal. An initial solution was found quickly, even for large instances. It took 32.27 s on average to find a solution for instances with 16 fields with a gap of 96.96% on average.

Table 3. Average makespan, gap and runtime for a given number of fields

#Fields	Makespan	Gap [%]	Runtime [s]	#Optimality
4	326.3	2.9	259.0	11/12
6	481.9	21.2	976.7	3/12
8	537.9	45.6	1200.0	0/12
10	609.3	52.7	1200.0	0/12
12	684.4	61.6	1200.0	0/12
14	808.7	69.2	1200.0	0/12
16	1042.3	78.0	1200.0	0/12

Table 3 shows the average of makespan, gap and runtime for all instances with the same number of fields. Although a feasible solution is usually found in

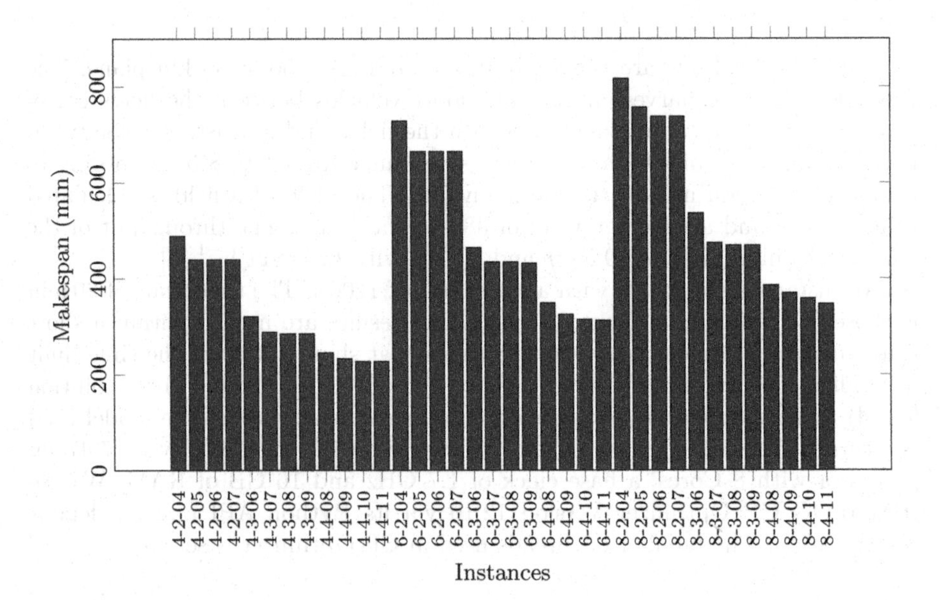

Fig. 2. Calculated makespan of instances with 4, 6 and 8 fields for different numbers of forage harvesters and transport vehicles

a short time, the computation time and gap increase rapidly with an increasing number of fields. No instance with 8 or more nodes could be solved optimally within the given time limit.

The influence of additional forage harvesters and transport vehicles on the makespan for instances with 4, 6 and 8 fields is presented in Fig. 2. Here, 4-2-05 describes an instance of four fields, two forage harvesters and five transport vehicles. The figure illustrates that additional transport vehicles shorten the makespan until the maximum utilization of the harvesters is reached. At this point, an additional transport vehicle cannot shorten the makespan any further. Similarly, the effect of an additional forage harvester on the makespan is affected by the number of transport vehicles. When the number of transport vehicles is larger, the benefit of an additional forage harvester is greater than when the number of transport vehicles is smaller. In general, it can be concluded that the reduction of makespan ist greater, when relatively scarce resources are added. This pattern can be observed for optimal and non-optimal solutions with up to 12 fields. For instances with 14 and more fields, deviations occur because the best solutions found within the time limit of some instances deviate strongly from the optimal solution.

3.3 Analysis of Model Variants

Even though for the objective of minimizing the makespan CP usually manages to find solutions faster than MIP, it often fails to prove optimality, terminating with a great gap. Thus, we introduce two variants in addition to the standard variant to address this issue. In the analysis of the variants, the changes in makespan and computation time are investigated. In the first variant, we reduce the amount of modes to two, only allowing forage harvesters to work in both of their most efficient modes (*dual mode*). By this, the minimum number of transport vehicles required to harvest a field is increased. In the second variant activity pre-emption and fast-tracking are removed, so each field is visited once by exactly one harvester. As this prevents harvest activities on the fields to be split, this will be called the variant with *no split* for the remainder of the paper. The number of optimal solutions and the average makespan, gap and computation time are summarized in Table 4. The dual mode variant shows similar results as the original model. Of the 84 instances, 13 were solved to optimality. On average, the makespan is slightly shorter in the dual mode variant compared to the standard variant. In terms of gap and runtime however, the standard variant shows slightly better results. The small differences in computation time suggest that a high mode was already frequently selected in the original model and lower modes were very quickly excluded by the solution algorithm.

In the tests of the variant without splits 56 of the 84 instances could be solved optimally. Since in this variant the service times are only determined by the number of transporters and can no longer be split arbitrarily between different forage harvesters, the solution space is greatly reduced. The standard model is a relaxation of the model without splits. Thus, the optimal solution of the standard model is at least as good as the optimal solution of the variant without splits. However, the solution space is so much larger that the simplified variant gives better results for 35 of the 36 instances with 12 or more fields.

Table 4. Comparison of the model variants regarding number of instances solved optimally, average makespan, gap and computation time

Variant	#Optimality	Average of		
		Makespan	Gap [%]	Runtime [s]
Standard	14/84	641.6	47.3	1033.7
Dual mode	13/84	639.1	47.9	1037.3
No split	56/84	622.3	19.9	444.8

Figure 3 shows the relative difference in makespan of the two additional variants compared to the standard variant per field and harvester combination. While the standard and dual mode variant have very similar results on average, the makespan of the model without splits is longer for smaller instances. On the contrary, for instances with more fields this variant results in shorter makespans. For larger instances, not using splits leads to better results in terms of makespan. Some smaller, non-optimal solutions have a shorter makespan in the standard variant compared to optimal solutions in the variant without splits, This could indicate that the solution is the optimal solution or close to it.

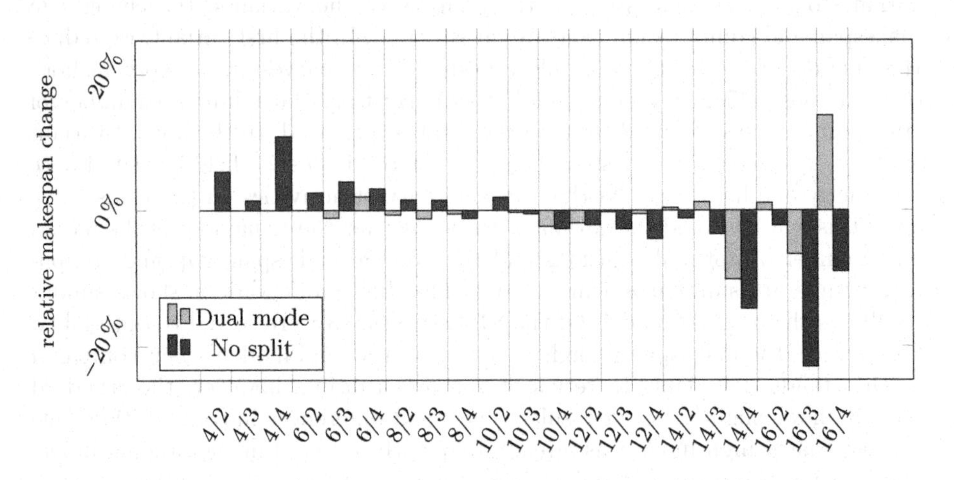

Fig. 3. Relative change in average makespan for the additional variants compared to the standard variant

4 Conclusion

In this paper, we present a CP formulation of the pre-emptive MRCPSP with fast-tracking, sequence-dependent time lags and synchronization in the field of harvest logistics. The focus lies on the utilization rate of forage harvesters influenced by the number of transport vehicles supporting it. Instances with various numbers of vehicles are examined to investigate the effect of additional forage harvesters or transport vehicles. We show that even for larger instances, good solutions could be found with CP in reasonable computation times. However, for most instances optimality could not be proved, terminating with a great gap. Therefore, two variants were introduced with the goal to shrink the solution space and obtain shorter computation times.

In the first variant, we increase the minimum number of transport vehicles needed to process a field. This reduces the number of modes in which a field can be processed by a forage harvester. However, our computational experiments show that reducing the modes to this extent does not significantly affect the computation time. In the second variant, in which pre-emption and fast-tracking are omitted, i.e. each field is visited once by exactly one forage harvester, a much larger part of the instances is solved optimally. Future work could include further simplified variants of the models presented in this paper to generate a starting point for instances with higher field numbers.

Experiments with greater computation time lead to slightly shorter makespan. But as short runtimes are a requirement to enable quick replanning due to uncertainties in terms of weather conditions or machine failure, the use of a computer with higher computing power is recommended. In additional experiments a simple symmetry breaking constraint (assignment of the furthest field to one of the transport vehicles) helped to find slightly better solutions in the given maximum runtime. Therefore, further research on more sophisticated symmetry breaking constraints is encouraged.

For practical application, additional constraints that take into account, e.g., the place and time for a lunch break or a preference in the order of fields to be harvested, should be considered.

Acknowledgments. The work was carried out as a part of the IGF research project 19509 BR "Landwirtschaftslogistik" (Logistikdienstleistungen in der Landwirtschaft - Aufträge sicherer und ressourcenoptimal planen) and supported by "Bundesvereinigung Logistik (BVL) e.V.".

A Appendix

(See Table 5).

Table 5. Overview of the objective value (OV), computation time (Time), and gap at termination (Gap) for a given number of fields $|N|$, number of forage harvesters $|K|$ and number of transport vehicles $|L|$.

$\|N\|$	$\|K\|$	$\|L\|$	Standard			Dual mode			No split		
			OV	Time [s]	Gap [%]	OV	Time [s]	Gap [%]	OV	Time [s]	Gap [%]
4	2	4	490	11.4	0.0	490	0.3	0.0	490	15.9	0.0
		5	440	1.7	0.0	474	0.2	0.0	440	1.3	0.0
		6	440	1.6	0.0	474	0.2	0.0	440	1.2	0.0
		7	440	1.7	0.0	474	0.2	0.0	440	1.7	0.0
	3	6	321	774.6	0.0	321	0.7	0.0	321	727.1	0.0
		7	288	18.1	0.0	305	0.3	0.0	288	19.6	0.0
		8	285	2.9	0.0	285	0.1	0.0	285	3.0	0.0
		9	285	2.5	0.0	285	0.2	0.0	285	3.4	0.0
	4	8	245	1200.0	34.7	269	2.1	0.0	245	1200.0	34.7
		9	233	891.7	0.0	253	0.2	0.0	233	1200.0	29.6
		10	227	191.2	0.0	253	0.2	0.0	227	195.4	0.0
		11	222	11.0	0.0	253	0.3	0.0	222	30.1	0.0
6	2	4	728	1200.0	41.8	753	1.3	0.0	728	1200.0	37.8
		5	664	287.9	0.0	679	0.7	0.0	664	296.5	0.0
		6	664	289.9	0.0	679	0.7	0.0	664	298.0	0.0
		7	664	342.2	0.0	679	0.8	0.0	664	339.0	0.0
	3	6	464	1200.0	35.1	478	6.0	0.0	458	1200.0	35.8
		7	434	1200.0	23.3	453	2.1	0.0	431	1200.0	17.6
		8	434	1200.0	21.4	453	2.2	0.0	427	1200.0	18.3
		9	431	1200.0	22.3	453	2.5	0.0	427	1200.0	21.6
	4	8	349	1200.0	49.3	351	7.9	0.0	352	1200.0	49.7
		9	325	1200.0	24.3	340	6.5	0.0	333	1200.0	37.8
		10	314	1200.0	17.5	325	0.6	0.0	329	1200.0	31.9
		11	312	1200.0	19.2	325	0.5	0.0	312	1200.0	18.6
8	2	4	817	1200.0	52.8	832	15.4	0.0	817	1200.0	40.3
		5	757	1200.0	50.6	754	1.9	0.0	737	1200.0	44.4
		6	737	1200.0	40.0	754	1.8	0.0	737	1200.0	30.4
		7	737	1200.0	34.7	754	5.1	0.0	737	1200.0	36.6
	3	6	536	1200.0	51.9	510	48.3	0.0	512	1200.0	49.4
		7	474	120000	44.5	489	2.7	0.0	474	1200.0	44.3
		8	469	1200.0	37.1	489	2.7	0.0	469	1200.0	39.9
		9	469	1200.0	30.1	489	5.6	0.0	469	1200.0	43.3
	4	8	385	1200.0	54.0	390	1200.0	54.6	385	1200.0	54.0
		9	369	1200.0	52.0	366	189.1	0.0	367	1200.0	51.8
		10	358	1200.0	50.6	343	11.9	0.0	340	1200.0	47.9
		11	347	1200.0	49.0	343	3.8	0.0	359	1200.0	50.7
10	2	4	939	1200.0	52.7	939	384.4	0.0	939	1200.0	57.4
		5	835	1200.0	42.4	856	3.9	0.0	834	1200.0	53.5
		6	834	1200.0	41.7	856	3.0	0.0	835	1200.0	46.4
		7	834	1200.0	41.9	856	3.1	0.0	834	1200.0	44.8
	3	6	593	1200.0	63.7	591	1200.0	60.9	593	1200.0	63.7
		7	550	1200.0	60.9	538	41.7	0.0	535	1200.0	59.8
		8	531	1200.0	50.1	533	1200.0	11.6	544	1200.0	52.0
		9	533	1200.0	49.7	533	7.5	0.0	528	1200.0	48.3
	4	8	442	1200.0	60.0	447	1200.0	60.4	437	1200.0	59.5
		9	425	1200.0	58.4	410	1200.0	56.8	417	1200.0	57.6
		10	401	1200.0	55.9	384	1200.0	53.9	396	1200.0	55.3
		11	395	1200.0	55.2	377	352.7	0.0	374	1200.0	52.7

(continued)

Table 5. (*continued*)

$\|N\|$	$\|K\|$	$\|L\|$	Standard			Dual mode			No split		
			OV	Time [s]	Gap [%]	OV	Time [s]	Gap [%]	OV	Time [s]	Gap [%]
12	2	4	1046	1200.0	69.3	1045	1200.0	35.3	1056	1200.0	70.1
		5	960	1200.0	52.9	923	4.0	0.0	923	1200.0	61.5
		6	946	1200.0	53.8	923	4.2	0.0	923	1200.0	51.3
		7	945	1200.0	56.4	923	6.4	0.0	923	1200.0	56.5
	3	6	677	1200.0	68.2	645	1200.0	65.4	672	1200.0	68.0
		7	601	1200.0	64.2	593	1200.0	54.5	601	1200.0	64.2
		8	601	1200.0	64.2	577	18.0	0.0	604	1200.0	64.4
		9	581	1200.0	63.0	577	15.9	0.0	577	1200.0	62.7
	4	8	512	1200.0	65.4	490	1200.0	63.9	520	1200.0	66.0
		9	474	1200.0	62.7	449	1200.0	60.6	472	1200.0	62.5
		10	453	1200.0	60.9	426	1200.0	58.5	434	1200.0	59.2
		11	417	1200.0	57.6	413	197.7	0.0	420	1200.0	57.9
14	2	4	1219	1200.0	72.0	1217	1200.0	46.8	1209	1200.0	73.1
		5	1073	1200.0	64.8	1069	37.0	0.0	1069	1200.0	64.0
		6	1093	1200.0	63.7	1069	137.2	0.0	1122	1200.0	68.1
		7	1093	1200.0	70.3	1069	194.1	0.0	1093	1200.0	70.3
	3	6	757	1200.0	71.6	767	1200.0	70.9	819	1200.0	73.8
		7	699	1200.0	69.2	676	1200.0	65.5	702	1200.0	69.4
		8	669	1200.0	67.9	657	228.9	0.0	702	1200.0	69.4
		9	732	1200.0	70.6	657	195.8	0.0	668	1200.0	67.8
	4	8	637	1200.0	72.2	544	1200.0	67.5	593	1200.0	70.2
		9	621	1200.0	71.5	525	1200.0	66.3	551	1200.0	67.9
		10	534	1200.0	66.9	491	1200.0	64.0	500	1200.0	64.6
		11	577	1200.0	69.3	466	285.8	0.0	484	1200.0	63.4
16	2	4	1546	1200.0	78.9	1490	1200.0	59.6	1646	1200.0	80.1
		5	1320	1200.0	74.4	1289	1200.0	27.4	1315	1200.0	73.6
		6	1272	1200.0	72.7	1259	607.9	0.0	1271	1200.0	74.3
		7	1280	1200.0	74.0	1259	709.5	0.0	1243	1200.0	66.2
	3	6	1434	1200.0	95.8	949	1200.0	76.5	1038	1200.0	79.0
		7	910	1200.0	76.0	860	1200.0	74.1	901	1200.0	75.8
		8	865	1200.0	74.8	775	1200.0	67.4	945	1200.0	76.9
		9	1121	1200.0	94.6	749	1200.0	63.4	1168	1200.0	94.8
	4	8	780	1200.0	77.3	685	1200.0	74.2	1092	1200.0	96.2
		9	793	1200.0	77.7	647	1200.0	72.6	695	1200.0	74.5
		10	603	1200.0	70.7	602	1200.0	70.6	734	1200.0	75.9
		11	583	1200.0	69.6	577	1200.0	69.3	623	1200.0	71.6

References

1. Aguayo, M.M., Sarin, S.C., Cundiff, J.S., Comer, K., Clark, T.: A corn-stover harvest scheduling problem arising in cellulosic ethanol production. Biomass Bioenergy **107**, 102–112 (2017). https://doi.org/10.1016/j.biombioe.2017.09.013
2. Basnet, C.B., Foulds, L.R., Wilson, J.M.: Scheduling contractors' farm-to-farm crop harvesting operations. Int. Trans. Oper. Res. **13**(1), 1–15 (2006). https://doi.org/10.1111/j.1475-3995.2006.00530.x
3. Bochtis, D., Dogoulis, P., Busato, P., Sørensen, C., Berruto, R., Gemtos, T.: A flow-shop problem formulation of biomass handling operations scheduling. Comput. Electron. Agric. **91**, 49–56 (2013). https://doi.org/10.1016/j.compag.2012.11.015

4. Bochtis, D.D., Sørensen, C.G., Busato, P.: Advances in agricultural machinery management: a review. Biosyst. Eng. **126**, 69–81 (2014). https://doi.org/10.1016/j.biosystemseng.2014.07.012
5. Döring, G., Schilcher, A., Strobl, M., Schleicher, R., Seidl, M., Mitterleitner, J.: Verfahren zum Transport von Biomasse. Tech. rep, Biogas Forum Bayern, Freising (2010)
6. El Hachemi, N., Gendreau, M., Rousseau, L.M.: A hybrid constraint programming approach to the log-truck scheduling problem. Ann. Oper. Res. **184**(1), 163–178 (2011). https://doi.org/10.1007/s10479-010-0698-x
7. Guan, S., Nakamura, M., Shikanai, T., Okazaki, T.: Resource assignment and scheduling based on a two-phase metaheuristic for cropping system. Comput. Electron. Agric. **66**(2), 181–190 (2009). https://doi.org/10.1016/j.compag.2009.01.011
8. He, P., Li, J., Wang, X.: Wheat harvest schedule model for agricultural machinery cooperatives considering fragmental farmlands. Comput. Electron. Agric. **145**, 226–234 (2018). https://doi.org/10.1016/j.compag.2017.12.042
9. Laborie, P.: An update on the comparison of MIP, CP and hybrid approaches for mixed resource allocation and scheduling. In: van Hoeve, W.J. (ed.) Integration of Constraint Programming, Artificial Intelligence, and Operations Research, vol. 10848, pp. 403–411. Springer, Heidelberg (2018). https://doi.org/10.1007/978-3-319-93031-2_29
10. Laborie, P., Rogerie, J., Shaw, P., Vilím, P.: IBM ILOG CP optimizer for scheduling. Constraints **23**(2), 210–250 (2018). https://doi.org/10.1007/s10601-018-9281-x
11. Lin, X., Negenborn, R.R., Duinkerken, M.B., Lodewijks, G.: Quality-aware modeling and optimal scheduling for perishable good distribution networks: the case of banana logistics. In: Bekta?, T., Coniglio, S., Martinez-Sykora, A., Voß, S. (eds.) Computational Logistics, vol. 10572, pp. 483–497. Springer, Heidelberg (2017). https://doi.org/10.1007/978-3-319-68496-3_32
12. Maleck, C., Nieke, G., Bock, K., Pabst, D., Stehli, M.: A comparison of an CP and MIP approach for scheduling jobs in product areas with time constraints and uncertainties. In: 2018 Winter Simulation Conference (WSC), pp. 3526–3537. IEEE (2018). https://doi.org/10.1109/WSC.2018.8632404
13. Orfanou, A., et al.: Scheduling for machinery fleets in biomass multiple-field operations. Comput. Electron. Agric. **94**, 12–19 (2013). https://doi.org/10.1016/j.compag.2013.03.002
14. Statistisches Bundesamt: Anbauflächen, Hektarerträge und Erntemengen ausgewählter Anbaukulturen im Zeitvergleich. https://www.destatis.de/DE/Themen/Branchen-Unternehmen/Landwirtschaft-Forstwirtschaft-Fischerei/Feldfruechte-Gruenland/Tabellen/liste-feldfruechte-zeitreihe.html
15. Talbot, F.B.: Resource-constrained project scheduling with time-resource tradeoffs: the nonpreemptive case. Manag. Sci. **28**(10), 1197–1210 (1982). https://doi.org/10.1287/mnsc.28.10.1197
16. Vanhoucke, M., Debels, D.: The impact of various activity assumptions on the lead time and resource utilization of resource-constrained projects. Comput. Ind. Eng **54**(1), 140–154 (2008). https://doi.org/10.1016/j.cie.2007.07.001
17. Vilím, P.: Re: solucion gap (2020). https://community.ibm.com/community/user/datascience/communities/community-home/digestviewer/viewthread?MessageKey=7fba1844-472c-4704-9500-f60737306084&CommunityKey=ab7de0fd-6f43-47a9-8261-33578a231bb7&tab=digestviewer#bm7fba1844-472c-4704-9500-f60737306084

18. Wang, T., Meskens, N., Duvivier, D.: Scheduling operating theatres: mixed integer programming vs. constraint programming. Eur. J. Oper. Res. **247**(2), 401–413 (2015). https://doi.org/10.1016/j.ejor.2015.06.008
19. Zenger, X., Friebe, R.: Agrarstrukturentwicklung in Bayern, IBA - Agrarstrukturbericht 2014. Tech. rep, Bayerische Landesanstalt für Landwirtschaft (LfL), Freising-Weihenstephan (2015)
20. Zhai, Z., Martínez, F.J., Beltran, V., Martínez, N.L.: Decision support systems for agriculture 4.0: survey and challenges. Comput. Electron. Agric. **170**, 105256 (2020). https://doi.org/10.1016/j.compag.2020.105256

Tackling Uncertainty in Online Multimodal Transportation Planning Using Deep Reinforcement Learning

Amirreza Farahani(✉), Laura Genga, and Remco Dijkman

School of Industrial Engineering, Eindhoven University of Technology, Eindhoven 5612, AZ, Netherlands
{A.Farahani,L.Genga,R.M.Dijkman}@tue.nl

Abstract. In this paper we tackle the container allocation problem in multimodal transportation planning under uncertainty in container arrival times, using Deep Reinforcement Learning. The proposed approach can take real-time decisions on allocating individual containers to a truck or to trains, while a transportation plan is being executed. We evaluated our method using data that reflect a realistic scenario, designed on the basis of a case study at a logistics company with three different uncertainty levels based on the probability of delays in container arrivals. The experiments show that Deep Reinforcement Learning methods outperform heuristics, a stochastic programming method, and methods that use periodic re-planning, in terms of total transportation costs at all levels of uncertainty, obtaining an average cost difference with the optimal solution within 0.37% and 0.63%.

Keywords: Optimization · Deep Reinforcement Learning · Online planning under uncertainty · Multimodal transport

1 Introduction

This paper introduces an online planning algorithm based on Deep Reinforcement Learning (DRL) in presence of uncertainty in container arrival times that we developed for a transportation company for their container allocation decision support system in the multimodal transportation planning domain. The problem can be formulated as a sequential decision-making problem (i.e., allocation decisions are taken for each individual container sequentially) for a particular transportation corridor between two locations (i.e., the set of available transportation options between two specific locations). Given a set of containers, each with its arrival time and due date, and a set of available vehicle options, each with its transportation costs, and arrival and departure time, the goal is to allocate each container to one of the available options, in such a way that

The work leading up to this paper is partly funded by the European Commission under the FENIX project (grant nr. INEA/CEF/TRAN/M2018/1793401).

M. Mes et al. (Eds.): ICCL 2021, LNCS 13004, pp. 578–593, 2021.
https://doi.org/10.1007/978-3-030-87672-2_38

the total cost is minimized. As trains have a lower cost than trucks, solving the planning problem corresponds to allocating as many containers to trains as possible. One of the crucial challenges within this context consists in dealing with *unexpected* events, i.e., events that can hinder the feasibility of the container allocation plan. These events can be classified in two groups: (1) 'Dynamicity', i.e., changes in demand to the logistics company, which is not predictable, and (2) 'Uncertainty', i.e., delays in containers availability for departure, due to, for instance, delays in container arrival time, which make the transportation environment non-deterministic [14].

Traditional offline planning methods are not suitable for dealing with these unexpected events. These methods allocate a batch of containers to available vehicles in one go. However, if an unexpected event happens after the allocation plan has been decided, it might make the plan not feasible for the new situation. In this case, the planner needs to re-plan only a single container or the few containers that are affected by such event. We refer to this as online (re)planning. Currently, there is little support for online planning. It is usually carried out by means of heuristics (e.g., re-optimization approaches and combinations of offline methods and greedy algorithms) whose outputs may be far from the optimal.

For online (re)planning in deterministic environments with dynamicity, our previous work [8] introduced a DRL-based online planning algorithm for multimodal transportation, which is able to learn rules to allocate individual containers to available vehicles while the plan is being executed. In this paper, we extend our previous approach, by introducing a novel DRL-based online planner able to deal with uncertainty due to container delays.

We tested our method using data representing a realistic scenario, designed on the basis of a real case study at a transportation company. Our results demonstrate that our algorithm can learn rules to effectively allocate containers to trains and trucks in the presence of uncertainty. Furthermore, it outperformed the tested competitors in terms of total transportation costs, generating a solution close to the theoretical optimal one.

The rest of this paper is organized as follows. Section 2 introduces relevant related literature. Section 3 provides a formal definition of the problem under investigation. Section 4 introduces our approach. Section 5 discusses our experimental evaluation. Section 6 draws conclusions and delineates future work.

2 Related Work

Previous work dealing with uncertainty in multimodal transportation usually rely on some a priori assumptions on probability distributions for the shipments, orders, and travel times, then integrating the unexpected events via offline model-driven methods such as stochastic programming or robust optimization [5,18]. To the best of our knowledge, there are only few exceptions, reported in Table 1. We categorize these online approaches based on modality, uncertainty measure, proposed method and problem type. Most of these

Table 1. Online methods in multimodal transportation (re)planning under uncertainty

Reference	Modality	Uncertainty	Problem	Method
[20]	Rail	Demand, Travel Time	Empty vehicle re-positioning	Approximate DP
[6]	Unspecified	Transit time	Re-routing	Genetic Algorithm
[21]	Rail, road, barge	Demand	Container re-allocation	LP-Decision Tree
[15]	Road, barge	Demand	Freight selection	Approximate DP

Table 2. Application of DRL in decision making under uncertainty

Reference	Domain	Uncertainty	RL Method
[4]	Cloud computing scheduling	Task, resource	Value based
[11]	Water management	Geological	Value based, Policy based
[13]	Supply chain	Demand	Policy based
[23]	ITS energy management	Environment	Policy based
[9]	EV Battery swapping	Electricity prices	Policy based
[7]	Internet of energy	Trajectory	Value based
[16]	Self-driving car	Driving scenario	Value based
[1]	Wind power producing	Wind generation, electricity price	Policy based
[24]	Wind power producing	Wind generation, electricity price	Value based
[22]	Unmanned aerial vehicle	Movement	Value based, Policy based
[25]	Pavement systems life-cycle	Traffic, cost, price indices, etc.	Value based
[17]	Operational management	Environment	Value based

approaches focus on demand uncertainty, while no approaches have been proposed to deal with delays in containers availability. This paper aims to fill that gap.

Reinforcement Learning applications have been studied for optimizing sequential decision-making problems under uncertainty in various domains. Table 2 categorizes previous approaches with respect to: (1) domain, (2) Uncertainty aspects, and (3) RL method. Current works use two main groups of RL methods: Value-based Reinforcement Learning methods, such as Deep Q-network (DQN), double DQN (DDQN), and dueling DDQN [4,7,16,24,25] and Policy-based approaches such as Deep Deterministic Policy Gradient (DDPG), and Proximal Policy Optimization (PPO) [1,9,13,23]. In our work, we also apply a value-based DRL method, which is mostly used in literature [4,7,16,24,25]. However, we apply the method in a new domain,i.e., to sequential container planning under uncertainty.

3 Problem Definition

In this section, first we define the problem in an offline setting using Integer Linear Programming (ILP) and Stochastic Programming (SP). Then, we define the problem in an online setting in the form of a Markov Decision Process (MDP).

3.1 Offline Planning Problem Definition

We create two different offline planning problem formulations. First, we formulate an ILP problem with two variants: (1) using actual arrivals with perfect knowledge of delays, to be used as a theoretical benchmark; (2) using estimated or expected arrivals without knowledge of delays, to be used by methods that do re-planning when a delay occurs. Second, we provide an SP formulation. Variables for both formulations are explained in Table 3.

Table 3. List of integer linear and stochastic programming elements

Sets	
I	Set of containers
T	Set of trains
Ω	Set of generated scenarios, $\omega \in \{0, 1, 2, ..., \lvert\Omega\rvert - 1\}$
Decision variables ILP, SP	
$X_i^t \in \{0, 1\}$	Put container i on train t
$B_i \in \{0, 1\}$	Put container i on truck
Decision variables SP	
$R_i^{t,\omega} \in \{0, 1\}$	Container i is removed from train t
$A_i^{t,\omega} \in \{0, 1\}$	Container i is re-planned to train t
$D_i^{\omega} \in \{0, 1\}$	Container i is re-planned to a truck
Parameters	
e_i	Estimated earliest day on which container i is available for transport
l_i	Delivery due date for container i
d_t	Day on which train t departs the origin
ar_t	Day on which train t arrives at its destination
cap_t	Number of spaces available on train t
C_t	Costs of transporting a container with train t
C	Costs of transporting a container with a truck
$late_i^{\omega}$	Number of days container i arrives late in scenario ω
$late_i$	Number of days container i arrives late

Integer Linear Program Problem Definition. Given a list of containers and vehicles, the goal is to determine an allocation of containers to vehicles such that the total cost of transportation is minimal. Formally, this is expressed by Eq. 1. This minimization problem has to fulfill the following constraints. First, each container must be allocated to exactly one train or truck (Eq. 2). Furthermore, a container can be allocated only to a train which departs on or after the earliest availability day of the container (Eq. 3), and arrives at the latest on the container due date (Eq. 4). Finally, the maximum capacity of a train can not be exceeded (Eq. 5).

$$Minimize \sum_{i \in I} \sum_{t \in T} C_t \cdot X_i^t + \sum_{i \in I} C \cdot B_i \tag{1}$$

Subject to:

$$\sum_{t \in T} X_i^t + B_i = 1, \qquad \forall i \in I \tag{2}$$

$$X_i^t \cdot d_t \geq X_i^t \cdot (e_i + late_i), \qquad \forall t \in T, i \in I \tag{3}$$

$$X_i^t \cdot ar_t \geq X_i^t \cdot l_i, \qquad \forall t \in T, i \in I \tag{4}$$

$$\sum_{i \in I} {X_i}^t \leq cap_t, \qquad \forall t \in T \tag{5}$$

It is worth noting that this program knows the actual schedules of the containers, because Eq. 3 takes the delay of the container into account. The ILP formulated in this setting acts as a benchmark, because it has perfect knowledge of the moment of arrival of the container. However, this method is not applicable in real-life, since it is unrealistic to know all possible delays in advance. In this paper, we also consider an ILP where we use the estimated arrivals. The formulation for this problem is identical except for Eq. 3, where only e_i is used. In this setting, when a delay occurs and one or more containers miss the allocated train, a re-planing is needed. We refer to this formulation as ILP-based re-optimization or re-planning, depending on whether another ILP solver or allocation heuristics are used to re-allocate the delayed containers.

Stochastic Program Problem Definition. Equation 6 formalizes the objective function to be minimized by our stochastic program. Again the total cost of transportation must be minimized. The first two terms correspond to the first-stage component, i.e., the summation of the original train costs and the original truck costs. The other elements belong to the second stage component: the costs of the recourse actions of re-planning containers. They involve the negative costs of containers that were removed from trains, cost for containers that were added to trains, and cost for containers that were added to trucks over all scenarios. This minimization problem is subject to the following set of constraints: the constraints for the ILP problem still apply (Equation 2–5), if a container is removed from a train, it must be added to another train or to a truck (Eq. 7), re-planned containers planned on a train should depart on or after their earliest availability day (note that the earliest leave day is delayed for these containers) (Eq. 8), re-planned containers planned on a train should arrive at the latest on their latest arrival day (Eq. 9), capacity constraints should also be met after re-planning (Eq. 10), a container can only be removed from a train if it was originally planned on that train (Eq. 11), a container should only be re-planned, if it is late (Eq. 12), a container must be re-planned, if it is too late for the train on which it was originally planned (Eq. 13).

$$Minimize \sum_{i \in I} \sum_{t \in T} C_t \cdot X_i^t + \sum_{i \in I} C \cdot B_i - \frac{1}{|\Omega|} \sum_{\omega \in \Omega} \sum_{i \in I} \sum_{t \in T} C_t \cdot R_i^{t,\omega}$$
$$+ \frac{1}{|\Omega|} \sum_{\omega \in \Omega} \sum_{i \in I} \sum_{t \in T} C_t \cdot A_i^{t,\omega} + \frac{1}{|\Omega|} \sum_{\omega \in \Omega} \sum_{i \in I} C \cdot D_i^{\omega} \tag{6}$$

Subject to:

$$\sum_{t \in T} R_i^{t,\omega} = \sum_{t \in T} A_i^{t,\omega} + D_i^{\omega}, \qquad \forall i \in I, \omega \in \Omega \tag{7}$$

$$A_i^{t,\omega} \cdot d_t \geq A_i^{t,\omega} \cdot (e_i + late_i^{\omega}), \qquad \forall t \in T, i \in I, \omega \in \Omega \tag{8}$$

$$A_i^{t,\omega} \cdot ar_t \leq A_i^{t,\omega} \cdot l_i, \qquad \forall t \in T, i \in I, \omega \in \Omega \tag{9}$$

$$\sum_{i \in I} X_i^t - \sum_{i \in I} R_i^{t,\omega} + \sum_{i \in I} A_i^{t,\omega} \leq cap_t, \qquad \forall t \in T, \omega \in \Omega \tag{10}$$

$$R_i^{t,\omega} \leq X_i^t, \qquad \forall t \in T, i \in I, \omega \in \Omega \tag{11}$$

$$\sum_{t \in T} R_i^{t,\omega} \leq late_i^{\omega}, \qquad \forall i \in I, \omega \in \Omega \tag{12}$$

$$X_i^t \cdot (d_t - e_i - late_i^{\omega}) + late_i^{\omega} \cdot R_i^{t,\omega} \geq 0, \qquad \forall t \in T, i \in I, \omega \in \Omega \tag{13}$$

3.2 Online Planning Under Uncertainty Problem Definition

We formulate the online version of our problem as a Markov Decision Process (MDP), defined by the following elements.

The set of **states** S, where each state has two components. The first component is a list of train capacities, $\{cap_1, cap_2, \ldots, cap_{|T|}\}$, where each cap_j represents the number of slots available on the train or trains that correspond to a particular train schedule. More precisely, we store in the environment a list of all train schedules $P = \{p_1, \ldots, p_{|T|}\}$, where each p_t corresponds to the pair (departure time, arrival time) for one or more trains. As a consequence, the arrival and departure times of trains are implicitly encoded in the state; each cap_j is equal to the sum of the capacities of all the train corresponding to the schedule p_j. The second component of our states is the information about the next container that must be allocated, where a container i is represented by the estimated earliest day on which this container is available for transportation e_i and the due delivery day l_i. The next container to allocate is selected using a heuristic, as explained later in the text. It should be noted that we do not take into account delays as a part of state components and we use estimated schedules.

The set of **actions** A, consisting of all possible train options T and an option 'Truck' that is assumed to be always available and uncapacitated. Note that not all actions are possible in each state, because of the constraints (see Sect. 3.1). For example, a train could have no more slots available, or its scheduled departure time could not meet the due delivery date of the container.

The **reward function** $R(s, a)$, which is the negative cost associated with selecting an action a from the list of eligible actions (Eq. 14). This reward function is defined as follows:

- If the selected action is a truck, it is never affected by possible delays and the reward is the negative cost of transportation by this truck.
- If the selected action is a train and if the train is eligible (considering the delay), the reward for this action is the negative cost of transportation by this train. If the selected train is ineligible, we penalize the selected action by setting the reward equal to the negative cost of transportation by truck (i.e., a cost much higher than any train cost).

$$R(s, a, late) = \begin{cases} -C, & \text{if } a = Truck, \\ -C_a, & \text{if } a \in T, d_t \geq e + late, \\ -C, & \text{if } a \in T, d_t < e + late \end{cases} \tag{14}$$

The **objective**, which is maximizing the expected cumulative reward of the selected actions. Note that this is equal to minimizing the expected cumulative cost of transportation. We use the Bellman Eq. [2] to calculate this.

4 Planning Under Uncertainty Using Deep Reinforcement Learning

In this paper, we extend our previous DRL method [8] to deal with uncertainty in online multimodal transportation planning problems. In the following, we provide a high-level overview of the Deep Q-Learning approach (described by Algorithm 1).

4.1 Multimodal Transportation Problem Environment

The DRL algorithm learns by performing a number of episodes E. During each episode a set of containers is planned either on a train or on a truck. The *environment* has the information on the trains, containers and occurrence of delays. It keeps a current state, and can be given actions to perform that will result in a reward and a new state (see Sect. 3.2). To this end, the environment has two main functions, discussed in the following.

- **Environment initialization**. At the beginning of each episode a new environment is generated by launching the data generator, to ensure that the starting point of each new episode is different from other episodes. The data generator creates a set of trains with their temporal features and initial capacities, a set of containers, with their temporal features and delay information, and transportation costs for each vehicle option (line 4, Algorithm 1).

Algorithm 1. Deep Q-Learning for Online Multimodal Transportation Planning under Uncertainty

1: Initialize Deep Q-Network Q
2: Initialize replay memory D
3: **for** episode = 1 to E **do**
4: Generate new containers, containers delays and trains
5: Set current state s with random capacity for all trains
6: **while** there is an unallocated container $i \in I$ **do**
7: $A' \leftarrow mask(s)$ forbidden actions (Eq. 15)
8: With probability ε select a random action $a \in A'$
9: Otherwise select $a = argmax_{a' \in A'} Q(s, a')$
10: $eligibility \leftarrow check(s, a, late)$
11: **if** $eligibility = False$ **then**
12: Allocate container to the truck option
13: Create new state s' from s without updating train capacity used by a
14: **else**
15: Allocate container to the selected action
16: Create new state s' from s by updating train capacity used by a
17: **end if**
18: Update new state s' with new container arrival
19: Calculate reward $r = R(s, a, late)$ (Eq. 14)
20: Record experience (s, a, r, s') in replay memory D
21: $s \leftarrow s'$
22: **if** every M iterations **then**
23: Sample random minibatch of experience from replay memory D
24: **for** (s, a, r, s') in minibatch **do**
25: $y \leftarrow$ Bellman Equation over $(s, a, r, s'), Q$
26: Update Deep Q-Network $Q(s, a) = y$
27: **end for**
28: **end if**
29: **end while**
30: **end for**

- **Interaction with the agent.** Once an agent selects an action, we update the environment, calculate the next state and calculate reward of this action. Updating the environment means updating the capacity of trains based on the selected action. Note that if the selected action is ineligible for allocation, then we do not change capacities (lines 11–13, Algorithm 1). Then, a new state is generated using the updated train capacities and selecting the next container to plan (line 18, Algorithm 1).
  We test four different allocation heuristics or policies for selecting the next container for decision making, (1) Earliest arrival first (or First In First Out - FIFO) with random allocation of containers arrived on the same day, (2) Earliest due date first (EDF) with random allocation of containers arrived on the same day, (3) FIFO with EDF allocation of containers arrived on the same day, (FIFO-EDF) (4) EDF with FIFO allocation of containers arrived on the same day (EDF-FIFO).

4.2 Feature Engineering and Deep Q-Network Architecture

The algorithm learns through a Deep Q-Network, which learns the Q values for state and action combinations. For the input features, we use a vector of size $|T|+2$, which consists of the list of train capacities defined in Sect. 3.2 and both the temporal features e_i, l_i of container i. The vector of the output nodes is equal to the size of the vehicle options (A), since we use a separate output unit for each action. For the output layer, we use a Softmax layer over the actions. Hence, the outputs of our Deep Q-Network correspond to the predicted $Q(a, s)$ of the individual action a for the input state s, and we select an action with highest Q-value. The network is fully-connected, with k hidden layers.

4.3 Action Selection Methods and Masking Approach

The list of eligible actions can be different for each state s. However, the use of a dynamic set of actions increases significantly the complexity of the problem, up to the point where the computation is not feasible. To deal with this challenge, we determine a static action list of all possible actions and then use we use a customized epsilon-greedy method with a masking approach to determine which actions are enabled at each state s as follows (Algorithm 1, line 7):

$$mask(\{cap_1, \ldots, cap_{|T|}, e, l\}) = \{t \in T \mid d_t \geq e, ar_t \leq l, cap_t \geq 1\} \cup \{Truck\} \tag{15}$$

The agent selects a random eligible action with a fixed probability, $0 \geq \varepsilon \geq 1$, or the action that is optimal with respect to the learned Q-function otherwise [19] (Algorithm 1, lines 8–9).

4.4 Replay Memory and Minibatch

We use a replay memory [12] method, which records the experiences of our agent into a replay memory D at each step (s, a, r, s') of each episode (Algorithm line 20). Every M steps, we then update the network. The main advantage of this method consists in decreasing the variance of the updates. Lines 22 to 26 show how we apply Q-learning updates, or minibatch updates, by first sampling experiences randomly from the replay memory, calculating the expected cumulative reward for each experience using the Bellman equation and then updating the Deep Q-Network for each experience with the expected cumulative reward.

5 Experiments and Results

This section discusses the experiments that we carried out to test the performance of our method. Section 5.1, introduces the experimental settings and the tested competitors, while Sect. 5.2 discusses the obtained results.

5.1 Experimental Settings

Dataset. We generated data with properties that are based on the long-haul transportation planning problem of a logistics company for a particular transportation corridor. These data include the following features (see Sect. 4): the number of trains, with their capacity and temporal properties, transportation costs, and containers with their temporal features and their delays. To simulate container delays, we follow the approach proposed by previous work on uncertainty handling [10], where delays are generated randomly based on the three different levels of uncertainty, which is the probability that the container is delayed by a given number of days x (Table 4): (1) Low, (2) Average, (3) High. We indicate this probability as $p(late = x)$. Time windows of this experiment are weekly. We assume that trucks are always available and uncapacitated.

Table 4. Probability of occurrence of estimated delay in each uncertainty level

Scenario	$p(late = 0)$	$p(late = 1)$	$p(late = 2)$
Low uncertainty	0.9	0.08	0.02
Average uncertainty	0.6	0.3	0.1
High uncertainty	0.3334	0.333	0.333

Training Parameters. We did hyperparameter tuning on: the number of episodes (with options 4000, 5000, 6000), learning rate (0.01, 0.1), number of hidden layers (2, 4), discount factor (0.5, 0.99), number of nodes per hidden layer (100, 150, 200), and mini batch size (5, 10, 15). The algorithm worked best and learning converged using $E = 6,000$ episodes of 7 days. In each episode 100 containers must be planned, i.e. 100 steps must be performed. The number of containers is chosen proportional to the train capacity over the week, in line with the properties of the planning problem at the logistics company. Each container has an estimated earliest availability day and a due date that are uniformly distributed over the week. Delays are generated based on Table 4 in three different levels of uncertainty. There are 28 train schedules per week. For the capacity of trains in each train schedule we test 7 different settings, i.e.: 6 different settings in which each train schedule $(1,1),(1,2),\ldots$ has the same capacity 1 through to 6; and one setting in which each schedule has a random number of available slots that is uniformly distributed over 0 to 6 spaces. The goal of using these different settings is to investigate the effect of uncertainty levels and available capacity on the planners performance in a realistic scenario.

We initialize a fully-connected feedforward neural network with backpropagation with 2 hidden layers of 100 nodes, ReLU activator, and Adam optimizer. We use a replay memory of size 10,000 and retrain the Deep Q-Network based on minibatches 5 times per epoch. The discount factor, used in the Bellman equation, is $\gamma = 0.99$, which means that future rewards are of high importance in the learning process. Remaining parameters are initialized according to PyTorch's default parameters. The probability ϵ with which a random action is chosen starts at 0.95 and is decreased after each episode in steps of 0.1 until it reaches 0.05. The agent and the simulation model are executed on a machine with an Intel(R) Core(TM) i7 Processor CPU @ 2.80 GHz and 16 GB of RAM, no graphics module is used for training the neural network.

Tested Competitors. We use the offline ILP solver based on actual schedules as a benchmark. Furthermore, we compare the performance of our method against (1) ILP-based re-optimization, and (2) ILP and SP joint Greedy heuristics as re-planner. These methods are inspired by the literature and discussions with the logistics company on how their (re) planner currently works. For the ILP-based (re) optimization, we run once per week an ILP planner based on estimated data; to update the plan in presence of delays, we also run a daily ILP based (re) optimization, which has information of delayed containers until the current day. ILP and SP joint Greedy heuristics are commonly used in practice as re-planner. These methods also run ILP and SP planner for estimated data once per week; however, they apply a greedy heuristic to allocate delayed containers separately to a train or, if no eligible trains are available, a truck. We refer to these combinations as '*ILP + First train*', '*SP + First train*', which re-plan a container on the first available train, and '*ILP + Cheapest train*', '*SP + Cheapest train*', which re-plan containers on the cheapest available option.

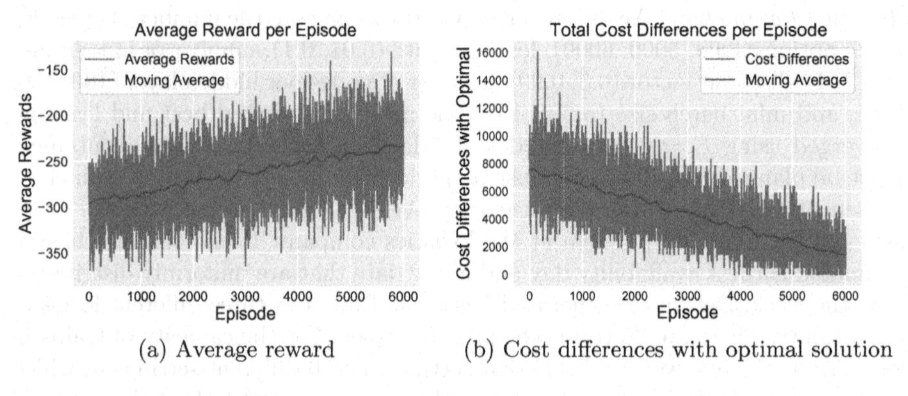

(a) Average reward (b) Cost differences with optimal solution

Fig. 1. Average Reward (a) and cost differences with ILP (b) per episode during training

5.2 Results

Training and Stability Analysis. We assess the performance of the learning process by using the total reward the agent collects in each episode during training [3]. We tested four different allocation heuristics discussed in Sect. 4.1 in 21 different settings (3 levels of uncertainty and 7 different available capacity settings). For the sake of space, here we discuss only the results of the most challenging (and realistic) setting, i.e., the one corresponding to random train capacity and high uncertainty level. As allocation heuristic, here we discuss EDF, since it turned out to be the best performing one in our previous work [8]. However, the results obtained for all the other tested settings show similar trends. Figure 1 shows the changes in the average reward (Fig. 1a) and in the cost differences with the 'benchmark' method (Fig. 1b) per episode. The red line is a moving average line highlighting the behavior of the model during these episodes. Figure 1a shows a smooth improvement of the average rewards, which demonstrates that we did not experience any divergence issues. Figure 1b shows that the cost differences with the benchmark solutions converge to zero, which proves that our agent is able to learn container allocation patterns, getting closer to the optimal solution as the training goes on.

Methods Comparison. We tested the above discussed methods in different experiments with seven different capacity settings and three different levels of uncertainty, obtaining in total 21 different settings. We measure their performance in terms of the average per week over 20 weeks of the total costs of transportation (i.e., the costs deriving from the actual allocation of each container to a train or to a truck). Figure 2 shows the average transportation costs of each method in the different capacity settings. For the sake of space, only the results obtained at high uncertainty level, i.e., the most challenging scenario, are plotted.

Table 5. Average cost differences (%) w.r.t. the optimal solution over 20 weeks.

Method	Low	Average	High
DRL-(FIFO)	27.87	21.61	17.01
DRL-(FIFO-EDF)	4.76	10.67	11.97
DRL-(EDF-FIFO)	0.63	0.53	0.37
DRL-(EDF)	0.65	0.78	0.66
Re-ILP	7.18	20.40	20.05
ILP+cheapest train	9.27	31.44	34.97
ILP+first train	9.28	31.38	34.96
SP+cheapest train	7.68	26.61	33.37
SP+first train	7.54	26.13	33.07

DRL-(EDF-FIFO), DRL-(EDF) are consistently better than other competitors and very close to the benchmark optimal solution for all tested capacities. Performance of DRL-(FIFO) and DRL-(FIFO-EDF), instead, decrease with increasing train capacity. Nevertheless, all DRL-based methods always perform better than the competitors.

For the other uncertainty levels, DRL-(EDF-FIFO), DRL-(EDF) achieved in all the settings performance very close to the optimal ones. However, in the low uncertainty settings differences with not DRL methods are less evident. In particular, in the most competitive scenario (i.e., with train capacity equal to 1), all the tested methods obtain results close to the optimal solution. This was expected, since few containers need re-planning and few options are available. Differences between DRL-based methods and other competitors become more evident with the increasing of the train capacity, even though the performance gap is less pronounced than in the other uncertainty levels. This suggests that uncertainty does have a significant impact on the performance of not DRL-based methods. Another interesting observation in the low uncertainty setting, is that the performance of the DRL-(FIFO) are overall worse than other competitors.

Table 5 reports differences of the average of the costs of each method with respect to the optimal solution. For each method, for each setting, we computed the average total cost of transportation over 20 weeks. Then, for each method we computed in each setting difference between its obtained cost and the cost corresponding to the optimal solution; finally, we compute the average of these differences over the 21 different settings considered in the experiments. All the DRL methods with EDF policy obtained on average values very close to the optimal solver in all the uncertainty levels. In particular, DRL-(EDF-FIFO) obtained the best results. All the other methods show a significant increase in costs. For example, the re-ILP method, which is the best one among the not DRL-based competitors, obtains on average increasing of the total costs of 7.18%, 20.40% and 20.05% for the low, average and high setting respectively. Some methods appear to perform better in higher uncertainty level; DRL-(FIFO) is the most evident example. This behavior may seem counter-intuitive. However, recall that this is a relative measure; while the absolute cost of EDF-(FIFO) increases moving from low to high uncertainty, the benchmark costs increases as well and at a faster pace, with the result that the relative cost decreases.

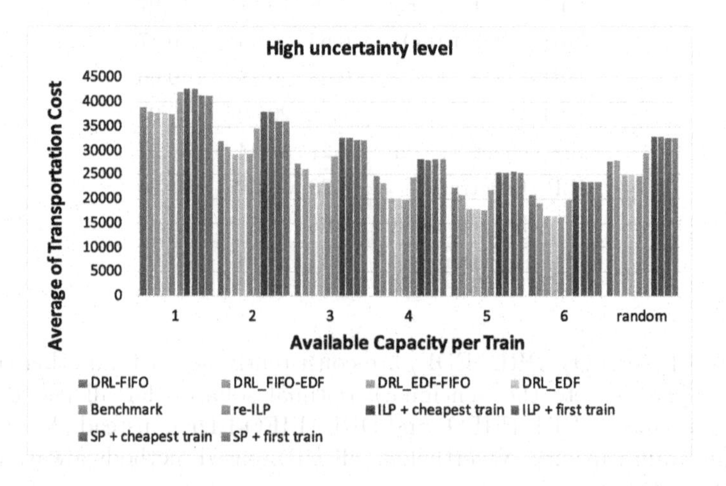

Fig. 2. Average transportation costs for high uncertainty levels in different capacity settings

Summing up the obtained results show that the use of ILP-based replanner methods with limited knowledge on future events leads in general to poor performance. SP re-planning methods partly remedy to this lack of knowledge using an a priori probability distribution for exploring different scenarios. However, when container arrivals do not fit this distribution, the obtained solutions are often quite far from the benchmark. In contrast, the proposed DRL-based methods can learn the container allocation rules under uncertainty, thus being able to take more informed (and efficient) allocation decisions.

6 Conclusions and Future Work

This paper investigated the application of DRL in tackling uncertainty in online container allocation in the multimodal transportation domain. The experimental results shown that the DRL based proposed approach can learn patterns of containers allocation under uncertainty, which allows it to consistently outperform the tested competitors, obtaining solutions close to the theoretical optimal benchmark solution. In particular, the Deep Q-Networks planner that uses an EDF heuristic to determine which container must be planned first, outperformed the heuristic, periodic re-planning, and stochastic programming competitors, obtaining an average cost difference with the optimal solution of only 0.63%, 0.53%, and 0.37% in the low, average, and high uncertainty levels respectively.

Overall, these results show how the use of Deep Reinforcement Learning can significantly decrease costs associated with container re-planning under uncertainty for logistics companies, thus suggesting that the use of these techniques can indeed bring significant practical advantages in the logistic domain with dynamic and non-deterministic environments.

Nevertheless, our method presents some limitations. The current version of the method supports only the allocation of containers to a single vehicle, rather than to a combination of vehicles. In future work, we plan to extend our model to incorporate this aspect, thus increasing the generality of the method. Also, we intend to take locations into account as a planning factor. Finally, we plan to investigate the application of different DRL approaches and reward functions.

References

1. Alves, J.C., Mateus, G.R.: Deep reinforcement learning and optimization approach for multi-echelon supply chain with uncertain demands. In: Lalla-Ruiz, E., Mes, M., Voß, S. (eds.) ICCL 2020. LNCS, vol. 12433, pp. 584–599. Springer, Cham (2020). https://doi.org/10.1007/978-3-030-59747-4_38
2. Barron, E., Ishii, H.: The Bellman equation for minimizing the maximum cost. Nonlinear Anal. Theory Methods Appl. **13**(9), 1067–1090 (1989)
3. Bellemare, M.G., Naddaf, Y., Veness, J., Bowling, M.: The arcade learning environment: an evaluation platform for general agents. J. Artif. Intell. Res. **47**, 253–279 (2013)

4. Bhargavi, K., Babu, B.S.: Soft-set based DDQ scheduler for optimal task scheduling under uncertainty in the cloud. In: 2017 2nd International Conference On Emerging Computation and Information Technologies (ICECIT), pp. 1–6. IEEE (2017)
5. Delbart, T., Molenbruch, Y., Braekers, K., Caris, A.: Uncertainty in intermodal and synchromodal transport: Review and future research directions. Sustainability **13**(7), 3980 (2021)
6. Escudero, A., Muñuzuri, J., Guadix, J., Arango, C.: Dynamic approach to solve the daily drayage problem with transit time uncertainty. Comput. Ind **64**(2), 165–175 (2013)
7. Fang, D., Guan, X., Peng, Y., Chen, H., Ohtsuki, T., Han, Z.: Distributed deep reinforcement learning for renewable energy accommodation assessment with communication uncertainty in Internet of Energy. IEEE Internet Things J. **8**, 8557–8569 (2020)
8. Farahani, A., Genga, L., Dijkman, R.: Online multimodal transportation planning using deep reinforcement learning. arXiv preprint arXiv:2105.08374 (2021)
9. Gao, Y., Yang, J., Yang, M., Li, Z.: Deep reinforcement learning based optimal schedule for a battery swapping station considering uncertainties. IEEE Trans. Ind. Appl. **56**(5), 5775–5784 (2020)
10. Gumuskaya, V., van Jaarsveld, W., Dijkman, R., Grefen, P., Veenstra, A.: Dynamic barge planning with stochastic container arrivals. Transp. Res. Part E Logist. Transp. Rev. **144**, 102161 (2020)
11. Ma, H., Yu, G., She, Y., Gu, Y., et al.: Waterflooding optimization under geological uncertainties by using deep reinforcement learning algorithms. In: SPE Annual Technical Conference and Exhibition (2019). Society of Petroleum Engineers
12. Mnih, V., et al.: Playing Atari with deep reinforcement learning. arXiv preprint arXiv:1312.5602 (2013)
13. Peng, Z., Zhang, Y., Feng, Y., Zhang, T., Wu, Z., Su, H.: Deep reinforcement learning approach for capacitated supply chain optimization under demand uncertainty. In: 2019 Chinese Automation Congress (CAC), pp. 3512–3517. IEEE (2019)
14. Powell, W.B., Jaillet, P., Odoni, A.: Stochastic and dynamic networks and routing. Handb. Oper. Res. Manag. Sci. **8**, 141–295 (1995)
15. Rivera, A.P., Mes, M.R.: Anticipatory scheduling of freight in a synchromodal transportation network. Transp. Res. Part E Logist. Transp. Rev. **105**, 176–194 (2017)
16. Sakib, N.: Highway lane change under uncertainty with deep reinforcement learning based motion planner (2020)
17. Shyalika, C., Silva, T.: Reinforcement learning based an integrated approach for uncertainty scheduling in adaptive environments using MARL. In: 2021 6th International Conference on Inventive Computation Technologies (ICICT), pp. 1204–1211. IEEE (2021)
18. SteadieSeifi, M., Dellaert, N.P., Nuijten, W., Van Woensel, T., Raoufi, R.: Multimodal freight transportation planning: a literature review. Eur. J. Oper. Res. **233**(1), 1–15 (2014)
19. Tokic, M., Palm, G.: Value-difference based exploration: adaptive control between epsilon-greedy and softmax. In: Bach, J., Edelkamp, S. (eds.) KI 2011. LNCS (LNAI), vol. 7006, pp. 335–346. Springer, Heidelberg (2011). https://doi.org/10.1007/978-3-642-24455-1_33
20. Topaloglu, H.: A parallelizable and approximate dynamic programming-based dynamic fleet management model with random travel times and multiple vehicle types. In: Dynamic Fleet Management, pp. 65–93. Springer, Heidelberg (2007). https://doi.org/10.1007/978-0-387-71722-7_4

21. van Riessen, B., Negenborn, R.R., Dekker, R.: Real-time container transport planning with decision trees based on offline obtained optimal solutions. Decis. Supp. Syst. **89**, 1–16 (2016)
22. Wan, K., Gao, X., Hu, Z., Wu, G.: Robust motion control for UAV in dynamic uncertain environments using deep reinforcement learning. Remote Sens. **12**(4), 640 (2020)
23. Wang, P., Li, Y., Shekhar, S., Northrop, W.F.: Uncertainty estimation with distributional reinforcement learning for applications in intelligent transportation systems: a case study. In: 2019 IEEE Intelligent Transportation Systems Conference (ITSC), pp. 3822–3827. IEEE (2019)
24. Yang, J., Yang, M., Wang, M., Du, P., Yu, Y.: A deep reinforcement learning method for managing wind farm uncertainties through energy storage system control and external reserve purchasing. Int. J. Electric. Power Energy Syst. **119**, 105928 (2020)
25. Yehia, A.: Understanding uncertainty: a reinforcement learning approach for project-level pavement management systems. PhD thesis, University of British Columbia (2020)

Robust Multi-Objective Gate Scheduling at Hub Airports Considering Flight Delays: A Hybrid Metaheuristic Approach

Abtin Nourmohammadzadeh(✉) and Stefan Voß

Institute of Information Systems (IWI), University of Hamburg, Hamburg, Germany
{abtin.nourmohammadzadeh,stefan.voss}@uni-hamburg.de

Abstract. Regarding the large number of flights that a hub airport usually has to serve and the competitiveness in the aviation industry, optimal scheduling of limited and expensive airport resources such as gates is really vital. This work focuses on the efficient scheduling of airport gates to achieve a balance between three important goals, namely reducing the walking distance of passengers, decreasing the number of flights assigned to the gates different from their reference gates as well as widening the total shopping area passed by passengers while walking to, from or between the gates. A set of different scenarios is considered for the arrival of flights regarding the possible delays. Robust multi-objective optimisation is followed through an exact solution approach according to the weighted sum method by the Baron solver as well as a metaheuristic method consisting of the hybridisation of multi-objective particle swarm optimisation (MOPSO) and the multi-objective simulated annealing (MOSA). The sets of Pareto-optimal solutions obtained by these two methods along with those of the pure MOPSO, MOSA and a tabu search algorithm from the literature are compared based on some evaluation metrics and with the aid of a statistical test.

Keywords: Airport gate scheduling · Robust optimisation · Multi-objective optimisation · Multi-objective particle swarm optimisation (MOPSO) · Multi-objective simulated annealing (MOSA)

1 Introduction

The world has witnessed tremendous growth in air traffic up to the hard era of the Covid-19 outbreak and there is great hope that the aviation industry can return to its flourishing days after controlling the pandemic. The European Commission's report on the Air Transport Market states that a wide increase happened in the total worldwide number of operated flights in the period 2010 to 2016, namely from 27.8 million to 36.8 million. The statistics on the volume of carried air passengers also show an increase of 219.28 million to reach 3.7

M. Mes et al. (Eds.): ICCL 2021, LNCS 13004, pp. 594–610, 2021.
https://doi.org/10.1007/978-3-030-87672-2_39

billion in 2016 [8]. Therefore, hub airports must handle a huge number of flights within short periods with their limited resources. Regarding the very competitive environment of the air transportation sector, efficient usage of resources to increase revenue and customer satisfaction is extremely important. Furthermore, in terms of the total revenues that airports gain, the average commercial revenues constitute about half of them [7]. Hence, airports are not only considered as the points for passenger transportation between origins and destinations but also as centres for shopping and leisure [13,15].

This necessitates investigating many optimisation problems which emerge at airports. Gate scheduling is one important resource assignment problem, where a set including a limited number of gates is assigned to a large set of arriving or departing flights. There are usually two kinds of gates, the ones connected to the terminal and equipped with air bridges, which are used by passengers to directly and conveniently get on and off the airplanes, and the ones existing as ramp spaces away from the terminal. The embarkation and disembarkation to/from airplanes that stay at the ramp gates are usually done by special buses and require more time. The passengers that a hub airport is involved with are from three categories. The departure passengers, who start their journey from the airport and traverse the distance between the entrance and their departure gate. The arrival passengers, who end their air travel at the airport and have to walk from the gates to the airport's exits. The last but most important ones are the transit passengers, who change their flights at the airport to continue travelling to their final destination. This group has to walk from their arrival to their departure gate. It is common that some gates are rent by specific airlines for their flights. Therefore, it is tried to assign those flights to their associated or reference gates and the assignment of them to any other gate is extremely undesirable. In addition, each flight can be served only by a specific set of gates with regard to the size and other specifications of the aircraft which operates it.

Three main groups of stakeholders exist in the gate scheduling problem (GSP), which are airport operators, airlines, and passengers. Each of them has its own interests and wills; however, there are also many common interests. Airport operators seek to gain more profits by giving a higher level of service to airlines and passengers while efficiently using the related resources and minimising the incurred costs by reducing the congestions, needed resources, interruptions, delays etc. On the other hand, passengers want convenient and smooth boarding and deboarding, short walking distances as well as good access to airport facilities such as shops, restaurants, or entertainment areas. Finally, airline operators prefer easy terminal access and short ground times for their aircraft [8].

These different goals and perspectives lead to the considerations of a variety of objectives for the GSP. The conflict between the objectives shows that the problem is multi-objective in nature and has to be modelled and tackled in this respect. Therefore, our research presents a model for the GSP with three objectives of minimising the total walking distances of passengers within the terminal, minimising the number of flights which are not assigned to their reference gate, and maximising the total shopping area that passengers go through within the terminal. The first objective is passenger-oriented, whereas the second objective is more in favour of airlines and the third one targets passengers and airports. Moreover, the stochastic nature of various influencing factors make it necessary to consider a robust version of the GSP. Hence, the model is later converted to its robust version.

We choose an exact and an evolutionary metaheuristic solution approach to tackle the problem. The application of the metaheuristic algorithm is due to the complexity of the problem and the demand for obtaining efficient solutions in short times. For the exact solution, the weighted sum method is chosen. The formulated model is programmed in GAMS [2] and the Baron solver [1] is called. The proposed metaheuristic approach comprises an integrative hybridisation of the multi-objective particle swarm optimisation (MOPSO) and the multi-objective Simulated Annealing (MOSA) to boost the search ability. The approaches are tested on some synthetic instances.

It is worth noting that the problem of determining efficient gates for flights is also called the gate assignment problem (GAP). However, due to providing each gate with the time plan for serving the assigned flights, in this work, we use the name GSP for the problem addressed.

The organisation of the remainder of this work is as follows: In Sect. 2, some previous related works are introduced and categorised. Section 3 explains the proposed model. Subsequently, Sect. 4 covers the solution approaches. Computational results are given and discussed in Sect. 5. At last, Sect. 6 draws the overall conclusions of this paper and recommends some potential subjects for future research.

2 Related Work

The GSP (or the GAP which is used interchangeably) has been under consideration for many years and numerous modelling approaches and solution methodologies are proposed for it. A very good literature review of this subject is [8], which categorises a vast collection of the GAP works based on the criteria such as the orientation of their objectives, and besides, a variety of mathematical models is introduced. [22] is another paper that addresses the state-of-art but with the focus on the presented mathematical models.

According to [8], the GAP objectives are of three groups: namely, passenger-oriented, which is the most common, e.g. [3,6,7,9–11,18,20,27–29], airline/airport-oriented, e.g. [3,7,9–12,14,16–18,20,25,28,29], or robustness-oriented, e.g. [7,10,12,17,18,20,29], which considers the possible changes in the problem inputs. Its other comparison aspects are if the models are single, e.g. [6,11,14,25, 27], or multi-objective, e.g. [7,9,10,12,16–18,20,28,29], as well as which objectives are optimised. Among the objectives considered in the GAP, the main ones are: minimising the total walking distance of passengers, e.g. in [3,6,7,9–11,18,22,28, 29], minimising total passenger waiting time, e.g. [27], minimising the number of flights assigned to remote gates, e.g. [9], minimising the number of towing moves, i.e. the assignment of two consecutive flights operated by one aircraft to different gates, e.g. in [11,12,20], minimising towing costs, e.g. in [18,28,29], minimising the sum of waiting times of aircraft for gates, e.g. in [16], maximising total flight-gate preferences, e.g. in [12,20], maximising the number of passengers at gates, e.g. [7,11], maximising the number of passengers at gates close to shopping facilities, e.g. [7], maximising potential airport commercial revenue, e.g. [11], minimising variance of idle times, e.g. in [7,10], minimising the number of conflicts of any two adjacent aircraft assigned to the same gate, e.g. in [18], minimising the expected gate conflict duration, e.g. in [28], minimising the expected gate conflict cost, e.g. in [29], and minimising the absolute deviation of new gate assignment from a reference schedule, e.g. in [20].

[6] examines the performance of a genetic algorithm (GA), a tabu search (TS), a simulated annealing (SA), and a hybrid approach based on the SA and TS on the GAP using flight data from the Incheon International Airport. [3] applies a kind of the TS algorithm where a probabilistic approach as an aspiration criterion is embedded. [7] aims at raising the shopping revenues through assigning passengers to specific gates near shopping facilities. [9] develops a metaheuristic based on the fuzzy bee colony optimisation to deal with its GAP. An improved adaptive particle swarm optimisation algorithm is used in [10]. [11] proposes an innovative approach to the tactical planning of the assignments of flights to terminal-connected and remote gates and uses the concept of recoverable robustness. [12] addresses the general case that the aircraft, which operate the flights, have arrival, parking, and departure tasks that can be each assigned to a different gate. It models the GAP with a graph-theoretical approach based on the clique partitioning problem.

[14] develops a method that makes use of the benefits of heuristics with a stochastic method rather than applying a purely probabilistic approach to find fast efficient solutions to the problem. The authors apply their approach to real data from the Istanbul Airport. [16] utilises CPLEX as well as an evolutionary multi-objective optimisation algorithm. A methodology that captures decision-makers' preferences in multi-objective environments is employed. [18] solves a version of the multi-objective GAP by the second version of the non-dominated sorting genetic algorithm (NSGA-II). The authors compare the performance of their approach with some other metaheuristics. A Pareto simulated annealing (PSA) approach is adapted for the GSP in [20] to get a representative approximation of the Pareto front and the uncertainty of inputs is treated by means of fuzzy numbers. The mathematical modelling of flight-to-gate reassignment with passenger connections is explored in [22] and a number of cases of various sizes and schedule scenarios, as well as a set based on a real European airport, are used as the test instances. [17] addresses a multi-criteria gate assignment problem with the objectives of maximising passenger connection revenues and gate plan robustness as well as minimising zone usage costs.

[23] forecasts and optimises movements of the passengers inside airport terminal buildings. [25] proposes a method to improve the robustness of solutions to reduce the need for gate re-planning regarding disturbances in the flight schedules. The method replaces the deterministic with stochastic gate constraints that

incorporate the inherent stochastic flight delays to reduce the gate conflicts. [28] considers a GAP, in which the traditional costs and the robustness are simultaneously considered. An adaptive large neighborhood search algorithm is designed with local search operators to efficiently tackle the problem.

3 Modelling

The GSP model which we present in this section includes the three mentioned objectives, while it ensures the assignment of all flights to gates and the separation time between two consecutive usages of an identical gate. Our tri-objective mathematical model for the GSP is presented in this section in three parts of introducing the notation, formulation and adding robustness.

3.1 Notation

The notations used in our model are as follows:

Parameters

H	The planning horizon
F	The set of flights including arrivals and departures, f is the index for a single flight
FR	The set of flights which have a reference gate, $FR \in F$
G	The set of gates connected to the terminal
RG	Represents the ramp or remote gates
PG_f	The set of gates which can serve flight f
R_f	The reference gate of flight f
$D_{g,g'}$	The walking distance between gates g and g'
$D_{0,g}$	The walking distance between the entrance and gate g
$D_{g,0}$	The walking distance between gate g and the exit
$S_{g,g'}$	The shopping area existing between gates g and g'
$S_{0,g}$	The shopping area existing between the terminal entrance and gate g
$S_{g,0}$	The shopping area existing between gate g and the terminal exit
$P_{f,f'}$	The number of transit passengers which get off flight f and then board flight f'
$P_{0,f}$	The number of departure passengers which enter the airport and board flight f
$P_{f,0}$	The number of arrival passengers which get off flight f and then exit the airport
ST_f	The arrival time of flight f at the gate, i.e. gating time of the flight
FT_f	The finish time of flight f at the gate
$SPT_{f,f'}$	The separation time required between the consecutive leaving of flight f from and gating time of flight f' at the same gate

Variables

$x_{f,g}$	Binary variable which is equal to 1 if flight f is assigned to gate g, otherwise it is zero
$y_{f,f'}$	Binary variable which is equal to 1 if flights f and f' are assigned to the same gate

3.2 Formulation

The mathematical formulations of our model are as follows:
Objectives

$$Min\ Z_1 = \sum_{f \in F} \sum_{f' \in F, f' \neq f} \sum_{g \in PG_f} \sum_{g' \in PG_{f'}, g' \neq g} x_{f,g} x_{f',g'} D_{g,g'} P_{f,f'} + \sum_{f \in F} \sum_{g \in PG_f} x_{f,g} (D_{0,g} P_{0,f} + D_{g,0} P_{f,0}) \quad (1)$$

$$Min\ Z_2 = \sum_{f \in FR} \sum_{g \neq R_f} x_{f,g} \quad (2)$$

$$Max\ Z_3 = \sum_{f \in F} \sum_{f' \in F, f' \neq f} \sum_{g \in PG_f} \sum_{g' \in PG_{f'}, g' \neq g} x_{f,g} x_{f',g'} S_{g,g'} P_{f,f'} + \sum_{f \in F} \sum_{g \in PG_f} x_{f,g} (S_{0,g} P_{0,f} + D_{g,0} S_{f,0}) \quad (3)$$

Constraints

$$\sum_{g \in PG_f} x_{f,g} = 1 \qquad f \in F \quad (4)$$

$$x_{f,g} x_{f',g} = y_{f,f'} \qquad f, f' \in F; g \in PG_f \cup PG_{f'} \quad (5)$$

$$ST_{f'} - FT_f - SPT_{f,f'} \leq H(1 - y_{f,f'}) \quad f, f' \in F; ST_{f'} > ST_f; g \in PG_f \cup PG_{f'} \quad (6)$$

Equation (1) is the first objective to be minimised, which sums up the total distances walked by all passengers between the gates, from the terminal entrance to the gates and from the gates to the terminal exit. Equation (2) is the second objective, which calculates the number of total non-reference assigned gates and we seek to minimise the resulted value. The third objective is represented by Eq.

(3), which is very similar to (1) with the difference that here the existing shopping area between the points is taken into account and it has to be maximised. It can also be transformed into a minimisation objective by replacing $S_{g,g'}$ with $D_{g,g'} - S_{g,g'}$, $S_{0,g}$ with $D_{0,g} - S_{0,g}$ and $S_{g,0}$ with $D_{g,0} - S_{g,0}$. In other words, we seek to reduce the passed non-shopping area. This is done to have the same optimisation direction for all objectives in order to make the solution approaches and evaluation metrics better applicable.

Constraint (4) enforces each flight to be assigned to exactly one gate. Constraints (5) set $y_{f,f'} = 1$ if both $x_{f,g}$ and $x_{f',g}$ are together one for any gate g. The separation time between any two flights dwelling consecutively at the same gate is ensured by constraint (6).

3.3 Considering Robustness

To insert robustness into our model, a set of scenarios is considered for the arrival times of flights at gates, which can be due to possible delays. Therefore, we have a set including the possible gating times for each flight f called PST_f. These make together numerous scenarios, which each consists of a combination of the possible gating times of flights. The set of probable scenarios is named $Scen$. In our robust optimisation process, for each solution, the worst value of each objective based on all the existing scenarios is regarded as the value of that objective. So even the worst scenario is taken into account in the solution evaluation and this is in accordance with the definition of robustness presented in [4]. It is worth noting that a feasible solution to the model must satisfy the constraints with regard to all scenarios according to our assumptions. Based on the given explanations, the robust version of our model can be expressed as:

$$Min(\max_{s \in Scen} Z_1) \tag{7}$$

$$Min(\max_{s \in Scen} Z_2) \tag{8}$$

$$Max(\min_{s \in Scen} Z_3) \tag{9}$$

$$Constraint\,(4) \tag{10}$$

$$Constraint\,(5) \tag{11}$$

$$Constraint\ (6) \qquad s \in Scen \tag{12}$$

The new objectives, i.e. (7–9) seek to optimise the worst objective value of all scenarios. While Constraints (10) and (11) are exactly the same as their equivalents in the non-robust version, Constraint (12) ensures the separation time of two consecutive flights at an identical gate based on all possible gating times of them included in the scenarios.

4 Solution Methodologies

4.1 Exact Approach with the Weighted Sum of Objectives

In multi-objective optimisation, each solution represents multiple objective values. Therefore, solutions can not be easily sorted based on only one criterion, which is the case in the single-objective optimisation. So we have to consider a trade-off between the objectives instead with the aid of a concept called dominance. It is said that in the presence of n objectives, a solution x dominates another solution y, $x \prec_d y$, if the following conditions are true:

1) $z_i(x) \leq z_i(y), \forall i \in 1, 2, ..., n$
2) $\exists j \in 1, ..., n : z_j(x) < z_j(y)$

In multi-objective optimisation, the goal is to discover a set of globally non-dominated solutions, which is known as Pareto-optimal. A set containing the Pareto-optimal solutions constitutes a Pareto front. An easily applicable method to tackle a multi-objective optimisation problem by any exact solver is to replace the multiple objectives with a single objective containing a weighted sum of the original objectives as below:

$$Z = \sum_{i=1}^{n} c_i Z_i \tag{13}$$

$$\sum_{i=1}^{n} c_i = 1 \tag{14}$$

where the problem has n objectives and c_i is the weight associated with the objective i. By manipulating these weights, i.e. the vector $C = (c_1, c_2, ..., c_n)$, different solutions can be obtained. Thus, each time the problem is solved based on a C, there is the possibility of finding one non-dominated or Pareto-optimal solution.

4.2 Hybrid Metaheuristic

There are some considerable weaknesses in exact methods in dealing with multi-objective problems such as long computational times, the necessity of numerous runs or difficulty in locating the non-convex part of the Pareto front. Therefore, metaheuristic multi-objective methods can be employed as alternative approaches to provide good results within much shorter computational times. Hence, here we devise a metaheuristic approach consisting of the combination of two famous evolutionary concepts, namely the MOPSO and the MOSA.

MOPSO. In the MOPSO, see [21], a population of candidate solutions is initialised as the particles and a random velocity is associated with each. The particles are moved in each iteration according to their velocities, which are steadily updated based on the best position that the particle has experienced as well as the global best position of all particles up to that point. The corresponding formulas are as follows:

$$v_{paticle} = \omega v_{particle} + \varphi_p(BP_{particle} - p_{particle}) + \varphi_g(BG - p_{particle}) \tag{15}$$

$$p_{particle} = p_{particle} + v_{particle} \tag{16}$$

where $v_{particle}$ is the velocity of the particle, $p_{particle}$ is the position of the particle, ω is the inertia coefficient that applies the effect of the previous positions, $BP_{particle}$ is the personal best position of the particle, BG is the global best position of all particles, φ_p and φ_g are random coefficients, which determine the influence of the personal and global best positions. These coefficients are separately chosen for each particle and changed in each iteration. A major difference in the MOPSO in comparison to the PSO is in the evaluation of the solutions to find the personal and global best positions. Different metrics can be considered for this sake. One of them is the distance from the nearest non-dominated solution found until then. The non-dominated solutions are saved in a set REP. The metric can be mathematically expressed as:

$$M_1(Particle) = \frac{1}{n} \min_{Particle' \in REP} \sum_{i=1}^{n} (Z_i(Particle') - Z_i(Particle))^2 \tag{17}$$

REP is updated each time that a new non-dominated solution (sol_{new}) is found. This update includes adding sol_{new} to REP and eliminating the solutions in REP which are dominated by sol_{new}. The global best (GB) is chosen within REP according to the diversity aspect. For this goal, each solution of REP is evaluated based on the distance of its two nearest neighbours which are also existing in REP. We call this metric M_2. The distance between any two solutions, sol and sol', is calculated as:

$$D_{sol,sol'} = \sum_{i=1}^{n} (Z_i(sol) - Z_i(sol'))^2 \tag{18}$$

The larger is M_2, the better is the solution because it shows that there is a larger gap between this solution and others. Thus, it is more likely that this gap can be filled with some new non-dominated solutions, which we try to find next. So the GB is a particle in REP corresponding to the largest M_2.

This algorithm continues iteration by iteration and REP is completed and corrected by population-based searches done by the movements of particles until a pre-defined termination criterion is met. The latest REP is considered as the output of this algorithm.

MOSA. The MOSA, originally presented in [26], applies neighbourhood searches by a probabilistic acceptance rule to increase the probability of landing on non-dominated solutions. For a problem with n objectives, a random weight is assigned to each objective considering $\sum_{i=1}^{n} \lambda_i = 1$. The acceptance probability of a neighbouring solution sol' of the solution sol is calculated as follows:

$$P(sol, sol', \lambda, T) = min(1, exp(-\max_{i=1,\ldots,n} \frac{\lambda_i[(z_i(sol') - z_i(sol)]}{T})) \quad (19)$$

where T is the current temperature. At the beginning of the algorithm, we have a high T that leads to larger acceptance probabilities for dominated solutions. By going further in the optimisation process, we decrease T, which means that a new dominated solution can be harder accepted instead of the current solution. This is analogous to the annealing process in reality that the shape of an object can be easily changed when it is melted but by cooling, it gets more stability and changing it is less possible.

We aim at using the concept of a population-based MOSA, which tries a number of neighbourhood searches for each solution of the population and replaces it with a neighbouring solution probabilistically. Here again, a set ρ is considered to keep the Pareto-optimal solutions. Upon finding any new non-dominated solution, ρ is updated. After the execution of *nbn* neighbourhood searches, the algorithm starts a new iteration by decreasing the temperature T. This algorithm continues iteratively while the stoppage condition does not hold.

Hybridisation. Hybridisation can be regarded as an approach to use the advantages of multiple metaheuristics at the same time. As the MOPSO and the MOSA have shown good capability and conform well with the structure of our problem, we decided to develop an integrative hybrid of them. In our hybrid algorithm, an initial population of particles is randomly generated. Then, in each iteration, firstly, the concept of the presented MOPSO is used and the particles move to their new positions. Henceforth, the mechanism of MOSA is applied by trying the neighbouring solutions of the positions with the acceptance probability. In other words, an MOSA is run at the end of each MOPSO iteration to improve the solutions. The termination condition of the MOSA is passing a pre-defined number of iterations $Maxit$, whereas the termination criterion of the MOPSO or the whole hybrid algorithm is stagnation over a number of consecutive iterations MUI. The pseudocode of this hybrid metaheuristic is given in Algorithm 1. This algorithm has some important parameters related to the MOPSO and the MOSA such as $|S|$, ω, MUI, T_0, *nbn* and $Maxit$, which have to be efficiently set.

Application for the GSP. An important part of the implementation of a meta-heuristic is how to encode a solution in a form that can be conveniently used in the algorithm. Two different solution representations are defined for our hybrid algorithm because it applies the concepts of two metaheuristics. The

Algorithm 1: The Proposed Hybrid Metaheuristic

Data: The problems' inputs and the algorithm's parameters
Result: A set of high-quality non-dominated solutions

```
1  - Generate initial solutions S, or initial positions of particles.
2  - Generate a random velocity for each particle.
3  - Set REP = ∅
4  - q = 0. q is the current number of consecutive unsuccessful iterations.
5  while q ≤ MUI do
6      for s ∈ S do
7          - Calculate the new velocity of the particle according to 15.
8          - Move the particle s to its new position according to 16.
9          - Evaluate the particle by the calculation of M_1 according to 17.
10         - Start the MOSA by setting T = T_0.
11         for it = 1:Maxit do
12             for neighbour = 1 : nbn do
13                 - Construct a neighbouring solution of the particle position.
14                 - Accept the neighbouring solution with the probability 19.
15             end
16             - Decrease T.
17         end
18         - Update the personal best position of the particle BP.
19         - Update the REP if necessary.
20         - Update GP if necessary based calculation of M_2.
21     end
22     - Set q = q + 1 if no non-dominated solution is found within the iteration;
         otherwise set q = 0.
23 end
24 - Report the REP.
```

MOSA is applicable to problems with discrete variables, which is the case in our GSP. So our first solution representation is a string of cells, where each cell corresponds to one flight and contains the id of the gate chosen to serve it. Figure 1 illustrates an encoded solution which can be used in the MOSA part.

3	5	2	2	5	1	4	2	4	1

Fig. 1. Solution representation for the MOSA: There are ten flights and five gates, according to this solution, the first flight is assigned to gate 3, the second to gate 5, and so on.

However, since the MOPSO works with continuous variables, the whole assignment is converted to only one real value by considering the mentioned representation as being a number in the base-B numeral system and B = *the number of gates plus one.* Consequently, this number can be converted to its decimal equivalent or one in the base-ten. For example, the structure shown in Fig. 1 is regarded as a number in the base-six numeral system (because there are five gates), i.e. $(3522514241)_6$, and it is converted to a decimal number as follows: $3(6^0)+4(6^1)+2(6^2)+4(6^3)+1(6^4)+5(6^5)+2(6^6)+2(6^7)+5(6^8)+3(6^9) = 39325491$.

The MOPSO can use it as the particle position. Therefore, in the process of the hybrid algorithm, these two representations are constantly turned to each other in order to be usable by the MOPSO and the MOSA. Finally, the constraints of the problem are handled by adding a penalty function to the objectives. The penalty functions are specific for each objective and expressed as $Pen_{z_i} = CP \times NSV \times Max(z_i)$, where CP is a constant factor, NSV is the normalised sum of violations in all constraints and Max_{z_i} is the considered maximum of the objective i.

5 Computational Experiments

The computational efforts done in this work begin with generating the test instances, then the considered methodologies have to be well parameterised, subsequently, the instances can be tackled with the approaches to compare their merit. Our experiments are run on machines with a Core(TM) i7 processor, 3.10 GHz CPU, and 16 GB of RAM.

5.1 Test Instances

The test instances are generated based on real data about the arrival and departure flights at five major airports in the world, which are obtained online. However, we consider a simple airport layout as shown in Fig. 2. This is an example where the airport has altogether 10 gates, the distance between two adjacent gates is one unit, the distance from the entrance/exit to the nearest gate is assumed to be 5 units, there are three shopping areas, one for every four gates, which cover 2 units each. An unlimited number of remote (non-connected) gates at the apron are available, which are with the consideration of required bus transfer efforts 10 units away from the terminal. We assume that the number of gates is directly proportional to the number of flights as $|G| = \frac{|F|}{20}$ or one for every 20 flights. This terminal layout is extended or shrunk according to the number of flights existing in each instance. It means that the distances between the entities remain the same but the number of them is increased or decreased. A random reference gate is considered for 20% of the flights and it is due to the fact that only a proportion of flights tend to be operated at a fixed gate in the real world. Three possible delays are assumed for 5%, 10%, and 20% of the flights. The set of possible delays is $\{0, 10, 30, 45, 60\}$ in minutes and the scenarios are built accordingly. The considered planning horizon is 12 h. The number of flights included in the instances is from 100 to 1000.

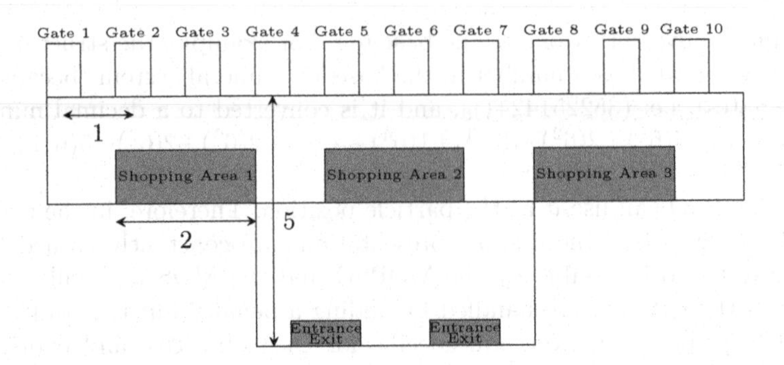

Fig. 2. The considered airport layout

5.2 Parameterisation of Algorithms

Setting the parameters at their efficient values plays a vital role in the successful performance of the optimisation algorithms. In the application of the weighted sum methods, the weights used in each run of the exact solver have to be correctly set and changed in order to increase the ability in finding Pareto-optimal solutions. In our implementation, the weights of the three objectives can take the values 0,0.1,0.2,0.3,...,1 regarding that their summation must be one. So the Baron solver processes the problem with the aggregated objective $10^3 = 1000$ times.

The tuning of parameters of our hybrid metaheuristic is done by the Taguchi method (see [19]), which works according to the design of experiment. We give the method three levels for each parameter, which are regarded as low, middle and high values, to choose from. The experiments are done on middle-sized instances. According to the applied parameter setting, the chosen parameter values are as follows: $MUI = 25$ from $\{10, 25, 50\}$, $|S|$=200 from $\{100, 200, 300\}$, $\omega = 0.6$ from $\{0.3, 0.6, 0.8\}$, $T_0 = 10^5$ from $\{10^3, 10^5, 10^7\}$, $nbn = 50$ from $\{10, 50, 100\}$, $Maxit = 100$ from $\{100, 200, 300\}$ and $CP = 100$ from $\{10, 100, 1000\}$.

5.3 Results and Comparison

The weighted sum method is programmed in GAMS, while the proposed hybrid metaheuristic along with a pure MOPSO and MOSA and a probabilistic TS approach based on the solution methodology presented in [3] are programmed in Python. The test instances are tackled by the method based on the presented robust model. Different metrics can be used to evaluate the performance of multi-objective optimisation approaches. Two evaluation metrics are here used. The first one is the quality metric (QM) that shows the share of each method in the set of overall non-dominated solutions obtained by all methods. It is calculated by the accumulation of Pareto-optimal sets of all methods in one pool and eliminating those which are dominated by any other solution within this pool; then the number of solutions provided by each method is divided by the size of the pool. Since duplicate answers are removed, one solution may be related to more than one method. The second one is Hypervolume (HV) that is used the most in the multi-objective literature [24] and measures the size of the objective space covered by the set of non-dominated solutions found by the method. A very significant advantage of this metric is that it indicates both accuracy and diversity. The metaheuristic methods run for each instance ten times and the average metrics of the replications are considered to have indicators which are based on wider experiments. Table 1 contains the QM and the HV based on the objective values normalised in [0,1] as well as the execution times of the applied solution methodologies. The solution time limit is set to 600 s for each run of the weighted sum method, so it is 6×10^5 in total.

Table 1. The (average) values of the multi-objective evaluation metrics and the (average) execution times in seconds of the methods in robust optimisation of the problem

$\|S\|$	Weighted sum			TS			MOPSO			MOSA			Hybrid		
	QM	HV	Time	QM	HV	Time	QM	HV	Time	QM	HV	Time	QM	HV	Time
100	0.15	0.65	6×10^5	0.54	0.71	25	0.56	0.72	18	0.54	0.68	25	0.82	0.83	41
200	0	0.53	6×10^5	0.56	0.70	30	0.52	0.68	26	0.53	0.50	33	0.80	0.78	48
300	0	0.48	6×10^5	0.48	0.63	34	0.50	0.67	32	0.51	0.48	41	0.76	0.75	56
400	0	0.42	6×10^5	0.52	0.56	45	0.51	0.64	36	0.48	0.47	49	0.78	0.73	63
500	0	0.41	6×10^5	0.51	0.42	52	0.48	0.62	42	0.45	0.50	48	0.72	0.79	73
600	0	0.38	6×10^5	0.48	0.58	51	0.47	0.65	47	0.46	0.45	53	0.67	0.76	82
700	0	0.36	6×10^5	0.47	0.55	64	0.45	0.62	55	0.40	0.46	61	0.71	0.74	88
800	0	0.35	6×10^5	0.49	0.58	70	0.46	0.62	62	0.41	0.43	68	0.73	0.78	97
900	0	0.31	6×10^5	0.45	0.62	107	0.42	0.60	87	0.42	0.41	95	0.68	0.75	111
1000	0	0.30	6×10^5	0.42	0.51	121	0.40	0.58	98	0.38	0.37	118	0.65	0.70	128

As it is observed in the results, the proposed hybrid metaheuristic outperforms the rest of the methods in terms of both metrics. Except for the smallest instance, where the weighted sum method can provide only one overall non-dominated solution, it is totally unable to find any in other cases within the very long considered time limit. The hybridisation has enhanced the ability in searching for better solutions. This can be perceived by comparing the results of the pure MOPSO and MOSA with those of the hybrid approach of these two. The results of the TS approach are quite similar to the pure MOPSO. In some cases, it provides partially better solutions, whereas in other cases its outcomes are slightly weaker. In terms of the execution time, the hybrid method is the slowest because two algorithms have to be run numerous times in every implementation of it. Nevertheless, the required execution times are not considerably longer and the method can be used in practice to deal with real-sized instances. So it provides high-quality solutions for the largest problem with 1000 flights only in 128 s, while the exact solution approach is unsuccessful even after almost one week of execution.

To have statistical comparisons, all the methods are compared through a non-parametric test (the Friedman test with the Bergmann-Hommel post hoc procedure [5]). The obtained p-values for the comparison of the hybrid method with others are near zero in all cases, which indicate a significant difference in their averages.

In the end, we conduct some complementary experiments to detect how much the objectives are deteriorated because of considering the robustness in our model. In these experiments, the non-robust model presented in Sect. 3.2 is solved for the same instances and the differences or the costs of robustness are analysed. The hybrid metaheuristic method, which is found to be the most efficient, is also used for the non-robust optimisation and the results are compared with those already presented in the three last columns of Table 1 in Table 2.

Table 2. The average metrics' values and the average execution times in seconds of the robust and non-robust approach and their average differences

$\|S\|$	Robust			Non-robust			Difference in percent (%)		
	QM	HV	Time	QM	HV	Time	QM	HV	Time
100	0.82	0.83	41	0.85	0.85	33	3.6	2.4	19.5
200	0.80	0.78	48	0.86	0.84	36	8.5	7.6	25
300	0.76	0.75	56	0.84	0.81	44	10.5	7.4	21.4
400	0.78	0.73	63	0.85	0.82	51	8.9	12.3	19
500	0.72	0.79	73	0.80	0.87	62	11.1	17.7	15.1
600	0.67	0.76	82	0.79	0.82	70	17.9	7.8	14.6
700	0.71	0.74	88	0.82	0.80	75	15.5	8.1	14.8
800	0.73	0.78	97	0.80	0.86	81	8.7	10.3	16.5
900	0.68	0.75	111	0.77	0.86	90	11.7	14.6	18.9
1000	0.65	0.70	128	0.74	0.81	106	13.8	15.7	17.1

As it is evident, the metrics' values related to the non-robust approach are better and the solutions are achieved in shorter execution times. This shows that we have to always accept a cost in terms of both the solution quality and the computation time if we aim at embedding robustness in the optimisation. It can be observed that the cost of robustness is more considerable in the large instances. However, regarding the important fact that the problem inputs do not have a deterministic nature, it can be worthwhile in many cases to bear this robustness cost in order to have more reliable solutions, which remain good in different scenarios.

6 Conclusions and Future Outlook

In this work, the GSP is addressed and a multi-objective model with the consideration of non-deterministic arrival times of flights at gates is presented for it. Some real-based instances of various sizes are created and solved with five different solution methodologies. The analysis of the methods' performances indicates that our proposed hybrid metaheuristic is superior to the exact solution method with the considered time limit, both its pure constituent metaheuristics and also another probabilistic approach based on a metaheuristic presented in a previous paper. Hence, we can provide fast non-dominated solutions for GSP instances of real sizes with several hundreds of flights. The Pareto-optimal solutions can be given to decision-makers to choose from based on their criteria and preferences.

An interesting future direction can be investigating the influence of the terminal layout or increasing the number of gates. Besides, other real facts such as breakdowns of gates and cancellation of flights can be taken into account. Finally, the application and development of other metaheuristics or matheuristic methods can be followed as a future subject with a lot of room for research.

References

1. https://www.gams.com/latest/docs/s_baron.html
2. GAMS Development Corporation, General Algebraic Modeling System (GAMS) Release 24.2.1. Washington, DC, USA (2013)
3. Aktel, A., Yagmahan, B., Özcan, T., Yenisey, M.M., Sansarcı, E.: The comparison of the metaheuristic algorithms performances on airport gate assignment problem. Transp. Res. Procedia **22**, 469–478 (2017). https://doi.org/10.1016/j.trpro.2017.03.061
4. Ben-Tal, A., Nemirovski, A.: Robust solutions of uncertain linear programs. Oper. Res. Lett. **25**(1), 1–13 (1999). https://doi.org/10.1016/s0167-6377(99)00016-4
5. Bergmann, B., Hommel, G.: Improvements of general multiple test procedures for redundant systems of hypotheses. In: Multiple Hypothesenprüfung / Multiple Hypotheses Testing, pp. 100–115. Springer, Heidelberg (1988). https://doi.org/10.1007/978-3-642-52307-6_8
6. Cheng, C.H., Ho, S.C., Kwan, C.L.: The use of meta-heuristics for airport gate assignment. Expert Syst. Appl. **39**(16), 12430–12437 (2012). https://doi.org/10.1016/j.eswa.2012.04.071
7. Daş, G.S.: New multi objective models for the gate assignment problem. Comput. Ind. Eng **109**, 347–356 (2017). https://doi.org/10.1016/j.cie.2017.04.042
8. Daş, G.S., Gzara, F., Stützle, T.: A review on airport gate assignment problems: single versus multi objective approaches. Omega **92**, 102146 (2020). https://doi.org/10.1016/j.omega.2019.102146
9. Dell'Orco, M., Marinelli, M., Altieri, M.G.: Solving the gate assignment problem through the fuzzy bee colony optimization. Transp. Res. Part C Emerg. Technol **80**, 424–438 (2017). https://doi.org/10.1016/j.trc.2017.03.019
10. Deng, W., Zhao, H., Yang, X., Xiong, J., Sun, M., Li, B.: Study on an improved adaptive PSO algorithm for solving multi-objective gate assignment. Appl. Soft Comput. **59**, 288–302 (2017). https://doi.org/10.1016/j.asoc.2017.06.004
11. Dijk, B., Santos, B.F., Pita, J.P.: The recoverable robust stand allocation problem: a GRU airport case study. OR Spectr. **41**(3), 615–639 (2018). https://doi.org/10.1007/s00291-018-0525-3
12. Dorndorf, U., Jaehn, F., Pesch, E.: Flight gate assignment and recovery strategies with stochastic arrival and departure times. OR Spectr. **39**(1), 65–93 (2016). https://doi.org/10.1007/s00291-016-0443-1
13. Freathy, P., O'Connell, F.: Planning for profit: the commercialization of European airports. Long Range Plan. **32**(6), 587–597 (1999). https://doi.org/10.1016/s0024-6301(99)00075-8

14. Genç, H.M., Erol, O.K., Eksin, İ, Berber, M.F., Güleryüz, B.O.: A stochastic neighborhood search approach for airport gate assignment problem. Expert Syst. Appl. **39**(1), 316–327 (2012). https://doi.org/10.1016/j.eswa.2011.07.021
15. Geuens, M., Vantomme, D., Brengman, M.: Developing a typology of airport shoppers. Tour. Manag. **25**(5), 615–622 (2004). https://doi.org/10.1016/j.tourman.2003.07.003
16. Kaliszewski, I., Miroforidis, J., Stańczak, J.: The airport gate assignment problem multi-objective optimization versus evolutionary multi-objective optimization. Comput. Sci. **18**(1), 41–52 (2017). https://doi.org/10.7494/csci.2017.18.1.41
17. Kumar, V.P., Bierlaire, M.: Multi-objective airport gate assignment problem in planning and operations. J. Adv. Transp **48**(7), 902–926 (2013). https://doi.org/10.1002/atr.1235
18. Mokhtarimousavi, S., Talebi, D., Asgari, H.: A non-dominated sorting genetic algorithm approach for optimization of multi-objective airport gate assignment problem. Transp. Res. Rec. J. Transp. Res. Board **2672**(23), 59–70 (2018). https://doi.org/10.1177/0361198118781386
19. Montgomery, D.C.: Design and Analysis of Experiments. John Wiley & Sons, Inc., Hoboken (2006)
20. Nikulin, Y., Drexl, A.: Theoretical aspects of multicriteria flight gate scheduling: deterministic and fuzzy models. J. Schedul. **13**(3), 261–280 (2009). https://doi.org/10.1007/s10951-009-0112-1
21. Parsopoulos, K.E., Vrahatis, M.N.: Particle swarm optimization method in multiobjective problems. In: Proceedings of the 2002 ACM symposium on Applied computing (SAC). ACM Press (2002). https://doi.org/10.1145/508791.508907
22. Pternea, M., Haghani, A.: Mathematical models for flight-to-gate reassignment with passenger flows: state-of-the-art comparative analysis, formulation improvement, and a new multidimensional assignment model. Comput. Ind. Eng. **123**, 103–118 (2018). https://doi.org/10.1016/j.cie.2018.05.038
23. Richter, S., Voss, S., Wulf, J.: A passenger movement forecast and optimisation system for airport terminals. Int. J. Aviat. Manag. **1**(1/2), 58 (2011). https://doi.org/10.1504/ijam.2011.038293
24. Riquelme, N., Von Lücken, C., Baran, B.: Performance metrics in multi-objective optimization. In: 2015 Latin American Computing Conference (CLEI), pp. 1–11 (2015). https://doi.org/10.1109/CLEI.2015.7360024
25. van Schaijk, O.R.P., Visser, H.G.: Robust flight-to-gate assignment using flight presence probabilities. Transp. Plan. Technol **40**(8), 928–945 (2017). https://doi.org/10.1080/03081060.2017.1355887
26. Serafini, P.: Simulated annealing for multi objective optimization problems. In: Multiple Criteria Decision Making, pp. 283–292. Springer, New York (1994). https://doi.org/10.1007/978-1-4612-2666-6_29
27. Yan, S., Tang, C.H.: A heuristic approach for airport gate assignments for stochastic flight delays. Eur. J. Oper. Res. **180**(2), 547–567 (2007). https://doi.org/10.1016/j.ejor.2006.05.002
28. Yu, C., Zhang, D., Lau, H.Y.: An adaptive large neighborhood search heuristic for solving a robust gate assignment problem. Expert Syst. Appl. **84**, 143–154 (2017). https://doi.org/10.1016/j.eswa.2017.04.050
29. Yu, C., Zhang, D., Lau, H.: MIP-based heuristics for solving robust gate assignment problems. Comput. Ind. Eng. **93**, 171–191 (2016). https://doi.org/10.1016/j.cie.2015.12.013

A Branch-and-Cut Algorithm for Aircraft Routing with Crew Assignment for On-Demand Air Transportation

Rafael Ajudarte de Campos(✉), Thiago Vieira, and Pedro Munari

Federal University of São Carlos, São Carlos, SP, Brazil
rafael.ajudarte@estudante.ufscar.br, munari@dep.ufscar.br

Abstract. We address the aircraft routing problem with crew assignment in the context of on-demand air transportation. This problem involves the design of least-cost routes for an aircraft set in order to service private flight requests, considering the customer preferences, fleet characteristics and maintenance events. Additionally, a crew team has to be assigned to each route while satisfying the crew legislation, including duty time limitations and minimum rest times. Despite its practical relevance, integrated aircraft routing and crew assignment has been barely explored in the literature addressing on-demand air transportation. In this paper, we propose a tailored branch-and-cut algorithm to effectively solve the addressed problem, which resorts to a strategy based on dynamic programming to separate cuts that guarantee the feasibility regarding crew legislation. In computational experiments carried out using real-life data provided by a company, the method obtained optimal solutions for all instances in less than five minutes. Moreover, these solutions indicate a potential improvement of around 23% in the operational cost when compared to the routes designed by the company, which highlights the benefits of using the proposed approach in practice.

Keywords: Aircraft routing · Crew assignment · On-demand air transportation

1 Introduction

The airline industry is known for operating in a dynamic environment where every decision may have a strong impact on operational costs. In a context like this, in which poorly-optimized decisions can easily undermine the company's already small profit margin, the Operations Research (OR) tools become of utmost importance. Optimization models and algorithms have been successfully

Supported by São Paulo Research Foundation (FAPESP) [grant numbers 19/22235-6, 19/23596-2, 20/11602-5, 16/01860-1], Coordenação de Aperfeiçoamento de Pessoal de Nível Superior - Brasil (CAPES) [Finance Code 001] and the National Council for Scientific and Technological Development (CNPq) [grant number 313220/2020-4].

M. Mes et al. (Eds.): ICCL 2021, LNCS 13004, pp. 611–626, 2021.
https://doi.org/10.1007/978-3-030-87672-2_40

used to assist the decision-making process, providing tools capable of generating efficient solutions within a reasonable time [1,10,13].

In this paper, we are interested in two important processes in the air transportation sector, namely aircraft routing, which involves the decision of which flights to assign to each aircraft [7,8,10], and crew assignment, which consists of assigning crew members to particular flight legs [4,6]. We consider these operations in the context of companies that offer private flights as part of fractional ownership programs. In such type of service, a customer partially owns an aircraft by paying a fraction of its price, giving them the right to fly for a certain number of hours per year [15]. These companies operate differently from traditional commercial airlines because the customer has decision-making power regarding the flight starting times, the route and the type of the requested aircraft. As customer's requests are mandatory by contract and generate revenue, this type of company usually focuses on minimizing the operating costs particularly related to *positioning* flights [14]. These are flights in which the aircraft flies without customers to the departure airport of a request, without providing any profit for the company. This type of flight usually represents approximately 35% of the time an aircraft is in the air, and thus it is an important cost component for the company [15].

Another relevant source of operating costs in this type of company is the *upgrade* costs, which arises when customers are serviced by an aircraft that is better, and more expensive, than the one they hired. Despite the additional cost, an upgrade can be used strategically by the company to obtain savings with respect to positioning a farther away aircraft of the type chosen by the customer. This approach allows the company to reduce their total costs and meet requests that would be impossible otherwise [10]. The downside is that this freedom considerably increases the complexity of the decision-making process. It should be noted that the companies' policy usually does not allow customers to be serviced by aircraft that are inferior to the one requested (downgrade).

Moreover, the company is responsible for the fleet maintenance. Periodically, each aircraft must go through a planned checking and maintenance process, becoming unavailable until it is finished. Although the start time of a maintenance event is pre-scheduled, the company is typically allowed to advance or delay this time within a relatively large margin of 24 h. Thus, maintenance can be seen as a request in which the aircraft must be stationary at a single airport for a certain period of time and which presents a comprehensive time window, allowing greater flexibility.

Another important point the company must consider while planning the aircraft routes is the various regulations related to the crew's working and resting time. Some of the most relevant are the maximum time allowed in a duty, the maximum accumulated flight time in a duty, the minimum time of rest between two duties, and the maximum time that crew members can be away from their base (the place where they are usually hosted when not working) [4,11].

Considering routing and crew requirements simultaneously in the planning stage, albeit a more complex activity, brings considerable economic advantages to the solution [2,9,12]. This is compatible with the company's focus on minimizing operating costs and can be interesting to apply in practice. However, it is worth noting that due to the high dynamics of this sector, it is essential not only to generate an efficient solution, but also to obtain it in a short time span. This is particularly more relevant in on-demand air transportation companies, such as the one studied in this paper, because, by contract, customers can request a flight as little as four hours in advance.

In the literature addressing models and solution approaches for aiding decision-making in the described context, authors typically resort to specialized algorithms to effectively obtain solutions, such as decomposition techniques [9], branch-and-price methods [14] and heuristic approaches [2]. We emphasize that the majority of works addressing integrated routing and crew requirements were developed for traditional companies [2,9]. Furthermore, we are not aware of any other study that considers all aspects of the situation addressed in this paper, in the context of on-demand air transportation.

In this paper, we aim to obtain economical solutions that consider requests' requirements and crew rules, based on the real case of an air transportation company that offers fractional ownership management services and operates primarily in Europe and Asia. We propose a branch-and-cut (B&C) algorithm that dynamically inserts cuts into a recent compact model from the literature [10], as a way to enforce the crew's requirements. Cut separation is done based on the labeling algorithm, a dynamic programming procedure that proved to be efficient in practice. In addition to the characteristics already considered in the literature, such as routing, maintenance request and possibility of service upgrade, we further ensure that the maximum duration of a duty and the minimum rest time between two duties are respected. Furthermore, we inserted into the formulation the possibility of outsourcing a customer requests. Computational experiments were carried out using real-life data provided by a company, and optimal solutions were quickly obtained for all instances.

The remainder of this paper is structured as follows. Section 2 presents the base model for aircraft routing that is used in the B&C algorithm. Section 3 describes the B&C algorithm and the cuts generated to enforce crew requirements. The results of computational experiments are shown in Sect. 4 and, finally, the concluding remarks and next steps are presented in Sect. 5.

2 Aircraft Routing Model

The B&C algorithm is based on an aircraft routing model that is initially created without considering crew requirements. This model was originally proposed in [10] for the same context of a fractional ownership management company. A feature that makes this model stand out among other compact formulations [5,14] is that instead of using a problem representation based on the traditional network in which nodes represent the airports, the authors considered an alternative network in which the nodes represent customer requests, as depicted in Fig. 1. Thus, the decision variables select the sequence of requests that each aircraft will perform. This typically leads to a more efficient optimization model than traditional formulations with the standard representation [10].

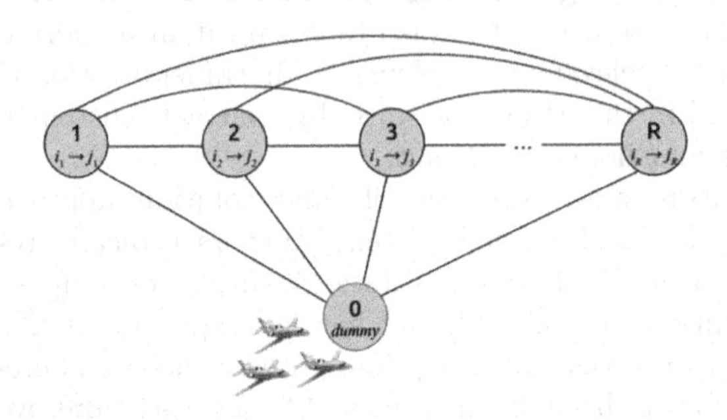

Fig. 1. Representation of the flow network through requests. Source: [10]

Consider the aircraft set V, which can be partitioned according to an ordered set of types P. For each type $p \in P$, we create subsets V_p that lists the aircraft of that type. The requests are represented by the set R which is composed of two subsets: L, which contains the customer requests; and M, which includes the maintenance events. Additionally, set R also contains a dummy request 0. The parameters considered in the model are: the positioning costs of aircraft v from the request r to the request s, C_{vrs}; the travel time between airports i and j for an aircraft of type p, T_{ij}^{p}; the moment when the aircraft $v \in V$ becomes available for its first flight in the planning horizon, AV_v; the airport at which an aircraft v is parked at the start of the planning horizon, k_v; the type of aircraft v, $t_v \in P$; the time the aircraft takes at an airport such as boarding and taxiing, TAT_k^r; the expected starting time of request $r \in R$, ST_r, which can be delayed by an amount of Δ_L minutes if it is a customer request, or advanced and delayed by Δ_M minutes if it is a maintenance event; we also need the departure and destination airports of request r, represented by i_r and j_r, respectively; the type of aircraft required for a request r is represented by parameter p_r; and TL_r is the maintenance time of a request $r \in M$.

Considering the sets and parameters previously presented, the model developed in [10] uses two groups of decision variables. The first one is composed by the binary variables y_{vrs}, which assume the value 1 if, and only if, aircraft v services request s immediately after servicing request r. The second is composed of the continuous variables w_r, which represent the earliest time that request r can be executed. Thus, the model is given by:

$$\min \quad \sum_{v \in V} \sum_{r \in R} \sum_{s \in R} C_{vrs} y_{vrs} + \sum_{r \in L} \sum_{\substack{s \in R \\ s \neq r}} \sum_{\substack{p \in P \\ p > p_r}} \sum_{v \in V_p} (c_{t_v} T_{i_r j_r}^{t_v} - c_{p_r} T_{i_r j_r}^{p_r}) y_{vrs} \tag{1}$$

$$\text{s.t.} \quad \sum_{\substack{p \in P \\ p \geq p_r}} \sum_{v \in V_p} \sum_{\substack{s \in R \\ s \neq r}} y_{vrs} = 1, \qquad r \in L, \tag{2}$$

$$\sum_{\substack{s \in R \\ s \neq r}} y_{vrs} = 1, \qquad r \in M, v \in V, \tag{3}$$

$$\sum_{\substack{s\in R\\ s\neq r}} y_{vrs} = \sum_{\substack{s\in R\\ s\neq r}} y_{vsr}, \qquad v \in V, r \in R, r > 0, \quad (4)$$

$$\sum_{s\in R} y_{v0s} = 1 = \sum_{r\in R} y_{vr0}, \qquad v \in V, \quad (5)$$

$$ST_r \leq w_r \leq ST_r + \Delta_L, \qquad r \in L, \quad (6)$$

$$ST_r - \Delta_M \leq w_r \leq ST_r + \Delta_M, \qquad r \in M, \quad (7)$$

$$w_s \geq w_r + \sum_{v\in V}(T^{t_v}_{i_r j_r} + TAT^s_{jr} + T^{t_v}_{j_r i_s} + TAT^s_{i_s})y_{vrs} + M^1_{rs}(\sum_{v\in V} y_{vrs} - 1),$$
$$r \in L, s \in R, r \neq s, s > 0, j_r \neq i_s, \quad (8)$$

$$w_s \geq w_r + \sum_{v\in V}(T^{t_v}_{i_r j_r} + TAT^s_{is})y_{vrs} + M^2_{rs}(\sum_{v\in V} y_{vrs} - 1),$$
$$r \in L, s \in R, r \neq s, s > 0, j_r = i_s, \quad (9)$$

$$w_s \geq (AV_v + T^{t_v}_{k_v i_s} + TAT^s_{i_s})y_{v0s}, s \in L, v \in V, k_v \neq i_s, \quad (10)$$

$$w_s \geq (AV_v + TAT^s_{i_s})y_{v0s}, \quad s \in L, v \in V, k_v = i_s, \quad (11)$$

$$w_s \geq w_r + TL_r + T^{p_r}_{j_r i_s} + TAT^s_{i_s} + M^3_{rs}(y_{v_r rs} - 1),$$
$$r \in M, s \in R, r \neq s, s > 0, j_r \neq i_s, \quad (12)$$

$$w_s \geq w_r + TL_r + TAT^s_{i_s} + M^4_{rs}(y_{v_r rs} - 1),$$
$$r \in M, s \in R, r \neq s, s > 0, j_r = i_s, \quad (13)$$

$$w_s \geq (AV_{v_s} + T^{p_s}_{k_{v_s} i_s})y_{v_s 0s}, \qquad s \in M, k_{v_s} \neq i_s, \quad (14)$$

$$w_s \geq AV_{v_s} y_{v_s 0s}, \qquad s \in M, k_{v_s} = i_s, \quad (15)$$

$$w_r \geq 0, \qquad r \in R, \quad (16)$$

$$y_{vrs} \in \{0, 1\}, \qquad v \in V, s, r \in R. \quad (17)$$

The objective function (1) consists of minimizing the operational costs. These costs are composed by the costs of aircraft positioning, which arise in trips that aircraft fly alone, represented by the first term, and upgrade cost, the increase in cost when servicing a request with an aircraft better than the one contracted, represented by the second term. Constraints (2)–(4) ensure that every request is fulfilled once and the correct flow of aircraft. Constraints (5) enforce the balance in the dummy request, where every aircraft must depart from and return to. The time windows for customer and maintenance requests are determined by constraints (6) and (7), respectively. The minimum time to start a customer's request s is computed by constraints (8) and (9). The first set is activated in the case in which positioning is needed to service request s after r while the second is used when the destination of request r is the same as the departure airport of s. Constraints (10) and (11) are used to calculate the time each aircraft will be ready to service the first request of their planning horizon, with the first set of constraints being used when positioning between the aircraft's starting airport and the request's departure airport is necessary and the second one is used when these airports are the same. Constraints (12)–(15) are analogous to (8)–(11), but for maintenance requests. The main difference between this type of request and customer ones is that instead of requiring a flight from i_r to j_r to execute the request, the aircraft must stay on the ground at the i_r airport during the whole duration (TL_r) of the maintenance process. Finally, the domain of the variables is defined in constraints (16) and (17).

We extend this formulation to consider the possibility of outsourcing a customer request to another company, if it is not possible to service them with the current fleet. Let Co_r be the cost of outsourcing request $r \in L$, and out_r be the binary decision variable that indicates whether request r should be serviced by other company. With this new information, we extend the objective function by adding the following term:

$$\sum_{r \in L} Co_r out_r. \tag{18}$$

It indicates the total cost of outsourced requests. We set Co_r as twice the maximum positioning costs between all existing requests using the best aircraft possible. Thus, Co_r is considerably bigger than any positioning costs, and should be used as little as possible. We also need to change constraints to allow the outsourcing to fulfill some requests if needed. To do this, we simply need to replace constraints (2) and (6), in this order, by the following expressions:

$$\sum_{\substack{p \in P \\ p \geq p_r}} \sum_{v \in V_p} \sum_{\substack{s \in R \\ s \neq r}} y_{vrs} + out_r = 1, \qquad r \in L, \tag{19}$$

$$ST_r(1 - out_r) \leq w_r \leq (ST_r + \Delta_L)(1 - out_r), \qquad r \in L. \tag{20}$$

With the base routing model finished, we can now describe the B&C algorithm developed to dynamically insert the crew rules.

3 Branch-and-Cut Algorithm

To incorporate crew assignment to the models described in the previous section, we develop a B&C algorithm that dynamically generates cuts when a candidate solution is infeasible regarding crew rules. Before detailing the complete algorithm, we describe the crew requirements that are evaluated in the separation routines.

The first requirement we consider is the maximum allowed time without rest in a single duty (*MaxDuty*), which is typically defined as 13 h in the studied company. This is guaranteed by international regulations due to the risks associated with crew fatigue. Moreover, the company must ensure the crew has at least 10 h (*MinRest*) of uninterrupted rest between two consecutive duties. Every time a complete rest occurs, the accumulated *duty* is reset. A particularity on maintenance requests is that, since the aircraft is parked during the entire process, the crew can rest during the event and, hence, the company can take advantage of the maintenance time. If maintenance lasts longer than *MinRest*, it is interesting to extend the crew's free time until the end of the request, as there is no reason to keep the pilots on stand-by if the aircraft is not ready yet. Conversely, if the duration of maintenance is less than *MinRest*, the crew still needs to rest for the minimum time, even if the aircraft is available earlier. Anyway, it is interesting to take advantage of maintenance to cover part of, or completely, the rest, as this allows for time saving and feasibility of solutions that would be impossible otherwise.

Other important elements considered by the company are the crew presentation times to prepare the plane and analyze the weather conditions and itinerary. The first one occurs at the beginning of a duty (*PRE*), usually taking 40 min, being counted within the crew's duty. The second one happens at the ending of a duty (*POS*), lasting 30 min and is neither counted in the duty nor the rest time. Thus, whenever there is a rest in the planning horizon we must also insert the presentation time in the start and end of the duty. Figure 2 presents a visual representation of these concepts, in which we created two examples to facilitate the understanding of the different behaviors of resting. In both examples we have a route that, during the planning horizon, starts from a rest, then the aircraft services a customer request, position itself, services another request, takes a rest and finally services a last customer request. The only difference between them is that in Example 1 the second request serviced by the aircraft is a customer's while in Example 2 we have a maintenance. We can note that in Example 2, the crew is allowed to start its rest right after positioning and does not need to be available during the whole maintenance (represented by the green outline in the rest period).

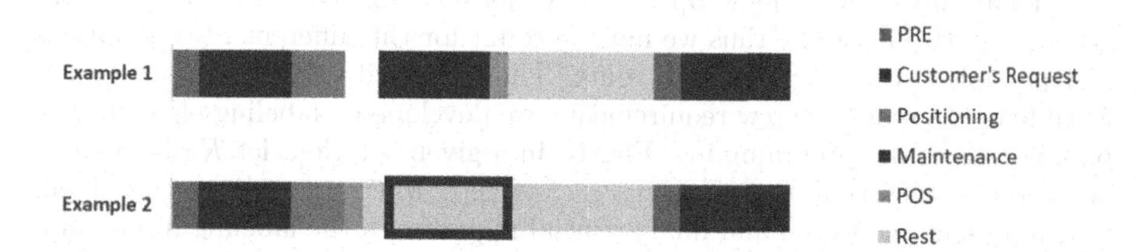

Fig. 2. Examples of rest behaviour for two vehicle routes. (Color figure online)

The B&C method, depicted in Fig. 3, incorporates these requirements by adding cuts that eliminate infeasible solutions regarding crew constraints. The separation algorithm is called inside CPLEX's enumeration tree, using callback procedures. Every node of this tree consists of a linear programming (LP) relaxation of model (1)–(17), possibly with additional branching constraints and cuts generated on previously solved nodes. If the solution of the LP relaxation in a given node is not integer, the solver automatically branches based on the fractional variables, and thus create two new nodes in the tree. Otherwise, if the solution is integer, the solver invokes the cut separation algorithm (which is detailed in Algorithm 1) to check if the solution is feasible considering crew requirements. If all the considered crew requirements are satisfied, we accept it as a candidate for optimal solution. Conversely, if the solution is infeasible regarding the crew constraints, we reject the solution and add cuts to the model, to avoid obtaining the same solution again.

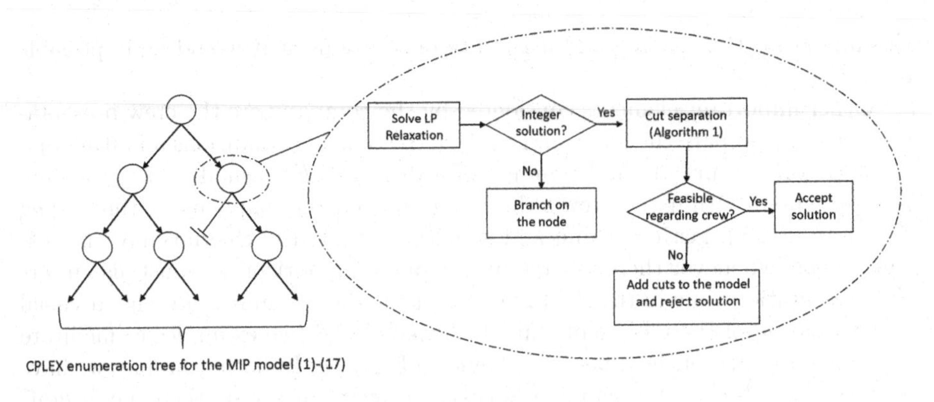

Fig. 3. General structure of the B&C algorithm.

Regarding the cut separation, there are different options that should be evaluated when considering the possibility of resting between two requests, and it is not possible to define the correct alternative when analyzing the route only up to a specific node. We also need to ascertain this decision's impact on the remaining nodes. For example, resting before the n-th node in a route might not violate any time window up to this node, but this extra time might make the $(n+1)$-th infeasible, thus we must account for the different resting options and possible future impacts in a route. Thus, to verify whether a solution is feasible in the scope of crew requirements, we developed a labeling algorithm [3] based on dynamic programming. Firstly, in a given solution, let $\mathcal{R}_v$ be a route of aircraft v such that $\mathcal{R}_v = (r_0, r_1, r_2, \ldots, r_{n+1})$, where r_0 and r_{n+1} are nodes that represent the initial and final artificial requests. In the labeling framework, we assign a *bucket* to each node and, inside each bucket, there are the labels related to this node. Labels are data structures, responsible to carry a series of essential information through the buckets for calculating accumulated resources and checking the feasibility condition. Regarding the crew requirements, we consider two resources, the total elapsed time ($Elap_l$) and the total duty time since the last rest (DT_l), both accumulated up to label l of a given bucket. In addition, each label keeps the information of the parent (previous) label, useful for backtracking in post-processing. This procedure is summarized in Algorithm 1.

The first operation in Algorithm 1, after getting the route R_v and creating its respective *buckets,* is defining a label at the bucket of node r_0 with the initial characteristics of the aircraft, such as the time it is available and the accumulated duty time since the last rest at the beginning of the planning horizon. Accumulated duty values are non-zero when the company's planning starts in the middle of a duty of a specific aircraft. After creating the label for node r_0, the algorithm extends it to the bucket of the first request in the route, r_1. Likewise, each generated label will be extended to the next nodes of the route, or at least an attempt will be made. This process is repeated until the last node in the route is reached or there is no way to create any label for the next node in the route, in which case the solution is deemed infeasible. As pointed out in line

Algorithm 1: Separation algorithm

```
   Input: Solution candidate.
   Output: Feasibility solution for the crew assignments.
 1 foreach aircraft v ∈ V do
 2     Get R_v and create a bucket for each node in it;
 3     Create label l in the bucket of node r_0 with the initial properties of v;
 4     for i ← 0 to n, step + 1 do
 5         foreach label l ∈ bucket of node r_i do
 6             Extend the current label l to all five possible cases and create child
               labels for those that are feasible in r_{i+1};
 7         end
 8         Check whether the created child labels dominate or are dominated by
           other existing labels in the bucket of node r_i or by themselves;
 9         Delete the dominated labels;
10         if there is no label in the current node then
11             Infeasible route: Add no-good cuts to eliminate solution;
12             break;
13         end
14     end
15 end
16 if cuts were not inserted in the model then
17     return Feasible solution for the crew requirements;
18 else
19     return Flag the solution candidate as infeasible for the crew requirements;
20 end
```

6 of the algorithm, when extending a label, there are five different possible cases related to rest decisions and whenever a verified case is feasible, a child label is generated using its strategy. Thus, each label can have up to five children. The options related to rest that should be checked are represented in Fig. 4.

The first case is the option of executing request r_i and immediately prepare to start request r_{i+1}, without any rest. Case 2 occurs when the crew takes a rest between requests r_i and r_{i+1} at the departure airport of request r_{i+1}. In this case the aircraft will start request r_{i+1} immediately after ending the rest. If a positioning is required, Case 2 can be interpreted as taking a rest after the positioning flight. Conversely, if there is no positioning the rest starts immediately after the end of request r_i, if it is a customer request, or at the start of it, if it is a maintenance request. Case 3 only exists if a positioning between requests r_i and r_{i+1} is required and represents the possibility to rest before the positioning. This option is usually taken when positioning would exceed the duty limit. Case 4 is a combination of Cases 2 and 3, and represents the attempt to take a complete rest before and after positioning, this one is especially useful on particularly long positioning trips. Finally, Case 5 can be chosen in a particular situation where two consecutive maintenance requests on an aircraft happen at the same airport. This allows the second maintenance to start while the crew

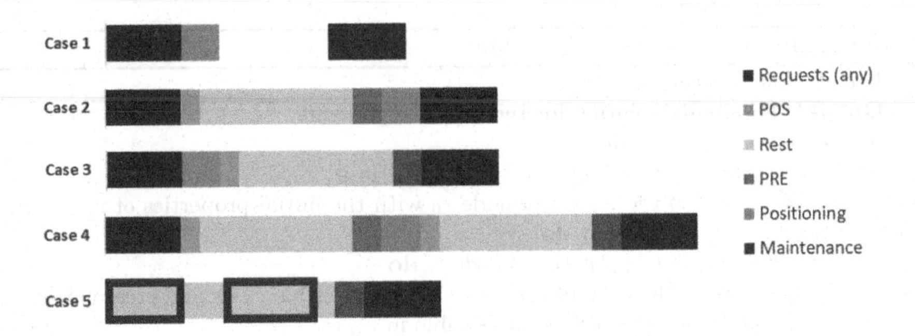

Fig. 4. Visual representation of all cases checked when extending a label to the next node.

is still resting, instead of forcing the crew to return to work to start the second maintenance like in Case 2. Note that the label generated in Case 5 is not inserted in the next node of the route (r_{i+1}), but in the following one (r_{i+2}).

When a label is created, the algorithm verifies if it dominates or is dominated by any other existing label in the evaluated node. A label l_1 dominates a label l_2 in the same node or bucket if and only if: $Elap_{l_1} \leq Elap_{l_2}$ and $DT_{l_1} \leq DT_{l_2}$. In other words, label l_1 dominates l_2 if, and only if, it has lower or equal elapsed time and less accumulated duty time at the start of the evaluated node. This can happen, for example, when it is possible to take a rest between two requests and the total elapsed time is still lower than the opening time windows, and thus Case 1 will be dominated by the ones that allow resting. Another situation in which this usually happens is when we compare Cases 2 and 3. The former usually dominates the latter, since its accumulated duty time at the start of the next request is lower, except in the situation Case 2 is infeasible. A label in the bucket of r_i is considered infeasible, and thus not allowed to be inserted, if it violates the time window of request r_i ($Elaps_l > ST_{r_i} + \Delta$) or violates the maximum duty time ($DT_l > MaxDuty$).

Figure 5 presents a numerical and visual example on how the separation algorithm checks the feasibility of a route. In this example, we want to verify if route $\mathcal{R} = (r_0, r_1, r_2, r_3, r_0)$ is feasible. This particular route is composed by the artificial node r_0 (from where the aircraft departs and returns) and three customer requests. The time windows for each request is found just below their respective bucket and the duration of each request is found above it, i.e., 120 min for r_1, 150 for r_2 and 180 for r_3. The positioning times are found over the arrow connecting two consecutive requests, thus 50 min between r_0 and r_1, 60 between r_1 and r_2 and 90 min between r_2 and r_3. We start by generating a label in node r_0, this label has a total elapsed time (E) and accumulated duty time (D) equal to 0. We then extend the generated label with the five possible cases, not forgetting to add the presentation time (PRE) in this first extension. Since the time window from request r_1 closes very early, Cases 2 to 4 are all infeasible because if the crew takes a rest the elapsed time will surpass the closing of the time windows. Furthermore, since there is no maintenance event in this example, Case 5 will never be used. The remaining label, generated by Case 1, is then extended to node r_2 and a similar situation happens, in which only the label created by Case 1 is feasible due to the time window constraints. However, this time the aircraft arrived before the request starting time, thus it needs to wait before starting the service, accumulating duty in the process. We now extend the label generated to the bucket of r_3. Unlike in the previous nodes, opting to not rest (Case 1) is not feasible, due to the maximum duty duration constraints, while Cases 2 and 3, in which we try to rest after and before the positioning, respectively, are. However, Case 3 is dominated by Case 2 because the accumulated duty from the latter is lower than the former, while both have the same elapsed time. Finally, we can easily extend the generated label to the remaining node (r_0) without any problem. Thus, we were able to successfully generate labels for all nodes in the route and confirm that this route is feasible.

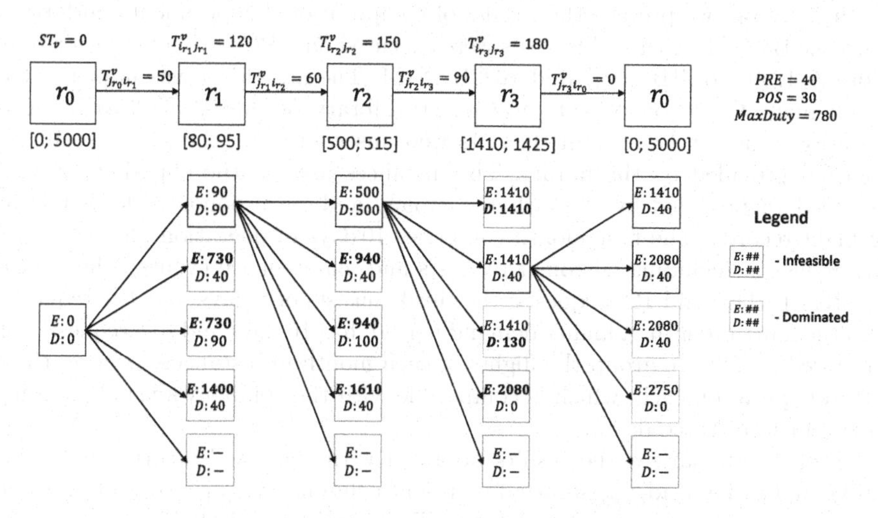

Fig. 5. Visual representation of a labeling structure.

If it is not possible to generate any label in some node, we consider the route infeasible and cut-off the solution. We consider two different types of cuts for this purpose. The first is the *feasibility cut*, in which we cut the route up to the node with no labels, limiting the sum of the binary variables. Suppose route $\mathcal{R}_v = (r_0, r_1, \ldots, r_k \ldots, r_n)$ is assigned to vehicle v and it becomes infeasible at node r_k. The algorithm would generate the following cut:

$$y_{vr_0r_1} + y_{vr_1r_2} + y_{vr_2r_3} + \cdots + y_{vr_{k-1}r_k} \leq k - 1; \tag{21}$$

This inequality prevents vehicle v to take this sequence of nodes up to node r_k, and thus cutting not only this route, but any other that would try this combination in v up to node r_k.

The second type of cut is used to define a lower bound for the arrival time in a customer if a route similar to the one evaluated is taken in future solution candidates. Let $MinElap_i$ be the smallest elapsed time among all the labels of node r_i. For the same infeasible route exemplified above, we insert for each node up to node r_k cuts of type:

$$MinElap_i(y_{vr_0r_1} + y_{vr_1r_2} + y_{vr_2r_3} + \cdots + y_{vr_{i-1}r_i} - i + 1) \leq w_{r_i}, \qquad i < k; \quad (22)$$

In the worst case, $MinElap_i$ will be equivalent to the opening of the time window constraints. However, if $MinElap_i$ is greater than the opening of the time windows, these cuts can prevent infeasible solutions for crew constraints without the need to call the separation algorithm again.

4 Computational Results

In this section, we present the results of computational experiments performed with the B&C algorithm. The tests were executed on a PC with a processor Intel Core i7-4790 3.6 GHz CPU and 16 GB RAM. The algorithm was implemented in language C++, on top of the Concert Library of the IBM CPLEX Optimization Studio v.12.10, and the cuts were inserted using the generic callback routines provided by the library. The instances used in the experiments were actual data provided by the airline company and corresponds to four months of flight records. The first month comprises 10 days of operation and a total of 112 requests (including customer requests and maintenance events); the second involves 10 days and 129 requests; the third consists of 8 days and 107 requests; and the fourth month, a higher demand period, has 16 days and 578 requests. As proposed in [10], we group the flights of each month in instances covering three days of operation each, which is compatible with the company's usual planning horizon (up to three days).

Table 1 summarizes the results obtained with the B&C algorithm. In this table, under the header *Instance*, we present columns *ID*, $|L|$, $|M|$ and n, which details the analyzed instances. Column *ID* identifies each individual instance in the format Mx_$y$to$z$, where x indicates the month, and y and z represent the first and last days covered by the instance, respectively. The three remaining columns present, respectively, the number of customer requests, the number of maintenance requests and their sum. Furthermore, we present the positioning (*Cpos*), upgrade (*Cup*), outsourcing (*Cout*) and total (*Ctot*) costs of the solutions provided by, the model without the crew's cuts and the full B&C algorithm. The company costs (*Ctot*) in particular are composed exclusively by positioning, since is the only information we were able to extract from its logs. We also present the computational times (*CPUt*) in seconds to obtain the optimal solution for both the model and the full algorithm. Finally, under the header "Comparison" we compare the B&C results against the solution provided by the company and the compact model by analyzing how much our algorithm improved the company's solution (*Ipv Co*), how much a solution with crew constraints is more expensive than the ones without these requirements (*Diff Crw*) and the increased computational time to obtain an optimal solution with the B&C algorithm against the simplified compact model (*Diff CPUt*). Note that, since we used an artificial big-M parameter to define the outsourcing costs, we compare only the positioning and upgrade costs against the company's solution for parameter *Ipv Co.*

A first point we identified is that, in general, our B&C algorithm not only obtained an optimal solution quickly, averaging 12.20 s and taking 239.43 s at most, but also improved the company's solution in most instances. On average, our solutions were 23.1% cheaper than the ones provided by the company. As expected, the instances of the fourth month, the one with higher demand and 110 requests on average, were the hardest to solve. Nevertheless, the B&C obtained optimal solutions for them with an average time of 30.94 s, and only two of them took more than 30 s to be solved.

Note that in some instances of months 3 and 4, our algorithm provided more expensive solutions than the company. This mostly happened because, in practice, the company can negotiate with the pilots to work overtime in exchange for longer rests or higher payment. For example, in instance M3_6to8, our solution was more expensive than the company's because it positioned an aircraft to service a specific customer, while the company actually negotiated with the crew members of an aircraft positioned in the requested airport to extend their duty over the limit to service that customer. Another factor that resulted in these differences is the *split duty rule*, a more complex strategy a company can employ in which the crew can take a break, a pause period shorter than a normal rest, and extend the duty duration by a fraction of the break time. Even so, our proposed algorithm proved to be a very useful tool for the studied company, being able to generate good solutions far quicker than the manual method previously used by them, which could take hours to design all routes from scratch.

When comparing the solutions of the B&C and the compact model, we noted that considering crew requirements had some negative impact in the optimal solution, increasing it by 13.0% on average. This was expected since we are now dealing with a problem with more, and tighter, constraints. This difference was more noticeable in months with higher customer demand, such as months 3 and 4, since the crew have less opportunities to rest than months with proportionally more maintenance events. Remarkably, the majority of instances in months 1 and 2 actually had the same solution for the problem with and without crew requirements. Moreover, as expected, using the B&C algorithm to insert these constraints iteratively resulted in longer solution times than simply running the compact model with no crew constraints. However, this increase was relatively small, with an average of only 10 s among all the instances. Notably, in instances of the first three months, the B&C algorithm increased the total computational time by less than one second.

Table 1. Computational results of all company's instances

Instance				Company solution	Compact model solution					Full algorithm solution					Comparison		
ID	|L|	|M|	n	Ctot	Cpos	Cup	Cout	Ctot	CPUt(s)	Cpos	Cup	Cout	Ctot	CPUt(s)	Ipv Co (%)	Diff Crw (%)	Diff CPUt
M1_1to3	14	22	37	87,527.54	64,413.69	3,860.80	0.00	68,274.49	0.11	71,708.87	1,046.86	0.00	72,755.73	0.19	16.9%	6.2%	0.08
M1_2to4	17	17	35	130,818.20	101,415.89	1,820.29	0.00	103,236.18	0.10	101,415.89	1,820.29	0.00	103,236.18	0.16	21.1%	0.0%	0.06
M1_3to5	24	16	41	127,816.68	81,364.02	2,893.93	0.00	84,257.95	0.23	81,364.02	2,893.93	0.00	84,257.95	0.22	34.1%	0.0%	-0.01
M1_4to6	24	18	43	93,718.44	61,739.88	1,937.22	0.00	63,677.10	0.28	61,739.88	1,937.22	0.00	63,677.10	0.26	32.1%	0.0%	-0.02
M1_5to7	22	17	40	89,792.18	66,465.02	5,167.71	0.00	71,632.73	0.19	66,465.02	5,167.71	0.00	71,632.73	0.25	20.2%	0.0%	0.05
M1_6to8	15	16	32	81,559.91	70,756.46	2,170.40	0.00	72,926.86	0.08	70,756.46	2,170.40	0.00	72,926.86	0.08	10.6%	0.0%	0.00
M1_7to9	14	14	29	74,651.34	58,959.51	2,912.28	0.00	61,871.79	0.05	58,959.51	2,912.28	0.00	61,871.79	0.05	17.1%	0.0%	0.00
M1_8to10	11	12	24	77,680.91	69,947.47	1,990.35	0.00	71,937.82	0.04	69,947.47	1,990.35	0.00	71,937.82	0.04	7.4%	0.0%	0.00
Avg M1	**17.63**	**16.50**	**35.13**	**95,445.65**	**71,882.74**	**2,844.12**	**0.00**	**74,726.87**	**0.13**	**72,794.64**	**2,492.38**	**0.00**	**75,287.02**	**0.16**	**21.1%**	**0.7%**	**0.02**
M2_1to3	8	19	28	166,836.06	99,768.30	3,112.39	0.00	102,880.69	0.04	99,768.30	3,112.39	0.00	102,880.69	0.04	38.3%	0.0%	0.00
M2_2to4	11	22	34	168,477.81	119,755.98	6,575.92	0.00	126,331.90	0.06	131,356.15	5,722.93	0.00	137,079.08	0.24	18.6%	7.8%	0.18
M2_3to5	13	24	38	156,857.73	119,703.68	6,654.25	0.00	126,357.93	0.12	124,633.85	6,654.25	0.00	131,288.10	0.15	16.3%	3.8%	0.03
M2_4to6	17	33	51	262,145.59	228,120.40	4,021.96	0.00	232,142.36	0.27	228,120.40	4,021.96	0.00	232,142.36	0.29	11.4%	0.0%	0.01
M2_5to7	18	31	50	179,238.59	158,376.20	0.00	0.00	158,376.20	0.31	158,376.20	0.00	0.00	158,376.20	0.33	11.6%	0.0%	0.02
M2_6to8	14	29	44	249,730.98	241,022.33	0.00	0.00	241,022.33	0.18	241,022.33	0.00	0.00	241,022.33	0.24	3.5%	0.0%	0.06
M2_7to9	10	28	39	181,115.44	127,410.97	2,802.51	0.00	130,213.48	0.13	127,410.97	2,802.51	0.00	130,213.48	0.11	28.1%	0.0%	-0.02
M2_8to10	12	29	42	226,630.02	187,461.89	13,517.26	0.00	200,979.15	0.14	202,445.74	7,125.80	0.00	209,571.54	0.25	7.5%	4.1%	0.11
Avg M2	**12.88**	**26.88**	**40.75**	**198,879.03**	**160,202.47**	**4,585.54**	**0.00**	**164,788.01**	**0.16**	**164,141.74**	**3,679.98**	**0.00**	**167,821.72**	**0.21**	**15.6%**	**1.8%**	**0.05**
M3_1to3	49	11	61	277,902.02	219,390.97	9,301.96	0.00	228,692.93	0.73	235,117.04	7,696.72	0.00	242,813.76	1.36	12.6%	5.8%	0.63
M3_2to4	44	7	52	241,087.32	182,774.91	0.00	279,726.34	462,501.25	0.44	183,661.83	4,387.92	279,726.34	467,776.09	0.67	22.0%	1.1%	0.23
M3_3to5	36	5	42	180,030.30	118,716.58	2,072.76	0.00	120,789.34	0.23	127,382.09	4,333.36	0.00	131,715.45	0.34	26.8%	8.3%	0.11
M3_4to6	26	2	29	164,932.85	155,934.44	7,005.75	290,126.09	453,066.28	0.07	158,134.52	7,005.75	290,126.09	455,266.36	0.07	-0.1%	0.5%	0.00
M3_5to7	28	2	31	230,174.19	175,763.32	11,005.47	0.00	186,768.79	0.10	216,373.13	11,005.47	0.00	227,378.60	0.28	1.2%	17.9%	0.18
M3_6to8	24	1	26	27,000.16	15,160.16	0.00	0.00	15,160.16	0.05	104,576.69	845.01	0.00	105,421.70	0.39	-290.4%	85.6%	0.34
Avg M3	**34.50**	**4.67**	**40.17**	**186,854.47**	**144,623.40**	**4,897.66**	**94,975.41**	**244,496.46**	**0.27**	**170,874.22**	**5,879.04**	**94,975.41**	**271,728.66**	**0.52**	**5.4%**	**10.0%**	**0.25**
M4_1to3	67	41	109	311,877.44	252,724.87	3,792.75	204,528.22	461,045.84	4.38	252,051.53	5,553.53	204,528.22	462,133.28	4.64	17.4%	0.2%	0.26
M4_2to4	58	39	98	350,029.82	205,585.46	15,705.56	0.00	221,291.02	5.24	209,718.69	15,705.56	0.00	225,424.25	12.30	35.6%	1.8%	7.06
M4_3to5	62	36	99	341,188.47	202,816.86	9,256.82	0.00	212,073.68	5.03	223,749.67	9,256.82	0.00	233,006.49	12.90	31.7%	9.0%	7.87
M4_4to6	68	35	104	350,093.88	215,138.24	5,854.93	0.00	220,993.17	5.18	215,138.24	5,854.93	0.00	220,993.17	5.08	36.9%	0.0%	-0.10
M4_5to7	70	39	110	389,702.09	253,584.96	3,909.45	0.00	257,494.41	4.54	322,249.91	3,909.45	0.00	326,159.36	14.15	16.3%	21.1%	9.61
M4_6to8	63	40	104	338,398.91	194,775.15	6,690.24	0.00	201,465.39	3.00	257,280.34	3,010.49	0.00	260,290.83	5.32	23.1%	22.6%	2.33
M4_7to9	72	41	114	518,951.14	243,357.56	3,664.01	0.00	247,021.57	4.13	247,997.56	3,664.01	0.00	251,661.57	4.29	51.5%	1.8%	0.17
M4_8to10	68	33	102	614,517.94	260,620.44	2,959.86	0.00	263,580.30	3.08	261,820.41	2,959.86	0.00	264,780.27	4.16	56.9%	0.5%	1.08
M4_9to11	76	34	111	569,620.72	312,402.20	6,838.90	0.00	319,241.10	4.46	329,398.85	6,838.90	297,325.91	633,563.66	16.40	41.0%	49.6%	11.94
M4_10to12	80	37	118	513,440.96	376,997.71	21,350.54	0.00	398,348.25	10.32	408,164.89	17,928.90	0.00	426,093.79	93.87	17.0%	6.5%	83.55
M4_11to13	82	42	125	476,176.69	350,013.98	11,733.17	0.00	361,747.15	6.82	465,611.11	11,733.17	0.00	477,344.28	239.43	-0.2%	24.2%	232.61
M4_12to14	71	49	121	417,695.68	307,678.68	5,862.01	0.00	313,540.69	6.61	358,477.41	5,862.01	284,259.56	648,598.98	7.15	12.8%	51.7%	0.54
M4_13to15	62	50	113	415,337.41	326,447.29	19,518.61	208,926.22	554,892.12	4.78	328,127.29	19,518.61	208,926.22	556,572.12	9.74	16.3%	0.3%	4.96
M4_14to16	56	48	105	373,520.17	265,439.80	10,862.69	0.00	276,302.49	2.47	262,773.20	16,062.32	0.00	278,835.52	3.71	25.3%	0.9%	1.24
Avg M4	**68.21**	**40.29**	**109.50**	**427,182.24**	**269,113.09**	**9,142.82**	**29,532.46**	**307,788.37**	**5.00**	**295,897.08**	**9,132.75**	**71,074.28**	**376,104.11**	**30.94**	**28.6%**	**18.2%**	**25.94**
Total	**39.06**	**26.08**	**66.14**	**262,674.32**	**180,333.48**	**6,022.85**	**27,314.08**	**213,670.41**	**2.05**	**196,202.65**	**5,903.10**	**43,469.23**	**245,574.99**	**12.20**	**23.1%**	**13.0%**	**10.14**

5 Conclusion

We studied the aircraft routing problem with crew assignment in the context of on-demand air transportation, based on the real case of a company that operates in the fractional ownership management sector. We proposed a tailored branch-and-cut algorithm that dynamically inserts constraints related to crew requirements into an aircraft routing model. The algorithm uses a labeling-based strategy to effectively separate and insert, if needed, feasibility and optimality cuts into the problem. In computational experiments using real-life data provided by the company, the algorithm obtained optimal solutions for all instances within reasonably short computational times, taking an average of 30.94 s to solve the largest instances in the set. Moreover, the solution costs were considerably smaller than the costs of the routes employed by the company, on average.

For next steps, we intend to consider additional crew rules, such as allowing overtime and split duties to further improve the solutions given by the algorithm. Furthermore, we plan to incorporate uncertainty on travel times to obtain solutions that are robust to this kind of variability. Finally, we intend to extend the problem to separate the crew from the aircraft, allowing them to take different aircraft in the planning horizon. This extension would also require to consider positioning the crew, to get to the new vehicle of their planning horizon, via the fleet or scheduled airlines to enable trips. We can also extend our formulation to consider not only the departure limits, but also the arrival time windows and allow slower, albeit cheaper, aircraft to service a request, as long as it respects both time windows. Finally, there is also the possibility to incorporate to the model the decision on where the maintenance events of each aircraft will take place, considering the availability and costs of maintenance services at different airports.

References

1. Barnhart, C., Cohn, A.M., Johnson, E.L., Klabjan, D., Nemhauser, G.L., Vance, P.H.: Airline Crew Scheduling, pp. 517–560. Springer, Boston (2003). https://doi.org/10.1007/0-306-48058-1_14
2. Dunbar, M., Froyland, G., Wu, C.L.: An integrated scenario-based approach for robust aircraft routing, crew pairing and re-timing. Comput. Oper. Res. **45**, 68–86 (2014)
3. Feillet, D.: A tutorial on column generation and branch-and-price for vehicle routing problems. 4OR Q. J. Oper. Res. **8**, 407–424 (2010)
4. Haouari, M., Zeghal Mansour, F., Sherali, H.D.: A new compact formulation for the daily crew pairing problem. Transp. Sci. **53**(3), 811–828 (2019)
5. Jamili, A.: A robust mathematical model and heuristic algorithms for integrated aircraft routing and scheduling, with consideration of fleet assignment problem. J. Air Transp. Manag. **58**, 21–30 (2017)
6. Kasirzadeh, A., Saddoune, M., Soumis, F.: Airline crew scheduling: models, algorithms, and data sets. EURO J. Transp. Logistics **6**(2), 111–137 (2015). https://doi.org/10.1007/s13676-015-0080-x

7. Khaled, O., Minoux, M., Mousseau, V., Michel, S., Ceugniet, X.: A compact optimization model for the tail assignment problem. Eur. J. Oper. Res. **264**, 548–557 (2017)
8. Liang, Z., Feng, Y., Zhang, X., Wu, T., Chaovalitwongse, W.A.: Robust weekly aircraft maintenance routing problem and the extension to the tail assignment problem. Transp. Res. Part B Methodol. **78**, 238–259 (2015)
9. Mercier, A., Soumis, F.: An integrated aircraft routing, crew scheduling and flight retiming model. Comput. Oper. Res. **34**, 2251–2265 (2007)
10. Munari, P., Alvarez, A.: Aircraft routing for on-demand air transportation with service upgrade and maintenance events: compact model and case study. J. Air Transp. Manag. **75**, 75–84 (2019)
11. Shebalov, S., Klabjan, D.: Robust airline crew pairing: move-up crews. Transp. Sci. **40**(3), 300–312 (2006)
12. Sherali, H.D., Bish, E.K., Zhu, X.: Airline fleet assignment concepts, models, and algorithms. Eur. J. Oper. Res. **172**(1), 1–30 (2006)
13. Vieira, T., et al.: Exact and heuristic approaches to reschedule helicopter flights for personnel transportation in the oil industry. Transp. Res. Part E Logist. Transp. Rev. **151**, 102322 (2021)
14. Yang, W., Karaesmen, I.Z., Keskinocak, P., Tayur, S.: Aircraft and crew scheduling for fractional ownership programs. Ann. Oper. Res. **159**(1), 415–431 (2008)
15. Yao, Y., Ergun, Ö., Johnson, E., Schultz, W., Singleton, J.M.: Strategic planning in fractional aircraft ownership programs. Eur. J. Oper. Res. **189**, 526–539 (2008)

Designing a Physical Packing Sequence Algorithm with Static Stability for Pallet Loading Problems in Air Cargo

Philipp Gabriel Mazur(✉), No-San Lee, Detlef Schoder, and Tabea Janssen

University of Cologne, Cologne Institute for Information Systems, Pohligstr. 1, 50969 Cologne, Germany
{mazur,lee,schoder}@wim.uni-koeln.de

Abstract. Large amounts of airfreight are loaded on pallets and containers for transport every day. Especially in the air cargo sector, fast and efficient pallet loading is crucial for smooth operations. Recently, scholars have proposed AI-optimized solutions for the pallet loading problem that include strongly heterogenous cargo. However, finding packing sequences that determine item loading order, receive scant attention in literature. In this research, we develop a design to solve the physical packing sequence problem that comprises requirements, features, and fitness criteria to equip an algorithm that automatically finds a physical packing sequence for a given cargo arrangement. We derive our algorithm based on previous findings and practical insights from a collaboration with a major cargo carrier. Also, we provide an integration design in combination with optimization heuristics. Our approach is implemented in a prototype, demonstrated, and evaluated on a set of real-world cargo data. Our findings reveal both the ability to find packing sequences in reasonable time and the ability to identify improvement potential with respect to stability.

Keywords: Pallet loading problem · Loadability · Physical packing sequence · Genetic algorithm

1 Introduction

In the air cargo sector, palletizers load freight into containers or onto pallets either manually or by using a forklift truck [1]. Since storage space of an airplane is limited, a high space utilization is desirable. Further, loaded pallets or container must comply to various constraints and aviation safety regulations [2, 3].

During palletizing, workers determine cargo positions and sequentially load cargo. Currently, the build-up process lacks IT support [2]. Due to strictly scheduled time windows, palletizers frequently load pallets under time pressure. Moreover, a rising proportion of heterogeneous cargo is complicating the build-up. Together, time pressure, loading process complexity, and absent support can lead to non-optimal space utilization or, in the worst case, repeated repacking of an arrangement.

M. Mes et al. (Eds.): ICCL 2021, LNCS 13004, pp. 627–641, 2021.
https://doi.org/10.1007/978-3-030-87672-2_41

In research, the problem of generating optimal arrangements of cargo is called pallet loading problem (PLP) [4] or container loading problem (CLP) [3]. A cargo arrangement defines the set of items to be packed, their orientation and final positions. A large body of literature exists that faces the optimization of cargo arrangements based on heuristics and mathematical programming [3]. On the other hand, one can distinguish between the problems of finding cargo arrangements and sequences, in which arrangements are assembled (physical packing sequence problem, PPSP) [5]. The PPSP opts to find an optimal packing sequence for an already calculated arrangement. A packing sequence provides step-by-step instructions on how to build up a specific arrangement that meets a set of real-world conditions, such as static stability of the cargo or reduction of physical strain for palletizers. Also, different modes of packing (e.g., manual, forklift) imply different characteristics that must be considered. Compared to the generation of cargo arrangements, research on packing sequence problems is scarce [5].

This work tackles the PPSP in the case of air cargo palletizing. To the best of our knowledge, no comparable artifact is described in related studies. Thus, our objective is to develop a packing sequence algorithm for a given, already optimized cargo arrangement that meets air cargo requirements.

Our approach is based on a genetic algorithm (GA) and must comply with a multitude of real-world conditions, such as item-related constraints, ULD-related constraints, packing device-related constraints, static stability and comprehensiveness. GA are well established in CLP research and can balance out multiple, potentially conflicting objectives by integrating several fitness dimensions in the objective function [6]. Problem-related constraints can be included in the GA in several ways, for example to avoid infeasible solutions. GA capabilities fit the requirements of PPSP due to high combinatorial complexity, strict deadlines and multiple optimization goals.

Since our goal is to solve the PPSP on a conceptual level and to achieve solutions that are directly applicable in practice, we develop both a design and an artifact that implements our design, thereby following a design science approach guided by Peffers et al. [7] that comprises six phases. After the problem identification and motivation, we derive solution objectives. Then, we design and implement the artifact. We demonstrate its functionality by deploying the algorithm with a set of cargo data from practice. For the evaluation, we collect and discuss performance data.

The remainder of the work is structured as follows: In the next chapter, we set out related work and constraints concerning the PPSP. Afterwards, we provide a description of our approach, which is followed by the presentation of results. In the following section, our results are discussed. Finally, we present a conclusion of this work.

2 State of the Art

Although the CLP is a widely researched topic, few studies explicitely considered the distinction between CLP and PPSP and the calculation of packing sequences for obtained cargo arrangements [5]. Some authors mention a separate algorithm for packing sequence generation but provide no details on design or procedure [8, 9]. A different picture is painted, when (un)loading is part of the problem, for example in multidrop scenarios or in combinations of CLP with vehicle routing problems [3]. The goal is to unload all boxes for one destination without having unload boxes for later destinations. Packing sequences might be implicitely present through placement heuristics, which iteratively place cargo in walls or layers. When considering the difference between cargo arrangements and packing sequences, a feasible and efficient packing sequence should be computed separately [5, 8, 9].

Different packing devices have been distinguished [5]. The authors differenciate between manual handling, handling equipment and automated systems. Manual handling is carried out by workers that lift and place cargo. Handling equipment comprises the usage of forklifts, or other mechanical equipment, while automated systems, robot arms or other systems carry out the packing.

The packing sequence defines the "sequence by which each box is placed inside the container in a specific location determined by the CLP algorithm" [5]. As such, the packing sequence should operationalize the caculated cargo arrangement into an actionable sequence of instructions. Every item must occur in the sequence exactly once [10]. The position of an item must not be blocked by an already loaded item [10] and it is not possible to lift an item over an already placed item. Following Ramos et al. [5], the PPSP can be formally described as: For a given cargo arrangement M that comprises items $b_i (i = 1, \ldots, n)$ with depth d_i, width w_i, and height and h_i and their final positions in the layout (x_i, y_i, z_i), find a sequence that fulfills a set of practical constraints. Practical constraints can be categorized into (1) item-related constraints, (2) ULD-related constraints, (3) packing device-related constraints, (4) static stability, and (5) comprehensiveness:

Item-related constraints: This category focuses on the possible packing sides and devices with respect to the cargo. Some items might be loaded from every side, others only from two sides. This situation frequently occurs when cargo arrives pre-palletized on wooden pallets that have holes for the forks [11]. In the air cargo sector, cargo is strongly heterogenous, i.e. cargo varies in shape, size and characteristics [1, 11]. Furthermore, packing device information on an item-level specify, if an item can be loaded manually or by a forklift.

ULD-related constraints: Unit loading devices (ULD) are employed in air cargo logistics to consolidate cargo and comprise standardized pallets and containers [12]. For containers, the constraint restricts the number of packing sides due to the presence of rigid walls. Containers can only be loaded through the container door, which might be congruent to the container wall or might be smaller. Nascimento et al. [13] permit packinf sides only to the container rear. Pallets can have a variable size and contour [1, 14]. Additionally, the ULD can restrict packing devices (i.e., some containers must be loaded manually).

Packing device-related constraints: This set of constraint focuses on the specifications of the device packing the items. When it comes to manual packing, stress and physical exertion plays a major role [5], especially when items are heavy, large or irregularly shaped. The packing sequence must be designed such that ergonomic factors are taken into account and should minimize lifting and movements especially under weight as far as possible to safe loading time and prevent physical fatigue. Lifting items with increasing distance to the human body increases the leverarm pressing on the palletizer's back, favoring the chances of injuries. Moreover, loading heavy items to high positions (e.g., overhead reaching) increases muscular exertion [5].

Furthermore, the packing sequence generation must consider body heights and arm lengths. For manual packing, some studies include a maximum reach value that represents the maximum distance between walls of the same destination that are allowed to surpass, or the maximum reach of the worker's arm or packing device's length [5, 15]. Ramos et al. [5] employ maximum reach for the calculation of arm's reach, which is the distance between a box and the nearest box to the palletizer. Nascimento et al. [13] include manual packing constraints to calculate CLP solutions based on an exact algorithm. Manual packing is expressed through both a maximum loadable height and a maximum horizontal distance between an item's end for each pair of items lying on top of each other.

For other packing devices, constraints exist that affect the generation of packing sequences [8]. Khan and Masood [16] present a method specificially designed to generate robot-packable patterns. Four placement strategies are developed along which the penta-block strategy performs best. Due to the length of the fork, a maximum reach might also be necessary for forklift packing. For cartesian robots, similar physical limitations exists, when items are loaded through a fork attached to the robot's arm. For humanoid robots, similar physical restrictions exists as for human palletizers.

Static stability: This constraint ensures that items maintain their positions during packing and prevents items from falling down on the container floor [17]. Altering the packing sequence heavily impacts static stability [5]. Many approaches exist, such as full base support, partial base support or static mechanical equilibrium calculations [18]. Ramos et al. [5] present a static stability algorithm that is based on the idea of force and moment equilibria and calculates the supporting polygon for every subset of boxes. In the air cargo sector with high item and shape heterogeneities, calculating static stability remains a major challenge, since most approaches impose assumptions (e.g., constant density, cuboidal shapes) on the cargo, which hardly reflects practical operation's complexity. To achieve realistic stability approximations, physical simulations with physics engines are chosen for dynamic [19] and static stability [11].

Comprehensiveness: This set of constraints applies only to manual packing or forklift packing, where a human steers the forklift. It contributes to the requirement that palletizers must understand the packing sequence, final positions and orientations. Brandt and Nickel [1] introduce the requirement of instructive visualizations to guide palletizers and load planners.

3 Solution Approach

Our goal was to solve the PPSP on a conceptual level and to achieve solutions that are directly applicable in practice. As mentioned previously, little is known about the solutions of PPSPs. Although previous studies provide useful insights, the entire complexity is uncovered during the process leading up to the implementation of the artifact. Therefore, we followed an design-oriented approach that builds an artefact for solving the PPSP based on requirements and constraints deduced from existing theory and practice. We iteratively improved our solution until it reached a maturity to be deployed in practical PPSP scenarios. The design science research (DSR) approach intends to create and evaluate IT artifacts with the purpose of solving organizational problems [20]. This work applies the design science research methodology (DSRM) by Peffers et al. [7] who structure a DSR project into the six phases (1) problem identification and motivation, (2) solution objectives definition, (3) design and development, (4) demonstration, (5) evaluation, and (6) communication [7]. All phases can be iterated repeatedly. From a practical viewpoint, the problem originated during a multi-year cooperation with a large airfreight handling company. Although an existing algorithm already calculated optimized cargo arrangements, practitioners were unable to evaluate the feasibility of obtained solutions since no practical packing sequence existed. Consequently, obtained solutions cannot directly be applied. We raised requirements for our solution based on previous findings from literature and team discussions. Consulted experts provided new input during multiple workshops. Further, we interviewed an experienced palletizer to deduct requirements for the PPSP. To demonstrate our artifact's problem-solving capabilities, we developed an interactive, web-based 3D-visualization that shows calculated packing sequences and provides packing instructions for palletizers. Further, the airfreight carrier provided us real world cargo data from operations, from which we extracted multiple test cases. Based on the test cases, we both demonstrated our design's feasibility to solve the problem and evaluated its performance. In this work, evaluation investigated development of fitness criteria and genetic operators scores during GA iterations. Afterwards, we compared results for varying algorithmic configurations. Particularly, we examined the impact of fitness criteria weight changes on the resulting best solution.

A genetic algorithm (GA) simulates the evolutionary process by representing solutions to an optimization problem with a DNA-like structure. For the purpose of finding optimal solutions, the solutions candidates go through an artificial evolution such that the best characteristics prevail. The general structure of a basic GA based on Kramer [6] is depicted in the following:

```
Initialize population
repeat
  repeat
    crossover
    mutation
    phenotype mapping
    fitness computation
  until population complete
  selection of parental population
until termination condition
```

The GA starts with the creation of an initial set of solutions (population). Every solution is represented by a genotype that contains all necessary information. The genotype for a solution to a combinatorial problem typically is a list of values from a set of symbols or bit strings. The recombinator (crossover) combines genotypes of two or more population members to generate new solutions (childs). Afterwards, child solutions undergo a mutation step. In most cases, the evaluation of solutions generated by recombination and mutation relies on a phenotype, a problem context specific representation. According to Rothlauf [21], to quantify solution quality, a fitness function $f : \Phi_g \rightarrow \mathbb{R}$ assigns every solution $x \in \Phi_g$ in the search space Φ_g a numerical score. The numerical fitness score allows comparison between two solutions$x_1, x_2 \in \Phi_g$. If $f(x_1) > f(x_2)$, x_1 is being superior with respect to solution fitness.The optimization goal is therefore to find a solution, which maximizes the fitness [21]:

$$\hat{x} = \max_{x \in \Phi_g} f(x)$$

When optimizing for multiple, potentially conflicting objectives, the fitness function can aggregate multiple sub-fitness scores, for example using a weighted sum. To get the desired solution quality convergence, a selector picks the best solutions in a population to become the next parental generation. In most optimization problems, constraints reduce the solution space. Problem-related constraints can be included in the GA in several ways, for example to avoid infeasible solutions. Genotype and genetic operators can be designed in a way that constraints are automatically fulfilled (1). A death penalty (2) causes infeasible solutions to enter a cycle of crossover and mutation until a feasible solution is found. Using penalty functions (3), the fitness score of invalid solutions might be reduced.

With respect to the PPSP, we observe a set of problem characteristics that well suit solution capabilities of GA. In CLP contexts, metaheuristics like GA are well established for problems with high combinatorial complexity [22]. On average a single flight segment contains around 350 items, with 3,8% of flight segments carrying over 1000 items [1], which impact packing sequence's complexity. Furthermore, feasibility of packing sequences is restricted through a set of constraints. GA are problem-agnostic and can balance out multiple, potentially conflicting objectives in their fitness function. Moreover, strict flight schedules and corresponding deadlines [11] combined with uncertain delivery times of input cargo implies the need to always find a good solution, even if terminated early. For every point in time, heuristic-based approaches guarantee a good solution exists.

3.1 Solution Objectives

Our artifact's objective is to find a packing sequence for a pre-defined cargo arrangement that contains strongly heterogenous cargo. Our artifact should incorporate item-related, ULD-related, packing device-related, static stability and comprehensiveness constraints. Further, due to the close nature of PPSP and PLP, we put special attention on integration design between both problems. With respect to the air cargo context, it should cope with the complexity of regular and irregular shapes and be configureable to adapt to varying inputs. To meet strict time windows, we opt for efficient calculations.

3.2 Design and Development

Our deducted design requirements (DR) are depicted in Table 1. We deducted our requirements based on the solution objectives and assigned corresponding system components and design features (DF). Design features remark implemented key characteristics of our proposed design. We implemented all design features in our instantiation.

Table 1. Design requirements, system compontens, and design features

Design requirement	System component	Design feature
DR.01: Single occurance	Genotype	DF.01: Sequence permutation
DR.02: Corridor	Genotype	DF.02: Corridor
DR.03: Include regular and irregular shapes	Genotype	DF.03: Box, cylinder, polygon prism
DR.04: Item-related constraints	Genotype, phenotype	DF.04: Possible and preffered packing sides on an item-level
DR.05: ULD-related constraints	Genotype	DF.05: Possible packing sides on an ULD-level
DR.06: Packing device-related constraints	Genotype	DF.06: Manual and forklift packing
	Fitness	DF.07: Ease of loading
		DF.08: Heavy first
		DF.09: Runway
		DF.10: Edge distance
		DF.11: Non-overhang
DR.07: Static stability	Fitness	DF.12: Static stability
DR.08: Comprehensiveness	Fitness	DF.13: Consecutive neighbors
		DF.14: Easy positioning
	Visualization	DF.15: Animated packing
		DF.16: Possible and prefereed packing sides
DR.09: Configureable algorithm	Architecture	DF.17: Job processing
DR.10: Close PLP and PPSP integration	Architecture	DF.18: Pre-assessment and stand-alone interface
DR.11: Reduced fitness calls	Populator, recombinator, mutator	DF.19: Create only feasible solutions

Architecture and Workflow. Our PPSP algorithm is integrated into the flow of finding palletizing solutions in two distinct ways: (1) As a pre-assessment and (2) as stand-alone (2) [DF.18]. Figure 1 shows a flowchart of our integration design. White boxes belong PLP optimization, grey boxes to PPSP calculation. The pre-assessment (1) estimates a packing sequence based on a subset of fitness criteria and provides an interface to PLP metaheuristics. With an estimation of packing sequence quality, the PLP metaheuristic can optimize for solutions that adhere to the requirements of packing sequences. The pre-assessment starts the populator of the PPSP algorithm and finishes if it finds a packing sequence. Fitness evaluation is conducted using a subset of fitness criteria. We modeled the stand-alone approach (2) as a full version, which is triggered after PLP optimization. The full algorithm employs a GA to find a packing sequence for an already optimized set of PLP solutions.

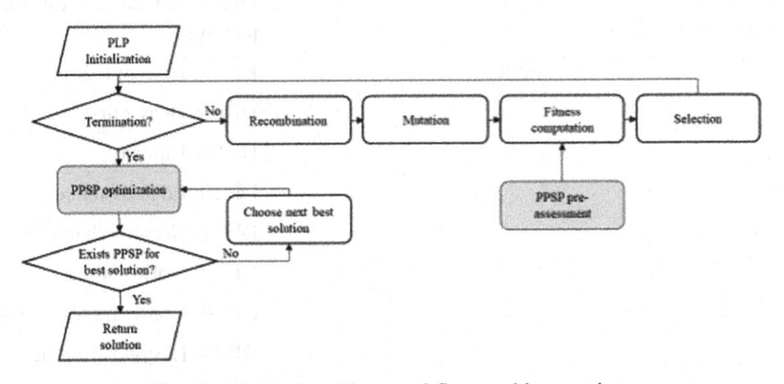

Fig. 1. PPSP algorithm workflow and integration.

Instantiation. Our PPSP algorithm supports four distinct shape types: Boxes, cylinders, and polygon prisms. Boxes are defined through a depth d_i,, width w_i, and height h_i. Cylinders are represented using radius r_i, and height h_i. Polygon prisms are represented using a two-dimensional vertex list and a fixed height h_i. Every item can be approximated using its bounding box, defined as minimal cuboid that entirely contains the item. The former two shapes are regular, the latter can become irregular. In air cargo logistics, most items are boxes [DF.03]. To meet the requirement of varying inputs (e.g., cargo characteristics, termination conditions, ULD), we employed job-processing. A job contains sufficient information to be executed as-is but is flexible enough to react to varying application circumstances [DF.17].

Genotype and Phenotype. Our algorithm uses order-based encoding as described in Raidl and Kodydek [23], so the genotype comprises a sequence of cargo item labels representing the packing sequence. Therefore, the overall solution space spans all permutations of items on the ULD [DF.01]. This set is reduced to permutations that represent *feasible* sequences [DF.19]. We opt to avoid infeasible solutions, since for every solution candidate, its fitness is evaluated. However, fitness evaluations are computationally expensive and should be minimized [6]. Therefore, we discard infeasible solution candidates by design. For this work, a *feasible* sequence must fulfill two conditions: (1)

The sequence contains items that are either placed on the pallet ground or if at least one supporting item is already placed, i.e., the supporting item is positioned earlier in the packing sequence. (2) For every item, there must be at least one packing device that can load the item from at least one side [DF.04]. Further, a free corridor between packing device and item's final position in the cargo arrangement must be present [DF.02]. Both, free corridor, and blocked corridor situations are depicted in Fig. 2 and Fig. 3. For irregular items, we employed the projected bounding box as free corridor estimation [DF.03].

Additionally, we consider individual packing device characteristics [DF.06]. Based on our observations from air cargo practice, we decided to support forklift packing and manual packing, as both are predominantly employed and no robot packing is possible in foreseeable future. Specifically, input consists of the device's maximum reach and item weight and size limits that constrain the usage of a specific packing device (e.g., work safety regulations limit the maximum lifting weight for a palletizer). For *manual* packing situations, the item's weight and longest side must not exceed weight and size limits. Further, we calculate the arm's reach to determine if the palletizer can reach the item's position and extend it to 360-degree. For *forklift* packing, items are primarily pre-palletized on transport pallet, which contain wholes for the forks. We specify a property on an axes-level that determines if and from which sides a forklift can load the item.

Similar to packing sides on an item-level, we include ULD-related packing side restrictions [DF.05]. Every ULD contains information about possible packing sides for every packing device specified (e.g., front, back, left, right for manual packing, front for forklift packing).

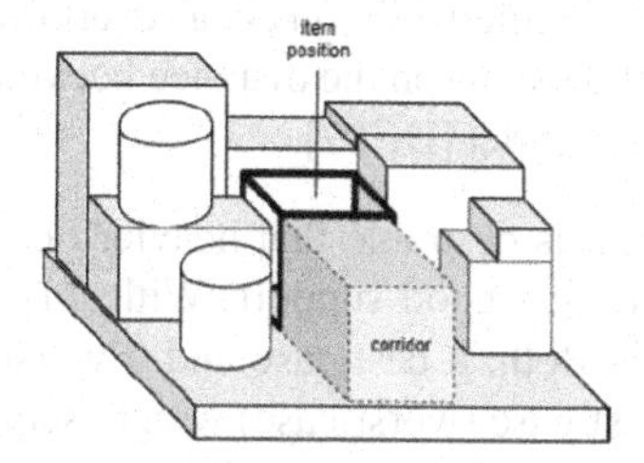

Fig. 2. Free corridor

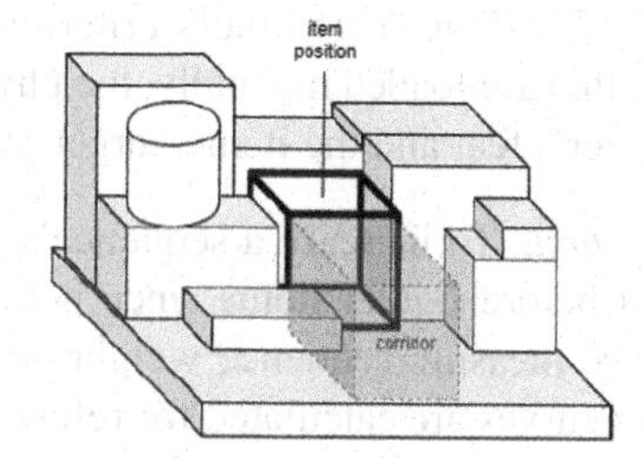

Fig. 3. Blocked corridor

Genetic Operators. For the initial population, we employ a random *populator*, which generates arbitrary start solutions that fulfill our *feasibility* conditions [DF.19]. The populator chooses on candidates that are currently supported and discard placements, in which candidate item would block unplaced items. With this check, we assert that all candidate items have at least one unblocked packing side [DF.02]. Due to our *feasibility* requirement, the populator might fail if no such sequence exists. Therefore, the population operator is limited to fixed number of iterations. Solutions are recombined by randomly choosing two parent solutions from our populations. The selection follows a uniform distribution between 0 and the population size. We implemented (1) a modified PMX *recombinator* (MPMX), inspired by Marian et al. [24] that copies the first part of one parent until a randomly chosen split point, and (2) an order keeping crossover (OKX). The modified OKX selects the first part of one parent until a randomly chosen index and then adds the unused items in the order they appear in the second parent's sequence. Again, the feasibility of the item is checked before insertion. In terms of *mutators*, we decided to implement (1) neighbor mutator swaps the position of two random subsequent items in the sequence if the result is a feasible sequence. (2) The swap mutator swaps the position of two random non-subsequent items in the sequence if the result is a feasible sequence. (3) The cut-and-paste mutator, which was inspired by Smith and Smith [25], sets both a random cut index and a random paste index. Then, the algorithm extracts the longest sequence of items including the item at the cut index that are consecutive neighbors and shifts subsequences to the paste index. We implemented two types of *selectors*: (1) The elitist selection operator chooses the solutions with the best score from the set of parents, children, and mutated children. (2) The tournament selection applies the tournament selection, where the solution with the higher score wins the battle. Furthermore, we added a check for duplicates check prior to adding a solution to the population. The *termination criteria* can be specified in the algorithm configuration. One can choose upon runtime, iterations, and solution quality, or a combination.

Fitness Criteria and Objective Function. Although previous studies provide useful insights to tackle the PPSP, few specify evaluation criteria for packing sequences. Thus, one main task in the design of the PPSP algorithm in this work was to determine characteristics of good packing sequences and to find fitness metrics that operationalize them. A solution's overall fitness score comprises the weighted sum of all individual fitness criteria scores. Our resulting fitness criteria contain:

Ease of Loading. This fitness criterion quantifies a palletizer's physical exertion. For items, that are loaded manually, the physical effort depends on the distance between the palletizer's feet and the item's target position on the pallet [DF.07].

Heavy-first. To increase a sequence's robustness, it is desirable that heavier items are placed before lighter items since heavy items provide good support. With this criterion, we measure sequence weight developments. Both, a best-case and a worst-case weight curves are calculated for reference. The best-case (worst-case) weight sequence orders items by weight in descending (ascending) order. Finally, our criterion incentives sequences that rather belong to the best-case [DF.08].

Runway. This criterion rewards sequences that suppress palletizer movements to save loading time. It adds up corners between the preferred packing sides of two consecutive items. Thereby, a perfect sequence comprises packing from one side, a cumbersome sequence enforces palletizers to switch between opposite ULD sides [DF.09].

Edge Distance. To achieve an easier handling of items and to reduce stress, this criterion incentives sequences with minimal edge distances between item and ULD [DF.10].

Non-overhang. During design cycle iterations, we frequently observed sequences that slid items below already placed items, which, in real-world contexts, is hard to realize. With this criterion, we count the number of items above other items that are located earlier in the sequence and penalize sequences with high number of overhangs [DF.11].

Static Stability. In this work, we implemented a physical simulation to meet our goal of real-world stability assessments that is based on the physics engine Bullet. We modelled cargo items as rigid bodies, i.e., they cannot be deformed. We assumed constant density. However, changing an items density (e.g., displacing its center of gravity) can be easily integrated when present. We implemented the *Sim2* approach described in Mazur et al. [11] that iteratively places items and triggers gravitational force acceleration. If at least one items moves, the sequence is accordingly marked as unstable and receives a penalty. Due to numerical simulation errors, a small quantity of delta movement is allowed [DF.12].

Consecutive Neighbors. This criterion incentives sequences that combine spatial proximity of two items with proximity in the sequence. Placing items that are located next to each other in both sequence and arrangement facilitates sequence understanding, as multiple information task requirements can be cognitively linked and palletizers can work from one area of the cargo arrangement onwards and recognize already placed items [DF.13].

Easy Positioning. Our algorithm incentives sequences that place items into corners, either in ULD corners or between already placed items and ULD edges, which simplifies the search for the correct position [DF.14].

Visualization. We implemented a web frontend with a visualization of packing sequences. Further, we displayed labeled arrows with the item's packing sequence that point to the item's final position from its preferred packing side and further displayed arrows that indicate possible alternative packing sides [DF.16]. Also, we implemented an animation that sequentially slid items on the ULD [DF.15]. We depict an exemplary visualization demonstration in Fig. 4.

Output. The PPSP output remarks a feasible packing sequence containing item sequences, packing devices, and preferred and possible packing side on an item-level.

4 Demonstration and Evaluation

For evaluation purposes, a large airfreight carrier provided us real-world cargo data. The data contains booking data from several flights with cargo of various sizes but is limited to outer bounding boxes. from the data, we created several sample jobs that comprise three degrees of shape heterogeneity (boxes only (BO), 15% non-cuboidal cargo (15H), and 40% non-cuboidal cargo (40H)) combined with three input item set sizes. This process led to a sample set of nine jobs. We enriched the jobs by adding further information and PPSP algorithm configuration, namely forklift capabilities and ULD packing sides. The forklift capable axes on an item-level were assigned at random. With respect to packing sides at an ULD-Level, we prohibit forklift packing from the ULD back, where cargo is frequently stored temporarily for the next station. For manual packing, all sides are reachable. Containers are loaded through the front side.

The algorithm runs through 100 iterations with a population size of 100. We used all implemented fitness criteria, recombinators, and mutators. The weightings resulted from observations during DSR cycles. We employed elitist selection since it resulted in higher scores. The recombinators each created 50% of a new generation. Both mutators were applied to 33% of the children. We ran 10 demonstration instances on three problem scenarios (BO, 15H, 40H) each. Our results are depicted in Table 2. Runtime average on 1–2 min. As the static stability criterion accounts for most of the runtime, we also conducted 10 runs runs without this criterion. The demonstration instance 40H was almost twice as fast in both cases due to a smaller item set. The mean best score (weighted sum of fitness criteria) exceeds 0.9 for all instances. Since resulting best sequences were very similar in all tests, rendered sequences range among the best for this configuration. All scenarios have been run 10 times, displayed are the mean values.

Table 2. Overview of test instances (N = 10)

Job scenario	# items	Runtime	Runtime w/o static stability	Best score
BO	20	106 s	2.16 s	0.936
15H	22	130 s	2.32 s	0.906
40H	13	55 s	1.68 s	0.904

For illustration, we depict the resulting best sequence in Fig. 4. As mirrored in our design requirements, the algorithm mostly proceeds from one area to another. In conclusion, the calculated sequences for demonstration seem applicable for practice.

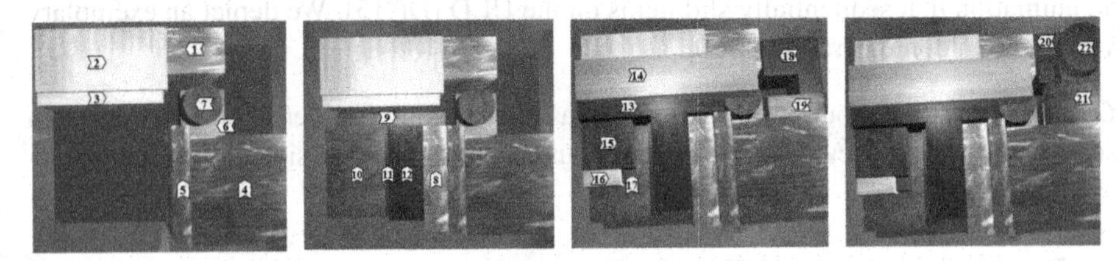

Fig. 4. Exemplary visualized packing sequence of 15H

We evaluate our instantiation with respect to overall and components' (fitness criteria, genetic operators) performance for sample problems. A unique solution is a sequence that has not appeared in any generation during algorithm run. We observe many new unique solutions at algorithm start, while for later generations, this number decreases. Afterwards, new solutions seem to supersede the current best solution only due to mutations. With increasing item heterogeneity, the number of unique solutions decreases. This observation might be due to more complex cargo arrangements as they impose more limitations to the packing sequence. Table 3 comprises information on the number of created and invalid solutions and duplicates for each genetic operator.

The OKX has a higher chance of producing valid and duplicate solutions. This might be due to its re-usage of already feasible sequences. If the split point is near sequence start or end, it has a high chance of duplicating one of the parents. It is noticeable that all

Table 3. Performance of genetic operators

	Populator	Recombinator		Mutator		
Operator	Random	OKX	PMX	cut-and-paste	Swap	Neighbor
Created solutions	100	1589	1803	1651	1586	2309
Invalid solutions	203	52	496	2283	3000	661
Duplicate solutions	0	3350	2701	175	140	330
Mean age	1.96	5.30	5.51	4.04	3.82	4.65
Variance age	1.92	65.70	82.94	21.32	14.48	43.82

mutators produce only few duplicates. The neighbor mutator has the highest mutation success rate, which reflects that it only makes a slight change to the sequence and the result would only be invalid if the second neighbor is placed on top of the first or the first item would be blocked by the second neighbor. The cut-and-paste and the swap mutator generate many invalid solutions. This is comprehensible because the chance that one or more items fit into the sequence at a randomly chosen position might be low. We designed recombination and mutation operators to complement each other. The recombination operators rather work on the sequence's overall structure. They define the sequences' start areas and where to proceed. In contrast, the mutation operators change sequences on a smaller scale. The neighbor- and swap mutation can create longer sequences of consecutive neighbors. However, the neighbor mutation has little impact on the overall packing strategy. The swap mutator has a higher impact but primarily fails because switching only one item frequently results in infeasible sequences. This disadvantage is tackled with the cut-and-paste operation. It incorporates the chance of removing "jumps" from one ULD area to another inside the sequence and a higher chance of creating feasible sequences than the swap mutator. Nevertheless, the swap mutator is crucial to remove single items from sequence positions where they do not really fit.

5 Discussion and Conclusion

This work contributes to solving ULD-loading problems with practical constraints as the availability of a packing sequence is an important part transition from cargo arrangement generation to physical assembly. Moreover, it addresses the PPSP as a problem class by itself. Despite its importance in practice, only few studies explicitly treat PPSPs. This work summarizes and illustrates requirements for the PPSP from related literature and practical viewpoints. We identified elven design requirements and 19 design features for PPSP algorithm design, which we implemented in our artifact. The requirements and features provide guidance to future research on how to model packing sequence algorithms in real-world applications. A further refinement might lead to abstractions in the form of design principles or theories. We laid the ground for the consideration of non-cuboidal cargo shapes, as frequently met in air cargo practice, and proposed a dual integration design to combine PPSP and PLP. Although we specifically designed our algorithm to meet practical requirements in the air cargo sector, applications in other transport modes such as truck or ship are conceivable. Our artifact supports pallets and containers, both of which play a major role in aviation industry. Depending on the problems' characteristics (e.g., available time, constraints) our algorithm can be adapted to run only few iterations or be stopped during execution. On the other hand, our work encounters limitations, which include, but are not limited to the following. Only few studies exist on packing sequence problems. Hence, only few findings could be reconceptualized for our deduction of requirements and features. Likewise, our study and specifically our evaluation lacks field tests. Although we evaluated on real-world data from operations of a large cargo carrier, requirements in terms of runtime, functionality and performance must be raised closely related to practical operations at a cargo hub. Further, feedback from palletizers is paramount when evaluating comprehensiveness and useability in terms of our 3D visualization. Finally, our algorithm's performance highly depends on its configuration (e.g., population size, used operators, fitness criteria weights). Although we developed a sense for algorithm configuration during DSR cycles, an optimization of configuration (e.g., using hyperparameter optimization) helps to find the best set ups for varying problem scenarios.

References

1. Brandt, F., Nickel, S.: The air cargo load planning problem - a consolidated problem definition and literature review on related problems. Eur. J. Oper. Res. **275**, 399–410 (2019). https://doi.org/10.1016/j.ejor.2018.07.013
2. Lee, N.-S., Mazur, P.G., Bittner, M., Schoder, D.: An intelligent decision support system for air cargo palletizing. In: Bui, T. (ed.) Proceedings of the 54st Hawaii International Conference on System Sciences. Proceedings of the Annual Hawaii International Conference on System Sciences. Hawaii International Conference on System Sciences (2021). https://doi.org/10.24251/HICSS.2021.170
3. Bortfeldt, A., Wäscher, G.: Constraints in container loading – A state-of-the-art review. Eur. J. Oper. Res. **229**, 1–20 (2013). https://doi.org/10.1016/j.ejor.2012.12.006
4. Dowsland, K.A.: An exact algorithm for the pallet loading problem. Eur. J. Oper. Res. **31**, 78–84 (1987). https://doi.org/10.1016/0377-2217(87)90140-8
5. Ramos, A.G., Oliveira, J.F., Lopes, M.P.: A physical packing sequence algorithm for the container loading problem with static mechanical equilibrium conditions. Int. Trans. Oper. Res. **23**, 215–238 (2016). https://doi.org/10.1111/itor.12124
6. Kramer, O.: Genetic Algorithm Essentials. Springer, Cham (2017). https://doi.org/10.1007/978-3-319-52156-5
7. Peffers, K., Tuunanen, T., Rothenberger, M.A., Chatterjee, S.: A design science research methodology for information systems research. J. Manag. Inf. Syst. **24**, 45–77 (2007). https://doi.org/10.2753/MIS0742-1222240302
8. Bischoff, E.E., Ratcliff, M.: Issues in the development of approaches to container loading. Omega **23**, 377–390 (1995). https://doi.org/10.1016/0305-0483(95)00015-G
9. Ngoi, B.K.A., Tay, M.L., Chua, E.S.: Applying spatial representation techniques to the container packing problem. Int. J. Prod. Res. **32**, 111–123 (1994). https://doi.org/10.1080/00207549408956919
10. Liu, W.-Y., Lin, C.-C., Yu, C.-S.: On the three-dimensional container packing problem under home delivery service. Asia Pac. J. Oper. Res. **28**, 601–621 (2011). https://doi.org/10.1142/S0217595911003466

11. Mazur, P.G., Lee, N.-S., Schoder, D.: Integration of physical simulations in static stability assessments for pallet loading in air cargo. In: Winter Simulation Conference (WSC). IEEE (2020). https://doi.org/10.1109/WSC48552.2020.9383878
12. Paquay, C., Schyns, M., Limbourg, S.: A mixed integer programming formulation for the three-dimensional bin packing problem deriving from an air cargo application. Int. Trans. Oper. Res. **23**, 187–213 (2014). https://doi.org/10.1111/itor.12111
13. Nascimento, O.X.d., Alves de Queiroz, T., Junqueira, L.: Practical constraints in the container loading problem: comprehensive formulations and exact algorithm. Comput. Oper. Res. **128**, 105186 (2021). https://doi.org/10.1016/j.cor.2020.105186
14. Chan, F.T., Bhagwat, R., Kumar, N., Tiwari, M.K., Lam, P.: Development of a decision support system for air-cargo pallets loading problem: a case study. Expert Syst. Appl. **31**, 472–485 (2006). https://doi.org/10.1016/j.eswa.2005.09.057
15. Junqueira, L., Morabito, R., Sato Yamashita, D.: MIP-based approaches for the container loading problem with multi-drop constraints. Ann. Oper. Res. **199**, 51–75 (2012). https://doi.org/10.1007/s10479-011-0942-z
16. Masood, S.H., A. Khan, H.: Development of pallet pattern placement strategies in robotic palletisation. Assembly Autom. **34**, 151–159 (2014). doi: https://doi.org/10.1108/AA-12-2012-092
17. Junqueira, L., Morabito, R., Sato Yamashita, D.: Three-dimensional container loading models with cargo stability and load bearing constraints. Comput. Oper. Res. **39**, 74–85 (2012). https://doi.org/10.1016/j.cor.2010.07.017
18. Ramos, A.G., Oliveira, J.F.: Cargo stability in the container loading problem - state-of-the-art and future research directions. In: Vaz, A.I.F., Almeida, J.P., Oliveira, J.F., Pinto, A.A. (eds.) APDIO 2017. SPMS, vol. 223, pp. 339–350. Springer, Cham (2018). https://doi.org/10.1007/978-3-319-71583-4_23
19. Bracht, E.C., Queiroz, T.A. de, Schouery, R.C.S., Miyazawa, F.K.: Dynamic cargo stability in loading and transportation of containers. In: IEEE International Conference on Automation Science and Engineering (CASE), pp. 227–232 (2016). https://doi.org/10.1109/COASE.2016.7743385
20. Hevner, A.R., March, S.T., Park, J., Ram, S.: Design science in information systems research. Manag. Inf. Syst. Q. **28**, 75–105 (2004). https://doi.org/10.2307/25148625
21. Rothlauf, F.: Representations for Genetic and Evolutionary Algorithms. Springer, Heidelberg (2006). https://doi.org/10.1007/3-540-32444-5
22. Zhao, X., Bennell, J.A., Bektaş, T., Dowsland, K.: A comparative review of 3D container loading algorithms. Int. Trans. Oper. Res. **23**, 287–320 (2016). https://doi.org/10.1111/itor.12094
23. Raidl, G.R., Kodydek, G.: Genetic algorithms for the multiple container packing problem. In: Goos, G., et al. (eds.) Parallel Problem Solving from Nature -- PPSN V. Lecture Notes in Computer Science, vol. 1498, pp. 875–884. Springer, Heidelberg (1998). https://doi.org/10.1007/BFb0056929
24. Marian, R.M., Luong, L.H., Abhary, K.: A genetic algorithm for the optimisation of assembly sequences. Comput. Ind. Eng. **50**, 503–527 (2006). https://doi.org/10.1016/j.cie.2005.07.007
25. Smith, G.C., Smith, S.S.-F.: An enhanced genetic algorithm for automated assembly planning. Robot. Comput-Integr. Manuf. **18**, 355–364 (2002). https://doi.org/10.1016/S0736-5845(02)00029-7

Intermodal Competition in Freight Transport - Political Impacts and Technical Developments

Joachim R. Daduna(✉)

Berlin School of Economics and Law, Badensche Str. 52, 10825 Berlin, Germany
daduna@hwr-berlin.de

Abstract. The competition between the various transport modes is characterized by the question of economic efficiency of transport services on the one hand and (transport) policy objectives on the other hand. A comparison shows that road freight transport dominates the terrestrial transport market, while the other transport modes in this segment generally rely on this due to their restrictions with regard to the provision of a widespread infrastructure. Despite of political prioritization and massive subsidies, the intended changes in modal split are not be achieved, due to significant changes in demand structures and their spatial distribution over the last years. By technological development the importance of road freight transport will significantly increase and also of River-sea and Short Sea Shipping, while rail freight transport and also freight transport on inland waterways will significantly lose market share.

Keywords: Competition in freight transport · Mono-modal and multimodal transport · Rail freight transport

1 Mobility and Economic Development

Intermodal competition between different modes of transport has been a controversially discussed problem for decades in the context of the design of regional, national as well as global transport processes. The main reason for this is that economic competition cannot exist and cannot develop if there are no efficient mobility services in freight (and passenger) transport. [83: 161] already stated that,…the improvement of the transport modes […] is a powerful incentive for efforts and objectives aimed at reducing the manufacturer price and improving the quality of goods…. This means that market structures and functioning competition are decisively influenced by the attained level of mobility. This is also indicated by the influence of transport on economic growth, for example for Germany from 1950 to 1990, which is given as 48.8% overall (and 26.1% for road transport) (see [10]). The importance of freight transport is ultimately to be seen in the fact that it forms mandatory basis structures for division of labor in economies.

[138] also comes to the result that there is a direct connection between industrialization and the available transport modes, whereby these are the dynamic factors in the development process. He says in this regard… the transport modes enforce on industrialization a certain independent influence, which otherwise industrialization would not

M. Mes et al. (Eds.): ICCL 2021, LNCS 13004, pp. 642–660, 2021.
https://doi.org/10.1007/978-3-030-87672-2_42

have made.... This means that industrial development does not determine the (transport-) logistics service structures, as is often claimed (see [67]), but quite simply formulated we can say: Logistics is not everything, but without (transport)logistics everything will be nothing.

The mobility of people and goods was already the basis of far-reaching trade relations in early ancient time. However, at that time, quasi-monopolistic structures existed due to the available transport alternatives based on very simple land vehicles, smaller inland barges and seagoing vessels for coastal areas. These were associated with considerable restrictions, especially with regard to the realizable transport quantities and trip durations as well as the transportability of goods. For a few thousand years, these were the (very limited) basis of worldwide mobility provision. It was not before the late 18th century that fundamental technical developments in the transport sector and a targeted construction of the necessary transport infrastructure came up (see [58: 186–188], [14]). With the availability of larger transport capacities and time-related plannable transport processes, it was possible to develop new industrial location structures and forms of manufacturing which were the basis of industrialization of European and North American economy.

For the *First Industrial Revolution* (see [28]) were improvements in the road network as well as a considerable expansion of the existing inland waterway network through its extension and the construction of canals (see [14]) the infrastructural basis. With the emergence of steamships, inland waterway transport (as well as maritime transport) became independent of weather conditions and to a certain extent of topographical influences, which led to disruptive changes in the transport of (bulk) goods at that time.

The *Second Industrial Revolution* (see [29]) created a completely new situation in freight transport. In the first phase, the focus was still on inland waterway transport, which reached almost its greatest network expansion during this period. However, it then lost its hitherto dominant position to rail freight transport (see [58: 186–188], [12]), which enabled completely new options in the design of network structures. This was in fact the first time that *intermodal competition* emerged, with a focus on industrial bulk transport.

In the middle of the 20th century, however, rail freight transport also lost importance due to significant improvement and wide-ranging expansion of the road transport infrastructure and the development of road vehicle technology in key transport market segments. Due to a higher flexibility, faster transport processes as well as direct origin-destination connections, changing industrial structures could emerge within the framework of small and medium-sized enterprises. The dependence of industrial sites on railroad sidings and inland ports decreased considerably, so that dislocated manufacturing structures could emerge. The *Third* and *Fourth Industrialization*, is on the other hand, largely characterized by improvements in transport modes in the context of digitalization and automation (see [56, 131]).

In the following chapters, first the existing transport modes are compared and evaluated. Next, the view of the competitive structure is modified due to operational restrictions of most transport modes by including multimodal transport services. Based on this, the impact of political influence and technological progress is viewed and an outlook on future developments is given. With regard to the lack of relevant data, mainly transport data from the *European Union* (EU) are the basis for the following analysis.

2 Comparison and Evaluation of Transport Modes

When discussing the available transport modes, it has to be taken into account that their possible applications must be viewed in a differentiated manner due to mode-related framework conditions. This concerns in particular the necessary infrastructure as well as the technical, capacitive and operational characteristics of the vehicles used. Simplified (one-dimensional) comparisons based only on the relation of transport volumes and costs for the three terrestrial transport modes (see Fig. 1), as used, among others, by [108: 113], are not very helpful. This is because, even if this form of comparison is repeatedly taken up in political discussions, they cannot provide a useful decision-making basis, as they are neither relevant from a theoretical nor an operational point of view.

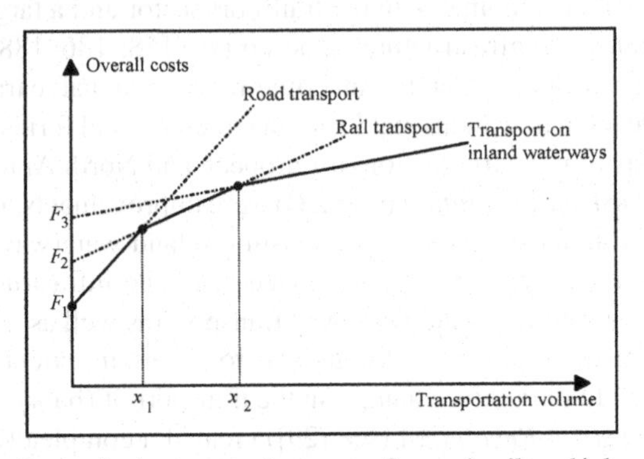

Fig. 1. Comparison of quantitative cost performances for road, rail, and inland waterway transport

The system-related advantages and disadvantages of the individual transport modes must be in the foreground of the analysis (see [138]) as well as the question of the fundamental suitability for the respective transport operation and their actual (spatial) availability. A basis for the qualitative rating is provided by the characteristics of *transport valuabilities* defined by [139: 80–92] (see also [18: 58–62], [39]).

- *Time required for transport execution* (*delivery time*): Sum of travel times and waiting times (due to technical or organizational reasons).
- *Ability for bulk good shipping*: Available capacity of a transport mode used as well as the cost-level for the transport of a weight- or space-unit of specified (bulk) goods.
- *Transport network design*: Possibility of forming direct transport connections between shippers and customers.
- *Time-related calculability of transport operations*: Adhere to travel times and delivery deadlines for transport operations.
- *Service frequency*: Level of transport service on a link in a defined time period.
- *Security and failure-free processes*: Avoidance of external influences on transport operations that would hinder or prevent the continuity of the processes.
- *Ease of access*: Extent of access to the transport mode in question.

In view of the changes in the framework conditions in recent decades, the flexibility of the possible applications and, in connection with the politically forced decarbonization of transport (see [55, 124]), the environmental friendliness of transport modes must be included as additional criteria. It is also possible that the question of the extent to which automated or autonomous driving will become established and will play a role in the future, also under the aspect of completely new design options in operational use and changed cost structures (see [25, 88, 120]).

Furthermore, due to structural developments, it makes sense to include *Short Sea Shipping* (SSS) (as well as *River-sea Shipping* (RSS)) as an independent transport mode in addition to the classic modes (road and rail freight transport, freight transport on inland waterways, sea and air freight transport, and transport via (pipe) lines). These are sea transport services that are largely limited to regional areas which can be seen here as an environmentally friendly alternative to terrestrial transport modes and as a capacity relief for them (see [33, 37, 99, 107, 125]). Under certain conditions, there is also the possibility of complementing existing terrestrial transport infrastructures (see [100: 162–186], [114], [133]). The growing importance of these transport modes in recent years is also reflected in their share of transport performance (in tkm) in the modal split, which was 29.2% in the EU in 2018 (see [42]: 38). The following figure (Fig. 2) shows the evaluation of the seven transport modes considered based on the criteria described. This is a present state, which, however, may lead to changed evaluations with a stronger market penetration of new vehicle technologies and qualified transport infrastructures in the coming years.

Transport mode Mode characteristics	Speed	Mass capability	Network integration	Calculabi-lity	Service frequency	Security	Conveni-ence	Flexibi-lity	Eco-friend-liness
Road	+2	-2	+2	+1	+2	0	+2	+2	0
Rail	+1	+2	-1	+1	0	+2	-1	-2	+1
Inland waterways	-1	+2	-2	0	0	+2	-1	-2	+1
Pipelines	+1	+2	-2	+1	+2	+2	-1	-2	+1
RSS / SSS	-1	+2	-2	+1	0	+2	-1	-2	+1
Sea	-1	+2	-2	+1	0	+2	-1	-2	+1
Air	+2	-2	-2	+2	0	+2	-1	-2	-2
The ratings are based on the comments by [139: 8092] and [39] as well as own estimations of the current situation.									

Fig. 2. Comparison of transport modes regarding their functional characteristics

The qualitative evaluation of the modes of transport described here represents the possibility of use, but it does not provide any information regarding the existing modal split, that means, the quantification of the realized market shares in intermodal competition. For the EU, Fig. 3 shows the shares of transport performance (in tkm) and their distribution in the modal split for the years 1995 and 2018, as well as their percentage change during this period (see, [42: 38]). However, these are aggregated values that do not reflect the significant differences within the individual member states (see, [42: 39]), which mainly result from the different geographical and topographical conditions and settlement structures.

Transport modes Mrd. tkm / %	1995	2018	Increase 1995 - 2018
Road	1127.2. / 47.0	1708.9 / 51.0	51.6
Rail	374.8 / 15.6	423.3 / 12.6	12.9
Inland waterways	121.9 / 5.1	135.0 / 4.0	10.7
Pipelines	103.8 / 4.3	104.0 / 3.1	0.2
RSS / SSS	671.4 / 28.0	979.2 / 29.2	45.8
Air	1.4 / 0.1	2.2 / 0.1	56.2
Total	2400.5	3352.6	39.7

Fig. 3. Comparison of freight transport performance (in tkm) and modal split in the EU by transport modes for 1995 and 2018

However, this is not only a phenomenon within the EU, but there are very differently structured transport markets worldwide with regard to the different used transport modes in the various regions (see, [42: 38]). The comparison of the situation in the EU, USA, Japan, China and Russia in Fig. 4 underlines this very clearly.

Transport mode Mrd. tkm / %	EU-27 2018	USA 2018	Japan 2017	China 2018	Russia 2018
Road	1708.9 / 51.2	2959.5 / 38,8	210.8 / 50.1	7124.9 / 39.9	241.2 / 5.7
Rail [1]	412.7 / 12.0	2525.2 / 33.1	21.7 / 5.2	288.1 / 1.6	2579.8 / 61.2
Inland waterways	135.0 / 4.1	463.0 / 6.1	---	---	32.7 / 0.8
Oil pipelines [2]	104.0 / 3.1	1429.8 / 18.7	---	530.1 / 3.0	1331.6 / 31.6
Sea [3,4,5]	979.2 / 29.3	253.5 / 3.3	180.9 / 43.6	9905.2 / 55.5	28.3 / 0.7
Total	3339.8	7631.0	413.4	17848.3	4213.6

[1] USA: Class I rail; [2] China: Oil and gas pipelines; [3] Domestic / Intra EU; [4] USA: Coastal shipping
[5] China: Coastal and inland waterway shipping

Fig. 4. Comparison of freight transport performance (in tkm) and modal split for the EU, USA, Japan, China, and Russia for 2018 (or 2017).

The data for the EU (see Fig. 3) show a dominance of road freight transport and SSS/RSS, also with regard to the development in recent years. Air freight transport has grown the most, but this is of minor importance due to its marginal share in transport performance. Critical, however, is the under-proportional growth in freight transport on inland waterways and especially in rail freight transport, which has been politically prioritized within the EU for decades. Although this sector has been subsidized to a significant extent, without any lasting success. It has not been possible to achieve any overall shifting effects to the disadvantage of road freight transport.

In order to understand this situation, the data on road freight transport volume must be analyzed in more detail (see Fig. 5). Since the relevant data are not available at the EU level, data on traffic volumes from Germany are used (see [20: 240–241]).

The key point is the distribution of distance-related volumes among *local* (< 50 km), *regional* (51–150 km) and *long-distance* (> 150 km) transport. The share of local and regional transport amounts to 79.6% in total in 2018 that means the major part of this market segment does not offer any relevant potential for modal shift in favor of the other terrestrial transport modes (see [70]). Furthermore, in the case of rail freight transport as well as freight transport on inland waterways, direct origin-destination links are normally

Road transport specification Mio, t	2008		2018		Percent change
Road freight transport	3438.0		3753.1		9.2
Foreign trucks	392.0	11.4	568.0	15.1%	44.9
German trucks	3046.0	88,6	3185.1	84.9%	4.6
Haulage	1889.8	62.0	2474.0	77.7	30.9
Transport by own trucks	1156.2	38.0	711.1	22.3	-39.0
< 50 km (near)	1749.1	57.4	1776.1	55.8%	1.5
51 - 150 (regional)	623.7	20.5	758.3	23.8%	21.6
>150 (long distance)	673.2	22.1	650.7	20.4%	-3.3

Fig. 5. Freight transport volume (in million tons) in road freight transport in Germany 2008 and 2018

not available due to the low network density, so that their use is only possible in the context of multimodal (or bi-modal) transport operations (see [30, 89, 111]), with pre-carriage and on-carriage services almost ultimately taking place by road (see Sect. 3). Transports in local and regional transport as well as in long-distance transport in the lower distances are definitely not affine to rail and to inland waterways transport. This means, that there is in this segment only an *intra-modal market.*

The explanations make it clear that the (operational and technical) utilization options can be described based on the transport valuabilities, but not the logistical deployment options or the market and competitive structures. This is because road freight transport is the only *mono-modal transport mode* that means, apart from a few exceptions, it is the only transport mode for which a direct origin-destination link is available. This means that a view of intermodal competition cannot be limited to the level of the individual transport modes, but due to the mandatory necessity of multimodal transport links, not only the included transport modes but also the necessary transshipment facilities must be considered when changing the transport mode.

3 Incorporation of Multimodal Transport Flows

Multimodal (or *combined*) *transport* is described (see [30, 38]) by the fact that (in their original definition) at least two different transport modes are used in freight forwarding. A special variant is the multimodal load unit transport, at which only (closed) and standardized load units (e.g. containers or swop bodies) are shipped continuously from the origin to the destination (see [59, 109]). This differentiation is not of fundamental importance for a comparison of multimodal transport processes, but they differ significantly in terms of the handling times and costs occurring in transshipment operations.

Since the economic and logistical importance of *load unit freight transport* is clearly in the foreground, the following comments are based on this. Accompanied bi-modal freight transport, such as Ro/Ro and ferry transport in the SSS (see [27, 80]) and ferry transport for river crossings (see [48]), is not included, as these have a de facto monopoly position largely due to geographical restrictions. The same applies to rolling roads (road/rail transport), which are often used in difficult topographical regions (for example for alp crossings) due to administrative restrictions on road freight transport (see [95, 110]), including to reduce high emission levels of road freight transport in ecologically sensitive areas (see [64]).

The (quantifiable) costs of multimodal transport include the fixed costs for their use and the variable costs (per km) of the respective transport modes as well as the handling costs (and, if necessary, also the additional costs for intermediate storage) in the transshipment facilities used. The following figure (see Fig. 6) shows a comparison of the resulting cost function for (a mono-modal) road freight transport with that for bi-modal road/rail freight transport (see [78, 113, 148]). An analogous structure also emerges in a comparison with the inclusion of freight transport on inland waterways (see [85]) in bi-modal freight transport solutions.

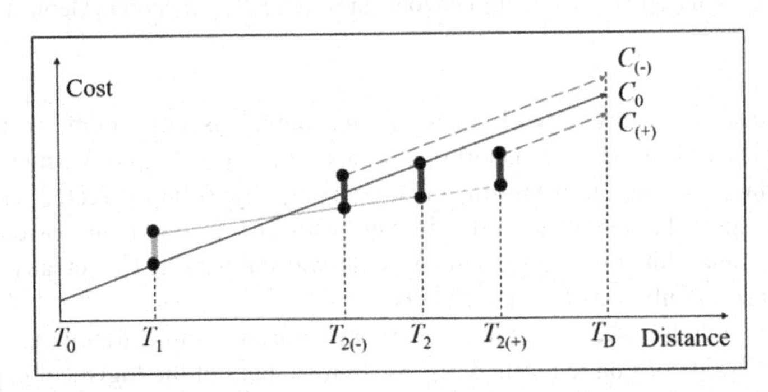

Fig. 6. Comparison of mono-modal and a bi-modal (road / rail) transport cost functions

A key issue in this comparison is the question of the break-even distance, the minimum transport distance that must occur in rail freight transport (or freight transport on inland waterways) for a bi-modal transport operation to achieve at least the same cost level (see [21, 90, 141]). The often stated 300 km for the share of rail freight transport, which is questioned by [90], seems to be plausibly based on data from the *Union Internationale Pour le Transport Combine Rail-Route* (UIRR) (see [42: 73]). For 2018, the percentage share of transport distances is for up to 300 km (1%), between 300 and 900 km (51%), and over 900 km (48%).

In the example presented, it can be seen that the break-even distance is at T_2. (with of costs C_0).This means that in the case when no rail/road transshipment is possible at point T_2 or between T_2 and T_D, that mono-modal road freight is the more cost-effective solution. Only in the case that a transshipment between T_2 and T_D is possible, the bi-modal transport is more cost effective.

However, this is only a rough estimate, because only on basis of a detailed (and case-by-case) comparative calculation (see [90]) can a sufficiently accurate decision be made. This is due to a number of parameters (see [78]), such as the capacity (= length) of the train, the wagon types used and the resulting loading options for different container types and lengths, and the availability and performance of handling facilities. However, there are also quality differences, which essentially result from reliability, transport flexibility, the time required for transport operations, and transport security.

With a view to sustainability of transport services to be provided, there is also an increasing interest considering external cost effects (see [9, 21, 64, 65, 85, 93]), which cannot be quantified or only to a limited extent. However, these are included by target-oriented administrative measures (including e.g. road tolls). The priority objective here is to make road freight transport more expensive.

When comparing transport processes, it also must be taken into account that there is usually more than one connection between an origin and destination, whereby the transport modes can be used in different combinations (see [35, 121]). These are determined by various parameters (including transport costs and time, type of goods and value, access options to the transport modes) (see [69]). Key objectives are, for example, to reduce the logistics costs occurring, the transport time, and the external costs that arise (including through transport-induced emissions). These problems are multi-criteria decision problems with non-congruent objective structures, where no relevant results for operational planning can be determined based on the calculated values. In order to enable a multi-criteria comparison, a standardization of the values is required (see [35]), which enables a multi-criterial comparison of alternative connections and form a useful basis for decision-making.

4 Impacts from Political Influence

In addition to measures in the field of research funding, political influence in the transport sector is extended essentially on two levels: the qualification and expansion of the transport infrastructure and regulatory measures. The objective here is, on the one hand, a targeted influence on competition to restrict road freight transport, among other by promoting bi-modal freight transport (see [115, 129]) and, on the other hand, reducing transport-induced environmental impacts by decarbonizing transport operations, especially regarding road transport (see [73, 92, 122]). For many years, these developments have been strongly pushed by subsidies on the EU level, among others with the *Trans-European Networks-Transport* (TEN-T) *projects* (see [100: 118–161], [106]), as well as on the national level of the member countries. The main objectives are the expansion and standardization of transport systems and infrastructure within the EU, including neighboring countries of Eastern Europe.

The focus of transport policy measures is on rail transport, which has been expected to play a much stronger role in European freight transport for years, although to date this has only been achieved to a very limited extent (see [105, 106]). So the eight *Rail Freight Corridors* (RFC) for providing more efficient seaport hinterland transport (see [79]) show not any significant effects (see [42]: 44) with regard to the overall share of rail freight transport. This is shown even though rail freight transport performance and transport volumes in bi-modal freight transport have increased somewhat in recent years.

The efforts of realizing the transport policy objectives regarding a prioritization of rail transport in the EU have been (and continue to be) extensively subsidized. Thus, subsidies amounting to almost 40 Mrd. € per year were paid out in the EU-15 area from 1998 to 2008 (see [31]). With growing subsidies and the enlargement of the EU, there has even been an increase to 50.6 Mrd. € in 2019 (see [43]:30–31). In fact, the subsidies are likely to be even higher, since financial resources can also flow into rail transport from other funding areas (e.g. environmental protection, sectoral and regional development) (see [43: 24]). Overall, however, a precise allocation of subsidies in the rail sector is only possible to a limited extent, since a differentiated view with a separation of passenger and freight transport is not possible (see [22]).

The introduction of distance-based charges on (heavy) trucks in various European countries with the aim of generating modal shift effects by making road freight transport more expensive (see [19, 52]) and halting the decline in single wagonload transport (see [60]) was and still is a failure. The reasons for this are multi-faceted, although from client's point of view, the focus is on inadequate performance in comparison with road freight transport (which are also to be seen as *quality differences*), which causes *asymmetric demand elasticity* (see [119]).

Furthermore, the still largely nationally oriented rail transport structures are an additional obstacle. There is a lack of consistent monitoring and control as well as safety technology in the various network areas (regional, national, international) (see [118, 127, 144]), which should already be available with the *European Train Control System* (ETCS). This has a significant negative impact on the actual core area of rail freight transport, the (cross-border) long-distance transport. In addition, there are problems due to a multi-layered user structure, the lack of vehicle standards at the international level (see [106]), as well as diverging national framework conditions and political interests.

If we also look at the developments in Germany in recent years, it is clear that the politically desired changes in the modal split of freight transport could not be achieved, even with extensive subsidies (see [144]). The currently planned high expenditures for qualification and expansion of the rail transport infrastructure will also not have any effect in this respect. The main reasons for this are the lengthy planning and approval procedures and the excessively long time period required for realization. Since continuity of industrial site structures has decreased significantly in the EU as well as in many regions worldwide in recent years and will continue to decrease, it must be assumed that if infrastructure becomes available, the expected demand for transport will no longer be met, even in an international context (see [46, 128]). A key factor is, among other things, the decarbonization of the energy sector, which will inevitably have a significant negative impact on bulk transport (see [41: 23]).

Even the rail freight transport between Europe and China (*Eurasian land bridge*), which has been pushed for several years within the framework of the Chinese *Belt and Road Initiative* (BRI) (see [82, 140, 146, 147]), cannot reach any significant market shares in container transport. So the available capacities for additional freight trains on the relevant routes must be seen, because the tracks are also used for passenger transport and national freight transport. If we look at the development of transport volumes between Asia and Europe in recent years, it also shows that the share of rail freight transport was only 1.4% in 2018 (see [17]). Apart from these facts, the expansion of the links is not uncontroversial from political and ecological view (see [130]).

As before, transport on the *Suez Canal Route* (SCR) is still dominant, which is also reflected in the container throughput of 42.255 million TEU in the EU ports of the North Range (see [41: 72]). However, depending on climate change developments, a competing (and especially shorter) link could become available for SCR with the *Northern Sea Route* (NSR) (see [145, 147]). So ultimately, rail freight can only provide a niche solution between air and sea transport in this market.

As already mentioned in Sect. 2, the SSS and RSS are competing with the terrestrial transport modes, which have reached a modal split share of 29.2% in the EU in 2018 (see Fig. 3), that means, significantly more than in rail freight transport and freight transport on inland waterways with an aggregated share of 16.6%.

Regarding to transport policy objectives, (not only) in the EU an (even) greater use of SSS and RSS is aimed at (see [57, 84, 112], [100: 162–186]). Influencing the modal split in (intra-European) freight transport to the disadvantages of road freight transport is the dominant objective (see [126, 132]). In the foreground here is the TEN-T project 21 (*Motorways of the Sea*) (see [8, 101, 125], [100: 118–161]). Possible examples of shifting effects to reduce road freight transport have been discussed increasingly (see [26, 107, 136]) and partially realized up to now. However, if we look at the transport valuabilities (see Sect. 2), it becomes clear that, due to similar performance parameters, a strong growing competition for rail freight transport has emerged with SSS/RSS.

5 Impacts of Technological Developments

One field of the technical development that will strongly influence the competitive structures in the future is automated and autonomous driving (see [25, 88, 120]). These technological changes must be seen in connection with an increasing digitalization in the transport sector and the developments in the area of low-emission and zero-emission drive systems (see [16, 143]). These fields will have a significant impact on future structures in competition of the transport sector, as key framework conditions will change in terms of performance and possible applications.

In road freight transport, autonomous and connected driving will lead to disruptive developments and new market opportunities (see [4, 40]). In the foreground are considerable (personnel) cost savings and completely new options for service design due to the elimination of legal working restrictions which currently affect the operational planning. This will significantly reduce the increase in the cost function of road freight transport (see Fig. 6), that means, the break-even distance in bi-modal road/rail (as also road/inland waterway) transport will inevitably increase unless there are corresponding developments for the other two transport modes. In the case of a successful market penetration of trucks with battery or fuel cell drive (see [98, 104]), the argument of ecological advantageousness of rail freight transport also disappears.

In the discussion about electric drives in the field of commercial vehicles in road transport, three main forms are in the foreground (see [16, 143]). These are battery-based drives (tending toward vans and light trucks) (see [68]), fuel-cell-based drives (tending toward trucks and heavy vehicles) (see [75]), and trucks with a power supply via catenaries (see [87]). At present, however, there is a controversial debate about which path is the right one; the decisive factor here will be the operational and economic evaluation of the respective overall system (costs).

Various technical developments are also available or about to be introduced in rail freight transport. However, autonomous driving will be limited to only a few applications due to safety regulations (see [52, 81]). The focus here is on automated (remote-controlled) driving (see [49, 118, 135]), which has actually been possible with self-driving units for about 20 years (see [49, 86, 118]). However, this technology is not been applied in the EU up to now, even though such solutions could have improved the marketability of regional freight transport services and operations on sidings (see [118]).

Fuel cell drives are also under discussion for the rail sector (see [1, 47, 123]). However, it is questionable whether their use in electrified core networks can bring relevant competitive advantages, especially in view of the necessary financial expense for restructuring the tracks. However, such solutions could be very interesting in the area of automated vehicle units (as well as in local passenger rail transport in non-electrified secondary networks).

Developments are also underway in inland navigation with regard to autonomous (see [102, 117]) and automated (remote-controlled) inland vessels (see [15]). However, competitiveness is sometimes considerably limited, also in the European area, as the possibilities of use are subject to widely differing capacity restrictions within the inland waterway networks (see [134]). The crucial question here is over which distances continuous transports are possible without the need for transshipment to smaller ship sizes, and to what extent do transshipment operations affect the time of transport processes and the economic efficiency.

In addition to the political prioritization of the SSS/RSS, there are now also significant technical developments. In the foreground are autonomously operating vessels (see [23, 44, 45, 50, 77]), which are currently under development and through which the fields of application will expand considerably. The associated concepts for sea-sea transshipment on feeder ships (see [3, 66]) can, for example, significantly reduce terrestrial seaport hinterland transport by distributing the cargo to smaller ports with regional catchment areas.

6 Outlook

The development of competition in freight transport, especially in the EU, will not meet the propagated transport policy objectives. This applies to a considerable extent to the intended modal shift from road to rail freight transport. In the EU, for example, this transport mode has developed at an under proportional low rate compared with overall freight transport in recent years, despite all the measures that have been taken. A modal split share of 38% predicted for 2050 in Germany (see [41: 26]) must be regarded as completely unrealistic and much more as a result of political wishful thinking (see Fig. 5), even taking into account increasing shares in bi-modal freight transport. The main reasons for this situation show a very complex background.

First, the in-depth changes in the freight structure must be seen. Rail freight-related bulk good transport will be significantly reduced in the future, due to the politically enforced decarbonization of power generation. In addition, there will be changes in transport demand due to dislocated (and also increasingly volatile) location structures. The disruptive developments associated with this are result from changed manufacturing processes based on *Additive Manufacturing* (AM) (see, [7, 11, 33, 74, 116]) with increasing on-demand and customer-based manufacturing. In this respect, the question must also be asked to what extent a qualification and expansion of the rail network structures the rural regions can be justified economically and also politically in the case of a very limited demand for freight transport due to the remarkable financial requirements.

The extent to which container transport can compensate for possible losses in demand is questionable, since there is also a powerful competition in this market segment from transport on inland waterways and in the RSS/SSS. In addition, a long-term decline in container traffic must be assumed based for example on AM developments (see the forecast of [97: 176] with decline of 38% in 2040), growing reshoring activities (see [6, 51, 103, 142]) as well as structural changes of the international supply chains under the aspect of (national) supply security (see [2, 35]).

Due to the significantly higher transport capacity of barges within bi- or multimodal transport connections, inland navigation can have considerable advantages over rail freight transport on certain routes. However, at this point, the question of possible impacts from climatic changes and resulting reductions in the depth of navigability on inland waterways must be considered, which can have a negative impacts on the performance of inland navigation and thus on competition in the medium and long term view (see [13, 71]). The inclusion of transport services in RSS/SSS, on the other hand, results in considerable competitive advantages over rail freight transport due to larger vehicle capacities and a future use of autonomously operating vessels as well as innovative forms of service (see [3, 66]).

Compared with road freight transport, rail freight transport will continue to lose importance due to its inherent disadvantages in the service provision processes. The previously propagated ecological disadvantages will no longer exist in the future due to the ongoing development of electric drives (see [54, 98, 104]) as well as the no longer existing volume-related (personnel) cost disadvantages with the use of autonomously driving trucks. Ultimately, this development will lead in road freight transport to more significant competitive advantages, especially in local and regional transport as well as on shorter distances of long-distance transport operations.

It is doubtful whether new transport modes be able to enter the terrestrial transport sector in the coming years. The repeatedly propagated use of cargo drones (see [63, 91]) will remain a niche application due to the lack of performance in terms of transport volume, transport weight and operating range. Larger vehicle versions will face aviation law problems as well as safety issues that will set limits. Also underground transport systems (see [137]), such as the *SwissMetro project* (see [72, 94]), which has been discussed for almost 50 years, or more recently the *Hyperloop project* (see [24, 62, 96]). These transport systems do not involve fundamentally new solutions, but they would be an additional competitor to rail freight transport, with a high transport speed but also with extremely high infrastructure costs. Locally oriented systems for inner-city transport as an alternative to (local) road freight transport, such as the *CargoCap* project (see [32, 76]), have also not yet come to implementation for certain reasons. However, in the case of the construction of new settlements, the inclusion of such systems may be possible as also useful (see [61]).

The data from the past years as well as the emerging developments show that ultimately there is no modal competition in the true sense but only between possible (mono- and multimodal) transport links. The crucial point is the dependence of almost all modes (rail, inland waterway, SSS/RSS as well as sea and air) on pre- and on-carriage links based on other modes in order to connect points of origin and destinations. These facts must be integrated more strongly into transport policy planning and decision-making, also to ensure an efficient use of resources and to overcome the current lack in congruence of transport policy objectives.

References

1. Al-Hamed, K.H., Dincer, I.: Development and optimization of a novel solid oxide fuel cell-engine powering system for cleaner locomotives. Appl. Therm. Eng. **183**, 116150 (2021)
2. Althaf, S., Babbitt, C.W.: Disruption risks to material supply chains in the electronics sector. Resour., Conserv. Recycl. **167**, 105248 (2021)
3. Akbar, A., Aasen, A.K., Msakni, M.K., Fagerholt, K., Lindstad, E., Meisel, F.: An economic analysis of introducing autonomous ships in a short-sea liner shipping network. Int. Trans. Oper. Res. **28**(4), 1740–1764 (2021)
4. Alonso Raposo, M., Grosso, M., Mourtzouchou, A., Krause, J., Duboz, A., Ciuffo, B.: Economic implications of a connected and automated mobility in Europe. Res. Transp. Econ. (in press) (2021)
5. Aperte, X.G., Baird, A.J.: Motorways of the sea policy in Europe. Marit. Policy Manag. **40**(1), 10–26 (2013)
6. Ashby, A.: From global to local - reshoring for sustainability. Oper. Manag. Res. **9**(3), 75–88 (2016)
7. Attaran, M.: The rise of 3-D printing - the advantages of additive manufacturing over traditional manufacturing. Bus. Horiz. **60**(5), 677–688 (2017)
8. Baird, A.J.: The economics of Motorways of the Sea. Marit. Policy Manag. **34**, 287–310 (2007)
9. Bask, A., Rajahonka, M.: The role of environmental sustainability in the freight transport mode choice. Int. J. Phys. Distrib. Logist. Manag. **47**(7), 560–602 (2017)
10. Baum, H.: Der volkswirtschaftliche Nutzen des Verkehrs. Zeitschrift für Verkehrswissenschaft **68**(1), 27–51 (1997)
11. Ben-Ner, A., Siemsen, E.: Decentralization and localization of production - the organizational and economic consequences of additive manufacturing (3d printing). Calif. Manage. Rev. **59**(2), 5–23 (2017)
12. Berger, T.: Railroads and rural industrialization - evidence from a historical policy experiment. Explor. Econ. Hist. **74**, 101277 (2019)
13. Beuthe, M., Jourquin, B., Urbain, N., Lingemann, I., Ubbels, B.: Climate change impacts on transport on the Rhine and Danube - a multimodal approach. Transp. Res. Part D **27**, 6–11 (2014)
14. Bogart, D.: Inter-modal network externalities and transport development - evidence from roads, canals, and ports during the english industrial revolution. Netw. Spat. Econ. **9**(3), 309–338 (2009)
15. Bratić, K., Pavić, I., Vukša, S., Stazić, L.: A review of autonomous and remotely controlled ships in maritime sector. Trans. Maritime Sci. **8**(2), 253–265 (2019)
16. Breuer, J.L., Samsun, R.C., Stolten, D., Peters, R.: How to reduce the greenhouse gas emissions and air pollution caused by light and heavy duty vehicles with battery-electric, fuel cell-electric and catenary trucks. Environ. Int. **152**, 106474 (2021)
17. Bucsky, P.: The iron silk road - how important is it? Area Dev. Policy **5**(2), 146–166 (2020)
18. Bühler, G.: Verkehrsmittelwahl im Güterverkehr. Physika, Heidelberg (2006)
19. Bulheller, M.: Verlagerung auf die Schiene muss teuer erkauft werden. Internationales Verkehrswesen **58**(7–8), 353–355 (2006)
20. Bundesministerium für Verkehr und digitale Infrastruktur (BMVI) (Hrsg.): Verkehr in Zahlen 2020/2021. Kraftfahrtbundesamt, Flensburg (2020)

21. Carboni, A., Dalla Chiara, B.: Range of technical-economic competitiveness of rail-road combined transport. Eur. Transp. Res. Rev. **10**(2), 1–17 (2018). https://doi.org/10.1186/s12544-018-0319-3
22. Catalano, G., Daraio, C., Diana, M., Gregori, M., Matteucci, G.: Efficiency, effectiveness, and impacts assessment in the rail transport sector - a state-of-the-art critical analysis of current research. Int. Trans. Oper. Res. **26**(1), 5–40 (2019)
23. Chaal, M., Banda, O.A.V., Glomsrud, J.A., Basnet, S., Hirdaris, S., Kujala, P.: A framework to model the STPA hierarchical control structure of an autonomous ship. Saf. Sci. **132**, 104939 (2020)
24. Chaidez, E., Bhattacharyya, S.P., Karpetis, A.N.: Levitation methods for use in the Hyperloop high-speed transportation system. Energies **12**, 4190 (2019)
25. Chan, C.-Y.: Advancements, prospects, and impacts of automated driving systems. Int. J. Transp. Sci. Technol. **6**, 208–216 (2017)
26. Chandra, S., Christiansen, M., Fagerholt, K.: Analysing the modal shift from road-based to coastal shipping-based distribution - a case study of outbound automotive logistics in India. Marit. Policy Manag. **47**(2), 273–286 (2020)
27. Christodoulou, A., Woxenius, J.: Short-distance maritime geographies - Short sea shipping, RoRo, feeder and inter-island transport. In: Wilmsmeier, G., Monios, J. (eds.) Geographies of Maritime Transport, pp. 134–148, Edward Elgar, Cheltenham, UK (2020)
28. Coluccia, D.: The first industrial revolution (c1760–c1870). In: Zanda, G. (ed.) Corporate Management in a Knowledge-Based Economy, pp. 41–51. Palgrave Macmillan UK, London (2012). https://doi.org/10.1057/9780230355453_3
29. Coluccia, D.: The second industrial revolution (late 1800s and early 1900s). In: Zanda, G. (ed.) Corporate Management in a Knowledge-Based Economy, pp. 52–64. Palgrave Macmillan UK, London (2012). https://doi.org/10.1057/9780230355453_4
30. Crainic, T.G., Kim, K.H.: Intermodal transportation. In: Barnhart, C, Laporte, G.V.(eds.) Handbooks in Operations Research and Management Science, vol. 14, pp. 467–537, North-Holland, Amsterdam et al. (2007)
31. Crössmann, K., Mause, K.: Rail subsidisation in the European Union - an issue beyond left and right? Comp. Eur. Polit. **13**(4), 471–492 (2015)
32. Cui, J., Nelson, J.D.: Underground transport - an overview. Tunn. Undergr. Space Technol. **87**, 122–126 (2019)
33. Daduna, J.R.: Short sea shipping and river-sea shipping in the multi-modal transport of containers. Int. J. Ind. Eng. **20**(1/2), 225–240 (2013)
34. Daduna, J.R.: Disruptive effects on logistics processes by additive manufacturing. In: Ivanov, D., Dolgui, A., Yalaoui, F. (eds.) IFAC Conference on Manufacturing Modelling, Management and Control. IFAC-PapersOnLine, vol. 13, pp. 2770–2775 (2019)
35. Daduna, J.R., Prause, G.: The baltic sea as a maritime highway in international multimodal transport. In: Doerner, K.F., Ljubic, I., Pflug, G., Tragler, G. (eds.) Operations Research Proceedings 2015. ORP, pp. 189–194. Springer, Cham (2017). https://doi.org/10.1007/978-3-319-42902-1_25
36. Dempsey, P.: The supply chain in a world of Covid-19. Eng. Technol. **15**(4), 44–47 (2020)
37. Douet, M., Cappuccilli, J.F.: A review of short sea shipping policy in the European union. J. Transp. Geogr. **19**(4), 968–976 (2011)
38. Dua, A., Sinha, D.: Quality of multimodal freight transportation - a systematic literature review. World Rev. Intermodal Transp. Res. **8**(2), 167–194 (2019)
39. Eisenkopf, A., Hahn, K., Schnöbel, C.: Intermodale Wettbewerbsbeziehungen im Verkehr und Wettbewerbsverzerrungen. In: Eisenkopf, A., Knorr, A. (eds.) Neue Entwicklungen in der Eisenbahnpoliti, pp. 9–138, Duncker & Humblot, Berlin, (2008)

40. Engholm, A., Björkman, A., Joelsson, Y., Kristoffersson, I., Pernestål, A.: The emerging technological innovation system of driverless trucks. Transp. Res. Procedia **49**, 145–159 (2020)
41. Erhardt, J., Reh, E., Treber, M., Oelinger, D, Müller-Görnert, M.: Klimafreundlicher Verkehr in Deutschland - Weichenstellungen bis 2050. Berlin/Bonn (2014)
42. European Commission (EC): EU transport in figures. Luxembourg, Publications Office of the European Union (2020)
43. European Commission (EC): State aid Scoreboard 2020. European Commission - DG Competition (2021)
44. Fan, C., Wróbel, K., Montewka, J., Gil, M., Wan, C., Zhang, D.: A framework to identify factors influencing navigational risk for maritime autonomous surface ships. Ocean Eng. **202**, 107188 (2020)
45. Felski, A., Zwolak, K.: The ocean-going autonomous ship - challenges and threats. J. Marine Sci. Eng. **8**(1), 41 (2020)
46. Fisch, J.H., Zschoche, M.: The effect of operational flexibility on decisions to withdraw from foreign production locations. Int. Bus. Rev. **21**(5), 806–816 (2012)
47. Fragiacomo, P., Piraino, F.: Fuel cell hybrid powertrains for use in Southern Italian railways. Int. J. Hydrogen Energy **44**(51), 27930–27946 (2019)
48. Gagatsi, E., Estrup, T., Halatsis, A.: Exploring the potentials of electrical waterborne transport in Europe - The E-ferry concept. Transp. Res. Procedia **14**, 1571–1580 (2016)
49. Gattuso, D., Cassone, G.C., Lucisano, A., Lucisano, M., Lucisano, F.: Automated rail wagon for new freight transport opportunities. In: IEEE International Conference on Models and Technologies for Intelligent Transportation Systems (MT-ITS), pp. 57–62 (2017)
50. Ghaderi, H.: Autonomous technologies in short sea shipping - Trends, feasibility and implications. Transp. Rev. **39**(1), 152–173 (2019)
51. Gharleghi, B., Jahanshahi, A.A., Thoene, T.: Locational factors and the reindustrialisation process in the USA - reshoring from China. Int. J. Bus. Globalisation **24**(2), 275–292 (2020)
52. Gleichauf, J., Vollet, J., Pfitzner, C., Koch, P., May, S.: Sensor fusion approach for an autonomous shunting locomotive. In: Gusikhin, O., Madani, K. (eds.) ICINCO 2017. LNEE, vol. 495, pp. 603–624. Springer, Cham (2020). https://doi.org/10.1007/978-3-030-11292-9_30
53. Gomez, J., Vassallo, J.M.: Has heavy vehicle tolling in Europe been effective in reducing road freight transport and promoting modal shift? Transportation **47**(2), 865–892 (2020)
54. González Palencia, J.C., Nguyen, V.T., Araki, M., Shiga, S.: The role of powertrain electrification in achieving deep decarbonization in road freight transport. Energies **13**, 2459 (2020)
55. Gota, S., Huizenga, C., Peet, K., Medimorec, N., Bakker, S.: Decarbonising transport to achieve Paris agreement targets. Energ. Effi. **12**(2), 363–386 (2018). https://doi.org/10.1007/s12053-018-9671-3
56. Greenwood, J.: The third industrial revolution - technology, productivity, and income equality. Econ. Rev. **35**(2), 2–12 (1999)
57. Grosso, M., Lynce, A.-R., Silla, A., Vaggelas, G.K.: Short sea shipping, intermodality and parameters influencing pricing policies - the Mediterranean case. NETNOMICS **11**, 47–67 (2010)
58. Grübler, A.: The rise and fall of infrastructures. Physica, Heidelberg (1990)
59. Guerrero, D., Rodrigue, J.P.: The waves of containerization - shifts in global maritime transportation. J. Transp. Geogr. **34**, 151–164 (2014)
60. Guglielminetti, P., Piccioni, C., Fusco, G., Licciardello, R., Musso, A.: Rail freight network in Europe - opportunities provided by re-launching the single wagonload system. Transp. Res. Procedia **25**, 5185–5204 (2017)

61. Guo, D. Yicun Chen, Y., Yang, J. Tan, Y.H., Zhang, C., Chen, Z.: Planning and application of underground logistics systems in new cities and districts in China. Tunn. Undergr. Space Technol. **113**, 10347 (2021)
62. Hansen, I.A.: Hyperloop transport technology assessment and system analysis. Transp. Plan. Technol. **43**(8), 803–820 (2020)
63. Hassanalian, M., Abdelkefi, A.: Classifications, applications, and design challenges of drones - a review. Prog. Aerosp. Sci. **91**, 99–131 (2017)
64. Heinold, A., Meisel, F.: Emission rates of intermodal rail/road and road-only transportation in Europe - A comprehensive simulation study. Transp. Res. Part D **65**, 421–437 (2018)
65. Heinold, A., Meisel, F.: Emission limits and emission allocation schemes in intermodal freight transportation. Transp. Res. Part E **141**, 101963 (2020)
66. Holm, M.B., Medbøen, C.A.B., Fagerholt, K., Schütz, P.: Shortsea liner network design with transshipments at sea - a case study from Western Norway. Flex. Serv. Manuf. J. **31**(3), 598–619 (2019)
67. Isenhardt, I., Solvay, A.F., Otte, T., Henke, C., Haberstroh, M.: Rolle und Einfluss der Industrie 4.0 auf die Gestaltung autonomer Mobilität. In: Frenz, W. (ed.) Handbuch Industrie 4.0: Recht, Technik, Gesellschaft, pp. 681–696. Springer, Heidelberg (2020). https://doi.org/10.1007/978-3-662-58474-3_35
68. Jahangir Samet, M., Liimatainen, H.; van Vliet, O.P.R., Pöllänen, M.: Road freight transport electrification potential by using battery electric trucks in Finland and Switzerland. Energies **14**, 823 (2021)
69. Jensen, A.F., et al.: A disaggregate freight transport chain choice model for Europe. Transp. Res. Part E **121**, 43–62 (2019)
70. Jonkeren, O., Francke, J., Visser, J.: A shift-share based tool for assessing the contribution of a modal shift to the decarbonisation of inland freight transport. Eur. Transp. Res. Rev. **11**(1), 1–15 (2019). https://doi.org/10.1186/s12544-019-0344-x
71. Jonkeren, O., Jourquin, B., Rietveld, P.: Modal-split effects of climate change - the effect of low water levels on the competitive position of inland waterway transport in the river Rhine area. Transp. Res. Part A **45**(10), 1007–1019 (2011)
72. Jufer, M., Perret, F.L., Descoeudres, F., Trottet, Y.: Swissmetro, an efficient intercity subway system. Struct. Eng. Int. **3**(3), 184–189 (1993)
73. Kaack, L.H., Vaishnav, P., Morgan, M.G., Azevedo, I. L., Rai, S.: Decarbonizing intraregional freight systems with a focus on modal shift. Environ. Res. Lett. **13**(8), 083001 (2018)
74. Kadir, A.Z.A., Yusof, Y., Wahab, M.S.: Additive manufacturing cost estimation models - a classification review. Int. J. Adv. Manuf. Technol. **107**(9), 4033–4053 (2020)
75. Kast, J., Vijayagopal, R., Gangloff, J.J., Jr., Marcinkoski, J.: Clean commercial transportation - medium and heavy duty fuel cell electric trucks. Int. J. Hydrogen Energy **42**(7), 4508–4517 (2017)
76. Kersting, M., Klemmer, P., Stein, D.: CargoCap - Wirtschaftliche Transportalternative im Ballungsraum. Internationales Verkehrswesen **56**(11), 493–498 (2004)
77. Kim, M., Joung, T.H., Jeong, B., Park, H.S.: Autonomous shipping and its impact on regulations, technologies, and industries. J. Int. Maritime Saf., Environ. Affairs, Shipping **4**(2), 17–25 (2020)
78. Kim, N.S., van Wee, B.: The relative importance of factors that influence the break-even distance of intermodal freight transport systems. J. Transp. Geogr. **19**(4), 859–875 (2011)
79. Knapcikova, L., Konings, R.: European railway infrastructure - a review. Acta logistica **5**(3), 71–77 (2018)
80. Kotowska, I.: The role of ferry and Ro-Ro shipping in sustainable development of transport. Rev. Econ. Persp. **15**(1), 35–48 (2015)
81. Krämer, I.: Shunt-E 4.0 - autonomous zero emission shunting processes in port and hinterland railway operations. J. Traffic Transp. Eng. **7**, 157–164 (2019)

82. Lasserre, F., Huang, L., Mottet, É.: The emergence of trans-Asian rail freight traffic as part of the belt and road initiative - development and limits. China Perspect. **2020**(2), 43–52 (2020)
83. Launhardt, W.: Mathematische Begründung der Volkswirtschaftslehre. Engelmann, Leipzig, Neudruck, Scientia, Aalen (1963)
84. Le, Y., Ieda, H.: Evolution dynamics of container port systems with a geo-economic concentration index - A comparison of Japan, China and Korea. Asian Transp. Stud. **1**(1), 47–62 (2010)
85. Lu, C., Yan, X.: The break-even distance of road and inland waterway freight transportation systems. Maritime Econ. Logistics **17**(2), 246–263 (2015)
86. Mairhofer, F.: CargoMover - an innovative mode of automated freight transport. Rail Eng. Int. **33**(1), 10–12 (2004)
87. Mareev, I., Sauer, D.U.: Energy consumption and life cycle costs of overhead catenary heavy-duty trucks for long-haul transportation. Energies **11**(12), 3446 (2018)
88. Martínez-Díaz, M., Soriguera, F.: Autonomous vehicles - theoretical and practical challenges. Transp. Res. Procedia **33**, 275–282 (2018)
89. Mathisen, T.A., Sandberg Hanssen, T.E.: The academic literature on intermodal freight transport. Transp. Res. Procedia **3**, 611–620 (2014)
90. Meers, D., Vermeiren, T., Macharis, C.: Intermodal break-even distances - a fetish of 300 kilometres? In: Macharis, C., Melo, S., Woxenius, J., van Lier, T. (eds.) Sustainable Logistics - Transport and Sustainability, Emerald, Bingley, vol. 6, pp. 217–243 (2014)
91. Merkert, R., Bushell, J.: Managing the drone revolution - a systematic literature review into the current use of airborne drones and future strategic directions for their effective control. J. Air Transp. Manage. **89**, 101929 (2020)
92. Meyer, T.: Decarbonizing road freight transportation - a bibliometric and network analysis. Transp. Res. Part D **89**, 102619 (2020)
93. Mostert, M., Limbourg, S.: External costs as competitiveness factors for freight transport - a state of the art. Transp. Rev. **36**(6), 692–712 (2016)
94. Nash, A., Weidmann, U., Buchmueller, S., Rieder, M.: Assessing feasibility of transport megaprojects - Swissmetro European market study. Transp. Res. Rec. **1995**(1), 17–26 (2007)
95. Nocera, S., Cavallaro, F., Irranca Galati, O.: Options for reducing external costs from freight transport along the Brenner corridor. Eur. Transp. Res. Rev. **10**(2), 1–18 (2018). https://doi.org/10.1186/s12544-018-0323-7
96. Nøland, J.K.: Prospects and challenges of the Hyperloop transportation system - a systematic technology review. IEEE Access **9**, 28439–28458 (2021)
97. OECD/International Transport Forum (ITF): ITF Transport Outlook 2019. OECD Publishing, Paris (2019). https://doi.org/10.1787/transp_outlook-en-2019-en
98. Olabi, A.G., Wilberforce, T., Abdelkareem, M.A.: Fuel cell application in the automotive industry and future perspective. Energy **214**, 118955 (2021)
99. Paixão Casaca, A.C., Marlow, P.B.: Logistics strategies for short sea shipping operating as part of multi-modal transport chains. Marit. Policy Manag. **36**, 1–19 (2009)
100. Papadimitriou, S., Lyridis, D.V., Koliousis, I.G., Tsioumas, V., Sdoukopoulos, E., Stavroulakis, P.J.: The dynamics of short sea shipping - New practices and trends. Palgrave Macmillan, Cham (2018)
101. Parantainen, J., Meriläinen, A.: The Baltic Sea motorway - recent developments and outlook for the future. J. Maritime Res. **4**, 21–30 (2007)
102. Peeters, G., et al.: An unmanned inland cargo vessel - design, build, and experiments. Ocean Eng. **201**, 107056 (2020)
103. Pegoraro, D., Propris, L.D., Chidlow, A.: De-globalisation, value chains and reshoring. Industry **4**, 152–175 (2020)
104. Peters, R., et al.: Future power train solutions for long-haul trucks. Sustainability **13**(4), 2225 (2021)

105. Pittman, R., Jandová, M., Król, M., Nekrasenko, L., Paleta, T.: The effectiveness of EC policies to move freight from road to rail - Evidence from CEE grain markets. Res. Transp. Bus. Manage. **37**, 100482 (2020)
106. Raitasuo, P., Bask, A., Rajahonka, M.: Sustainable intermodal train transport. In: de Boer, L., Houman Andersen, P. (eds.) Operations Management and Sustainability, pp. 195–222, Palgrave Macmillan, Cham (2019)
107. Raza, Z., Svanberg, M., Wiegmans, B.: Modal shift from road haulage to short sea shipping - a systematic literature review and research directions. Transp. Rev. **40**(3), 382–406 (2020)
108. Rodrigue, J.-P., Comtois, C., Slack, B.: The geography of transport systems. Routledge, London / New York (2006)
109. Rodrigue, J.P., Notteboom, T.: The geography of containerization - half a century of revolution, adaptation and diffusion. Geo J. **74**(1), 1–5 (2009)
110. Rothengatter, W.: Environmental charges levied on heavy goods vehicles in the EU. In: Hayashi, Y, Morisugi, M., Iwamatsu, S. (eds.) Balancing nature and civilization - Alternative sustainability perspectives from philosophy to practice, pp. 77–91, Springer, Cham (2020)
111. Saeedi, H., Wiegmans, B., Behdani, B., Zuidwijk, R.: Analyzing competition in intermodal freight transport networks - the market implication of business consolidation strategies. Res. Transp. Bus. Manag. **23**, 12–20 (2017)
112. Sánchez, R.J., Wilmsmeier, G.: Short-sea shipping potentials in central America to bridge infrastructural gaps. Marit. Policy Manag. **32**, 227–244 (2005)
113. Sandberg Hanssen, T.E., Mathisen, T.A., Jørgensen, F.: Generalized transport costs in intermodal freight transport. Procedia Soc. Behav. Sci. **54**, 189–200 (2012)
114. Santos, T.A., Guedes Soares, C.: Modeling transportation demand in short sea shipping. Maritime Econ. Logistics **19**(4), 695–722 (2017)
115. Santos, B.F., Limbourg, S., Carreira, J.S.: The impact of transport policies on railroad intermodal freight competitiveness - the case of Belgium. Transp. Res. Part D **34**, 230–244 (2015)
116. Savolainen, J., Collan, M.: How additive manufacturing technology changes business models? Review of literature. Addit. Manuf. **32**, 101070 (2020)
117. Schiaretti, M., Chen, L., Negenborn, R.R.: Survey on autonomous surface vessels - Part I - a new detailed definition of autonomy levels. In: Bektaş, T., Coniglio, S., Martinez-Sykora, A., Voß, S. (eds.) Computational Logistics, pp. 219–233, Springer, Cham (2017). https://doi.org/10.1007/978-3-319-68496-3_15
118. Schindler, C.: Schienenverkehrstechnik 4.0. In: Frenz, W. (ed.) Handbuch Industrie 4.0: Recht, Technik, Gesellschaft, pp. 719–757. Springer, Heidelberg (2020). https://doi.org/10.1007/978-3-662-58474-3_38
119. Schulz, W.: Industrieökonomik und Transportsektor - Marktdynamik und Marktanpassungen im Güterverkehr. Kölner Wissenschaftsverlag, Köln (2004)
120. Schwarting, W., Alonso-Mora, J., Rus, D.: Planning and decision-making for autonomous vehicles. Ann. Rev. Control, Robotics, Auton. Syst. **1**, 187–210 (2018)
121. Seo, Y.J., Chen, F., Roh, S.Y.: Multimodal transportation - the case of laptop from Chongqing in China to Rotterdam in Europe. Asian J. Shipping Logistics **33**(3), 155–165 (2017)
122. Shankar, R., Pathak, D.K., Choudhary, D.: Decarbonizing freight transportation - an integrated EFA-TISM approach to model enablers of dedicated freight corridors. Technol. Forecast. Soc. Change **143**, 85–100 (2019)
123. Siddiqui, O., Dincer, I.: A review on fuel cell-based locomotive powering options for sustainable transportation. Arab. J. Sci. Eng. **44**, 677–693 (2019)
124. Skjærseth, J.B.: Towards a European green deal - the evolution of EU climate and energy policy mixes. Int. Environ. Agreements - Politics, Law Econ. **21**(1), 25–41 (2021)
125. Suárez-Alemán, A.: Short sea shipping in today's Europe - a critical review of maritime transport policy. Maritime Econ. Logistics **18**(3), 331–351 (2016)

126. Svindland, M., Hjelle, H.M.: The comparative CO_2 efficiency of short sea container transport. Transp. Res. Part D **77**, 11–20 (2019)
127. Tasler, G., Knollmann, V.: The introduction of highly automatic operation - Towards fully automatic train operation. Signalling + Datacommunication **110**(6), 6–14 (2018)
128. Tate, W.L., Ellram, L.M., Schoenherr, T., Petersen, K.J.: Global competitive conditions driving the manufacturing location decision. Bus. Horiz. **57**(3), 381–390 (2014)
129. Torres de Miranda Pinto, J., Mistage, O., Bilotta, P., Helmers, E.: Road-rail intermodal freight transport as a strategy for climate change mitigation. Environ. Dev. **25**, 100–110 (2018)
130. Tracy, E.F., Shvarts, E., Simonov, E., Babenko, M.: China's new Eurasian ambitions - The environmental risks of the silk road economic belt. Eurasian Geogr. Econ. **58**(1), 56–88 (2017)
131. Ullrich, G.: Fahrerlose Transportsysteme. Springer, Wiesbaden (2014)
132. Vallejo-Pinto, J.A., Garcia-Alonso, L., Fernández, R.Á., Mateo-Mantecón, I.: Iso-emission map - a proposal to compare the environmental friendliness of short sea shipping vs road transport. Transp. Res. Part D **67**, 596–609 (2019)
133. van den Bos, G., Wiegmans, B.: Short sea shipping - a statistical analysis of influencing factors on SSS in European countries. J. Shipping Trade **3**(1), 1–20 (2018)
134. van Dorsser, C.: Existing waterway infrastructures ad future need. In: Wiegmans, B., Konings, R. (eds.) Inland Waterway Transport, pp. 99–124, Routledge, New York (2016)
135. Venkateswaran, K.G., Nicholson, G.L., Roberts, C., Stone, R.: Impact of automation on the capacity of a mainline railway. In: IEEE International Conference on Intelligent Transportation Systems, pp. 2097–2102 (2015)
136. Vierth, I., Sowa, V., Cullinane, K.: Evaluating the external costs of trailer transport - a comparison of sea and road. Maritime Econ. Logistics **21**(1), 61–78 (2019)
137. Visser, J.G.S.N.: The development of underground freight transport - an overview. Tunn. Undergr. Space Technol. **80**, 123–127 (2018)
138. Voigt, F.: Verkehr und Industrialisierung. Zeitschrift für die gesamte Staatswissenschaft **109**(2), 193–239 (1953)
139. Voigt, F.: Verkehr - Die Theorie der Verkehrswirtschaft (Band 1, Teil 1). Duncker & Humblot, Berlin (1973)
140. Wen, X., Ma, H.L., Choi, T.M., Sheu, J.B.: Impacts of the belt and road initiative on the China-Europe trading route selections. Transp. Res. Part E **122**, 581–604 (2019)
141. Wiegmans, B., Konings, R.: Intermodal inland waterway transport - Modelling conditions influencing its cost competitiveness. Asian J. Shipping Logistics **31**(2), 273–294 (2015)
142. Wiesmann, B., Snoei, J.R., Hilletofth, P., Eriksson, D.: Drivers and barriers to reshoring - a literature review on offshoring in reverse. Eur. Bus. Rev. **29**(1), 15–42 (2017)
143. Wolff, S., Fries, M., Lienkamp, M.: Technoecological analysis of energy carriers for long-haul transportation. J. Ind. Ecol. **24**(1), 165–177 (2020)
144. Wissenschaftlicher Beirat beim BMVI: Gutachten des Wissenschaftlichen Beirats beim Bundesminister für Verkehr und digitale Infrastruktur November 2020. BMVI, Bonn (2021)
145. Xu, H., Yang, D., Weng, J.: Economic feasibility of an NSR/SCR-combined container service on the Asia-Europe lane - a new approach dynamically considering sea ice extent. Marit. Policy Manag. **45**(4), 514–529 (2018)
146. Yang, D., Jiang, L., Ng, A.K.: One belt one road, but several routes - a case study of new emerging trade corridors connecting the far East to Europe. Transp. Res. Part A **117**, 190–220 (2018)
147. Zeng, Q., Lu, T., Lin, K.C., Yuen, K.F., Li, K.X.: The competitiveness of Arctic shipping over Suez Canal and China-Europe railway. Transp. Policy **86**, 34–43 (2020)
148. Zgonc, B., Tekavčič, M., Jakšič, M.: The impact of distance on mode choice in freight transport. Eur. Transp. Res. Rev. **11**(1), 1–18 (2019). https://doi.org/10.1186/s12544-019-0346-8

Author Index

Printed in the United States
by Baker & Taylor Publisher Services